ASTRONOMY IN FOCUS

AS PRESENTED AT THE IAU XXX GENERAL ASSEMBLY, 2018

COVER ILLUSTRATION:

Nuzzled in the chest of the constellation Virgo (the Virgin) lies a beautiful cosmic gem—the galaxy Messier 61. This glittering spiral galaxy is aligned face-on towards Earth, thus presenting us with a breathtaking view of its structure. The gas and dust of the intricate spiral arms are studded with billions of stars. This galaxy is a bustling hub of activity with a rapid rate of star formation, and both a massive nuclear star cluster and a supermassive black hole buried at its heart.

Messier 61 is one of the largest members of the Virgo Cluster, which is made up of more than a thousand galaxies, and is itself at the centre of the Virgo Supercluster—to which our Milky Way also belongs. This dazzling beauty was first discovered in 1779, and it has been capturing astronomers' interest ever since. Set against a dark sky littered with galaxies, this image shows the awe-inspiring M61 in its full glory—even at its distance of over 50 million light-years.

This image was taken as part of ESO's Cosmic Gems Programme, an outreach initiative to produce images of interesting, intriguing or visually attractive objects using ESO telescopes, for the purposes of education and public outreach. The programme makes use of telescope time that cannot be used for science observations. In case the data collected could be useful for future scientific purposes, these observations are saved and made available to astronomers through ESO's Science Archive.

Credit: ESO

IAU SYMPOSIUM PROCEEDINGS SERIES

INTERNATIONAL ASTRONOMICAL UNION

UNION ASTRONOMIQUE INTERNATIONALE

International Astronomical Union

IAU

ASTRONOMY IN FOCUS XXX

AS PRESENTED AT THE IAU XXX GENERAL ASSEMBLY, VIENNA, AUSTRIA 2018

Edited by

MARIA TERESA LAGO
General Secretary

CAMBRIDGE UNIVERSITY PRESS
University Printing House, Cambridge CB2 8BS, United Kingdom
1 Liberty Plaza, Floor 20, New York, NY 10006, USA
10 Stamford Road, Oakleigh, Melbourne 3166, Australia

First published 2020

Printed in the UK by Bell & Bain, Glasgow, UK

Typeset in System LaTeX 2ε

A catalogue record for this book is available from the British Library Library of Congress Cataloguing in Publication data

This journal issue has been printed on FSC™-certified paper and cover board. FSC is an independent, non-governmental, not-for-profit organization established to promote the responsible management of the world's forests. Please see www.fsc.org for information.

ISBN 9781108488730 hardback
ISSN 1743-9213

Table of Contents

FM3 – Radio Galaxies: Resolving the AGN Phenomenon

FM4 – Magnetic Fields along the Star-Formation Sequence

FM5 – Understanding Historical Observations to Study Transient Phenomena

FM6 – Galactic Angular Momentum

FM7 – Radial Metallicity Gradients in Star Forming Galaxies

FM8 – New Insights in Extragalactic Magnetic Fields

FM9 – Solar Irradiance: Physics-Based Advances

FM10 – Nano Dust in Space and Astrophysics

FM11 – JWST: Launch, Commissioning, and Cycle 1 Science

FM12 – Calibration and Standardization Issues in UV-VIS-IR Astronomy

FM15 – Astronomy for Development

Astronomy in Focus

Preface

The scientific programme of the XXXth IAU General Assembly held in Vienna, Austria, from August 20th to 31st, 2018, included the usual six Scientific Symposia, plus one extra Symposium, the "Centennial Symposium" focused on the 100 years of IAU, and fifteen Focus Meetings.

The Focus Meetings were introduced for the first time in the programme of the XXIXth IAU General Assembly in 2015 aimed at discussing specific themes of great relevance in the current astronomical research scenario. Learning from that experience, the IAU Executive Committee took in 2016 the decision that the scientific programme of the XXXth IAU General Assembly should include only twelve scientific Focus Meetings plus three other of different character and duration: one addressing the activities of the Executive Committee Working Group on Global Coordination of Ground and Space Astrophysics (FM13), and the other two focused on the mission and activities carried out by two IAU Offices: the Office of Astronomy Outreach (FM14) and the Office of Astronomy for Development (FM15).

The Focus Meetings were selected from a set of thirty two Focus Meetings Proposals submitted by the deadline, December 15th 2016. The high quality and diversity of the proposals made the task of the Evaluation Panel particularly difficult.

The quality of the invited and contributed talks, as well as the attendance at all the Focus Meetings testify the success of the overall GA Scientific Programme: the XXXth General Assembly was not only a very participated astronomical gathering - the largest so far in terms of participants - but also a very stimulating one.

Besides the traditional two "Transaction" Volumes produced as outcome of the IAU General Assembly: Transaction-A, containing the reports by the Divisions, Commissions, Working Groups and a summary of the SpS and JD meetings, and Transaction-B presenting the more IAU "business" oriented reports, official speeches and documents, as well as the updated list of IAU Members, the "Astronomy in Focus" Volume introduces the most relevant contributions presented in the various Focus Meeting. Regrettably two of the Focus Meetings (FM2 and FM13) did not in the end produce the expected Proceedings which are essential to keep a full record of the meeting.

"Astronomy in Focus" is published in paper and electronic versions, allowing for each Focus Meeting up to 40 pages plus an introduction by the respective Chair and Editor. Furthermore, it was also offered the possibility to include extra supplementary material, in electronic version only.

Teresa Lago
IAU General Secretary

Table 1. Astronomy in Focus

FM #	Chair	Title
FM 1	Joseph Masiero	A century of Asteroid Families
FM 2	Wei Cui	Warm and Hot Baryonic Matter in the Cosmos
FM 3	Volker Beckmann	Radio Galaxies: Resolving the AGN phenomenon
FM 4	Analle Maury	Magnetic Fields along the Star-Formation Sequence
FM 5	Ralph Neuhaeuser	Understanding historical observations to study transient phenomena
FM 6	Danail Obreschkow	Galactic Angular Momentum
FM 7	Letizia Stanghellini	Radial metallicity gradients in star forming galaxies
FM 8	Luigina Feretti	New Insights in Extragalactic Magnetic Fields
FM 9	Greg Kopp	Solar Irradiance: Physics-Based Advances
FM 10	Ingrid Mann	Nano Dust in Space and Astrophysics
FM 11	Nikole Lewis	JWST: Launch, Commissioning, and Cycle 1 Science
FM 12	J. Allyn Smith	Calibration and Standardization Issues in UV-VIS-IR Astronomy
FM 13	David Spergel	Global Coord. of Intern. Astrophysics and Heliophysics Activities from Space and Ground
FM 14	Sze-leung Cheung	IAU's role on global astronomy outreach, the latest challenges and bridging different communities
FM 15	Kevindran Govender	Astronomy for Development

FM1 – A Century of Asteroid Families

Introduction

In 1918, Kiyotsugu Hirayama published a manuscript in the Astronomical Journal titled "Groups of asteroids probably of common origin", opening the field study of asteroid families. The IAU 2018 General Assembly was held on the 100th anniversary of this seminal work. In the intervening century, Hirayama's initial insight that these groups of objects were not random has been supported by countless studies spanning orbital elements, colors, spectral taxonomy, albedo, rotation, impact physics, and dynamical evolution. This wealth of information has become a key aspect in family identification and interpretation, expanding our understanding of family-forming processes.

Asteroid families have provided unique insights into the forces shaping both our Solar System and other planetary systems. Families provide us observable evidence of large-scale catastrophic impacts in the Solar System, giving us the tools to test impact physics on planetary scales. Families also have been critical in revealing the fingerprints left on the Solar System by gravitational mean motion and secular resonances as well as the Yarkovsky effect, the most important non-gravitational force for sculpting the Main Belt of asteroids and supplying new objects into near-Earth space. The quantification and simulation of the Yarkovsky force has enabled age dating of asteroid families, providing us a chronology of the impacts in the inner Solar System. Families allow us to probe the heterogeneity of the protoplanetary disk in the terrestrial planet region, and homogeneity of the parent bodies prior to the family-forming collision. And recently, collisional families have been identified beyond the Main Belt in populations as diverse as the Jovian irregular satellites and the Trans-Neptunian Objects.

Asteroid families are currently undergoing a revolution in understanding due to the combination of a rapid increase in asteroid survey data, and improved computational resources and techniques allowing for detailed simulation of family formation and evolution. Some of the avenues explored in this Focus Meeting included:

- improvements to family classification routines that detect family halos as well as core members;
- simulations probing the internal structure of $D < 30$km scale family members, formed either from fracturing during impact or reaccretion post-impact;
- the lack of differentiation signatures in the set of known families, and the apparent conflict with the presence of differentiated meteorites in our collections;
- asteroid families formed through mechanisms other than catastrophic disruption (e.g. YORP-induced rotational fission);

This meeting gave the international community the chance to review the history of asteroid family science; highlight some of the major results as well as watershed moments in the field; discuss new work being done; provide predictions for the future of the field in light of the new techniques and data sets that are currently being developed; and celebrate the centennial of the birth of this field. Asteroid families will continue to be a touchstone for Solar System science in their next century, providing insights and test populations for planetary formation models not available in any other way.

Joe Masiero, SOC chair
on behalf of the SOC

Astronomy in Focus - XXX
Proceedings IAU Symposium No. XXX, 2018
M. T. Lago, ed.

doi:10.1017/S1743921319003211

The History of Asteroid Family Identification

Zoran Knežević

Serbian Academy of Sciences and Arts
Kneza Mihaila 35, 11000 Belgrade, Serbia
email: zoran.knezevic@sanu.ac.rs

Abstract. In this paper the early history of search for asteroid groupings is briefly reviewed. Starting from the first attempts by Kirkwood, who managed to identify a number of asteroid pairs and triples with adjacent orbits, via the similar contributions of Tisserand and Mascart, we arrive to Hirayama and his discovery of asteroid families, marking the beginning of modern asteroid science.

Asteroid families were discovered a century ago by K. Hirayama. The seminal paper of his: "Groups of asteroids probably of common origin" (Hirayama 1918), as well as a series of subsequent papers devoted to asteroid families (Hirayama 1919, 1920, 1922, 1927, 1933), received a lot of attention in the past, and have been thoroughly described in a recent review (Knežević 2016). In the present paper, we shall therefore pay more attention to several almost forgotten attempts to find asteroid groupings that precede the Hirayama's discovery.

The first attempts to find groups of asteroids with similar orbits date back to 1877, when D. Kirkwood, in an article entitled "The asteroids between Mars and Jupiter" (Kirkwood 1877), presented "evidence of a similarity more than accidental between adjacent orbits of the asteroidal group". Among only 172 asteroids known at the time, he found 4 pairs (Fortuna and Eurynome, Fides and Maia, Clotho and Juno, Sirona and Ceres) with striking similarity in the magnitude, form and position of orbits. For an additional pair (Urda and Gerda) he states that the discoverer of these bodies, Dr. Peters, in the American Journal of Science for February 1877, calls attention to coincidences between the elements of their orbits. Kirkwood does not venture (yet) to speculate on the origin of such pairs, stating simply that they must be of common origin.

In 1888 Kirkwood publishes a booklet (Kirkwood 1888) devoted entirely to asteroids, in which he again mentions "adjacent orbits", drawing, however, a special attention to the pair Hilda and Ismene, for which he finds distances, periods, inclinations and eccentricities nearly identical, but longitudes of perihelia different by almost exactly 180°. Wondering whether this indicates that the two objects detached at about the same time from the opposite sides of the solar nebula, he suggests that in general the asteroid pairs with similar orbits originate from single asteroids, which, like comets, "may have been separated by the sun's unequal attraction on their parts".

Finally, in a couple of papers published in astronomical journals (Kirkwood 1890, 1891) he again returns to the topic of similarity of asteroid orbits, proposing in the former paper the three most significant pairs (including a new one – Vera and Semele) and again discussing possible formation by separation/dismemberment; in the latter paper, however, starting from the catalog with about 300 asteroids, he reports on finding of 10 groups (4 triples and 6 pairs) of bodies with similar orbits. Assuming certain characteristics of the solar nebula, he concludes that the disturbing influence of Jupiter could have been sufficient not only to detach the masses of asteroids from the central body, but also to subdivide by unequal attraction on the different parts the newly-formed nebulous planets, until the fragments finally resulted in the existing asteroids.

In his article "La question des petites planètes", F. Tisserand (1891) reports "sans explication" on the three pairs of asteroids with similar orbital elements: (106) Dione and (245) Vera (forming with (86) Semele one of the triples proposed by Kirkwood), (218) Bianea and (246) Asporina, and (84) Klio and (249) Ilse (again a pair which makes part of the Kirkwood's triple with (115) Thyra).

Finally, J. Mascart published two studies (Mascart 1899, 1902) in which he considers pairs of asteroids with similar orbits. In the former paper, he first computes Tisserand invariants for a total of 417 asteroids, identifying subsequently 20 pairs of asteroids with very close values of the invariants and with more or less similar orbital elements. In the latter paper he then analyses in detail the probabilities of these coincidences.

As pointed out by Hirayama (1922), all these attempts to find significant groupings of asteroids failed. Hirayama first rightly cautions against having only two or three asteroids with similar orbits, because this gives rise to a considerable probability that they may be accidentally coincident. Then, explaining the failure to find more asteroids with similar orbits, he attributes it to the "use of actual orbits ... for comparison, whereas these orbits are varied remarkably by the action of the planets", and correctly states that such attempts could be successful only if the separation had occurred very recently. To identify families originated at remoter age, Hirayama asserts that some kind of invariable elements must be used to identify families. Indeed, in the same paper he formally introduces the *proper elements* as such invariable parameters and computes them by means of the Lagrangian linear theory of secular perturbations, stating: "... if there may be noticed a group of the asteroids with the elements approximately common to all, and if at the same time, their number is sufficiently large to be insured from the effect of chance, then we can conclude that they have probably originated from the breaking up of a single asteroid". His success in identifying asteroid families with tens of members marks the end of the pioneer era of the search for asteroid groupings, and the beginning of the modern asteroid science.

References

Hirayama, K. 1918, *Astron. J.*, 31, 185
Hirayama, K. 1919, *Proc. Phys. Mat. Soc. Japan*, III, 1, 52
Hirayama, K. 1920, *Proc. Phys. Mat. Soc. Japan*, III, 2, 236
Hirayama, K. 1922, *Japanese J. Astron. Geo.*, 1, 55
Hirayama, K. 1927, *Japanese J. Astron. Geo.*, 5, 137
Hirayama, K. 1933, *Proc. Imp. Acad. Japan*, 9, 482
Kirkwood, D. 1877, *The Annual Review of the Board of Regents of the Smithsonian Institution for 1876* (Washington: Gov. Print. Office) p. 358
Kirkwood, D. 1888, *The Asteroids or Minor Planets Between Mars and Jupiter* (Philadelphia: J.B.: Lippincot Comp.)
Kirkwood, D. 1890, *Publ. Astron. Soc. Pacific*, 2, 48
Kirkwood, D. 1891, *Publ. Astron. Soc. Pacific*, 3, 95
Knežević, Z. 2016, in: S. R. Chesley, A. Morbidelli, R. Jedicke & D. Farnocchia, (eds.) *Proceedings IAU Symposium 318: Asteroids: New Observations, New Models* (Cambridge: Cambridge Univ. Press), p. 16
Mascart, J. 1899, *Bull. Astron.*, 16, 369
Mascart, J. 1902, *Annales Obs. Paris*, XXIII, F
Tisserand, F. 1891, *Annuaire Pour L'An 1891 publie par Bureau des Longitudes* (Paris: Gauthier-Villars et files) p. B.18

Astronomy in Focus - XXX
Proceedings IAU Symposium No. XXX, 2018
M. T. Lago, ed.

doi:10.1017/S1743921319003223

Japanese asteroid studies in the century after the discovery of the Hirayama families

Tsuko Nakamura[1] and Fumi Yoshida[2]

[1]Daito-bunka University, Tokyo, Japan
email: tsukonk@yahoo.co.jp

[2]Planetary Exploration Research Center, Chiba Institute of Technology, Chiba, Japan
email: fumi.yoshida@perc.it-chiba.ac.jp

A number of studies relating to asteroids have been conducted by Japanese astronomers in the century since the discovery of asteroid families by Kiyotsugu Hirayama in 1918. The true concept of the "asteroid family" was recognized correctly by a few eminent astronomers like Brouwer (1950) in the astronomical community worldwide by the 1970s. The Hirayama's monumental discovery likely stimulated research activities on the dynamics of asteroids after WWII in Japan, as represented by the Kozai mechanism (Kozai (1962)).

Kitamura (1959) performed the pioneering work on the physical nature of asteroids, in 1959. He obtained the 2-band colors of 42 asteroids by using a photomultiplier and compared to the proper orbital elements of asteroid families, the reflectance spectra of nine meteorites, and some rock minerals. He concluded that the observed asteroid colors showed no correlation with their proper elements, and both the colors of asteroids and meteorites were quite similar as a whole.

Starting in the 1920s, Suzuki & Nagashima (1938) had started primitive impact experiments of rock samples, and in 1975, Fujiwara *et al.* (1977) initiated modern impact experiments, which soon became an important mean for investigating the origin of asteroid families and collisional phenomena in the Solar System.

Recent research outcomes are listed as follows.

- the discovery of a great number of meteorites in Antarctica by the Japanese expedition team (Yanai & Kojima (1986)),
- the asteroid discovery race by Japanese amateur asteroid hunters during the 1980s – 2000s, especially, T. Kobayashi who independently discovered more than 2000 new asteroids,
- an experimental solution for the space weathering in asteroid reflectance spectra (Sasaki *et al.* (2001))
- the establishment of the Institute of Space and Astronautical Science (ISAS) and its achievements in the asteroid mission to the asteroid Itokawa by the Hayabusa spacecraft in 2005 (e.g., Fujiwara *et al.* (2006)), and
- survey observations of sub-km-sized asteroids and faint Trojan asteroids using the 8.2m Subaru telescope equipped with a wide-field CCD camera (e.g.,Yoshida & Nakamura (2005), Yoshida & Nakamura (2007)).

References

Brouwer, D. 1950, *AJ*, 55, 162
Fujiwara, A., Kamimoto, G., & Tsukamoto, A. 1977, *Icarus*, 31, 277
Fujiwara, A. *et al.* 2006, *Science*, 312, 1330

Hirayama, K. 1918, *AJ*, 31, 185
Kitamura, M. 1959, *Pub. Astron, Soc. Japan*, 11, 79
Kozai, Y. 1962, *AJ*, 67, 591
Sasaki, S. *et al.* 2001, *Nature*, 410, 555
Suzuki, S. & Nagashima, H. 1938, *Proc. of the Physico-Mathematical Society of Japan, 3rd Series*, 20, 517
Yanai, K. & Kojima, H. 1986, *Lunar and Planetary Institute Technical Report*, 86-01, Houston, Lunar and Planetary Institute.
Yoshida, F. & Nakamura, T. 2005, *AJ*, 130, 2900
Yoshida, F. & Nakamura, T. 2007, *Planetary and Space Science*, 55, 1113

Astronomy in Focus - XXX
Proceedings IAU Symposium No. XXX, 2018
M. T. Lago, ed.

doi:10.1017/S1743921319003235

Review of asteroid-family and meteorite-type links

Peter Jenniskens

SETI Institute,
189 Bernardo Ave., Mountain View, CA 94043, U.S.A.
email: Petrus.M.Jenniskens@nasa.gov

Abstract. The materials of large asteroids and asteroid families are sampled by meteorites that fall to Earth. The cosmic ray exposure age of the meteorite identifies the collision event from which that meteorite originated. The inclination of the orbit on which the meteoroid impacted Earth measures the inclination of the source region, while the semi-major axis of the orbit points to the delivery resonance, but only in a statistical sense. To isolate the sources of our meteorites requires multiple documented falls for each cosmic ray exposure peak. So far, only 36 meteorites have been recovered from observed falls. Despite these low numbers, some patterns are emerging that suggest CM chondrites originated from near the 3:1 resonance from a low-inclined source (perhaps the Sulamitis family), LL chondrites came to us from the ν_6 resonance (perhaps the Flora family), there is an H chondrite source at high inclination (Phocaea?), and one group of low shock-stage L chondrites originates from the inner main belt. Other possible links are discussed.

1. Introduction

Our meteorites are the remains of meteoroids that were created during a significant cratering event on a large asteroid or during the disruption of a small asteroid. Those are discrete events in time.

The cosmic ray exposure (CRE) age of a meteorite measures the time since that event took place. It is the time since the meteoroid was no longer shielded from exposure to cosmic rays by several meters of overlaying burden. CRE ages have been measured for many meteorite types (Eugster *et al.* 2006), and they show distinct peaks (Fig. 1).

Peaks are broadened by measurement error, by two-phase exposure (where material was pre-exposed near the surface before being excavated), and by later disruptions into smaller meteoroids. Even so, it appears that most of our meteorite classes originated in more than one collision event (tentatively labeled A, B, C, ... in Fig. 1, in order of their contribution to the influx on Earth).

The CRE diagrams are dominated by the biggest collision events in the past 10s of millions of years (Ma). Some classes have dominant events that must have been impact craters of at least about 10-km in size, or disrupted asteroids that were at least about 1-km in size, in order to create sufficient debris to impact Earth. The influx does not seem to be dominated by a background of smaller collision events, which would create a continuous age-distribution with a shape determined by the timescales on which 0.5 – 5 m sized meteoroids diffuse from their source region into resonances, and subsequently evolve dynamically to reach Earth.

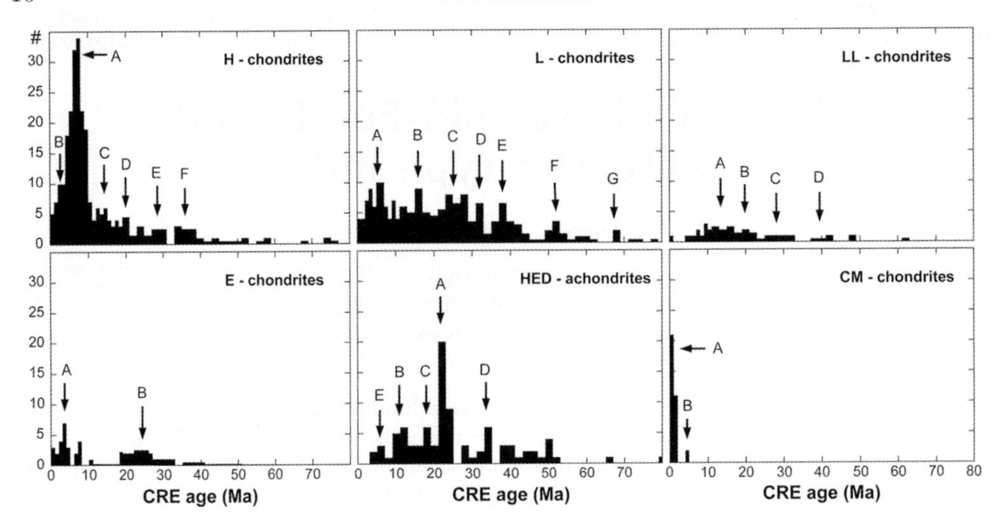

Figure 1. Cosmic ray exposure age distributions for six meteorite classes. Data from: Eugster *et al.* 2006.

2. Methods

When evolving for 5 – 50 Ma, most meteoroids will impact Earth on orbits that still have a semi-major axis close to the delivery resonance (Fig. 2). They will also mostly still have an inclination close to that of the source asteroid. Asteroid families, as well as large asteroids, have the most surface area for collisions, more so than the background asteroids (Dermott *et al.* 2018). That identifies a discrete number of potential source regions for the individual CRE peaks in Fig. 1. The spectroscopic properties of those asteroids can further constrain which asteroids or families are good candidates.

To find the source asteroids and asteroid families, we need to measure a statistically relevant sample of impact orbits of observed meteorite falls to identify the delivery resonance and inclination of the source for each meteorite type with given CRE age peak. Figure 1 tentatively identifies at least 25 collision events for just these six meteorite

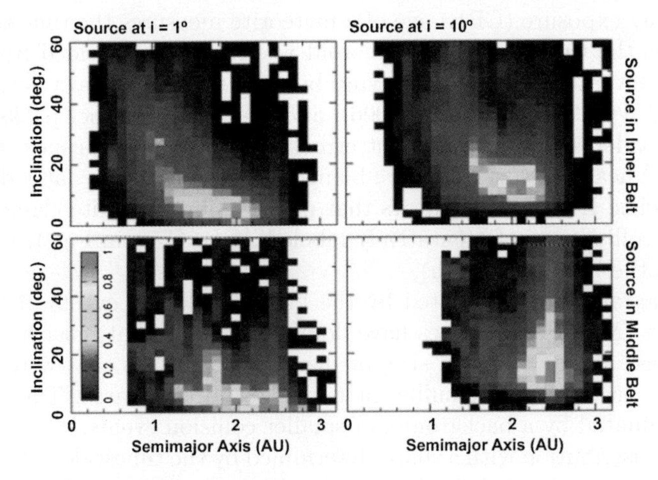

Figure 2. Relative density of Earth impact orbital elements for 20-cm meteoroids that were ejected from the Inner Main Belt (2.1-2.4 AU) or Central Main Belt (a = 2.55-2.85 AU) at initial inclinations of 1 and 10 degrees, respectively. Meteoroids are followed to perihelion distance q < 1.1 AU for 50 Ma, in calculations by David Nesvorny (priv. corr.), using methods by (Nesvorny *et al.* 2009).

classes. Figure 2 shows that while the semi-major axis tends to cluster near the delivery resonance (a = 2.0 AU for the ν_6 secular resonance and a = 2.5 AU for the 3:1 mean-motion resonance), there is a wide tail to each distribution following close encounters with the terrestrial planets.

3. Results

As of today, only 36 meteorite falls have been observed well enough to derive a pre-impact orbit. Several case studies are ongoing. Even so, some patterns are already emerging (Fig. 3). The two *CM chondites* Sutters's Mill and Maribo show low-inclined eccentric orbits that appear to originate in the 3:1 mean motion resonance (Jenniskens *et al.* 2012). Recently, two *LL chondrites* were observed to fall, Stubenberg and Hradek Kralove, that have a similar short orbit as Chelyabinsk pointing to an origin in the ν_6 resonance (Popova *et al.* 2013). Dingle Dell is classified as an L/LL5 chondrite (Eugster *et al.* 2006), but has an L-like orbit.

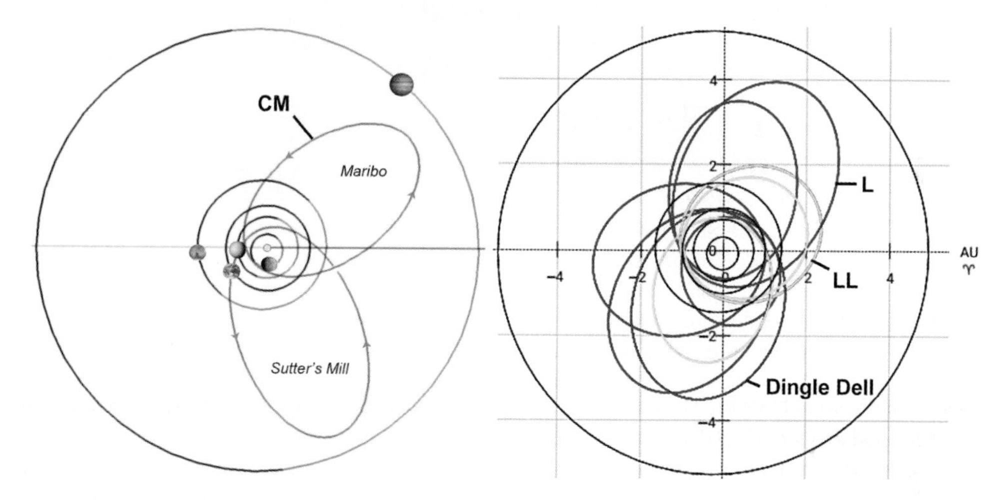

Figure 3. Orbits of CM chondrites (left) and both LL and L ordinary chondrites (right).

Most observed *L chondrites* appear to come from longer orbits, arriving from both the ν_6 and 3:1 resonances (Fig. 3). Recent work on Creston showed that there is a source of L chondrites of moderate shock stage that originated in the inner asteroid belt. This group lacks the 470 Ma K-Ar age found in many shock blackened L chondrites (Jenniskens *et al.* 2018).

H chondrites stand out because so far three have been found on steeply inclined orbits (24 - 32 degrees), some on short orbits. These may be the source of the 7 Ma CRE peak (source A in Fig. 1), because the only observed H chondrite fall so far with a 7-Ma CRE age had such a high inclined orbit.

4. Discussion

Figure 4 provides an overview of where we are now. Since my previous review (Jenniskens 2013), it has become clear that the Eulelia (and Polana) family lacks water bands in spectra (de Leon *et al.* 2018), which makes the Sulamitis family, in my opinion, a more likely candidate for CM chondrites among the current eight known primitive collisional families located in the inner belt. Like Eulalia, this C-class family also hugs the 3:1 mean motion resonance and is relatively low inclined (de Leon *et al.* 2018). The Eulelia and Polana families may, instead, be the source of our ureilites. The optical

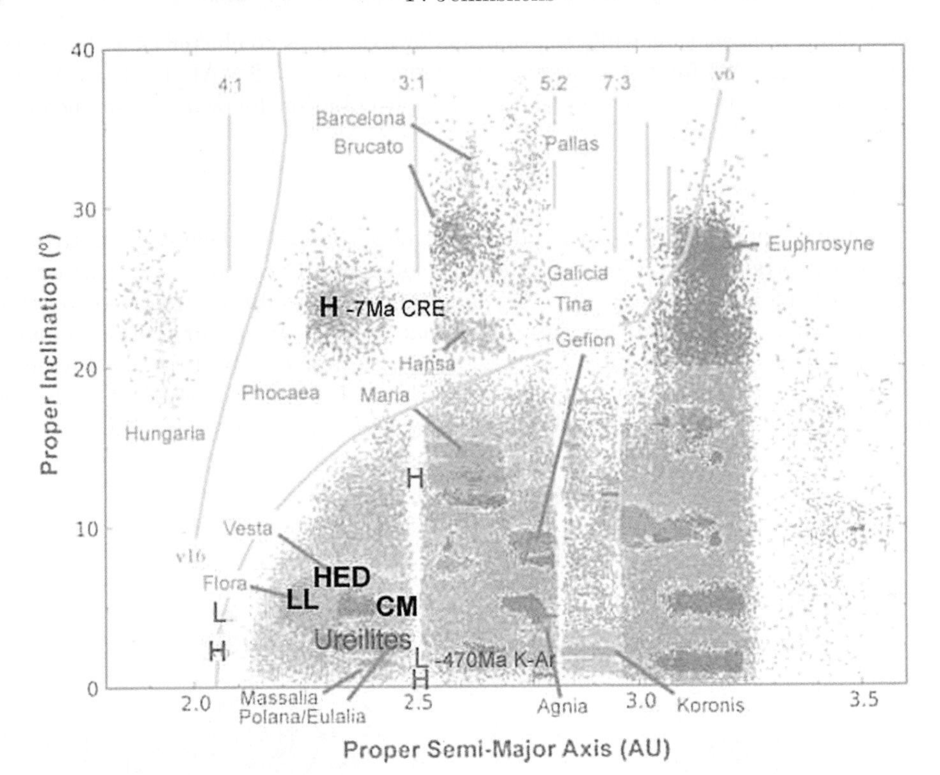

Figure 4. Possible source asteroids and families based on observed meteorite falls. Background image is based on Wise data (Masiero *et al.* 2013)

spectra of Polana-Eulalia B-types (de Leon *et al.* 2018) resemble that of asteroid 2008 TC3 (Jenniskens *et al.* 2010).

Recent work by Granvik & Brown 2018 provides more nuanced insight into the most likely delivery resonance for the first 25 observed meteorite falls. Based on the tabulated results for the high-inclined H chondrites, they suggest these come to Earth from the Hungaria family, or from an inner main belt source via the ν_6 resonance. However, the Hungaria family is poor in S class asteroids. Further dynamical calculations may show that they originated from the Phocaea family via the 4:1 and 3:1 resonances, instead.

References

de Len, J., Campins, H., Morate, D., *et al.* 2018, *Icarus*, 313, 25

Dermott, S. F., Apostolos, C. A., Li, D., *et al.* 2018, *Nature Astronomy*, 2, 549

Devillepoix H. A. R., Sansom, E. K., Bland, P. A., *et al.* 2018, *MAPS*, (in press)

Eugster, O., Herzog, G. F., & Marti, K. 2006, in: D. S. Lauretta & H. Y. McSeen Jr. (eds.), *Meteorites and the Early Solar System II* (Tucson: University of Arizona Press), p. 829

Granvik, M. & Brown, P. 2018, *Icarus*, 311, 271

Jenniskens, P., Vaubaillon, J., Binzel, R. P., *et al.* 2010, *MAPS*, 45, 1590

Jenniskens, P., Fries, M., Yin Q.-Z., *et al.* 2012, *Science*, 338, 1583

Jenniskens, P. 2013, in: T. J. Jopek, F. J. M. Rietmeijer, J. Watanabe & I. P. Williams (eds.), *Meteoroids 2013* (Poznan, Poland: Adam Mickiewicz University Press), p. 57

Jenniskens, P., Utas, J., Yin, Q.-Y., *et al.* 2018, *MAPS*, 54, 699

Masiero, J. R., Mainzer, A. K., Bauer, J. M., *et al.* 2013, *ApJ*, 770, 7

Nesvorny, D., Vokrouhlicky, D., Morbidelli, A., & Bottke, W. F. 2009, *Icarus*, 200, 698

Popova, P, Jenniskens, P., Emel'yanenko, V., *et al.* 2013, *Science*, 342, 1069

Astronomy in Focus - XXX
Proceedings IAU Symposium No. XXX, 2018
M. T. Lago, ed.

doi:10.1017/S1743921319003247

Numerical Simulations of Catastrophic Impacts Resolving Shapes of Remnants

Keisuke Sugiura, Hiroshi Kobayashi and Shu-ichiro Inutsuka

Graduate School of Science, Nagoya University,
464-8602, Furo-cho, Chikusa-ku, Nagoya, Aichi, Japan
email: sugiura.keisuke@a.mbox.nagoya-u.ac.jp

Asteroids have a variety of shapes that are probably formed through asteroidal collisions. Recent high-velocity impacts mainly result in catastrophic disruption and also the formation of asteroid families. Therefore, if we clarify asteroidal shapes formed through catastrophic disruption and compare them with those of family asteroids, we may extract valuable information on the past collisional events.

We conduct numerical simulations of catastrophic impacts using Smoothed Particle Elastic Dynamics (SPED: Sugiura & Inutsuka 2017) method with the self-gravity and the models of fracture of rock (Benz & Asphaug 1994) and friction of damaged rock (Jutzi 2015). We use basaltic spheres as impacting asteroids. The radius of target asteroids, the friction angle of damaged rock, the total number of SPED particles for each simulation, and the impact angle are set to 50 km, 40°, 4 million, and 15°, respectively. We conduct the three simulations with $v_{\mathrm{imp}} = 350\,\mathrm{m/s}$ and $M_i/M_t = 1$, $v_{\mathrm{imp}} = 700\,\mathrm{m/s}$ and $M_i/M_t = 1/4$, and $v_{\mathrm{imp}} = 1.7\,\mathrm{km/s}$ and $M_i/M_t = 1/16$, where v_{imp} is the impact velocity, M_t and M_i show the mass of targets and impactors, respectively.

We select the collisional remnants composed of more than 5,000 SPED particles resulting from the simulations, and analyze their shapes. As a result, we find that the remnants produced through catastrophic impacts mainly have spherical and bilobed shapes (Fig. 1). Spherical remnants are firstly formed through reaccumulation of fragments, and then coalescence of two spherical remnants produces bilobed remnants. However, flat shapes are difficult to form (hereafter, "flat" means the shapes with the ratio of the minor to major axis lengths less than 0.6 and the ratio of the intermediate to major axis lengths larger than 0.8): There are only two flat remnants among 106 remnants produced in the simulations.

We also analyze the shape models of asteroids with diameters larger than 10 km obtained from DAMIT database (Ďurech *et al.* 2010), and compare the shape distribution of non-family asteroids with that of asteroids belonging to families produced through catastrophic disruption. Here, we use AstDyS-2 database (http://hamilton.dm.unipi.it/astdys/) to distinguish between family and non-family asteroids. For non-family asteroids, there are 49 flat asteroids among 573 asteroids; the fraction of flat shapes is 8.6%. For family asteroids, there are 9 flat asteroids among 237 asteroids; the fraction of flat shapes is 3.8%. Thus the fraction of flat shapes for non-family asteroids is more than twice as large as that for family asteroids. This implies that catastrophic disruption is difficult to produce flat asteroids and supports the results of our simulations.

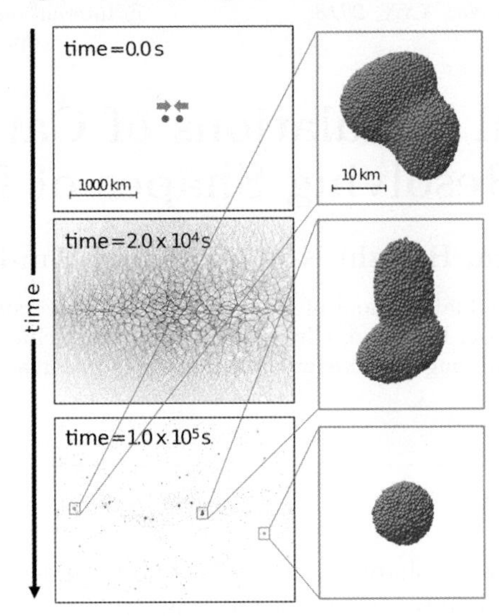

Figure 1. Snapshots of the impact simulation with $v_{\rm imp} = 350\,{\rm m/s}$ and $M_i/M_t = 1$.

References

Benz, W. & Asphaug, E. 1994, *Icarus*, 107, 198
Ďurech, J., Sidorin, V., & Kaasalainen, M. 2010, *Astron. & Astrop.*, 513, 46
Jutzi, M. 2015, *Plan. & Space Sci.*, 107, 3
Sugiura, K. & Inutsuka, S. 2017, *Jour. of Comp. Phys.*, 333, 78

Astronomy in Focus - XXX
Proceedings IAU Symposium No. XXX, 2018
M. T. Lago, ed.

doi:10.1017/S1743921319003259

YORP equilibria: ways out of YORP cycles

Oleksiy Golubov[1,2,3], Daniel J. Scheeres[1] and Yurij N. Krugly[3]

[1]Department of Aerospace Engineering Sciences, University of Colorado at Boulder, 429 UCB, Boulder, CO, 80309, USA

[2]V. N. Karazin Kharkiv National University, 4 Svobody Sq., Kharkiv, 61022, Ukraine
email: oleksiy.golubov@karazin.ua

[3]Institute of Astronomy of V. N. Karazin Kharkiv National University, 35 Sumska Str., Kharkiv, 61022, Ukraine

Abstract. Here, we discuss the YORP equilibria, to which asteroids can come as the result of their evolution due to the YORP effect.

The evolution of small asteroids is governed by the YORP effect, which can include several different varieties: normal YORP (NYORP), tangential YORP (TYORP), and binary YORP (BYORP) (Vokrouhlický *et al.*, 2015). The conventional wisdom says that asteroids evolve according to YORP cycles: an asteroid is accelerated by the YORP effect to the disruption limit, forms a binary, then the binary decays, and the asteroid starts a new YORP cycle, with a decreased mass, an altered shape and a small angular momentum.

Still, there are several mechanisms, by which YORP cycles can be interrupted and YORP evolution of asteroids stopped:

(*a*) Equilibrium between TYORP and NYORP, which can cancel each other at a certain rotation rate (Golubov & Krugly, 2012; Golubov *et al.*, 2016).

(*b*) For a singly-synchronous binary, equilibrium between TYORP, NYORP and tides acting on the primary, and between BYORP and tides acting on the secondary (Golubov *et al.*, 2018).

(*c*) For a doubly-synchronous binary, equilibrium between NYORP and BYORP, which are equal at a certain distance between the primary and the secondary (Golubov & Scheeres, 2016).

For each of the three kinds of equilibria, their probabilities are of the order of 10 percent, implying that each asteroid should be caught into an equilibrium after several YORP cycles. Stable asteroids survive, while unstable ones are disrupted and recreated in YORP cycles, until they perhaps become stable, in a process similar to natural selection. Thus the YORP effect can check itself and minimize its own importance for asteroids evolution.

References

Golubov O. & Krugly Yu. N. 2012, *ApJL*, 752, 11

Golubov O., Lipatova V., & Scheeres D. J. 2016, Modelling evolution of asteroid's rotation due to the YORP effect. In AAS/Division of Dynamical Astronomy Meeting, Vol. 47

Golubov O. & Scheeres D. J. 2016, *ApJL*, 833, L23

Golubov O., Unukovytch V., & Scheeres D. J. A. 2018, *ApJL*, 857, L5

Vokrouhlický, D., Bottke, W. F., Chesley, S. R., Scheeres, D. J., & Statler, T. S. 2015, Asteroids IV, 509

Astronomy in Focus - XXX
Proceedings IAU Symposium No. XXX, 2018
M. T. Lago, ed.

doi:10.1017/S1743921319003260

Dynamical evolution of asteroid pairs on close orbits

Eduard D. Kuznetsov†, Dmitrij V. Glamazda, Galina T. Kaiser, Vadim V. Krushinsky, Aleksandr A. Popov, Viktoriya S. Safronova, Andrej A. Shagabutdinov, Pavel V. Skripnichenko and Yuliya S. Vibe

Kourovka Astronomical Observatory, Ural Federal University,
Lenina Avenue, 51, Yekaterinburg, Russia 620000
email: eduard.kuznetsov@urfu.ru

Abstract. We apply Kholshevnikov metrics defined in the space of Keplerian orbits to search for asteroids in close orbits. We showed that the Yarkovsky effect was required to take into account accurately to carry out precise simulation of dynamical evolution of the asteroid pairs. Determination of physical and rotational parameters of asteroids is needed to solve this problem.

We apply natural metrics (Kholshevnikov metrics) defined in the space of Keplerian orbits to search for asteroids in close orbits. We use as a metric ϱ_2 the distance between two orbits in the five-dimensional space of Keplerian orbits.

We have used 582 120 orbits of asteroids (using both numbered and multi-opposition objects – 22.06.2018 release) from the Asteroids Dynamic Site – AstDyS. There are 105 asteroid pairs that have $\varrho_2 < 0.002$ (au)$^{1/2}$ ($\varrho_2^2 < 600$ km). Of these, 33 asteroid pairs were identified within known families of asteroids using AstDyS:

16 pairs are members of fragmentation asteroid families: (158) Koronis, (434) Hungaria;

10 pairs are members of cratering asteroid families: (4) Vesta, (15) Eunomia, (20) Massalia;

2 pairs are members of young asteroid families: (396) Aeolia, (1547) Nele;

1 pair is member of one-sided asteroid family: (93) Minerva;

4 pairs are members of unclassified asteroid families: (135) Hertha, (298) Baptistina, (1338) Duponta.

There are 72 asteroid pairs that are not identified with known families of asteroids.

We consider orbital evolution of two the tightest pairs: (63440) 2001 MD30 and (331933) 2004 TV14, and (355258) 2007 LY4 and (404118) 2013 AF40. To carry out a high accuracy numerical simulation it is necessary to take the Yarkovsky effect into account. In total, we used 7 test orbits for each of the two paired orbits that were assigned different values of the secular semimajor axis drift rate da/dt. The test values were chosen as $da/dt = 0$, $\pm 10^{-5}$, $\pm 10^{-4}$, $\pm 10^{-3}$ au/Myr. We numerically integrated the orbits of these pairs backward in time (a time span of 20 kyr) with the code known as Orbit9.

We showed that the Yarkovsky effect was required to accurately carry out precise simulations of dynamical evolution of the asteroid pairs. Determination of physical and rotational parameters of asteroids is needed to solve this problem.

† The reported study was funded by RFBR according to the research project no. 18-02-00015.

Astronomy in Focus - XXX
Proceedings IAU Symposium No. XXX, 2018
M. T. Lago, ed.

doi:10.1017/S1743921319003272

Mapping of the asteroid families

Nataša Todorović

Belgrade Astronomical Observatory
Volgina 7, P.O.Box 74 11060 Belgrade, Serbia
email: ntodorovic@aob.rs

Abstract. In this work we map one of the most populated regions in the main belt - the asteroid family Themis. Computed with a good choice of parameters, the map enables us to get a refined picture of the dynamics in the family, to reexamine the role of resonances therein, to understand better the distribution of asteroids inside the region and to identify dynamical pathways along which particles can drift away.

1. Map of the Themis asteroid family

Asteroid family Themis, discovered one century ago by Hirayama (1918), is one of the most populous (close to 6000 identified members) and most studied asteroid families. Themis family is located at low eccentricity $(0.11 < e < 0.184)$ and low inclination $(i < 3)$ in the semi-major axis range $a \in [3.14, 3.22]$ AU, on the outer edge of the main belt. It is bordered by the strong 2A:1J mean motion resonance with Jupiter and intersected by numerous weaker resonances, which all have some contribution in sculpting the family.

For a better understanding of the evolution of its members, it is very helpful to observe dynamical structures in the region with good precision. Therefore, we compute clear stability maps, that are produced using a realistic Solar System model, a high precision integrator - ORBIT9 and a sensitive short term numerical tool for chaos detection - FLI (Froeschlé *et al.* (1997)).

We map the family in the (a, e) plane in the domain that slightly exceeds the boundaries of the family i.e. for $[a \times e] = [3, 3.25] AU \times [0.11, 0.2]$. We divide this segment in a grid of equidistant points and for each point on the grid we calculate the corresponding FLI value for 200 Kyrs. The map is colorized according to FLI values, in a sense that larger chaoticity implies larger FLIs. The remaining four orbital elements: inclination i, longitude of the node Ω, argument of the pericenter ω, and mean anomaly M are fixed to $(i, \Omega, \omega, M) \sim (0.75, 36, 107, 280)$ which are the corresponding angles for the asteroid 24 Themis, the parent body and largest member of the family. In all calculations, we use osculating orbital elements for the epoch 2456200.5 MJD, which is another argument to have a realistic situation on the map. In this way we obtain exact structures of the resonances in the region, their separatrix borderlines and stability domains. This will contribute to an easier identification of places with higher diffusion abilities, i.e. to localize those regions along which particles drift apart in an efficient way.

References

Froeschlé, Cl., Lega, E., & Gonczi, R. 1997, *CeMDA*, 67, 41
Hirayama, K. 1918, *ApJ*, 31, 743

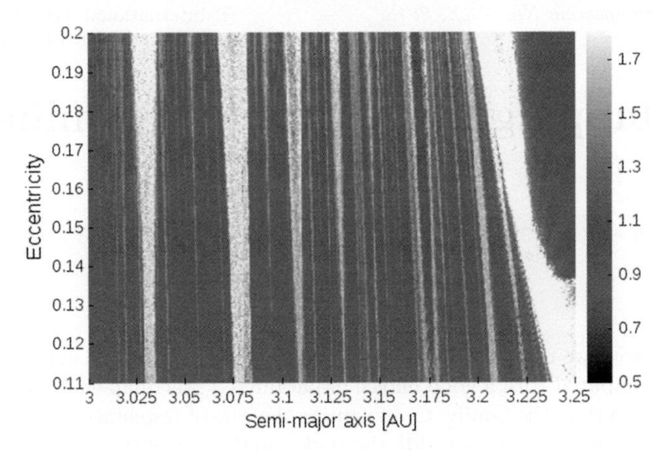

Figure 1. FLI map of the Themis family region.

Astronomy in Focus - XXX
Proceedings IAU Symposium No. XXX, 2018
M. T. Lago, ed.

doi:10.1017/S1743921319003284

Recent results in family identification

Anatolii Kazantsev

Astronomical observatory of Kyiv National Taras Shevchenko University
Observatorna Str. 3, Kyiv, Ukraine, 04053
email: ankaz@observ.univ.kiev.ua

Abstract. To select asteroid families, the D(a) distribution of asteroid sizes by their semimajor axes and the N(p) distribution of the number of asteroids by their albedo values for individual families were used. A statistically significant reduction in the mean albedo with increasing semimajor axis is observed for almost all correctly identified families that are not truncated by resonances. This points on an action of a specific nongravitational effect (NGE) in the asteroid belt, and results in the spatial separation of asteroids with different albedos.

---

1. Introduction

Usually the H(a) absolute magnitude distribution by proper semimajor axes are used to select families. The D(a) distribution gives clearer picture on the correctness of the family selection, since the D values are obtained with taking into account the albedo of separate bodies. The D(a) distribution for a family should have a central maximum with descending wings on either side. Each parent body belongs to a certain taxonomic type. Therefore, the albedos of all members of the family should not greatly differ from the albedo typical for this type. Thus the N(p) distribution for a family should have a single primary maximum with a certain scatter on either side. This distribution should not have a pair (or more) of such maxima.

2. The albedo distributions for separate families

The p(a) distribution of a family asteroid albedos by their semimajor axes right after the parent body disintegration should be close to a uniform one. Gravitational disturbances in the process of evolution of orbits cant affect this distribution qualitatively as well. So the p(a) distributions for separate families at present can show an influence of a NGE on the families. Of course, the families should be correctly identified and not truncated by resonances. There were selected 21 such families using the D(a) and N(p) distributions. The D and p values were taken from the WISE database (Masiero *et al.* 2011). The averaged linear p(a) dependences were plotted for these 21 families: p = b1 a + b0. A possible NGE influence exerted on the asteroids from a certain family is detected by the value and significance of b1 coefficient. If this coefficient differs from zero at a sufficiently high statistical significance, one may conclude that a NGE influence on the bodies of a family is actually manifested. The b1 coefficients are negative for 15 of 21 families. The b1 values are significant at a level of 2-sigma and higher for 8 of 15. There are no significant positive b1 values. The b1 value for 221 Eos family is significant at a level higher than 13-sigma. The accuracy of proper semimajor axes and albedos are similar for all families. Thus we may superpose the centers of all families and compare the average albedos of the left wing and the right one for that of the combined family. We renormalize the center of each family to be equal to 3.00 AU, while keeping unchanged the difference of proper semimajor axes of every family from the center. The combined family includes more than 7500 bodies of 20 families (without the 221 family). The b1 coefficient in the p(a) dependence for the family is equal -0.11 at a significance of higher than 18 sigma. Thus one may be sure that the average albedo decreases along the proper semimajor axes inside the asteroid families.

3. Conclusions

Using the asteroid sizes and albedos gives a possibility to identify the families more correctly. The average albedo decreases along the proper semimajor axes inside the asteroid families. This points on an action of a specific NGE in the asteroid belt results in the spatial separation of asteroids with different albedos.

References

Masiero, Joseph R., Mainzer, A. K., Grav, T., *et al.* 2011, *ApJ*, 741, 20

Astronomy in Focus - XXX
Proceedings IAU Symposium No. XXX, 2018
M. T. Lago, ed.

doi:10.1017/S1743921319003296

Are Large Families in The Main Belt Homogeneous?

Ivan G. Slyusarev

Department of Astronomy and Space Informatics,
V.N. Karazin Kharkiv National University,
61022, Kharkiv, Ukraine
email: i.slyusarev@karazin.ua

Abstract. Using color indexes from SDSS and albedos from WISE we tested the homogeneity of 56 large Main belt families from Nesvorny list using the "color - albedo" plots. 25% of the analyzed families are non-homogeneous in terms of albedos and colors. Only two families (Flora and Vesta) contain low, moderate and high albedo asteroids, that are separated in a "color-albedo" plot. The fraction of the low albedo asteroids in bimodal families is not negligible ($10 - 30\%$). Seven bimodal families may contain members from two overlapping families.

---

1. Introduction

Asteroid families were formed during collisional disruptions and their physical properties provide unique information about the internal material of the parent bodies. From analysis of data on physical properties of the family members we can also find possible interlopers in homogeneous families or we can distinguish overlapping families in the (ap,ip,ep) space. We perform an analysis of the physical homogeneity of 56 large Main belt asteroid families based on the color and albedo data. Main belt asteroid families were taken from Nesvorny (2015). We have considered only numbered asteroids in our analysis. We use WISE albedos from Mainzer *et al.* (2016), and a* which is the first principal component in the r-i versus g-r SDSS color-color plane (Parker *et al.* 2008) to plot albedo distributions and "albedo - color" diagram for each family.

2. Results

Using data of albedo (p) and color index (a*) from WISE and SDSS databases for large main belt asteroids families we find that all points on a* - albedo plots for all families can be separated in to three subgroups: I ($p < 0.1$; $a^* < -0.05$); II ($0.1 < p < 0.25$; $a^*<0.05$) and III ($p>0.15$; $a^*>0.05$). In all bimodal families with some exceptions are present the dark subgroup I and high albedo subgroup III. Only two families (Vesta and Flora) include all three subgroups. Analysis of taxonomic interpretations of these three subgroups gives a clear result only for subgroup III - these are S-type asteroids (also may include V and E-type). Subgroup I is a mixture of dark asteroids that belong to B, F, C, P, D types. Subgroup II probably can be consistent with the M-type asteroids. Presence of subgroup II is a characteristic only for the Flora region (2.2 a.u.) and Karin region (2.8 a.u.). Outside these two regions this subgroup is absent. Analyzed families are divided into homogeneous: (37 families); bimodal (13 families); and trimodal families (2 families Vesta and Flora). Families Aeolia, Xizang, Aeria and (15477) have not shown bimodality in their color and albedo distributions, but they contain asteroids that are intermediate between low (I) and middle albedo (II) subgroups. All these families are located very near to 2.7 A.U. More deep analysis of the distribution of proper elements for subgroups and V-shape plots for each bimodal family show that several families may consists of two overlapping families as in the case of Nysa-Polana. In the Flora family it is clearly seen that the core of the family is characterised by prominent V-shape which belong to the moderate albedo subgroup which are consistent with M-type objects.

3. Conclusions

A significant fraction (25%) of the analyzed families are inhomogeneous in terms of albedos and colours. A fraction of the dark subgroup (I) in bimodal families is not negligible ($10-30\%$). In seven bimodal families asteroids from different subgroups have a slightly different proper elements distributions.

References

Nesvorny, D. 2015, *Nesvorny HCM Asteroid Families V3.0.*, NASA Planetary Data System

Mainzer, A. K., Bauer, J. M., Cutri, R. M., Grav, T., Kramer, E. A., Masiero, J. R., Nugent, C. R., Sonnett, S. M., Stevenson, R. A., & Wright, E. L. 2016, *NEOWISE Diameters and Albedos V1.0*, NASA Planetary Data System

Parker, A. H., Ivezic, Z., Juric, M., Lupton, R., Sekora, M. D., & Kowalski, A. 2000, *Icarus*, 198, 138

Astronomy in Focus - XXX
Proceedings IAU Symposium No. XXX, 2018
M. T. Lago, ed.

doi:10.1017/S1743921319003302

The common origin of family and non-family asteroids: Implications

Stanley F. Dermott[1], Apostolos Christou[2] and Dan Li[3]

[1]University of Florida, Department of Astronomy, Gainesville, FL 32611,USA
email: sdermott@ufl.edu

[2]Armagh Observatory and Planetarium, College Hill, Armagh BT61 9DG, UK
email: apostolos.christou@armagh.ac.uk

[3]National Optical Astronomical Observatory, Tucson, AZ, USA
email: dli@noao.edu

Abstract. Because the number of asteroids in the IMB with absolute magnitude H<16.5 is effectively complete, the distributions of the sizes and the orbital elements of these asteroids must be devoid of observational selection effects. This allows us to state that the observed size-frequency distributions (SFDs) of the five major asteroid families in the IMB, defined by Nesvorný (2015) using the Hierarchical Clustering Method (Zappala *et al.* 1990), are distinctly different and deviate significantly from the linear log-log relation described by Dohnanyi (1969). The existence of these differences in the SFDs, and the fact that the precursor bodies of the major families have distinctly different eccentricities and inclinations, provides a simple explanation for the observations that the mean sizes of the family asteroids, *taken as a whole*, are correlated with their mean proper eccentricities and anti-correlated with their mean proper inclinations. While the latter observations do have a simple explanation, we observe that the mean sizes of the non-family asteroids in the IMB are also correlated with their mean proper eccentricities and anti-correlated with their mean proper inclinations. We deduce from this, and from the fact that the SFDs of the non-family and the family asteroids (again *taken as a whole)* are almost identical, that the family and most of the non-family and asteroids have a common origin. We estimate that ~85% of all the asteroids in the IMB with H<16.5 originate from the Flora, Vesta, Nysa, Polana and Eulalia families with the remaining ~15% originating from either the same families or, more likely, a few ghost families (Dermott *et al.* 2018).

More information is available online.

References

Dermott, S. F., Christou, A. A., Li, D., Kehoe, T. J. J., & Robinson, J. M. 2018, *Nat. Astron.*, 2, 549

Dohnanyi, J. S. 1969, *J. Geophys. R.* 74, 2531

Nesvorný, D. 2015, EAR-A-VARGBDET-5-NESVORNYFAM-V3.0. NASA Planetary Data System

Zappala, V., Cellino, A., Farinella, P., & Knežević, Z. 1990, *Astron. J.*, 100, 2030

Astronomy in Focus - XXX
Proceedings IAU Symposium No. XXX, 2018
M. T. Lago, ed.

doi:10.1017/S1743921319003314

Families among the Hildas and Trojans

Tamara A. Vinogradova

Institute of Applied Astronomy of Russian Academy of Science,
nab. Kutuzova 10, St. Petersburg 191187, Russia
email: vta@iaaras.ru

Abstract. A search for asteroid families among the Hildas and Jupiter Trojans was performed with the use of a new set of proper elements. The proper elements were calculated by the empirical method. Besides well known families, several new probable families were found in addition.

---

1. The Hildas and Trojans resonant regions

A search for asteroid families among the Hildas and Trojans by analytical methods is especially difficult because these asteroids move in resonance regions. The Hildas are in the 3:2 and Trojans in the 1:1 mean motion orbital resonances with Jupiter. To find asteroid families, proper elements should be calculated. A concept of proper elements has been introduced by Hirayama (Hirayama 1918). There are different methods of proper elements calculation: analytical, numerical, and empirical which uses distributions of orbital elements.

2. Hirayama (1918) work

What method did Hirayama use? Hirayama selected some condensations in distributions of osculating elements. For each of the selected groups he studied its distribution in the $(\tan i \cos \Omega, \tan i \sin \Omega)$ and $(e \cos \varpi, e \sin \varpi)$ planes. Here i is inclination, Ω - longitude of ascending node, e - eccentricity, ϖ - longitude of perihelion. Hirayama noticed that these distributions have the form of a circle. For each of selected group he found visually the position of a center of a circle. This point corresponds to forced elements. A distance of a dot from the center is a proper element (i_p or e_p). Also he applied the secular theory and defined that such distribution is due to secular perturbations by Jupiter. So, Hirayama used two methods: first empirical, then analytical.

3. The Empirical method

After Hirayama's work the empirical method was not used. The new empirical method described in (Vinogradova (2015)) allows us to calculate forced elements. If forced elements are known, proper ones can be calculated with the use of a coordinate transformation formula. The method is simple and not time-consuming. It allows us to use all available asteroids for a family identification. The MPC catalogue, version May. 2018, was used as a source of initial osculating elements.

4. Asteroid families in the Hilda-group

The number of multi-opposition Hildas now exceeds 3000. Two forced elements derived for the Hildas, $i_f = 1.20^o \pm 0.05^o$ and $\Omega_f = 99^o \pm 1^o$, are in a good agreement with results of the secular theory (Brouwer & van Woerkom (1950)), but two other elements, $e_f = 0.069 \pm 0.002$ and $\varpi_f = 20^o \pm 2^o$, differ significantly from these results (0.043 and 12°, respectively). One of the features of the motion in these resonant regions is the libration of the semimajor axis. It was accepted to use a value $a_p = 3.97au + da$ as a proper semimajor

Table 1. Families among the Hildas and Jupiter Trojans.

Region	Name	FIN	Diam (km)	N	d_{cut} 10^{-4}	Tax	p_v	Δa_p (au)	Δe_p	Δi_p (deg)
Hilda-group	1911 Schubart	002	80+66	658	30-110	CX-7:1(36)	0.04	3.98-4.04	0.15-0.23	2.6- 3.2
	153 Hilda	001	167+100	433	90-200	CX-6:2(59)	0.06	3.98-4.03	0.11-0.25	7.8-10.4
	1212 Francette	-	83+ 21	42	110-210	P -(1)	0.06	3.98-4.01	0.22-0.24	6.9- 7.6
	51874 2001 PZ28	-	13+ 28	52	130-250	DS-8:1(6)	-	3.98-4.02	0.20-0.25	10.3-11.8
	2483 Guinevere	-	43+ 38	81	140-230	DC-5:2(4)	0.07	3.98-4.03	0.20-0.27	4.7-6.3
	4757 Liselotte	-	18+ 24	17	180-230	-	-	3.99-4.02	0.14-0.16	1.1-1.8
	5661 Hildebrand	-	38+ 28	29	220-370	-	-	3.98-4.01	0.19-0.23	13.7-15.0
L4-Trojans	3548 Eurybates	005	66+ 80	317	50-120	CX-5:4(10)	0.06	5.28-5.33	0.03-0.07	7.1- 7.8
	2148 Epeios	008	39+ 55	104	100-160	-	-	5.23-5.33	0.02-0.05	8.6-9.2
	624 Hektor	004	207+121	100	110-290	DL-9:1(11)	0.05	5.23-5.35	0.03-0.09	17.8-19.4
	9799 1996 RJ	006	65+ 30	19	140-390	-	0.04	5.23-5.24	0.03-0.05	31.4-31.8
	9713 Oceax	-	31+ 66	122	160-210	DX-5:5(2)	-	5.25-5.38	0.02-0.05	3.4-5.0
	2797 Teucer	-	112+113	41	160-280	D (2)	0.07	5.23-5.32	0.06-0.08	20.3-21.5
	1583 Antilochus	-	117+ 79	41	260-490	D (2)	0.09	5.23-5.33	0.01-0.08	28.3-29.5
L5-Trojans	1172 Aneas	-	137+109	55	190-300	DC-7:1(9)	0.05	5.22-5.33	0.03-0.06	16.6-18.8
	1867 Deiphobus	009	128+ 91	139	200-270	DC-7:1(6)	0.06	5.23-5.34	0.02-0.05	26.8-31.2
	11487 1988 RG10	-	28+ 48	37	210-270	DX-7:1(7)	-	5.24-5.37	0.02-0.04	3.8-4.8
	37519 Amphios	010	35+ 27	14	210-270	-	-	5.22-5.23	0.04-0.05	24.4-25.1

Notes: Diam - diameter of the main asteroid + diameter of debris; N - number of members; d_{cut} - dimensionless d_{cut}-distance; Tax - shares of taxonomical types and a number of members with known taxonomy (in brackets); p_v - albedo; Δa_p, Δe_p, Δi_p - intervals of proper elements.

axis for the Hildas (*da* is the libration amplitude). We adopt an approach similar to the hierarchical clustering method for identification of asteroid families. As a result, two large families were found here: (1911) Schubart and (153) Hilda. Both these families were found earlier by other authors (Schubart (1982)), (Brož & Vokrouhlický (2008)). In Asteroids IV these families were assigned FIN - family identification number (Nesvorný, Brož, Carruba (2015)). In addition, our new set of proper elements enables five probable families to be identified (see Tab.1).

5. Asteroid families in the Trojans

About 5000 multi-opposition asteroids are known in the Trojans now (3300 in L4 + 1700 in L5). Calculating the proper elements of the Trojans is easy, because orbital elements of Jupiter can be used as forced elements: $i_f = 1.3^o$, $\Omega_f = 100.5^o$, $e_f = 0.049$, $\varpi_{jup} = 14.2^o$. For L4-Trojans, $\varpi_f = \varpi_{jup} + 60^o = 74.2^o$. For L5-Trojans, $\varpi_f = \varpi_{jup} - 60^o = 314.2^o$. A proper semimajor axis $a_p = 5.20 + da$, where da is a libration amplitude. Large discrepancies take place in the Trojans family lists obtained by different authors. The number of families found differs from one (Brož & Rozehnal 2011) to about 20 (Beaugé & Roig 2001). With our new set of proper elements seven families were found among the L4-Trojans and four among the L5-Trojans. All additional families among the Trojans were previously found by different authors, but at the same time a large number of families published earlier were not confirmed.

References

Beaugé C. & Roig F. 2001, *Icarus* 153, 391
Brouwer D. & van Woerkom A. J. J. 1950, *Astron. Papers Amer. Ephem.*
Brož M. & Rozehnal J. 2011, *MNRAS*, 414, 565
Brož M. & Vokrouhlický D. 2008, *MNRAS*, 390, 715
Hirayama K. 1918, *AJ*, 31, 185
Nesvorný D., Brož M., & Carruba V. 2015, *Asteroid IV* chapter
Schubart J. 1982, *Cel. Mech. Dynam. Astron.* 28, 189
Vinogradova T. A. 2015, *MNRAS*, 454, 2436

Astronomy in Focus - XXX
Proceedings IAU Symposium No. XXX, 2018
M. T. Lago, ed.

doi:10.1017/S1743921319003326

The Near Earth Asteroid associations

Tadeusz J. Jopek

Institute Astronomical Observatory, Faculty of Physics, A. M. University, Poznań, Poland
email: jopek@amu.edu.pl

Abstract. We searched for associations (not for families) amongst the near Earth asteroids (NEAs) and, similarly as in our previous studies (Jopek 2011; Jopek 2015), a dozen groups of 10 or more members was found with high statistical reliability. We present some details of our most numerous finding: association (2061) Anza which, at the moment, incorporates 191 members.

---

1. Data, method, result, implication

Applying the cluster analysis technique we made an extensive search amongst the 18291 NEA's retrieved in June 2018 from the NEODyS-2 database (http://newton.dm.unipi.it/neodys/). The most numerous finding was the (2061) Anza association which consists of 191 members. The probability that such a grouping is a statistical fluctuation is less than 1%. In Figure 1 the orbits of the Anza association are illustrated. In Table 1 the designations of its selected members are given. To quantify the orbital similarity we used D_H function (see Jopek 1993) and the similarity threshold was determined by a statistical approach. Regardless of their origin, existence of such associations increase the NEA collisional probability with the Earth. Analogously to meteoroid streams, each year the Earth almost crosses the orbits of such associations. This implies that searching for grouping amongst the NEAs is an important issue.

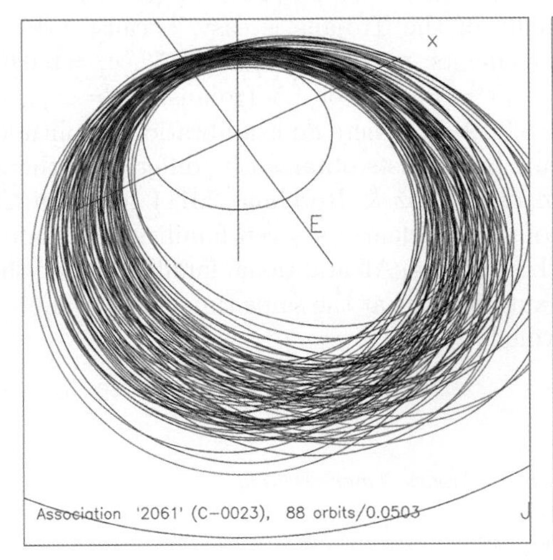

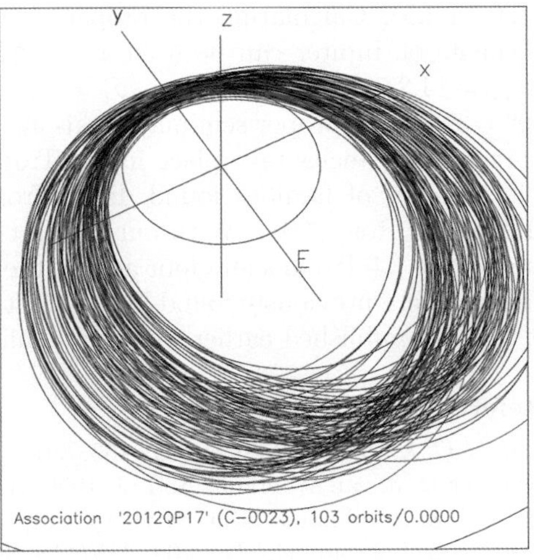

Figure 1. Association (2061) Anza. 191 NEAs of this group were identified: before 2012, 88 orbit on the left panel, and after 2012, 103 orbits on the right. As this example shows, we do not know really how many members are in the Anza association.

Table 1. Designations of 27 selected members of (2061) Anza association.

2061	52760	100085	100085	162183	222008	354030	363831	446826
455322	477719	481989	1991XA	1993UA	1998SS4	1999RB32	1999RK33	1999TM13
2013PD39	2013RO21	2014SE145	2014ST1	2015MT96	2016EU85	2016PS26	2017QW32	2017WT1

Acknowledgements

TJJ work was supported by 2016/21/B/ST9/01479 grant of the National Science Centre in Poland. We used NASA's ADS Bibliographic Services.

References

Jopek, T. J., 1993, *Icarus*, 106, 603
Jopek, T. J. 2011, *EPSC-DPS Joint Meeting 2011*, 2-7 Oct. 2011 in Nantes, France. Vol. 6, 15
Jopek, T. J. 2015, *Highlights of Astronomy*, Volume 16, 474

Astronomy in Focus - XXX
Proceedings IAU Symposium No. XXX, 2018
M. T. Lago, ed.

doi:10.1017/S1743921319003338

Trojans distribution in the Solar system

Anatolii Kazantsev and Lilia Kazantseva

Astronomical observatory of Kyiv National Taras Shevchenko University,
Observatorna Str. 3, Kyiv 04053, Ukraine
email: ankaz@observ.univ.kiev.ua

Abstract. Orbits of potential Trojans of different planets in the solar system were selected from the MPC catalog on February, 2017. The evolution of those orbits was calculated. The bodies on librating orbits around the points L4 and L5 were determined. The quantities of real Trojans in the MPC catalog are as follows: Mars - 5, Jupiter - over 4500, Saturn - none, Uranus - 2, Neptune - 15. A reasoned explanation of such distribution of Trojans in the solar system is proposed.

---

1. Introduction

A decade ago a specific non-gravitational effect (NGE) in the asteroid belt was revealed (Kazantsev, 2007). More detail description of the NGE is presented in (Kazantsev & Kazantseva, 2017). This NGE causes an increase in the semimajor axes of orbits of low-albedo asteroids relative to the semimajor axes of orbits of high-albedo bodies. The NGE existence is confirmed by the distributions of albedo values in separate asteroid families. The physical mechanism behind this NGE should be close in nature to the NGE seen in comets. It is possible to propose a reasoned explanation for the origin of Trojans in the Solar system using the NGE.

2. Origin of Planet Trojans under the NGE action

Planetary perturbations can move a small body from the inner or outer zone to become Trojans. But the initial orbit eccentricity of the body should be pretty high. Therefore such body cant remain in the 1:1 commensurability with the planet for a long time. We carried out numerical calculations of asteroid orbital evolution taking into account not only planet perturbations but also a model NGE. The model NGE in every step of integration gives the asteroid an additional impulse either in the direction of its orbital velocity, or in the opposite direction. In the first case the NGE causes an increase in the semimajor axis, in the latter case decrease in the semimajor axis. During the evolution under the NGE influence, the semimajor axis increases or decreases in response to small changes of the eccentricity. Therefore such orbit can remains in the commensurability 1:1 with the planet for a long time. If the NGE is several times more powerful than for Jupiter family comets (JFC), an asteroid from the outer edge of the asteroid belt can move into the 1:1 resonance with Jupiter. The bodies that do not fall into 1:1 resonance with Jupiter, behind its orbit pass into orbits typical for Centaurs. They will never be able to become stable Trojans of other planets. Neptune Trojans can pass from the inner edge of Kuiper belt under the NGE action. In that case the NGE power may be less than for JFC. The semimajor axes of orbits that do not fall into 1:1 resonance with Neptune continue to decrease smoothly at small eccentricities. Some bodies on such orbits can become Uranus Trojans. It is clear the quantity of such bodies should be much less than

Neptune Trojans. And sizes of Uranus Trojans should be noticeable smaller than sizes of Neptune Trojans. It is difficult to say if some bodies from Kuiper belt can become Saturn Trojans. In any case, such bodies should be very small (if the bodies exist generally). This conclusion corresponds to the available data as well. Origin of Mars Trojans can be explained by transition of bodies from the inner edge of MBA under the NGE action. But Mars cant hold at its Lagrange points these bodies, which undergo the NGE action like for JFC. If the NGE is three orders of magnitude lesser than for JFC, the bodies have a possibility to become Mars Trojans.

3. Conclusions

The origin of the planetary Trojans in the Solar system can be explained by the NGE action. There exists a simple method to verify the presented explanation: 1) Jupiter Trojans should mainly have prograde rotation; 2) Mars Trojans should mostly have retrograde rotation.

References

Kazantsev, A. 2007, *Kinem. Phys. Celest. Bodies*, 23, 258
Kazantsev, A. & Kazantseva, L. 2017, *Sol. Syst. Res.*, 51, 527

Astronomy in Focus - XXX
Proceedings IAU Symposium No. XXX, 2018
M. T. Lago, ed.

doi:10.1017/S174392131900334X

Studies of physical parameters of kilometer sized NEA by the RTT-150 telescope.

Irek M. Khamitov[1,2], Rustem I. Gumerov[2,3], Ilfan F. Bikmaev[2,3], Selcuk Helhel [4], Eldar N. Irtuganov[2,3], Sergey S. Melnikov[2,3], Gizem Okuyan[1,4] and Oğuzhan Okuyan[1,4]

[1]TÜBİTAK National Observatory, Antalya, Turkey
[2]Kazan Federal University, Kazan, Russia
[3]Academy of Sciences of Tatarstan, Kazan, Russia
[4]Akdeniz University, Antalya, Turkey
email: `irek_khamitov@hotmail.com`

We are carrying out a program of experimental studies of an as-large-as-possible sample of kilometer sized NEAs with the 1.5m Russian-Turkish telescope RTT150. We present the results of polarimetric and photometric observations in V bandpass of 32 NEAs (23 objects with diameter larger than 1 km and 9 of them only a few hundred meters in diameter) with high proper motion near their close approaches performed at RTT150 between the August 2014 and June 2018 (regular observations had begun only in the spring 2018). The estimation of albedos and diameters of asteroids were made by means of polarimetric and photometric data. Because sources were observed at large phase angles from 40° to 100° for different asteroids, the measured linear polarization degree makes it possible to differentiate the taxonomy classes of observed sources even when using single shot observations. The highest polarization 27.1% was found for (164201) at a phase angle of ∼75°. Taking into account the high inclination of the object (∼35°), (164201) has a possibility to be a cometary nuclei. All objects smaller than 1 km were eliminated from analysis and the remaining 23 NEAs were separated into two groups according to the obtained albedo: 0.10$<$MODERATE (55%)$<$ 0.35, LOW (40%)$<$0.10. For each group the probability density function of NEA perihelion was built as a fraction of asteroid number inside of given interval of perihelion parameter (Δq=0.3). It seems to be that there is increase of moderate albedo NEAs with perihelion parameter. However, this may be due to the incompleteness of sample. In addition to polarimetric determination, the spectral class of NEA (333888) was estimated as Sq using the low resolution spectra (R∼600, covered whole visual range from 4000Å to 9000Å) obtained with RTT150 facilities. Complex observations in the frame of this project will be continued. To avoid the loss of darker kilometer sized NEAs, the object observational list will be broadened with 416 more NEAs which meet the conditions: $q<0.6$, $18.0^m < H < 19.5^m$. To estimate photometric slope (G) and shape of the target NEAs, additional long-term photometric observations in V-band will be performed at phase angles of 10° – 60° on T100 and T60 telescopes of the TUG Observatory.

Acknowledgement

This work was supported by the Russian Foundation for Basic Research (project no. 18-02-00105 A) and funded by TÜBİTAK (Grand No. 113F263). Authors thank TÜBİTAK, KFU, AST, and IKI for partial support in using RTT150 (the Russian-Turkish 1.5-m telescope in Antalya). This work was partially funded by the subsidy 3.6714.2017/8.9 allocated to KFU for the state assignment in the sphere of scientific activities.

Astronomy in Focus - XXX
Proceedings IAU Symposium No. XXX, 2018
M. T. Lago, ed.

doi:10.1017/S1743921319003351

Photometry and spin rates of 4 NEAs recently observed by the Mexican Asteroid Photometry Campaign

J. C. Saucedo[1], S. A. Ayala-Gómez[2], M. E. Contreras[1], S. A. R. Haro-Corzo[3], P. A. Loera-González[1], L. Olguín[1] and P. A. Valdés-Sada[4]

[1] Departamento de Investigación en Física, Universidad de Sonora
Blvd. Rosales y Colosio, Ed. 3H, 83190 Hermosillo, Sonora, México
email: jsaucedo@cifus.uson.mx

[2]FCFM–Universidad Autónoma de Nuevo León

[3]ENES–Universidad Nacional Autónoma de México

[4]DFM–Universidad de Monterrey

Abstract. We present photometric observations of (4055) Magellan, (143404) 2003 BD_{44}, 2014 JO_{25} and (3122) Florence, four potentially hazardous Near Earth Asteroids (NEAs). The data were taken near their approaches to Earth by 3 observatories participating in the Mexican Asteroid Photometry Campaign (CMFA). The results obtained: light curves, spin rates, amplitudes and errors, are in general agreement with those obtained by others. During the day of a NEAs maximum approach to our planet, its light curve may present significant changes. In the spin rate, however, only minute changes are observed. 2014 JO_{25} is briefly discussed in this regard.

---

1. Introduction

NEAs have perihelion distances $\leqslant$ 1.3 AU, and if their Earth Minimum Orbit Intersection Distance $\leqslant$0.05 AU and their size are $\geqslant$140 m, they are considered "potentially hazardous". According to their orbits NEAs are divided into 4 classes: the Amors, the Apollos, the Atens and the Atiras/Apohele. Over 18,500 NEAs have been discovered so far and the number keeps growing at a rate unmatched to that at which they are characterized. One way of characterizing asteroids is by obtaining and analyzing their photometric light curves, from which physical properties such as the spin rate can be obtained. According to the Asteroid Lightcurve Photometry Database (ALCDEF) spin rates are known for ~6000 minor planets, roughly representing ~7% of NEAs, and <1% of the general asteroid known population (over 750,000). With this in mind, in 2015 a network was implemented to carry out photometric observations of asteroids: the CMFA (Campaña Mexicana de Fotometría de Asteroides). The CMFA has published spin rates for several main belt asteroids, see Schuster *et al.* in this IAU proceedings. Here we concentrate on the results for NEAs of the CMFA. The spin rate is related to an asteroids cohesion strength. In addition, 3D asteroid shapes can be inferred from light curves obtained at different epochs, since objects are observed from different view angles (Durech *et al.* (2010)). Finally, images used to generate light curves are also used to derive astrometric information that allow us to improve asteroid's orbital parameters.

2. Data and reductions

Consecutive images of the asteroid are taken throughout the night and usually during several nights; to obtain a complete light curve, typically requires from 5 to 10 observing nights for each asteroid. Due to instability issues of telescope mounts, exposure times for individual images are in the range of 30 to 240 seconds. This restriction imposes a severe limitation for the follow-up of interesting faint objects. Filters are not used, to facilitate the observation of fainter objects and to compare and combine the results of the 3 participating observatories: 0.84 m telescope at the Observatorio Astronómico Nacional at Sierra San Pedro Martir (OAN-SPM); 0.40 m telescope at Observatorio Astronómico Carl Sagan (OACS) located in Hermosillo and 0.36 m telescope of the Universidad de Monterrey Astronomical Observatory, Monterrey.

Basic steps of data reduction are performed using IRAF or MaximDL4 software. For light curve extractions and period determinations, MPO Canopus software is used. For crowded stellar fields, first DAOPHOT (Stetson 1987) is used to obtain the light curve and then MPO Canopus is used for the spin rate determination.

Basic information on the 4 NEAs reported here: (4055) Magellan (named in honor of Fernando de Magallanes) is an amor NEA discovered by Eleanor Francis Helin on February 24 1985 at Palomar Observatory. Observations by the Keck Observatory and the NEOWISE mission permit to estimate a diameter of 2.2 - 2.8 km. (143404) 2003 BD_{44} is an Apollo-type NEA discovered by Brian Skiff at Lowell Observatory, Anderson Mesa Station (LONEOS survey). It is considered to be potentially hazardous. Its diameter is estimated to be $\sim$1.3 km. On April 2017 it came within 0.1 AU of the Earth. Radar observations at Arecibo Observatory were carried out during its approach. (3122) Florence (named in honor of Florence Nightingale) is an amor NEA discovered by Schelte John Bus at Siding Spring Observatory in 1981. It is one of the largest NEAs (4.5 km). It had its closest encounter (0.047 AU = 706,106 km) to the Earth (until the year 2500) on September 1 2017. During this flyby, radar observations at Arecibo Observatory and at the Goldstone Deep Space Communications Complex revealed it has two satellites, probably formed due to the YORP effect. The inner moon has a diameter 180 - 240 m, and the outer 300 - 360 m. 2014 JO_{25} is an Apollo NEA discovered by Al Grauer on May 5 2014 at the Mount Lemmon Survey in Arizona, United States, in the NEOS program (NASA and University of Arizona collaboration). Its closest approach on April 19 2017 (1.8×10^6 km) was the closest by any known asteroid of this size, or larger, since asteroid Toutatis in 2004. From NEOWISE spacecraft observations in 2014 Masiero estimated a size of $\sim$650 m. Arecibo and Goldstone radar observations during this approach found an irregular shape with 2 components connected by a narrow neck (it resembles a peanut or perhaps the nucleus of Comet 67P/Churyumov-Gerasimenko). Its long axis is $\sim$1 km, its short axis is $\sim$600 m, and the neck $\sim$200 m). It is potentially hazardous.

3. Results

In Figures 1 and 2 we present the lightcurves obtained. Figure 1 shows 4055 Magellan (left side) and (143404) 2003 BD_{44} (right side) lightcurves; in Figure 2 we show 2014 JO_{25} lightcurve (left side) including the night of the maximum approach, while the right side is also for 2014 JO_{25} where data at the maximum approach were excluded. Notice that the form of the light curves in Figure 2 are very different, but that the amplitudes do not present a significant change. Warner *et al.* 2017b found a similar behavior in the striking difference of the light curves of 2014 JO_{25} during the same approach to Earth. He proposes these abrupt changes are due to this NEAs shape. Table 1 presents a summary of the results, along with comparisons with data for these objects in the literature. Column 1 is the name of the NEA, columns 2 to 4 are the period and error,

Table 1. NEAs periods and lightcurve amplitudes.

Name		Period (h)	Amp (mag)	N	Literature Period	Reference
4055	Magellan	7.479 ± 0.001[1]	0.50	3	7.4805 ± 0.0013	Waszczak *et al.* 2015
143404	2003 BD_{44}	78.617 ± 0.009[2]	0.66	9	78.633 ± 0.004	Warner *et al.* 2017
3122	Florence	2.354 ± 0.003	0.22	4	2.357 ± 0.0002	Franco *et al.* 2018
	2014 JO_{25}^{3}	4.527 ± 0.0002	0.65	1	...	
	2014 JO_{25}	4.527 ± 0.007	0.20	5	4.533 ± 0.02	Warner *et al.* 2017b

Notes:
[1] Sada *et al.* 2016
[2] Sada *et al.* 2017
[3] For day of maximum approach.

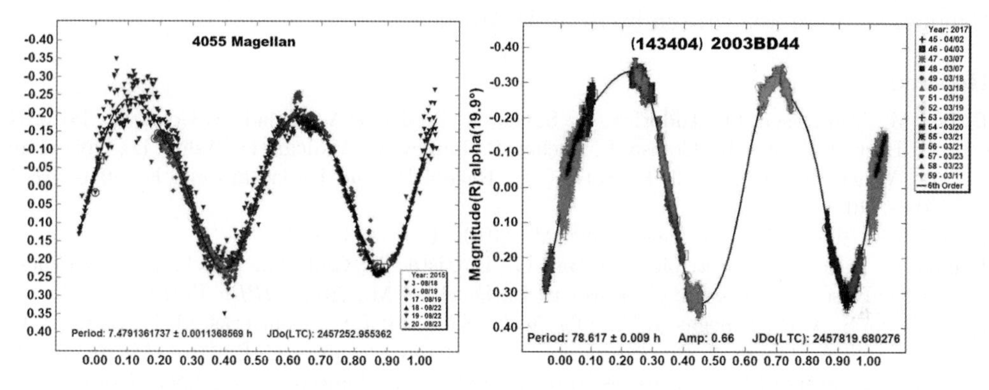

Figure 1. Lightcurves for (4055) Magellan and (143404) 2003 BD_{44}.

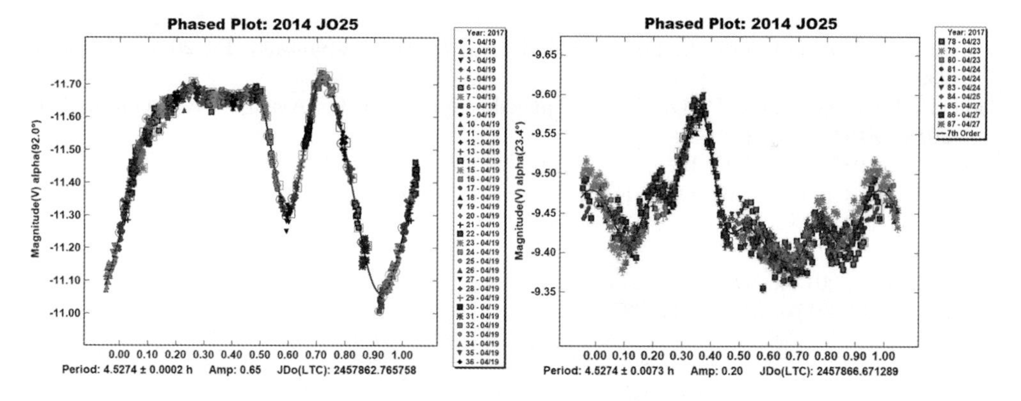

Figure 2. Lightcurves for 2014 JO_{25}

amplitude and number of nights measured in this work. Column 5 present the periods and errors observed by authors shown in column 6.

4. Concluding remarks and future work

Relatively small telescopes can contribute to narrow the gap between asteroid discoveries and characterization. Observing networks, both national and international, such as CMFA, EURONEAR (Birlan *et al.* 2010) and others, ought to be encouraged in order to reach a wider range of longitudes and latitudes to allow continuous observations of individual NEAs, while at the same they can be used to train observers throughout the world. From our own perspective, we plan to start taxonomical classification of bright objects (V<15), to obtain astrometry, to carry out infrared observations, to estimate more precisely the asteroids' size and albedo, as well as undertake theoretical studies of asteroid.

5. Acknowledgements

M.E.C. acknowledges the financial support by CONACyT Fellowship C-841/2018 (México). IRAF is distributed by the National Optical Astronomy Observatory, which is operated by the Association of Universities for Research in Astronomy (AURA) under a cooperative agreement with the National Science Foundation. Collaborative Asteroids Lightcurve Link (CALL) can be accesed at www.minorplanet.info/call.html. Asteroid Lightcurve Photometry Database (ALCDEF) can be accesed at http://alcdef.org. MPO Canopus software was written by Brian D. Warner.

References

Birlan, M., Vaduvescu, O., Tudorica, A., Sonka, A., Nedelcu, A., Galad, A. Colas, F., Pozo, N. F., Barr, A., Toma, R., Comsa, I., Rocher, P., Lainey, V., Vidican, D., Asher, D., Opriseanu, C., Vancea, C., Colque, J. P., Soto, C.P., Rekola R., and Unda-Sanzana, E., 2010, *A&A*, 511, A40

Durech J.D, Sidorin V., and Kaasalainen M., 2010, *A&A*, 513, A46

Franco, L., Bacci, P., Maestripieri, M.,Baj, G., Battista, G., Galli, G., Marchini, A., Noschese, A., Valvasori, A., Caselli, C., Barbieri, L., Facchini, M., 2017, *MPBu* 45, 174

Haro-Corzo, S. A. R., Villegas, L. A., Olguín, L., Saucedo, J. C., Contreras, M. E., Sada, P. V., Ayala, S. A., Garza, J. R., Segura-Sosa, J. & Benítez-Benítez, C. P., 2018, *MPBu*, 45, 233

Sada, P. V., Navarro-Meza, S., Reyes-Ruiz, M., Olguín, L., Saucedo, J. C., Loera-González, P., 2016, *MPBu*, 43, 154

Sada, P. V., Olguín, L., Saucedo, J. C., Loera-González, P., Cantú-Sánchez, L., Garza, J. R., Ayala-Gómez, S. A., Avilés, A., Pérez-Tijerina, E., Navarro-Meza, S., Silva, J.S., Reyes-Ruiz, M., Segura-Sosa, J., López-Valdivia, R., & Álvarez-Santana, F., 2017, *MPBu*, 44, 239

Sada, P., Loera-González, P., Olguín, L., Saucedo-Morales, J. C., Ayala, S. A.& Garza, J. R., 2018, *MPBu*, 45, 122

Stetson, P., 1987, *PASP*, 99, 191

Waszczak, A., Chang, C. K., Ofek, E. O., Laher, R., Masci, F., Levitan, D., Surace, J., Cheng, Y. C., Huen, W., Kinoshita, D., Helou, G., Prince, T. A., and Kulkarniet, S., 2015, *AJ*, 150, 35

Vaduvescu, O., 2017a, *MPBu*, 44, 335

Warner, B. D., 2017b, *MPBu*, 44, 327

Astronomy in Focus - XXX
Proceedings IAU Symposium No. XXX, 2018
M. T. Lago, ed.

doi:10.1017/S1743921319003363

The Mexican Asteroid Photometry Campaigns: Aiming for Asteroids' Rotation Periods

W. J. Schuster[1], S. A. Ayala-Gómez[2], A. Avilés[2], M. E. Contreras[3], S. A. R. Haro-Corzo[4], P. A. Loera-González[3], S. Navarro-Meza[1], L. Olguín[3], E. Pérez-Tijerina[2], M. Reyes-Ruiz[1], J. C. Saucedo[3], J. Segura-Sosa[5], P. A. Valdés-Sada[6], I. L. Fuentes-Carrera[7], C. Chávez-Pech[2], M. Rodríguez-Martínez[4] and R. Vázquez[1]

[1]IA–Ensenada, Universidad Nacional Autónoma de México,
Km 107 Carretera Tijuana-Ensenada, Ensenada, B.C., Mexico
email: schuster@astro.unam.mx

[2]FCFM–Universidad Autónoma de Nuevo León

[3]DIF–Universidad de Sonora

[4]ENES–Universidad Nacional Autónoma de México

[5]FCFM–Universidad Autónoma de Coahuila

[6]DFM–Universidad de Monterrey

[7]ESFM–Instituto Politécnico Nacional

Abstract. Thousands of new asteroids are discovered every year and the rate of discovery is by far larger than the determination rate of their physical properties. In 2015 a group of researchers and students of several Mexican institutions have established an observational program to study asteroids photometrically. The program, named Mexican Asteroid Photometry Campaign, is aiming to derive rotation periods of asteroids based on optical photometric observations. Since then four campaigns have been carried out. The results obtained throughout these campaigns, as well as future work, are presented.

---

1. Overview

Currently, more than 760,000 asteroids are known and every year thousands of new ones are discovered. Nevertheless, most of their physical parameters: size, shape, albedo, rotation, taxonomical classification, etc. are unknown. Thus, we need to make a greater effort to increase our knowledge about these objects. Some asteroid properties, such as their rotation period, can be derived from their lightcurve. However, according to the Asteroid Lightcurve Photometry Database (ALCDEF), only a few thousand (∼6000) asteroids have reliable rotation periods. These periods are important because they are related to the cohesion strength of the material from which an asteroid has been made. Figure 1 shows the relation between the asteroid rotation frequency and its size. We can see that fast rotators have sizes smaller than ∼300 m. In addition, 3D asteroid shapes can be inferred from lightcurves obtained at different epochs, since objects are observed from different viewing angles.

In order to contribute to asteroid characterization, in 2015 a group of researchers and students of several Mexican institutions, have established an observational program to study asteroids photometrically. The program, named Mexican Asteroid Photometry

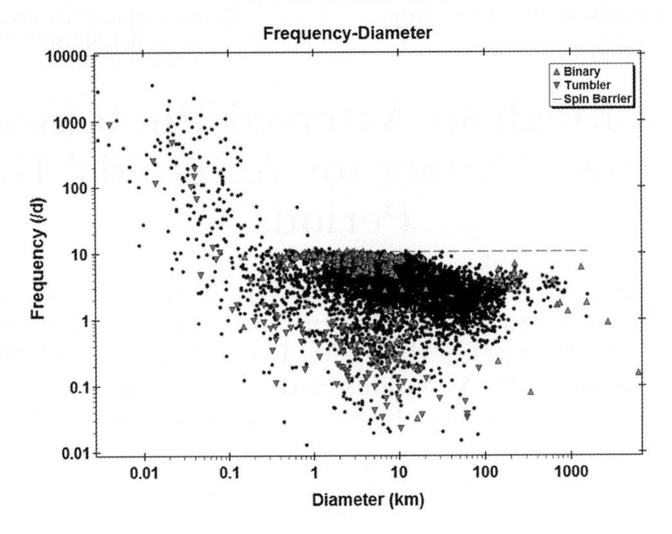

Figure 1. Relation between the asteroid rotation frequency and its size. Figure taken from the ALCDEF web site.

Campaign (CMFA, from the Spanish), is aiming to derive rotation periods of asteroids based on optical photometric observations. This program has been established at an introductory stage to generate the abilities and basic knowledge related to asteroids, allowing us to carry out broader and deeper studies in the near future. The first campaign started during the second half of 2015. Since then, three campaigns have been completed and the fourth is ongoing.

2. Observations, data reduction, analysis and results

The observatories and telescopes involved in the campaigns are the following: i) 0.84-m telescope at the Observatorio Astronómico Nacional at Sierra San Pedro Mártir (OAN-SPM), operated by Universidad Nacional Autónoma de México, at Baja California, ii) 0.40-m telescope at Observatorio Astronómico Carl Sagan (OACS), Universidad de Sonora, Hermosillo, Sonora, and iii) 0.36-m telescope of the Universidad de Monterrey Astronomical Observatory, Monterrey, Nuevo León.

Typically, to obtain complete lightcurves, five to ten nights of observation are dedicated to each asteroid. Exposure times for individual images are in the range of 30 to 240 seconds, depending on asteroid's visual magnitude. Generally, a few hundred images per object are taken during each observing night.

Basic steps in data reduction are performed using IRAF or MaximDL software. For lightcurve extractions and period determinations, MPO Canopus software is used. For crowded stellar fields, first DAOPHOT (Stetson 1987) is used to obtain the photometry and then MPO Canopus is used for deriving the lightcurve and rotation period. Figure 2 shows examples of lightcurves derived using these procedures.

The asteroid sample has been obtained from the Collaborative Asteroids Lightcurve Link (CALL). In general, asteroids with poorly, or unknown, rotation periods were chosen. Only those with declinations $> -30^o$ were observed. In three years more than fifty objects have been observed in four campaigns from which a dozen lightcurves and rotation periods have been published (Sada *et al.* 2016; Sada *et al.* 2017; Sada *et al.* 2018; Haro-Corzo *et al.* 2018). Some data are still under analysis. A sample of derived rotation periods and relevant information are presented in Table 1. Near Earth Asteroids' (NEAs') information is not included because it has been presented in a separate contribution by J.C. Saucedo *et al.* at this same IAU Focus Meeting.

Table 1. Asteroids with reliable period determinations.

Name	Period (h)	Error (h)	Amplitude (mag)	Nights	Group
703 Noemi	11.108	0.014	0.28	11	Flora
1084 Tamariwa	6.195	0.001	0.30	5	Outer Main Belt
1218 Aster	3.1581	0.0002	0.35	7	Flora
1305 Pongola	4.349	0.0003	0.19	8	Outer Main Belt
1491 Balduinus	15.3044	0.0057	0.45	4	Outer Main Belt
1856 Ruzena	5.957	0.001	0.68	11	Main Belt
2022 West	14.1385	0.0031	0.54	7	Outer Main Belt
2535 Hameenlinna	3.2311	0.0001	0.11	10	Flora
2733 Hamina	93.23	0.02	0.36	13	Inner Main Belt
3394 Banno	7.3249	0.0008	0.21	5	Main Belt
3887 Braes	5.81	0.01	0.60	6	Main Belt
4775 Hansen	3.1187	0.0001	0.16	5	Mars Crossing
8433 Svecica	20.9905	0.0015	0.65	10	Outer Main Belt
18301 Konyukhov	2.6667	0.0003	0.15	6	Outer Main Belt

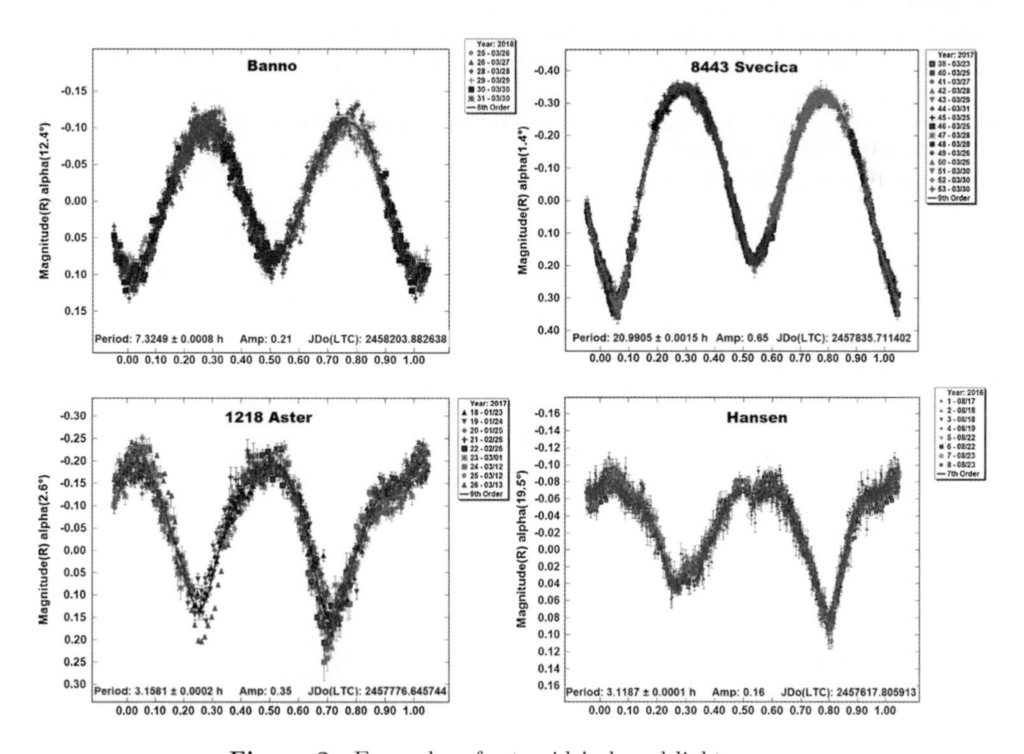

Figure 2. Examples of asteroids' phased lightcurves.

3. Concluding remarks and future work

After three years we have refined observational and analysis methods to derive reliable asteroids' lightcurves. The pace at which we are obtaining and processing photometric data is increasing. However, due to the shortage of trained people, we are reaching our maximum rate of studied objects to about 12 objects per year. Therefore, we are planning to coach a new generation of students and researchers in the observation and analysis techniques. In addition, we plan to improve and increase our observation facilities. In the meantime, we will fulfill the following tasks: i) carry out infrared observations to estimate precise asteroid sizes and albedos, ii) start taxonomical classification of bright objects (V<15), and iii) undertake theoretical studies of asteroids' dynamical behavior.

4. Acknowledgments

WJS gratefully acknowledges financial support from the Universidad Nacional Autónoma de México, grant DGAPA-PAPIIT, project IN100918. M.E.C. acknowledges financial support from CONACyT Fellowship C-841/2018 (México). IRAF is distributed by the National Optical Astronomy Observatory, which is operated by the Association of Universities for Research in Astronomy (AURA) under a cooperative agreement with the National Science Foundation. Collaborative Asteroids Lightcurve Link (CALL) can be accessed at www.minorplanet.info/ call.html. Asteroid Lightcurve Photometry Database (ALCDEF) can be accessed at alcdef.org.

References

Haro-Corzo, S. A. R., Villegas, L. A., Olguín, L., Saucedo, J. C., Contreras, M. E., Sada, P. V., Ayala, S. A., Garza, J. R., Segura-Sosa, J., & Benítez-Benítez, C. P. 2018, *MPBu*, 45, 233

Sada, P. V., Navarro-Meza, S., Reyes-Ruiz, M., Olguín, L., Saucedo, J. C., & Loera-González, P. 2016, *MPBu*, 43, 154

Sada, P. V., Olguín, L., Saucedo, J. C., Loera-González, P., Cantú-Sánchez, L., Garza, J. R., Ayala-Gómez, S. A., Avilés, A., Pérez-Tijerina, E., Navarro-Meza, S., Silva, J.S., Reyes-Ruiz, M., Segura-Sosa, J., López-Valdivia, R., & Álvarez-Santana, F. 2017, *MPBu*, 44, 239

Sada, P., Loera-González, P., Olguín, L., Saucedo-Morales, J. C., Ayala-Gómez, S. A., & Garza, J. R. 2018, *MPBu*, 45, 122

Stetson, P. B. 1987, *PASP*, 99, 191

Astronomy in Focus - XXX
Proceedings IAU Symposium No. XXX, 2018
M. T. Lago, ed.

doi:10.1017/S1743921319003375

Lightcurve Photometry of (2525) O'Steen with the new "Milanković" 1.4 m telescope

N. Todorović[1], G. Apostolovska[2] and E. V. Bebekovska[2]

[1]Belgrade Astronomical Observatory, Belgrade, Serbia email: ntodorovic@aob.rs

[2]Institute of Physics, Faculty of Science Ss. Cyril and Methodius University
Skopje, Republic of Macedonia

The Astronomical Station (AS) Vidojevica, located on the Vidojevica mountain, has been established by the Astronomical Observatory of Belgrade. A new ASA AZ1400 "Milanković" Ritchey - Chrétien 1.4m telescope was mounted in April 2016 and in September 2018 the telescope was placed in the newly constructed automated dome. The asteroid (2525) O'Steen, a member of the Themis asteroid family, was observed with the "Milanković" telescope on 28 March 2017. Recently, it was showed that O'Steen and some other Themis members might have a low-level cometary activity. Our main aim is to use the light-curve of O'Steen obtained at the AS Vidojevica, in combination with light-curves from other observatories obtained during previous apparitions at various geometric conditions, to calculate the asteroid's spin vector, rotational properties and estimation of its preliminary shape model. We acknowledge the support of the European Commission through the project BELISSIMA and the support from the Ministry of Education, Science and Technological Development of the Republic of Serbia through the project 176011.

Astronomy in Focus - XXX
Proceedings IAU Symposium No. XXX, 2018
M. T. Lago, ed.

doi:10.1017/S1743921319003387

Observation of 44(Nysa) Asteroid by IST60 Telescope

Remziye Canbay and Fulin Gursoy

Istanbul University, Graduate School of Science and Engineering, Department of Astronomy and Space Sciences, 34116 Beyazıt, Istanbul, Turkey
email: rmzycnby@gmail.com and fulingursoy@gmail.com

Abstract. In this study, the main belt asteroid (44) Nysa, which is also a known member of the Nysian asteroid family, was observed by IST60 telescope. The orbital elements were compared with MPO(Minor Planet Center) and NASA Horizons Web-Interface results.

---

1. Introduction-Method

Asteroid studies enable us to learn about the physical properties and chemical composition of asteroid surfaces through the analysis of the sunlight reflected off the asteroid and the calculation of asteroid orbit. Orbital measurements can show us if an asteroid poses a threat the world or not. Examples of how asteroid impacts on the world have created a disaster have already been seen. Nysa was observed at Ulupnar Observatory located in anakkale Onsekiz Mart University. The observations were carried out using a CCD camera which is installed in the IST60 telescope. The CCD camera has 2048x2048 pixels, a pixel size of 13.5 microns, and the focal length of the telescope is 480 cm.

2. Results

The orbital elements were calculated using the equatorial coordinates of Nysa obtained from the 1220 observations made from the IST60 telescope at the 13th, 14th, 15th, 25th, 26th of September 2013 and the 5th, 6th of May 2015. We compared the results of these data with orbit parameters from NASA's JPL Small-Body Database Browser and Minor Planet Center, and the results showed very good agreement. The results obtained are $\Omega=131^\circ.562$, $\omega=343^\circ.295$, $i=3^\circ.7071$, $a=2.423143$ AU, $e=0.14870$ and $\zeta=2456906.100345$ JD. We compared our results with the Minor Planet Center and NASA Horizons Web-Interface results and found that they are consistent with an accuracy of $0^\circ.003$ for Ω, $0^\circ.029$ for ω, $0^\circ.0005$ for i, 0.0002 AU for a, and 0.0004 for e. Similar differences occur in the third or fourth digits of the trajectory reductions made by adding new observations at different dates. This shows that Nysa's orbit is still exposed to perturbations of Jupiter and Mars. If not only the Nysa asteroid, but also the positions of the asteroids in the system are determined and new orbital elements are calculated from these positions, at least the results obtained on how the dynamic structure of the solar system in the inner region occurred can provide important clues to the theoretical researchers.

3. Acknowledgements

The observations in this study were taken from project 21183 of the Executive Secretary of Scientific Research Projects of Istanbul University.

Reference

Canbay, R. 2017, DETERMINATION OF THE ORBITAL ELEMENTS OF ASTEROID (44) NYSA, M.Sc. THESIS, Istanbul University

Astronomy in Focus - XXX
Proceedings IAU Symposium No. XXX, 2018
M. T. Lago, ed.

doi:10.1017/S1743921319003399

Detection of Longitudinal Albedo and Metallicity Variations of Asteroids with Ground-Based, Part-Per-Million Polarimetry

Sloane J. Wiktorowcz[1] and Joseph Masiero[2]

[1]The Aerospace Corporation,
2310 E. El Segundo Blvd., El Segundo, CA 90245, USA
email: sloane.j.wiktorowicz@aero.org

[2]Jet Propulsion Laboratory,
M/S 183-601, 4800 Oak Grove Drive, Pasadena, CA 91109, USA
email: Joseph.Masiero@jpl.nasa.gov

The linear polarization and albedo of rocky solar system bodies is anticorrelated: dark surfaces, dominated by single scattering, are strongly polarized, but multiple scattering in bright surfaces randomizes the electric field orientation and reduces polarization. As an asteroid rotates, both shape changes and surface albedo variations affect reflected light flux, causing difficulty in the identification of albedo variations. Polarimetry, however, is insensitive to shape changes: as total flux varies with instantaneous cross-sectional area, fractional polarization does not. Thus, rotational variability in linear polarization is a hallmark of albedo inhomogeneity, and it cannot be identified with photometry alone.

Until now, polarimeters have only discovered high significance rotational variation of linear polarization for (4) Vesta. We report on Lick 3-m observations of Main Belt asteroids with the POLISH2 polarimeter, which utilizes photoelastic modulators instead of a waveplate. We have not only confirmed rotational variations in (4) Vesta, but we have also discovered variations in (1) Ceres and (7) Iris. The amplitude of variations for both (4) Vesta and (7) Iris are stronger than those of (1) Ceres due to the latter's relatively homogeneous surface.

Circular polarization, which may originate from multiple scattering or from the phase retardance introduced by a metalliferous surface, has been observed in nearly all Solar System bodies except for asteroids. POLISH2 simultaneously measures linear and circular polarization, and we report the discovery of non-zero circular polarization from (2) Pallas, (7) Iris, and the metalliferous (16) Psyche and (216) Kleopatra. This is in stark contrast to the non-detection of time-averaged circular polarization from the differentiated (1) Ceres and (4) Vesta. Therefore, optical circular polarization may provide a new way to identify metalliferous asteroidal surfaces.

Astronomy in Focus - XXX
Proceedings IAU Symposium No. XXX, 2018
M. T. Lago, ed.

doi:10.1017/S1743921319003405

Multi-color observation of active comets

Liu Jinzhong and Zhang Xuan

Xinjiang Astronomical Observatory, Chinese Academy of Sciences, Urumqi, 830011, China
email: liujinzh@xao.ac.cn

In general, due to a high temperature and a presence of gas sublimation leading to coma emission, a comet is supposed to be active when it is close to the Sun. We start to investigate photometric information from some special active comets. The data were obtained with the 1-m wide field optical telescope (hereafter NOWT) at Nanshan Observatory in Xinjiang. This project was started in 2017. By multi-color observation and cometary morphological studies, we can estimate the surface brightness profiles, Afρ parameters, mass production rates, the sizes of nuclei radii and coma colors. Meanwhile the physical driving mechanism of activity in comets needs to be discussed. Here we will give a brief description of our photometric observations of comet C/2015 O1 with NOWT. Object C/2015 O1 was discovered on 2015 July 19 by the Pan-STARRS 1 telescope. The followup observations with NOWT were carried out starting 2017 August 31. In total, the number of observations collected of C/2015 O1 for B, V, R bands are 44, 36, and 37, respectively. The standard IRAF packages were used to measure the core magnitude of coma. We also gave a rough estimate of the possible outside structure of C/2015 O1. Finallywe conclude that C/2015 O1 has a lower activity than expected. And a solar radiation pressure model can be used to explain the surface brightness profiles of C/2015 O1.

Acknowledgements

This work is supported by the program of the Light in China's Western Region (LCWR, grant No. 2015-XBQN-A-02).

Astronomy in Focus - XXX
Proceedings IAU Symposium No. XXX, 2018
M. T. Lago, ed.

doi:10.1017/S1743921319003417

A new pipeline for astrometry and photometry of asteroids

Y. Kılıç[1,2] M. Kaplan[2] and Z. Eker[2]

[1]TÜBİTAK National Observatory, Antalya, Turkey
email: `yucel.kilic@tubitak.gov.tr`

[2]Department of Space Sciences and Technologies, Akdeniz University, Antalya, Turkey

Abstract. Using Python 3, astropy and astrometry.net, we have developed a pipeline to obtain photometric light curves of asteroids automatically queried by the SkyBoT database from sequential FITS images. The pipeline provides: pre-reduction of data, astrometry, standard differential photometry and light curves by auto-selecting multiple comparison stars (maximum user-defined) from NOMAD catalog via VizieR. The code is an open source, free and hosted on GitHub with the GNU GPL v3 license.

The code

The pipeline can be downloaded from the link: https://github.com/yucelkilic/astrolib.

Acknowledgement

We acknowledge TÜBİTAK National Observatory for partial support in using T100 telescope with project number 13BT100-464. This research has made use of IMCCE's SkyBoT VO tool.

Astronomy in Focus - XXX
Proceedings IAU Symposium No. XXX, 2018
M. T. Lago, ed.

doi:10.1017/S1743921319003429

Testing the Yarkovsky-driven evolution of the Eureka cluster with LSST

Apostolos A. Christou

Armagh Observatory and Planetarium, College Hill, Armagh BT61 9DG, UK
email: apostolos.christou@armagh.ac.uk

The Trojan clouds of Mars are occupied by a handful of asteroids, the orbits of which, rather than being random, form a tight cluster around the 2-km body (5261) Eureka (Christou 2013, see Fig. 1). Of all clusters known to exist, this genetic association of asteroids is the closest to the Sun. How this family formed is still under investigation (Ćuk *et al.* 2015; Christou *et al.* 2017) but its proximity to the Sun, family member physical properties and existence of similar clusters of asteroids (Pravec *et al.* 2010; Pravec *et al.* 2018) thought to have arisen from YORP-induced fission of a parent body, all point to this latter mechanism as the culprit.

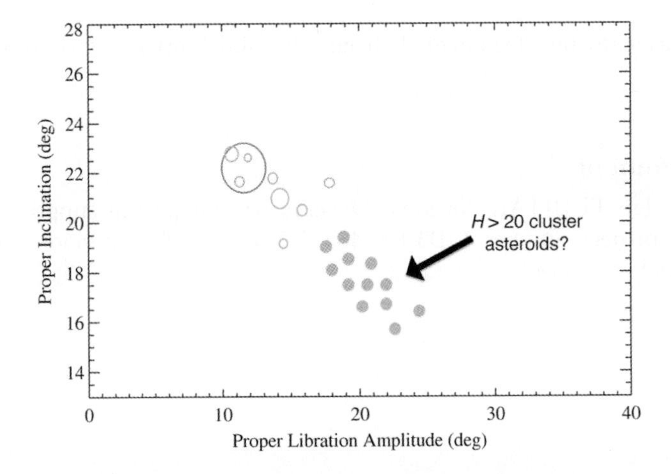

Figure 1. Eureka (red circle) and known cluster members (amber circles) in proper inclination vs proper libration amplitude space. Circle size is proportional to asteroid size. Solid amber symbols indicate the likely location of fainter, as-yet-undiscovered, members of this cluster if Yarkovsky efficiently modifies their orbits as per Ćuk *et al.* (2015).

The question of family age is closely related to its post-formation dynamical history. Christou (2013) and Ćuk *et al.* (2015) showed that, over a period $>10^8$ yr, the orbits gradually disperse due to the Yarkovsky effect. In particular, Ćuk *et al.* found that family member orbits plot along the locus of negative Yarkovsky acceleration suggesting that the observed distribution may reflect a long-term dominance of seasonal over diurnal Yarkovsky and proposed a $\sim$1 Gyr age for the family. If correct, smaller objects evolve faster than larger ones; this can be used to confirm the role of Yarkovsky. We estimate that the Large Synoptic Survey Telescope will push observational completeness of the population up to $H{\sim}23$ and discover a few hundred additional members. The location of the smaller objects relative to Eureka and the other Trojans will test the Ćuk *et al.* hypothesis (Fig. 1). Also, because the path in the space of libration amplitude vs inclination is largely deterministic, an approach similar to Milani *et al.* (2017) may constrain the family age.

References

Christou, A. A. 2013, *Icarus*, 224, 144
Christou, A. A., Borisov, G., Dell'Oro, A., Cellino, A., & Bagnulo, S. 2017, *Icarus*, 293, 243
Ćuk, M., Christou, A. A., & Hamilton D. P. 2015, *Icarus*, 252, 243
Milani, A. & 5 co-authors 2015, *Icarus*, 288, 240
Pravec, P. & 25 co-authors 2010, *Nature*, 466, 1085
Pravec, P. & 28 co-authors 2018, *Icarus*, 304, 110

Astronomy in Focus - XXX
Proceedings IAU Symposium No. XXX, 2018
M. T. Lago, ed.

doi:10.1017/S1743921319003430

Asteroid Families and the Next Generation Surveys

Joseph Masiero

NASA Jet Propulsion Laboratory, Caltech
4800 Oak Grove Dr, MS 183-301, Pasadena, CA 91109, USA
email: joseph.masiero@jpl.nasa.gov

Asteroid families represent unique sets of data for testing the physical and mechanical properties of protoplanets and other small bodies. One of the main sources of data on the physical properties for these bodies has been large-scale astronomical surveys that have collected incidental observations of a large number of minor planets. Surveys such as the Sloan Digital Sky Survey (SDSS; Ivezić *et al.* 2001; Parker *et al.* 2008) and the Wide-field Infrared Survey Explorer (WISE; Mainzer *et al.* 2011; Masiero *et al.* 2011) have provided griz colors (SDSS) and albedos and diameters (WISE) for tens of thousands of members of asteroid families.

Given the utility of these astronomical surveys to understanding Solar system objects, we naturally look forward to the next generation of surveys to further our knowledge of families, and provide us unique insights. The assets that will have the nearest-term impact on our understanding of families are those that are currently either ongoing, or will become operational in the next few years. While there are many groups conducting a wide range of astronomical surveys across the electromagnetic spectrum, two in particular stand out as likely to have the biggest impact on the study of families: Gaia and LSST.

Gaia is ESA's all-sky astrometric survey, which has already provided milliarcsecond astrometry for thousands of asteroid family members and will soon release measurements for the full moving object data set containing $\sim 120,000$ family asteroids (Spoto *et al.* 2018). Future data releases will also increase the baseline of these measurements, which will enable constraints on the masses for hundreds of family members. When combined with known diameters, density measurements as a function of size will test predictions of rubble pile reaccumulation from family formation simulations. Future Gaia data releases will also contain low resolution spectra of tens of thousands of family members, providing improved taxonomic classifications.

The Large Synoptic Survey Telescope (LSST, *lsst.org*), is currently under construction in Chile, with a planned first light in 2021 and a formal beginning of survey in late 2022. With an 8.4 m diameter primary mirror and a 3.2 Gpix camera, LSST will conduct a 10 year survey of the sky in 5 optical bandpasses, repeating coverage every 3 days. These survey data would provide optical colors for possibly millions of asteroid family members, greatly expanding on the results from SDSS.

Beyond the immediate future, the surveys that are currently planned or proposed offer us a view of what new data may become available for asteroid families. These designs and debates highlight where the interest lays in the broader astronomical community for future survey strategies, and what the probable next-steps are going to be.

Two upcoming missions focus on high precision photometric surveys in the near-infrared. ESA's Euclid space telescope (http://sci.esa.int/euclid/) with planned launch date in 2021, and NASA's Wide Field InfarRed Survey Telescope (WFIRST, https://www.nasa.gov/wfirst) with launch date targeting the mid 2020's, would focus on conducting a photometric survey from the visible to 2 μm to study dark energy. Incidental observations of asteroid family members would enable NIR photometric taxonomies similar to those developed by e.g. Sykes *et al.* (2000), Popescu *et al.* (2018). When combined with data from LSST, these would provide full visible-to-NIR taxonomic classifications for hundreds-of-thousands of asteroid family members.

SphereX (http://spherex.caltech.edu/) has been selected by NASA for Phase B development, targeting a 2023 launch date. The goal of this mission is to perform an all-sky spectroscopic survey from $0.5 - 5$ μm, by sweeping a series of low resolution dispersion elements over the sky during the telescope's low-Earth polar orbit. Spectra would be built piece-wise, so moving objects would only receive partial spectral coverage or would need to be reassembled after accounting for rotation, but tens of thousands of family members would receive some spectral characterization, including of the water absorption features near 3 μm.

The Near-Earth Object Camera (NEOCam, https://neocam.ipac.caltech.edu/) is a proposed mid-infrared asteroid survey mission, with a launch date targeting the mid 2020's. NEOCam would collect observations simultaneously at $4 - 5.2$ μm and $6 - 10$ μm, detecting millions of Main Belt asteroids and enabling thermal modeling to characterize their diameters. When combined with data from LSST, albedo measurements for these millions of asteroids would also be possible. While it is specifically designed for observations of near-Earth objects, incidental observations of family members at multiple epochs would also support detailed thermophysical modeling of the surface.

Beyond the next few decades, it becomes harder to guess what kinds of surveys and data streams will be needed and desired by astronomers, or how those data will be used in light of the discoveries that precede them. Looking to the past, we see that new detector technologies often mature into pioneering surveys, while successful surveys often are repeated when order-of-magnitude improvements become possible.

One obvious area of future exploration will be deeper, all-sky moderate resolution spectral surveys. The rapid growth of integral field unit (IFU) spectroscopy will enable a spectral survey of all objects in the sky, even asteroids, to be carried out. This could provide taxonomic classifications for a large fraction of the known members of asteroid families. Another probable future survey is a next-generation version of Gaia after a few decades have passed that would expand our measurements of masses via gravitation perturbations and Yarkovsky drift measurements to many thousands of family members, or more.

In addition to new instrumentation, new processing of survey data will also allow us to expand our knowledge of families and family-forming impacts. In particular, we can expect that ongoing surveys will continue to detect asteroids that have recently undergone catastrophic impacts, and that the rate of these events will increase as new surveys come online. But it is also probable that as our catalog of objects increases and our knowledge of their orbits improves with new astrometric surveys we will soon be able to not just witness, but to predict asteroid-asteroid collisions before they occur. This will enable us to study the objects pre- and post-collision, and thereby observe the details of catastrophic impacts on a protoplanetary-scale, possibly even with *in situ* spacecraft.

The future of asteroid family studies already promises to be one of revolutionary ideas and understandings. I for one look forward with anticipation to what we will learn in the second century of asteroid families.

References

Ivezić, Ž., Tabachnik, S., Rafikov, R. *et al.* 2001, *AJ*, 122, 2749
Mainzer, A., Bauer, J., Grav, T. *et al.* 2011, *ApJ*, 731, 53
Masiero, J., Mainzer, A., Grav, t., *et al.* 2011, *ApJ*, 741, 68
Parker, A., Ivezić Ž., Jurić, M. *et al.* 2008, *Icarus*, 198, 138
Popescu, M., Licandro, J., Carvano, J.M., *et al.*, 2018, *A&A*, 617, A12
Gaia Collaboration, Spoto, F., Tanga, P., Mignard, F., *et al.* 2018, *A&A*, 616, A13
Sykes, M., Cutri, R., Fowler, J., *et al.*, 2000, *Icarus*, 146, 161

FM3 – Radio Galaxies: Resolving the AGN Phenomenon

Astronomy in Focus - XXX
Proceedings IAU Symposium No. XXX, 2018
M. T. Lago, ed.

doi:10.1017/S1743921319003442

Focus Meeting #3: Radio Galaxies – Resolving the AGN Phenomenon

Volker Beckmann

CNRS/IN2P3, 3 rue Michel Ange, 75016 Paris, France
email: beckmann@in2p3.fr

Radio galaxies provide excellent laboratories for investigating the physical aspects, unification and cosmic evolution of active galactic nuclei (AGN). Thanks to recent multi-wavelength observations, we are now able to separate many different physical components of radio galaxies through imaging and spectroscopy. Observations from radio through X-ray wavelengths can probe the ejection of matter into jets and monitor decades of jet evolution. Gamma-ray observations have shown that radio galaxies are detectable up to the very-high-energy range despite unfavorable jet alignment.

We observe radio galaxies out to redshifts greater than $z = 5$, which makes them important cosmological probes. Planck maps have provided us with new insights into the populations of radio galaxies and their distributions in space in the 30 – 900 GHz range. NuSTAR provides high-quality spectra in the hard X-ray range. The Event Horizon Telescope (EHT) has begun mapping close to the event horizon of the Milky Way's central black hole, and the Square Kilometre Array (SKA), the next generation of extremely large telescopes (ELTs) and other future telescopes will open up a new and vast discovery space.

Focus Meeting #3 brought together multiwavelength observers and theorists to synthesize progress made over the last three years and to define future directions. In order to channel the presentations and discussions, the meeting was organized into five sessions:

• During the first session, starting on August 22nd, we discussed the radio-galaxy structures that are produced on kiloparsec scales and beyond, with an emphasis on plasma composition, and sought to address their underlying causes. Daniel Schwartz (Harvard-Smithsonian Center for Astrophysics, USA) talked about high-resolution studies of 100 kpc jets based on data from the Chandra X-ray Observatory. For example, a new Chandra survey for jets in radio quasars at $z > 3$ has revealed X-ray jets and lobes extending beyond the region of detected radio emission (Schwartz 2018).

• The second session focused on the central engine and production of the jet(s). Alexander Tchekhovskoy (University of California, Berkeley, USA) introduced the topic with a presentation about how numerical simulations incorporating general relativity and magnetism allow us to use black hole accretion phenomena to quantitatively probe strong-field gravity and constrain black hole physics in various astrophysical contexts.

• On the second day of the meeting we first discussed populations and statistics of radio galaxies, motivated by a review from Elaine Sadler (University of Sydney, Australia) focusing on results from multi-wavelength radio surveys.

• This was followed by a session about future prospects. Lindy Blackburn (Harvard-Smithsonian Center for Astrophysics, USA) started the discussion with a presentation on the goals and status of the EHT, which has already revealed structure on the scale of the Schwarzschild radius in Sagittarius A*, the supermassive black hole at the centre of our galaxy, and in Messier 87 in the Virgo Cluster.

• A fifth session dealt with the interaction of radio galaxies with their environments. Andy Fabian (Institute of Astronomy, Cambridge University, UK) discussed AGN feedback in clusters of galaxies and explained how energy can be transported and dissipated throughout the cluster core.

In addition to the invited reviews, 20 high-profile contributed talks reviewed the state of the art within the field, and 20 posters had been selected for one-minute/one-slide presentations. The sessions were accompanied by dedicated discussions, giving the audience ample opportunities to help us obtain a wide and complete view of what is driving radio galaxies and what we can learn from them about physical processes in the universe.

Finally, Annalisa Celotti (International School for Advanced Studies (SISSA), Italy) wrapped up FM3 by giving a summary of the results presented during the 1.5 days of the meeting.

In the following the proceedings present the review article of Dan Schwartz, followed by eight contributions, which are representative of the presentations given at the focus meeting: Birkinshaw, Rawes & Worrall (2019), Bruni *et al.* (2019), Jarvis (2019), Morganti *et al.* 2019, Nyland (2019), Shaballa (2019), Whittam (2019), and Worrall, Duffy & Birkinshaw (2019). These articles are sorted in alphabetical order. Additional articles can be found in the supplements of these proceedings. Photos of the focus meeting can be found on Twitter under https://twitter.com/iau18radiogalax

Acknowledgments

We would like to thank the local organisers of the IAU general assembly and the IAU Executive Committee for their efficient support! The Scientific Organizing Committee put together the program, gave the meeting its scientific direction, and edited these proceedings: Loredana Bassani (IASF / INAF, Italy), Volker Beckmann (chair, CNRS / IN2P3, France), Markus Böttcher (North-West University, South Africa), Chris Done (Durham University, UK), Melanie Johnston-Hollitt (Victoria University, New Zealand), Anne Lähteenmäki (Aalto University, Finland), Raffaella Morganti (ASTRON, The Netherlands), Rodrigo Nemmen (Sao Paolo University, Brazil), Paolo Padovani (ESO, Germany), Claudio Ricci (co-chair, UDP, Chile), Yoshihiro Ueda (Kyoto University, Japan), Sylvain Veilleux (University of Maryland, USA), and Diana Worrall (co-chair, Bristol University, UK).

References

Birkinshaw, M., Rawes, J., & Worrall, D. M. 2019, *Astronomy in Focus* 1
Bruni, G., Ursini, F., Panessa, F., *et al.* 2019, *Astronomy in Focus* 1
Jarvis, M. E. 2019, *Astronomy in Focus* 1
Morganti, R., Schulz, R., Nyland, K., et al. 2019, *Astronomy in Focus* 1
Nyland, K. 2019, *Astronomy in Focus* 1
Schwartz, D. 2019, *Astronomy in Focus* 1
Shabala, S. S. 2019, *Astronomy in Focus* 1
Whittam, I. H. 2019, *Astronomy in Focus* 1
Worrall, D. M., Duffy, R. T., & Birkinshaw, M. 2019, *Astronomy in Focus* 1

Astronomy in Focus - XXX
Proceedings IAU Symposium No. XXX, 2018
M. T. Lago, ed.

doi:10.1017/S1743921319003454

The High-energy emission of jetted AGN

Daniel A. Schwartz

Smithsonian Astrophysical Observatory,
60 Garden St.,Cambridge, MA 02138, USA
email: das@cfa.harvard.edu

Abstract. Quasars with flat radio spectra and one-sided, arc-second scale, ≈ 100 mJy GHz radio jets are found to have similar scale X-ray jets in about 60% of such objects, even in short 5 to 10 ks *Chandra* observations. Jets emit in the GHz band via synchrotron radiation, as known from polarization measurements. The X-ray emission is explained most simply, i.e. with the fewest additional parameters, as inverse Compton (iC) scattering of cosmic microwave background (cmb) photons by the relativistic electrons in the jet. With physics based assumptions, one can estimate enthalpy fluxes upwards of 10^{46} erg s^{-1}, sufficient to reverse cooling flows in clusters of galaxies, and play a significant role in the feedback process which correlates the masses of black holes and their host galaxy bulges. On a quasar-by-quasar basis, we can show that the total energy to power these jets can be supplied by the rotational energy of black holes with spin parameters as low as $a = 0.3$. For a few bright jets at redshifts less than 1, the *Fermi* gamma ray observatory shows upper limits at 10 Gev which fall below the fluxes predicted by the iC/cmb mechanism, proving the existence of multiple relativistic particle populations. At large redshifts, the cmb energy density is enhanced by a factor $(1+z)^4$, so that iC/cmb must be the dominant mechanism for relativistic jets unless their rest frame magnetic field strength is hundreds of micro-Gauss.

Keywords. (galaxies:) quasars: general, galaxies: jets, X-rays: galaxies, radio continuum: galaxies, black hole physics,(cosmology:) cosmic microwave background

---

1. Introduction

Radio astronomy is intimately related to high energy astronomy. Estimates of magnetic fields in the lobes of radio sources led to the recognition that TeV electrons must be present to radiate via the synchrotron mechanism. This motivated early suggestions (Morrison (1958), Savedoff (1959)) that direct detection of celestial gamma rays might be feasible. Estimates of the energy contents of these radio lobes were extremely large, posing difficulties for explaining the origin and possible relation to the associated galaxy or quasar. The dilemma was solved by theoretical explanations of how collimated beams of particles and fields, i.e. jets, could carry energy to the lobes (Rees (1971), Longair *et al.* (1973), Scheuer (1974), Blandford & Rees (1974), Begelman, Blandford, & Rees), and by direct imaging of these radio jets (Turland (1975), Waggett *et al.* (1977), Readhead *et al.* (1978), Perley *et al.* (1979), Bridle & Perley (1984)). The existence of X-ray emission from the nearest, brightest jets in Cen A (Feigelson *et al.* (1981)), 3C273 (Willingale (1981)), and M87 (Schreier *et al.* (1982)) resulted from observations by the *Rosat* and *Einstein* X-ray telescopes, each with about 5$''$ angular resolution. The *Chandra* X-ray observatory (Weisskopf *et al.* (2002), Weisskopf *et al.* (2003), Schwartz (2014)) with its 0$''$.5 resolution telescope gives a 100-fold increase in 2-dimensional imaging capability. This has led to the discovery of X-ray jets in a wide variety of astronomical systems (Schwartz (2010)), and in particular has exploded the study of X-ray emission from extra-galactic radio jets.

1.1. *Importance of jets*

In their 1984 review, Begelman, Blandford, & Rees wrote "...the concept of a jet is crucial to understanding all active nuclei, ..." Jets can carry significant amounts of energy, and since that power is not subject to the Eddington radiation limit jets may allow super-Eddington accretion rates. Such accretion may be relevant to the growth of super-massive black holes in the early universe. High energy observations using the *Chandra* Observatory revealed the effects of jets on the gas filling clusters of galaxies Fabian *et al.* (2000). This solved a long standing problem that the cooling time of gas in clusters of galaxies was much less than the Hubble time, implying that the cluster gas should collapse catastrophically. Jets on parsec scales in the nuclei of galaxies explain the blazar phenomena of rapid variability and apparent superluminal expansion. It is now known from direct imaging that regions within X-ray jets may be variable even 10's of kpc from the black hole (Marshall *et al.* (2010), Hardcastle *et al.* (2016)). From gravitational lensing observations (Barnacka (2018)) of γ-ray blazars it was found that variability could originate from regions many kpc from the black hole (Barnacka *et al.* (2015)) and that γ-ray flares occur at locations distinct from the radio core (Barnacka *et al.* (2016)). These X-ray and γ-ray variability cases constrain the mechanisms of particle acceleration.

1.2. *Outline of this review*

This review considers non-thermal jets from extra-galactic sources, and will emphasize X-ray observations of powerful quasars. In the case of FR-I radio sources, the X-ray jet emission can generally be interpreted as an extension of the radio synchrotron spectrum (Worrall (2005), Harris & Krawczynski (2006), Harris & Krawczynski (2007), Worrall (2009)). The subject of this review will be FR-II quasars, in which case an optical flux or upper limit generally shows that the X-rays can not be an extension of the radio synchrotron emission. In section 3 we will see how this gives information on the relativistic parameters of the jet, and/or on multiple distinct populations of relativistic electrons.

2. Application of the minimum energy assumption

The intensity of synchrotron radiation from a power law distribution of relativistic electrons, $dN/d\gamma = \kappa\gamma^{-(2\alpha+1)}$ is proportional to the product of κ and the magnetic field strength B. Here, α is the energy index of the observed radiation, γmc^2 is the electron energy, and the spectrum is usually assumed to extend from a minimum γ_1 to a maximum γ_2. From the radio synchrotron flux density alone, one cannot determine either the magnetic field strength or the relativistic particle density. Another relation between B and κ can be obtained by assuming minimum total energy in particles and fields, which is nearly equivalent to assuming equipartition of energy between those two channels (Burbidge (1956)). This assumption is now widely used for the interpretation of sources emitting synchrotron radiation.

Miley (1980) has previously discussed the assumptions necessary to apply the minimum energy condition. We update that discussion by considering the relativistic particle spectrum to extend from γ_1 to γ_2 instead of fine tuning the γ to extend from the limits of observed radio frequencies, and by considering the application to X-ray jet measurements. This picture allows the equation for the minimum energy magnetic field strength to be written in the form given by Worrall (2009):

$$\boldsymbol{B_{min}} = f_{min}[\boldsymbol{G(\alpha)}(\boldsymbol{1} + k)\boldsymbol{L_\nu \nu^\alpha}(\boldsymbol{\gamma_1}^{\boldsymbol{1-2\alpha}} - \boldsymbol{\gamma_2}^{\boldsymbol{1-2\alpha}})/(\phi \boldsymbol{lrt})]^{\boldsymbol{1/(\alpha+3)}},$$

where G(α) is a combination of physical constants and functions of α.

In the on-line version we have color-coded the symbols as follows: green symbols for quantities which can be measured directly, namely the spectral index α, the luminosity density L_ν at frequency ν, and the length of the jet element l. In turn, the luminosity is determined from the flux density and the length is determined from the angular extent by using the measured redshift and a cosmological model. Here we use H_0=67.3 km s^{-1} Mpc^{-1} and Ω_m=0.315 from the Planck Collaboration *et al.* (2017) results, in a flat universe. Blue symbols are not directly measured, but have some observational limits. These include the extremes of the power law electron spectrum, γ_1 and γ_2, and the width of the jet r. We note that γ_2 has a negligible effect on the calculation, since it is typically orders of magnitude larger than γ_1. Red symbols are based on intuitive or simplifying assumptions, such as the ratio, k, of energy in relativistic protons to the energy in electrons (including positrons), the filling factor, ϕ, of particles and fields in the jet, the line of sight thickness, t, through the jet, and the assumption, f_{min} that the magnetic field strength corresponds exactly to the minimum energy condition. Here we will take the values $k = \phi = f_{min} = 1$ and t=r. In addition to assuming minimum energy, we must transform observed quantities to the jet rest frame using the Doppler factor $\delta = 1/(\Gamma(1 - \beta\cos(\theta)))$ where θ is the angle of the jet to our line of sight and Γ is the Lorentz factor of the jet relative to the cmb frame. The above discussion shows that we must consider possible magnetic field strength uncertainty of a factor of 2 or 3 in any modeling.

The *Lynx* X-ray observatory (Gaskin *et al.* (2017), Gaskin *et al.* (2018), Özel (2018), Vikhlinin (2018)), which is being studied for submission to the 2020 Decadal Committee for Astronomy and Astrophysics, would have $0\farcs5$ angular resolution and 2 m^2 effective area. The *Lynx* capabilities will allow significant improvements in the determination of some of the quantities needed to calculate B_{min} for X-ray jets. Currently, the modest photon statistics from *Chandra* observations allow only an upper limit of $\approx 0\farcs5$ for the width, r, and assumed thickness, t, of X-ray jets. With the 30-fold increase in collecting area, *Lynx* should allow measurements of widths down to $0\farcs1$, or of order 700 pc even for the most distant jets. At 200—300 eV, the *Lynx* throughput will be 100 times that of *Chandra*. This should allow direct determination of γ_1 via measurement of the soft X-ray turn-over at those energies. With *Chandra* this measurement could only be applied to PKS 0637-752 (Mueller & Schwartz (2009)) due to the build-up of contamination on the filter of the ACIS camera. The improved statistics will also allow precise measurement of the jet X-ray spectral index α. The iC/cmb mechanism assumes this is the same index as the GHz spectrum, although it could be flatter if the radiative lifetime of the GHz emitting electrons is comparable to or less than the age of the jet.

3. The iC/cmb interpretation

The X-ray emission of the luminous jet in PKS 0637-752 could not be explained by reasonable models of synchrotron, inverse Compton, or thermal mechanisms (Schwartz *et al.* (2000), Chartas *et al.* (2000)). Tavecchio *et al.* (2000) and Celotti *et al.* (2001) provided the insight of invoking relativistic bulk motion of the jet with Lorentz factor Γ, and using the result from Dermer and Schlickeiser (1994) that this increased the energy density of the cosmic microwave background in the rest frame of the jet by the factor Γ^2. Subsequently the iC/cmb model has been widely used to interpret the X-ray emission from the jets of powerful quasars (Siemiginowska *et al.* (2002), Sambruna *et al.* (2002), Sambruna *et al.* (2004), Sambruna *et al.* (2006), Marshall *et al.* (2005), Schwartz (2005), Marshall *et al.* (2011), Marshall *et al.* (2018), Schwartz *et al.* (2006), Schwartz *et al.* (2006b), Worrall (2009), Perlman *et al.* (2011), Massaro *et al.* (2011)).

Arguments supporting the iC/cmb model were originally based on the similarity of the X-ray and the radio jet profiles over extended angular distances, with the interpretation that they therefore originated from a single, broad relativistic electron population.

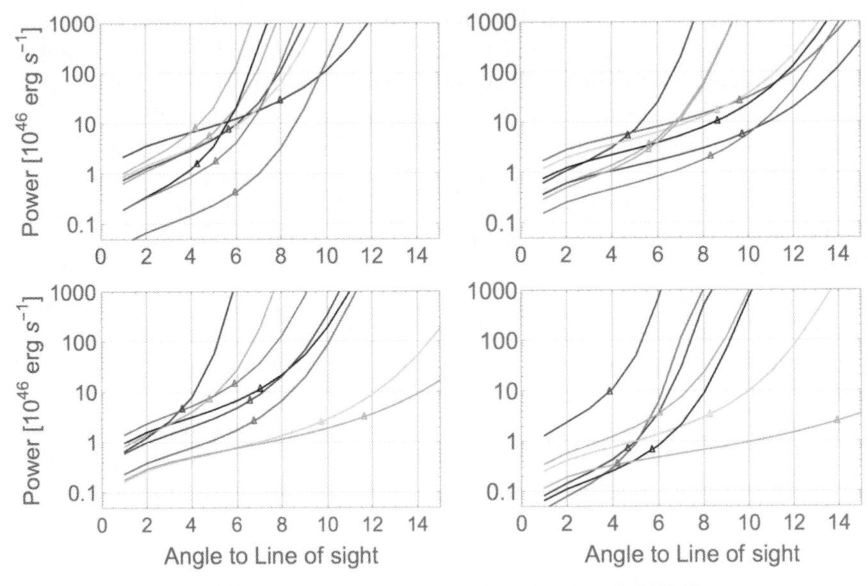

Figure 1. Each curve in one of the panels gives what would be the enthalpy flow in one of the 31 detected jets, calculated as a function of the unknown angle of that jet to our line of sight, and using the iC/cmb model. The triangle on each curve marks the point for which $\Gamma = \delta$, and gives reasonable values for the distribution of parameters of the jet sample.

(Schwartz *et al.* (2000), Schwartz (2005b), Harris *et al.* (2017)), Further evidence is suggested by the termination of X-ray emission just where the radio jet goes through a large change of direction (Schwartz *et al.* (2003), Schwartz (2010)). This is naturally explained by the X-ray to radio ratio of $\delta^{1+\alpha}$ discussed by Dermer (1995). However, powerful evidence against the iC/cmb mechanism in the case of PKS 0637-752 was presented by Meyer *et al.* (2015), and Meyer *et al.* (2017), while Breiding *et al.* (2017) presented similar evidence for 3 jets for which the iC/cmb model had not been indicated. They showed *Fermi* upper limits to GeV γ-ray emission was below the level expected *if* the electron spectrum producing the GHz radio emission also produced the jet emission observed in the *ALMA* and IR/optical regions. In any event, there must be two distinct populations of relativistic electrons, and the X-rays could possibly arise from iC/cmb if the second population produced the *ALMA* and optical emission. The existence of two populations immediately shows that the assumption on the filling factor, $\phi = 1$, is not correct.

In this review we will continue to interpret the X-ray emission as iC/cmb, which we note involves no new parameters for these relativistic jets. In particular, we discuss the results from the systematic survey of Marshall *et al.* (2005), Marshall *et al.* (2011), Marshall *et al.* (2018) and previously presented by Schwartz *et al.* (2015). Figure 1 shows the deduced enthalpy power (Bicknell (1994)) carried by the 31 detected jets, parameterized as a function of the unknown angle to our line of sight. Qualitatively it is apparent that very small angles imply a very large population of unrecognized sources which lie at larger angles. Large angles, greater than 6° to 12° for most of the objects, imply unrealistically large powers; namely, greater than 10^{49} ergs s^{-1}. An assumption is often made that $\Gamma = \delta$. This corresponds to the jet being at the largest possible angle to our line of sight for the given value of δ. Those points are indicated by the triangles on each curve, and we expect they give a reasonable estimate of the distribution of results from the sample.

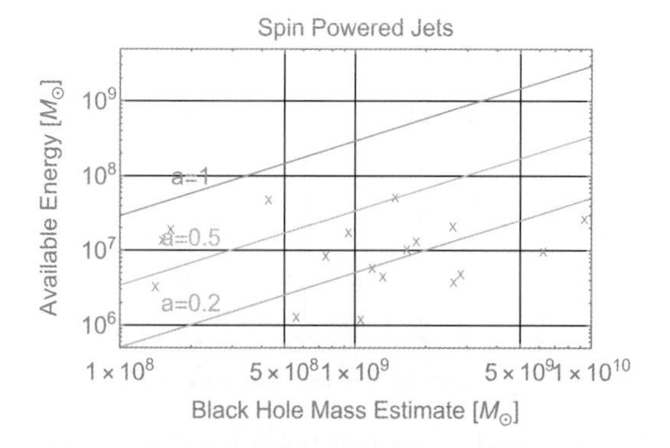

Figure 2. The solid lines show the available rotational energy, in terms of rest mass, which can be recovered by optimal spin-down of a black hole as a function of initial mass. The different spin parameters a=1,0.3,0.2, respectively allow 29%, 1%, and 0.5% of the mass-energy of the black hole to be recovered in principle. The x's plot the amount of energy required to sustain the power of the jets for 10^7 years. The black hole mass estimate here is taken as a median value of estimates in the literature tabulated by Shen *et al.* (2011), and by Xiong & Zhang (2014) and references therein. Even black holes with only 0.2 to 0.3 of the maximum possible angular momentum, can power the jets we measured for millions of years.

4. Connection to the super-massive black hole

The rotational energy, E_r of a Kerr black hole manifests as a contribution $E_r/c^2=M_r$ to the total mass of the black hole. The relation is non-linear, with the fraction

$$M_r/M = 1 - \sqrt{0.5(1+\sqrt{(1-a^2/M^2)})}$$

being available in principle to be expelled. Here, a is the spin parameter in units of the total black hole mass M. The solid lines in Figure 2 show the available mass that can, in principle, be released as a function of the total black hole mass. The $x's$ plot the mass-energy equivalent required to energy to power the jets for 10^7 years, as determined from the $\Gamma = \delta$ point in Figure 1. We plot those masses against the estimated mass of the given black hole. That black hole mass is taken from the median of values given in the literature for 18 objects where it is available. We see that maximally spinning black holes, $a \approx 1$ can provide more than the required energy, and super-massive black holes with spin parameters even as low as 0.3 could in principle power the observed jets for millions of years. Of course, this consideration does not address the dynamics of actually extracting such energy in a collimated flow.

Numerical calculations of the magnetically arrested disk (MAD) model (Narayan *et al.* (2003), Igumenshchev (2008), Sądowski *et al.* (2014), Tchekhovskoy *et al.* (2011), Zamaninasab *et al.* (2014)) for a rapidly spinning black hole have had some success showing that jets can be formed, and can extract more than the potential energy of the accreting matter. A magnetic field is advected in with the accreting matter, and is compressed and amplified until its pressure balances the gravitation pull (Narayan *et al.* (2003)). This results in chaotic variability of the accretion, and the rapidly spinning black hole wraps the magnetic field lines into a tight spiral about the spin axis. Observations of rotation measure gradients across pc-scale radio jets provides evidence for such field geometry (Gabuzda (2014), Gabuzda *et al.* (2015), Gabuzda *et al.* (2017)). The magnetic pressure then causes ejection along the spin axis. If we assume that the initial ejection right at the gravitational radius of an extreme Kerr black hole is a pure Poynting flux, then we can obtain a lower limit to that initial magnetic field by assuming 100% efficiency

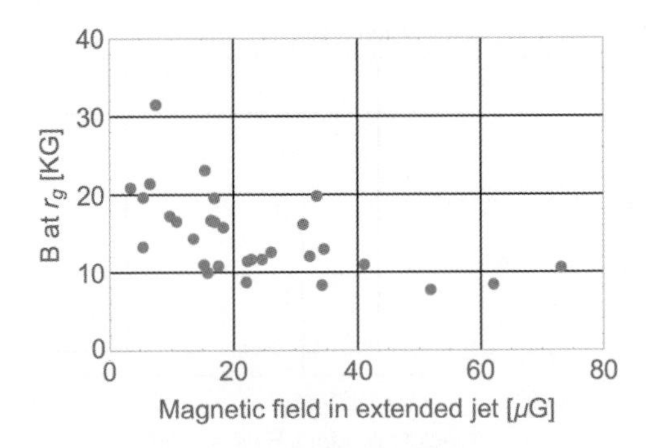

Figure 3. Ordinate: Magnetic field strength at the gravitational radius such that the Poynting vector equals the power deduced for the kpc-scale jet. Abscissa: Mean magnetic field strength in the rest frame of the kpc scale jet, according to the iC/cmb model. Points are the 31 jets detected in the survey discussed in section 3. (In view of numerous uncertainties, we do not consider the possible anti-correlation to be significant.)

for converting that Poynting flux into the energy flux deduced for the kpc-scale jet. Such initial field strengths are shown in Figure 3. We have taken the mass from fundamental plane relation of Gültekin *et al.* (2009) to deduce the gravitational radius r_g of each quasar.

5. Summary

We use Chandra X-ray observations to estimate the power of quasar jets, by observing the jet itself. We tie this to the central black hole mass on an individual object basis. The rotational energy of super-massive black holes can power these quasar jets, even with spin parameters as low as a=0.2, for lifetimes longer than millions of years. If the power we observe originates as a pure Poynting flux, we derive initial magnetic field strengths of a few 10's of kilo-Gauss. For models of magnetically arrested disks this inferred magnetic flux is of the order of magnitude of predictions, for Eddington limited accretion onto maximally spinning super-massive black holes.

References

Barnacka, A., Geller, M. J., Dell'Antonio, I. P., & Benbow, W. 2015, *ApJ*, 809, 100
Barnacka, A., Geller, M. J., Dell'Antonio, I. P., & Zitrin, A. 2016, *ApJ*, 821, 58
Barnacka, A. 2018, *Physics Reports*, 778, 1
Begelman, M. C., Blandford, R. D., & Rees, M. J. 1984, *Reviews of Modern Physics*, 56, 255
Bicknell, G. V. 1994, *ApJ*, 422, 542
Blandford, R. D., & Rees, M. J. 1974, *MNRAS*, 169, 395
Breiding, P., Meyer, E. T., Georganopoulos, M., *et al.* 2017, *ApJ*, 849, 95
Bridle, A. H., & Perley, R. A. 1984, *AAR&A*, 22, 319
Burbidge, G. R. 1956, *ApJ*, 124, 416
Celotti, A., Ghisellini, G., & Chiaberge, M. 2001, *MNRAS*, 321, L1
Chartas, G., *et al.* 2000, *ApJ*, 542, 655
Dermer, C. D. 1995, *ApJL*, 446, L63
Dermer, C. D. & Schlickeiser, R. 1994, *ApJS*, 90, 945
Fabian, A. C., Sanders, J. S., Ettori, S., *et al.* 2000, *MNRAS*, 318, L65
Feigelson, E. D., Schreier, E. J., Delvaille, *et al.* 1981, *ApJ.*, 251, 31
Gaskin, J. A., Allured, R., Bandler, S. R., *et al.* 2017, *SPIE Conference Series*, 10397, 103970S
Gaskin, J. A., Dominguez, A., Gelmis, K., *et al.* 2018, *SPIE Conference Series*, 10699, 106990N
Gabuzda, D. 2014, *Nature*, 510, 42
Gabuzda, D. C., Knuettel, S., & Reardon, B. 2015, *MNRAS*, 450, 2441
Gabuzda, D. C., Roche, N., Kirwan, A., *et al.* 2017, *MNRAS*, 472, 1792

Gültekin, K., Cackett, E. M., Miller, J. M., *et al.* 2009, *ApJ*, 706, 404
Hardcastle, M. J., Lenc, E., Birkinshaw, M., *et al.* 2016, *MNRAS*, 455, 3526
Harris, D. E., & Krawczynski, H. 2006, *ARAA*, 44, 463
Harris, D. E., & Krawczynski, H. 2007, *Revista Mexicana de Astronomia y Astrofisica*, vol. 27, 27, 188
Harris, D. E., Lee, N. P., Schwartz, D. A., *et al.* 2017, *ApJ*, 846, 119
Igumenshchev, I. V. 2008, *ApJ*, 677, 317
Longair, M. S., Ryle, M., & Scheuer, P. A. G. 1973, *MNRAS*, 164, 243
Marshall, H. L., Schwartz, D. A., Lovell, J. E. J., *et al.* 2005, *ApJS*, 156, 13
Marshall, H. L., Hardcastle, M. J., Birkinshaw, M., *et al.* 2010, *ApJL*, 714, L213
Marshall, H. L., Gelbord, J. M., Schwartz, D. A., *et al.* 2011, *ApJS*, 193, 15
Marshall, H. L., Gelbord, J. M., Worrall, D. M., *et al.* 2018, *ApJ*, 856, 66
Massaro, F., Harris, D. E., & Cheung, C. C. 2011, *ApJS*, 197, 24
Meyer, E. T., Georganopoulos, M., Sparks, W. B., *et al.* 2015, *ApJ*, 805, 154
Meyer, E. T., Breiding, P., Georganopoulos, M., *et al.* 2017, *ApJL*, 835, L35
Miley, G. 1980, *ARAA*, 18, 165
Morrison, P. 1958, *Nuovo Cimento*, 7, 858
Mueller, M., & Schwartz, D. A. 2009, *ApJ*, 693, 648
Narayan, R., Igumenshchev, I. V., & Abramowicz, M. A. 2003, *PASJ*, 55, L69
Narayan, R., McClintock, J. E., & Tchekhovskoy, A. 2014, in: J. Bičák & T. Ledvinka (eds.), *General Relativity, Cosmology and Astrophysics* (Springer) p. 523
Özel, F. 2018, *Nature Astronomy*, 2, 608
Perley, R. A., Willis, A. G., & Scott, J. S. 1979, *Nature*, 281, 437
Perlman, E. S., Georganopoulos, M., Marshall, H. L., *et al.* 2011, *ApJ*, 739, 65
Planck Collaboration, Aghanim, N., Akrami, Y., *et al.* 2017, *A&Ap*, 607, A95
Readhead, A. C. S., Cohen, M. H., & Blandford, R. D. 1978, *Nature*, 272, 131
Rees, M. J. 1971, *Nature*, 229, 312
Sądowski, A., Narayan, R., McKinney, J. C., & Tchekhovskoy, A. 2014, *MNRAS*, 439, 503
Sambruna, R. M., Maraschi, L., Tavecchio, F., Urry, C. M., Cheung, C. C., Chartas, G., Scarpa, R., & Gambill, J. K. 2002, *ApJ*, 571, 20
Sambruna, R. M., Gambill, J.K., Maraschi, L., Tavecchio, F., Cerutti, R., Cheung, C. C., Urry, C. M., & Chartas, G., 2004, *ApJ*, 608, 698
Sambruna, R. M., Gliozzi, M., Donato, D., *et al.* 2006, *ApJ*, 641, 717
Savedoff, M. P. 1959, *Nuovo Cimento*, 13, 12
Scheuer, P. A. G. 1974, *MNRAS*, 166, 513
Schreier, E. J., Gorenstein, P., & Feigelson, E. D., 1982, *ApJ*, 261, 42
Schwartz, D. A., *et al.* 2000, *ApJL*, 540, L69
Schwartz, D. A., Marshall, H. L., Miller, B. P., *et al.* 2003, in: S. Collin, F. Combes and I. Shlosman (eds.) *ASP Conference Series*, 290, 359
Schwartz, D. A. 2005, *EAS Publications Series*, 15, 353
Schwartz, D. A. 2005, in: Pisin Chen, Elliott Bloom, Greg Madejski, and Vahe Patrosian (eds.), *22nd Texas Symposium on Relativistic Astrophysics*, p.38
Schwartz, D. A., Marshall, H. L., Lovell, J. E. J., *et al.* 2006, *ApJ*, 640, 592
Schwartz, D. A., Marshall, H. L., Lovell, J. E. J., *et al.* 2006b, *ApJL*, 647, L107
Schwartz, D. 2010, *PNAS*, 107, 7190
Schwartz, D. A. 2014, *Rev. Sci. Inst.*, 85, 061101
Schwartz, D. A., Marshall, H. L., Worrall, D. M., *et al.* 2015, *IAU Symposium Extragalactic Jets from Every Angle*, 313, 219
Shen, Y., Richards, G. T., Strauss, M. A., *et al.* 2011, *ApJS*, 194, 45
Siemiginowska, A, Bechtold, J., Aldcroft, T. L., Elvis, M., Harris, D. E., & Dobrzycki, A. 2002, *ApJ*, 570, 543
Tavecchio, F.,Maraschi, L., Sambruna, R. M., & Urry, C. M. 2000, *ApJL*, 544, L23
Tchekhovskoy, A., Narayan, R., & McKinney, J. C. 2011, *MNRAS*, 418, L79
Turland, B. D. 1975, *MNRAS*, 172, 181

Vikhlinin, A. 2018, *AAS Meeting* #231, 231, 103.04
Waggett, P. C., Warner, P. J., & Baldwin, J. E. 1977, *MNRAS*, 181, 465
Weisskopf, M. C., Brinkman, C., Canizares, C., *et al.* 2002, *PASP*, 114, 1
Weisskopf, M. C., Aldcroft, T. L., Bautz, M., *et al.* 2003, *Exp. Astron.*, 16, 1
Willingale, R. 1981 *MNRAS*, 194, 359
Worrall, D. M. 2005, *Highlights of Astronomy*, 13, 685
Worrall, D. M. 2009, *A&ARv*, 17, 1
Xiong, D. R., & Zhang, X. 2014, *MNRAS*, 441, 3375
Zamaninasab, M., Clausen-Brown, E., Savolainen, T., & Tchekhovskoy, A. 2014, *Nature*, 510, 126

Astronomy in Focus - XXX
Proceedings IAU Symposium No. XXX, 2018
M. T. Lago, ed.

doi:10.1017/S1743921319003466

Extreme jet distortions in low-z radio galaxies

Mark Birkinshaw[1] Josie Rawes[2] and Diana Worrall[3]

[1]HH Wills Physics Laboratory, University of Bristol,
Tyndall Avenue, Bristol BS8 1TL, U.K.
email: Mark.Birkinshaw@bristol.ac.uk

[2]HH Wills Physics Laboratory, University of Bristol,
Tyndall Avenue, Bristol BS8 1TL, U.K.
email: J.Rawes@bristol.ac.uk

[3]HH Wills Physics Laboratory, University of Bristol,
Tyndall Avenue, Bristol BS8 1TL, U.K.
email: D.Worrall@bristol.ac.uk

Abstract. Jets often display bends and knots at which the flows change character. Extreme distortions have implications for the nature of jet flows and their interactions. We present the results of three radio mapping campaigns. The distortion of 3CRR radio galaxy NGC 7385 is caused by a collision with a foreground magnetised gas cloud which causes Faraday rotation and free-free absorption, and is triggered into star formation. For NGC 6109 the distortion is more extreme, creating a ring-shaped structure, but no deflector can be identified in cold or hot gas. Similar distortions in NGC 7016 are apparently associated with an X-ray gas cavity, and the adjacent NGC 7018 shows filaments drawn out beyond 100 kpc. Encounters with substructures in low-density, magnetised, intergalactic gas are likely causes of many of these features.

Keywords. Galaxies:active, galaxies:jets, radio continuum:galaxies, intergalactic medium

---

1. NGC 7385

NGC 7385 is radio-bright and a member of the complete 3CRR sample (Laing, Riley & Longair 1983) despite failing to be in the 3C catalogue because of its low redshift ($z = 0.0243$) and consequent large angular size. In low-resolution radio maps (e.g., Schilizzi & Ekers 1975) NGC 7385 appears as a radio trail source, with two distinct tails appearing about 300 kpc southwest of the core. The higher-resolution study by Simkin & Ekers (1979) found that the northeastern side of the source encounters an optically-bright, line-emitting cloud about 10 kpc from the nucleus of NGC 7385. The radio structure seems to disrupt at this point.

We have undertaken new radio, optical, and X-ray imaging of NGC 7385. An L-band radio map constructed from archival VLA data is shown in Fig. 1 superimposed on a galaxy-subtracted HST image (Rawes, Worrall & Birkinshaw 2015). The main jet extends to the southwest and is seen in the optical and X-ray in its inner regions. The wide bandwidth of our new data (Rawes, Birkinshaw & Worrall 2018b), allows spectral indices and Faraday rotation to be mapped at high angular resolution. Prominent features are associated with the gas cloud, showing that it lies on the near side of the counter-jet plume. The free-free optical depth and rotation measure through the cloud can be reconciled with a gas density of about 5 cm^{-3} and a magnetic field of order 1 μG provided that the cloud is strongly clumped.

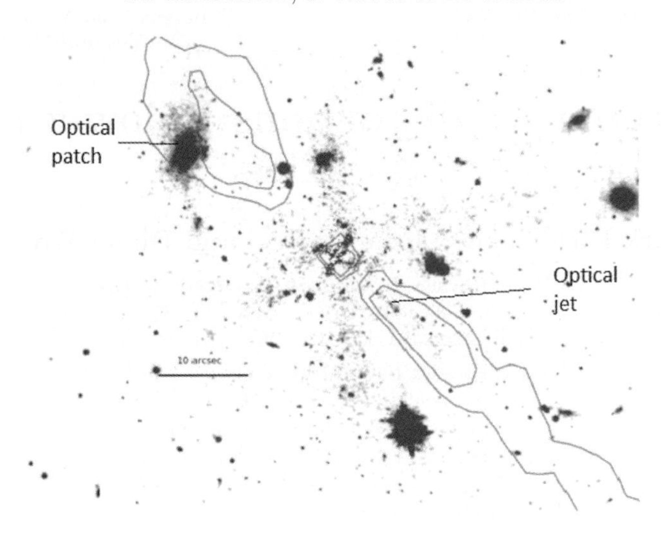

Figure 1. The centre of NGC 7385. At $z = 0.0243$ 10 arcsec corresponds to a projected scale of 4.9 kpc. Galaxy-subtracted HST F160W image with superimposed radio contours, from Rawes, Worrall & Birkinshaw (2015). An optical/X-ray jet extends southwest from the core. To the northeast the counter-jet is deflected and disrupted by the optical patch of Simkin & Ekers (1979), but retains enough integrity to flow back behind the main jet. We detect free-free absorption and Faraday rotation from the optical patch (Rawes, Birkinshaw & Worrall 2018b).

The strength of the cloud/counter-jet interaction causes the flow to turn through about 180 degrees, with the expanded, post-encounter, flow running back almost along the projected main jet, so that low-resolution maps show only a one-sided flow until different environmental conditions cause the projected structure to separate into two tails. This is likely due to gas flows and substructures on 100-kpc scales in the intergalactic medium of Zwicky cluster Zw 2247.3+1107, of which NGC 7385 is a member.

A high signal/noise IFU study of the cloud interaction might provide information about the momentum flux down the jet, and so potentially the jet composition, in the same way as has been possible for other jet/cloud interactions (e.g., PKS 2152-699; Worrall *et al.* 2012, Smith *et al.* 2018). While such interactions are rare, they can provide the kinematic information that cannot be obtained from jet synchrotron (or synchrotron plus inverse-Compton) emission alone.

2. NGC 6109

NGC 6109, another neglected member of the 3CRR sample, exhibits a striking example of a radio ring. This low-redshift ($z = 0.0296$) galaxy appears, at low angular resolution, to be a radio trail source. O'Dea & Owen (1985) remarked that the source exhibited a small circular component to one side of the core, and a long tail to the other, and noted that this might indicate that the jet on one side was stopped by external gas, to create an FR II-like lobe, while it was drawn out into a long tail on the other.

Our high-resolution radio map (Fig. 2), shows that the southeastern feature is not a lobe but a circular loop (Rawes, Birkinshaw & Worrall 2018a). No excess or deficit of X-ray emission is seen in the neighbourhood of the loop, and there is no optical or infra-red emission, so the mechanism causing the loop is not like the interaction in NGC 7385.

The brightness contrast between the jet and counter-jet on sub-kpc scales can be interpreted in terms of a relativistic Doppler factor that associates the loop with the jet brightening 12-24 arcsec northwest of the core, so that both could arise from a single event in the AGN. However, relativistic flow and simple precession models cannot explain

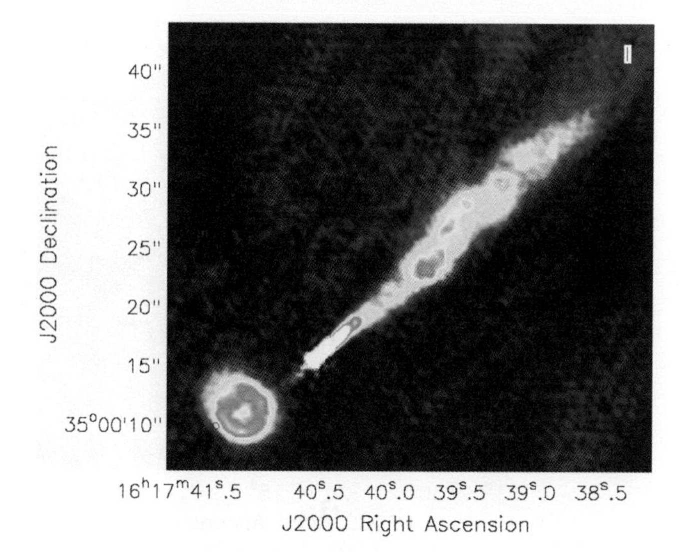

Figure 2. The centre of NGC 6109. At $z = 0.0296$ 5 arcsec corresponds to a projected scale of 3.0 kpc. S-band radio surface brightness image from Rawes, Birkinshaw & Worrall (2018a). The main jet extends northwest of the core for about 5 arcsec before fading, and then re-brightening in a less collinear structure 12-24 arcsec from the core. The counter-jet is faintly detected to about 5 arcsec, then develops a bright loop structure with radius 5 arcsec that lies outside the flattened X-ray emitting atmosphere detected by *Chandra*.

the structure: there must be asymmetrical acceleration/deceleration of the jet/counter-jet flow by the external medium. Although *Chandra* imaging shows little structured gas in the neighbourhood of NGC 6109, the galaxy is embedded in the large-scale atmosphere of a poor cluster, and our new polarisation data shows a Faraday rotation feature across the loop that could suggest interaction with a magnetic substructure such as a galactic wake in the intracluster medium.

3. Abell 3744

NGC 7016 and 7018 lie in Abell 3744, a moderately rich ($kT \approx 3$ keV) cluster of galaxies at $z = 0.0381$. Both galaxies host bright radio sources. The heating effects of the sources should combine to cause significant changes in the intracluster medium. Strong temperature and density structures are, indeed, seen across the cluster though these may be a consequence of a merger (Worrall & Birkinshaw 2014). Both radio sources are highly distorted (Cameron 1988; Bicknell, Cameron & Gingold 1990; Fig. 3), confirming that their interactions with the intracluster medium are strong.

The brightest parts of NGC 7018 show it to be of FR II character, with prominent hot spots at the ends of the lobes, a compact core, and a jet extending from the core towards the NE hot spot. Faint filamentary structures extend from both lobes and lie *around* a prominent cavity in the cluster's X-ray emission (Worrall & Birkinshaw 2014). One filament, < 2 kpc in diameter, can be traced to > 150 kpc, where it seems to end in a faint radio source that might have punctured one lobe of NGC 7018 (Birkinshaw, Worrall & Rawes 2018).

NGC 7016 shows a bent, twin-sided, FR I structure in its inner parts. However, to the north the jet appears to reverse direction and expand, and then to create a radio ring (Worrall & Birkinshaw 2014), before disrupting into a faint array of radio filaments and flowing back parallel to the southern jet (Birkinshaw, Worrall & Rawes 2018). This is seen as a broad structure around the jet in low-resolution maps. The sequence of structures,

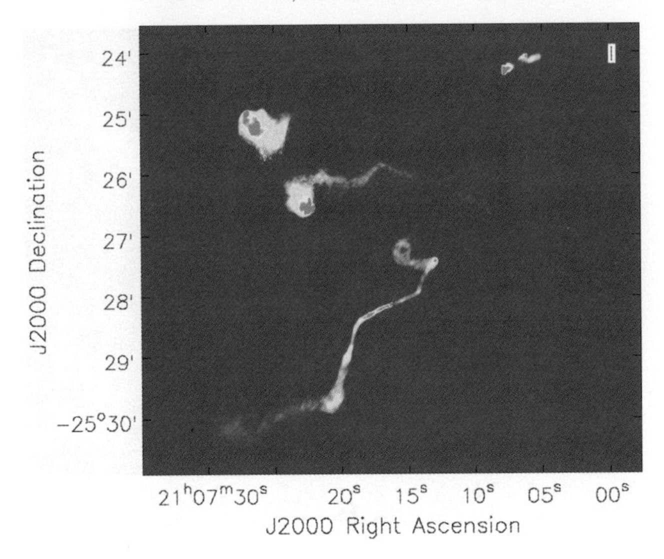

Figure 3. Abell 3744 from Birkinshaw, Worrall & Rawes (2018). At $z = 0.0381$ 5 arcmin corresponds to a projected scale of 230 kpc. The image is dominated by NGC 7016 to the southwest and NGC 7018 to the notheast. An unusual (background?) radio source lies to the northeast.

with the ring downstream of the reversal, might indicate that the reversal is a younger version of the ring, and will develop into a full ring in a few Myr.

The X-ray cavity would collapse rapidly without internal pressure support. This can arise from the repository of electrons within the cavity that produce the diffuse radio emission mapped by Cameron (1988). The irregular X-ray structure suggests that Abell 3744 is dynamically young (Worrall & Birkinshaw 2014), so that rapid gas motions are likely and could cause some of the radio distortions, though the ring and reversal features seem surprisingly compact unless they are short-lived.

4. Generalisation

The extreme distortions in Figs. 1–3 were detected in arcsec-resolution, high dynamic-range, low-frequency mapping. High resolution is necessary, even at low redshift, to detect the kpc-scale substructures we are reporting. Extreme distortions cannot be rare, since we have two examples (NGC 6109 and NGC 7385) in the low-redshift 3CRR sample of only 35 radio galaxies, so the SKA should find many such structures in the future, as its sub-arcsec capabilities reveal them at redshifts beyond $z = 1$.

While some distortions arise from interactions with optically-emitting gas, where there is the potential for obtaining important dynamical information about the jet flows (e.g., NGC 7385, PKS 2152-699), sometimes there is no clear deflecting gas (NGC 7016, NGC 6109). The evolution of dynamical activity in clusters and groups around low-thrust radio sources could be studied by future radio mapping.

References

Bicknell, G. V., Cameron, R. A., & Gingold, R. A. 1990, *ApJ*, 357, 373
Birkinshaw, M., Worrall, D. M., & Rawes, J. 2018, in preparation
Cameron, R. A. 1988, PhD thesis, ANU
Laing, R. A., Riley, J. M., & Longair, M. S. 1979, *MNRAS*, 204, 151
O'Dea, C., & Owen, F. 1985, *AJ*, 90, 927
Rawes, J., Worrall, D. M., & Birkinshaw, M. 2015, *MNRAS*, 452, 3064
Rawes, J., Birkinshaw, M., & Worrall, D. M. 2018a, *MNRAS*, 480, 3644

Rawes, J., Birkinshaw, M., & Worrall, D. M. 2018b, *MNRAS*, in preparation
Schilizzi, R., & Ekers, R. 1975, *A&A*, 40, 221
Simkin, S. M., & Ekers, R. D. 1979, *AJ*, 84, 56
Smith, D. P., Young, A. J., Worrall, D. M. & Birkinshaw, M. 2018, *MNRAS*, submitted
Worrall, D. M., & Birkinshaw, M. 2014, *ApJ*, 784, 36
Worrall, D. M., Birkinshaw, M., Young, A. J, *et al.* 2012, *MNRAS*, 424, 1346

Astronomy in Focus - XXX
Proceedings IAU Symposium No. XXX, 2018
M. T. Lago, ed.

doi:10.1017/S1743921319003478

Probing restarting activity in hard X-ray selected giant radio galaxies

G. Bruni[1]†, F. Ursini[2], F. Panessa[1], L. Bassani[2], A. Bazzano[1], A. J. Bird[3], E. Chiaraluce[1], D. Dallacasa[4,5], M. Fiocchi[1], M. Giroletti[5], L. Hernández-García[6], A. Malizia[2], M. Molina[2], L. Saripalli[7], P. Ubertini[1] and T. Venturi[5]

[1]INAF - Istituto di Astrofisica e Planetologia Spaziali
via del Fosso del Cavaliere 100, 00133 Roma, Italy
email: gabriele.bruni@inaf.it

[2]INAF - Osservatorio di Astrofisica e Scienza dello Spazio
via Piero Gobetti 93/3, 40129 Bologna, Italy
email: francesco.ursini@inaf.it

[3]School of Physics and Astronomy, University of Southampton, SO17 1BJ, UK

[4]DIFA - Dipartimento di Fisica e Astronomia
Università di Bologna, via Gobetti 93/2, 40129 Bologna, Italy

[5]INAF - Istituto di Radioastronomia, via Piero Gobetti 101, 40129 Bologna, Italy

[6]IFA - Instituto de Física y Astronomía
Universidad de Valparaíso, Gran Bretaña 1111, Playa Ancha, Valparaíso, Chile

[7]Raman Research Institute, C. V. Raman Avenue, Sadashivanagar, Bangalore 560080, India

Abstract. With their sizes larger than 0.7 Mpc, Giant Radio Galaxies (GRGs) are the largest individual objects in the Universe. To date, the reason why they reach such enormous extensions is still unclear. One of the proposed scenarios suggests that they are the result of multiple episodes of jet activity. Cross-correlating the INTEGRAL+Swift AGN population with radio catalogues (NVSS, FIRST, SUMSS), we found that 22% of the sources are GRG (a factor four higher than those selected from radio catalogues). Remarkably, all of the sources in the sample show signs of restarting radio activity. The X-ray properties are consistent with this scenario, the sources being in a high-accretion, high-luminosity state with respect to the previous activity responsible for the radio lobes.

Keywords. galaxies: active, galaxies: evolution, galaxies: jets, radio continuum: galaxies, X-rays: galaxies

---

1. Introduction

A relatively small fraction of powerful radio galaxies (~6% in the 3CR catalogue, Ishwara-Chandra & Saikia 1999) exhibits rather large linear extents, i.e. above 0.7 Mpc, making them the largest individual objects in the Universe. These sources are usually referred to as Giant Radio Galaxies (GRG), and can exhibit both Fanaroff-Riley type I and type II radio galaxies (FRI and FRII respectively, Fanaroff & Riley 1974). While FRI GRGs are associated with early type galaxies, those with FRII morphology are hosted both in early type galaxies and quasars. The samples of GRGs available in the literature, mainly drawn from all sky radio surveys such as NVSS, SUMSS, WENSS, have been used to test models for radio galaxy evolution and investigate the origin of

† website: http://gral.iaps.inaf.it

such incredibly extended structures (i.e. Blundell *et al.* 1999). Despite the dynamical ages typically overestimate the radiative age of the radio source by a factor 2-4 (see Fig. 5 in Parma *et al.* 1999), the general correlation observed between size and age in radio galaxies (Fig. 6 in Parma *et al.* 1999) suggests that GRGs represent the oldest tail of the age distribution for radio galaxies. Beyond the source age, the main intrinsic parameters that allow a radio galaxy to reach a linear size of the order of a Mpc during its lifetime are still unclear. The medium must play a role in the overall jet expansion, but its effects remains difficult to evaluate, not to mention that the density of the medium explored by the radio jet during its life/development may change considerably over the large scales considered here. Some GRGs are associated with the dominant member of a galaxy group (e.g. the FRI-GRG NGC 315, Giacintucci *et al.* 2011), while others have been detected at high redshift in a likely less dense environment (Machalski *et al.* 2004). Those authors also concluded that the jet power and the central density of the galaxy nucleus seem to correlate with the size of radio galaxies. Yet another study, based on optical spectroscopy of galaxies in a large-scale environment around the hosts of 19 GRGs (Malarecki *et al.* 2004) finds a tendency for their lobes to grow to giant sizes in directions that avoid dense galaxy on both small and large scales. Finally, other authors suggested that GRG could reach their size thanks to more than one activity episode, being restarting radio sources (Subrahmanyan *et al.* 1996). More recently, progress in the study of this class of sources has been achieved thanks to the use of low frequency facilities - such as LOFAR, JVLA and GMRT - where old relativistic plasma is better seen (e.g. Orrù *et al.* 2010; Clarke *et al.* 2017; Sebastian *et al.* 2018). All in all, however, the origin and evolution of GRGs remains until now very much unconstrained.

2. Hints of restarting activity

Starting from 2002, the hard X-ray and γ-ray sky has been surveyed by *INTEGRAL*/IBIS and *Swift*/BAT in the spectral range from 10-200 keV. Up to now many catalogues have been released, the most recent ones comprising more than 1000 high energy sources (Bird *et al.* 2016; Oh *et al.* 2018), with a large fraction of objects unambiguously associated with AGN. Our group is carrying out a multi-wavelength study of a sample of hard X-ray selected GRG extracted from these high energy catalogues. Bassani *et al.* (2016) undertook a radio/γ-ray study of the combined *INTEGRAL+Swift* AGN populations, and found 64 sources associated with extended radio galaxies with measured redshift. They belong to both the FRI and FRII morphological classes. Interestingly, inspection of NVSS and SUMSS revealed that 14 of them are GRG, i.e. $\sim$22% of the sample. Considered the classical fraction of giant sources in radio-selected samples of radio galaxies (1-6%), this fraction is impressive, and suggests a tight link between the nuclear/accretion properties of the AGN and the radio source size.

In order to better characterize the GRG in this sample and to study their evolutionary history, we are collecting multi-band high sensitivity observations for the fraction of sample observable from the Northern hemisphere (12 targets). Remarkably, all of the sources in the sample show signs of restarting radio activity from previous studies in the literature or our radio campaign (Bruni *et al.* 2019, Bruni *et al.* in prep.), with one even showing an extreme reorientation from radio-galaxy to BL Lac (Hernández-García *et al.* 2017). This large fraction suggests that multiple radio phases could justify the large size, and be responsible for the strong hard X-ray emission coming from a possible refueled radio core - as suggested from X-ray data (Ursini *et al.* 2018). Our aim is to confirm the restarting activity for these sources, and possibly understand whether this scenario is a distinctive property of hard X-ray selected GRGs, or rather a general property of GRGs. In 2014, we have collected GMRT deep observations at 325 and 610 MHz for a pilot sample of 4 sources from our GRG sample. Those GMRT data allowed our group to

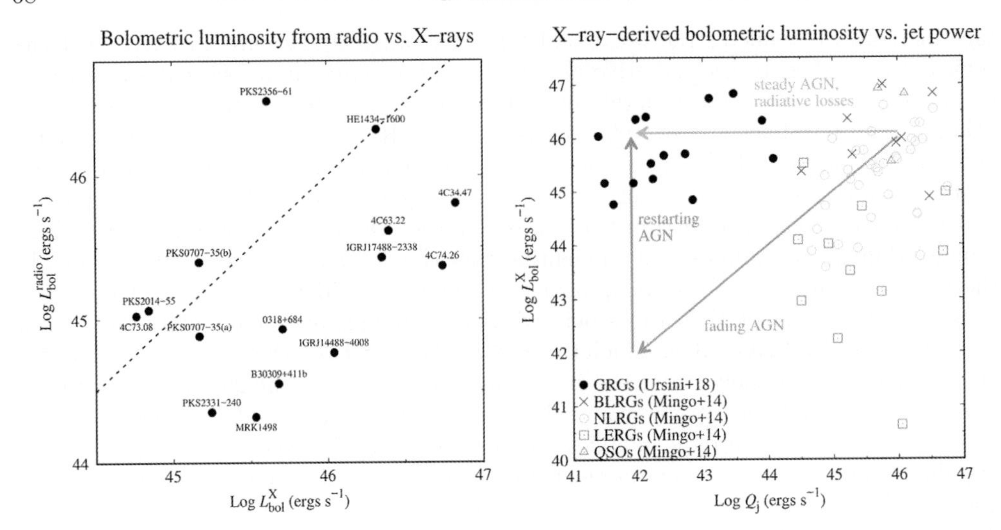

Figure 1. *Left panel:* Bolometric luminosity estimated from the radio luminosity of the lobes versus that estimated from the 2–10 keV luminosity. The dashed line represents the identity $y = x$. *Right panel:* Bolometric luminosity estimated from the 2–10 keV luminosity versus jet power estimated from the relation of Willott *et al.* (1999). Black dots denote the GRGs of our sample, overlayed in the plot of Mingo *et al.* (2014). The colored arrows represent two putative evolutionary paths of radio galaxies. Both panels are adapted from Ursini *et al.* (2018).

identify the second known giant X-shaped radio galaxy (IGR J14488-4008: Molina *et al.* 2015), and a newly discovered GRG (IGR J17488-2338: Molina *et al.* 2014).

3. X-ray properties

The bulk of the X-ray emission of GRGs in our sample is consistent with originating from a Comptonizing corona coupled to a radiatively efficient accretion flow (Eddington ratio > 0.02) (Ursini *et al.* 2018), like in normal-size FR II radio galaxies. This indicates that the nuclei are currently active, despite the likely old age of the radio lobes. The peculiar morphology makes it possible to study the relation between the X-ray emission and the radio emission in detail, separating the contribution from the core and from the lobes. We find that:

- The X-ray luminosity $L_{2\text{–}10\text{keV}}$ correlates with the radio core luminosity $L_{1.4\text{GHz}}^{\text{core}}$, as expected from the so-called fundamental plane of black hole activity (Merloni *et al.* 2003). The slope of the correlation is consistent with the 'radiatively efficient' branch of the fundamental plane rather than the 'standard/inefficient' branch (Coriat *et al.* 2011).

- In most sources, the X-ray luminosity yields an estimate of the bolometric luminosity ($L_{\text{bol}}^{\text{X}}$) an order of magnitude larger than the corresponding estimate from the radio lobes luminosity ($L_{\text{bol}}^{\text{radio}}$), from the relation of van Velzen *et al.* (2015) (Fig. 1, left panel).

- The time-averaged kinetic power of the jets, as estimated from the radio luminosity using the relation of Willot *et al.* (1999), is much lower than in the radio luminous AGNs studied by Mingo *et al.* (2014). This discrepancy is up to 3 orders of magnitude, while the bolometric luminosity is perfectly consistent with high-excitation radio galaxies (Fig. 1, right panel).

These results are consistent with a restarting activity scenario, i.e. the sources are currently highly accreting and in a high-luminosity state compared with the past activity that produced the old and extended radio lobes. GRGs could start their life with high nuclear luminosities and high jet powers; with time, the central engine gradually fades while the radio lobes expand. Eventually, the nuclear activity can be triggered again following a new accretion episode, resulting in a strong increase of the core luminosity. Alternatively, the nuclei would need to sustain a steady activity during their lifetime, with a nearly constant accretion rate and core luminosity; radiative losses would produce the observed dimming of the radio lobes as they grow in size and interact with the environment. In this case, however, the nuclei are required to stay active for at least 100–250 Myrs (Machalski *et al.* 2004).

4. Future work

Triggered by these results, we are collecting more data both in radio and X-ray bands, in order to connect the nuclear accretion status to the Mpc-scale structure of these objects. The cores have been observed in single-dish mode (Effelsberg-100m telescope) in June 2018, with the aim of reconstructing the radio SED and test the fraction of young radio components. In the X-ray band, a *Swift*/XRT campaign has been planned to build a complete comparison sample of radio-selected GRG (Schoenmakers *et al.* 2000), and highlight the different properties with respect to the hard X-ray selected ones. Indeed, radio selected GRGs are not necessarily X-ray bright, thus allowing us to explore new portions of the parameter space in luminosity-luminosity diagrams such as those in Fig. 1.

References

Bassani, L., Venturi, T., Molina, A., *et al.* 2016, *MNRAS*, 461, 3165
Bird, A. J., Bazzano, A., Malizia, A., *et al.* 2016, *MNRAS*, *ApJS*, 223, 15
Blundell, K. M., Rawlings, S., & Willott, C. J. 1999, *AJ*, 117, 677
Bruni, G., Panessa, F., Bassani, G., *et al.* 2019, *ApJ*, 875, 88B
Clarke, A. O., Heald, G., Jarrett, T., *et al.* 2017, *A&A*, 601, A25
Coriat M., Corbel, S., Prat, L., *et al.* 2011, *MNRAS*, 414, 677
Fanaroff, B. L. & Riley, J. M. 1974, *MNRAS*, 167, 31
Giacintucci, S., O'Sullivan, E., Vrtilek, J., *et al.* 2011, *ApJ*, 732, 95
Hernández-García, L., Panessa, F., Giroletti, M., *et al.* 2017, *A&A*, 603, A131
Ishwara-Chandra, C. H. & Saikia, D. J. 1999, *MNRAS*, 309, 100
Machalski, J., Chyzy, K. T. & Jamrozy, M. 2004, *Acta Astronomica*, 54, 249
Malarecki, J. M., Jones, D. H., Saripalli, L., *et al.* 2015, *MNRAS*, 449, 955
Merloni A., Heinz S., di Matteo T., *et al.* 2003, *MNRAS*, 345, 1057
Mingo, B., Hardcastle, M. J., Croston, J. H., *et al.* 2014, *MNRAS*, 440, 269
Molina, M., Bassani, L., Malizia, A., *et al.* 2014, *A&A*, 565, A2
Molina, M., Venturi, T., Malizia, A., *et al.* 2015, *MNRAS*, 451, 3
Oh, K., Koss, M., Markwardt, C. B., *et al.* 2018 *ApJS*, 234, 4
Orrù, E., Murgia, M., Feretti, L., *et al.* 2010, *A&A*, 515, A50
Parma, P., Murgia, M., Morganti, R., *et al.* 1999, *A&A*, 344, 7
Schoenmakers, A. P., Mack, K.-H., de Bruyn, A. G., *et al.* 2000, *A&AS*, 146, 293
Sebastian, B., Ishwara-Chandra, C. H., Joshi, R., *et al.* 2018, *MNRAS*, 473, 4
Subrahmanyan, R., Saripalli, L., & Hunstead, R. W. 1996, *MNRAS* 279, 257
Ursini, F., Bassani, L., Panessa, F., *et al.* 2018, *MNRAS*, in press
van Velzen S., Falcke H., Körding E., *et al.* 2015, *MNRAS*, 446, 2985
Willott, C. J., Rawlings, S., Blundell, K. M., *et al.* 1999, *MNRAS*, 309, 1017

Astronomy in Focus - XXX
Proceedings IAU Symposium No. XXX, 2018
M. T. Lago, ed.

doi:10.1017/S174392131900348X

Prevalence of radio jets associated with quasar outflows and feedback

Miranda E. Jarvis[1,2]

[1]Max-Planck Institut fûr Astrophysik, Karl-Schwarzschild-Str. 1, 85741 Garching, Germany
email: mjarvis@eso.org

[2]European Southern Observatory, Karl-Schwarzschild-Str. 2, 85748 Garching, Germany

Abstract. We have identified that radio jets are commonly associated with "radiative mode" feedback in quasars. By performing a systematic multi-wavelength study of $z < 0.2$ quasars, we have found that 70–80% of our sample of 'radio-quiet' type 2 quasars, which host kpc-scale ionized gas outflows, exhibit radio jet structures. Here, we discuss our results on the pilot sample of 10 objects that combine high resolution (~0.25-1 arcsec) radio imaging at 1-7 GHz with optical IFU observations. Our results demonstrate that it is extremely common for jets to be spatially and kinematically linked to kpc-scale ionized gas kinematics in such quasars. Therefore, radio jets may be an important driver of outflows during 'radiative mode' feedback, apparently blurring the lines between the traditional divisions of feedback modes.

Keywords. galaxies: active –galaxies: evolution – galaxies: jets – quasars: emission lines

1. Introduction

The mechanisms by which AGN influence the gas of their host galaxies (feedback) are often split into two categories: the 'quasar' or 'radiative' mode, where radiation pressure dominates the feedback, and the 'jet' or 'radio' mode, where the majority of the AGN's energy is in a relativistic jet (see e.g., Heckman & Best 2014; Fabian 2012). In particular, the exact mechanism by which 'quasar' mode AGN interact with their host galaxies has been widely studied with little consensus, with radiative winds, compact jets and star formation all suggested (see e.g. Condon *et al.* 2013; Mullaney *et al.* 2013; Zakamska & Greene 2014). However, much of the previous observational work has lacked the spatial resolution to unambiguously distinguish between these possibilities. Our current work seeks to shed additional light on this problem by using high spatial resolution (~1kpc) radio images of 10 'radio-quiet' quasars, and in so doing we have found that the division between quasar and jet mode feedback might not be as sharp as has previously been believed. This work is presented in full in Jarvis *et al.* (2019).

2. Sample properties

The ten quasars discussed here were selected from our parent sample of 24 264 $z < 0.4$ AGN, that were spectroscopically identified as AGN from SDSS (DR7; Abazajian *et al.* 2009). We selected type 2 ('obscured') AGN with $z < 0.2$, with 'quasar' like luminosities ($\log L_{\rm AGN} = 45$–$46[\,{\rm erg\,s^{-1}}]$), and a luminous broad [O III] component, indicative of a powerful ionised outflow (i.e., a broad component that contributes $\geq$30% of the total flux and has a full width at half-maximum FWHM>700 $\rm km\,s^{-1}$; Mullaney *et al.* 2013). In Harrison *et al.* (2014), we found large scale ($\gtrsim$kpc) ionised outflows in all 10 AGN, using Gemini-South GMOS (Gemini Multi-Object Spectrograph) IFS data.

All of our targets were detected by the FIRST survey (Becker *et al.* 1995) with moderate radio luminosities ($\log L_{1.4\,\rm GHz}$ =23.2–24.4[$\rm W\,Hz^{-1}$]) and are either unresolved or

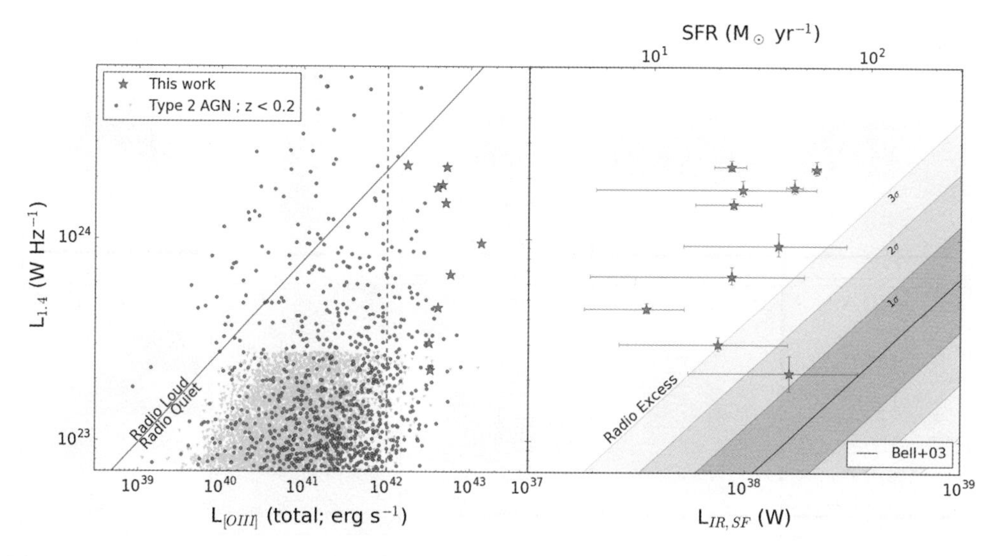

Figure 1. Left: Radio versus [O III] luminosity for our galaxies (red stars), with the division between "radio-loud" and "radio-quiet" from Xu *et al.* (1999) (blue line) and our parent sample of type 2 AGN with $z < 0.2$ (black points, grey for upper limits; Mullaney *et al.* 2013). Our selection criterion of $L_{[\text{OIII}]} > 10^{42}$ erg s^{-1} is shown by a dashed green line. Right: The FIR-radio correlation of Bell (2003) compared to the values for our primary sample. We take $L_{IR,SF}$ from the luminosity of the dust component from UV-FIR SED fitting. The solid black line is the average correlation from Bell (2003), with the cyan regions marking 1, 2 and 3σ regions respectively. Although all of our targets are classified as radio-quiet by most traditional methods, all but one appear to have excess radio emission above what is predicted from star formation.

marginally resolved at FIRST's $\sim 5''$ resolution (see Harrison *et al.* 2014). They are all classified as 'radio-quiet' using the separation criteria of Xu *et al.* (1999) (see Figure 1) and 'star formation dominated' by the separation criteria of Best & Heckman (2012). However, we find that all but one of our targets have a radio excess from what would be expected from their star formation alone. We determined this by using the infrared–radio correlation for normal star forming galaxies from Bell (2003) and UV-FIR SED fitting to isolate the infrared luminosity due to star formation from the AGN contribution (see Figure 1; Jarvis *et al.* 2019). This provides initial evidence for non star formation processes (likely AGN) producing the majority of the radio emission in our sources.

3. Observations

We present radio images from our observing campaigns with the Karl G. Jansky Very Large Array (VLA) in the L and C-bands with A and B configurations and the enchanced Multi-Element Radio Linked Interferometer Network (*e*MERLIN) that were designed to obtain high resolution (i.e., $\approx$0.3–1$''$; $\approx$600 pc at a representative redshift of $z = 0.1$) 1–7 GHz radio images of these targets. We compared the radio images to the ionized gas outflow properties traced by [O III] in our GMOS (Harrison *et al.* 2014) and VIMOS observations (Harrison *et al.* 2015 and Jarvis *et al.* 2019). We fit the [O III] line profile with multiple Gaussians to reduce the effect of noise and measure the kinematic properties of the ionized gas in all pixels (Harrison *et al.* 2014). Specifically, in Figure 2 we show the signal-to-noise ratio (S/N) of [O III] with radio contours from our VLA C-band A configuration data (0.3$''$ beam) and where needed to show additional morphological features the *e*MERLIN (0.3$''$ beam) and VLA C-band B-configuration data (1$''$ beam)†.

† For J1338+1503 we only have VLA C-band B configuration data

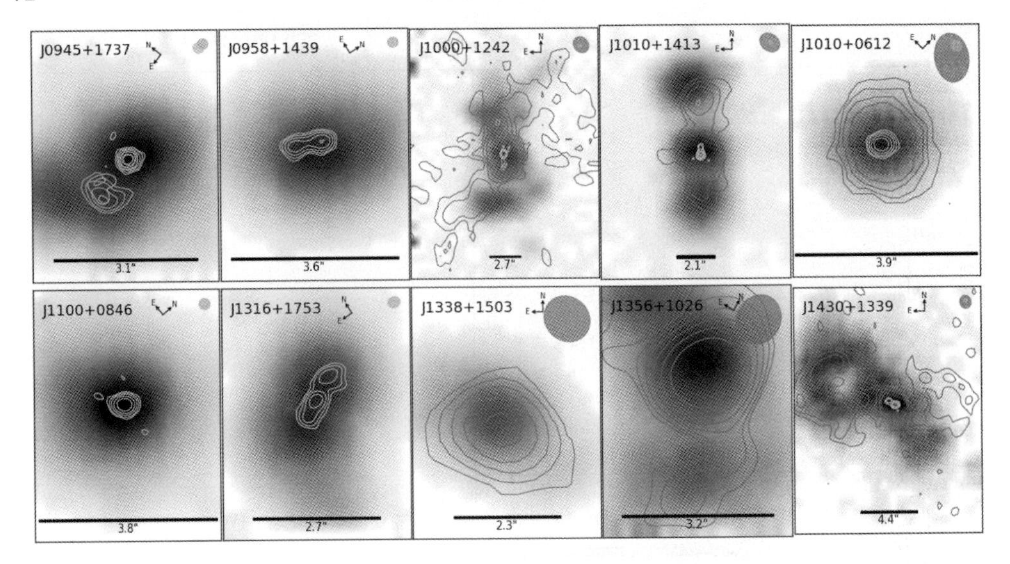

Figure 2. The distribution of [O III] emission (S/N maps) with contours overlaid from our radio images. Specifically, our VLA C-band B-configuration data is shown in green, VLA C-band A-configuration data in cyan and *e*MERLIN in orange. The beam for each radio image is shown as an appropriately coloured ellipse in the top right corner. The scale bar in each is 7kpc long. We observe a close connection between the radio and ionised gas morphologies.

4. Results and conclusions

Perhaps surprisingly for a 'radio-quiet' sample, 70–80% show distinct, collimated morphological features in the radio on 1–25 kpc scales (for one target, J1338+1503, we cannot rule out the presence of ~kpc scale features due to the lack of high resolution radio images). For three sources in particular (J0945+1737, J1000+1242 and J1010+1413) the high resolution images reveal elongated jet like components and flat spectrum ($\alpha > -0.2$) compact components. These latter structures, identified in the high-resolution images, are embedded in low resolution lobes, and we suggest that the most viable interpretation is that they are hot-spots from a jet. Furthermore, the sizes and lumionsities of our sources in the radio are comparable to the low luminosity population of compact steep spectrum (CSS) and Fanaroff-Riley type I (FRI) radio galaxies (see e.g. An & Baan 2012), suggesting a possible continuity between radio-loud and radio-quiet populations and supporting our interpretation that the collimated structures we identify are due to jets. Finally, these jets are in all cases coincident with kinematically disturbed ionised gas, an example of which can be seen in Figure 3. The combination of ionized gas kinematics and radio morphologies suggests that the jets and gas are interacting with each other, with the jet possibly driving outflows and the gas possibly deflecting the jets (see e.g. Leipski & Bennert 2006; Jarvis *et al.* 2019). We note that shallow, low-resolution radio images such as those from the FIRST survey are insufficient to unambiguously determine the origin of the radio emission or the driving mechanism of the outflows in typical quasars, even at these low redshifts. Our observations support a scenario where compact radio jets, with modest radio luminosities, are a crucial feedback mechanism for massive galaxies during the quasar phase. In our future work, we will explore the impact of these jets on the molecular (star forming) gas through a recently accepted ALMA proposal and extend the sample to 42 objects, through newly acquired VLA images.

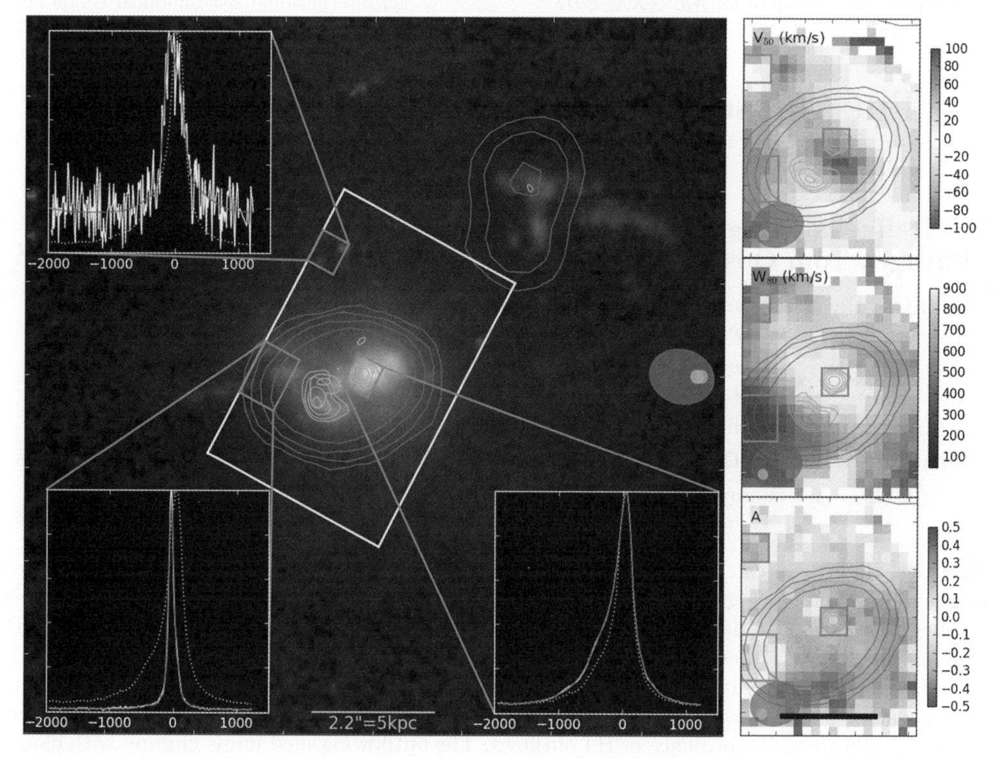

Figure 3. A comparison of the ionized gas and the radio features for J0945+1737. Left: contours from our radio data colour coded as they are in Figure 2 with the beams shown in the middle right, over-layed on HST data with continuum emission in red, [O III] and Hβ in green and H-alpha in blue (proposal id.13741; Cui *et al.* 2001). The FOV of the GMOS IFS is shown in white. [O III]5007 emission line profiles from the grey regions are shown each corner. The weighted average in the box is shown in white, the fit in magenta and the total over the cube as a dotted white line, all normalised to 1 and with the velocity given in km/s. Right: maps of the median velocity (v_{50}), line width (w_{80}) and asymmetry (A; see Harrison *et al.* 2014) with the boxes from the main panel overlayed. The scalebar is the same as in the main panel. The interactions between the radio jet features and the ionised gas are clearly visible.

References

Abazajian, K. N., Adelman-McCarthy, J. K., Agueros, M. A., *et al.* 2009, *ApJS*, 182, 543

An, T. & Baan, W. A. 2012, *ApJ*, 760, 77

Becker, R. H., White, R. L., & Helfand, D. J. 1995, *ApJ*, 450, 559

Bell, E. F. 2003, *ApJ*, 586, 794

Best, P. N. & Heckman, T. M. 2012, *MNRAS*, 421, 1569

Condon, J. J., Kellermann, K. I., Kimball, A. E., Ivezic, ., & Perley, R. A. 2013, *ApJ*, 768, 37

Cui, J., Xia, X.-Y., Deng, Z.-G., Mao, S., & Zou, Z.-L. 2001, *AJ*, 122, 63

Fabian, A. C. 2012, *ARAA*, 50, 455

Harrison, C. M., Alexander, D. M., Mullaney, J. R., & Swinbank, A. M. 2014, *MNRAS*, 441, 3306

Harrison, C. M., Thomson, A. P., Alexander, D. M., *et al.* 2015, *ApJ*, 800, 45

Heckman, T. M., & Best, P. N. 2014, *ARAA*, 52, 589

Jarvis, M. E. *et al.* 2019, *MNRAS*, 485, 2710

Leipski, C. & Bennert, N. 2006, *A&A*, 448, 165

Mullaney, J. R., Alexander, D. M., Fine, S., *et al.* 2013, *MNRAS*, 433, 622

Noll, S., Burgarella, D., Giovannoli, E., *et al.* 2009, *A&A*, 507, 1793

Xu, C., Livio, M., & Baum, S. 1999, *AJ*, 118, 1169

Zakamska, N. L. & Greene, J. E. 2014, *MNRAS*, 442, 784

Astronomy in Focus - XXX
Proceedings IAU Symposium No. XXX, 2018
M. T. Lago, ed.

doi:10.1017/S1743921319003491

The parsec-scale structure of jet-driven H I out ows in radio galaxies

Raffaella Morganti[1,2], Robert Schulz[1], Kristina Nyland[3], Zsolt Paragi[4], Tom Oosterloo[1,2], Elizabeth Mahony[5] and Suma Murthy[1,2]

[1]ASTRON, the Netherlands Institute for Radio Astronomy, Oude Hoogeveensedijk 4, 7991 PD wingeloo, The Netherlands. email: `morganti@astron.nl`

[2]Kapteyn Astronomical Institute, University of Groningen, P.O. Box 800, 9700 AV Groningen, The Netherlands;

[3]National Radio Astronomy Observatory, Charlottesville, VA 22903, USA;

[4]Joint Institute for VLBI ERIC, Oude Hoogeveensedijk 4, 7991 PD Dwingeloo, Netherlands;

[5]CSIRO Astronomy and Space Science, PO Box 76, Epping NSW 1710, Australia

Abstract. Radio jets can play multiple roles in the feedback loop by regulating the accretion of the gas, by enhancing gas turbulence, and by driving gas outflows. Numerical simulations are beginning to make detailed predictions about these processes. Using high resolution VLBI observations we test these predictions by studying how radio jets of different power and in different phases of evolution affect the properties and kinematics of the surrounding H I gas. Consistent with predictions, we find that young (or recently restarted) radio jets have stronger impact as shown by the presence of H I outflows. The outflowing medium is clumpy with clouds of with sizes up to a few tens of pc and mass $\sim 10^4 M_\odot$) already in the region close to the nucleus (< 100 pc), making the jet interact strongly and shock the surrounding gas. We present a case of a low-power jet where, as suggested by the simulations, the injection of energy may produce an increase in the turbulence of the medium instead of an outflow.

Keywords. ISM: jets and outflows, radio lines: galaxies, galaxies: active

---

The impact of active galactic nuclei (AGN) on the surrounding medium can be due to either winds and radiation from the nuclear region, or to plasma jets. Both these mechanisms are known to play a role and, depending on the situations and on the physical conditions, one can dominate via a strong coupling with the surrounding medium (e.g. Cielo *et al.* 2018). However, quantifying the actual impact of these phenomena is still a challenging task. Radio jets play a particularly important role in the feedback process, providing the best examples of AGN-driven feedback seen in action by preventing the cooling of the X-ray gas on cluster scales. However, radio jets can also provide an effective mechanism on *galaxy scales*. Their impact manifests itself in different ways: by counterbalancing the cooling of the hot coronae present around even isolated galaxies (e.g. Croston *et al.* 2008; Ogorzalek *et al.* 2017); by driving fast outflows traced by different gas phases or by injecting turbulence in the ISM, e.g. Alatalo *et al.* (2015); Guillard *et al.* (2015). All these mechanisms are relevant and need to be quantified.

Particularly interesting are the results from recent numerical simulations (Wagner *et al.* 2012, Mukherjee *et al.* 2018a, 2018b, Cielo *et al.* 2018). These show that radio jets can couple more strongly to the ISM if that is modelled to be clumpy (Wagner *et al.* 2012). Furthermore, a connection is expected with the cycle of activity of radio galaxies: given their small size, young (and recently restarted) radio jets have the highest impact on the gas. A dependence is also expected with jet power: powerful jets can drive faster outflows while low power jets can be "trapped" for longer times and induce more turbulence in the galactic ISM (Mukherjee *et al.* 2018b). Finally, the orientation of the jet with respect to the distribution of gas in the host galaxy is also relevant for the impact of the jet.

AGN-driven and jet-driven outflows are known to be multi-phase and can be traced also by atomic neutral hydrogen. This has opened the possibility to test the impact of plasma jets using radio data and, in particular, 21-cm H I observed in absorption (see Morganti & Oosterloo 2018 for an overview). The advantage of this is that the gas can be traced down to very small scales and the location of the outflow and their properties can be studied. This can be done using (sub-)arcsec down to milli-arcsec data (i.e. down to pc scales) as shown by the global Very Long Baseline Interferometry (VLBI) data described below and obtained by arrays including telescopes from the European VLBI Network, the Very Long Baseline Array (VLBA), as well as Arecibo. The results allow us to investigate not only the impact of radio jets, but also whether the predictions from the simulations are confirmed.

1. Where and how often do we see jet-driven outflows?

Jet-driven outflows are long known from ionised gas, but more recent work has not only shown that also atomic neutral hydrogen (H I) and molecular gas can be associated with these outflows and also that they may carry a significant (possibly the largest) fraction of the outflowing gas mass. The jet-driven origin of outflows of cold gas has been confirmed in a number of cases traced by molecular gas (see e.g. Alatalo *et al.* 2011, Dasyra & Combes 2012, Combes *et al.* 2013, Morganti *et al.* 2015, García-Burillo *et al.* 2014, Oosterloo *et al.* 2017, Runnoe *et al.* 2018) and by HI, see Morganti & Oosterloo (2018) for a review. The H I outflows have typically velocities between a few hundred to $\sim$1300 km s^{-1}.

In addition to this, a relation between the occurrence of H I outflows and the evolutionary status of the radio jet has been found by observations of a relatively large sample (248 objects) presented in Geréb *et al.* (2015). They find that at least 5% of all sources (15% of H I detections) show H I outflows. These numbers represent lower limits, given that absorption measurements are sensitive only to gas (and outflows) located in front of the radio continuum. Particularly interesting is that the vast majority of the H I outflows are detected in sources with newly born (or reborn) radio jets. This supports the idea that these phases in the evolution are those where the jet has most of its impact on the surrounding medium. This is in agreement with predictions from the simulations of Wagner *et al.* (2012) and Cielo *et al.* (2018). The recurrent nature of radio sources (see e.g. Morganti 2017) would ensure that this impact is repeated during the life of the host galaxy. A similar effect was also seen in the ionised gas (see Holt *et al.* 2008). However, this phase of the gas was found to show mass outflow rates reaching at most 1 $M_\odot$ yr^{-1}, while mass outflow rates up to 50 $M_\odot$ yr^{-1} have been found for the H I outflows.

2. Do we see the interaction of the jet with a clumpy medium?

For a small number of objects we have used VLBI observations to trace the properties of the H I outflows down to pc scales. The results show that a clumpy distribution of the gas is seen in all observed objects. Figure 1 illustrates the distribution of H I absorption in the central region of the restarted radio galaxy 3C 236, Schulz *et al.* (2018). Interestingly, in this and other targets observed so far, fast outflowing clouds (many hundred km s^{-1}) are detected already in the very inner region, at distances $< 50 - 100$ pc from the core. The clouds have masses of a few $\times 10^4$ $M_\odot$, and are unresolved on VLBI scales (< 40 pc). The presence of a clumpy medium (see also Oosterloo *et al.* 2017) is of key importance and confirms the prediction of the numerical simulations. A clumpy medium can make the impact of the jet much larger than previously considered: because of the clumpiness of the medium, the jet is meandering through the ISM to find the path of minimum

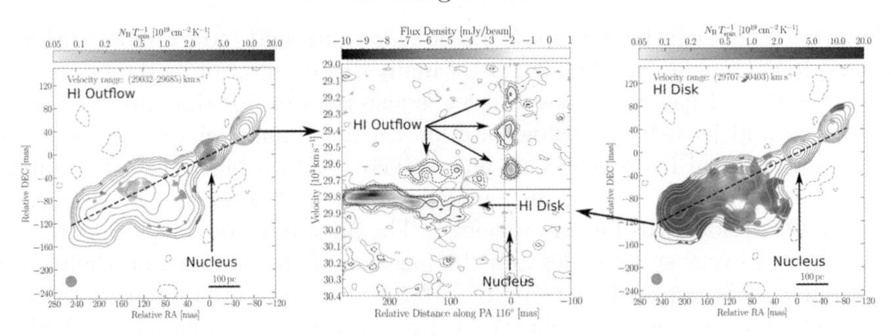

Figure 1. Radio continuum superposed on the H I absorption column density of the H I outflow (left) and of the H I disk (right) of the radio galaxy 3C236 obtained with VLBI. Position-velocity plot (centre) along the jet showing the outflowing clouds. From Schulz *et al.* (2018).

resistance and so creating an overpressured cocoon of outflowing and shocked gas, as suggested by Wagner *et al.* (2012); Mukherjee *et al.* (2018a).

Furthermore, for the smaller (and perhaps younger) sources in the sample (4C 12.50, Morganti *et al.* 2013 and 4C 52.37, Schulz *et al.* in prep) the VLBI observations not only spatially resolve the outflows, but they also recover all the H I flux observed at low resolution. This suggests that these outflows are mostly made up by relatively compact structures, easy to be detected at the very high resolution of VLBI observations. In the largest (and likely more evolved) sources, like 3C 293 (Schulz *et al.* in prep) and 3C 236 (Schulz *et al.* 2018), we also find evidence of a clumpy structure but the H I outflows in these sources are only partly recovered by VLBI. This suggests the additional presence of a diffuse component in which the clumps are embedded. This could be due to the expansion of the jet in the medium changing the structure of the outflows, the fraction of diffuse component increasing with time.

3. Do the low power jets have also an impact?

Interestingly, a growing number of cases (among which some listed above e.g. NGC1433, IC 5063, NGC1068, PG1700+518) show that, despite being classified as *radio quiet*, the power of the jet is sufficient to be the driving mechanism of their outflows. However, the simulations also show that an other effect expected from low power jets (Mukherjee *et al.* 2018b) is to increase the turbulence of the gas. Figure 2 shows the H I absorption detected against the kpc-scale jet of the low-luminosity radio source B2 0258+35. The width of the absorption (~ 400 km s^{-1}) is too large to be explained by the rotation of the large scale H I disk known to be present in this object. The most likely hypothesis is that the jet enters the disk and, being trapped there, disturbs the kinematics of the gas without being able to produce a fast outflow, but injects energy increases the turbulence of the gas (Murthy *et al.* 2019). As already suggested for the low-power radio source NGC 1266 (Alatalo *et al.* 2015), jet-induced turbulence may play a role in preventing star formation despite the large reservoir of cold gas observed in these objects.

4. Implications

The observations are showing evidence - in a growing number of sources - of interaction between the radio jets and the surrounding ISM. The properties appear to be, to first order, consistent with the predictions from some of the recent numerical simulations. This supports the idea that also on galactic scales the role of radio jets should not be neglected. However, the impact of outflows may not always be as large as required by models of galaxy evolution. A relatively small fraction of the outflowing gas may actually

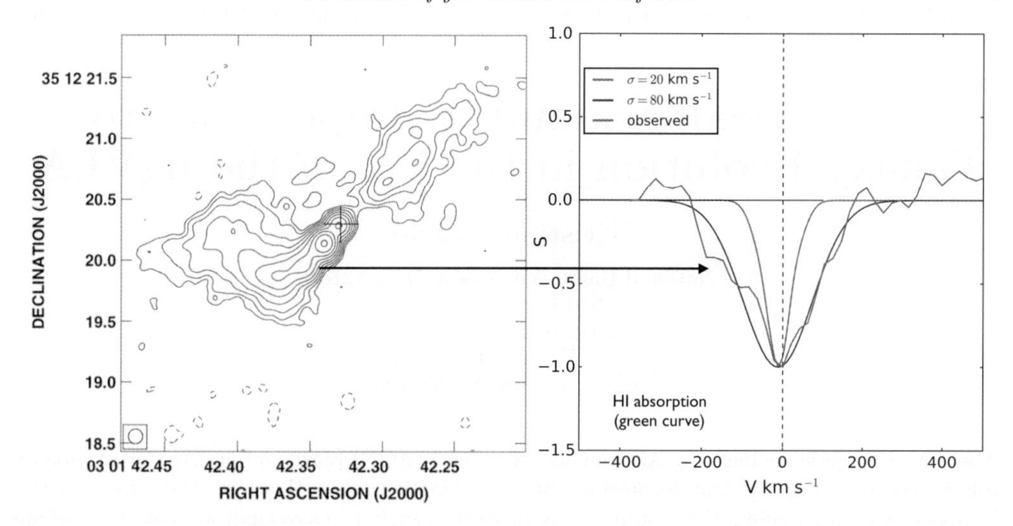

Figure 2. Left: Radio continuum image from Giroletti *et al.* (2005). **Right:** H I absorption profile (green) from VLA observations with superposed (red) the model from the H I disc observed in emission in this object by Struve *et al.* (2010) and in blue the model adding a component of turbulence due to the interaction of the jet with the H I in the disk (Murthy *et al.* 2019).

leave the galaxy. This is also seen in more AGN-driven outflows, i.e. those driven by winds or radiation. Thus, the likely main effect of jet-ISM interactions and their injection of energy could be on the redistributing the gas and possibly in keeping it turbulent for longer periods of time. This has been now seen in particular for (much more common) low luminosity radio jets. Thus, in addition to the search for violent processes like outflows, other more subtle effects needs to be searched for and investigated.

References

Alatalo, K., Blitz, L., Young, L. M., *et al.* 2011, *ApJ*, 735, 88
Alatalo, K., Lacy, M., Lanz, L., *et al.* 2015, *ApJ*, 798, 31
Cielo, S., Bieri, R., Volonteri, M., Wagner, A. Y., & Dubois, Y. 2018, *MNRAS*, 477, 1336
Combes, F., García-Burillo, S., Casasola, V., *et al.* 2013, *A&A*, 558, A124
Croston, J. H., Hardcastle, M. J., Kharb, P., Kraft, R. P., & Hota, A. 2008, *ApJ*, 688, 190
Dasyra, K. M., & Combes, F. 2012, *A&A*, 541, L7
García-Burillo, S., Combes, F., Usero, A., *et al.* 2014, *A&A*, 567, A125
Geréb, K., Maccagni, F. M., Morganti, R., & Oosterloo, T. A. 2015, *A&A*, 575, A44
Giroletti, M., Giovannini, G., & Taylor, G. B. 2005, *A&A*, 441, 89
Guillard, P., Boulanger, F., Lehnert, M. D., *et al.* 2015, *A&A*, 574, A32
Holt, J., Tadhunter, C. N., & Morganti, R. 2008, *MNRAS*, 387, 639
Morganti, R., Fogasy, J., Paragi, Z., Oosterloo, T., & Orienti, M. 2013, Science, 341, 1082
Morganti, R., Oosterloo, T., Oonk *et al.* 2015, *A&A*, 580, A1
Morganti, R. 2017, Nature Astronomy, 1, 596
Morganti, R. & Oosterloo, T. 2018, *A&ARev.* in press, arXiv:1807.01475
Mukherjee, D., Wagner, A. Y., Bicknell, G. V., *et al.* 2018a, *MNRAS*, 476, 80
Mukherjee, D., Bicknell, G. V., Wagner, A. Y. *et al.* 2018b, *MNRAS*, 479, 5544
Murthy, S., Morganti, R., Oosterloo, T., *et al.* 2019, *A&A*, 629, A58
Ogorzalek, A., Zhuravleva, I., Allen, S. W., *et al.* 2017, *MNRAS*, 472, 1659
Oosterloo, T., Raymond Oonk, J. B., Morganti, R., *et al.* 2017, *A&A*, 608, A38
Runnoe, J. C., Gültekin, K., & Rupke, D. S. N. 2018, *ApJ*, 852, 8
Schulz, R., Morganti, R., Nyland, K., *et al.* 2018, *A&A* in press, arXiv:1806.06653
Struve, C., Oosterloo, T., Sancisi, R., Morganti, R., & Emonts, B. H. C. 2010, *A&A*, 523, A75
Wagner A. Y., Bicknell G. V., & Umemura M. 2012, *ApJ*, 757, 136

Astronomy in Focus - XXX
Proceedings IAU Symposium No. XXX, 2018
M. T. Lago, ed.

doi:10.1017/S1743921319003508

AGN Feedback and its Importance to Galaxy Evolution in the Era of the ngVLA

Kristina Nyland

National Radio Astronomy Observatory
520 Edgemont Rd.
Charlottesville, VA 22903
United States
email: knyland@nrao.edu

Abstract. Energetic feedback by Active Galactic Nuclei (AGN) plays an important evolutionary role in the regulation of star formation (SF) on galactic scales. However, the effects of this feedback as a function of redshift and galaxy properties such as mass, environment and cold gas content remain poorly understood. The broad frequency coverage (1 to 116 GHz), high collecting area (about ten times higher than the Karl G. Jansky Very Large Array), and superb angular resolution (maximum baselines of at least a few hundred km) of the proposed next-generation Very Large Array (ngVLA) are uniquely poised to revolutionize our understanding of AGN and their role in galaxy evolution.

Keywords. galaxies: active, galaxies: jets, galaxies: evolution

1. Introduction

A key missing element in our understanding of cosmic assembly is the nature of energetic feedback from supermassive black holes (SMHBs) and the impact of active galactic nuclei (AGN) on galaxy evolution. Energetic feedback produced by Active Galactic Nuclei (AGN) is believed to play an important role in galaxy evolution through the regulatory effect it may have on the star formation rate and efficiency of the host galaxy. Despite its importance, identifying AGN-driven feedback in action is observationally challenging and requires high sensitivity, high angular resolution, and broad frequency coverage. As a result, large-scale studies of jet-driven AGN feedback using existing radio telescopes, such as the Karl G. Jansky Very Large Array (VLA) and Atacama Large Millimeter and Submillimeter Array (ALMA), have not been feasible. With its unprecedented current reference design consisting of $\sim$214 $\times$ 18m antennas operating from 1 to 116 GHz with baselines out to several hundred km, the next-generation Very Large Array (ngVLA) will overcome current observational limitations and enable significant advancements in our understanding of the impact of jet-driven feedback on galaxy evolution.

2. Observing Jet-ISM feedback with the ngVLA

The most well-studied population of sources exhibiting radio jet-driven feedback consists of massive elliptical galaxies residing at the centers of galaxy clusters with powerful jets capable of influencing galaxy evolution through the regulation of cooling flows. Jet-ISM feedback may also occur on sub-galactic scales in lower-mass and/or gas-rich galaxies, which typically have less massive SMBHs and much weaker radio jets. However, the prevalence of low-power ($L_{1.4\,\mathrm{GHz}} < 10^{24}$ W Hz^{-1}), $\sim$kpc-scale radio jets, and their impact on galaxy evolution, remain poorly constrained. The ngVLA will be an ideal

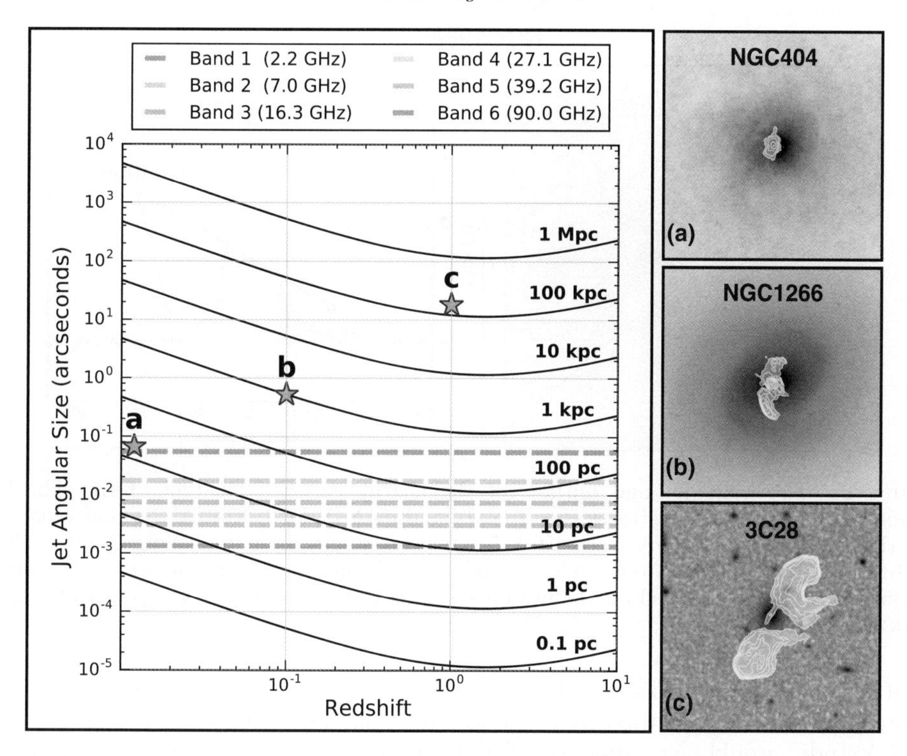

Figure 1. Jet angular size as a function of redshift. The black solid lines trace the redshift dependence of the angular extent of a jetted AGN for intrinsic jet sizes (measured from end to end along the major axis of the jet) from 0.1 pc to 1 Mpc. The maximum angular resolution of the ngVLA at the center of each of the ngVLA bands as defined in (Selina *et al.* 2017) is denoted by the dashed colored lines. The magenta stars and thumbnails to the right of the main figure indicate three jetted radio AGN representing a wide range of jet size scales: **a)** the dwarf galaxy NGC 404 with a jet extent of 10 pc, **b)** the jet-driven feedback host NGC 1266 with a jet extent of 1 kpc, and **c)** the radio galaxy 3C28 with a jet extent of 150 kpc. The redshifts of the representative sources correspond to simulated ngVLA maps from Nyland *et al.* (2018) at $z \approx 0$ ($D = 10$ Mpc), $z = 0.1$, and $z = 1.0$, respectively.

instrument for studying sub-galactic-scale radio jets and their impact on the interstellar medium (ISM), particularly for sources with extents of a few pc to a few kpc such as young radio AGN, jetted AGN hosted by low-mass galaxies, and radio jets that are interacting strongly with the interstellar medium of the host galaxy. Figure 1 illustrates the redshift dependence of the angular jet extent as observed by the ngVLA for a wide range of radio jet size scales ranging from sub-parsec jets to giant radio galaxies with Mpc-scale lobes. Thus, future ngVLA studies of radio jets with intrinsic extents of a few pc to a few kpc will be able to fully utilize the unique combination of angular resolution, collecting area, and frequency coverage of the ngVLA over a wide range of redshifts. *ISM Content and Conditions*. The ngVLA will link source morphologies and energetics from deep, high-resolution continuum observations with spectral line data that encode information on the ISM content and conditions. The combination of broadband continuum and spectral line imaging will allow the ngVLA to uniquely probe the energetic impact of radio jets on the cold gas reservoirs of their hosts. In particular, spectral line measurements of both the molecular (Figure 2) and atomic gas may be used to identify AGN-driven outflows, perform detailed kinematic studies to estimate the amount of energy injected into the ISM via feedback, and address the future evolutionary impact on the star formation efficiency on local/global scales caused by the AGN feedback (e.g.,

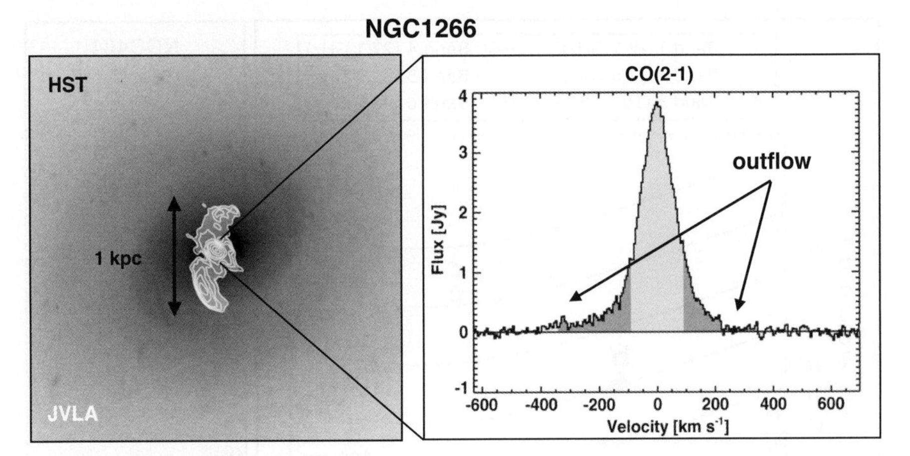

Figure 2. The radio jet and outflow of the nearby galaxy NGC 1266. **Left:** The background colorscale image illustrates the *HST J*-band data (WFC3, F140W; Nyland *et al.* 2013) and the filled white contours trace VLA 5 GHz continuum data from Nyland *et al.* (2016). **Right:** The CO(2–1) data from CARMA highlights the molecular outflow originally identified in Alatalo *et al.* (2011) based on the presence of excess emission in the wings of the spectrum (dark blue).

negative vs. positive feedback; Gaibler *et al.* 2012). Through comparisons with state-of-the-art jet-feedback simulations (e.g., Mukherjee *et al.* 2016), these continuum + cold gas ngVLA studies would place direct constraints on the prevalence and energetic importance of jet-ISM feedback as a function of redshift, galaxy mass, and environment.

Radio Spectral Ages. The ngVLA will uniquely excel at studies of radio AGN spanning a wide range of ages at low redshift, as well as radio AGN that are young or embedded in dense environments at higher redshifts (Figure 3). As shown by results from the Australia Telescope 20 GHz (AT20G) survey, continuum measurements in the tens of GHz range are needed to adequately model the radio spectral energy distributions (Sadler *et al.* 2006). This is particularly important for modeling the ages of young, low-redshift sources less than 10 Myrs old (Patil *et al.* 2018). Lower frequency continuum data in the MHz range will be important for constraining the ages of high-z sources; however, the inclusion of the lowest-frequency ngVLA bands down to ∼1 GHz would provide sufficient frequency coverage for measuring ages of sources as old as 30–40 Myrs at $z \sim 1$.

AGN Hosted by Low-mass Galaxies. Recent studies suggest that accreting SMBHs with masses in the range of $10^3 \lesssim M_{\rm BH} \lesssim 10^6$ M$_\odot$ may commonly reside in the nuclei of nearby low-mass ($M_* < 10^{10}\, M_\odot$) galaxies (Mezcua 2017, and references therein), thus motivating deep searches for their radio continuum signatures. However, identifying SMBHs in this population of galaxies is inherently difficult due to their faint accretion signatures. An ngVLA survey of accreting SMBHs hosted by nearby low-mass galaxies (e.g., analogs to NGC 404; Nyland *et al.* 2017) would offer new insights into the occupation fraction of SMBHs analogous to the SMBH seeds that formed at high redshift, thus profoundly impacting our understanding of the origin of SMBHs. Deep, high-angular-resolution observations with the ngVLA will both help constrain the SMBH seed mass distribution and also provide new constraints on the energetic impact of AGN feedback associated with sub-million-solar-mass SMBHs.

3. Opportunities for multiwavelength synergy

The unique capabilities of the ngVLA will facilitate exciting advancements in our understanding of AGN feedback and its broader connection to galaxy evolution, particularly when combined with multiwavelength data from other current and next-generation

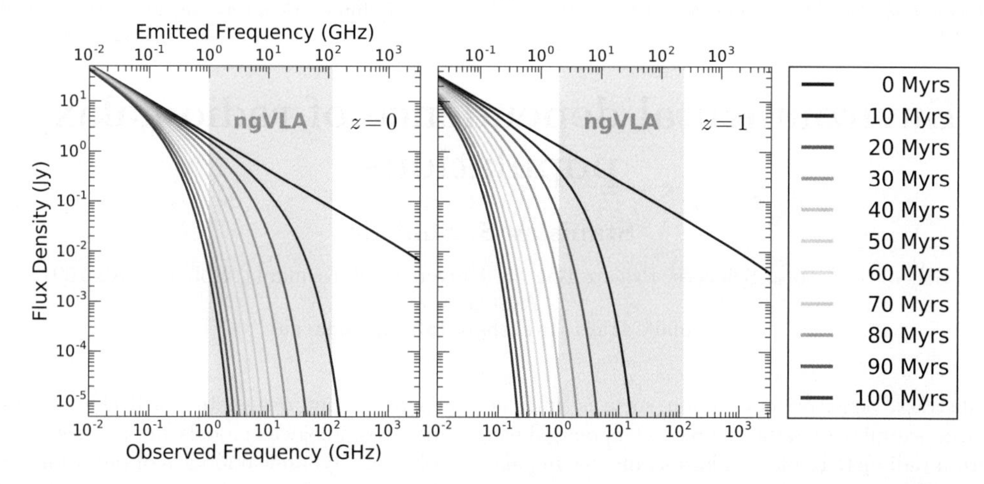

Figure 3. Example of JP model (Jaffe & Perola 1973) spectral ages calculated using the BRATS software (Harwood *et al.* 2013) demonstrating the need for ngVLA observations spanning a wide range of frequencies. The left and center panels correspond to redshifts of 0 and 1, respectively. The flux density values shown on the y-axis have been arbitrarily scaled. Because of its advantages of wide frequency range and angular resolution compared to the SKA, the ngVLA will uniquely excel in studies of low-redshift radio AGN that are young, or higher-redshift AGN that are embedded in dense environments.

instruments. In terms of synergy with current radio telescopes, observations with ALMA at frequencies above the ngVLA's limit of 116 GHz will provide key insights into the energetic and chemical impact of jet-driven feedback on the dense gas phase of the ISM. In the low radio frequency regime, the Square Kilometre Array (SKA) and its pathfinders will probe the 21 cm line out to higher redshifts (though at lower spatial resolution) than the ngVLA (Morganti 2017). The combination of constraints on both the atomic and molecular gas conditions from ngVLA and SKA observations will be important for studying the full impact of energetic jet-driven feedback on the ISM.

References

Alatalo, K., Blitz, L., Young, L. M., Davis, T. A., *et al.* 2011, *ApJ*, 735, 88
Mukherjee, D., Bicknell, G. V., Sutherland, R., & Wagner, A. 2016, *MNRAS*, 461, 967
Harwood, J. J., Hardcastle, M. J., Csoton, J. H., & Goodger, J. L. 2013, *MNRAS*, 435, 3353
Jaffe, W. J. & Perola, G. C. 1973, *A&A*, 26, 423
Mezcua, M. 2017, *IJMPD*, 26, 1730021
Morganti, R. 2017, *AN*, 338, 165
Mukherjee, D., Bicknell, G. V., Sutherland, R., & Wagner, A. 2016, *MNRAS*, 461, 967
Nyland, K., Alatalo, K., Wrobel,J. M., Young, L. M., *et al.* 2013, *ApJ*, 779, 173
Nyland, K., Young, L. M., Wrobel, J. M., Sarzi, M., *et al.* 2016, *MNRAS*, 859, 23
Nyland, K., Davis, T. A., Nguyen, Dieu D., Seth, A., *et al.* 2017, *ApJ*, 845, 50
Nyland, K., Harwood, J. J., Mukherjee, D., Jagannathan, P., *et al.* 2018, *ApJ*, 458, 2221
Patil, P., Nyland, K., Whittle, M., Lonsdale, C., & Lacy, M. 2018, *Astronomy in Focus*, 1
Sadler, E. M., Ricci, R., Ekers, R. D., Ekers, J. A., *et al.* 2006, *MNRAS*, 371, 898

Astronomy in Focus - XXX
Proceedings IAU Symposium No. XXX, 2018
M. T. Lago, ed.

doi:10.1017/S174392131900351X

Environmental dependence of radio galaxy populations

Stanislav S. Shabala

School of Natural Sciences, Private Bag 37, University of Tasmania, Hobart, TAS 7001, Australia
email: stanislav.shabala@utas.edu.au

Abstract. Sensitive continuum surveys with next-generation interferometers will characterise large samples of radio sources at epochs during which cosmological models predict feedback from radio jets to play an important role in galaxy evolution. Dynamical models of radio sources provide a framework for deriving from observations the radio jet duty cycles and energetics, and hence the energy budget available for feedback. Environment plays a crucial role in determining observable radio source properties, and I briefly summarise recent efforts to combine galaxy formation and jet models in a self-consistent framework. Galaxy clustering estimates from deep optical and NIR observations will provide environment measures needed to interpret the observed radio populations.

Keywords. galaxies: jets, galaxies: active, hydrodynamics, radio: continuum

1. Introduction

Properties of radio Active Galactic Nuclei (AGN), their host galaxies and larger-scale environment are closely linked. On the one hand, radio sources are now widely accepted to be responsible for the bulk of the feedback required to restrict gas cooling and star formation in the most massive galaxies and clusters since $z \sim 1$ (e.g. Silk & Rees 1998, Croton *et al.* 2006). These objects are prevalent in environments where cooling needs to be suppressed, namely massive elliptical galaxies (Sadler *et al.* 1989) and clusters with short cooling times (Mittal *et al.* 2009); and there appears to be approximate heating/cooling equilibrium in the hot haloes (Best *et al.* 2007, Shabala *et al.* 2008). There is also strong evidence for an environmental dependence of the AGN triggering mechanisms (Sabater *et al.* 2013, Pimbblet *et al.* 2013, Poggianti *et al.* 2017, Marshall *et al.* 2018), related to two quite different accretion modes: direct cold gas fuelling of the central engine in strong line radio galaxies, and chaotic accretion of gas cooled out of the hot phase in weak line radio galaxies (Hardcastle 2018).

On the other hand, the morphologies of radio galaxies are also strongly influenced by environment. On sub-kpc scales, the jets may be mass-loaded, ultimately determining whether a core (Fanaroff-Riley type I; FR) or edge-brightnened (FR-II) structure emerges. On group and cluster scales of tens and hundreds of kpc, the observed properties of radio lobes and efficiency with which they impart feedback onto the surrounding gas are shaped by their environment (Hardcastle & Krause 2013, Yates *et al.* 2018).

Dynamical models of radio galaxies provide the framework within which observed radio galaxy properties can be interpreted. In this contribution, I describe recent development of environment-sensitive jet models, and future prospects for studying the energetics and duty cycles of radio galaxies in large continuum surveys.

2. Dynamical modeling of radio sources

Dynamical models of powerful FR-II radio sources date back to work by Scheuer (1974). In the basic picture, the momentum flux of relativistic jets drives expansion along the jet axis, while transverse growth is due to the expansion of an overpressured cocoon of radio plasma inflated by backflow from the jet termination shock. The radio cocoon expands supersonically through the surrounding gas, driving strong bow shocks. The dynamics of lobed FR-Is is similar to FR-IIs, despite differences in jet morphology and lobe particle content (Croston *et al.* 2018). Jetted FR-Is, on the other hand, are dominated by velocity structure in the sheared jet forward flow, and the surface brightness of the uncollimated jets decreases with distance from the core. Radio emissivity is calculated by making some assumptions about jet particle content and lobe magnetic field strength, the latter typically parametrised in terms of lobe pressure.

A key feature of all these models is their sensitivity to the pressure profile of the atmosphere into which the jets expand. While X-ray observations are the "gold standard" for quantifying jet environments, these are only available for a relatively small fraction of systems, and are biased towards gas-rich haloes. The alternative approach, adopted by Turner & Shabala (2015) in developing the **R**adio **A**GN **i**n **S**emi-analytic **E**nvironments (RAiSE), is to use statistical properties of galaxy haloes inferred from cosmological galaxy formation models. The RAiSE model successfully reproduces a number of key observables, including departure from self-similarity, the relationship between radio luminosity, morphology, and host galaxy mass (Turner & Shabala 2015); and by including broadband radio spectra, lobe magnetic field strengths estimated from Inverse Compton observations (Turner *et al.* 2018b). Recently, Turner *et al.* (2018a) combined the RAiSE model with numerical simulations of backflow in FR-IIs, and showed that the discrepancy between spectral and dynamical ages in the powerful radio galaxy 3C436 can be explained by the mixing of electron populations of different ages; this effect is most pronounced in regions far from the hotspots, where a relatively small number of young electrons can skew the broadband radio spectra to younger spectral ages. Combining jet and galaxy formation models within a self-consistent framework also places strong constraints on AGN feedback models: requiring observations of both galaxy *and* radio AGN populations to be matched simultaneously can rule out some feedback models (Raouf *et al.* 2017).

3. Radio source populations

Application of dynamical models to well-defined radio AGN samples can in principle yield a census of jet energetics and duty cycles. Figure 1 shows the results of applying the RAiSE model (Turner & Shabala 2015) to a volume-limited sample of local ($z < 0.1$) radio AGN. Best-fitting jet power and age are estimated for each observed point in size – luminosity space, after marginalising over all other parameters. The derived mass scaling of jet powers suggests the hot haloes are in heating – cooling equilibrium, and jet generation efficiencies are consistent with predictions of jet production models. The derived jet kinetic luminosity function shows that most of the kinetic energy budget is provided by the low number of bright radio sources.

Figure 1 clearly shows selection effects. The observed dearth of low-luminosity, large sources is simply a consequence of survey surface brightness sensitivity limit. By extrapolating inferred properties of sources just above the detection limit, it is possible to quantify the fraction of sources missed in the survey (see right panel).

The overabundance of compact, low-luminosity sources is more puzzling. Sometimes referred to as FR-0 radio sources (e.g. Baldi *et al.* 2015), these have been variously proposed to be young radio sources, or "frustrated" older jets whose expansion has been impeded by a dense atmosphere. The other possibility is that at least some of these objects

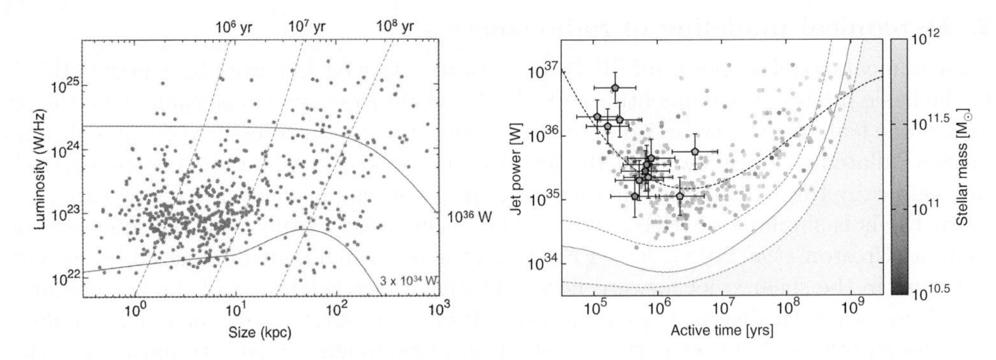

Figure 1. Application of the RAiSE model to a local ($z < 0.1$) sample of radio AGN from Shabala *et al.* (2008). *Left*: size-luminosity distributions at 1.4 GHz, with two representative tracks. *Right*: inferred jet kinetic power - age distributions (from Turner & Shabala 2015). Lines are detection limits for 50 (solid), 16 and 84 (dashed) percent of low surface brightness sources.

are cores whose diffuse lobes are below the surface brightness detection threshold. This scenario is supported by the modelling of Shabala *et al.* (2017), who showed that lobes inflated by VLBI-detected jets will rapidly become too diffuse for detection if the AGN host galaxy is located in a poor environment, consistent with observational evidence for an increased fraction of compact radio AGN in lower-mass haloes (Shabala 2018).

The potential implications of such a population of radio sources could be important. If jets can form lobes which rapidly fade below the detection limit, existing models would be underestimating both the lifetimes and kinetic powers of jets, and hence the energy available for feedback on their hot haloes. If there is sufficient pressure to collimate the jets and form well-defined lobes (Krause *et al.* 2012), more sensitive observations may constrain jet parameters through measurements of lobe volume and/or spectral ageing. Jetted FR-Is are a more challenging proposition, since the lack of backflow in these objects ensures that the oldest electrons will be located in the most diffuse regions; here, observations sensitive to low surface brightness emission are the best hope, as spectacularly shown by Heesen *et al.* (2018) for the archetypal jetted radio galaxy 3C31. Environment-sensitive simulations connecting galaxy and hot halo scales will provide the theoretical framework within which observations can be interpreted.

4. Probing environment through asymmetric radio sources

Jet production models predict that the two anti-parallel jets should be intrinsically identical. Hence asymmetric radio sources provide the ideal test bed for both models of jet – environment interaction, and metrics used to quantify environment. Rodman *et al.* (2019) used data from the Radio Galaxy Zoo citizen science project (Banfield *et al.* 2015) to test the hypothesis that any observed asymmetry in the radio continuum emission associated with the two jets is due to interaction with the environment. To minimize the effects of projection and model uncertainties, this analysis was restricted to a sample of straight FR-II lobes; environment associated with each lobe was quantified by counting SDSS galaxies with redshifts consistent with the AGN host. These authors found that the observed correlation between lobe length and galaxy clustering† is in excellent quantitative agreement with model predictions. Galaxy clustering therefore provides an environmental metric which can be used in conjunction with radio source models to extract physical parameters of radio source populations.

† Correlations with lobe luminosity are much weaker, in agreement with model predictions of highly non-linear relations with age and environment (e.g. Shabala & Godfrey 2013).

5. Conclusions and future prospects

We are fast entering an era of large radio surveys. Dynamical models provide a mechanism for interpreting radio source energetics and lifetimes, yet both these models and inferences about radio source population properties are sensitive to assumptions about the environments into which the jets expand. We have developed a dynamical model in which the adopted measure of environment is halo mass. Studies of asymmetric radio sources suggest that halo masses measured through galaxy clustering will provide an excellent description of jet environments. A combination of sensitive radio surveys with ancillary galaxy clustering data therefore holds much promise. The in-progress GAMA Legacy ATCA Southern Survey (GLASS; Huynh *et al.* in prep.) will survey 60 deg^2 of the GAMA G23 field; on much larger scales, over the next few years SKA surveys will be complemented by optical catalogues from next-generation optical/IR instruments.

Acknowledgements

This research was supported by the Australian Research Council (grant DE130101399), and an Endeavour Research Fellowship. I am grateful to all my collaborators, past and present, for their insights into the physics of radio galaxies.

References

Baldi, R., Capetti, A., & Giovannini, G., 2015, *A&A*, 576A, 38
Banfield, J. K. *et al.* 2015, *MNRAS*, 453, 2326
Best, P. N. *et al.* 2007, *MNRAS*, 379, 894
Croston, J. H., Ineson, J., & Hardcastle, M J. 2018, *MNRAS*, 476, 1614
Croton, D. J. *et al.* 2006, *MNRAS*, 365, 11
Hardcastle, M. J. & Krause, M. G H. 2013, *MNRAS*, 430, 174
Hardcastle, M. J. 2018, Nature Astronomy, 2, 273
Heesen, V. *et al.* 2018, *MNRAS*, 474, 5049
Krause, M. G. H., Alexander, P., Riley, J. M., & Hopton, D. 2012, *MNRAS*, 427, 3196
Marshall, M. A. *et al.* 2018, *MNRAS*, 474, 3615
Mittal, R., Hudson, D. S., Reiprich, T. H., & Clarke, T. 2009, *A&A*, 501, 835
Pimbblet, K. A. *et al.* 2013, *MNRAS*, 429, 1827
Poggianti, B. *et al.* 2017, *Nature* 548, 304
Raouf, M. *et al.* 2017, *MNRAS*, 471, 658
Rodman, P. E. *et al.* 2019, *MNRAS*, 482, 5625
Sabater, J., Best, P. N., & Argudo-Fernández, M. 2013, *MNRAS*, 430, 638
Sadler, Elaine M.; Jenkins, C. R., & Kotanyi, C. G. 1989, *MNRAS*, 240, 591
Scheuer, P. A. G. 1974, *MNRAS*, 166, 513
Shabala, S. S. & Godfrey, L. E. H. 2013, *ApJ*, 769, 129
Shabala, S. S. 2018, *MNRAS*, 478, 5074
Shabala, S. S., Ash, S., Alexander, P., & Riley, J. M. 2008, *MNRAS*, 388, 625
Shabala, S. S. *et al.* 2017, *MNRAS*, 464, 4706
Silk, J. & Rees, M. J. 1998, *A&A Lett.*, 331, 1
Turner, R. J. & Shabala, S. S. 2015, *ApJ*, 806, 59
Turner, R. J., Rogers, J. G., Shabala, S. S., & Krause, M. G. H. 2018, *MNRAS*, 473, 4179
Turner, R. J., Shabala, S. S., & Krause, M. G. H. 2018, *MNRAS*, 474, 3361
Yates, P. M., Shabala, S. S., & Krause, M. G. H. 2018, *MNRAS*, 480, 5286

Astronomy in Focus - XXX
Proceedings IAU Symposium No. XXX, 2018
M. T. Lago, ed.

doi:10.1017/S1743921319003521

Understanding mechanical feedback from HERGs and LERGs

Imogen H. Whittam

Department of Physics and Astronomy, University of the Western Cape,
Robert Sobukwe Road, Bellville 7535, South Africa
email: iwhittam@uwc.ac.za

Abstract. The properties of ~ 1000 high-excitation and low-excitation radio galaxies (HERGs and LERGs) selected from the Heywood *et al.* (2016) $1-2$ GHz VLA survey of Stripe 82 are investigated. The HERGs in this sample are generally found in host galaxies with younger stellar populations than LERGs, consistent with other work. The HERGs tend to accrete at a faster rate than the LERGs, but there is more overlap in the accretion rates of the two classes than has been found previously. We find evidence that mechanical feedback may be significantly underestimated in hydrodynamical simulations of galaxy evolution; 84 % of this sample release more than 10 % of their energy in mechanical form. Mechanical feedback is significant for many of the HERGs in this sample as well as the LERGs; nearly 50 % of the HERGs release more than 10 % of their energy in their radio jets.

Keywords. radio continuum: galaxies, galaxies: active, galaxies: evolution

---

1. Introduction

One of the key unknowns in galaxy evolution is how star-formation in galaxies becomes quenched; it is widely thought that feedback from active galactic nuclei (AGN) is responsible for this, but the mechanisms are not well understood. Observational evidence (e.g. Best & Heckman (2012)) suggests that AGN can be split into two distinct classes; high-excitation radio galaxies (HERGs; also known as cold mode, quasar mode or radiative mode sources) which radiate efficiently across the electromagnetic spectrum and posses the typical AGN accretion-related structures such as an accretion disk and a dusty torus, and low-excitation radio galaxies (LERGs; also known as hot mode, radio mode or jet mode sources) which radiate inefficiently and emit the bulk of their energy in mechanical form as powerful radio jets (see e.g. Heckman & Best (2014)).

It is thought that these two AGN accretion modes have different feedback effects on the host galaxy (see review by Fabian (2012)) and lead to the two different feedback paths in semi-analytic and hydrodynamic simulations, but despite being widely studied over the last decade (e.g. Hardcastle *et al.* (2007); Cattaneo *et al.* (2009); Heckman & Best (2014)) these processes are not well understood. In these proceedings I use a sample of ~ 1000 HERGs and LERGs to investigate the host galaxy properties and accretion rates of the two classes, and explore the implications of these results for AGN feedback.

2. Data used and source classification

This work is based on a $1-2$ GHz Karl G. Jansky Very Large Array (VLA) survey covering 100 deg^2 in SDSS Stripe 82 which has a 1σ rms noise of 88 μJy beam^{-1} and a resolution of 16×10 arcsec; full details of the radio survey are given in Heywood *et al.* (2016). This radio catalogue was matched to the SDSS DR14 optical catalogue (Abolfathi *et al.*

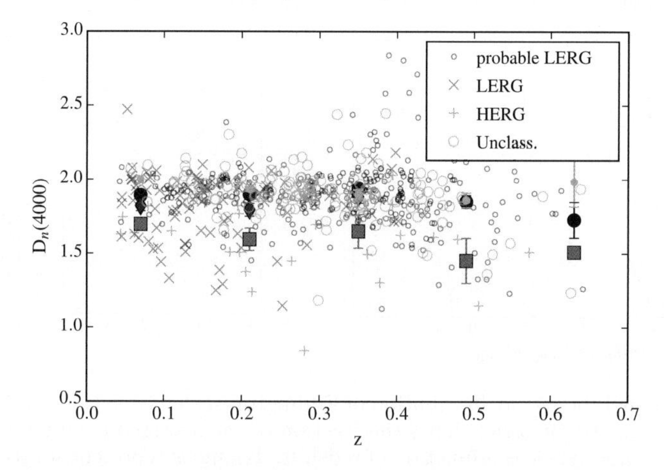

Figure 1. 4000 Å break strength as a function of redshift with the HERGs, LERGs, probable LERGs and unclassified sources shown separately. The filled shapes show the mean values in each luminosity bin for the different samples. From Whittam *et al.* (2018)

(2018)) by eye; details of the matching process are described in Prescott *et al.* (2018). We restrict our analysis to sources with a counterpart in the spectroscopic catalogue with $z < 0.7$, our sample therefore has 1501 sources which cover the range $0.01 < z < 0.7$ and $10^{21} < L_{1.4\ \mathrm{GHz}}/\mathrm{W\ Hz}^{-1} < 10^{27}$. We use the the value-added spectroscopic catalogues described in Thomas *et al.* (2013).

Sources are classified as either AGN or star-forming galaxies using information from their optical spectra, full details of this process are given in Prescott *et al.* (2018). The AGN in the sample are then classified as HERGs or LERGs using the criteria given in Best & Heckman (2012), which use a combination of emission line ratios and [OIII] equivalent width. This is explained in detail in Whittam *et al.* (2018). Additionally to the Best and Heckman classification scheme, we introduce a 'probable LERG' class for sources which cannot be classified according to the full criteria but which have an [OIII] equivalent width < 5 Å. The total number of sources in each category is as follows; HERGs = 60, LERGs = 149, probable LERGs = 600, QSOs = 81 and unclassified sources = 271, with 340 star-forming galaxies.

3. Host galaxy properties

Using the wealth of multi-wavelength data available in the field, we can compare the properties of the host galaxies of the HERGs and LERGs. 4000 Å break strength, which traces stellar age, is shown as a function of redshift in Fig 1. This shows that HERGs tend to be found in host galaxies with younger stellar populations than LERGs across the redshift range probed here. This is agrees with other results in the literature (e.g. Best & Heckman (2012)) and is consistent with the idea that HERGs have a supply of cold gas which provides the fuel for both star-formation and AGN activity. We refer the reader to Whittam *et al.* (2018) for further discussion of this and other host galaxy properties.

4. Accretion rates

There is a scenario building up in the literature that there are two distinct accretion modes which are responsible for HERGs and LERGs respectively; in this scenario there is a dichotomy in accretion rates between the two classes, relating to the two different modes. The radiative accretion rates of the AGN in this sample are estimated from their

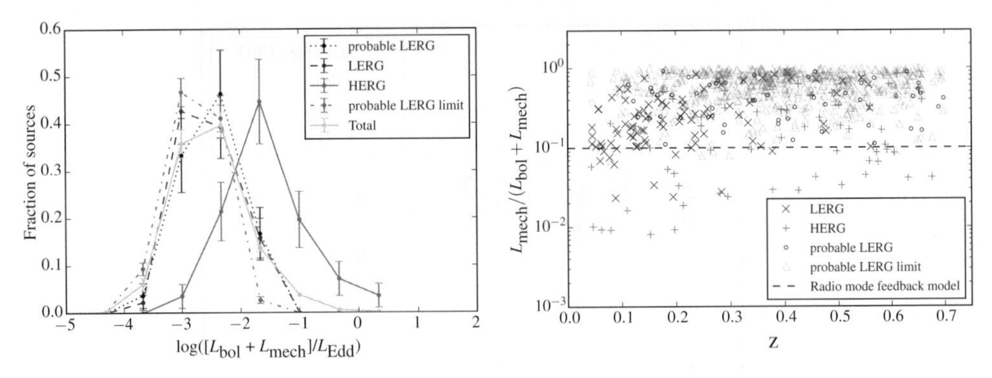

Figure 2. Left panel shows the distribution of Eddington-scaled accretion rates for the different source classifications. Right panel shows the fraction of the accreted energy released in the jets for the different source types as a function of redshift. Triangles represent sources with an upper limit on their radiative accretion rate, so the fraction of energy released in the jet is a lower limit. The dashed line is the radio mode feedback model used in Horizon-AGN from Dubois *et al.* (2014). The uncertainties in the scaling relations used to estimate $L_{\rm bol}$ and $L_{\rm mech}$ are 0.4 and 0.7 dex respectively. From Whittam *et al.* (2018).

[OIII] 5007 line luminosity and the mechanical accretion rates are estimated from the 1.4-GHz radio luminosity using the Cavagnolo *et al.* (2010) relationship. Black hole masses are estimated from the local black hole mass - bulge mass relation, allowing Eddington-scaled accretion rates to be calculated as follows: $\lambda = (L_{\rm bol} + L_{\rm mech})/L_{\rm Edd}$. The left panel of Fig. 2 shows the distribution of Eddington-scaled accretion rates for the HERGs and LERGs in this sample. It is clear from this figure that the HERGs generally accrete at a faster Eddington-scaled rate than the LERGs, with a distribution that peaks just below 0.1 compared to 0.01. However, there is a significant overlap in accretion rates between the two classes, with HERGs found across nearly the full range of accretion rates.

The dichotomy in accretion rates between HERGs and LERGs is therefore less clear in this study than it is in other studies in the literature; for example Best & Heckman (2012) and Mingo *et al.* (2014) both find almost no overlap in accretion rates between the two classes. In contrast, our sample seems to suggest a more continuous range of accretion rates. Note the our sample probes fainter radio luminosities ($10^{21} < L_{1.4\ \rm GHz}/{\rm W\ Hz}^{-1} < 10^{27}$) than other results in the literature; this could be part of the reason for the difference in our results, although we see some overlap in the accretion rates of the HERGs and LERGs across the luminosity range sampled here. We also do not observe any dichotomy in the [OIII] equivalent width or Excitation Index distributions, the two main parameters used to classify the HERGs and LERGs, suggesting that any dividing value chosen in these parameters is perhaps arbitrary for our sample.

5. Implications for AGN feedback

AGN feedback is required in all leading hydrodynamical simulations of galaxy evolution to quench star-formation. Some simulations implement mechanical and radiative feedback (assumed to relate to LERGs and HERGs respectively) separately (e.g. Horizon-AGN; Dubois *et al.* (2014)) while others do not (e.g. MUFASA; Davé *et al.* (2016)).

The right panel of Fig. 2 shows $L_{\rm mech}/(L_{\rm bol} + L_{\rm mech})$, which provides an estimate of the fraction of the total accreted energy deposited back into the interstellar medium in mechanical form. The dashed line shows the mechanical feedback efficiency of 10 % assumed in Horizon-AGN; it is clear that this is a significant underestimate for the sources in this sample, with 84 % of the sample depositing more than 10 % of their energy in mechanical form. This plot also demonstrates that mechanical feedback can be significant for HERGs as well as for LERGs; nearly 50 % (29/60) of the HERGs in this sample release more than 10 % of their accreted energy in mechanical form.

There is a scatter of ~ 2 dex in $L_{\mathrm{mech}}/(L_{\mathrm{bol}} + L_{\mathrm{mech}})$, which shows that the assumption that there is a direct scaling between accretion rate and mechanical feedback which is used in most hydrodynamical simulations does not necessarily hold. This may be because environment plays a significant role.

6. Conclusions and future perspectives

We have used the Heywood *et al.* (2016) VLA 1-2 GHz radio survey covering 100 deg^2 in Stripe 82 along with optical spectroscopy to probe the properties of ~ 1000 high- and low-excitation radio galaxies. They key results of this work are:

- HERGs tend to be found in host galaxies with younger stellar populations than LERGs, consistent with other results in the literature.
- While the HERGs in our sample tend to have higher accretion rates than the LERGs, we find considerable overlap in the accretion rates of the two samples.
- Mechanical feedback can be significant for HERGs as well as for LERGs, and may be underestimated for both populations in hydrodynamical simulations.

The advent of new radio telescopes, such as MeerKAT, LOFAR and ASKAP, means there is potential to make a large step forward in our understanding of radio galaxies and their mechanical feedback effects in the next few years. One example of a survey planned with a new instrument is the MeerKAT MIGHTEE survey (Jarvis *et al.* (2016)) which has just started to collect data and will survey 10 deg^2 to a depth of 1 μJy at 800 - 1600 MHz in four different fields. The unique combination of deep radio images over a significant cosmological volume along with excellent multi-wavelength coverage means we will be able to, amongst other things, extend the study described in this proceedings to significantly fainter luminosities and probe whether or not there is an accretion mode dichotomy, particularly at lower luminosities.

Acknowledgements

The author thanks Matthew Prescott, Matt Jarvis, Kim McAlpine and Ian Heywood for their significant contributions to this work. This research was supported by the South African Radio Astronomy Observatory, which is a facility of the National Research Foundation, an agency of the Department of Science and Technology.

References

Abolfathi B. *et al.* 2018, *ApJS*, 235, 42
Best P. N. & Heckman T. M. 2012, *MNRAS*, 421, 1569
Cattaneo A. *et al.* 2009, *Nature*, 460, 213
Cavagnolo K. W. *et al.* 2010, *ApJ*, 720, 1066
Davé R., Thompson R., & Hopkins P. F. 2016, *MNRAS*, 462, 3265
Dubois Y. *et al.* 2014, *MNRAS*, 444, 1453
Fabian A. C. 2012, *ARA&A*, 50, 455
Hardcastle M. J., Evans D. A., & Croston J. H. 2007, *MNRAS*, 376, 1849
Heckman T. M. & Best P. N. 2014, *ARA&A*, 52, 589
Heywood I. *et al.* 2016, *MNRAS*, 460, 4433
Jarvis M. *et al.* 2016, *Proceedings of MeerKAT Science: On the Pathway to the SKA. 25-27 May, 2016 Stellenbosch, South Africa*, 6
Mingo B. *et al.* 2014, *MNRAS*, 440, 269
Prescott M. *et al.* 2018, *MNRAS*, 480, 707
Thomas D. *et al.* 2013, *MNRAS*, 431, 1383
Whittam I. H., Prescott M., McAlpine K., Jarvis M. J., & Heywood I. 2018, *MNRAS*, 480, 358

Astronomy in Focus - XXX
Proceedings IAU Symposium No. XXX, 2018
M. T. Lago, ed.

doi:10.1017/S1743921319003533

The intermediate-power population of radio galaxies: morphologies and interactions

Diana M. Worrall[1], Ryan T. Duffy[2] and Mark Birkinshaw[3]

[1]HH Wills Physics Laboratory, University of Bristol,
Tyndall Avenue, Bristol BS8 1TL, U.K.
email: D.Worrall@bristol.ac.uk

[2]HH Wills Physics Laboratory, University of Bristol,
Tyndall Avenue, Bristol BS8 1TL, U.K.
email: R.Duffy@bristol.ac.uk

[3]HH Wills Physics Laboratory, University of Bristol,
Tyndall Avenue, Bristol BS8 1TL, U.K.
email: Mark.Birkinshaw@bristol.ac.uk

Abstract. Radio galaxies of intermediate power dominate the radio-power injection in the Universe as a whole, due to the break in the radio luminosity function, and so are of special interest. The population spans FR I, FR II, and hybrid morphologies, resides in a full range of environmental richness, and sources of all ages are amenable to study. We describe structures and interactions, with emphasis on sources with deep high-resolution *Chandra* X-ray data. As compared with low-power sources there is evidence that the physics changes, and the work done in driving shocks can exceed that in evacuating cavities. A range of morphologies and phenomena is identified.

Keywords. galaxies:active, galaxies:jets, radio continuum:galaxies, intergalactic medium, X-rays:galaxies

---

1. The intermediate-power population

Our motivation for studying intermediate-power radio galaxies is that the radio luminosity function breaks at intermediate powers (e.g., Best *et al.* 2005), where sources from the lower-power FR I and higher-power FR II populations (Fanaroff & Riley 1974) overlap (hereafter called the FR I/II boundary zone). This means that in weighting power by density it is the intermediate-power radio galaxies that dominate the overall radio power in the Universe. If radio power is a proxy for jet power, then these sources should dominate radio-galaxy heating in the Universe.

For radio galaxies in significant atmospheres in the local Universe, jet powers have been measured by combining the enthalpy of cavities evacuated of X-ray-emitting gas by the radio lobes with their ages (e.g., Bîrzan *et al.* 2004). Such work has led to correlations between jet and radio power (e.g., Bîrzan *et al.* 2008, Cavagnolo *et al.* 2010, O'Sullivan *et al.* 2011). Godfrey & Shabala (2016) have argued that common distance spreading, forcing correlations to slopes approaching unity, may be a strong factor driving the results, but more recently Ineson *et al.* 2017 have estimated jet powers for samples at higher radio power from measurements of lobe overpressuring with respect to the intergalactic medium (IGM). The conclusions are that correlations broadly hold, but that different radio-source morphology and composition contribute a large scatter (Croston,

Ineson & Hardcastle 2018). We find the sources in the FR I/II boundary zone to be intermediate both in radio power and estimated jet power, so supporting the hypothesis that their environmental heating is of particular importance.

FR I/II boundary sources span a wide range of atmosphere type. For example, the boundary-zone sources in the sample of radio galaxies with low excitation emission lines of Ineson *et al.* (2015) span three orders of magnitude in the X-ray luminosity of their environments.

In our work we concentrate on radio galaxies at redshift less than 0.1 for the best spatial resolution. This has given us two redshift-selected subsamples from two primary samples. One is 3CRR (Laing, Riley & Longair 1983) where there are 15 low-redshift boundary-zone sources, all of which are *Chandra* observed. Our matched southern-hemisphere sample is selected from MS4 (Burgess 1998, Burgess & Hunstead 2006a, Burgess & Hunstead 2006b), and is only partially observed with *Chandra.* Additionally we've observed other FR I/II boundary-zone sources falling outside those complete samples. As a rough approximation, about 30 per cent of the radio galaxies lie in significant group or cluster atmospheres.

2. A new rich-cluster atmosphere

One of our southern-sample sources, PKS B1416-493, led to a recent discovery with *Chandra* of a $z < 0.1$ 4.6-keV cluster (Worrall & Birkinshaw 2017). Finding new nearby rich clusters has become rare. The radio source lies central to the cluster emission which, unusually for such situations, is void of a strong cool core. Detection of lobe inverse Compton emission and an X-ray lobe cavity have enabled good estimates of source energetics, and this FR I/II boundary source is consistent with jet-radio-power correlations and so should be a fair representative of a source providing heating in a rich cluster environment.

3. Two sources in a poorer cluster

The poor cluster Abell 3744 at $z = 0.038$ contains two FR I/II boundary sources, NGC 7016 and 7018 (Worrall & Birkinshaw 2014, see also Birkinshaw *et al.*, these proceedings). A pronounced cavity overlaps lobe emission from each source. The cluster temperature, at 3.5 keV, is much hotter than expected based on cluster temperature-luminosity correlations, but since it would take the energy from 85 cavities to provide such heating, a recent merger is the likely case of the excess heat. It is extraordinary that the large-scale plumes from the two radio galaxies seem to be hugging X-ray-emitting gas of different temperature. The radio plasma appears to be acting as a thermal barrier, perhaps due to magnetic-field compression and a smaller gyroradius inside the radio plasma than in the external gas.

4. Shocks

While earlier Chandra work argued for weak or absent shocks around radio galaxies (see McNamara & Nulsen 2007), the situation changes for FR I/II boundary-zone sources, perhaps due to their more representative range of atmospheres. Moderate Mach 2 shocks are seen in 3C 305 and 3C 310 from the 3CRR sample (Hardcastle *et al.* 2012, Kraft *et al.* 2012). In PKS B2152-699 from our southern sample, we measure shocks of roughly Mach 2.7 (Worrall *et al.* 2012). Here the detection of lobe inverse Compton emission coupled with *Chandra* measurements of the gas properties enables a secure estimate of the source energetics, leading to the important conclusion that the work in driving shocks dominates that in producing cavities, so jet power can be sorely underestimated from cavity estimates alone.

5. Protons not contributing to minimum energy

Our studies have led to an important conclusion concerning minimum energy. The argument is as follows. Measurements of X-ray inverse Compton emission from the lobes of FR II radio galaxies have found magnetic fields that are typically about a third of the minimum-energy value calculated using the radiating leptons as the only significant contributer to the particle energy density (e.g., Croston *et al.* 2005). More recently we find the same situation for the lobes/plumes of FR I radio galaxies (Duffy, Worrall & Birkinshaw 2018). It's known from pressure-balance considerations that FR II lobes cannot contain significant proton pressure (Croston *et al.* 2005, Ineson *et al.* 2017, Croston, Ineson & Hardcastle 2018) while FR Is do (e.g., Morganti *et al.* 1988, Worrall & Birkinshaw 2000), for which the best candidate is re-energized entrained particles (Croston *et al.* 2008). So, similar departures from minimum energy are seen in FR Is and IIs, but only FR Is contain a significant proton pressure. The logical physics conclusion is that only electrons are relevant to the state of minimum energy, even in the presence of protons. This is perhaps because electrons are light, and so should react quicker to magnetic field irregularities. A further conclusion is that if relativistic protons enter lobes from FR II jets (including quasars) in similar numbers to electrons, they need a lower minimum Lorentz factor than the electrons so as not to increase lobe pressure significantly.

6. Local versus large-scale heating

A relatively small fraction of FR I/II boundary sources appears to be interacting with their environments on a scale that may provide large-scale heating. In some cases, significant environments appear already to have been lost, with heating complete. However, local heating is rather common.

In this regard, study of our 3CRR subsample of FR I/II boundary-zone sources finds almost half to show evidence of enhanced central belt-like gas structures, seen between and roughly orthogonal to the lobes (Duffy, Worrall & Birkinshaw 2018). Such gas can have originated from mergers, fossil groups or cool cores. Less clear is whether or not the radio plasma is driving the gas towards the AGN nucleus, as might assist fuelling, since measurements of temperature structure in the belt gas are generally insufficient for firm conclusions. The low-redshift radio galaxy best supporting inflow is 3C 386 (Duffy *et al.* 2016), but this source is underpowered for the FR I/II boundary zone by a factor of two to three.

On an even more localized scale, FR I/II boundary-zone sources show cases where the jets have bent appreciably in interactions with large gas clouds. A well-studied case is PKS 2152-699 at $z = 0.0282$, where the jet shows at least two distinct bends on its path to the northern hotspot (Worrall *et al.* 2012). The outer bend lies adjacent to a bright high ionization cloud (Tadhunter *et al.* 1987), and a recent kinematic study of the gas argues for a proton contribution to the jet to avoid ram-pressure deflection exceeding that observed (Smith *et al.* 2018). The gas cloud, of mass of order 10^8 $M_\odot$, has been heated to X-ray temperatures by the jet's passage (Worrall *et al.* 2012). The inner jet bend is embedded within galaxy emission and less amenable to study, but lies adjacent to a bright feature seen in HST data and which in a similar way is likely to be responsible for the jet deflection here. A further example is the FR I/II boundary-zone radio galaxy 3C 277.3 (Coma A) at $z = 0.0853$. Despite the source lying closer to the plane of the sky than PKS 2152-699, the projected deflection is greater, at 40°, and again the gas cloud adjacent to the bend is heated to X-ray temperatures (Worrall, Birkinshaw & Young 2016).

Both PKS 2152-699 and 3C 277.3 also show interfaces for heating on hundred-kpc scales. PKS 2152-699, which fills much of its 1-keV group atmosphere, is strongly shocking gas around its lobes, as mentioned above. At the outer extremities of the lobes of 3C 277.3 lie Hα-emitting filaments believed to be merger remnants (Tadhunter *et al.* 2000). An anti-correlation between the locations of arms of enhanced X-ray emission and the filaments suggests that shocks advancing around the lobe are inhibited by the dense colder material (Worrall, Birkinshaw & Young 2016).

References

Best, P. N., Kauffmann, G., Heckman, T. M., & Ivezić, Ž 2005, *MNRAS*, 362, 9

Bîrzan, L., Rafferty, D. A., McNamara, B. R., Wise, M. W., & Nulsen, P. E. J. 2004, *ApJ*, 607, 800

Bîrzan, L., McNamara, B. R., Nulsen, P. E. J., Carilli, C. L., & Wise, M. W. 2008, *ApJ*, 686, 859

Burgess, A. M. 1998, Ph.D. Thesis, University of Sydney

Burgess, A. M. & Hunstead R. W. 2006a, *AJ*, 131, 100

Burgess, A. M. & Hunstead R. W. 2006b, *AJ*, 131, 114

Cavagnolo, K. W., McNamara, B. R., Nulsen, P. E. J., Carilli, C. L., Jones, C., & Bîrzan, L. 2010, *ApJ*, 720, 1066

Croston, J. H., Hardcastle, M. J., Harris, D. E., Belsole, E., Birkinshaw, M., & Worrall, D. M. 2005, *ApJ*, 626, 733

Croston, J. H., Hardcastle, M. J., Birkinshaw, M., Worrall, D. M., & Laing, R. A. 2008, *MNRAS*, 386, 1709

Croston, J. H., Ineson, J., & Hardcastle, M. J. 2018, *MNRAS*, 476, 1614

Duffy R. T., Worrall, D. M., Birkinshaw, M., & Kraft, R. P. 2016, *MNRAS*, 459, 4508

Duffy R. T., Worrall, D. M., & Birkinshaw, M. 2018, in preparation

Fanaroff, B. L., & Riley, J. M. 1974, *MNRAS*, 167, 31P

Godfrey, L. E. H., & Shabala, S. S. 2016, *MNRAS*, 456, 1172

Hardcastle, M. J., Massaro, F., Harris, D. E. *et al.* 2012, *MNRAS*, 424, 1774

Ineson, J., Croston, J. H., Hardcastle, M. J., Kraft, R. P., Evans, D. A, & Jarvis, M. 2015, *MNRAS*, 453, 2682

Ineson, J., Croston, J. H., Hardcastle, M. J., & Mingo, B. 2017, *MNRAS*, 467, 1586

Kraft, R. P., Birkinshaw, M., Nulsen, P. E. J. *et al.* 2012, *ApJ*, 748, 19

Laing, R. A., Riley, J. M., & Longair, M. S. 1983, *MNRAS*, 204, 151

McNamara B. R., & Nulsen P. E. J. 2007, *ARA&A*, 45, 117

Morganti, R., Fanti, R., Gioia, I. M., Harris, D. E., Parma, P., & de Ruiter, H. 1988, *A&A*, 189, 11

O'Sullivan, E., Giacintucci, S., David, L. P., Gitti, M., Vrtilek, J. M., Raychaudhury, S., & Ponman, T. J. 2011, *ApJ*, 735, 11

Smith, D. P., Young, A. J., Worrall, D. M., & Birkinshaw, M. 2018, *MNRAS*, submitted

Tadhunter, C. N., Fosbury, R. A. E., Binette, L., Danziger, I. J., & Robinson, A. 1987, *Nature*, 325, 504

Tadhunter, C. N., Villar-Martin, M., Morganti, R., Bland-Hawthorn, J., & Axon, D. 2000, *MNRAS*, 314, 849

Worrall, D. M. & Birkinshaw, M. 2000, *ApJ*, 530, 719

Worrall, D. M. & Birkinshaw, M. 2014, *ApJ*, 784, 36

Worrall D. M. & Birkinshaw, M. 2017, *MNRAS*, 467, 2903

Worrall D. M., Birkinshaw, M., Young, A. J., Momtahan, K., Fosbury, R. A. E., Morganti, R., Tadhunter, C. N., & Verdoes Kleijn, G. 2012, *MNRAS*, 424, 1346

Worrall, D. M., Birkinshaw, M., & Young, A. J. 2016, *MNRAS*, 458, 174

FM4 – Magnetic Fields along the Star-Formation Sequence

Astronomy in Focus - XXX
Proceedings IAU Symposium No. XXX, 2018
M. T. Lago, ed.

doi:10.1017/S1743921319003545

Magnetic fields along the star-formation sequence: bridging polarization-sensitive views

Anaëlle Maury[1], Swetlana Hubrig[2] and Chat Hull[3,4]

[1]AIM, CEA, CNRS, Université Paris-Saclay, Université Paris Diderot, Gif-sur-Yvette, France
email: anaelle.maury@cea.fr

[2]Leibniz-Institut für Astrophysik Potsdam (AIP), Potsdam, Germany

[3]National Astronomical Observatory of Japan, NAOJ Chile Observatory, Santiago, Chile

[4]Joint ALMA Observatory, Santiago, Chile

Abstract. It is believed that magnetic fields play important roles in the processes leading to the formation of stars and planets. Polarimetry from optical to centimeter wavelengths has been the most powerful observing technique to study magnetic fields: the development of polarimetric capabilities on a wide range of observational facilities now allows to probe the magnetic field properties in various objects along the star formation sequence, from star-forming molecular clouds to young stars and their protoplanetary disks. However, the complexity of combining results from different observational techniques and facilities emphasizes the need to transcend historical barriers and bring together the various communities working with polarimetric observations. This Focus Meeting was a first step to compare observations of magnetic fields at the various evolutionary stages and physical scales involved in star formation processes, such that we can establish a coherent view of their key role in the multi-scale process of star formation.

Keywords. magnetic fields, stars: formation, polarization, techniques: polarimetric

---

Starting with a subtle interplay between gravity, turbulence, and magnetic fields during the formation of molecular clouds, the complexity of the star formation process is rooted in the conditions of the gas and the effectiveness of processes that remove angular momentum and magnetic flux, not just at the onset of gravitational instability but throughout the evolution of the forming star toward the zero-age main sequence (ZAMS). During the $\sim$1 million years needed to form a star, gravity has to overcome two main barriers to transform a dense core into a star: the dense core's angular momentum and its magnetic flux. The outcome of the collapse (manifested in the stellar mass function, occurrence and properties of protoplanetary disks, stellar multiplicity, etc.) depends critically on how and when these two barriers are surmounted. While the gas dynamics at most of the relevant scales ($0.1 - 10000$ au) are now accessible to observations, a thorough understanding of the star formation process will not be achieved until we have characterized the role of the magnetic field across all of the spatial and timescales relevant to the process of star formation. High-sensitivity, high-resolution polarization observations by new and planned next-generation facilities (ALMA; BLAST-TNG; HAWC+ on SOFIA; SCUBA-2 on JCMT; NIKA2 on the IRAM-30m; PolKa at APEX; CanariCam at the GTC; SPIROU on the CFHT; CRIRES+ on the VLT; PEPSI at the LBT; MIMIR and many others) are ushering in a new era in the study of polarized light, allowing us to uncover the properties of the magnetic field over a broad range of wavelengths and physical scales. Our Focus Meeting was built to gather the different communities studying magnetic field properties in objects along the evolutionary sequence leading to star formation, from molecular clouds to young stars reaching the ZAMS, and to identify synergies to compare, combine, and synthesize knowledge of the end-to-end role of the magnetic fields in the formation of stars. Can we establish a coherent picture of the role of the magnetic field in the star-formation sequence across time and spatial scale, in spite of the diverse observational techniques and analysis tools used to observe magnetic fields in molecular clouds, filaments, cores, young stars and their disks? We briefly summarize below the models and observations of magnetic fields at the various stages of the star-formation process that were shown and discussed at the meeting.

The initial setting for the formation of stars takes place in dense filamentary structures within molecular clouds. Observations of the Zeeman effect from molecular lines emission have revealed magnetic fields at the ~parsec scale in nearly all star-forming clouds (which have typical densities $100 - 10^4$ cm^{-3}) . Recent results from the polarized dust emission observed with *Planck* have confirmed the ubiquity of magnetic fields at molecular-cloud scales, showing that the orientation of the magnetic field is mostly parallel to diffuse filamentary structures in the ISM, whereas the field tends to be perpendicular to the dense, self-gravitating structures where star formation takes place. The comparison of these observations with numerical simulations of magnetohydrodynamic (MHD) turbulence suggests a rough equipartition between magnetic and turbulent kinetic energies, and a key role of magnetic fields in the evolution and fate of star-forming material in molecular clouds. However, even along the same line of sight, different wavelengths probe different density regimes, and both line-of-sight and plane-of-sky depolarization effects can affect the determination of the magnetic field orientation.

Once ~0.1 pc dense cores (which have typical densities $10^3 - 10^6$ cm^{-3}) are formed from the fragmentation of molecular cloud material, they can remain gravitationally stable or undergo collapse to form one or several stellar embryos. Both starless cores and protostellar cores actively forming stars show polarized dust emission in the (sub)millimeter domain, suggesting magnetic field strengths of $\sim 10 - 1000\,\mu$G. New polarimetric capabilities at various observatories will allow us to estimate the energy budgets of gravity, turbulence, rotation and magnetic fields in star-forming cores, thereby characterizing the physical processes contributing to the ability (or inability) of dense cores to form stars. During the meeting, new sensitive observations covering multiple scales were presented; however it is still unclear whether the magnetic field is retained from the filamentary clouds during core formation, and how important the field is for physically supporting the cores. Large-scale surveys with future space observatories (SPICA, OST) that would probe the magnetic field structure with sensitivity and spatial dynamic ranges comparable to those achieved by, e.g., *Herschel* images of the cold ISM in unpolarized emission, will be key to address this question.

A stellar embryo evolves by accreting material from the parent core, and moves through the protostellar and pre-main-sequence stages until the resulting star finally arrives on the main sequence. One of the major challenges for the formation of stars is the "angular momentum problem": observations show that, over the course of the ~100,000 yr that it is being accreted onto the stellar embryo, the material in the protostellar core needs to reduce its specific angular momentum by 5 to 10 orders of magnitude. While the formation of multiple systems, disks, jets and outflows may help remove excess angular momentum, these processes seem unable to fully solve the problem during the short ($< 10^5$ yrs) main accretion phase. Analytical models and MHD numerical simulations of the evolution of star-forming cores show that the magnetic field is critical for transporting angular momentum during the protostellar phase: characterizing the efficiency of magnetic braking during the main accretion phase is therefore of utmost importance.

Moreover, MHD models suggest that the magnetic field is a key player in the formation of massive stars, because the field increases the gravitational stability of massive cores – the seeds of massive stars – and also sets the conditions for strongly anisotropic accretion of the surrounding material, ultimately permitting more mass to be accreted. While observations of the polarized dust continuum emission with both high dynamic range and angular resolution (e.g., with ALMA) have started providing constraints on magnetic models of protostellar collapse, new challenges are emerging to probe the magnetic field properties in the very high density regimes where the dust thermal emission at sub-mm wavelengths becomes optically thick. The meeting gathered communities working on polarization observations, as well as experts from dust modeling, who discussed how to best characterize the properties of magnetic fields at high densities where their effect on the angular momentum evolution during the protostellar stage might be crucial to describe the formation of stars and disks.

Regarding the late evolutionary stages of the star formation process, high-spectral-resolution circular spectropolarimetry of pre-main-sequence (PMS) stars and their circumstellar environments (which have typical densities $10^{10} - 10^{30}$ cm^{-3}) reveals strong kGauss fields in these T Tauri objects. Their properties suggest some of the parental core magnetic conditions ("fossil fields") remain, in addition to the fields generated by the stellar dynamo. Moreover, observations of young stars of intermediate mass (> 1.4 $M_{\odot}$: Herbig Ae/Be stars) have shown evidence for large-scale organized magnetic fields, suggesting that the magnetic structure of these stars could be partly due to compression of the pristine magnetic fields during the early phase of the star-formation process. Characterizing the connection of the magnetic field between the PMS star and the inner part of its disk is also key to understand the evolution of young stars. Indeed, not only are these magnetic field lines thought to be the dominant pathway for accretion during the PMS stage ("magnetospheric accretion"), but the role of the field in controlling the evolution of angular momentum of the star might be paramount for slowing the rotation of young stars prior to their arrival on the main sequence. Future studies should pave the way towards a global picture of magnetospheric accretion processes over the whole stellar-mass range. We discussed how the synergy of new observing techniques now makes it possible to investigate how much of the initial magnetic field survives the star formation process, and whether the resulting field in young stars strongly depends on the properties of the parent core. Finally, we briefly discussed how our efforts to account for the diversity of exoplanetary systems must comprise a deeper understanding of the interaction between stellar magnetic fields and the protoplanetary disks: if magnetic fields affect the structure and evolution of disks, then planet formation and migration could be greatly affected by magnetic conditions during star formation.

In summary, the magnetic field is omnipresent at all scales and stages of star formation, and thus drives several key processes during the formation of stars such as (1) removing angular momentum during the main accretion phase, (2) affecting disk initial conditions and subsequent planet formation, (3) allowing late-stage accretion onto PMS stars, and possibly even (4) braking the newly born stars themselves through star-disk interaction. However, magnetic fields remain difficult to observe since they can rarely be observed directly; rather, we mostly rely on observations of polarized emission from dust and spectral lines, both of which have significant limitations. Very diverse observational techniques and analysis tools must be used to probe magnetic fields in the different types of star-forming objects: only by bringing together communities studying magnetic fields in molecular clouds, star-forming cores, protostars, disks, and PMS stars, can we overcome the current limitations and ultimately move synergistically toward a coherent picture of the role and evolution of magnetic fields in the star formation process.

Astronomy in Focus - XXX
Proceedings IAU Symposium No. XXX, 2018
M. T. Lago, ed.

doi:10.1017/S1743921319003557

The role of magnetic field in the formation and evolution of filamentary molecular clouds

Shu-ichiro Inutsuka

Department of Physics, Graduate School of Science, Nagoya University, 464-8602, Nagoya, Japan
email: inutsuka@nagoya-u.jp

Abstract. Recent observations have emphasized the importance of the formation and evolution of magnetized filamentary molecular clouds in the process of star formation. Theoretical and observational investigations have provided convincing evidence for the formation of molecular cloud cores by the gravitational fragmentation of filamentary molecular clouds. In this review we summarize our current understanding of various processes that are required in describing the filamentary molecular clouds. Especially we can explain a robust formation mechanism of filamentary molecular clouds in a shock compressed layer, which is in analogy to the making of "Sushi." We also discuss the origin of the mass function of cores.

References

Inoue, T. *et al.* 2018, *PASJ*, 70, S53
Inutsuka, S. *et al.* 2015, *A&Ap*, 580, 49
Iwasaki, K. *et al.* 2018, submitted to *ApJ* (arXiv:1806.03824)

Astronomy in Focus - XXX
Proceedings IAU Symposium No. XXX, 2018
M. T. Lago, ed.

doi:10.1017/S1743921319003569

Observations of magnetic fields in star-forming clouds

Juan D. Soler

Max-Planck-Institute for Astronomy, Königstuhl 17, 69117, Heidelberg, Germany.
email: soler@mpia.de

Abstract. This review examines observations of magnetic fields in molecular clouds, that is, at spatial scales ranging from tens to tenths of parsecs and densities up to hundreds of particles per cubic centimetre. I will briefly summarize the techniques for observing and mapping magnetic fields in molecular clouds. I will review important examples of observational results obtained using each technique and their implications for our understanding of the role of the magnetic field in molecular cloud formation and evolution. Finally, I will briefly discuss the prospects for advances in our observational capabilities with telescopes and instruments now beginning operation or under construction.

Keywords. ISM: general / ISM: magnetic fields / ISM: clouds / ISM: structure / dust, extinction / submillimeter: ISM / infrared: ISM /magnetic fields / (magnetohydrodynamics:) MHD.

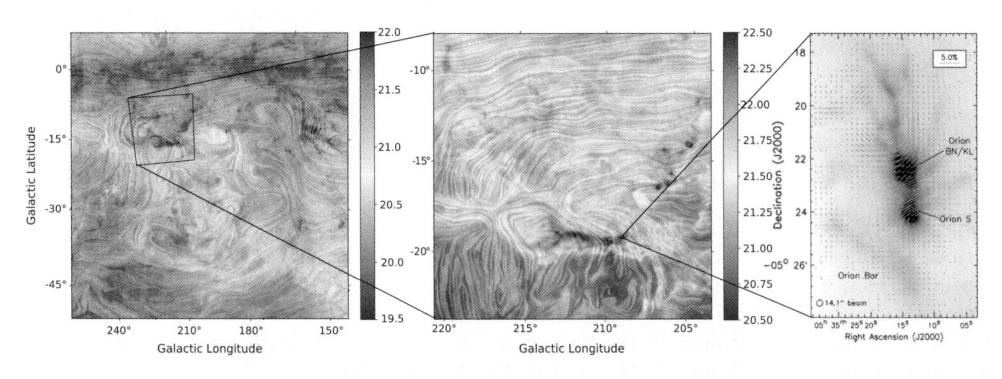

Figure 1. Magnetic field morphology in and around star-forming clouds as revealed by dust polarized thermal emission. The drapery pattern and vectors represent the direction of the magnetic field integrated along the line of sight and projected in the plane of the sky as inferred from submillimeter polarization observations by European Space Agency's *Planck* satellite and POL-2 at the East Asian Observatory James Clerk Maxwell Telescope. From left to right, the panels show: the Orion-Eridanus superbubble (Soler, Bracco & Pon, 2018), the Orion molecular clouds (*Planck* Collaboration I, 2016; *Planck* Collaboration XXXV, 2016), and the Orion A filament (Pattle *et al.*, 2017; Ward-Thompson *et al.*, 2017).

References

Pattle, K., *et al. ApJ*, 846 (2017) 112P
Planck 2015 results I, 2016. *A&A*, 594 (2016) A1
Planck Collaboration XXXV, 2016. *A&A*, 586 (2016) A138
Soler, J. D., Bracco, A. & Pon, A. *A&A*, 609L (2018) 3S
Ward-Thompson, D., *et al. ApJ*, 842 (2017) 66W

Astronomy in Focus - XXX
Proceedings IAU Symposium No. XXX, 2018
M. T. Lago, ed.

doi:10.1017/S1743921319003570

Magnetic field structures in star-forming regions revealed by imaging polarimetry at multi-wavelengths

Jungmi Kwon

Institute of Space & Astronautical Science (ISAS) / Japan Aerospace Exploration Agency (JAXA), Japan

Abstract. Magnetic fields are ubiquitous in various scales of astronomical objects, and they are considered as playing significant roles from star to galaxy formations. However, the role of the magnetic fields in star forming regions is less well understood because conventional optical polarimetry is hampered by heavy extinction by dust. We have been conducting extensive near-infrared polarization survey of various star-forming regions from low- and intermediate-mass to high-mass star-forming regions, using IRSF/SIRPOL in South Africa. Not only linear but also circular polarizations have been measured for more than a dozen of regions. Both linear and circular polarimetric observations at near-infrared wavelengths are useful tools to study the magnetic fields in star forming regions, although infrared circular polarimetry has been less explored so far. In this presentation, we summarize our results of the near-infrared polarization survey of star forming regions and its comparison with recent submillimeter polarimetry results. Such multi-wavelength approaches can be extended to the polarimetry using ALMA, SPICA in future, and others. We also present our recent results of the first near-infrared imaging polarimetry of young stellar objects in the Circinus molecular cloud, which has been less studied but a very intriguing cluster containing numerous signs of active low-mass star formation.

References

Kwon, J., Nakagawa, T., Tamura, M., *et al.* 2018, *AJ*, 156, 1
Kwon, J., Doi, Y., Tamura, M., *et al.* 2018, *ApJ*, 859, 4
Kwon, J., Nakagawa, T., Tamura, M., *et al.* 2018, *ApJS*, 234, 42
Kwon, J., Tamura, M., Hough, J. H., Nagata, T., & Kusakabe, N. 2016, *AJ*, 152, 67
Kwon, J., Tamura, M., Hough, J. H., *et al.* 2016, *ApJ*, 824, 95
Kwon, J., Tamura, M., Hough, J. H., *et al.* 2015, *ApJS*, 220, 17
Kwon, J., Tamura, M., Hough, J. H., *et al.* 2014, *ApJL*, 795, L16
Kwon, J., Tamura, M., Lucas, P. W., *et al.* 2013, *ApJL*, 765, L6

Astronomy in Focus - XXX
Proceedings IAU Symposium No. XXX, 2018
M. T. Lago, ed.

doi:10.1017/S1743921319003582

The Line-of-Sight Magnetic Field Structure in Filamentary Molecular Clouds

M. Tahani, R. Plume and J. Brown

Department of Physics & Astronomy, University of Calgary, Calgary, Alberta, Canada
email: `mtahani@ucalgary.ca`

Magnetic fields pervade in the interstellar medium (ISM) and are believed to be important in the process of star formation, yet probing magnetic fields in star formation regions is challenging. We present a new method to find the line-of-sight strength and direction of magnetic fields in star forming regions using Faraday rotation measurements. Tahani *et al.* (2018) describes the method, in details. In this technique, we use Taylor *et al.* (2009) rotation measure data and adopt a simple approach, based on relative measurements, to estimate the amount of rotation measure induced by the molecular clouds versus that from the rest of the Galaxy. We then use a chemical evolution code, along with Kainulainen *et al.* (2009) extinction maps of each cloud, to find the electron column density of the molecular cloud at the position of each rotation measure data point. Combining the rotation measures produced by the molecular clouds and the electron column density, we calculate the line-of-sight magnetic field strength and direction.

We applied this method to four relatively nearby regions of Orion A, Orion B, Perseus, and California. In the California cloud and Orion A, we found clear evidence that the magnetic fields at one side of these filamentary structures were pointing towards us and were pointing away from us at the other side. This behaviour is consistent with a helical magnetic field morphology. In the vicinity of available Zeeman measurements in Orion A, Orion B, and Perseus, we found magnetic field values of -23 ± 38 μG, -129 ± 28 μG, and 32 ± 101 μG, respectively, which are in agreement with the Zeeman measurements.

References

J. Kainulainen, H. Beuther, T. Henning and R. Plume, 2009, *A&A*, 508, L35
M. Tahani, R. Plume, J. C. Brown and J. Kainulainen, 2018, *A&A*, 614, A100
A. R. Taylor, J. M. Stil and c. Sunstrum, 2009, *ApJ*, 702, 1230

Astronomy in Focus - XXX
Proceedings IAU Symposium No. XXX, 2018
M. T. Lago, ed.

doi:10.1017/S1743921319003594

Statistics on the relative orientation between magnetic fields and filaments hosting Planck Galactic Cold Clumps

D. Alina[1], I. Ristorcelli[2], L. Montier[2] and M. Juvela[3]

[1]Department of Physics, School of Science and Technology, Nazarbayev University, Astana 010000, Kazakhstan

[2]IRAP, Université de Toulouse, CNRS, UPS, CNES, Toulouse, France

[3]Department of Physics, PO Box 64, University of Helsinki, 00014, Helsinki, Finland

Abstract. We present a statistical analysis of the relative orientation between the plane-of-sky magnetic field and the filaments associated with the Galactic Cold Clumps. We separated polarization parameters components of the filaments and their background using thin optical medium assumption, the filaments were detected using the Rolling Hough Transform algorithm and we separated the clump and the filament contributions in our maps. We found that in high column density environments the magnetic fields inside the filaments and in the background are less likely to be aligned with each other. This suggests a decoupling between the inner and background magnetic fields at some stage of filaments evolution. A preferential alignment between the filaments and their inferred magnetic fields is observed in the whole selection if the clumps contribution is subtracted. Interestingly, a bimodal distribution of relative orientation is observed between the filamentary structures of the clumps and the filaments' magnetic field. Similar results are seen in a subsample of nearby filaments. The relative orientation clearly shows a transition from parallel to no preferential and perpendicular alignment depending on the volume densities of both clumps and filaments. Our results confirm a strong interplay between the magnetic field and filamentary structures during their formation and evolutionary process.

Astronomy in Focus - XXX
Proceedings IAU Symposium No. XXX, 2018
M. T. Lago, ed.

doi:10.1017/S1743921319003600

Fragmentation of a Filamentary Cloud Threaded by Perpendicular Magnetic Field

Tomoyuki Hanawa[1], Takahiro Kudoh[2] and Kohji Tomisaka[3]

[1]Center for Frontier Science, Chiba University, 1-33 Yayoi-cho, Inage-ku, Chiba, Chiba 263-8522, Japan. email: hanawa@faculty.chiba-u.jp

[2]Faculty of Education, Nagasaki University, 1-14 Bonkyo-machi, Nagasaki, Nagasaki 852-8521, Japan. email: kudoh@nagasaki-u.ac.jp

[3]Division of Theoretical Astronomy, National Astronomical Observatory of Japan, Mitaka, Tokyo 181-8588, Japan. email: tomisaka@th.nao.ac.jp

Abstract. Filamentary molecular clouds are thought to fragment to form clumps and cores. However, the fragmentation may be suppressed by magnetic force if the magnetic fields run perpendicularly to the cloud axis. We evaluate the effect using a simple model. Our model cloud is assumed to have a Plummer like radial density distribution, $\rho = \rho_c \left[1 + r^2/(2pH^2)\right]^{2p}$, where r and H denote the radial distance from the cloud axis and the scale length, respectively. The symbols, ρ_c and p denote the density on the axis and radial density index, respectively. The initial magnetic field is assumed to be uniform and perpendicular to the cloud axis. The model cloud is assumed to be supported against the self gravity by gas pressure and turbulence. We have obtained the growth rate of the fragmentation instability as a function of the wavelength, according to the method of Hanawa, Kudoh & Tomisaka (2017). The instability depends crucially on the outer boundary. If the displacement vanishes in regions very far from the cloud axis, cloud fragmentation is suppressed by a moderate magnetic field. If the displacement is constant along the magnetic field in regions very far from the cloud, the cloud is unstable even when the magnetic field is infinitely strong. The wavelength of the most unstable mode is longer for smaller index, p.

Keywords. ISM: clouds, ISM: magnetic fields, magnetohydrodynamics: MHD

Reference

Hanawa, T., Kudoh, T., & K. Tomisaka 2017, *ApJ*, 30, 490

Astronomy in Focus - XXX
Proceedings IAU Symposium No. XXX, 2018
M. T. Lago, ed.

doi:10.1017/S1743921319003612

Dust and polarization of cold clumps

Mika Juvela

Department of physics, University of Helsinki, FI-00014 Finland
(on behalf of the Planck and Herschel Cold Cores projects and the TOP-SCOPE collaboration)

Abstract. The Planck catalogue of Galactic cold clumps, PGCC, contains sources of ongoing and future star formation. The data show clear variations also in their dust properties.

We use Planck polarization measurements to investigate the polarization fraction in PGCC clumps and the relative orientation of filamentary structures and magnetic fields (Alina *et al.* 2017). The decrease of polarization fraction as a function of column density can be related to the field geometry but also suggest some loss of grain alignment.

PGCCs have been studied with ground-based observations (Liu *et al.* 2018). The first SCUBA-2/POL-2 polarization studies have targeted the infrared dark cloud G35.39-0.33. The magnetic field is found to be mostly perpendicular to the main filament. The plane-of-the-sky field strength is $\sim$50 μG, a noticeable support against gravity. The polarization fraction decreases with increasing column density. This matches predictions of RAT grain alignment models but the relative contribution of the field morphology is hard to quantify (Juvela *et al.* 2018).

We continue to use MHD simulations to study the same phenomena, with synthetic observations of clumps and filaments.

References

Alina D., Ristorcelli I., Montier L., et al. 2017, submitted, arXiv1712.09325
Juvela M., Guillet V., Liu T., *et al.* 2018, *A&A*, in press, arXiv1809.00864
Liu T., Kim K.-T., Juvela M., *et al.* 2018, *ApJS* 234, 28

Astronomy in Focus - XXX
Proceedings IAU Symposium No. XXX, 2018
M. T. Lago, ed.

doi:10.1017/S1743921319003624

B-fields and gas motion in the L1689 region: an interpretation of Planck polarization data

Masafumi Matsumura and Team BISTRO Japan

Faculty of Education, Kagawa University, Takamatsu, Kagawa, 760-8522, Japan
email: `matsu@ed.kagawa-uac.jp`

Abstract. With using the Planck polarization data (PR2, Planck Collaboration *et al.* 2016), we investigate the magnetic fields in L1689 and associated clouds, and compare them with centroid velocities $V_{\rm LSR}$ of ^{12}CO and ^{13}CO from the COMPLETE survey (Ridge *et al.* 2006). We observe two components in this elongated region: in one component, the position angle of the magnetic field varies from -10 to 110 degrees in the galactic coordinate, while $V_{\rm LSR}$ is rather constant ($= 4 \pm 0.5$ km/s). In the other component with the position angle being constant ($= 110 \pm 15$ degrees), the velocity $V_{\rm LSR}$ shows a spatial gradient from 3 to 5 km/s, as one goes from west to east along the direction of elongation. If the east side of the component is more distant from us than the west, this gradient suggests that this component is stretching. This work is supported by JSPS KAKENHI Grant Number JP18H03720 (PI: Koji S. Kawabata).

References

Planck Collaboration, Aghanim, N., Ashdown, M., *et al.* 2016, *A&A*, 596, A109
Ridge, N. A., Di Francesco, J., Kirk, H., *et al.* 2006, *AJ*, 131, 2921

Astronomy in Focus - XXX
Proceedings IAU Symposium No. XXX, 2018
M. T. Lago, ed.

doi:10.1017/S1743921319003636

The role of the magnetic field in a translucent molecular cloud

Georgia Panopoulou

California Institute of Technology
1200 E. California Blvd, CA, USA 91125
email: panopg@caltech.edu

Abstract. Translucent molecular clouds represent a vastly underexplored regime of cloud evolution in terms of the effect of the magnetic field. Their pristine nature renders them ideal for investigating the initial properties of the magnetic field, prior to the onset of star formation. Using starlight polarimetry, we map the plane-of-sky magnetic field orientation throughout 10 sq. degrees of the Polaris Flare translucent molecular cloud. We provide the first quantitative estimate of the magnetic field strength in this type of system. By combining our measurements with the high-resolution Herschel dust emission map, we find a preferred alignment between filaments and the observed magnetic field. Our results support the presence of a strong magnetic field in this system (Panopoulou *et al.* 2016).

Reference

Panopoulou, G., Psaradaki, I., & Tassis, K. 2016, *MNRAS*, 462, 1517

Astronomy in Focus - XXX
Proceedings IAU Symposium No. XXX, 2018
M. T. Lago, ed.

doi:10.1017/S1743921319003648

Near-IR imaging polarimetry of the RCW 106 cloud complex

Shohei Tamaoki[1], Koji Sugitani[1] and Takayoshi Kusune[2]

[1]Nagoya City University, Graduate School of Natural Sciences
Mizuho-cho, Mizuho-ku, Nagoya 467-8501, Japan

[2]National Astronomical Observatory of Japan, 2-21-1 Osawa, Mikata, Tokyo 181-8588, Japan

Abstract. We have carried out near-IR imaging polarimetry toward RCW 106 with the JHK$_s$-simultaneous imaging polarimeter SIRPOL mounted on the IRSF 1.4m telescope at SAAO, in March and May, 2017 and January, 2018. We have observed 29 fields and covered mostly the southern part of the giant molecular cloud complex associated with the H$_{II}$ region RCW 106, which is located at a distance of 3.5 kpc (Moises *et al.* 2011) and is elongated approximately in the north-south direction with a size of $\sim$70$\times$15 pc. Our preliminary analysis indicates that the magnetic field seems to globally run along the complex elongation, unlike many other elongated clouds that are often reported to have their global elongations perpendicular to the magnetic fields. The RCW 106 complex consists of many small filaments or clumps. Some of such filaments seem to parallel to the magnetic fields, but some others perpendicular. Around the central part of the H$_{II}$ region RCW 106, the magnetic field appears to be influenced by the expansion of this H$_{II}$ region. Here, we present our preliminary results by comparing with the archival molecular line and far- to mid-IR data.

Reference

Moises, A. P., Damineli, A., Figueredo, E., Blum, R. D., Conti, P. S., Barbosa, C. L., *et al.* 2011, *MNRAS*, 411, 705M

Astronomy in Focus - XXX
Proceedings IAU Symposium No. XXX, 2018
M. T. Lago, ed.

doi:10.1017/S174392131900365X

Signposts of shock-induced magnetic field compression in star-forming environments

Helmut Wiesemeyer

Max Planck Institite for Radio Astronomy, Auf dem Hügel 69, 53121 Bonn, Germany
email: hwiese@mpifr.de

Abstract. In star-forming environments, shock-compressed magnetic fields occur in cloud-cloud collisions, in molecular clouds exposed to supernova remnants (SNRs), and in photo-dissociation regions (PDRs). Besides their dynamical role, they increase the cosmic ray flux above the Galactic average, and the trapped particles contribute to the heating of the shocked gas. The associated dust emission is polarized perpendicularly to the sky plane projection of the field, $B_{\rm sky}$. In edge-on viewed shock planes, highly ordered polarization patterns are expected. In search of such a signature, the dust emission from the Orion bar (a prototypical PDR) and from a molecular cloud/SNR interface (IC443-G) was studied with a $\lambda 870\,\mu$m polarimeter at the APEX (Wiesemeyer *et al.* 2014 and references therein). While our polarization map of OMC1 confirms the hourglass shape of $B_{\rm sky}$ (e.g., Schleuning 1998, Houde *et al.* 2004), a deep integration towards the Orion bar reveals an alignment of $B_{\rm sky}$ with the shock forming in response to the wind and to the ionizing radiation from the Trapezium cluster (Fig. 1). This structure suggests a compressed magnetic field accelerating cosmic-ray particles, a scenario proposed by Pellegrini *et al.* (2009) to explain the high excitation temperature of rotationally warm H_2 and CO (Shaw *et al.* 2009, Peng *et al.* 2012, respectively).

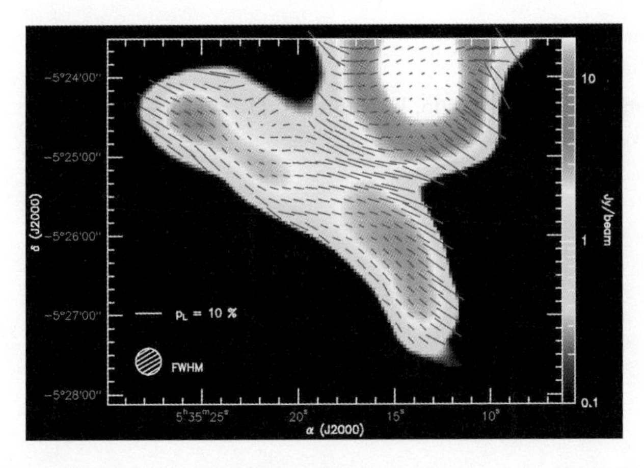

Figure 1. $B_{\rm sky}$ towards the Orion bar (corrected for instrumental polarization) with 345 GHz dust emission underneath. The Orion south core appears in the north-west.

References

Houde, M., Dowell, C.D., Hildebrand, R.H. *et al.* 2014, *ApJ*, 604, 717
Pellegrini, E.W., Baldwin, J.A., Ferland, G.J. *et al.* 2009, *ApJ*, 693, 285
Peng, T.-C., Zapata, L.A., Wyrowski, F. *et al.* 2012, *A&A*, 538, A12
Schleuning, D.A. 1998, *ApJ*, 493, 811
Shaw, G., Ferland, G.J., Henney, W.J. *et al.* 2009, *ApJ*, 701, 677
Wiesemeyer, H., Hezareh, T., Kreysa, E. *et al.* 2014, *PASP* 126, 1027

Astronomy in Focus - XXX
Proceedings IAU Symposium No. XXX, 2018
M. T. Lago, ed.

doi:10.1017/S1743921319003661

Revealing magnetic fields towards massive protostars: a multi-scale approach using masers and dust

Daria Dall'Olio, W. H. T. Vlemmings and M. V. Persson

Department of Space, Earth and Environment, Chalmers University of Technology,
Onsala Space Observatory,Observatorievägen 90, 43992 Onsala, Sweden
email: `daria.dallolio@chalmers.se`

Abstract. Magnetic fields play a significant role during star formation processes, hindering the fragmentation and the collapse of the parental cloud, and affecting the accretion mechanisms and feedback phenomena. However, several questions still need to be addressed to clarify the importance of magnetic fields at the onset of high-mass star formation, such as how strong they are and at what evolutionary stage and spatial scales their action becomes relevant. Furthermore, the magnetic field parameters are still poorly constrained especially at small scales, i.e. few astronomical units from the central object, where the accretion disc and the base of the outflow are located. Thus we need to probe magnetic fields at different scales, at different evolutionary steps and possibly with different tracers. We show that the magnetic field morphology around high-mass protostars can be successfully traced at different scales by observing maser and dust polarised emission. A confirmation that they are effective tools is indeed provided by our recent results from 6.7 GHz MERLIN observations of the massive protostar IRAS 18089-1732, where we find that the small-scale magnetic field probed by methanol masers is consistent with the large-scale magnetic field probed by dust (Dall'Olio *et al.* 2017 A&A 607, A111). Moreover we present results obtained from our ALMA Band 7 polarisation observations of G9.62+0.20, which is a massive star-forming region with a sequence of cores at different evolutionary stages (Dall'Olio *et al.* submitted to A&A). In this region we resolve several protostellar cores embedded in a bright and dusty filamentary structure. The magnetic field morphology and strength in different cores is related to the evolutionary sequence of the star formation process which is occurring across the filament.

Keywords. magnetic field – masers – stars: formation – stars: massive – dust – polarisation

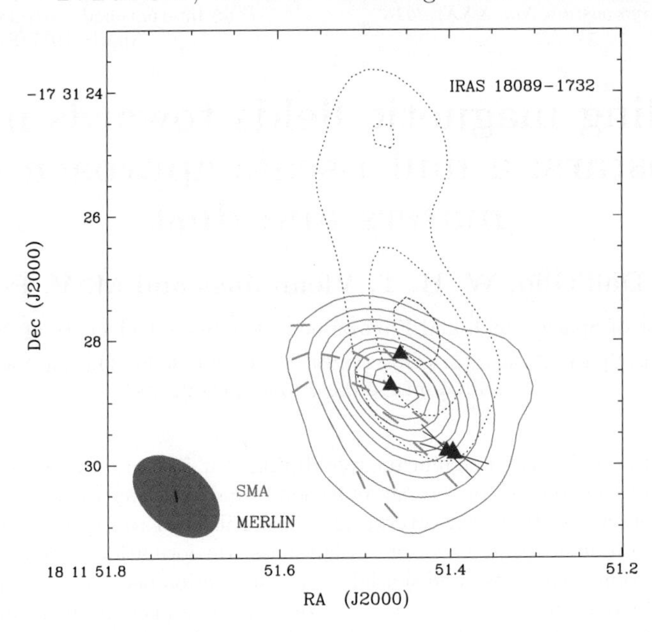

Figure 1. The red contours show the integrated Stokes I image of the continuum emission in 10σ steps. Dotted black contours show the SiO(5-4) emission, tracing the outflow, integrated from 30 to 40 km s^{-1} (5σ steps of 260 mJy $beam^{-1}$). The magenta line segments plot the orientation of the magnetic field derived from dust polarization (obtained with SMA, Beuther *et al.* 2010). The blue triangles mark the position of methanol masers and the blue lines indicate the direction of the magnetic field from methanol maser linear polarization (MERLIN observations discussed in this work).

Astronomy in Focus - XXX
Proceedings IAU Symposium No. XXX, 2018
M. T. Lago, ed.

doi:10.1017/S1743921319003673

The role of magnetic fields in the early stages of star formation

Maud Galametz, Anaëlle Maury and Valeska Valdivia

AIM, CEA, CNRS, Université Paris-Saclay, UniversitéParis Diderot, Sorbonne Paris Cité, F-91191 Gif-sur-Yvette, France
email: maud.galametz@cea.fr

Abstract. Magnetic fields are believed to redistribute part of the angular momentum during the collapse and could explain the order-of-magnitude difference between the angular momentum observed in protostellar envelopes and that of a typical main sequence star. The Class 0 phase is the main accretion phase during which most of the final stellar material is collected on the central embryo. To study the structure of the magnetic fields on 50-2000 au scales during that key stage, we acquired SMA polarization observations (870μm) of 12 low-mass Class 0 protostars. In spite of their low luminosity, we detect dust polarized emission in all of them. We observe depolarization effects toward high-density regions potentially due to variations in alignment efficiency or in the dust itself or geometrical effects. By comparing the misalignment between the magnetic field and the outflow orientation, we show that the B is either aligned or perpendicular to the outflow direction. We observe a coincidence between the misalignment and the presence of large perpendicular velocity gradients and fragmentation in the protostar (Galametz *et al.* 2018). Our team is using MHD simulations combined with the radiative transfer code POLARIS to produce synthetic maps of the polarized emission. This work is helping us understand how the magnetic field varies from the large-scale to the small-scales, quantify beam-averaging biases and study the variations of the polarization angles as a function of wavelength or the assumption made on the grain alignment (see poster by Valdivia).

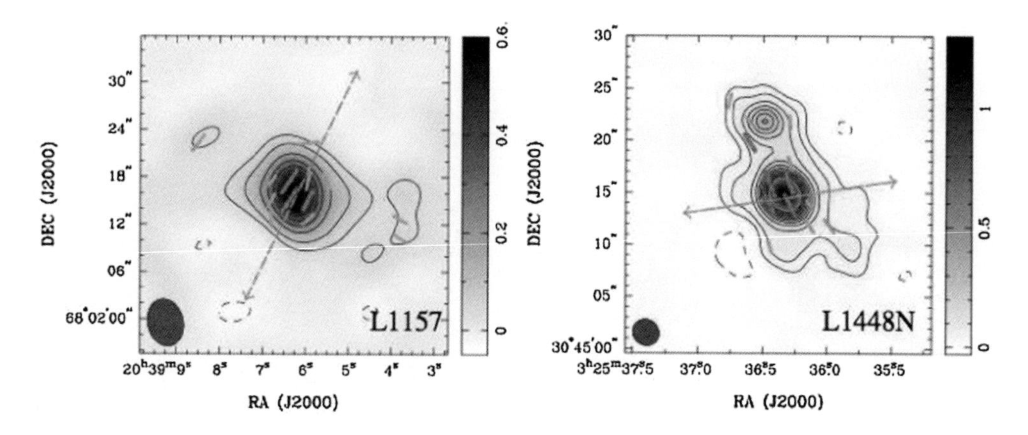

Figure 1. SMA 870μm dust emission of two Class 0 protostellar enveloppes. The blue arrows indicate the outflow direction. The B line segments are overlaid (>3σ in red, >2σ detections in orange).

References

Galametz, M., Maury, A. J., Girart, J. M., Rao, R., Zhang, Q., Gaudel, M., Valdivia, V., Keto, E. & Lai, S.-P. 2018, *A&A*, 616A, 139G

Maury, A. J., Girart, J. M., Zhang, Q., Hennebelle, P., Keto, E., Rao, R., Lai, S.-P., Ohashi, N., Galametz, M. 2018, *MNRAS*, 477, 2760

Astronomy in Focus - XXX
Proceedings IAU Symposium No. XXX, 2018
M. T. Lago, ed.

doi:10.1017/S1743921319003685

Trying to Make Sense of Polarization Patterns in Circumstellar Disks

Ian W. Stephens[1], Haifeng Yang[2,3], and Zhi-Yun Li[2]

[1]Harvard-Smithsonian Center for Astrophysics, 60 Garden Street, Cambridge, MA, USA
email: ian.stephens@cfa.harvard.edu

[2]Astronomy Department, University of Virginia, Charlottesville, VA 22904, USA

[3]Institute for Advanced Study, Tsinghua University, Beijing, 100084, People's Republic of China

Abstract. In the era of ALMA, we can now resolve polarization within circumstellar disks at (sub)millimeter wavelengths. While many initially hoped that these observations would map magnetic fields in disks, the observed polarization patterns indicate other possible polarization mechanisms. These alternative polarization mechanisms include Rayleigh self-scattering, grains aligning with the radiation anisotropy (k-RAT alignment), and mechanical alignment. Stephens *et al.* (2017) specifically showed that the polarization morphology in HL Tau changes rapidly with wavelength; the morphology is uniform at $870\,\mu$m, azimuthal at 3.1 mm, and ~50%/50% mix of the two at 1.3 mm. Although it has been suggested that the polarized emission at $870\,\mu$m is due to scattering and at 3.1 mm is due to k-RAT alignment, both mechanisms appear to have shortcomings. Specifically, Kataoka *et al.* (2017) showed that scattering requires much smaller grains (10s of μm) than that suggested by other studies, while k-RAT alignment suggest a significant decrease in polarization along the minor axis, which is not seen. Studies of other disks have suggested that polarization may come from grains aligned with the magnetic fields, but these studies are inconclusive. Understanding and extracting information about the polarized emission from disks requires multi-wavelength and high resolution observations.

Keywords. polarization, stars: protoplanetary disks, submillimeter

---

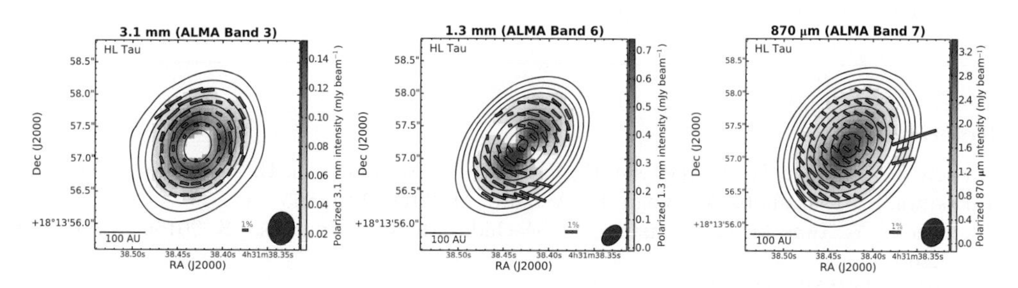

Figure 1. Images from Stephens *et al.* (2017), showing HL Tau ALMA polarization observations at 3 mm (left), 1.3 mm (middle), and $870\,\mu$m (right).

References

Kataoka, A., Tsukagoshi, T., Pohl, A., *et al.* 2017, *Ap. Lett.*, 844, L5
Stephens, I. W., Yang, H., Li, Z.-Y., *et al.* 2017, *ApJ*, 851, 55

Astronomy in Focus - XXX
Proceedings IAU Symposium No. XXX, 2018
M. T. Lago, ed.

doi:10.1017/S1743921319003697

The impact of non-ideal effects on the circumstellar disk evolution and their observational signatures

Y. Tsukamoto[1], S. Okuzumi[2], K. Iwasaki[3], M. N. Machida[4] and S. Inutsuka[5]

[1]Graduate Schools of Science and Engineering, Kagoshima University, Kagoshima, Japan
email: m.lugaro@phys.uu.nl

[2]Department of Earth and Planetary Sciences, Tokyo Institute of Technology, Tokyo, Japan

[3]Department of Earth and Space Science, Osaka University, Osaka, Japan

[4]Department of Earth and Planetary Sciences, Kyushu University, Fukuoka, Japan

[5]Department of Physics, Nagoya University, Aichi, Japan

Abstract. It has been recognized that non-ideal MHD effects (Ohmic diffusion, Hall effect, ambipolar diffusion) play crucial roles for the circumstellar disk formation and evolution. Ohmic and ambipolar diffusion decouple the gas and the magnetic field, and significantly reduces the magnetic torque in the disk, which enables the formation of the circumstellar disk (e.g., Tsukamoto *et al.* 2015b). They set an upper limit to the magnetic field strength of $\sim$0.1 G around the disk (Masson *et al.* 2016). The Hall effect notably changes the magnetic torques in the envelope around the disk, and strengthens or weakens the magnetic braking depending on the relative orientation of magnetic field and angular momentum. This suggests that the bimodal evolution of the disk size possibly occurs in the early disk evolutionary phase (Tsukamoto *et al.* 2015a, Tsukamoto *et al.* 2017). Hall effect and ambipolar diffusion imprint the possibly observable characteristic velocity structures in the envelope of Class 0/I YSOs. Hall effect forms a counter-rotating envelope around the disk. Our simulations show that counter rotating envelope has the size of 100–1000 au and a recent observation actually infers such a structure (Takakuwa*et al.* 2018). Ambipolar diffusion causes the significant ion-neutral drift in the envelopes. Our simulations show that the drift velocity of ion could become 100-1000 m s^{-1}.

Keywords. stars: formation, protoplanetary disks, magnetic fields

---

References

Masson, J., Chabrier, G., Hennebelle, P., Vaytet, N., & Commerçon, B. 2016, A&A, 587, A32
Takakuwa, S., Tsukamoto, Y., Saigo, K., & Saito, M. 2018, ApJ, 865, 51
Tsukamoto, Y., Iwasaki, K., Okuzumi, S., Machida, M. N., & Inutsuka, S. 2015a, ApJL, 810, L26
—. 2015b, MNRAS, 452, 278
Tsukamoto, Y., Okuzumi, S., Iwasaki, K., Machida, M. N., & Inutsuka, S.-i. 2017, PASJ, 69, 95

Astronomy in Focus - XXX
Proceedings IAU Symposium No. XXX, 2018
M. T. Lago, ed.

doi:10.1017/S1743921319003703

Magnetically regulated collapse in the B335 protostar ?

A. J. Maury[1], J. M. Girart[2] and Q. Zhang[3]

[1]AIM, CEA, CNRS, Université Paris-Saclay, UniversitéParis Diderot, Sorbonne Paris Cité, F-91191 Gif-sur-Yvette, France
email: anaelle.maury@cea.fr

[2]Institut de Ciències de l'Espai (ICE, CSIC), Cerdanyola del Vallès, Catalonia, Spain

[3]Harvard-Smithsonian Center for Astrophysics, Cambridge, MA 02138, USA

Abstract. The role of the magnetic field during protostellar collapse is still poorly constrained from an observational point of view, and only few constraints exist that shed light on the magnetic braking efficiency during the main accretion phase. I presented our ALMA polarimetric observations of the thermal dust continuum emission at 1.3 mm, towards the B335 Class 0 protostar (Maury *et al.* 2018a). Linearly polarized dust emission is detected at all scales probed by our observations (50 to 1000 au). The magnetic field structure has a very ordered topology in the inner envelope, with a transition from a large-scale poloidal magnetic field, in the outflow direction, to strongly pinched in the equatorial direction. We compared our data to a family of magnetized protostellar collapse models. We show that only models with an initial core mass-to-flux ratio $\mu \sim 5-6$ are able to reproduce the observed properties of B335, especially the upper-limits on its disk size, its large-scale envelope rotation β and the pronounced magnetic field lines pinching observed in our ALMA data. In these MHD models, the magnetic field is dynamically relevant to regulate the typical outcome of protostellar collapse, suggesting a magnetically-regulated disk formation scenarios is at work in B335.

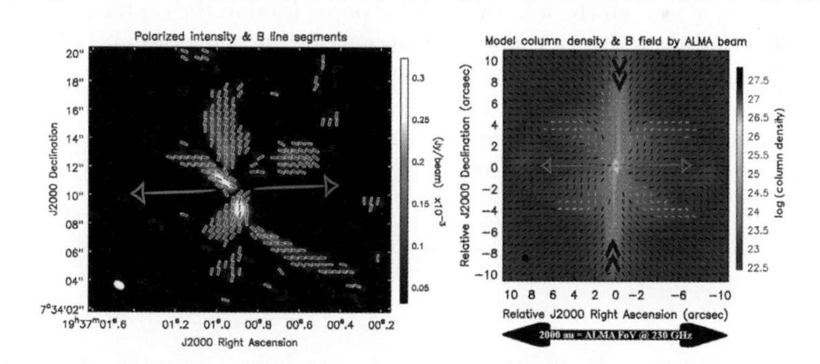

Figure 1. Left: the background image shows the ALMA polarized dust continuum emission, and the superimposed line segments show the B-field (polarization angle rotated by 90°) where the polarized dust continuum emission is detected at $> 3\sigma$. Right: Column density from a non-ideal MHD model of the protostellar collapse. The magnetic field topology in the core at scales 2000 au, integrated along the line-of-sight and convolved by the ALMA synthesized beam, are shown as black line segments, while the white line segments highlight the general areas where they are detected in our ALMA map of B335.

Reference

Maury, A.J., Girart, J.M., Zhang, Q., Hennebelle, P., Keto, E., Rao, R., Lai, S.P., Ohashi, N., & Galametz, M., *MNRAS*, 477, 2, 2760 (2018a)

Astronomy in Focus - XXX
Proceedings IAU Symposium No. XXX, 2018
M. T. Lago, ed.

doi:10.1017/S1743921319003715

Twinkle little stars: Massive stars are quenched in strong magnetic fields

Fatemeh S. Tabatabaei[1], M. Almudena Prieto[2] and Juan A. Fernández-Ontiveros[3]

[1]Instituto de Astrofísica de Canarias, Vía Láctea S/N, E-38205 La Laguna, Spain
School of Astronomy, Institute for Research in Fundamental Sciences, 19395-5531 Tehran, Iran
email: ftaba@ipm.ir

[2]Instituto de Astrofísica de Canarias, Vía Láctea S/N, E-38205 La Laguna, Spain
Departamento de Astrofísica, Universidad de La Laguna, E-38206 La Laguna, Spain
email: aprieto@iac.es

[3]Instituto de Astrofísica de Canarias, Vía Láctea S/N, E-38205 La Laguna, Spain
Departamento de Astrofísica, Universidad de La Laguna, E-38206 La Laguna, Spain
email: jafo@iac.es

Abstract. The role of the magnetic fields in the formation and quenching of stars with different mass is unknown. We studied the energy balance and the star formation efficiency in a sample of molecular clouds in the central kpc region of NGC 1097, known to be highly magnetized. Combining the full polarization VLA/radio continuum observations with the HST/Hα, Paα and the SMA/CO lines observations, we separated the thermal and non-thermal synchrotron emission and compared the magnetic, turbulent, and thermal pressures. Most of the molecular clouds are magnetically supported against gravitational collapse needed to form cores of massive stars. The massive star formation efficiency of the clouds also drops with the magnetic field strength, while it is uncorrelated with turbulence (Tabatabaei *et al.* 2018). The inefficiency of the massive star formation and the low-mass stellar population in the center of NGC 1097 can be explained in the following steps: I) Magnetic fields supporting the molecular clouds prevent the collapse of gas to densities needed to form massive stars. II) These clouds can then be fragmented into smaller pieces due to e.g., stellar feedback, non-linear perturbations and instabilities leading to local, small-scale diffusion of the magnetic fields. III) Self-gravity overcomes and the smaller clouds seed the cores of the low-mass stars.

Reference

Tabatabaei, F. S., Minguez, P., Prieto, M. A., & Fernández-Ontiveros, J. A. 2018, *Nature Astronomy*, 2, 83

Astronomy in Focus - XXX
Proceedings IAU Symposium No. XXX, 2018
M. T. Lago, ed.

doi:10.1017/S1743921319003727

Role of the magnetic field on the formation of solar type stars

Valeska Valdivia, Anaëlle J. Maury and Patrick Hennebelle

AIM, CEA, CNRS, Université Paris-Saclay, Université Paris Diderot, Sorbonne Paris Cité, F-91191 Gif-sur-Yvette, France

email: valeska.valdivia@cea.fr, anaelle.maury@cea.fr, patrick.hennebelle@cea.fr

Abstract. Magnetic fields play a key role during the gravitational collapse of dense protostellar cores. In recent years mm and sub-mm observations of dust polarized emission have been used to unveil the morphology of the magnetic field, but this method relies on the assumption that non-spherical dust grains are well aligned with the magnetic field.
Using non-ideal MHD numerical simulations, we study the evolution of the magnetic field during the gravitational collapse. We use the state-of-the-art radiative transfer code POLARIS to compute the Stokes parameters and produce synthetic observations of mm/submm polarized dust emission. We compare the results obtained using the radiative torques (RAT) mechanism to the results obtained by assuming that grains are perfectly aligned to constrain how well polarized dust emission traces the magnetic field orientation.
The complexity of the magnetic field produces a mild depolarization. The depolarization observed in the inner regions is rather caused by a decrease of the dust alignment efficiency and it cannot be reproduced by just scaling down the polarisation degree obtained for a uniform efficiency. We find that the magnetic field orientation is well constrained by the polarized dust emission as long as its 3D topology remains organized.

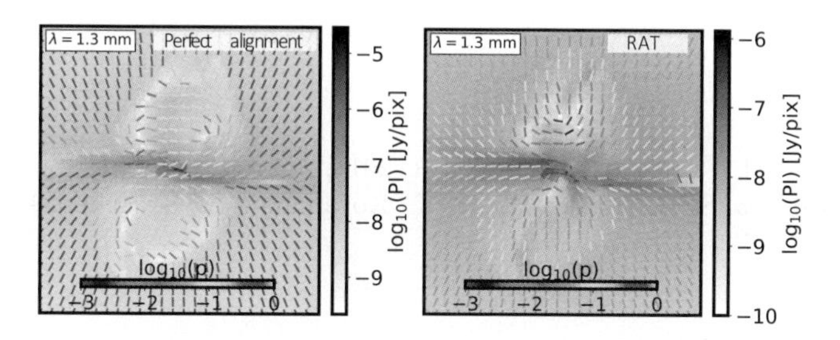

Figure 1. Total linearly polarized emission (background) and magnetic field orientation (vectors) for the perfect alignment case (*left*) and the RAT case (*right*). The color-scale of vectors shows the polarization fraction.

References

Galametz, M., Maury, A., Girart, J. M., *et al.* 2018, *A&A*, 616, A139
Maury, A. J., Girart, J. M., Zhang, Q., *et al.* 2018, *MNRAS*, 477, 2760
Reissl, S., Wolf, S., & Brauer, R. 2016, *A&A*, 593, A87
Teyssier, R. 2002, *A&A*, 385, 337

Astronomy in Focus - XXX
Proceedings IAU Symposium No. XXX, 2018
M. T. Lago, ed.

doi:10.1017/S1743921319003739

Kinematics of neutral and ionzied gas in the candidate protostar with efficient magnetic braking B335

Hsi-Wei Yen[1,3], Bo Zhao[2] and Patrick M. Koch[3] *et al.*

[1]European Southern Observatory, Karl-Schwarzschild-Str. 2, D-85748 Garching, Germany
email: `hwyen@asiaa.sinica.edu.tw`

[2]Max-Planck-Institut fur extraterrestrische Physik, Garching, Germany, 85748

[3]Academia Sinica Institute of Astronomy and Astrophysics, P.O. Box 23-141, Taipei, Taiwan

Abstract. Ambipolar diffusion can cause a velocity drift between ions and neutrals. This is one of the non-ideal MHD effects proposed to enable the formation of large Keplerian disks with sizes of tens of au (Zhao *et al.* 2018). To observationally study ambipolar diffusion in collapsing protostellar envelopes, we analyzed the ALMA $H^{13}CO^+$ (3–2) and $C^{18}O$ (2–1) data of the protostar B335, which is a candidate source with efficient magnetic braking (Yen *et al.* 2015). We constructed kinematical models to fit the velocity structures observed in $H^{13}CO^+$ and $C^{18}O$. With our kinematical models, the infalling velocities in $H^{13}CO^+$ and $C^{18}O$ are both measured to be 0.85 ± 0.2 km s^{-1} at a radius of 100 au, suggesting that the velocity drift between the ionized and neutral gas is at most 0.3 km s^{-1} at a radius of 100 au in B335. The Hall parameter for $H^{13}CO^+$ is estimated to be $\gg$1 on a 100 au scale in B335, so that $H^{13}CO^+$ is expected to be attached to the magnetic field. Our non-detection or upper limit of the velocity drift between the ionized and neutral gas could suggest that the magnetic field remains rather well coupled to the bulk neutral material on a 100 au scale in B335, and that any significant field-matter decoupling, if present, likely occurs only on a smaller scale, leading to an accumulation of magnetic flux and thus efficient magnetic braking in the inner envelope in B335.

References

Zhao, B., Caselli, P., Li, Z.-Y., Krasnopolsky, R. 2018, *MNRAS*, 473, 4868
Yen, H.-W., Takakuwa, S., Koch, P.M., Aso, Y., Koyamatsu, S., Krasnopolsky, R., Ohashi, N. 2015, *ApJ*, 812, 129

Astronomy in Focus - XXX
Proceedings IAU Symposium No. XXX, 2018
M. T. Lago, ed.

doi:10.1017/S1743921319003740

Magnetic fields of T Tauri stars and inner accretion discs

Jean-Francois Donati

Univ. de Toulouse, CNRS, IRAP, 14 avenue Belin, F-31400 Toulouse, France
email: jean-francois.donati@irap.omp.eu

Abstract. Magnetic fields play a key role in the early life of stars and their planets, as they form from collapsing dense cores that progressively flatten into large-scale accretion discs and eventually settle as young suns orbited by planetary systems. Pre-main-sequence phases, in which central protostars feed from surrounding planet-forming accretion discs, are especially crucial for understanding how worlds like our Solar System are born.

Magnetic fields of low-mass T Tauri stars (TTSs) are detected through high-resolution spectroscopy and spectropolarimetry (e.g., Johns Krull 2007), whereas their large-scale topologies can be inferred from time series of Zeeman signatures using tomographic techniques inspired from medical imaging (Donati & Landstreet 2009). Large-scale fields of TTSs are found to depend on the internal structure of the newborn star, allowing quantitative models of how TTSs magnetically interact with their inner accretion discs, and the impact of this interaction on the subsequent stellar evolution (e.g., Romanova *et al.* 2002, Zanni & Ferreira 2013).

With its high sensitivity to magnetic fields, SPIRou, the new near-infrared spectropolarimeter installed in 2018 at CFHT (Donati *et al.* 2018), should yield new advances in the field, especially for young embedded class-I protostars, thereby bridging the gap with radio observations.

References

Donati, J.-F., & Landstreet, J.D. 2009, *ARA&A*, 47, 333
Donati, J.-F., *et al.*, 2018, *Deeg H., Belmonte J. (eds) Handbook of Exoplanets. Springer*
Johns Krull C. 2007, *ApJ* 664, 975
Romanova M.M., Ustyugova G.V., Koldoba A.V., Lovelace R.V.E. 2002, *ApJ* 578, 420
Zanni, C., & Ferreira, J. 2013, *A&A* 550, 99

Astronomy in Focus - XXX
Proceedings IAU Symposium No. XXX, 2018
M. T. Lago, ed.

doi:10.1017/S1743921319003752

Formation and Structure of Magnetized Protoplanetary Disks

Susana Lizano[1] **and Carlos Tapia**[2]

[1]Instituto de Radioastronomía y Astrofísica, UNAM,
Apartado Postal 3-72, 58089 Morelia, Michoacán, México
email: s.lizano@irya.unam.mx

[2]Instituto Nacional de Astrofísica, Optica y Electrónica,
Luis Enrique Erro 1, CP 72840, Tonantzintla, Puebla, México
email: c.tapia@irya.unam.mx

Abstract. Protoplanetary disks are expected to form through the gravitational collapse of magnetized rotating dense cores. We discuss the structure and emission of models of accretion disks threaded by a poloidal magnetic field and irradiated by the central star, expected to form in this process (Shu *et al.* 2007; Lizano *et al.* 2016). The poloidal magnetic field produces sub-keplerian rotation of the gas which can accelerate planet migration (Adams *et al.* 2009). It can make the disk more stable against gravitational perturbations (Lizano *et al.* 2010). Also, the magnetic compression can reduce the disk scale height with respect to nonmagnetic disks. We find that the mass-to-flux ratio λ is a critical parameter: disks with a weaker magnetization (high values of λ) are denser and hotter and emit more at millimeter wavelengths than disks with a stronger magnetization (low values of λ). Applying these models to the millimeter observations of the disk around the young star HL Tau indicate the large grains are present at the external radii in order to reproduce the observed 7 mm emission that extends up to 100 AU (Tapia & Lizano 2017). In the near future, observations with ALMA and VLA will be able to determine the level of magnetization of protoplanetary disks, which will be important to understand their formation and evolution.

Keywords. accretion, accretion disks, ISM: magnetic fields, planetary systems: protoplanetary disks, stars: formation

---

Astronomy in Focus - XXX
Proceedings IAU Symposium No. XXX, 2018
M. T. Lago, ed.

doi:10.1017/S1743921319003764

Observations of magnetic fields in Herbig Ae/Be stars

Markus Schöller[1] **and Swetlana Hubrig**[2]

[1]European Southern Observatory, Karl-Schwarzschild-Str. 2, 85748 Garching, Germany,
email: `mschoell@eso.org`

[2]Leibniz-Institut für Astrophysik Potsdam (AIP), An der Sternwarte 16, 14482 Potsdam, Germany, email: `shubrig@aip.de`

Abstract. Models of magnetically driven accretion reproduce many observational properties of T Tauri stars. For the more massive Herbig Ae/Be stars, the corresponding picture has been questioned lately, in part driven by the fact that their magnetic fields are typically one order of magnitude weaker. Indeed, the search for magnetic fields in Herbig Ae/Be stars has been quite time consuming, with a detection rate of about 10% (e.g. Alecian *et al.* 2008), also limited by the current potential to detect weak magnetic fields. Over the last two decades, magnetic fields were found in about twenty objects (Hubrig *et al.* 2015) and for only two Herbig Ae/Be stars was the magnetic field geometry constrained. Ababakr, Oudmaijer & Vink (2017) studied magnetospheric accretion in 56 Herbig Ae/Be stars and found that the behavior of Herbig Ae stars is similar to T Tauri stars, while Herbig Be stars earlier than B7/B8 are clearly different. The origin of the magnetic fields in Herbig Ae/Be stars is still under debate. Potential scenarios include the concentration of the interstellar magnetic field under magnetic flux conservation, pre-main-sequence dynamos during convective phases, mergers, or common envelope developments. The next step in this line of research will be a dedicated observing campaign to monitor about two dozen HAeBes over their rotation cycle.

References

Ababakr, K.M., Oudmaijer, R.D., & Vink, J.S. 2017, *MNRAS*, 472, 854
Alecian, E., *et al.*, 2008, *Contr. of the Astr. Obs. Skalnate Pleso*, 38, 235
Hubrig, S., *et al.*, 2015, *MNRAS*, 449, L118

Astronomy in Focus - XXX
Proceedings IAU Symposium No. XXX, 2018
M. T. Lago, ed.

doi:10.1017/S1743921319003776

Magnetic field detections in Herbig Ae SB2 systems

S. P. Järvinen[1], S. Hubrig[1], T. A. Carroll[1], M. Schöller[2] and I. Ilyin[1]

[1]Leibniz-Institut für Astrophysik Potsdam (AIP), An der Sternwarte 16, 14482 Potsdam, Germany

email: sjarvinen@aip.de, shubrig@aip.de, tcarroll@aip.de, ilyin@aip.de

[2]European Southern Observatory, Karl-Schwarzschild-Str. 2, 85748 Garching, Germany

email: mschoell@eso.org

Abstract. Studies of the presence of magnetic fields in Herbig Ae/Be stars are extremely important because they enable us to improve our insight into how the magnetic fields of these stars are generated and how they interact with their environment, including their impact on the planet formation process and the planet-disk interaction. We report new detections of weak mean longitudinal magnetic fields in the close Herbig Ae double-lined spectroscopic binary AK Sco and in the presumed spectroscopic Herbig Ae binary HD 95881 (Järvinen *et al.* 2018) based on observations obtained with HARPSpol attached to ESO's 3.6 m telescope. Such studies are important because only very few close spectroscopic binaries with orbital periods below 20 d are known among Herbig Ae stars. Our detections favour the conclusion that the previously suggested low incidence (5-10%) of magnetic Herbig Ae stars can be explained by the weakness of these fields and the limited accuracy of the published measurements. The search for magnetic fields and the determination of their geometries in close binary systems will play an important role for understanding the mechanisms that are responsible for the magnetic field generation.

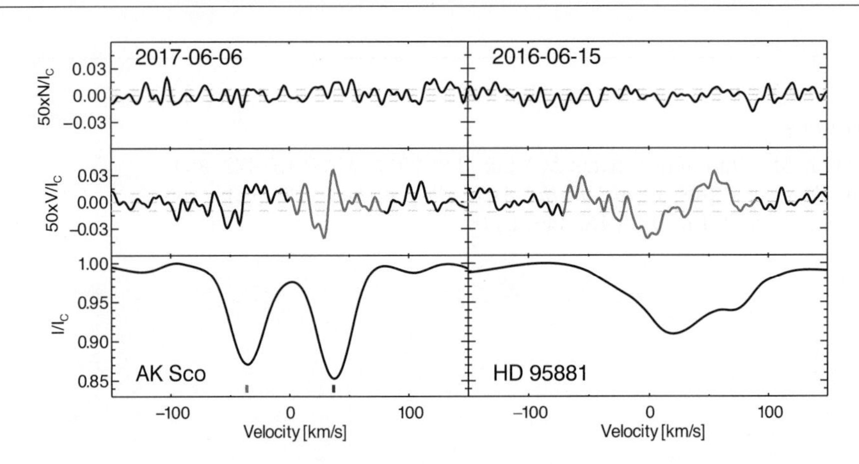

Figure 1. Examples of Singular Value Decomposition (SVD) Stokes I (bottom) and V (middle) as well as diagnostic null (N) profiles obtained for the Herbig SB2 systems AK Sco and HD 95881. The horizontal dashed lines indicate the average values and the $\pm 1\sigma$-ranges. The primary component of AK Sco is marked with a red and the secondary with a blue tick (Järvinen *et al.* 2018).

Reference

Järvinen, S.P., *et al.* 2018, *ApJL*, 858, 18

Astronomy in Focus - XXX
Proceedings IAU Symposium No. XXX, 2018
M. T. Lago, ed.

doi:10.1017/S1743921319003788

Magnetic field and accretion in EX Lup

Á. Kóspál[1,2], J.-F. Donati[3], J. Bouvier[4] and P. Ábrahám[1]

[1]Konkoly Observatory, Konkoly-Thege Miklós út 15-17, H-1121, Budapest, Hungary
[2]Max Planck Institute for Astronomy, Königstuhl 17, D-69117 Heidelberg, Germany
[3]Université de Toulouse, UPS-OMP, IRAP, F-31400 Toulouse, France
[4]Université Grenoble Alpes, CNRS, IPAG, F-38000, Grenoble, France

Abstract. While the Sun is a quiet and well-balanced star now, during its first few million years it possessed a strong magnetic field and actively accreted material from its circumstellar environment. Theoretical models predict that under certain circumstances the interaction of a strongly magnetic star and its circumstellar disk may lead to short bursts of increased accretion onto the star (D'Angelo & Spruit 2012). Examples for this phenomenon may be the members of a group of young eruptive stars called EXors. Their prototype, EX Lup, had its historically largest outburst in 2008. Spectroscopic evidence suggests that the mass accretion proceeds through the same magnetospheric accretion channels both in quiescence and in outburst but with different mass flux (Sicilia-Aguilar *et al.* 2012). To characterize for the first time EX Lup's magnetic field, we obtained spectropolarimetric monitoring for it with the CFHT/ESPaDOnS. We detected strong, poloidal magnetic field with a prominent cool polar cap and an accretion spot above it. We compared our results with numerical simulations, in order to check the applicability of the d'Angelo & Spruit model as an explanation of EX Lup's accretion outbursts. If EX Lup is a good proxy for the proto-Sun, similar magnetic field-disk interactions and outbursts might have happened during the early evolution of the Solar System as well.

References

D'Angelo, C. R., & Spruit, H.C. 2012, *MNRAS*, 420, 416
Sicilia-Aguilar, A., Kóspál, Á., Setiawan, J., Ábrahám, P., Dullemond, C., Eiroa, C., Goto, M., Henning, Th., & Juhász, A. 2012, *A&A*, 544, A93

Astronomy in Focus - XXX
Proceedings IAU Symposium No. XXX, 2018
M. T. Lago, ed.

doi:10.1017/S174392131900379X

How to identify magnetic field activity in young circumstellar disks

Mario Flock and Gesa H.-M. Bertrang

Max Planck Institute for Astronomy, Heidelberg, Germany
email: flock@mpia.de; bertrang@mpia.de

Abstract. Recent advanced simulations of protoplanetary disks allow us to search for observational constraints to identify the magnetic field activity in protoplanetary disks. With our 3D radiation non-ideal magneto-hydrodynamical (MHD) models including irradiation from an Herbig type star we are able to model the thermal and dynamical evolution in a so far never reached detail (Flock *et al.* 2017). The activity of the magneto-rotational instability in the inner hot ionized regions comes along with a magnetic dynamo. The oscillations in the mean toroidal magnetic field with a timescale of 10 local orbits can slightly bend the inner dust rim and so the irradiation surface. This causes a clear variability pattern in the near infrared (NIR) emission at the dust inner rim surface. Another way to identify the presence of magnetic fields are to search for polarization signatures. Using 3D non-ideal MHD simulations of the outer disk regions (Flock *et al.* 2015) we calculate synthetic images of the intrinsically polarized continuum emitted by aspherical grain aligned with the dominantly toroidal magnetic field (Bertrang *et al.* 2017). Our results show a clear radial polarization pattern for face-on observed disk, similar to recent observations by Ohashi *et al.* (2018). Additionally, we are even able to see the change of the polarization pattern inside the vortex as the poloidal magnetic field dominates therein.

Keywords. (stars:) planetary systems: protoplanetary disks, magnetic fields, (magnetohydrodynamics:) MHD, instabilities, polarization

---

References

Bertrang, G. H.-M., Flock, M. & Wolf, S. 2017, *MNRAS Letters*, 464, L61
Flock, M., Ruge, J. P., Dzyurkevich, N., Henning, Th., Klahr, H., Wolf, S. 2015, *A&A*, 574, 68
Flock, M., Fromang, S., Turner, N., Benisty, M. 2017, *ApJ*, 835, 230
Ohashi, S., Kataoka, A., Nagai, H., Momose, M., *et al.* 2018, *ApJ*, 864, 81

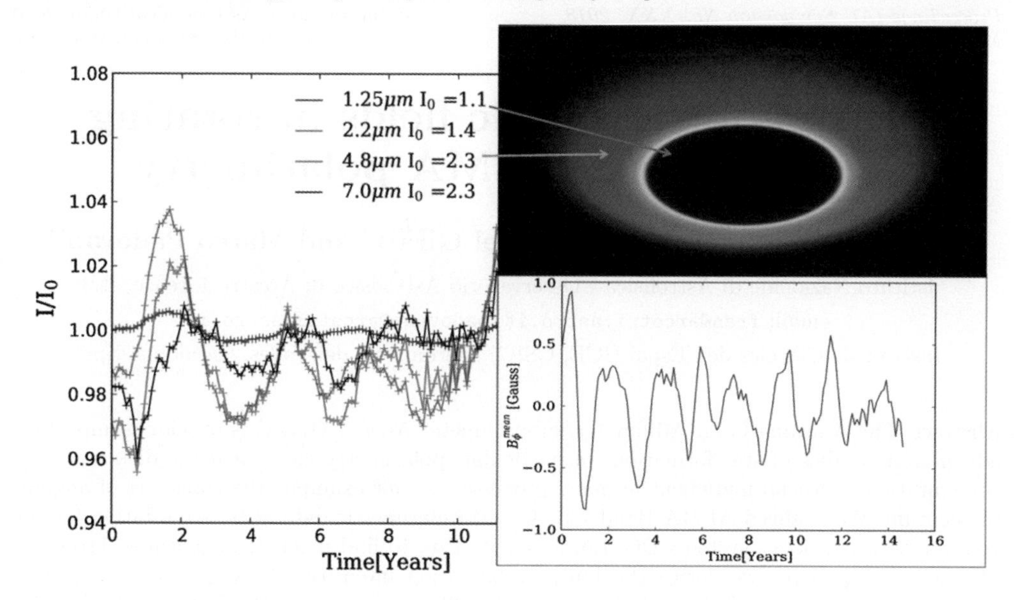

Figure 1. Left: Lightcurve of the simulation for different wavelengths. Top right: synthetic image for 3 different wavelength bands for the model. Bottom right: time evolution of the mean toroidal magnetic field at the inner dust rim.

Astronomy in Focus - XXX
Proceedings IAU Symposium No. XXX, 2018
M. T. Lago, ed.

doi:10.1017/S1743921319003806

Seeking for magnetic fields in rotating disk/jets with ALMA polarimetry

Francesca Bacciotti[1], Josep Miquel Girart[2] and Marco Padovani[1]

[1]Istituto Nazionale di Astrofisica - Osservatorio Astrofisico di Arcetri, Firenze, Italy
email: fran@arcetri.astro.it, padovani@arcetri.astro.it

[2]Institut de Ciències de l'Espai (ICE, CSIC), Cerdanyola del Valles, Catalonia, Spain

Abstract. The Atacama Large Millimeter/submillimeter Array (ALMA) is providing important advances in studies of star formation. In particular, polarimetry can reveal the disk magnetic configuration, a crucial ingredient in many processes, as, for example, the transport of angular momentum. We analized ALMA Band 7 (870 μm) polarimetric data at 0.″2 resolution for the young rotating disk/jet systems DG Tau and CW Tau, to find magnetic signatures. From the Stokes I, U, Q maps, we derive the linear polarization intensity, $P = \sqrt{Q^2 + U^2}$, the linear polarization fraction, and the polarization angle. The alignment of the latter with the disk minor axis (Fig. 1) shows that self-scattering of dust thermal emission rather than magnetic alignment dominates the polarization in both targets (Bacciotti *et al.* 2018). However, several dust properties can be diagnosed comparing the polarization data with the models of self-scattering (e.g. Kataoke *et al.* 2017, Yang *et al.* 2017). The maximum grain size turns out to be in the range 50 - 70 μm for DG Tau and 100 - 150 μm for CW Tau. The asymmetry of the polarized intensity in DG Tau, observed for the first time around a T Tauri star, indicates that the disk is flared. Moreover, the observed belt-like feature may betray the presence of a disk substructure. In contrast, the polarization maps of CW Tau indicate that here the grains have settled to the disk midplane. Polarimetry is thus very important in studies of the dust evolution.

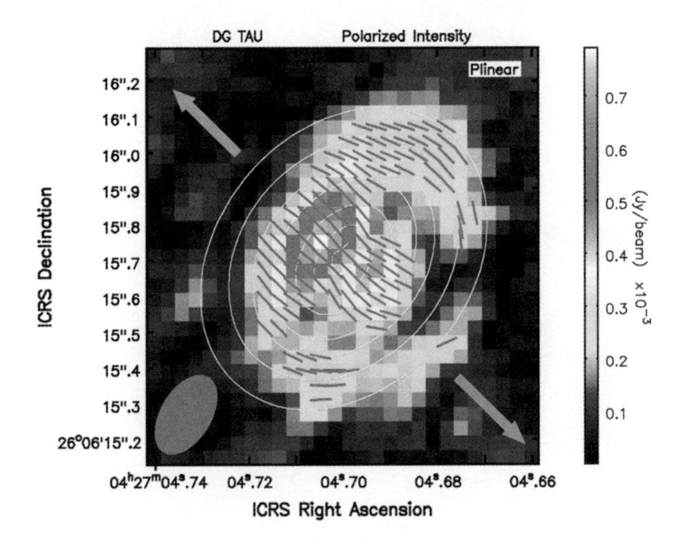

Figure 1. Linearly polarized intensity in the DG Tau disk, with superposed total intensity contours. The alignment of the polarization vectors (fixed length segments) along the disk minor axis supports self-scattering of the dust emission as the origin of the polarization. The higher polarized intensity toward the disk near-side (toward the receding jet lobe (red arrow)) suggests a flared disk geometry (Bacciotti *et al.* 2018).

References

Bacciotti, F., Girart, J. M., Padovani, M., *et al.* 2018, *Astroph. J. Lett.*, 865, L12
Kataoka, A., Tsukagoshi, T., Pohl, A., *et al.* 2017 *Astroph. J. Lett.*, 844, L5
Yang, H., Li, Z.-Y., Looney, L. W., Girart, J. M. & Stephens, I. W. 2017 *M.N.R.A.S.*, 472, 373

Astronomy in Focus - XXX
Proceedings IAU Symposium No. XXX, 2018
M. T. Lago, ed.

doi:10.1017/S1743921319003818

Magnetic fields in disks: A solution to an polarized ambiguity

Gesa H.-M. Bertrang[1], Paulo C. Cortés[2] and Mario Flock[1]

[1]Max Planck Institute for Astronomy, Königstuhl 17,
69117 Heidelberg, Germany,
email: bertrang@mpia.de, flock@mpia.de

[2]National Radio Astronomy Observatory, Joint ALMA Observatory,
Alonso de Cordova 3107, Santiago, Chile,
email: pcortes@alma.cl

Abstract. Numerous numerical studies suggest that magnetic fields influence the transport of dust and gas, the disk chemistry, the migration of planetesimals within the disk, and above all the accretion of matter onto the star. In short: Magnetic fields are crucial for the evolution of planet-forming disks. First indirect comparisons of theory and observations support this picture (Flock *et al.* 2017); however, profound observational constraints are still pending. Recent studies show that the intrinsically polarized continuum emission, the classical tracer of magnetic fields, might trace other physics as well (radiation field or dust grain size). The nearly face-on protoplanetary disk HD 142527 shows predominantly radial polarization vectors consistent with aspherical grains aligned by a toroidal magnetic field (Fig. 1; Bertrang *et al.* 2017a,b; Ohashi *et al.* 2018). However, the number of cutting-edge polarization observations presenting inconclusive data, for which these three different origins of polarization are not clearly distinguishable, increases continuously. We present a solution to this polarized ambiguity: observations

Toroidal Magnetic Field in Protoplanetary Disk

MHD based generic disk (Bertrang+2017a,b)

HD142527 - ALMA Band 7 (adapted from Ohashi+2018)

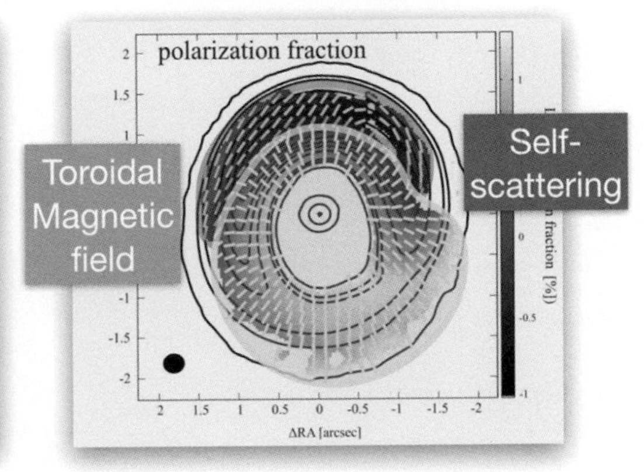

Figure 1. The predicted intrinsically polarized continuum emission of aspherical grains aligned by a toroidal magnetic field (*left*; Bertrang *et al.* 2017a,b) explains both the observed radial polarization vectors as well as the polarization fraction in the protoplanetary disk HD 142527 (*right*; green region). In the Northern part (blue region), the disk is optically thick and the polarization signal consistent with current models of self-scattering (Ohashi *et al.* 2018).

and simulations of the most direct tracer of magnetic fields, polarized gas emission, in combination with multi-wavelength continuum polarization observations will disentangle the sources of continuum polarimetry with ALMA (Bertrang *et al.* 2017a,b; Bertrang & Cortés in prep.).

Keywords. (stars:) planetary systems: protoplanetary disks, magnetic fields, polarization, radiative transfer, astrochemistry, radiation mechanisms: thermal, scattering, line: formation, (magnetohydrodynamics:) MHD, instabilities

---

References

Bertrang, G. H.-M., Flock, M. & Wolf, S. 2017, *MNRAS Letters*, 464, L61
Bertrang, G. H.-M. & Wolf, S. 2017, *MNRAS*, 469, 2869
Bertrang, G. H.-M. & Cortés, P. C., *in prep.*
Flock, M., Nelson, R. P., Turner, N. J., Bertrang, G. H.-M. *et al.* 2017, *ApJ*, 850, 131
Ohashi, S., Kataoka, A., Nagai, H., Momose, M. *et al.* 2018, *ApJ*, 864, 81

Astronomy in Focus - XXX
Proceedings IAU Symposium No. XXX, 2018
M. T. Lago, ed.

doi:10.1017/S174392131900382X

Magnetic massive stars in star forming regions

Swetlana Hubrig[1], Markus Schöller[2] and Silva P. Järvinen[1]

[1]Leibniz-Institut für Astrophysik Potsdam (AIP), An der Sternwarte 16, 14482 Potsdam, Germany
email: shubrig@aip.de, sjarvinen@aip.de

[2]European Southern Observatory, Karl-Schwarzschild-Str. 2, 85748 Garching, Germany
email: mschoell@eso.org

Abstract. One idea for the origin of magnetic fields in massive stars suggests that the magnetic field is the fossil remnant of the Galactic ISM magnetic field, amplified during the collapse of the magnetised gas cloud. A search for the presence of magnetic fields in massive stars located in active sites of star formation led to the detection of rather strong magnetic fields in a few young stars. Future spectropolarimetric observations are urgently needed to obtain insights into the mechanisms that drive the generation of kG magnetic fields during high-mass star formation.

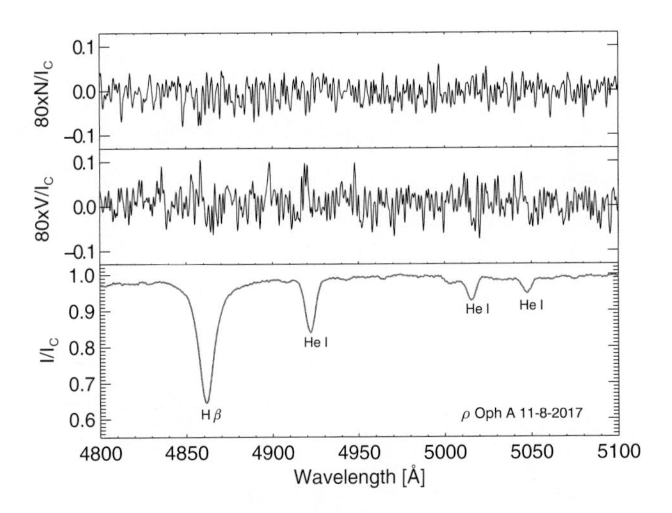

Figure 1. Stokes I, Stokes V, and diagnostic N spectra (from bottom to top) of ρ Oph A in the vicinity of the Hβ line. For better visibility, Stokes V and diagnostic N spectra are magnified by a factor of 80. Our study of the spectral variability indicates a behaviour similar to that observed in typical He-rich magnetic early-type Bp stars (Hubrig *et al.* 2018).

Reference

Hubrig, S., *et al.*, 2018, *AN*, 339, 72

Astronomy in Focus - XXX
Proceedings IAU Symposium No. XXX, 2018
M. T. Lago, ed.

doi:10.1017/S1743921319003831

An abundance analysis of AK Sco, a Herbig Ae SB2 system with a magnetic component

Swetlana Hubrig[1], Fiorella Castelli[2], and Silva P. Järvinen[1]

[1]Leibniz-Institut für Astrophysik Potsdam (AIP), An der Sternwarte 16, 14482 Potsdam, Germany

email: `shubrig@aip.de`, `sjarvinen@aip.de`

[2]Instituto Nazionale di Astrofisica, Osservatorio Astronomico di Trieste, via Tiepolo 11, 34143, Trieste, Italy

email: `castelli@oats.inaf.it`

Abstract. AK Sco is an SB2 system formed by two nearly identical Herbig Ae stars, with $T_{\rm eff} = 6500$ K and $\log g = 4.5$, surrounded by a circumbinary disk.This actively accreting system is of special interest among the pre-main-sequence binaries because of its prominent ultraviolet excess and the high eccentricity of its orbit. Moreover, recent spectropolarimetric observations using HARPSpol indicate the presence of a weak magnetic field in the secondary component (Järvinen *et al.* 2018). An abundance analysis of both components has shown that all elements have a solar abundance in the two stars, except for Li and Ba. These elements are enhanced by 2.2 and 0.5 dex, respectively, in the A component and by 2.4 and 0.5 dex, respectively, in the B component.

Reference

Järvinen, S.P., *et al.*, 2018, *ApJ*, 858, L18

Astronomy in Focus - XXX
Proceedings IAU Symposium No. XXX, 2018
M. T. Lago, ed.

doi:10.1017/S1743921319003843

Magnetic fields along the pre-main sequence: new magnetic field measurements of Herbig Ae/Be stars using high-resolution HARPS spectropolarimetry

S. P. Järvinen[1], S. Hubrig[1], M. Schöller[2] and I. Ilyin[1]

[1]Leibniz-Institut für Astrophysik Potsdam (AIP), An der Sternwarte 16, 14482 Potsdam, Germany

email: sjarvinen@aip.de, shubrig@aip.de, ilyin@aip.de

[2]European Southern Observatory, Karl-Schwarzschild-Str. 2, 85748 Garching, Germany

email: mschoell@eso.org

Abstract. Herbig Ae/Be-type stars are analogs of T Tauri stars at higher masses. Since the confirmation of magnetospheric accretion using Balmer and sodium line profiles in the Herbig Ae star UX Ori, a number of magnetic studies have been attempted, indicating that about 20 Herbig Ae/Be stars likely have globally organized magnetic fields. The low detection rate of magnetic fields in Herbig Ae stars can be explained by the weakness of these fields and rather large measurement uncertainties. The obtained density distribution of the root mean square longitudinal magnetic field values revealed that only a few stars have magnetic fields stronger than 200 G, and half of the sample possesses magnetic fields of about 100 G or less. We report on the results of our analysis of a sample of presumably single Herbig Ae/Be stars based on recent observations obtained with HARPSpol attached to ESO's 3.6m telescope. Knowledge of the magnetic field structure combined with the determination of the chemical composition are indispensable to constrain theories on star formation and magnetospheric accretion in intermediate-mass stars. As of today, magnetic phase curves have been obtained only for two Herbig Ae/Be stars, HD 101412 and V380 Ori.

Astronomy in Focus - XXX
Proceedings IAU Symposium No. XXX, 2018
M. T. Lago, ed.

doi:10.1017/S1743921319003855

Doppler images of V1358 Ori

Levente Kriskovics[1], Zsolt Kővári[1], Krisztián Vida[1], Katalin Oláh[1] and Thorsten A. Carroll[2]

[1]Konkoly Observatory of the Hungarian Academy of Sciences,
H-1121 Konkoly-Thege Street 15-17, Budapest, Hungary
email: kriskovics.levente@csfk.mta.hu

[2]Leibniz-Institut für Astrophysik Potsdam (AIP), An der Sternwarte 16,
14482 Potsdam, Germany
email: tcarroll@aip.de

Abstract. We present Doppler images of the active dwarf star V1358 Ori using high-resolution spectra from the NARVAL spectropolarimeter mounted on the Bernard Lyot Telescope. The spectra were taken between 09-20 Dec, 2013 with a resolution of R=80000. Doppler imaging was carried out with our new generation multi-line Dopper imaging code iMap (Carroll *et al.* 2012). 40 individual photospheric lines were selected by line depth, temperature sensitivity and blends. Two data subsets were formed to get two consecutive Doppler images. Prominent cool spots at lower latitudes are found on both maps. At 0.5 phase there is a prominent equatorial feature on both maps. Weaker polar features can be seen on the first map, which somewhat diminishes for the second map. On the first image there is a cool surface feature at 30 degrees latitude which seems to fade greatly on the second map. Around 0.75 phase, a new spot seems to form. These changes suggest a rapid surface evolution. Spot displacements may also indicate surface differential rotation, which was derived by cross-correlating the two subsequent Doppler images (see e.g. Kővári *et al.* 2012). We fit the latitudinal correlation peaks with a sine-squared law. The fit suggests solar-type surface differential rotation with a shear parameter of $\alpha = 0.02 \pm 0.02$. The shear parameter fits the $P_{\rm rot} - |\alpha|$ diagram in Kővári *et al.* (2017) quite well.

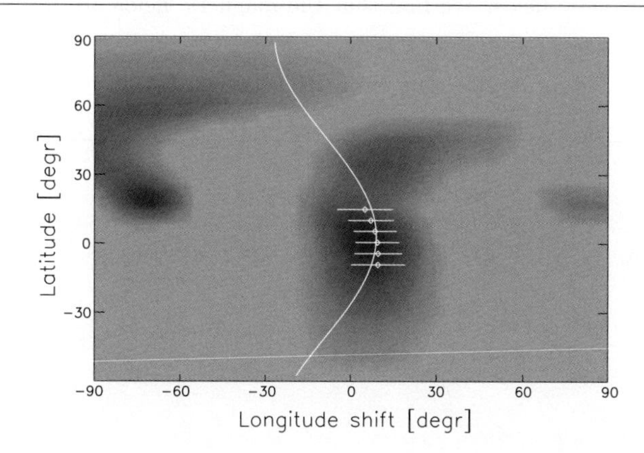

Figure 1. Cross correlation map derived with ACCORD. Dots indicate the best correlation on each 5 degree latitude strip. The minima are fitted with a sine-squared law, resulting in $\alpha = 0.02 \pm 0.02$.

References

Carroll, T. A., Strassmeier, K. G., Rice, J. B & Künstler, A 2012, *A&A*, 548, 95
Kővári, Zs., Korhonen, H., Kriskovics, L. *et al.* 2012, *A&A*, 539, 50
Kővári, Zs., Oláh, K., Kriskovics, L. *et al.* 2017, *AN*, 338, 903

Astronomy in Focus - XXX
Proceedings IAU Symposium No. XXX, 2018
M. T. Lago, ed.

doi:10.1017/S1743921319003879

Two Different Grain Distributions within the Protoplanetary Disk around HD 142527

Satoshi Ohashi[1], Akimasa Kataoka[2] and Hiroshi Nagai[2]

[1]RIKEN Cluster for Pioneering Research, 2-1, Hirosawa, Wako-shi, Saitama 351-0198, Japan
email: satoshi.ohashi@riken.jp

[2]National Astronomical Observatory of Japan, 2-21-1 Osawa, Mitaka, Tokyo 181-8588, Japan

Abstract. The origin of polarized emission from protoplanetary disks is uncertain. Three mechanisms have been proposed for such polarized emission so far, (1) grain alignment with magnetic fields, (2) grain alignment with radiation gradients, and (3) self-scattering of thermal dust emission. Aiming to observationally identify the polarization mechanisms, we present ALMA polarization observations of the 0.87 mm dust continuum emission toward the circumstellar disk around HD 142527 with a spatial resolution of ~ 0.2 arcsec as shown in Ohashi *et al.* (2018). We confirm that the polarization vectors in the northern region are consistent with self-scattering because of a flip of the polarization vectors. Furthermore, we show that the polarization vectors in the southern region are consistent with grain alignment by magnetic fields, although self-scattering cannot be ruled out. To understand these differences between the polarization mechanisms, we propose a simple grain size segregation model: small dust grains ($\lesssim 100$ microns) are dominant and aligned with magnetic fields in the southern region, and middle-sized (~ 100 microns) grains in the upper layer emit self-scattered polarized emission in the northern region. The grain size near the middle plane in the northern region cannot be measured because the emission at 0.87 mm is optically thick. However, it can be speculated that larger dust grains ($\gtrsim$ cm) may accumulate near this plane. These results are consistent with those of a previous analysis of the disk, in which large grain accumulation and optically thick emission from the northern region were found. This model is also consistent with theories where smaller dust grains are aligned with magnetic fields. We find that the magnetic fields are toroidal, at least in the southern region.

Keywords. polarization, protoplanetary disks, HD 142527

Reference

Ohashi, S., Kataoka, A., Nagai, H., *et al.* 2018, *ApJ*, 864, 81

Astronomy in Focus - XXX
Proceedings IAU Symposium No. XXX, 2018
M. T. Lago, ed.

doi:10.1017/S1743921319003880

The accretion process in the magnetic Herbig Ae star HD 104237

Markus Schöller[1] and Mikhail A. Pogodin[2]

[1]European Southern Observatory, Karl-Schwarzschild-Str. 2, 85748 Garching, Germany,
email: `mschoell@eso.org`

[2]Central Astronomical Observatory at Pulkovo, Pulkovskoye chaussee 65,
196140 Saint Petersburg, Russia,
email: `mikhailpogodin@mail.ru`

Abstract. After successfully retrieving the known rotation period $P = 42.076$ d in the Herbig Ae star HD 101412 using spectroscopic signatures of accretion tracers (Schöller *et al.* 2016), we have studied magnetospheric accretion in the Herbig Ae SB2 system HD 104237 using spectroscopic parameters of the He I 10830, Paγ, and He I 5876 lines, formed in the accretion region. Employing 21 spectra obtained with ISAAC and X-shooter, we found that the temporal behavior of these parameters can be explained by a variable amount of matter being accreted in the region between the star and the observer. Using a periodogram analysis, we examined the possible origin of the accretion flow in HD 104237 and considered the following four scenarios: matter flows from the circumbinary envelope, mass exchange between the system's components, magnetospheric accretion (MA) from the disk onto the star, and fast high-latitude accretion from a disk wind onto a weakly magnetized star. Based on a correlation analysis, we were able to show that the primary component is responsible for the observed emission line spectrum of the system. Since we do not find any correlation of the spectroscopic parameters with the phase of the orbital period ($P \approx 20$ d), we can reject the first two scenarios. We found a variation period of about 5 d, which likely represents the stellar rotation period of the primary and favors the MA scenario.

Reference

Schöller, M., *et al.*, 2016, *A&A*, 592, A50

Astronomy in Focus - XXX
Proceedings IAU Symposium No. XXX, 2018
M. T. Lago, ed.

doi:10.1017/S1743921319003892

3D simulations of accretion onto a star: Fast funnel-wall accretion

Shinsuke Takasao[1], Kengo Tomida[2], Kazunari Iwasaki[2] and Takeru K. Suzuki[3]

[1]Department of Physics, Nagoya University, Nagoya, Aichi, 464-8602, Japan
email: takasao@nagoya-u.jp

[2]Department of Earth and Space Science, Osaka University, Toyonaka, Osaka, 560-0043, Japan

[3]School of Arts & Sciences, University of Tokyo, 3-8-1, Komaba, Meguro, Tokyo, 153-8902, Japan

Abstract. We show the results of global 3D magnetohydrodynamics simulations of an accretion disk with a rotating, weakly magnetized central star (Takasao *et al.* 2018). The disk is threaded by a weak large-scale poloidal magnetic field. The central star has no strong stellar magnetosphere initially and is only weakly magnetized. We investigate the structure of the accretion flows from a turbulent accretion disk onto the star. Our simulations reveal that fast accretion onto the star at high latitudes is established even without a stellar magnetosphere. We find that the failed disk wind becomes the fast, high-latitude accretion as a result of angular momentum exchange mediated by magnetic fields. The rapid angular momentum exchange occurs well above the disk, where the Lorentz force that decelerates the rotational motion of gas can be comparable to the centrifugal force. Unlike the classical magnetospheric accretion model, fast accretion streams are not guided by magnetic fields of the stellar magnetosphere. Nevertheless, the accretion velocity reaches the free-fall velocity at the stellar surface owing to the efficient angular momentum loss at a distant place from the star. Our model can be applied to Herbig Ae/Be stars whose magnetic fields are generally not strong enough to form magnetospheres, and also provides a possible explanation why Herbig Ae/Be stars show indications of fast accretion.

Reference

Takasao, S.,Tomida, K., Iwasaki, K., & Suzuki, K.T. 2018, *ApJ*, 857, 4

Astronomy in Focus - XXX
Proceedings IAU Symposium No. XXX, 2018
M. T. Lago, ed.

doi:10.1017/S1743921319003909

Role of magnetic field in star formation

Gemechu Muleta Kumssa and Solomon Belay Tessema

Ethiopian Space Science and Technology Institute (ESSTI),Entoto Observatory and Research Center(EORC), Astronomy and Astrophysics Research Division, Addis Ababa, Ethiopia
email: gemechumk@gmail.com, tessemabelay@gmail.com

Abstract. Magnetic fields are a key component in star formation theories. Nevertheless, their exact role in the formation of stars is still a matter of debate. The process of angular momentum transportation by the disturbance caused during magnetic field reconnection still needs theoretical formulation in terms of the collapsing cloud's parameters. The purposes of this study are: to model the critical mass of a magnetized, gravitating and turbulent star forming molecular cloud (MC) and to formulate the momentum carried out by a magnetic field through magnetic field reconnection in terms of the MC's parameters. By applying theoretical modeling, we show how angular momentum transported via an Alfvén wave can be described in terms of mass, radius and dispersion velocity of a collapsing cloud core and a model equation of the critical mass for a gravitating, turbulent, and magnetized molecular cloud core. The outflow of angular momentum by magnetic fields facilitates the inflow of mass. On the other side, magnetic pressure prevents collapse. Therefore, magnetic fields have a dual purpose in the process of star formation. This momentum outflow triggers the inflow of mass to conserve angular momentum. The results show that Alfvén waves are like a machine that extracts angular momentum from a magnetized collapsing cloud core. Thus the total angular momentum transported by magnetic field at a distance R from the core's center depends on the size, mass and turbulent velocity dispersion of the collapsing cloud core.

Keywords. Star formation, molecular cloud, magnetic pressure, turbulence, angular momentum

Astronomy in Focus - XXX
Proceedings IAU Symposium No. XXX, 2018
M. T. Lago, ed.

doi:10.1017/S1743921319003910

Zeeman Effect Observations in Class I Methanol Masers

Emmanuel Momjian[1] and Anuj P. Sarma[2]

[1]National Radio Astronomy Observatory, P.O. Box O, Socorro, NM 87801, USA
email: `emomjian@nrao.edu`

[2]Physics Department, DePaul University, 2219 N. Kenmore Ave., Byrne Hall 211, Chicago, IL 60614, USA
email: `asarma@depaul.edu`

Abstract. We report the detection of the Zeeman effect in the 44 GHz Class I methanol maser line toward the star forming region DR21W. The 44 GHz methanol masers in this source occur in a $\sim 3''$ linear structure that runs from northwest to southeast, with the two dominant components at each end, and several weaker maser components in between. Toward a 93 Jy maser in the dominant northwestern component, we find a significant Zeeman detection of -23.4 ± 3.2 Hz. If we use the recently published result of Lankhaar *et al.* (2018) that the $F = 5 - 4$ hyperfine transition is responsible for the 44 GHz methanol maser line, then their value of $z = -0.92\,\mathrm{Hz\,mG^{-1}}$ yields a line-of-sight magnetic field of $B_{\rm los} = 25.4 \pm 3.5\,\mathrm{mG}$. If Class I methanol masers are pumped in high density regions with $n \sim 10^{7-8}\,\mathrm{cm^{-3}}$, then magnetic fields in these maser regions should be a few to several tens of mG. Therefore, our result in DR21W is certainly consistent with the expected values.

Using the above noted splitting factor in past Zeeman effect detections in Class I methanol masers reported by Sarma & Momjian (2011) and Momjian & Sarma (2017) in the star forming regions OMC-2 and DR21(OH) result in $B_{\rm los}$ values of $20.0 \pm 1.2\,\mathrm{mG}$ and $58.2 \pm 2.9\,\mathrm{mG}$, respectively. These are also consistent with the expected values.

Keywords. Masers, magnetic fields, techniques: interferometric, radio lines: ISM

References

Lankhaar, B., Vlemmings, W., Surcis, G., *et al.* 2018, *Nature Astronomy*, 2, 145
Momjian, E. & Sarma, A.P. 2017, *ApJ* 834, 168
Sarma, A.P. & Momjian, E. 2011, *ApJ* (Letters) 730, 5

Astronomy in Focus - XXX
Proceedings IAU Symposium No. XXX, 2018
M. T. Lago, ed.

doi:10.1017/S1743921319003922

Magnetic fields and massive star formation

Qizhou Zhang

Center for Astrophysics | Harvard & Smithsonian
email: qzhang@cfa.harvard.edu

Abstract. Massive stars (M > 8 $M_\odot$) often form in parsec-scale molecular clumps that collapse and fragment, leading to the birth of a cluster of stellar objects. The role of magnetic fields during the formation of massive dense cores is still not clear. The steady improvement in sensitivity of (sub)millimeter interferometers over the past decade enabled observations of dust polarization of large samples of massive star formation regions. We carried out a polarimetric survey with the Submillimeter Array of 14 massive star forming clumps in continuum emission at a wavelength of 0.89 mm. This unprecedentedly large sample of massive star forming regions observed by a submillimeter interferometer before the advent of ALMA revealed compelling evidence of strong magnetic influence on the gas dynamics from 1 pc to 0.1 pc scales. We found that the magnetic fields in dense cores tend to be either parallel or perpendicular to the mean magnetic fields in their parental molecular clumps. Furthermore, the main axis of protostellar outflows does not appear to be aligned with the mean magnetic fields in the dense core where outflows are launched. These findings suggest that from 1 pc to 0.1 pc scales, magnetic fields are dynamically important in the collapse of clumps and the formation of dense cores. From the dense core scale to the accretion disk scale of $\sim 10^2$ au, however, gravity and angular momentum appear to be more dominant relative to the magnetic field.

Reference

Zhang, Q., Qiu, K., Girart, J. M., *et al.* 2014, *ApJ*, 792, 116

FM5
Understanding Historical Observations to Study Transient Phenomena

Astronomy in Focus - XXX
Proceedings IAU Symposium No. XXX, 2018
M. T. Lago, ed.

doi:10.1017/S1743921319003934

Terra-Astronomy – Understanding historical observations to study transient phenomena

Ralph Neuhäuser[1], Dagmar L. Neuhäuser[2] and Thomas Posch[3]†

[1]Astrophysical Institute, University of Jena, 07745 Jena, Germany
email: ralph.neuhaeuser@uni-jena.de

[2]Independent scholar, Merano, Italy & Jena, Germany

[3]Department of Astrophysics, University of Vienna, 1180 Vienna, Austria

Abstract. We give an overview of Focus Meeting 5 on the new field of Terra-Astronomy.

Keywords. solar activity, supernovae, novae, comets, eclipses, history of astronomy

---

Focus Meeting 5. During the General Assembly 2018 of the International Astronomical Union in Vienna, Austria, Focus Meeting 5 (FM5) took place about *Understanding historical observations to study transient phenomena* (Terra-Astronomy). Scientific Organizing Committee: S.-H. Ahn, A. Ankay, D. Banerjee, J. Evans, L. Fletcher, R. Gautschy, D.L. Neuhäuser, R. Neuhäuser (co-chair), T. Posch (co-chair), B. Schaefer, J. Steele, F.R. Stephenson, J. Vaquero, N. Vogt, M. Werner, H.R.G. Yazdi. See www.astro.uni-jena.de/IAU for program and details. Editors of the proceedings for FM5 were Ralph Neuhäuser, Elizabeth Griffin, and Thomas Posch. FM5 covered 11 reviews, 16 contributed talks, and 18 poster papers, only some of them could be included here.

What is Terra-Astronomy? The study of historical observations motivated by contemporary astrophysical questions is valuable for different fields in astronomy. Terra-Astronomy concentrates on transient phenomena. Terra-Astronomy uses *terrestrial* archives: historical archives (text records from previous centuries) and natural archives (e.g. ^{14}C in trees and ^{10}Be in polar ice as solar activity proxies); and Terra-Astronomy studies phenomena which potentially can affect *Terra*:
(1) While we can study solar activity in great detail with satellites only since a few decades (too short for secular changes), telescopic sunspot observations are available for four centuries (Clette *et al.*, this issue) – yet, polar lights and sunspots were recorded since more than two millennia ago. While solar activity can also be studied with radioisotopes (solar wind being inversely proportional to cosmic-ray influx), they depend on the Earth magnetic field, whose reconstruction is also uncertain.
(2) Solar and lunar eclipses (and other conjunctions) which were recorded with a precise location (and being dated) are used to study secular variations in the Earth rotation and acceleration of the Moon (Stephenson *et al.*, this issue).
(3) Historical observations of comets (e.g., Zolotova *et al.*, this issue) and meteor streams facilitate orbit solutions and the study of their origin and dynamics.
(4) Pre-telescopic observations of novae and supernovae enable detailed studies of their remnants (Pagnotta *et al.* and Cosci, this issue); age and celestial location are well-known from the historical records, some light curves could be derived; historical supernovae are the most nearby stellar explosions, so that deep follow-up observations can reveal

† deceased April 2019

possible former donor stars (SN Ia), other previous companions, and neutron stars (SN II). New records on historical supernovae written in Arabic were uncovered recently (e.g. R. Neuhäuser *et al.* 2016), so that progress is possible also in this field – for a precise dating on the Muslim lunar calendar, one needs to know when the lunar crescent was first visible at that time and location (Gautschy & Thomann, this issue).
(5) More transient phenomena were possibly observed in history, such as meteoritic impacts on Earth or Moon, changes in brightness or color of stars, or optical transients in gamma-ray bursts; some may also be detectable on photographic plates.
Previous works have shown that applied historical astronomy can be very useful for astrophysics. The recently measured ^{14}C variation around AD 775 triggered a new interest in historical observations, so that, e.g., several teams have studied northern lights observed at around that time to reconstruct the solar activity state and to evaluate the possibility of a solar super-flare (e.g., R. Neuhäuser & D. L. Neuhäuser 2015). This Focus Meeting also covered observations after the invention of the telescope: e.g., we discussed telescopic (and naked-eye) sunspots and aurorae until around AD 1715, the end of the Maunder Minimum, as well as the Carrington flare in AD 1859.

Understanding historical records. In all such cases, it is essential to correctly understand the historical reports, which are texts in old to ancient languages: translation may work best in collaboration between astronomers and scholars of languages (articles by Chapman, Hunger, and Shylaja, this issue). Available competence in the field of History of Astronomy and cultural knowledge can significantly advance the correct understanding of historical observations. The widespread interpretation of transient phenomena as problematic omina often had a deep cultural impact.

E.g., most recently, a *red cross* observed in England around AD 775 was interpreted as absorbed supernova, as airglow after a gamma-ray burst, as aurora borealis indicating a solar activity maximum (all in connection to a strong ^{14}C variation around AD 775), and as an unrelated halo display – indeed, there are many meteorological phenomena recorded in, e.g., aurora catalogs (see D. L. Neuhäuser & R. Neuhäuser 2015).

We put forward eight recommendations:

- Historical records must not be used as quarry: we have to approach the problem unbiased, we have to be aware of our modern interests. E.g., what did our *ancestors* see, when they reported a *red cross* in their chronicle ? They wrote down, what was important for them, and maybe also to preserve it for us.
- Any such work should be based on critical editions, which contain variants of the different copies, dating corrections, sources, history of transmission etc.
- Any translation is an interpretation, and the meaning of words can evolve.
- No text without context – especially, it is not sufficient to search just digitally for a certain keyword in large text corpora; digital searches have to be conducted carefully (D. L. Neuhäuser *et al.* 2018).
- Consider author's intention and the ideological background – the zeitgeist: chronicles reflect the connection between (celestial) signs and following (terrestrial) events – understanding of the signs as portents.
- Today's terminology is defined by physics – historical descriptions are pheno-typical: a phenomenon can be, e.g., rainbow-like, but it is not necessarily a *rainbow* in our sense.
- Criteria help to identify the likely true physical character of the phenomena (e.g. polar lights, halo effects, meteor showers, comets), historical records provide information on up to five criteria: (i) timing, (ii) direction, (iii) color and form, (iv) dynamics/changes, and (v) duration/repetition, see D. L. Neuhäuser *et al.* (2018).
- One should provide references of previous catalogs considering the discussed events and a list of rejected false-positives.

Working on astronomical problems by interpreting historical texts is different in methodology from other fields of astrophysics.

Methodological and epistemological aspects. Terra-Astronomy is a study of historical astronomical events with respect to questions relevant to current astrophysics. It needs to use specific methods of the humanities (e.g., a hermeneutical approach to ancient texts), but also shares its methods and goals with the natural sciences.

Astronomy in general, much more so than, e.g., theoretical physics, includes elaborate studies of individual objects. But are we interested in such remote individual objects for their own sake – or rather in the context of general laws (e.g., those on the evolution of the solar system)? We also discussed epistemological aspects of Terra-Astronomy, where we would deduce general knowledge from individual events.

Terra-Astronomy deals both with *nature* and with *history*: our primary data are typically historical records – hence, the exceptional role of *understanding of texts*. The final purpose of our studies is typically the refinement of general laws of nature. In Terra-Astronomy, as in Kant and Rickert, *historical* means a *past event*, while *something regarded as a unique individual* can be termed *historic*, in German both *historisch*. Any event studied in Terra-Astronomy includes *facts in their uniqueness and individuality*. Thereby, it has the potential to expand the scientist's epistemological perspective.

As synthesis of the prototypical natural-scientific and the *individualising* approach to reality, *Astronomy is one of those disciplines in which the generalising way and the individualising way of concept formation are most intimately connected to each other* (Rickert 1986). The case to be made here is: this characteristic feature of astronomy is even more characteristic of Terra-Astronomy, it is indeed one of the defining features of the latter. It is the reason for the fact that Terra-Astronomy is truly trans-disciplinary.

Further aspects. This newly emerging field has a strong potential also in education and outreach (e.g., Zotti & Wolf, this issue), as the extraction of celestial observations from local chronicles can indeed be done by, e.g., school students and amateur historians, e.g. in a world-wide Citizen Science project; such a project can raise the public awareness for the variety of celestial phenomena and their relevance for Earth. Studying solar activity proxies is also a timely service to humanity, as the Sun currently shows a weak Schwabe cycle, so that a new Grand Minimum may be starting.

The vast and disparate amount of historical observations also calls for a world-wide and homogeneous catalogue, also implementing oral reports and drawings – a call to save this valuable heritage, so that we can understand the past and utilize it for solving forefront astrophysical problems. Resolution B3 recommends *that a concerted effort be made to ensure the preservation, digitization, and scientific exploration of all of astronomy's historical data, both analogue and primitive digital, and associated records.* Some celestial phenomena are also relevant for other natural science fields, e.g. atmospheric darkenings for geophysics (volcanos) and meteorology. Absolute dating of celestial observations (e.g. with comet orbits, eclipses, conjunctions) advances historical chronology. Many celestial phenomena were already found in annals, more old texts become available in scientific editions and are translated to modern languages, so that it is possible to advance the study of historical observations to more cultures (e.g. Arabic, Indian) and earlier epochs.

References

Neuhäuser, D. L. & Neuhäuser, R. 2015, *AN*, 336, 913
Neuhäuser, R. & Neuhäuser, D. L. 2015, *AN*, 336, 225
Neuhäuser, R., Rada, W., Kunitzsch, P., & Neuhäuser, D. L. 2016, *JHA*, 47, 359
Neuhäuser, D. L., Neuhäuser, R., & Harrak, A. 2018, *JCSSS*, 18, 67
Neuhäuser, D. L., Neuhäuser, R., & Chapman, J. 2018, *AN*, 339, 10
Rickert, H. 1986, The limits of concept formation in natural science, CUP Cambridge

Astronomy in Focus - XXX
Proceedings IAU Symposium No. XXX, 2018
M. T. Lago, ed.

doi:10.1017/S1743921319003946

Applying Historical Observations to Study Transient Phenomena

Elizabeth Griffin

Herzberg Astronomy & Astrophysics Research Centre, Victoria, BC, Canada
email: elizabeth.griffin@nrc-cnrc.gc.ca

Abstract. Astronomy has an enviable wealth of historical observations. Some verge on the archaeological, and display rare events such as novæ and supernovæ; others range up to 100 or more years in age, and bear unique information about events that will never repeat in detail. Yet most astronomers today know little of those resources and the scientific potential which they harbour, so rather infrequent use is made today of those historical data. The problem is that historical data were perforce obtained in analogue formats, and because of those formats the data too tend to be regarded as hailing from a culture whose scientific significance is passé. But *the medium is not the message!* Astronomy's archives of photographic observations constitute an irreplaceable resource. The change in technology from analogue to electronic recording in the late 20th Century was abrupt, and it left most of today's astronomers unable to handle and use photographic data, and led to a general skepticism of the value of photographic observations for present-day studies of variability in the cosmos. But that is precisely what older data can do; in particular, the older the data the more reliable the base-line against which one can measure new trends, refine orbital parameters, discern period modulations, etc.

Keywords. instrumentation: detectors – astronomical data bases – surveys – stars: variables – stars: long-term variability

---

1. Background

Astronomy has an enviable wealth of historical observations, mostly ranging up to 100 years or so in age. Some verge on the archaeological, and display evidence of rare events such as novæ and supernovæ. All bear unique information about transient events that may never repeat, or not exactly, in detail. Yet most astronomers today know little of those resources and the rich scientific potential which they could tap.

The problem is accessibility. Historical data were recorded in analogue formats, and in the present era of born-digital data they are regarded as hailing from a culture whose scientific significance has been completely overtaken by that of modern data. This is a serious misrepresentation of the facts; our archives of photographic observations are irreplaceable, and constitute an invaluable resource†. The technology change to electronic data in the latter half of the 20th Century was abrupt and in some ways cathartic, but it left most of today's astronomers unable to handle photographic data, let alone to appreciate the feasibility of including "old-fashioned" technological output in the service of research. True, photographic observations suffered from low DQE and relatively low time-resolution, and the application of electronic detectors, in particular the ability to observe fainter objects than before and (more recently) to resolve with sufficiently high time-resolution so as to detect very rapid flickerings, opened up new vistas in research, *but they cannot look backwards in time.* Not all astrophysics can be conducted with

† The fact that the whole of classical astrophysics was based upon photographic observations has been forgotten. There has not been a perceived need to re-do everything with CCD data

sufficient thoroughness, or even meaningfully, on the basis of observations that span less than 20–25 years.

2. The Need for an Action Plan

Our photographic observations have the potential to contribute unique information to *present-day* studies of variability in the cosmos. Furthermore, the older the data, the longer (for *longer*, read *more reliable*) the base-line against which one can measure new trends, determine orbital parameters, discern period modulations, assess changes and distributions of features on the Sun, and so on. But all analogue data are vulnerable to damage and destruction, and the older the data, the more will they *and their supporting documentation* have already been subjected to the ravages of time, vermin and natural hazards. Non-digital data must therefore be regarded as "data at risk", and should be an Action Item on every observatory's agenda.

This problem will not evaporate by our continuing to do nothing. Discussions have been under way for some time as to how best to rescue that historical information from becoming impaired as the plates age, or (in some cases) get discarded "for lack of interest", but it always comes back to the same road-block: the competition for funding between new hardware that promises to break new ground, versus allocating even relatively small amounts to re-visiting an outdated technology. What is needed first is a strong propaganda campaign that demonstrates the contributions that historical data have already made, and will continue to make, to modern studies of transient phenomena. In 2000 the IAU accepted a Resolution that expressed concern over the plight of its heritage data and the losses threatening to accrue to the science, and voiced the need to encourage digitizing programmes. In a fresh attempt to bring the urgency of the situation to the attention of astronomers worldwide, a new IAU Resolution (B3 2018) builds on the earlier one but emphasizes the dangers of continuing to do nothing.

A full digitizing programme of this scale will clearly require skills, hardware, and sufficient funds to see it through. Suitable variants of specialist hardware for scanning plates correctly can now be designed and built, so it is up to the community to garner the necessary support to set up major digitizing programmes, including the corresponding log-books or equivalent records for the all-important meta-data. That support will be nourished by rehearsing some of the triumphs achieved when information from historical data has been included alongside modern observations. Exemplary projects and some individual scientific cases are summarized below.

3. New Science Already Achieved

3.1. *Multi-million light-curves from DASCH*

What seemed originally an ambitious project to scan the entire collect of over >500,000 sky images in Harvard College Observatory's plate archive is now nearing completion, and is resulting in an enormous wealth of new knowledge. The project's website at dasch.rc.fas.harvard.edu/status.php reports 0.2 billion photometric estimates, from which 50 million light-curves have been generated. But in addition to those impressive statistics are many "odd balls", the objects that do not conform to any previously known pattern and whose discoveries are ripe for following up with high-resolution spectrographs. Here is new science in abundance!

The precedent set by the *DASCH* project at Harvard has triggered pilot studies elsewhere (e.g., Yu, Zhao, Tang, & Shang 2017), and will be an inspiration for others.

3.2. *Transient phenomena detected via the digitized Byurakan survey*

The objective of the Plate Archive Project (www.aras.am/PlateArchive) of the Byurakan Astrophysical Observatory in Armenia was to digitize all 37,000 plates obtained between 1947–1991, to derive astrometric solutions, create extraction and analysis software, and build an electronic database plus a webpage and an interactive sky map. The 1800 plates in the Markarian Survey (a.k.a. the First Byurakan Survey, FBS) were digitized in 2002–2007, from which the Digitized FBS (DFBS; www.aras.am/Dfbs/dfbs.html) was created. The archive is comprised of low-dispersion spectra, and is supporting new scientific projects. The database of the Armenian Virtual Observatory (ArVO; www.aras.am/Arvo/arvo.html) will accommodate these new data, and provide the standards and tools needed to use the scientific output efficiently and integrate it into international databases. Full details of the project are given in Mickaelian *et al.* (2016). The DFBS is a valuable resource in the search for transients, and can be conducted by comparing the same fields observed at different epochs, and by comparing BAO plates with POSS 1/2 (DSS 1/2) records. It also provides some cover for the gap years between POSS1 and POSS2, and may reveal new transients and variables. A separate study has revealed asteroids at positions not previously recorded, and is helping to correct their ephemerides and provide templates of their low-dispersion spectra. 7 extremely high-amplitude variables (differences of 7^m–8^m or more between two epochs) were also discovered by comparing BAO with DSS fields.

3.3. *The unsolved enigma of ϵ Aurigae*

ϵ Aurigae is no ordinary star; at least, it certainly doesn't seem to be. An early-F supergiant in a binary system with a period close to 27 years, it undergoes eclipses that last for nearly 2 years, but no-one has yet been able to fathom the nature of the companion that causes the eclipses. Moreover, ϵ Aur is not just a terribly faint smudge whose spectrum is hard to resolve; it is a *third magnitude* object. Its binary nature was recognized over a century ago.

ϵ Aurigae went through eclipse from 2009–2011. Communities of observers – some professionals, but mostly "backyard" amateurs – were galvanized into action, supplying high-quality series of ground-based digital spectra from the UV to the near IR. Our plate archives contain series of high-dispersion spectra recorded during both the 1955–57 and the 1982–84 eclipses. Eclipses are not quite total, but quantitative comparisons between the spectrum during eclipse, as recorded on those earlier occasions, and the most recent one had not been attempted, leaving unanswered a number of critical questions regarding the physical constancy of the (unknown) eclipsing body, or period modulations triggered by mass-loss, mass exchange, or flaring. Series of plates taken during the two previous eclipses from the Mount Wilson and the DAO plate archives were digitized with the DAO's PDS microdensitometer (PDS $\equiv$ Photometric Data Systems; no longer in business). Comparing the older spectra with new CCD observations (also recorded at the DAO) revealed fascinating features about the system that could not otherwise have been learned (Griffin & Stencel 2013), in particular that a stream of material rich in rare-earths (*Why??*) from the primary is constantly being accreted by the secondary.

3.4. *Variations in stratospheric ozone: a transdisciplinary challenge*

Ground-based observers and observatories know only too well that the Earth's atmosphere absorbs incoming radiation at wavelengths that permit only the optical window to be transmitted. The UV is blocked by bands of ozone that are totally opaque – and very necessarily so, from the point of view of the health of terrestrial bio-organisms.

For astronomy it means observing from space, and the concomitant expenses and limitations are well rehearsed. However, valuable science can be gained by inverting the problem and investigating the properties of the atmosphere rather than the properties of the stars. This is especially important in the case of the UV ozone. Ozone is now monitored externally from space, but for some 50 years the only routine monitoring was carried out by observing the Sun from the ground (actually mile-high ground) at Arosa in Switzerland. Unfortunately, the noise levels in the early data were not as low as in modern data, making it somewhat difficult to know if the concentrations that were typical during the 1920s, when monitoring commenced, held steady until freed chlorofluorocarbons (CFCs) started to destroy ozone catalytically in the 1970s (the Antarctic "ozone hole"), or whether they undergo natural variations. The way to discover is to examine historical data from other sources, such as tree rings or ice-cores. One such (untapped) source is astronomy's historical spectra of early-type stars, whose far-UV spectra are uncluttered.

Appropriate spectra, borrowed for the purpose from elsewhere, were digitized with the DAO's PDS, and the broad absorption bands of ozone in the wavelength region $\sim\lambda\,3050$–3350 Å were analyzed and modelled. A pilot study successfully validated the method of analysis that was developed (Griffin 2005), though progress was hindered by a total lack of *any* observatory plate inventory being on-line (and was halted by a serious equipment failure). But the potential remains, and represents a way of contributing indispensable information to a science that affects us all.

3.5. *Studying the dynamical evolution of trapezium systems*

Knowledge of the internal motions in multiple trapezium-like systems is necessary to understand their dynamical evolution. Despite relatively low accuracy, historical observations provide a long time-baseline. Carefully selected historical measures of the separations and position angles (selected from the best observers) enable the construction of dynamical evolution models of trapezium systems; detailed results for a dozen or so such systems have been described by Allen *et al.* (2018) and in papers referenced therein. Monte Carlo N-body integrations were performed, yielding extremely small dynamical lifetimes for these systems (10 to 40 thousand years). Literature searching turned up data for only 10 systems likely to be true trapezia, which indicates that trapezia are very scarce, as would be consistent with the short dynamical lifetimes that were found.

References

Allen, C., Ruelas-Mayorga, A., Sánchez. L. J., & Costero, R. 2018, MNRAS, 481, 3953

Griffin, R. E. M., 2005, *PASP*, 117, 885

Griffin, R. E. M., & Stencel, R. E. 2013, *PASP*, 125, 775

Mickaelian, A. M., *et al.* 2016, in: A. M. Mickaelian, A. Lawrence & T. Yu. Magakian. (Eds.) *Astronomical Surveys and Big Data*, *ASPCS*, 505, 262

Yu, Y., Zhao, J. H., Tang, Z.-H., & Shang, Z. J. 2017, *RAA*, 17, 28

Astronomy in Focus - XXX
Proceedings IAU Symposium No. XXX, 2018
M. T. Lago, ed.

doi:10.1017/S1743921319003958

Ambiguity, Scope, and Significance: Difficulties in Interpreting Celestial Phenomena in Chinese Records

J. J. Chapman

Dept. of East Asian Languages & Cultures, Univ. California, Berkeley, CA 97420, USA
email: `contactjchapman@gmail.com`

Abstract. Several problems contribute to difficulties in interpreting transient celestial phenomena as described in Chinese records. Frameworks are an overarching problem. *Tianwen*, the modern Chinese term for astronomy, in pre-modern times included meteorological phenonemena and was concerned with omenology. Manuscripts that include star charts and comets but also meteorological phenomena and omen reading texts were routinely reframed in modern scholarship to appear as if they included only astronomical content. The scope of pre-modern *tianwen*, however, was broader than its modern sense. Pre-modern celestial phenomena had political and religious significance. Apparent ambiguity arises from the presence of both meteorological and astronomical phenomena in a single category and from features of the classical Chinese language. Accounting for these problems is essential for research into transient phenomena using historical archives.

Keywords. history and philosophy of astronomy, sociology of astronomy, (Sun:) sunspots, comets: general

---

1. Misinterpretations and Category Errors

Misinterpretations of East Asian records of transient phenomena stem from three basic problems: (a) the broad scope of the celestial sciences in East Asia, (b) the omenological significance of celestial phenomena within broader historical narratives, and (c) aspects of the classical Chinese language, East Asia's pre-modern *lingua franca*.

Category errors lie at the heart of the problem. The modern Chinese word for astronomy is *tianwen*, but the same term had a broader scope in pre-modern times. Literally, it meant "celestial patterns." *Tianwen* included both astronomical and meteorological phenomena, encompassing omenological interpretation as well as observational practices. Nonetheless, modern scholarship sometimes treats *tianwen* as if it included only astronomical phenomena. Liu *et al.* (2014) misinterpreted a dust storm in *tianwen* records as a reference to "dust rain" from a comet, citing this as evidence that a comet's impact with the Earth's atmosphere had caused the global spike in levels of ^{14}C and ^{10}B in the late 8th century. Because this explanation is based on a faulty translation, it has been rejected (Chapman, Csikszentmihalyi, & Neuhäuser 2014; Stephenson 2015).

Understanding East Asian records of celestial phenomena requires considering how and why pre-modern observers recorded and categorized them. Modern readers do not expect to see human historical events mixed up with records of celestial ones. While Chinese models of celestial regularities became increasingly precise, accurate and predictive over time (Morgan 2017), the political significance of celestial phenomena persisted. This was especially true of irregular and unpredictable phenomena–transient phenomena–such as comets, novæ, and auroræ, as well as appearances of what modern researchers regard as meteorological events: strange clouds, rainbows, floods and droughts, and halos around the Sun and the Moon.

2. Frameworks

The category of astronomy is a framework that influences the way technical materials are represented in scholarly journals and monographs. The scholar writing on the history of astronomy faces a difficult task when working with pre-modern records and manuscripts. Excising seemingly irrelevant parts of technical manuscripts obscures differences between modern astronomy and pre-modern modes of observing the heavens.

Scholarship surrounding a complex omenological manuscript found in the tombs of the Marquises of Dai at Mawangdui (*terminus ad quem* 168 BC) provides an illustrative case. In his history of *tianwen* in China, Chen Meidong refers to the manuscript simply as a "chart of comets" (Chen 2008). Archaeologist Feng Shi likewise creates the sense that the manuscript primarily features comets, presenting in his *Archaeoastronomy in China* a photographic image including eight comets and a line-drawing including 29 comets (Feng 2001). However, the six horizontal registers of the full manuscript contain a variety of phenomena including rainbows, halos around the Sun and Moon, eclipses, clouds shaped like animals, trees, and in one case the Big Dipper, plus others more difficult to identify. The images are paired with prognostic statements predicting military victories and defeats, bumper crops and famines, murdered kings and political disorder (Qiu 2014).

A manuscript discovered at the Dunhuang caves (Or.8210/S.3326), billed as the world's earliest extant star chart, presents a better-known example. Non-astronomical portions of this manuscript have also been excised so that it appears more cleanly and unambiguously scientific. Like the Mawangdui manuscript, the so-called "Dunhuang Star Chart" or "Dunhuang Star Atlas" is more complex than its conventional name implies. Images of clouds paired with prognostic information make up about half of the total manuscript. Bonnet-Bidaud *et al.* (2009) largely elide the cloud divination text in their discussion of the manuscript, while Feng Shi (2001) and Sun & Kistemaker (1997), further removed from the manuscript, leave it out entirely.

3. Scope of the Celestial Sciences

The scope of phenomena that early observers tracked can be seen not only in rare manuscripts but also in treatises in the standard histories. Technical treatises reveal both the organization and the significance of actor categories. These categories are also manifest in the structures of the institutions responsible for observing and recording celestial phenomena.

The first of the standard histories in the Chinese tradition was compiled by Sima Qian (145–86). He served as *Taishi ling*, a title that has been rendered into English as "Senior Archivist" (Nylan 1998-1999), "Director of Astronomy" (Loewe 2000), and "Prefect Grand Astrologer" (Bielenstein 1980). Sima Qian included in his history a series of technical treatises. His treatise on the "Celestial Offices" described celestial bodies, explained their omenological significance, and argued that the observation and interpretation of celestial signs was a crucial component of sage governance.

Sima Qian's treatise compiles technical information concerning celestial bodies and omen reading. It enumerates some 412 stars scattered over 89 asterisms (Pankenier 2013), and gives detailed accounts of the movements of each of the five visible planets. It provides omenological information on halos around the Sun and Moon, eclipses, *qi* phenomena, and aberrant behaviour by the planets, such as unexpected retrograde motion or scintillation. Transient phenomena play a major role in the treatise. It describes various types of *xing* – a word usually translated as star – that appear, or are said to appear, as baleful or auspicious omens; meteors, comets, and novæ are identified as *xing* of this type. It also includes information on interpreting clouds and *qi*.

Eastern Han records (24–220) show institutions that map onto a scope of practice similar to that in Sima Qian's treatise. Under the *Taishi ling*, there were three major Assistants, one of whom acted as director of the Imperial Observatory. The observatory employed fourteen officials responsible for observing various types of *xing*. It also included twelve officials responsible for observing *qi* phenomena such as auroræ, halos, and strange clouds (Bielenstein 1980).

By the Eastern Han, Chinese approaches to the heavens would be divided into two major categories, *tianwen* and *lüli* (harmonics and mathematical astronomy), both of which received individual treatises in many of the Chinese dynastic histories. They played complementary roles. Modern scholars sometimes make a dichotomy between astrology or astro-omenology in *tianwen*, counter to mathematical astronomy in *lüli*. However, this division can be deceptive. Records of comets, auroræ and novæ are much more likely to appear in a *tianwen* treatise than in a *lüli* treatise, as transient phenomena usually cannot be assimilated by any mathematical description under pre-modern conditions. The key distinction between *tianwen* and *lüli* lies in irregularity *versus* regularity. *Lüli* establishes precise ratios to describe both harmonic intervals and the movements of celestial bodies.

Many of the standard dynastic histories include a general omenological treatise, called the *Wuxing* or "Five Phases" treatise. These include numerous records of celestial phenomena, as well as many phenomena (ranging from two-headed chickens to earthquakes) that have no obvious relationship with the heavens. The *Hanshu* (*History of the Han*) *Wuxing* treatise contains more records of celestial phenomena than does the treatise on *Tianwen*. It organizes each category of sign chronologically, gravitating toward the collapse of the Western Han dynasty.

4. Significance of Celestial Signs

Celestial signs had political and religious significance. The *Hanshu tianwen* treatise explains: *[Signs] originate in the earth and erupt into the heavens. When governance fails below, then aberrations appear above, just as shadows are the counterparts of their forms and echoes are responses to sounds. This is why the clear-sighted ruler sees them and awakens, putting himself in order and rectifying his affairs* (26.1273).

An early folk etymology for the Chinese graph for king demonstrates how early court figures thought about the relationship between human beings and the cosmos. It reads the three horizontal lines in the graph as referring to heaven above, the human realm in the middle, and the earth below. The emperor acts as a veritable *axis mundi*, the vertical line that binds together the whole of the cosmos (Lai 1984, 44.295). His political policies and ritual comportment explain the appearance of inauspicious transient phenomena.

5. The Problem of Ambiguity

The scope and significance of celestial phenomena in East Asian records creates substantial difficulties for researchers who mine them for astronomical data. This difficulty is compounded at times by ambiguities arising from particular words and features of classical Chinese syntax. *Qi*, like *xing*, presents problems. *Qi* operated in human bodies, the terrestrial world, and the cosmos. By the late 1st Century BC, *qi* had multiple processual aspects, *yin* and *yang* and the Five Phases, and it played a role in fields ranging from philosophy to medicine. *Qi* was a kind of "matter that incorporates vitality" (Lloyd & Sivin 2003, p. 198). *Qi* has been variously rendered into English as *material force*, *vapour*, *psychophysical stuff* (Gardner 2007), and *materia vitalis* (Pankenier 2013).

Syntax also presents problems. Classical Chinese does not distinguish between active and passive senses of verbs. The verb *guan*, often used to describe halos or shafts of light, may mean to penetrate or to be penetrated, to encircle or to be encircled. Prepositions are often ambiguous from a purely linguistic standpoint. The phrase *ri pang* could mean "next to the Sun" or "on the side of the Sun", depending on context. When black *qi* appears on the *ri pang* in a sunspot record, context means the latter form must be chosen (*contra* e.g., Abbott & Juhl 2016).

6. Conclusion

Three points should be kept in mind when using East Asian records to study transient phenomena. First, meteorological phenomena intermingle with (and can be mistaken for) astronomical phenomena in East Asian records. A light in the sky that is actually a backlit cloud might easily be mistaken for an aurora. Second, prepositions are always suspect. There is seldom a hard linguistic distinction between being next to an object or being on the side of an object, or between penetrating an object or encircling it. Finally, records are mediated by the large-scale textual projects in which they are included. Records are part of a broader historical narrative, and historical circumstances played a role in what was transmitted and what survived.

References

Abbott, D. & Juhl, R. 2016, *Adv. Sp. Res.*, 58, 2181

Ban G. (32-92) 1962, *Hanshu (History of the Han)* (Beijing: Zhonghua Shuju)

Bielenstein, H. 1980, *The bureaucracy of Han times* (Cambridge: CUP)

Bonnet-Bidaud, J.M., Praderie, F., & Whitfield, S. 2009, *J. Astron. Hist. Herit.*, 12, 39

Chapman, J., Csikszentmihalyi, M., Neuhäuser, R., 2014, *AN* 335, 964

Chen, M.D. 2008, *Zhongguo gudai tianwenxue sixiang* (Beijing: Zhongguo Kexue Jishu)

Feng, S. 2001, *Zhongguo tianwen kaoguxue* (Beijing: Shehui Kexue Wenxian)

Gardner, D. 2007 *The four books: The basic teachings of the later Confucian tradition* (Indianapolis: Hackett Pub. Co.)

Lai Y.Y. comm. 1984 *Chunqiu fanlu jinzhu jinyi (A modern commentary with translation to the Luxuriant Dew of the Annals)* (Taipei: Taiwan Shangwu)

Liu, Y., Zhang, Z., Peng, Z., *et al.*, 2014, *Nat SR* 4E3728

Lloyd, G. & Sivin, N. 2003, *The way and the word: Science and medicine in early China and Greece* (New Haven: Yale UP)

Loewe, M. 2000, *A biographical dictionary of the Qin, Former Han, and Xin periods: 221 BC-24 AD* (Leiden: Brill)

Morgan, D. P. 2017. *Astral sciences in early imperial China: Observation, sagehood and the individual* (Cambridge: CUP)

Nylan, M. 1998-1999, *Early China*, 23/24, 203

Pankenier, D. 2013, *Astrology and cosmology in early China* (Cambridge: CUP)

Qiu, X.G. 2014, *Changsha Mawangdui Hanmu jianbo jicheng 4* (Beijing: Shehui Kexue Wenxian)

Sima Q. (ca. 145-ca. 86) 1959, *Shiji (Records of the senior archivist)* (Beijing: Zhonghua Shuju)

Stephenson, F.R. 2015, *Adv. Sp. Res.*, 55, 1537

Sun, X.S. & Kistemaker, J. 1997, *The Chinese sky during the Han: Constellating the stars and society* (Leiden: Brill)

Astronomy in Focus - XXX
Proceedings IAU Symposium No. XXX, 2018
M. T. Lago, ed.

doi:10.1017/S174392131900396X

Sunspot and Group Number: Recent advances from historical data

Frédéric Clette[1], José M. Vaquero[2], María Cruz Gallego[3] and Laure Lefèvre[1]

[1]Royal Observatory of Belgium, 3, avenue Circulaire, 1180 Brussels, Belgium
email: `frederic.clette@oma.be`

[2]Departamento de Física, Universidad de Extremadura, 06071 Mérida, Spain

[3]Departamento de Física, Universidad de Extremadura, 06071 Badajoz, Spain

Abstract. Due to its unique 400-year duration, the sunspot number is a central reference for understanding the long-term evolution of solar activity and its influence on the Earth environment and climate. Here, we outline current data recovery work. For the sunspot number, we find historical evidence of a disruption in the source observers occurring in 1947–48. For the sunpot group number, recent data confirm the clear southern predominance of sunspots during the Maunder Minimum, while the umbra-penumbra ratio is similar to other epochs. For the Dalton minimum, newly recovered historical observations confirm a higher activity level than in a true Grand Minimum.

Keywords. Sun: sunspots, Sun: photosphere, Sun: activity, methods: data analysis, history of astronomy

---

1. Introduction

For multiple applications, ranging from the physical modeling of the solar magnetic cycle to the evolution of the Earth climate, the sunspot number S_N defined by Wolf (1859) and the sunspot group number G_N introduced more recently by Hoyt & Schatten (1998) provide a unique multi-century record of the long-term evolution of the solar cycle. The homogeneity of those sunspot data series over several past centuries is of crucial importance, but poses difficult challenges.

Over recent years, this motivated new investigations, which led to the release of the first official revision of the S_N series (cf. SILSO Web site: http://www.sidc.be/silso). The associated key advances were published in a topical issue of Solar Physics (Clette *et al.* 2016a). While several methodological issues must still be verified and settled, future progress definitely depends on the production of complete and fully verified databases of historical sunspot observations. In the case of G_N, such a fully updated database of raw observations was recently produced (Vaquero *et al.* 2016). However, the period of the Maunder minimum still poses a challenge, given the sparsity of data and ambiguity of spotless reports. For S_N, instead of the current corrections to the original series, a full reconstruction is now envisioned from the original observations. In this case, the construction of an entirely new database is needed. Hereafter, we are outlining some of this ongoing data-recovery work.

2. Rebuilding a database for the Zurich sunspot number

Between the creation of the sunspot number in 1849 and the closing of the Zurich observatory in 1981, the daily sunspot number rested on fairly simple principles. By default, on most days, the daily S_N was the number from the primary observer at the Zurich reference station. On missing days (bad weather), sunspot counts from a network of auxiliary stations were used as a replacement. Consequently, most of the data sent by external stations were not effectively used to produce the original S_N values (Version 1). Moreover, as all Zurich observers were considered equivalent, no distinction was made between the primary observer (i.e. the successive directors) and the assistants.

So, a lot of information was ignored, but fortunately, those data can now be recovered, as the base observations were published in yearly tables in the collection of *Mittheilungen* of the Zurich Observatory. Over 2017–2018, we fully encoded those tables to rebuild the first sunspot number database in digital form. Before making it accessible to the scientific community, we are now conducting a full quality control. However, this thorough data recovery already revealed a few important facts:

- All input data were published only until 1919. Then, following a steep growth of the network after World War I, only a minor part of the data were still published thereafter, up to 1945, when no data tables were published anymore (Figure 1). All published data have now been entirely digitized, but although the Zurich Observatory mentioned in 1919 that the remaining unpublished data were archived and could be provided on request, those archives are missing and still need to be recovered.
- When M. Waldmeier became Director and primary observer in 1945, the former team of W. Brunner (previous Director) and assistant Brunner actually continued observing until 1947. Only starting in 1948, new assistants are recruited, but most of them appear only during short durations. None of them observed in parallel with the aforementioned long-term observers active before 1948. This thus marks a sharp and unprecedented break in the composition of the Zurich team.
- A similar synthesis of external contributing stations show that all stations that contributed before World War II progressively stopped between 1938 and 1945. During the war, a temporary network of local amateur observers was set up by the Zurich Observatory. However, none of those observers contributes for a long duration. Only after the war, Waldmeier re-creates a full worldwide network. However, there is almost no overlap between this new network and the one that was active before the war. This thus marks another disruption in the input data for the sunspot number series.

It turns out that a sharp 18% jump was found in the scale of the original sunspot number precisely in 1947 (Clette & Lefèvre 2016, Clette *et al.* 2014, Lockwood, Owens & Barnard 2014). So far, such a jump and its timing were difficult to explain. Indeed, the main cause of this inhomogeneity was a methodological change (Zurich internal counts weighted as a function of sunspot size) that was introduced progressively and much earlier by A. Wolfer (Friedli 2016, Svalgaard, Cagnotti, & Cortesi 2017). Now, by their temporal coincidence with this 1947–1948 jump, the two above discontinuities in the source data of the sunspot number provide strong historical evidence that the long-term chain of primary and auxiliary observers was broken at that moment, making possible a shift in the scale of the resulting sunspot numbers.

3. The Maunder minimum (1645–1715)

During the last years, some important advances have been made regarding the Maunder minimum after the controversy generated by the publication of two studies with mismatched results (Zolotova & Ponyavin 2015, Usoskin *et al.* 2015).

First, a digital version of sunspot latitude data from previous studies is available to the international community (Vaquero *et al.* 2015a). These data have confirmed the butterfly diagram for the Maunder minimum as well as the strong hemispheric asymmetry, with predominance of spots in the southern hemisphere (Figure 2). In addition, another study has confirmed the presence of the solar cycle during the Maunder minimum using

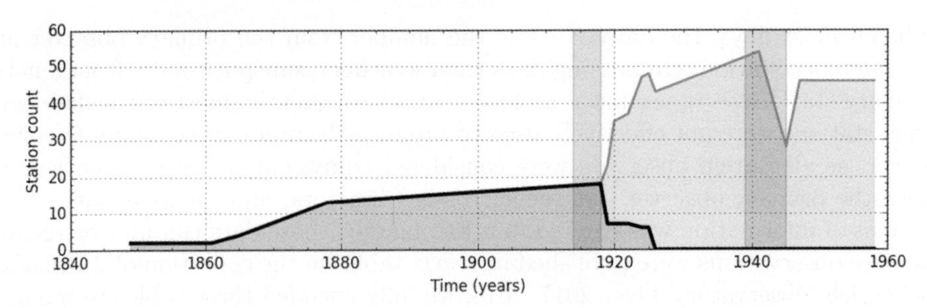

Figure 1. Evolution of the number of auxiliary stations contributing to the production of the Zurich sunspot number. From 1849 to 1919, all data were published in the Mittheilungen of the Zurich observatory and are now digitized (thick line), but thereafter, only part of them, when the network greatly expanded (thin line). So far, a large part of those more recent unpublished data has not been recovered. The vertical shaded bands mark the two World Wars, which had a clear influence on the sunspot observing network.

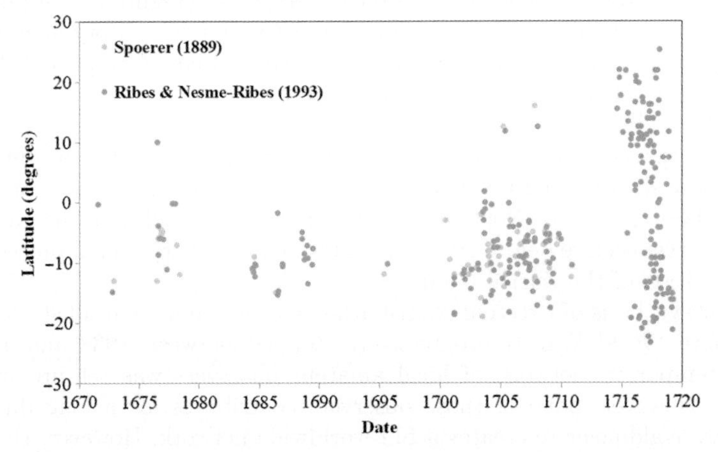

Figure 2. Butterfly diagram during the Maunder minimum using the data from Spoerer (orange circles) and Ribes and Nesme-Ribes (green circles).

subsets of observations from the Hoyt & Schatten (1998) database. These subsets were constructed using different quality criteria (Vaquero *et al.* 2015b). Several articles have appeared improving the Hoyt and Schatten database during the Maunder minimum (Carrasco *et al.* 2015, Carrasco & Vaquero 2016, Vaquero *et al.* 2016), and this effort to improve data availability continues (Neuhäuser *et al.* 2018).

Very recently, the umbra–penumbra area ratio during the Maunder minimum has been evaluated using 196 sunspot drawings published in scientific journals of that epoch (Carrasco *et al.* 2018a). They cover 48 different sunspots observed during the period 1660–1709. The mode value of the ratio obtained from the occurrence frequency distribution lies between 0.15 and 0.25, very similar to values found for other epochs.

4. The Dalton minimum (1793–1827)

Recent studies have confirmed that the Dalton minimum cannot be considered a Grand Minimum but a secular minimum (Vaquero *et al.* 2016, Ogurtsov 2018). Thus, for example, Vaquero *et al.* (2016) found that the expected value of active days for the Maunder minimum (9.94%) was significantly lower than for the Dalton Minimum (61.63%).

Several articles have been published recently adding new sunspot observations into the databases. Hayakawa *et al.* (2018) recovered a sunspot drawing made by Iwahashi Zenbei on 26 August 1793 and Denig & McVaugh (2017) analyzed a set of 25 drawings made by Jonathan Fisher in 1816 and 1817. Carrasco *et al.* (2018b) recovered the sunspot observations performed by Cassian Hallaschka in 1814 and 1816. These works have helped to improve our knowledge of solar activity for the Dalton minimum.

5. Acknowledgements

This work was supported by the Belgian Solar-Terrestrial Center of Excellence (STCE), funded by the Belgian Science Policy office. This work was also partly funded by FEDER-Junta de Extremadura (Research Group Grant GR15137 and project IB16127) and from the Ministerio de Economía y Competitividad of the Spanish Government (AYA2014-57556-P and CGL2017-87917-P).

The authors have benefited from the participation in the ISSI Sunspot Number Recalibration Working Team (*http://www.issibern.ch/teams/sunspotnoser/*).

References

Carrasco, V. M. S., Álvarez, J. V., & Vaquero, J. M. 2015, *Sol. Phys.*, 290, 2719
Carrasco, V. M. S., & Vaquero, J. M. 2016, *Sol. Phys.*, 291, 2493
Carrasco, V. M. S., García-Romero, J. M., Vaquero, J. M., *et al.* 2018a, *Astrophys. J.*, 865, 88
Carrasco, V. M. S., Vaquero, J. M., Arlt, R., & Gallego, M. C. 2018b, *Sol. Phys.*, 293, 102
Chatzistergos, T., Usoskin, I., Kovaltsov, G., Krivova, N.A., & Solanki, S.K. 2017, *Astron. & Astrophys.*, 602, A69
Clette, F., Svalgaard, L., Vaquero, J.M., & Cliver, E.W. 2014, *Space Sci. Rev.*, 186, 35
Clette, F., & Lefèvre, L. 2016, *Sol. Phys.*, 291, 2629
Clette, F., Cliver, E.W., Lefèvre, L., Svalgaard, L., Vaquero, J.M., & Leibacher, J.W. 2016a, *Sol. Phys.*, 291, 2479
Denig, W. F., & McVaugh, M. R. 2017, *Space Weather*, 15, 857
Friedli, T.K. 2016, *Sol. Phys.*, 291, 2505
Hayakawa, H., Iwahashi, K., Tamazawa, H., Toriumi, S., & Shibata, K. 2018, *Sol. Phys.*, 293, 8
Hoyt, D.V., & Schatten, K.H. 1998, *Sol. Phys.*, 181, 491
Lockwood, M., Owens, M.J., & Barnard, L. 2014 *J. Geophys. Res.*, 119(A7), 5193
Neuhäuser, R., Arlt, R., & Richter, S. 2018, *Astronomische Nachrichten*, 339, 219
Ogurtsov, M. G. 2018, *Astronomy Letters*, 44, 278
Svalgaard, L., Cagnotti, M., & Cortesi, S. 2017, *Sol. Phys.*, 292, 34
Usoskin, I. G., Arlt, R., Asvestari, E., *et al.* 2015, *Astron.& Astrophys.*, 581, A95
Vaquero, J. M., Nogales, J. M., & Sánchez-Bajo, F. 2015a, *Advances in Space Res.*, 55, 1546
Vaquero, J. M., Kovaltsov, G. A., Usoskin, I. G., Carrasco, V. M. S., & Gallego, M. C. 2015b, *Astron.& Astrophys.*, 577, A71
Vaquero, J.M., Svalgaard, L., Carrasco, V.M.S., Clette, F., Lefèvre, L., Gallego, M.C., Arlt, R., Aparicio, A.J.P., Richard, J.-G., & Howe, R. 2016 *Sol. Phys.*, 291, 3061
Wolf, R. 1859, *Astron. Mitt. Eidgnöss. Sternwarte Zürich*, I (VIII), 66
Zolotova, N. V., & Ponyavin, D. I. 2015, *Astrophys. J.*, 800, 42

Astronomy in Focus - XXX
Proceedings IAU Symposium No. XXX, 2018
M. T. Lago, ed.

doi:10.1017/S1743921319003971

Eclipses and the Earth's Rotation

F. R. Stephenson[1], L. V. Morrison[2] and C. Y. Hohenkerk[3]

[1]University of Durham, DH1 3LE, UK
email: f.r.stephenson@durham.ac.uk

[2]Formerly Royal Greenwich Observatory
28 Pevensey Park Road, Westham, BN24 5HW, UK

[3]Formerly HM Nautical Almanac Office, UK

Abstract. Analysis of historical records of eclipses of the Sun and Moon between 720 BC and AD 1600 gives a measure of the time difference, TT − UT = ΔT. The first derivative in time along a smooth curve fitted to the values of ΔT measures the changes in the length of the day (*lod*). The average rate of change of the *lod* is found to be significantly less than that expected on the basis of tidal friction. Fluctuations on a time-scale of centuries to millennia are mainly attributed to the effects of post-glacial uplift and core-mantle coupling.

Keywords. Eclipse, length of day

1. Introduction

Historical records of eclipses of the Sun and Moon between 720 BC–AD 1600 provide vital information on the rotation of the Earth. From several hundred records we derive an estimate of the difference, TT − UT = ΔT, which measures the cumulative discrepancy in UT due to fluctuations in the Earth's rate of rotation. The fluctuations are derived from the slope on a smooth curve fitted to ΔT, which are conveniently measured by changes in the length of the mean solar day (*lod*).

2. Historical Sources of Eclipses Observations

The extant records come from the ancient civilizations of Babylonia, China, and Greece, together with the medieval Arab Dominions and medieval Europe, and their provenance is described in Stephenson (1997). The calculated values of ΔT are listed in Stephenson *et al.* (2016), with a few additions in Morrison *et al.* (2019).

3. Observations

The eclipse observations are subdivided into two main categories: untimed and timed. These in turn are subdivided into solar and lunar.

Untimed solar −708 to +1567 (61 observations): The accuracy of the ΔT results from untimed observations depends on the width of the paths of totality parallel to the equator at the places of observation. We have used only observations where the place, date and description of the eclipse are unambiguous. In most cases, the observation gives a clear statement that the eclipse was total or that the Sun completely disappeared, along with details such as 'day turned to night' and/or 'stars appeared'. This gives a solution space for ΔT with sharp upper and lower bounds. We have also used a few observations of large solar eclipses where a description such as 'not total, like a hook' is given. This defines a solution space of ΔT outside the range of totality. Generally this is wide, but in some cases the observations give a useful boundary condition.

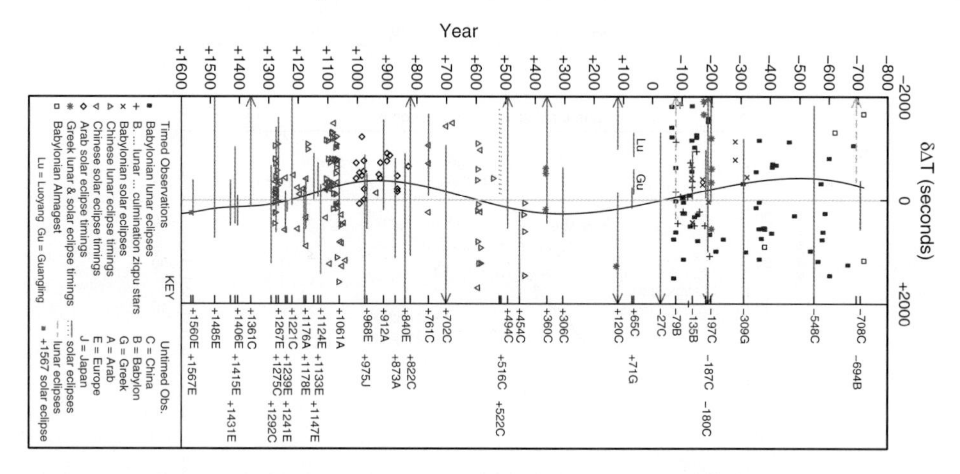

Figure 1. Residuals $\delta\Delta$T and spline plotted with respect to the parabola (Eqn. 4.1).

Untimed lunar −719 to −79 (14): Useful limits on ΔT can also be obtained from Babylonian untimed observations of lunar eclipses, where the Moon rose or set eclipsed, or the degree of obscuration at rising or setting is given.

Timed solar −356 to +1004 (100): The Babylonians timed the beginning or end of solar eclipses by measuring the time elapsed from sunrise or sunset to the nearest 4 minutes. The Chinese measured time as a fraction of the day length, typically to the nearest 1/100th (14 minutes). The Greek observations are given to a precision in the range 1/3rd to 1/6th hour (20-10 minutes). The Arab timings were made by astrolabe measurement of the Sun's altitude, giving a precision of 4 min.

Timed lunar −720 to +1277 (≈250): The Babylonian lunar observations were timed using the same method as the solar observations, and after the mid-6th century BC, with the same resolution (4 minutes). The precision of the Chinese observations ranged from 0.5 hours prior to +1050, to 14 minutes afterwards. The Greek observations were recorded with a precision of 1/3 h, typically. The Arab precision is similar to the solar observations (4 minutes), using astrolabe altitudes of the Moon or reference stars.

4. Analysis of ΔT −720 to +1600

The behaviour of ΔT over time is predominantly parabolic as a result of the deceleration in the Earth's rate of rotation caused by the action of tidal friction due to the Moon, and, to a lesser extent, the Sun. In our paper Stephenson *et al.* (2016), we find the average parabola fitted to ΔT to be

$$\Delta\mathrm{T(parabola)} = -320.0 + (32.5 \pm 0.6)((\mathrm{year} - 1825)/100)^2 \text{ s.} \tag{4.1}$$

This parabola is subtracted from the values of ΔT to obtain the residuals $\delta\Delta$T, which are plotted in Figure 1 together with the subset of critical observations.

There are clearly fluctuations of $\delta\Delta$T around this parabola. We weighted the data, both discrete points (mainly due to timing) and extended solutions (untimed), and fitted a smooth curve using cubic splines. That is plotted in Figure 1. Particular care was taken over the reliability of the critical limits of the untimed eclipses of −708C, −694B, −135B, +454C, +761C, +1239E, +1361C and +1567E. They are discussed in Stephenson *et al.* (2016), Morrison *et al.* (2019), and Stephenson *et al.* (2018).

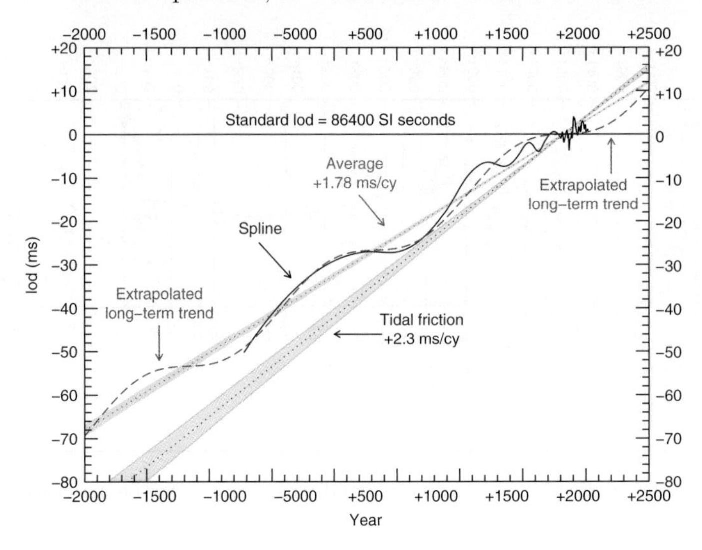

Figure 2. Change in the *lod* derived from the slope of the spline fitted to ΔT.

5. Change in *lod*: −720 to +1600

The slope along the spline curve in Figure 1 gives the change in *lod* plotted in Figure 2. The average long-term increase in the *lod* is found to be +1.78 milliseconds per century (ms/cy), which is significantly less than the increase of +2.3 ms/cy expected on the basis of tidal friction. This implies an accelerative component decreasing the *lod* by −0.5 ms/cy since 720 BC, which is mainly attributed to the effects of post-glacial uplift and core-mantle coupling. The decadal fluctuations resolved after +1600 are derived from timings of lunar occultations (Stephenson *et al.* 2016), which are not discussed here.

6. Acknowledgements

The authors acknowledge support by HM Nautical Almanac Office and the IAU's Standards for Fundamental Astronomy WG of Division A.

References

Stephenson, F. R. 1997, *Historical Eclipses and Earth's Rotation*, (CUP), ISBN 0-521-46194-4
Stephenson, F. R., Morrison, L. V., & Hohenkerk, C. Y. 2016, *Proc. Roy. Soc. A*, 472, 2016.0404
Morrison, L. V., Stephenson, F. R., & Hohenkerk, C. Y. 2019, *Proc. Journées 2017, Sysèmes de Référence et de la Rotation Terrestre*, (in press), See http://astro.ukho.gov.uk/nao/lvm/ for the data and spline fit from this paper and Stephenson *et al.* (2016)
Stephenson, F. R., Morrison, L. V., & Hohenkerk, C. Y. 2018, *J. Hist. Astr.*, 49, 425–471

Astronomy in Focus - XXX
Proceedings IAU Symposium No. XXX, 2018
M. T. Lago, ed.

doi:10.1017/S1743921319003983

Dating Historical Arabic Observations

Rita Gautschy[1] **and Johannes Thomann**[2]

[1]Dept. of Ancient Civilizations, University of Basel,
Petersgraben 51, CH-4051 Basel, Switzerland
email: rita.gautschy@unibas.ch

[2]Institute of Asian and Oriental Studies, University of Zurich,
Rämistrasse 59, CH-8001, Zurich, Switzerland
email: johannes.thomann@aoi.uzh.ch

Abstract. The first visibility of the lunar crescent signals the start of a new month in the Islamic calendar. The eminent astronomer Ḥabash al-Ḥāsib developed a method of uncompromising complexity for predicting the visibility of the lunar crescent. He derived his threshold value from a moonwatch carried out at different places in Iraq on November 17th, 860 CE. We will allude to a few modern visibility criteria as well and highlight the uncertainties of today's calculations when converting historical Arabic into Julian or Gregorian dates. Tables of first visibility of the lunar crescent for different locations are provided for the purpose of date conversions. Since the Islamic calendar was based on the observation of the lunar crescent, historical dates imply information on positive and negative sightings of the lunar crescent. From such information estimations of cloudiness in different regions of the Islamic world can be extracted.

Keywords. lunar crescent visibility, moonwatch, Islamic calendar, Ḥabash al-Ḥāsib, cloudiness

---

1. Introduction

The first or last visibility of the lunar crescent is the decisive phenomenon for the start of a new month in various calendars, e.g. the Islamic, Jewish, Babylonian or lunar Egyptian calendar. In order to convert correctly the dates of astronomical observations documented by these cultures into Julian or Gregorian dates, it is necessary to compute the beginnings of the lunar months. The Babylonians were the first to formulate visibility criteria in order to predict whether or not the crescent would be visible. In Arabic times, special moonwatch programmes were organised. We will discuss one extended moonwatch programme from 860 CE in the Middle East and also modern attempts, and discuss the uncertainties of modern calculations.

2. A Lunar Crescent Visibility Criterion and a Moon Watch in 9th Century Iraq

Many criteria for the visibility of the lunar crescent were developed by astronomers in the Islamic world. In early Abbasid time Ya^{c}qūb ibn Ṭāriq (active 778 CE) predicted the visibility of the lunar crescent, based on a very rough approximation of the thickness of the crescent (Hogendijk 1988). Other criteria were based on the time difference of sunset and moonset, or on the sum of the Moon's elongation and its latitude, or on the negative altitude of the Sun. Ibn Yūnus (d. 1009) established a list of several necessary conditions for visibility based on the difference in setting times, the thickness of the crescent and the velocity of the Moon. A very simple criterion, a 10° threshold for the negative altitude of the Sun at the time of the setting of the Moon, is ascribed to Ḥabash al-Ḥāsib (d. after 869 CE), a fact frequently referred to in scholarship (Yazdi 2018). However, it has gone unnoticed until recently that Ḥabash also proposed the most complicated description for predicting the visibility of the lunar crescent in any premodern astronomical work (Thomann 2017). Ḥabash's chapter on crescent visibility consists of two parts. It begins with a general step by step description of his solution, which consists of more than sixty steps of calculation. It leads to a critical value which is compared to a threshold value of 14° 29′. If the critical value is smaller, the Moon is supposed to be invisible, and if greater, it should be visible. After the description of the method in general, a fully calculated numerical example for November 17th, 860 CE at Samarra follows. Ḥabash finds a value of 14° 26′, which is too small for visibility, and indeed the crescent did not appear in the sky on that evening. It was also not seen at Baghdad, but it became visible in Kufa and Anbar, which are west of Baghdad. Ḥabash did not give an explanation for his threshold value of 14° 29′, but obviously he chose it with the intention to receive a correct prediction for the sightings of November 17th, 860 CE. For that purpose, he must have organized a moon watch.

3. Modern visibility criteria and calculations

Computations of first and last visibility of the lunar crescent for the distant past are subject to several uncertainties. The most important uncertainty is the decreasing rotation rate of the Earth, which causes the length of the day to increase with time. The resulting time difference due to the slowing down of Earth's rotation is referred to as ΔT. For details about ΔT see the contribution of Stephenson *et al.* (page 160). In our computations we use the long-period DE406-ephemeris of the Jet Propulsion Laboratory (Standish 1998) and for ΔT the values of Morrison & Stephenson (2004) with a small correction that accounts for the different values of the Moon's secular acceleration used by them and by the DE406-ephemeris. Morrison & Stephenson stressed that their ΔT -values should not be extrapolated further to the past than about 700 BCE because there are no ancient observations earlier on that can be used to derive a value of ΔT. The uncertainty of these values was estimated with the formula of Huber (2006) that is based on the analysis of the stochastic behaviour of the length of day process and a model comprising a global Brownian motion process with infinite relaxation time. All our calculations were made for three different values of ΔT: a mean value, a lower ΔT value that is equal to the mean ΔT value minus the error of ΔT, and an upper ΔT value that is equal to the mean ΔT value plus the error of ΔT.

The second important point is the visibility criterion. Various criteria for the first visibility of the lunar crescent have been proposed based on ancient as well as on modern observational data. The first modern attempts to formulate a crescent visibility criterion by Fotheringham (1910), Maunder (1911), Schoch (1927) and Neugebauer (1929) incorporated two parameters: the true lunar altitude and the azimuthal difference between Sun and Moon at sunset. The most commonly used criterion of this kind is the one of Carl Schoch published posthumously by Neugebauer (1929). Frans Bruin chose a different ansatz in 1977, which was later modified by Bradley Schaefer (Bruin 1977; Schaefer 1988). Bruin calculated the necessary minimal brightness of the Moon in order to be observable at a certain sky brightness. He provided a purely graphic solution of the problem. One has to know the width of the Moon and its altitude above the horizon at sunset to determine whether or not the Moon will be visible. In addition, Bruin gives the time for a best possible sighting.

Bernard Yallop tried to merge the Frans Bruin's approach with the criterion of Carl Schoch, from whom he adopted the values for the minimal altitude of the Moon at a certain difference in azimuth (Yallop 1997). At the same time he resorted to the moment of the best possible sighting – the so-called "best time" – and the width of the lunar crescent defined by Bruin. Yallop accounted for the lunar parallax and the topocentric width of the lunar crescent and introduced a parameter q that describes the threshold for a possible successful sighting. He distinguished in total six zones for q. For historical purposes, only the first three zones are relevant. For the computations, we use the criterion of Yallop which was adapted slightly for our topocentric calculations. This criterion makes use of the so-called "best time" of observation when the Moon is still some degrees above the horizon – hence, effects of the local horizon can be neglected as long as the horizon is not higher than about 3°. In the calculations standard values of refraction are accounted for.

A successful sighting of the lunar crescent is highly dependent on the prevailing seeing conditions: dust in the air or slight fog can easily cause delayed first sightings. But especially in cases where the visibility parameter is close to the critical threshold, a lunar crescent may be missed even if the seeing is good.

4. Tables of first and last visibility

According to the tables of Wüstenfeld *et al.* (1961) the Arabic era began on July 16th in 622 CE. These tables are widely used, and starting from the mentioned zero point a regular calendar scheme is usually applied. However, modern calculations for Medina show that the lunar crescent should easily have been visible on July 15th. It is well known that the calendar dates of Wüstenfeld *et al.* (1961) may be in error by 1–2 days in comparison to the calendar that was actually used. Additionally, there are two more points of concern: first, Sura 9, 36-37 of the Koran suggests that a lunar calendar without intercalation was applied only from year 10 of the Hijra onwards. Earlier, a lunisolar calendar was used. Secondly, in many countries the first actual sighting of the lunar crescent was decisive for the beginning of a new month until quite recent times and not a regular scheme.

While we follow Wüstenfeld *et al.* (1961) in extrapolating a pure lunar calendar backwards to the beginning of the Hijra and earlier, our tables of first and last visibility of the lunar crescent take into account the fact that actual observations were decisive for the beginning of a new lunar month (Gautschy 2018). First and last visibility of the lunar crescent and the corresponding Islamic calendar date between 600 and 2000 CE were calculated for the following locations: Medina, Damaskus, Baghdad, Samarra, Cairo, Cordoba, Aleppo, Isfahan, Bursa and Istanbul. The Islamic calendar date is given on one hand according to the scheme following the tables of Wüstenfeld *et al.* (1961), but also according to observation for the desired location. In addition, the day of the week is specified. If an observational report mentions the day of the week in addition to the Islamic calendar date, then – and only then – an unambiguous conversion into a Julian or Gregorian calendar date can be achieved.

5. Historical Dates in the Islamic Calendar as Documents of Astronomical Observations

If a "calendar" is defined as a set of rules for organizing time, one can speak of a single Islamic calendar. But if a "calendar" is defined as a certain nomenclature giving an individual name for each natural day, month or year, one must speak of a multitude of local calendars in the Islamic world that differed from town to town. Furthermore, it has to be considered that even in the same town calendars could differ according to the religious parties and legal schools living side by side in the same community. All these calendars (except that of the Isma'ilis) are based on observations of the lunar crescent, usually performed in the evening of the 29th day of the month. Therefore, a month of 29 days documents a positive sighting of the lunar crescent, and suggests favourable atmospheric conditions at that day. If modern calculations indicate that on the 29th of a month the lunar crescent must have been visible, but nevertheless the next month started after the 30th day of that month, on can conclude that it was cloudy in the evening of the 29th of that month. The historical records of the Islamic wold – an immense corpus of writings – therefore contain records of atmospheric data that are precise in space and time. They enable estimates to be made of cloudiness in a vast area and over a time period of more than a millennium. All that has yet to be explored. A data collection is about to be built, and a selection has already been published (Thomann 2017). Methods for estimating seasonal cloudiness values from the binary data (positive/negative sightings) have been successfully tested. The significance of the data referred to is due to the fact that on one hand historical cloudiness data are otherwise not available in great quantity, and on the other hand that cloudiness has become a major issue in climatology, which previously concentrated, rather on temperature and precipitation (Lehmann *et al.* 2016).

References

Bruin, F. 1977, Vistas in Astronomy 21, 331

Fotheringham, J. K. 1910, MNRAS 70, 1910, 527

Gautschy, R. 2018, *Islamic Calendar, http://www.gautschy.ch/~rita/archast/mond/arabcal.html*

Hogendijk, J. P. 1988, Journal of Near Eastern Studies 47, 95

Huber, P. J. 2006, Journal of Geodesy 80, 83

Lehmann, U. *et al.* 2016, *An Introduction to Clouds: From the Microscale to Climate*, CUP

Maunder, E. W. 1911, Journal of the British Astronomical Association, 21, 355

Morrison, L., & Stephenson, F. R. 2004, JHA, 35, 327

Neugebauer, P. V. 1929, *Astronomische Chronologie*, Table B17

Schaefer, B. E. 1988, Quarterly Journal of the Royal Astronomical Society, 29, 511

Schoch, C. 1927, *Planeten-Tafeln für Jedermann*

Standish, E. M. 1998, *JPL Planetary and Lunar Ephemerides, DE405/LE405, Jet Propulsion Laboratory Interoffice Memorandum 312.F*

Thomann, J. 2017, *JPL "Few things more perfect": Ḥabash al-Ḥāsib's Criterion for the Visibility of the Lunar Crescent and the Dustūr al-munajjimīn, in: E. Orthmann et al. (eds.), Science in the city of Fortune*, 2017, 137

Wüstenfeld, F., Mahler, E., Mayr, E., Spuler, B. 1961, *Wüstenfeld-Mahler'sche Vergleichungs-Tabellen zur muslimischen und iranischen Zeitrechnung*

Yallop, B. D. 1997, *A Method for Predicting the First Sighting of the New Crescent Moon, NAO Technical Note no. 69*

Yazdi, H.-R. G. 2018, Archive for History of Exact Sciences 72, 89

Astronomy in Focus - XXX
Proceedings IAU Symposium No. XXX, 2018
M. T. Lago, ed.

doi:10.1017/S1743921319003995

Cuneiform Descriptions of Transient Phenomena

H. Hunger

Institut für Orientalistik, University of Vienna, A-1090 Vienna, Austria
email: hermann.hunger@univie.ac.at

Abstract. Sources from Ancient Mesopotamia contain mention of transient astronomical phenomena in two contexts: in records of observations, many of which can be dated, and in collections of omens, which use the appearance of such phenomena to predict future events. These omens consider quite a range of phenomena, but only rarely can they be dated in a precise way. This paper describes how transient phenomena were handled in both kinds of context.

Keywords. History and philosophy of astronomy

---

1. The Babylonian Diaries

Transient *objects* like comets and meteors are individual objects in their own right, whereas transient *phenomena* like eclipses and occultations mostly happen to other objects such as the Sun, Moon and stars. Many eclipses and occultations are fairly rapid (start to finish may take a few hours or less), but that is not a general rule, particularly as concerns stars. Records of observations collected from archaeological times have mainly been preserved in the so-called "Astronomical Diaries", which contain day-by-day observations of events in the sky and on the Earth. Written on clay tablets, the Diaries bear records from the 7th to the 1st Centuries BC, but are unevenly distributed. Very few remain from the earlier half of this time, and only by the 3rd Century do they become more numerous. More than 1500 (mostly fragmentary) tablets have been preserved, and almost all come from the city of Babylon. Unfortunately they are far from complete: the texts may cover only about 5% of the time-span mentioned. Apart from the movement of the Moon and planets among the stars, the tablets list the first and last visibilities of the planets, and lunar and solar eclipses. They also report the weather, the water level in the river Euphrates, prices of basic commodities such as barley and dates, and events of local and regional interest. We also find mentions of meteors and comets.

2. Omens

While the number of Diaries may be impressive, they constitute only a minor section of Babylonian celestial lore. Most of the tablets concerning the sky are celestial omens. All Babylonian omens are in the form of conditional clauses: "If A (the sign) happens, then B (a future event) will happen". It was generally believed in Mesopotamia that if nothing was done, the predicted events would in fact happen. The omens were seen as messages from the gods. Favourable omens indicated the benevolence of the gods; bad ones were meant to warn humans of undesirable future events. Nevertheless, the predicted events were not inevitable; it was possible to prevent unfavourable predictions from happening by means of prayer, rites and sacrifices.

Celestial omens first appear in the first half of the 2nd millennium BC, and mostly concern lunar eclipses. Omens from the sky did not, however, play an important role in those early times; divinations from the liver of sheep were far more frequent. Only in the 1st millennium BC did celestial divinations become a dominant source of signs. However, interest in them had existed for centuries before that, and pertinent omens were also collected, as were the other signs. That finally led to the organization of celestial omens into an extensive edition comprised of about 70 tablets. The chief problem with omens is that they have no associated dates, and they need not be based on observations. While some of the descriptions of ominous signs sound plausible, others are hard to understand. The Babylonian omen collections have a tendency to strive for completeness in order to take care of every possible phenomenon, even if not yet observed. In the case of eclipses, impossible cases are sometimes included, such as lunar eclipses on day 20 or 25 of the lunar month. We therefore refer only occasionally to the omen literature.

3. Halley's Comet

In 1983 I was entrusted with the publication of the Diary tablets. When in 1984 the expected recurrence of Halley's Comet became a topic of general interest, I was asked whether the comet was mentioned in the Diaries. Since its approximate period is known, it was easy to look in the tablets concerning the years in question. What was not so easy was to read the pertinent passages, since the tablets were mostly somewhat damaged. The results were published by Stephenson & Walker (1985).

The earliest cometary observations in the clay tablets that could be identified with Halley's comet are from 164 BC. They are found on two tablets, with a slightly different wording. Both concern month VIII of the year 148 of the Seleucid era, corresponding to the span from October 20 to November 18 of 164 BC in the Julian calendar. Unfortunately, no day number was preserved for these observations; we know only the Babylonian month.

One passage reads: "The comet which previously had been seen in the east in the path of Anu in the area of Pleiades and Taurus, to the west (break).... and passed along in the path of Ea." The other reads: ".... in the path of Ea in the area of Sagittarius, 1 cubit in front of Jupiter, the comet being 3 cubits high to the north ..."

Both passages need some explanation. The Babylonians divided the sky into three zones called "paths", which were parallel to the equator and bore the names of three major gods. The "path of Anu" refers to a band from $\sim$+17 to –17 declination; the "path of Ea" is the region of the sky south of it. The words "The comet which previously had been seen in the east" must refer to an observation in an earlier month, probably VII or VI. Thus, while the comet had been further north, in the path of Anu when it was observed earlier, it now had moved into the southern band. The second passage refers to an approach of Jupiter and the comet. 1 cubit is $\sim 2°$. The relative position given was very useful for determining the orbit of the comet more precisely.

The appearance of Halley's Comet in 87 BC is documented in a Diary which is very fragmentary but which can at least be dated reliably. The observation of the comet is dated on month V, day 13, equivalent to August 24. The passage reads: "Night of the 13th, first part of the night, the comet .... (break) ... which in month IV day beyond day 1 cubit ... (break) ... between north and west, its tail 4 cubits ...."

Whatever the tablet said about the comet on the 13th of month V, is lost in a break. The observation reported was either the reappearance after perihelion, or the last visibility. The following partly broken sentences seem to concern only the preceding month, IV. The comet moved between 1° and 6° daily for approximately the duration of that month, as could be interpreted from the words, "day beyond day 1 cubit". When first visible, the comet was high in the eastern sky before dawn, so its tail would have pointed north-west. After perihelion the tail pointed more to the north-east. The remark about the tail therefore very likely refers to what was seen in month IV. Though fragmentary, these reports can be identified as referring to Halley's comet.

There are also mentions of other comets, similarly badly preserved. They cannot be connected to a later comet because we do not have reports from the times of antiquity or later that could bridge the gap from modern observations.

Reports of comets also crop up in letters sent by observers to the Assyrian kings in the 7th Century BC. One contains the following sentence:

"If a comet becomes visible in the path of Anu: there will be a downfall of Elam in battle". The construction using "If" identifies it as a quote from an omen, which are always expressed as conditional sentences. But the quote occurs in a report to the Assyrian king, so something must have been observed that "fitted" the "If"clause of the omen. The date of the letter can be determined as 675 BC, but unfortunately that does not match the period of Halley's Comet.

The Babylonian word used for comet, ṣallummû, may be used not only for comets. In fact, the observations of Halley's Comet are a proof of that translation. In other contexts the word could just as well refer to meteors, which is how the *Chicago Assyrian Dictionary* translated ṣallummû in 1962.

4. Meteors

Meteors were frequently called "big stars" (a designation which was unfortunately also used occasionally for planets), or simply "star"; only from the description of its behaviour is it apparent that a meteor must be meant.

An early report is again from 7th Century letters to the Assyrian kings: "One double-hour of night had passed: a big star flashed from north to south". That was considered a good omen by the writer of the letter.

The Astronomical Diaries contain observations of meteors in 419 BC: "in the middle part of the day, a big star which was like a torch flashed from south to north, and the land heard the noise of the sky". In 292 BC: "a big star which was very bright and had a train, flashed from north to south; its light was seen on the ground ...".

5. Occultations and Eclipses

Occultations of stars or planets by the Moon are also transient phenomena, and were regularly noted in the Diaries. The Babylonian expression was "star X entered the Moon", – which is actually how it looks.

A good example is found for 400 BC April 11: "First part of the night, Saturn entered the Moon. At 28° of night (i.e. 112 minutes after sunset), Saturn returned from its inside". Sometimes an occultation was expected but not observed. This is expressed by the words, "the star (or planet) was set to the Moon's inside", meaning that the star was on a track leading to an occultation.

Close approaches of planets to stars often bear similar visual effects to occultations, and are frequently mentioned in the Diaries, when the smallest distance between them is 1 finger or 2 fingers, corresponding to $5'$ or $10'$, respectively. In rare cases the two bodies are said to have become one star.

The most vivid transient phenomena are eclipses of the Sun and Moon. As early as the first half of the 2nd millennium BC we find omens deriving from lunar eclipses, though they cannot be identified with eclipses that actually happened. From the 1st millennium BC, 8 chapters in a large omen collection deal with lunar eclipses. The descriptions of eclipses in those omens are sometimes rather detailed, but in my opinion they cannot be connected to actual eclipses, although some scholars believe the converse.

Apart from omens, eclipses are occasionally included in texts referring to daily life. For instance, a lunar eclipse is reported in a letter from the city of Mari in Syria that dates from the 18th Century BC; its significance was considered unclear so an inspection of the liver of a sheep was ordered to find out. The Astronomical Diaries and their derivatives contain frequent reports of eclipses. The oldest are from the 8th Century BC, and became increasingly detailed in the course of time. Reports from the 4th Century BC onwards can contain the following information:
(a) Date (year, month, day). (b) Time between moonrise and sunset (like most of time measurements, it is given in degrees of time, equivalent to 4 minutes). (c) Entrance angle of the shadow (indicated by direction, e.g., "it began on the north-east side". (d) Time (from sunset) to maximal phase ("maximal phase" translates a Babylonian word that literally means "weeping, lamentation"; such lamentation was probably done during totality. The word is used both for total eclipses and for very long ones, hence the expression "maximal phase". (e) Magnitude of maximal phase (often expressed as part of the lunar disk, e.g., "a quarter of the disk", or measured in fingers, where 12 fingers correspond to totality). (f) Duration of maximal phase. (g) Time from the end of maximal phase to the end of the eclipse. (h) Direction in which the shadow leaves the disk (refers to the shadow covering increasingly little of the lunar disk, e.g., "from north-east to south-west it became clear"). (i) Total duration of the eclipse. (j) Weather (e.g., the wind blowing during the eclipse, and/or lightning and thunder, may be reported). (k) Presence or absence of planets (e.g., if planets - or Sirius - were visible during the eclipse, and/or if one of them rose or set during the eclipse). (l) Position of the Moon relative to a fixed star. (m) Time of start of the eclipse, relative to sunset or sunrise. (n) Time between sunrise and moonset. If an eclipse did not take place it is marked as "which passed by". In the case of lunar eclipses, this could be known in advance; if a lunar eclipse was expected during daylight, it would not be possible to observe it.

There are collections of eclipse reports, some arranged like a spreadsheet in 18-year groups, such that each eclipse is accompanied to the left or right by an eclipse from 18 years before or after. This arrangement reflects the Saros cycle of 223 lunar months, after which eclipses with similar characteristics occur on the same dates in the Babylonian calendar, only with $\sim$8 hours displacement. This period is therefore very good for predicting eclipses, and was probaby used by the Babylonians from the 7th Century BC. It is also the basis of arithmetic formulæ for calculating eclipses, as found in Babylonian mathematical texts for astronomy.

References

Bjorkman, J. K. 1973, *Meteoritics*, 8, 91

Huber, P. J. & De Meis, S. 2004, *Babylonian Eclipse Observations from 750 BC to 1 BC*, (Milan: Mimesis)

Hunger, H. 1992, *Astrological Reports to Assyrian Kings*, (Helsinki: HUP)

Sachs, A. J., & Hunger, H., 1988ff, *Astronomical Diaries and Related Texts from Babylonia*, (Vienna: ÖAW)

Stephenson, F. R., & Walker, C. B. F. (eds.) 1985, *Halley's Comet in History*, (London: British Museum Publications)

Astronomy in Focus - XXX
Proceedings IAU Symposium No. XXX, 2018
M. T. Lago, ed.

doi:10.1017/S1743921319004009

Historical Novæ and Supernovæ

A. Pagnotta[1], D. W. Hamacher[2], K. Tanabe[3], V. Trimble[4] and N. Vogt[5]

[1]Department of Physics and Astronomy, College of Charleston, SC 29424, USA
email: pagnotta@cofc.edu

[2]Department of Indigenous Studies, Macquarie University, NSW 2109, Australia

[3]Okayama University of Science, Okayama, Japan

[4]Department of Physics and Astronomy, University of California, Irvine, CA 92697-4575, USA

[5]Instituto de Física y Astronomía, Universidad de Valparaíso, Valparaíso, Chile

Abstract. This paper brings together the chief points raised during FM5 by astronomers, archaeologists, and historians whose research interests centred on novæ and supernovæ. The common focus was the use of historical observations to study transient astronomical phenomena. The presenters covered a wide variety of topics within that theme, and this report summarizes some of the aspects specific to historical novæ and supernovæ.

Keywords. (stars:) novae, cataclysmic variables, (stars:) supernovae: general

---

1. Fundamental Properties of Novæ and Supernovæ

Novæ and supernovæ are extreme, and usually unpredictable, brightenings of stars. They have been observed since at least two millennia ago, recorded by Chinese astronomers as "guest stars" (Clark & Stephenson 1982). A nova occurs when a white dwarf in a binary system accretes enough hydrogen onto its surface to trigger a runaway thermonuclear reaction, causing an eruption in the surface layer. Supernovæ come in two basic types: thermonuclear, or Type Ia (when either one white dwarf reaches the Chandrasekhar mass limit and explodes (single-degenerate) or when two white dwarfs in a binary (double-degenerate) merge and thereby exceed the Chandrasekhar mass limit), and core collapse (when a massive star runs out of fuel in its core and can no longer sustain fusion, causing a rapid collapse and explosion). Supernovæ and neutron star mergers produce and distribute the elements from CNO up to Th, U, and Pu; they heat and stir interstellar gas and dust, push gas and dust out of galaxies, trigger the formation of new stars, accelerate cosmic rays, and produce pulsars, X-ray binaries, neutrinos and (probably) gravitational waves. They aid the mapping of dark matter in galaxy clusters, measuring cosmological distances, and feeding magnetic fields into galaxies. Their records appear in many formats, from ancient written records to possible variations in ionic concentrations in geological records to (more recent) glass plates. The talks and posters presented on this theme reflected that variety.

Both novæ and supernovæ can simulate the appearance of temporary new stars, and can be bright enough to be seen with the naked eye from days to months, depending on the distance and type of event. Additionally, instabilities in the disks surrounding accreting white dwarfs can produce small outbursts known as dwarf novæ, which are not bright enough to appear in historical records but which can now be observed in systems that were previously identified by ancient astronomers as guest stars.

1.1. *Historical supernovæ as research tools*

Despite the danger of small number statistics, historical supernovæ (of which there are some 7–14) can tell us something about SN rates, types, parent populations, 3-D structures, what they put into the ISM, their formation, initial rotation and magnetic fields, and early evolution of neutron stars and black holes, late mass loss from massive stars, nucleosynthesis, and more. For some systems, such as CM Tau (also known as the Crab Pulsar, SN 1054, NGC 1952, 3C144, Tau X-1, NP0532) the observational evidence is still evolving, as publications find different mass values, detect a pulsar (with rather slow initial rotation), a jet off to one side, and a gamma ray source. The full ensemble includes 1572 (Tycho), 1604 (Kepler), Cas A, SN 1006, the progenitor of SNR 0519-76.5, S And, RCW86 = SN 185, and 1987A in addition to SN 1054. The statistics suggest that we are overdue for another Galactic event, and that classification can be difficult.

1.2. *Novæ and supernovæ in ancient records*

Pre-telescope ("historical") supernovæ events in the Milky Way were recorded by sky watchers of several different cultures (though not by all that could have done so). The ancient Egyptians expected new stars when the souls of pharaohs rose to the heavens (Trimble 1964), but we find no inscriptions or papyri confirming such events. The Babylonian tradition, as preserved in *Genesis*, puts all star births on the fourth day of creation, so no new ones are expected. Mayan records include astronomically-based calendars, but no supernova-like items (but it is important to note that many records were burned in 1540 CE). The Greek and Roman myths imply on-going star formation, but only recurrent events were thought important in their more serious writings. However, Chinese (and later Japanese and Korean) writings and star maps are highly valuable (Stephenson & Green 2002). Their authors kept track of planetary positions, eclipses, meteors, comets, and new stars of various types. The most firmly established examples happened in 1006 (also seen from Switzerland), 1054 (also one Arabic record), 1181, 1572 (also Tycho), 1604 (also Kepler), and 1987A in the Large Magellanic Cloud. There are possible or probable events from 70, 185, 369, 386, 393 and 837 CE; a couple of southern supernova remnants are less than 2000 years old. Yet it remains an abiding puzzle that there is no general agreement over records of the event (in about 1685) that gave birth to the radio source Cas A.

The known remnants lie close to the galactic plane (the Crab, SN 1054, is furthest, at $\sim 5^\circ$ south). What they teach us is that, first, they are all different (Branch & Wheeler 2017; Alsabti & Murdin 2017). Secondly, supernovæ really do add to the heavy element inventory of the Galaxy. The Suzuku X-ray spectrum of the 1006 event is bristling with emission features of O, Ne, Mg, Si, S, Ar, Ca and Fe. The same can be said of the XMM Newton X-Ray spectrum of Tycho's remnant. The 1054 remnant includes a solar mass or more of hydrogen, which would have been conspicuous in peak-light spectra, agreeing with its central neutron star, PSR0532, and identifying it as a core-collapse event. The peak-light spectra of Tycho (1572) and of the Cas A birth event can be studied through light echoes. Dust clouds near the SN events have reflected peak light to us with multi-century delays. Cas A bore a definite resemblance to the bright SNII 1993J, complete with helium and hydrogen, while SN 1572, devoid of hydrogen then as now, resembled SN 1994D and others (Krause 2008; Krause *et al.* 2008). A couple of the historical remnants are sources of Ti-44 decay gamma rays – additional evidence of nucleosyntheses in SN events – as have been the gamma rays and radioactive-decay powered light curve of SN 1987A.

Future large surveys are expected to yield thousands of events per year, and should yield much better statistics on rates, types and parent populations, examples of rare sub-and super-luminous types, double detonations, and unusual environmental effects.

We may also expect answers to the deepest ongoing SN questions: (1) Are the progenitors of nuclear explosions white-dwarf pairs that merge or white dwarfs that accrete from some other sort of companion? and (2) What is the mechanism that kicks off the outer layers of core-collapse events? A Type II event in the Milky Way would provide a flood of neutrinos (compared to about a few tens from SN 1987A), and an unpredictable flux of gravitational waves. The most likely nearby progenitors are Betelgeuse and Antares.

2. Transient Astronomical Phenomena in Australian Indigenous Oral Traditions: 65,000 years of Oral History

Historical novæ and supernovæ are well recorded in the written records of cultures across Eurasia (Stephenson & Green 2002). Such brightly visible objects would also have been observed by Indigenous peoples across the world and probably incorporated into their oral traditions and material culture. However, demonstrating this poses a number of significant challenges. Oral traditions encode knowledge in narrative forms, while motifs in material culture, such as paintings and rock art, are open to interpretation without the producers of that art telling us what it means. The prospect of finding Indigenous traditions of novæ and supernovæ is an attractive one, but also one that has succumbed to speculation, conjecture, and pseudoscience in the literature. The most famous example is of the alleged depiction of SN 1054 in Anasazi rock art in SW USA (Brandt *et al.* 1975). Although the evidence was sketchy, the idea caught on and has now become an established "fact" that is wildly popular with the public. The Proceedings in which that paper was published included a full refutation article on the next page (Ellis 1975), which systematically demonstrated that the supernova interpretation was incorrect. However, that second paper is almost never mentioned or cited and seems to have been ignored.

This poses a problem whose solution can be addressed by the development of rigorous selection criteria for determining the validity and support of novæ and supernovæ claims in oral traditions and material culture. Hamacher (2014) offered a detailed methodology showing that despite a number of claims published in the literature, not one had sufficient evidence to conclude that the tradition or motif was that of a known nova or supernova. The only example that was vindicated related to Boorong Aboriginal traditions of western Victoria in Australia. In the 1840s, William E. Stanbridge learned about the local astronomical traditions of the Boorong clan (Wergaia language group) near Lake Tyrell, Victoria (Hamacher & Frew 2010). The Aboriginals who taught him their knowledge claimed to be the most skilled astronomers in the region. His paper (Stanbridge 1858) listed around 40 astronomical objects and provided a brief description of each. One was of a bright red star in Robur Carolinium, and although it was not listed in his catalogue he provided a description of it, its location, and a catalogue number of what he thought it might be. It proved to be a description of the luminous blue variable star Eta Carinae during its Great Eruption in the late 1830s and early 1840s. Its identity was proved conclusively during its "supernova impostor" event thanks to the details provided by Stanbridge from his Aboriginal informants.

Other Aboriginal traditions from Australia describe, for example, the bright appearance of stars in the Milky Way (Wells 1973). The stars attributed to sky ancestors were located in the tail of Scorpius, where known historical supernovæ such as SN 393 appeared. The evidence is not conclusive and other novæ candidates could also explain it, but it does offer important information about possible novae and supernovæ observations by Indigenous people around the world. But although those traditions are important, we must also be vigilant that we do not plant false or misleading conclusions into the public sphere (Schaefer 2006). That can cause degradation of traditional knowledge, and fuel speculation that can easily become established "fact" in scientific discourse, despite evidence to the contrary. This requires careful and robust scholarship, and calls for the community (a) not to publish scholarship without strong supporting evidence, and (b) to reject unfounded or refuted claims.

3. Historical Observations of Novae: Guest Stars to Glass Plates

Two examples are documented in which modern observations have been matched with the "guest stars" mentioned in historical records. The first is Nova Sco 1437, recorded by Korean royal astronomers in the Sejong Sillok, a chronicle of King Sejong's reign from 1418–1464 CE. Visible for 14 days, the guest star was located "between the second and third stars of Wei. It was nearer to the third star, about half a chi away". Previous modern interpretations had placed that location just north of ζ Sco, but searches for nova shells or other signs of nova activity in the relevant region showed nothing. However, a new numbering of the stars in Wei, starting with μ Sco (the determinant of the mansion) places the guest star just *east* of ζ Sco, where a cataclysmic variable (CV) and nova shell are in fact located. However, there is no confirmation available that this numbering was indeed applied by Korean astronomers in historical times. In quiescence the CV is visible on scans of the Harvard College Observatory plate archives dating from the 1920s to the 1950s. An image of the system on plate A12425 (taken on 1923 June 10 in Arequipa, Peru) was used in conjunction with modern observations (made using the Swope 1-m telescope at Las Campanas in 2016 June) to obtain a long-baseline measurement of the proper motion of the CV, finding $\mu_\alpha = -12.74 \pm 1.79$ milliarcseconds per year and $\mu_\delta = -27.72 \pm 1.21$ milliarcseconds per year. Accounting for that motion shows that the location of the CV in 1437 CE coincided with the centre of the nova shell, providing strong evidence of a causal link between the two, and linking the CV to the guest star in the Korean records. The Harvard plates also show dwarf nova eruptions in this system in 1934, 1935, and 1942, providing data that can be used to test theories, such as Hibernation (Shara *et al.* 1986), that seek to explain the connections between the various types of events that occur on or near massive white dwarfs (novæ, dwarf novæ, and thermonuclear supernovæ).

The second example is BK Lyn, a nova-like CV identified in 1986 and since linked to a guest star that appeared on 101 December 30 and was listed in Chinese records. BK Lyn has been heavily monitored by members of the Centre for Backyard Astrophysics. It transitioned from a nova-like CV to an ER UMa-like dwarf nova in 2005, but now appears to have entered a standstill phase, possibly similar to that of Z Cam-type stars; no dwarf nova outbursts have been seen since 2012. Further details about Nova Sco 1437 can be found in Shara *et al.* 2017, and on BK Lyn in Patterson *et al.* (2013). The combination of data from ancient records, archived plates and modern observations is clearly a powerful tool for understanding the life cycles of accreting white-dwarf binaries.

4. Evidence of Galactic Supernovæ from Natural Terrestrial Archives?

Asian records of historical guest stars have been linked to Antarctic ice-core records. (The issue of nitrate concentrations was discussed in a poster paper by Tanabe). One suggestion is that a rise in terrestrial radiocarbon seen in CE 1009 could have been caused by SN 1006 (Damon *et al.* 1995): if a large flux of gamma rays had been provided by SN 1006, they could have produced terrestrial radiocarbon which was incorporated into tree rings a few years later (via the carbon cycle). However, it may also be possible that normal variations in solar activity (from a Schwabe cycle maximum in the first half of the first decade of the 11th century to a Schwabe cycle minimum a few years later) led to an increase in radiocarbon from around CE 1005 to 1010. This example shows that reconstructing solar activity is also important for the study of supernovæ.

5. The Expected Accuracy of Classical Nova Identifications among Historical Far-Eastern Guest Star Observations

It is clearly important to assess the probability that identifications of classical novæ among historical Far Eastern guest star observations are correct. The approach adopted by Vogt, Hoffmann, Neuhäuser *et al.* (poster paper) was to compare the coordinates of eight supernovæ derived by Stephenson (1976) and Clark & Stephenson (1977) from information given in old texts with those of the corresponding modern supernova remnants. It yielded a typical angular difference of the order of 0.3 to 7 degrees. That value could then be adopted for the expected deviation in coordinates between a classical nova observed as a guest star and its modern counterpart among known cataclysmic variables. However, there are considerable disagreements among modern authors over the interpretation of ancient Far Eastern texts, emphasizing the need to consult again the original sources in order to improve the positional reliability.

References

Alsabti, A. S., & P. Murdin (eds.). 2017, *Handbook of Supernovæ*, (Springer), Part II
Branch, D., & J. C. Wheeler. 2017, *Supernovae Explosions*, (Springer), sects. 7, 4 & 11
Brandt, J. C., *et al.* 1975, in: *Archaeoastronomy in Pre-Columbian America*, p. 45–58
Clark, D. H., & Stephenson, F. R. 1977, *The Historical Supernovæ* (Elsevier)
Clark, D. H., & Stephenson, F. R. 1982, in: M. J. Rees & R J. Stoneham (eds), *ASI Series C*, 90, 355
Damon, P.E., Kocharov, G.E., Peristykh, A.N., *et al.* 1995, in: N. Iucci & E. Lamanna (eds.), *24th Int. Cosmic Ray Conf.*, p. 311
Ellis, F. H. 1975, in: *Archaeoastronomy in Pre-Columbian America*, p. 59–87
Hamacher, D. W. 2014, *JAHH*, 17, 161
Hamacher, D. W., & Frew, D. J. 2010, *JAHH*, 13, 220
Krause, O. 2008, *Nature*, 456, 617
Krause, O., *et al.* 2008, *Science*, 320, 1195
Patterson, J., *et al.* 2013, *MNRAS*, 434, 1902
Schaefer, B., in: T. W. Bostwick, & B. Bates (eds.), 2006, *Viewing the Sky Through Past and Present Cultures* (Phoenix: Pueblo Grande Museum), p. 27–56
Shara, M. M., Livio, M., Moffat, A.J.F., & Orio, M. 1986, *ApJ*, 311, 163
Shara, M. M., *et al.* 2017, *Nature*, 548, 558
Stanbridge, W. E. 1858, Trans. Phil. Inst. Victoria, 2, 137
Stephenson, F. R. 1976, *QJRAS*, 17, 121
Stephenson, F. R., & Green, D. A. 2002, *Historical Supernovæ and their Remnants* (OUP), 5
Trimble, V. 1964, *Mitt. des Instituts Orientforschung*, 10, 183
Wells, A. E. 1973, Stars In The Sky (Rigby)

Astronomy in Focus - XXX
Proceedings IAU Symposium No. XXX, 2018
M. T. Lago, ed.

doi:10.1017/S1743921319004010

Stone Inscriptions from South Asia as Sources of Astronomical Records

B. S. Shylaja

Jawaharlal Nehru Planetarium, Bangalore, 560001, India
email: shylaja.jnp@gmail.com

Abstract. Stone inscriptions from all over India provide records of eclipses, solstices and planetary conjunctions. Extending the study to South Asia, to include Cambodia, Sri Lanka, Nepal and Thailand, threw light on many new aspects such as evolution of calendars independently from the influence of Indian system of time measurement as early as the 3rd Century BCE. Many interesting records of planetary conjunctions are available. One record from Cambodia hints at a possible sighting of the 1054 supernova, while another from Thailand suggests a pre-planetary nebula event.

Keywords. History of Astronomy, planets, eclipses

1. Introduction

Stone inscriptions were engraved to leave a permanent record of donations and grants given by the kings, their feudatories, chiefs, and village headmen. A good number of them record the heroic deeds of soldiers and commoners fighting enemies, or wild animals during hunting. In some cases they mark the self-immolation of ascetics, widows, and devotees. They also carry accurate records of the date in whatever local system, and show details of the positions of the Sun and Moon. Those details were written in different languages, and thus serve as important documents on celestial events.

The Archaeological Survey of India undertook the publication of these records (as Epigraphia Indica), and also the regional versions (as Epigraphia Carnatica), resulting in the *South Indian Inscriptions.* More than 40 volumes are now available, each averaging about 200 inscriptions (Shylaja & Geetha 2016a).

The most important aspect of this study for astronomers is the recordings of the visibility of the totality of solar eclipses. Five such records have been used, in order to place a limit on the path of totality and therefore on the range of ΔT, the variation of the rotation period of the Earth. In one case it has been possible to link a record with one from China (Tanikawa *et al.* 2019).

2. Inscriptions from South Asia

The influence of Indian culture and Sanskrit were widespread in S and SE Asia, so this type of documentation might well be expected outside India. Documentation and translations of inscriptions from several countries were carried out during the colonial era. This paper summarized the results, based on a limited number of records and references.

Sri Lanka. The majority of inscriptions from Sri Lanka were the edicts of Buddha. It is estimated (though no specific dates are mentioned) that they date back to 3rd Century BCE when King Ashoka spread the messages of the Buddha across SE Asia. The earliest dated record corresponds to 183 CE. The languages used were mostly Sanskrit and Pali. The method of reckoning the year is BE (Buddha Era); the name of the month and the lunar phases are mentioned, but there is a conspicuous absence of the name of the day of the week (Vara) (see Epigraphia Zeylanika, Vol II and Vol III).

The count of the year in many records corresponds to Shalivahana Saka (78 CE). Numerals are written down as words and numbers; for example, 327 is written as 300 027, a method also seen in contemporary records in India.

One record uses the word *nabhomarakayam*, which is interpreted by epigraphists as lunar eclipse. Using the information that it was the 13th year *trisahita dasame* (or 3 with 10) of King Shrisanghabhodi (982–1017 CE), we arrive at the lunar eclipse of CE 955 Jul 14. The month is given as Nikini (a translation of Nabhas, the Vedic name corresponding to Jul/Aug). There is another record of a solar eclipse in CE 982 Sep (Müller 1883).

Nepal. Most of the inscriptions in Nepal are edicts for administrative purposes; they cover a period from about the 3rd century BCE to the 12th century CE. While the lunar phase and year count are available, the absence of a specific choice of events such as eclipses is noticeable. The information is therefore insufficient to deduce any astronomical event that could have occurred. Again, there is no mention of the vara (day of the week), and the Saka year is written in numerals (Agarval 2010).

Cambodia. Studies by French epigraphists have been a great help in understanding the texts which are available in Khmer; many are bilingual, and Sanskrit is also widely used. The inscriptions are now being translated to English. The historian Majumdar (1953) produced direct translations of Sanskrit verses, though the ones in Khmer have not yet been decoded fully. Most of the inscriptions provide complete details of the positions of the Sun, Moon (with phases), planets, nodes and even the ascendant (lagna), so the time of an event can be fixed quite precisely. The planetary positions are verifiable and agree within a few degrees with the positions from modern calculations. Interestingly, there is no emphasis on events like eclipses. While the year count is similar to the one on the mainland, Shalivahana saka, in some places the Chinese method of counting is followed.

In an inscription of 605 Saka (equivalent of 683 CE), the date is written down as numerals – the first ever record of decimal notation. Generally the numbers are written in text, e.g., "five added to six hundreds". In the South Indian records, numerals are avoided by following the bhuta sankhya system, whereby objects are used to specify a number; for example, an eye or a hand refers to 2, a 'muni' or sage ('wise man') to 7.

The word *dhumakethu* is mentioned for a comet; however the date is not readable.

The names of the zodiacal constellations start to appear after about the 7th Century, e.g. *Bhouma* (son of the Earth, Mangala, Mars), *Induja* (son of the Moon, Budha, Mercury), and *Arka* (the Sun). *Guru* (Jupiter) is with the Moon in *apa* (water, Sagittarius). Venus, Shukra, is in *keeta* (a small creature, viz. Scorpius).

An inscription in Sanskrit and Khmer (Fig. 1) for worship by an ascetic includes a phrase praising Lord Shiva by *shukra tara prabhavaya*, and may be a record of SN 1054.

There are many Sanskrit inscriptions in and around Angkor Wat. Many of them have been translated into French, through the efforts of Cedes (1968). One of the inscriptions (Beer 1967) refers to the year 1217 of the Saka era (corresponding to 1295 CE), or the 12th day of the first half of the month of Vaishakha, on a Thursday, in the month Chitra, when the king erected two statues. The Sun and Saturn were in Taurus, Mars and Rahu in Gemini, the Moon in Libra, Jupiter in Scorpio, Mercury, Venus and *Ketu* in Aries, and the ascendant in Cancer. *Ketu* can mean either a comet or a descending node. However, in the absence of either of them at the specified location, we deduced a possible pre-planetary nebula event based on the rings around it. A Chinese record reports a very interesting case of two eruptions in the same month, but is dated CE 1297 Sep 9–18 (Shylaja & Geetha 2014).

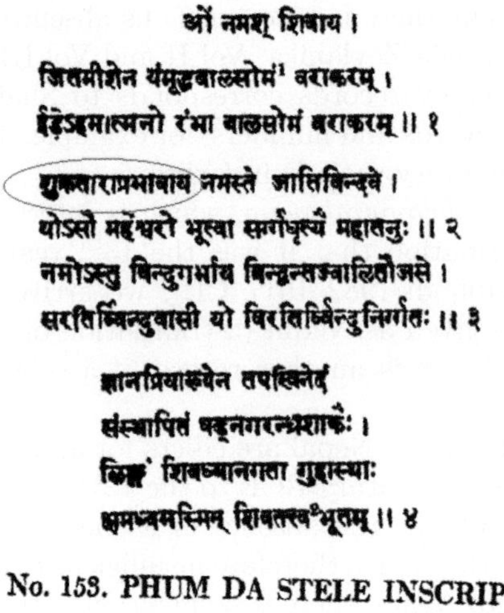

ओं नमश् शिवाय ।

जितमीशेन यंमूद्भवाल्सोमं[1] वराकरम् ।

ईड़ेऽहमात्मनो रंभा वाल्सोमं वराकरम् ॥ १

शुक्रताराप्रभावाय नमस्ते जातिबिन्दवे ।

योऽसौ महेश्वरो भूत्वा सर्गभृत्यै महातनुः ॥ २

नमोऽस्तु बिन्दुगर्भाय बिन्दुन्तज्वालितौजसे ।

सरतिर्ब्बिन्दुवासी यो विरतिर्ब्बिन्दुनिर्गतः ॥ ३

ज्ञानप्रियाख्येन तपस्विनेदं

संस्थापितं षड्नगरन्ध्रशाकैः ।

लिङ्गं शिवध्यानगता गुहास्थाः

क्षमध्वमस्मिन् शिवतत्त्व[2]भूतम् ॥ ४

No. 153. PHUM DA STELE INSCRIPTION.

Figure 1. The inscription of interest (number 153 on page 382 of Majumdar 1953) is from a small village called Phum Da in the Kampong Cham province, written partly in Sanskrit and partly in Khmer. The Sanskrit part has the *saka* year mentioned in the *bhuta sankhya* system as *sat* (6) *naga* (7) *randhra* (9). The phrase praising Lord Shiva include an adjective *sukra tara prabhavaya*, which means *"one who creates a star as bright as Venus"*, and could well suggest SN 1054 since the year (Saka 976) is equivalent to 1054 CE.

3. Conclusions

This study revealed that many interesting events are recorded in these inscriptions. It threw light on a different system of reckoning of time, called the 'Buddha era', which is not very common in India, and can help us understand the evolution of the calendars, e.g., through the introduction of the 12 zodiacal signs along with the 27 star system. It has also shown the importance of these records in the context of fixing the variation in the rotation period of the Earth (see Stephenson *et al.*, page 160).

The figurative description of CE 1054 is particularly interesting, as it could indicate the supernova. This possibility was missed by historians, who would be concentrating on the genealogy of kings, their coronations and the duration of their reigns.

References

Agarval, K. D. 2010, *Importance of Nepalese Sanskrit Inscriptions* (New Delhi: Rashtriya Sanskrit Sansthan)

Beer, A. 1967, *Vistas in Astronomy*, 9, 202

Cœdes, G. 1968, *The Indianized States of Southeast Asia* (Univ, Hawaii Press, Honolulu)

Majumdar, R. C., 1953, *Sanskrit Inscriptions of Kambuja, Asiatic Soc. Mon. Ser.*, 8

Müller, E. 1883, *Ancient Inscriptions of Ceylon* (London: Trubner & Co.)

Shylaja, B. S., & Geeta, K. G. 2014, *Current Science*, 107, 1751

Shylaja, B. S. & Geeta, K. G. 2016a, *History of the Sky – On Stones* (Bangalore: Infosys Found.)

Shylaja, B. S. & Geetha, K. G. 2016b, *IJHS*, 51, 206

Tanikawa, K., Sôma, M., Shylaja, B. S., & Vahia, M. 2019. The Reliability of the Records of Observed Solar Eclipses in India and Comparison with Contemporaneous Eclipse Data from Other Countries. In: Orchiston W., Sule A., Vahia M. (Eds.) *The Growth and Development of Astronomy and Astrophysics in India and the Asia-Pacific Region.* Astrophysics and Space Science Proceedings, vol 54. Springer, Singapore

Astronomy in Focus - XXX
Proceedings IAU Symposium No. XXX, 2018
M. T. Lago, ed.

doi:10.1017/S1743921319004022

Galileo's Account of Kepler's Supernova (SN 1604): A Copernican Assessment

M. Cosci

Universita Ca' Foscari Venezia
email: matteo.cosci@unive.it

Abstract. The name of Kepler is inseparably associated with the supernova of 1604 (SN 1604; V843 Ophiuchi), but there are reasons why Galileo Galilei might also claim to leave his name on that phenomenon, given the assiduousness of his observations.

Keywords. Supernova: SN 1604

Why is it that today the supernova of 1604 is known as "Kepler's supernova" and not as Galileo's? The simple answer is that, unlike the German astronomer, Galileo did not publish anything specific on it. However, that does not mean that he did not study it deeply, or that he did not publish anything at all on it. This note gave a summary of Galileo's account of Kepler's supernova, as recovered from surviving sources. The main challenge is that the surviving evidence is scattered, and in many texts of different types. The study of the "stella nova" was a life-long research objective for Galileo and he intended to utilize the observational opportunity offered by it for confirming the Copernican heliocentric hypothesis – his lifetime astronomical goal.

A few weeks after the supernova's visible outburst, Galileo was asked to give some public lessons on the unexpected and unsettling celestial novelity. Galileo accomplished this well, even though it was the first time he had had to handle an issue of observational, rather than theoretical, astronomy. The texts of his lectures survive, though largely incomplete and fragmentary. The main aim of his teaching was to demonstrate that the nova was located beyond the distance of the Moon, as proved by his parallactic measurements. By itself it was a provocative result, since it was contrary to the dominant Aristotelian model, according to which no substantial change could take place in the superlunary region of the heavens. The interpretation of the fragments is aided by a recently discovered external report on Galileo's lessons. Galileo's assessment of the superlunary position of the nova, placed at the distance of the fixed stars, finds confirmation there.

The main reference which Galileo consulted for the interpretation of the nova was the *Progymnasmata* by Tycho Brahe. Galileo intended to offer a comparative study of the nova of 1572 with the one now visible. Brahe included some positional measures taken by the astronomer Elias Camerarius and which he disregarded as wrong, but they could have been right (Galileo noted) if the Earth were moving while the star stayed now.

Another important source of information for Galileo was Ilario Altobelli, one of the first observers of the new star. Galileo received many letters from him on the subject. The correspondent described openly his anti-Aristotelian interpretation of the appearance, and provided the observational data he collected. (At the same time Galileo also received observational notes from others). However, shortly afterwards Galileo was attacked by a pro-Aristotelian, Antonio Lorenzini, the point of contention being the alleged superlunary position and the presumed applicability of a geometrical method, i.e. parallax, to matters within the domain of natural philosophy. Galileo pseudonimously replied *thrice*, also by the aid of one of his students, making plain the principles of parallax in an elementary way, defending the validity of its application for calculating relative astronomical distances, and cautiously sympathizing with Copernican heliocentrism.

18 months after its first appearance the nova had faded, and Kepler published his *De Stella nova*, hailed as the magnum opus on the subject. Galileo's only comment (written on a little slip of paper) was a consideration regarding the apparent centrality of the solar source of light as observed upon other heavenly bodies, stars included, an argument that he planned to discuss in his own account of the nova.

Another Aristotelian opponent, named Baldassar Capra, then attacked Galileo openly, discrediting his work because he did not attribute to him (Capra) the priority of having spotted the nova first – something that Galileo actually never pretended for himself. Galileo defended his professional respectability and his scientific results vigorously. From this dispute we learn that Galileo probably used a quadrant for his observations (the telescope had not then been invented), and that he provided precise alignment measures of the nova in respect to other surrounding stars in order to validate its fixity.

After the publication of his *Sidereus Nuncius*, Galileo was still reviewing the observations that he made in order to determine two unsettled key features of the nova: (1) whether it was in motion, and (2) what was its actual physical conformation. The observed decrease in luminosity suggested that the nova had an upward receding motion. From its progressive uniform fading he deduced that its estimated motion was constant, while from the lack of any detectable angular parallax he deduced that the motion occurred in a straight line. The "optical" nova was therefore considered to be in uniform rectilinear upward motion at constant velocity, far above the atmosphere.

In his *Dialogue Concerning the Two Chief World Systems* Galileo focused mainly on the earlier nova in Cassiopeia, but intended to extend his results to the one of 1604. He double-checked the measurements taken by others, and definitively corrected its position as superlunary (in opposition to the account provided earlier by Chiaramonti). He even took into account effects like atmospheric refraction and instrumental errors, but judged them irrelevant for the level of precision of the observations. Responding to his critics, he maintained that the "new star" had to be a sort of tailless, reflecting comet. 30 years later the nova was also a hot topic for discussion.

Among Galileo's papers we can also find a couple of undated sketches that turn out to be of paramount importance for assessing his interpretation of the object. One reveals how he intended to set his observations within a Copernican heliocentric framework. The other shows how he intended to use the expected parallactic shifts to confirm the Earth's revolution around the Sun: if the nova moved steadily on along a line, changes in the annual parallax could have falsified the geostatic hypothesis and confirmed the Copernican one. But no annual parallax could be detected (because of the nova's fading, and because of the insufficient measuring precision). Galileo later became fully aware that the actual distances to the stars were vastly greater than those commonly estimated. Nonetheless, even if the premises on which his hypothesis was built were wrong, his confirmation model was correct. Despite this, for almost 200 years after Galileo's death no-one was able to demonstrate the Earth's motion around the Sun by stellar parallax, as he brilliantly anticipated.

The cultural impact which the sudden appearance of the supernova had on the early-modern scientific mindset(s) was deep and unsettling, and proved key to the abandonment of the Aristotelian-Ptolemaic view of the world. Besides the distinguished contribution by Kepler, Galileo too (willing or not) played a star role in that historical transition.

Astronomy in Focus - XXX
Proceedings IAU Symposium No. XXX, 2018
M. T. Lago, ed.

doi:10.1017/S1743921319004034

What can Observations of Comets Tell Us about the Solar Wind at the Maunder Minimum?

N. V. Zolotova[1], Y. V. Sizonenko[2], M. V. Vokhmyanin[1] and I. S. Veselovsky[3,4,5]

[1]St. Petersburg State University, 198504 St. Petersburg, Russia
email: ned@geo.phys.spbu.ru
[2]Main Astronomical Observatory, National Academy of Sciences, 03143 Kyiv, Ukraine
[3]Lomonosov Moscow State University, 119991 Moscow, Russia
[4]Skobeltsyn Institute of Nuclear Physics, Lomonosov SU, 119991 Moscow, Russia
[5]Space Research Institute, Russian Academy of Sciences, 117997 Moscow, Russia

Abstract. This paper discussed whether 17th Century observers left historical records of the plasma tails of comets that would be adequate to enable us to extract the physical parameters of the solar wind. The size of the aberration angle between a comet's tail and its radius-vector defines the *type* of the tail: plasma or dust. We considered Bredikhin's calculations of the parameters for 10 comet tails observed during the Maunder minimum (1645 – 1715). For those comets the angle between the tail's axis and the radius-vector on average exceeded the value of 10° that is typical for dust tails. It was noted that visual observations of the ion tails of comets are very difficult to make owing to the spectral composition of their radiation, confirming the conclusion that observations of comet tails made in the 17th Century are not suitable for deriving past values of the physical parameters of the solar wind.

Keywords. history and philosophy of astronomy, (Sun:) solar wind, comets: general, comets

---

1. Introduction

The fact that historical reports of comets contain observations of plasma tails indicates the persistence of the solar wind phenomenon. Even before the discovery of the plasma tails, Bessel (1836) suggested that a repulsive force ejects particles from the cometary nucleus. The larger this force, the lighter the elements composing the cometary tail, and the closer the tail to the anti-solar direction. This idea was exploited by Bredikhin (1879a, 1879b, 1880, 1886), who developed a classification of cometary tails. Although the theory of the shape based on mechanical arguments is not valid for plasma tails, a qualitative description ("straight and narrow") matches their type I tail. In the Maunder minimum $\sim$ 20 comets were reported, most of them with nearly straight tails.

2. Orientation of cometary tails

Fig. 1 illustrates the trajectory of a comet with straight plasma and curved dust tails, in relation to its orbit around the Earth. ε is the angle between the plasma tail and the radius-vector passing through the Sun and the comet's nucleus, V_c is the orbital speed of the comet, and γ is the angle between V_c and the anti-solar direction. However, there are circumstances in which the dust tail may appear to the observer to be straight; those include a non-stationary outflow of cometary particles, a short dust

Table 1. Deviation ε of the cometary tail angle

Comet	Average ε	Comet	Average ε
C/1471 Y1	12° [1]	C/1664 W1	27°
C/1577 V1	12.5° – 33.5°	C/1665 F1	−2° – 12.9°
C/1580 T1	15°	C/1668 E1	Unknown
C/1582 J1	37.5°	C/1672 E1	Unknown
C/1618 W1	25°	C/1677 H1	Unknown
C/1652 Y1	11.7°	C/1680 V1	$\varepsilon > 10°$ [2]
C/1558 P1	10°	1P/1682 Q1	23° (Halley)
C/1661 C1	Unknown		

[1] 1472 Jan 20: $\varepsilon = 6°$, Feb 2: $\varepsilon = 18°$. [2] ε from 1° to 20°

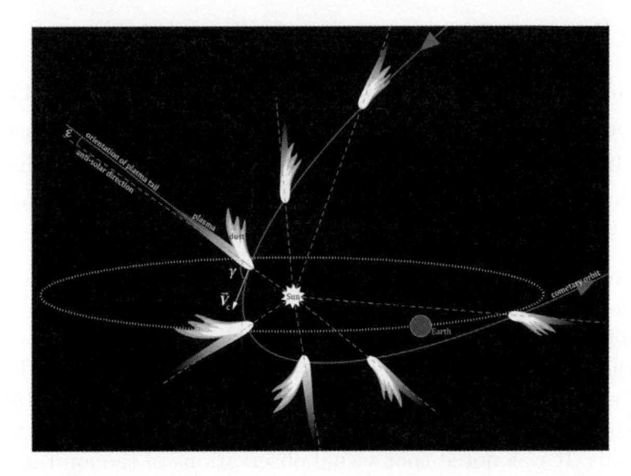

tail (less than $6° - 7°$), and the projection effect when an observer is close to the plane of the orbit. The latter also introduces an uncertainty in the calculation of the angle of aberration.

The reliable method (Mendis 2007) for defining the plasma tail is to evaluate ε, the ratio of the transverse component of the comet's orbital speed and the radial component of the solar wind flow: $\varepsilon \approx \arctan(|V_c| sin\gamma / |V_{sw}|)$. Typical values of the transverse velocity, ε, should be less than 6°. Larger deviations of the cometary tail from the anti-solar direction were recorded when an observer was able to report the dust tail. By applying the calculations by Bredikhin (1879a, 1879b, 1880, 1886), we can specify the type (dust or plasma) of the tails from the historical records. Tab. 1 lists the average angle of deviation of the tails from the radius-vector on dates before and during the Maunder minimum.

The orientation of tails is obviously not as accurate from drawings as from photographs. For four comets, the quality of observations proved to be too poor to determine the type of tail. For Great comet C/1665 F1, ε varied significantly, enough to prevent us from determining the type of its tail. According to Hevelius (1668), in 3.5 hours the orientation of the tail of Comet C/1652 Y1 changed by 18°, but a change of that amount is hardly possible even if the solar wind had storm-level gusts.

3. Conclusions

In spite of the poor quality of observations, the average deviation of tails from the anti-solar direction was found to exceed 10°, thereby suggesting that observers had actually reported dust tails rather than plasma tails. However, it is worth noting that plasma tails can be difficult to discern, because fluorescence of ionized gases is hard to detect by the unaided eye or even with a telescope.

Acknowledgments

This study received partial funding from the RFBR under project 16-02-00300-a. ISV is supported by grant RSCF 16-12-10062-a.

References

Bessel, F. W. 1836, Ann. Phys., 114, 498

Bredikhin, F. A. 1879a, *Ann, obs. Moscow*, 5, No. 2

Bredikhin, F. A. 1879b, *Ann. obs. Moscow*, 6, No. 1

Bredikhin, F. A. 1880, *Ann. obs. Moscow*, 7, No. 1

Bredikhin, F. A. 1886, *Ann. obs. Moscow*, *2nd* *ser.*, 1, No. 1

Hevelius, J. 1668, *Cometographia* (Danzig)

Mendis, D. A. 2007, in: Y. Kamide & A. C. -L. Chian (eds.), *Handbook of the Solar-Terrestrial Environment*, p. 494

Astronomy in Focus - XXX
Proceedings IAU Symposium No. XXX, 2018
M. T. Lago, ed.

doi:10.1017/S1743921319004046

Changes in the Unchangeable: Simulation of Transient Astronomical Phenomena with Stellarium

Georg Zotti[1] **and Alexander Wolf**[2]

[1]Ludwig Boltzmann Institute for Archaeological Prospection and Virtual Archaeology,
Hohe Warte 38, A-1190 Vienna, Austria
email: Georg.Zotti@univie.ac.at

[2]Altai State Pedagogical University, Barnaul, Russia
email: aw@altspu.ru

Abstract. The open-source desktop planetarium Stellarium has become very popular in astronomical education and outreach. Our recent changes aim for its applicability in historical and archaeoastronomical simulation contexts. Apart from visualizing the seemingly perpetual regular motions of the celestial bodies, it can be used to visualize and demonstrate historical solar and lunar eclipses, historical and present comets, meteors, and also novae and supernovae.

Keywords. Simulation, phenomenology, desktop planetarium, outreach

1. Introduction

One of the most popular desktop planetarium programs is the open-source Stellarium project started by Fabien Chéreau in 2001 (Zotti & Wolf 2018). The original purpose of the project was to visualise the current skies as realistically as possible. Although it was mostly geared towards contemporary amateur astronomers and laypeople, some of its original features, such as its easily exchangeable "sky cultures" (constellation patterns), as well as its simple usability and free availability, also made the program very attractive to researchers in cultural astronomy (i.e., ethnoastronomy and archaeoastronomy).

A powerful aspect of Stellarium is the possibility to extend the program with plug-ins for special — educational or scientific — purposes without deep changes in the core code. Several of those, including a 3D mode to explore potentially astronomically oriented buildings in their surrounding landscape, have been developed for purposes of visualisation and research in historical astronomy and archaeoastronomy (Zotti 2016).

In this paper we want to highlight several features which can be used to simulate for research, but also and especially for outreach, several transient astronomical phenomena: Solar and Lunar eclipses, comets, meteors, and Supernovae and Novae.

2. Simulation of Eclipses

Total Solar eclipses may well be described as nature's visually most impressive, yet still harmless, spectacle (in contrast to "true" disasters like earthquakes). An unprepared witness will easily be deeply impressed or even terrorized by the visually stunning phenomenon, which in contrast nowadays draws thousands of travellers around the globe.

The geometry between Moon and observer location on the surface of Earth (modelled as ellipsoid) automatically should cause eclipse geometry to be correct in the simulation. When the Moon covers the Sun, the amount of light illuminating the atmosphere is reduced and the sky becomes much darker. During totality, the brightest planets and stars are shown. We have also added a view of the solar corona. However, a complete simulation of the "circular twilight" during totality, or even of the approaching or receding "wall of darkness", has not been achieved so far.

A critical parameter for all eclipse simulations is the computation of ΔT, the irregular deceleration of Earth's rotation. In Stellarium, the user can select from more than 30 different models of ΔT, including a custom parabolic formula.

Lunar eclipses are likewise simulated by projecting Earth's shadow onto the Lunar sphere. During the eclipse, the sky brightening by the Full Moon is diminished and more stars are visible in the sky. The penumbra is indicated by a small minimum darkening which gets gradually darker, then the umbra border is represented by a steep darkening.

Currently, Stellarium does not compute particular numerical data for eclipses like Bessel elements, nor does it have an "eclipse finder". Contact times for the total solar eclipse of 2017-08-21 were within seconds of reality, and solar eclipses from Espenak & Meeus (2006) and Mucke & Meeus (1983) are well reproduced.

3. Simulation of Comets and Meteors

Bright comets used to surprise people all over the world, and most cultures saw them as bad *omina.* Their sudden appearance did not fit into the antique world view of heavenly bodies perpetually orbiting Earth on circular orbits or crystalline spheres, and only Tycho Brahe finally could show that they must be farther from Earth than the Moon.

Comets have been available in Stellarium pretty early, but they were only displayed as core with Coma. In 2014 we added a simple tail model based on Zotti (2001). It consists of two parabolic shells, one narrow and long which points straight away from the sun and represents the gas tail, and the other, which represents the dust tail, is wider and curved. The gas tail length and coma diameter follow the models from *Project Pluto* (n.d.). The curvature of the dust tail takes the comet's speed and distance from the Sun into account, and its strength relative to the gas tail can be parameterized in the orbital elements data file. The final appearance was crucially influenced by our own observations of the two Great Comets of the 1990s, C/1996 B2 Hyakutake and C/1995 O1 Hale-Bopp.

As is usual for this kind of software, position is computed from osculating orbital elements which are valid for a limited time only. Stellarium includes a collection of over 1000 element sets for historical comets. Orbital elements can be updated using the element service provided by the Minor Planetary Center (MPC).

Related to comets, Stellarium can also provide simulation of sporadic meteors and information (radiant location, activity) on meteor showers.

4. Historical Supernovae and Bright Novae

Only few bright supernovae exist in the historical records, yet for observers in earlier times, deeply rooted in the belief of the unchanging stellar sphere, a bright additional "guest star" must have been a very uncommon and miraculous sight. The idea of creating a tool for visualization of supernovae outbursts was born in a discussion between users and developers of Stellarium on a Russian astronomical forum in October 2010. This idea was obvious and yet none of the planetarium programs at that time had the ability to visualize supernovae outbursts. A first public release of the "Historical Supernovae" plugin was available near the end of the year 2010, and this version contained 10 bright historical supernovae observed by unaided eyes or with telescopes since 1006 (Tsvetkov & Bartunov 1993). By today (version 0.18.2 of August 2018) the list contains 47 supernovae brighter than 12.00^m at maximum. The JSON based format of our supernovae catalog can be updated through the Internet or modified by the user. Light curves of the historical supernovae cannot be obtained for obvious reasons, therefore we model the typical light curves for type I and type II supernovae. Those models are very simple, fast and limited to V-band observations of magnitudes (Utrobin 1986). For visual validation of models we used data from Green & Stephenson (2003) and Polcaro & Martocchia (2005).

Similar to the "Historical Supernovae" plugin, also bright Novae have been made available to simulation in Stellarium in 2013. We use a simple model for calculation of light curves based on decay time by N magnitudes from the maximum brightness value, where $N \in [2, 3, 6, 9]$. If a nova has no recorded values for decay of magnitude then this plugin will use generalized values for it. The list of novae is limited to magnitude 9.00^m and brighter at peak of brightness and currently (2018) contains 54 stars (Strope *et al.* 2010).

5. Conclusion and Further Work

In this short review we presented and discussed the possibilities of simulating transient astronomical phenomena for research and outreach in the popular open-source desktop planetarium Stellarium. Many more details are given in the User Guide which was also significantly enlarged in recent years (Zotti & Wolf 2018).

The application of Stellarium for simulation of historical skies has required several improvements in the accuracy of astronomical computation like including the long-time precession model by Vondrák *et al.* (2011) and accessing the DE430/431 solutions from Folkner *et al.* (2014). A known shortcoming is the current handling of stellar proper motion mentioned by De Lorenzis & Orofino (2018). Future developments should clearly include GAIA data and 3D proper motion.

Stellarium as a community project lives from voluntary contributions and donations. We invite contributions in code, data or just pointers to better models.

References

De Lorenzis, A. & Orofino, V. (2018), *Astron. and Comput.* **25**, 118–132

Espenak, F. & Meeus, J. (2006), *Five Mill. Can. of Sol. Eclipses: -1900 to +3000*, NASA

Folkner, W. M., Williams, J. G., Boggs, D. H., Park, R. S. & Kuchynka, P. (2014), The Planetary and Lunar Ephemerides DE430 and DE431, IPN Progress Report 42-196, JPL/NASA

Green, D. A. & Stephenson, F. R. (2003), Historical Supernovae, *in* Vol. 598 of *Lecture Notes in Physics*, Springer, Berlin, pp. 7–19

Mucke, H. & Meeus, J. (1983), *Canon of Solar Eclipses -2003 to +2526 / Canon der Sonnenfinsternisse -2003 bis +2526*, 2nd edn, Astron. Büro, Wien

Polcaro, F. & Martocchia, A. (2005), 'Supernovae astrophysics from Middle Age documents', *ArXiv Astrophysics e-prints*

Project Pluto (n.d.), online at https://www.projectpluto.com/update7b.htm#comet_tail_formula (Accessed Oct. 29, 2018)

Stellarium website (n.d.), online at https://stellarium.org (Accessed 2018-10-25)

Strope, R. J., Schaefer, B. E. & Henden, A. A. (2010), *Astron. J.* **140**, 34–62

Tsvetkov, D. Y. & Bartunov, O. S. (1993), 'Sternberg Astron. Inst. supernova catalogue', *Bull. d'Inf. du CDS* **42**, 17

Utrobin, V. P. (1986), Sverhnovye zvezdy, *in* R. A. Syunayev, ed., 'Fizika Cosmosa', Sovetskaya Encyclopediya, Moskva, pp. 600–607

Vondrák, J., Capitaine, N. & Wallace, P. (2011), *Astron. Astrophys.* **534**(A22), 1–19

Zotti, G. (2001), A Multi-Purpose Virtual Reality Model of the Solar System (VRMoSS), Master's thesis, TU Wien

Zotti, G. (2016), *Mediterranean Archaeology and Archaeometry* **16**(4), 17–24

Zotti, G. & Wolf, A. (2018), *Stellarium 0.18.2 User Guide*

Astronomy in Focus - XXX
Proceedings IAU Symposium No. XXX, 2018
M. T. Lago, ed.

doi:10.1017/S1743921319004058

Meteor Observations at Kazan Federal University (Russia)

D. V. Korotishkin[1], S. A. Kalabanov[1], O. N. Sherstyukov[1], F. S. Valiullin[1] and R. A. Ishmuratov[2]†

[1]Kazan Federal University, Kazan, Russia
e-mail: kazansergei@mail.ru

[2]Kazan State Power Engineering University, Kazan, Russia

Summary. This poster paper described initial results of recent meteor observations in Kazan obtained with a new meteor radar, SKiYMET. Significant improvements in the number of registrations are being recorded, enabling better statistics.

Radar observations of meteors have been carried out and archived since the mid-20th Century; at Kazan University (56°N, 49°E) they commenced in 1956, and have continued with only a few interruptions. A quasi-tomographic method (developed in-house in 2000) has been used to determine the coordinates of radiants of meteor showers, and to derive the orbital elements of small showers (microshowers).

The observations provide valuable information about the distribution of meteoric matter near the Earth's orbit. That information supports studies of stratospheric temperatures. A new radar, SKiYMET, was deployed in 2015 and measured parameters such as meteor speeds, but in 2016 its software was supplemented by an in-house package which performed more efficient pre-filtering of the primary data to eliminate non-meteor reflections from various sources. The new filtering algorithm has reduced substantially the threshold S/N ratio for detecting meteors (so fainter meteors can now be recorded), while the package also enables meteor velocities to be calculated much more efficiently than with SKiYMET, improving significantly the quality and statistical indicators of the processed data. By detecting a much larger number of meteors, it enables more detailed studies of the distribution of meteor velocities per day and per season.

Daily meteor counts for 2016–2017, extracted from the observations by the KFU programme, yielded significantly higher rates than with SKiYMET. These results support much better statistics, and also enable much finer details to be discerned. In particular, plots of meteor rates against speed are found to exhibit two maxima, one near 30 km s^{-1} and one near 55 km s^{-1}. This bifurcation in the relationship can also be discerned in the seasonal dependences of the speed distribution.

Improvements in the measurements of the speeds of meteor particles entering the atmosphere reveal an increase in the numbers with higher speeds ($> 50\,\text{km s}^{-1}$). The distributions of speeds and heights show a rather strong dependence of the meteor speed upon altitude; as predicted by meteor physics, those with higher speeds begin to burn at higher altitudes. A local increase in the number of meteors that have speeds of about 55 km s^{-1} has also been reported in the literature. The increase could be associated with heterogeneities in the distribution of meteoric matter in the vicinity of the Earth's orbit.

Keywords. Meteors and meteoroids, Meteor radar

---

† This work was funded by the subsidy allocated to Kazan Federal University for the state assignment in the sphere of scientific activities (project 3.7400.2017/8.9).

FM6
Galactic Angular Momentum

Astronomy in Focus - XXX
Proceedings IAU Symposium No. XXX, 2018
M. T. Lago, ed.

doi:10.1017/S174392131900406X

Focus Meeting: Galactic Angular Momentum

Danail Obreschkow[1,2]

[1]International Centre for Radio Astronomy Research (ICRAR), M468,
University of Western Australia, WA 6009, Australia

[2]ARC Centre of Excellence for All Sky Astrophysics in 3 Dimensions (ASTRO 3D)
email: danail.obreschkow@icrar.org

Abstract. The 6th Focus Meeting (FM6) at the XXXth IAU GA 2018 aimed at overviewing the rise in angular momentum (AM) science seen in the last 10 years and debating new emerging views on galaxy evolution. The foundational works on galaxy formation of the 1970s and 80s clearly exposed the fundamental role of AM, suspected since the time of Kant. However, quantitative progress on galactic AM remained hampered by observational and theoretical obstacles. Only in the last 10 years, numerical simulations began to produce galactic disks with realistic AM. Simultaneously, the fast rise of Integral Field Spectroscopy (IFS) and millimetre/radio interferometry have opened the door for systematic AM measurements, across representative samples and cosmic volumes. The FM bridged between cutting-edge observational programs and leading simulations in order to review, debate and resolve core issues on AM science, ranging from galactic substructure (e.g. gas fraction, turbulence, clumps) to global properties (e.g. size evolution, morphologies) and cosmology (spin alignment, cosmic origin of AM). The co-chairs and SOC members strived to assemble a representative selection of leading scientists in the field, while adhering to principles of equal opportunity and inclusivity.

Keywords. Galaxies: general – Galaxies: formation – Galaxies: evolution – Galaxies: galaxies: fundamental parameters – Galaxies: halos – Galaxies: bulges – Cosmology: theory.

---

1. Rationale: state of affairs motivating this meeting

In the standard model of galaxy formation, cold dark matter (CDM) haloes grow from primordial density fluctuations while acquiring AM through tidal torques. Galactic disks then condense at the halo centres by radiative dissipation of energy. The cooling baryons naturally exchange AM with their haloes, but the mass-size relation of local star-forming galaxies suggests that, on average, the specific AM of the baryons must remain approximately conserved. Explaining this apparent conservation has been a long-standing problem for theory: until recently (early 2010), hydro-gravitational simulations (using both particle-based and grid-based techniques) systematically failed at reproducing disks as large and thin as normal late-type galaxies, such as the Milky Way. The simulated galaxies were deficient in AM, making them too small and too bulgy – a problem so severe that it became known as the 'AM catastrophe'. Overcoming this catastrophe via increased computing power and refined feedback physics has been one of the major recent success stories of galaxy simulations. However, there is still debate about exactly how this is achieved: is outflowing gas torqued so that re-accreted gas has higher AM, is low AM material preferentially removed from galaxies, or do the winds prevent loss of AM by making inflows smoother?

A certain result from the AM catastrophe is that AM is one of the most critical quantities for explaining galaxy morphologies, opening a new bridge between theory and observation. The recent fast rise of IFS has enabled simultaneous measurements of the composition and Doppler velocity at every position in a 2D galaxy image, hence enabling a pixel-by-pixel integration of the AM. Such measurements of AM in early-type galaxies (ATLAS^{3D} survey, 2011) led to the surprising discovery that most of these seemingly featureless objects exhibit a rotational structure akin to that of normal spiral galaxies, thus containing more AM than previously suspected. The fewer actual 'slow-rotators' host up to ten times less AM at a fixed mass. AM thus offers a more fundamental, albeit harder to measure, classification of galaxy types than the classical Hubble sequence. This conclusion was cemented by recent AM measurements in spiral galaxies, again suggesting that the Hubble morphology sequence might be substituted for a more physical classification by AM. The precise form of this new AM-based classification scheme remains nonetheless a source of much argument. Many recent hydro-gravitational simulations (e.g. Illustris, EAGLE, Horizon, Magneticum, MAGICC, CLUES, NIHAO) contribute to this discussion, as do most major kinematic observing programs. Prominent examples include optical IFS/IFS-like surveys (e.g. ATLAS^{3D}, CALIFA, MaNGA, SLUGGS, PN.S, KROSS, SAMI Survey), interferometric radio surveys (e.g. THINGS on the VLA) and many other kinematic observations on modern and future instruments (e.g. KMOS, MUSE, SINFONI, HECTOR, ALMA, NOEMA, JWST, SKA and precursors).

The strong correlations between morphology and AM of local galaxies raises the question as to whether the cosmic evolution of morphologies is paralleled, or even driven, by the evolution of AM. Observationally, the Hubble Space Telescope's (HST) exquisite spatial resolution showed that star-forming galaxies at redshift $z > 1$ had very different structures to local grand-design spirals: The rapidly star-forming early galaxies showed a predominance of 'clumpy' and 'irregular' morphologies caused by super-giant (300 to 1000 pc) star-forming complexes. The physical origin of these clumpy morphologies and the processes that drive the large star formation rates are currently heavily debated. High-z IFS observations surprised with the finding that most of the clumpy star-forming galaxies have a regular, rotating disk structure. Interestingly, the emission line velocity dispersions appear to be about five times larger than in mass-matched local disks, which presents a major puzzle, because high velocity dispersions are predicted to stabilise the disks, preventing them from fragmenting into star-forming clumps. While high gas fractions could explain instabilities in spite of high dispersion, deep IFS studies (on Keck-OSIRIS, Gemini-GMOS) in rare nearby clumpy disks suggest that low AM is the dominant driver of instabilities. This motivates the arguable conjecture that the cosmic evolution of AM plays indeed a major role in the morphological transformation of the star-forming population – a hypothesis that was debated at this FM.

Answers to key questions regarding the cosmic evolution of AM are about to emerge from new high-z IFS observations on 8m-class telescopes (e.g. KMOS and MUSE on the VLT), as well as from an array of cosmological hydro-gravitational simulations (see above). Meanwhile multi-wavelength surveys are about to pile up evidence for strong correlations between AM and various baryonic processes (e.g. star formation rates, the transition from atomic to molecular gas and the growth of black holes). Moreover, ongoing and near-future surveys (see above) are about to expand AM science to smaller and larger scales: for the first time, enough spatially resolved velocity maps are available to systematically study the spatial distribution of the baryon AM in galaxies, which offers a nuanced test of different evolution models. On large scales, the number and spatial completeness of galaxies mapped using IFS are about to become sufficient to test the weak correlations between AM and cosmic large scale structure predicted by simulations.

In summary, observational and computational studies of AM have induced major progress in galaxy evolution theory over the last decade. The rich and fast evolving diversity of AM-related topics and the need for bringing observers and theoreticians together were the primary drivers for this FM at the XXXth IAU GA 2018.

2. Contributions: proceedings, talks, posters

2.1. *Proceedings*

All our proceedings are published and indexed online on Zenodo. A subset of forty pages of substantial reviews and summaries are additionally printed in Focus in Astronomy by Cambridge University Press.

Proceedings of oral presentations published on Zenodo and in Focus in Astronomy

- Danail Obreschkow, "Focus Meeting: Galactic Angular Momentum".
- Françoise Combes, "Angular Momentum – Conference Summary".
- P. J. E. Peebles, "On the History and Present Situation".
- Claudia del P. Lagos, "Angular Momentum Evolution of Galaxies: the Perspective of Hydrodynamical Simulations".
- Matthew Colless, "Emerging Angular Momentum Physics from Kinematic Surveys".
- Susana Pedrosa, "The Fundamental Physics of Angular Momentum Evolution in a ΛCDM Scenario".
- Filippo Fraternali and Gabriele Pezzulli, "Angular Momentum Accretion onto Disc Galaxies".

Proceedings of oral presentations published online on Zenodo

- Daniel DeFelippis, *et al.*, "Baryonic Angular Momentum in Simulated Disks: The CGM".
- Kareem El-Badry, "The Interplay between Galactic Angular Momentum and Morphology".
- Michael Fall & Aaron Romanowsky, "New Perspectives on Galactic Angular Momentum, Galaxy Formation, and the Hubble Sequence".
- Shy Genel, "A Lagrangian View on the Relation between Galaxy and Halo Angular Momentum".
- Chandrashekar Murugeshan, *et al.*, "Does Angular Momentum Regulate the Atomic Gas Content in H I-deficient Spirals?".
- Aura Obreja, "Galaxy Simulations after the Angular Momentum Catastrophe".
- Lorenzo Posti, "Angular Momentum-Mass Law for Discs in the Nearby Universe".
- Claudia Pulsoni, *et al.*, "The extended Planetary Nebula Spectrograph (ePN.S) Early Type Galaxy Survey: the Kinematic Diversity of Stellar Halos".
- Rhea-Silvia Remus, *et al.*, "Connecting Angular Momentum, Mass and Morphology: Insights from the Magneticum Simulations".
- Francesca Rizzo, "S0 Galaxies Are Faded Spirals: Clues from their Angular Momentum Content".
- Kanak Saha, "Angular Momentum Transport in Lopsided Galaxies".
- Sarah M. Sweet, *et al.*, "Spatially Resolved Galaxy Angular Momentum".
- Charlotte Welker, "Stellar Kinematics in the Cosmic Web: Lessons from the SAMI Survey and the Horizon-AGN Simulation".

2.2. *Oral presentations*

All oral presentations are available online at http://gam18.icrar.org. The abstracts of all oral presentations, other than summaries and addresses, are available online at https://astronomy2018.univie.ac.at/abstractsFM06.
Each proceeding listed in §2.1 corresponds to an oral presentation. The titles of the additional oral presentations are:

- Jenny Greene, "The role of AM in central and satellite ETGs".
- Xiaohu Yang, "Observing various alignment signals of galaxies ".

- James Bullock, "From halos to disks: the physics of AM profiles".
- Rhea-Silvia Remus, "Connecting AM, mass and morphology: insights from the Magneticum simulations".
- Michele Cappellari, "Surprises in kinematic observations of ETGs".
- Jean Brodie (presented by presented by R.-S. Remus, "Understanding the assembly histories of ETGs from the kinematics of stars and globular clusters".
- Jayaram Chengalur, "Expanding AM measurements to dwarf galaxies".
- Sarah Blyth, "Future high-redshift observations of H I kinematics".
- Lia Athanassoula, "AM and the evolution of disc galaxies".
- Susan Kassin, "The Assembly of Disk Galaxies: From Keck to JWST".
- Caroline Foster, "Spinning galaxies into shape".
- Rachel Somerville, "On the relationship between galaxy and halo size and spin".
- Martha Tabor, "Untangling galaxy components: the AM of bulges and disks in the MaNGA survey".
- Hoseung Choi, "Spin evolution of Horizon-AGN ETGs".
- Anelise Audibert, "Morphology and kinematics of the cold gas inside the central kiloparsec of nearby AGN with ALMA".
- Arianna Di Cintio, "Poster overview session".

2.3. *Posters*

The titles and abstracts of all posters are available online at `https://astronomy2018.univie.ac.at/PosterAbstracts/posterFM06` Posters were presented by Valentina Abril Melgarejo, Sung-Ho An, Aleksandra Antipova, Joan Font (for John Beckman), Sebastian Bustamante, Bernardo Cervantes Sodi, Horacio Dottori, Joan Font, Shy Genel, Jesus A. Gomez-Lopez, Katherine Harborne, Ivan Kacala, Eunbin Kim, Keiichi Kodaira, Baerbel Koribalski, Andrea Lapi, Jie Li, Katharina Lutz, Brisa Mancillas-Vaquera, Kyoko Onishi, Sol Rosito, Luis Enrique Prez Montao, Nicolas Peschken, Antonio J. Porras, Christoph Saulder, Felix Schulze, Yun-Kyeong Sheen, Shravan Shetty, Olga Silchenko, Matthias Steinmetz and Jolanta Zjupa.

3. People: organisers, presenters, attendees

The Scientific Organising Committee (SOC) consisted of 16 members assembled from 12 countries and 5 continents, including senior astronomers as well as earlier career researchers. Collectively, the SOC members led and won the original proposal to host this FM at the XXXth IAU GA, put forward the list of invited speakers, selected the contributed talks and posters, chaired most of the sessions during the meeting and edited the proceedings. The organisation of the local logistics of the entire General Assembly was handled by a dedicated LOC, headed by Gerhard Hensler, in conjunction with the IAU secretary and a national organising committee.

The SOC invited 20 high-level speakers for 20-min "review talks" and 14-min "feature talks" and selected 14 submitted proposals for 9-min "highlight talks". These talks were paralleled by a poster exhibition featuring the 31 highest-ranked additional submissions, out of 102 submissions in total (see Figure 2, right). According to NASA ADS, 10 of the 15 highest cited researchers on "Galactic AM" (and similar) of the last 20 years were present among the speakers and SOC members, making this meeting representative of the state-of-the-art.

Figures 1 and 2 (left) show a balanced mix of speakers in terms of gender and geographical distribution, bar a natural over-density in Europe, the site of many world-leading academic institutions and the location of the meeting. Figure 1 also reveals a fairly balanced mix of theoretical and observational expertise among the SOC members and speakers. Furthermore, the list of presenters included a wide range of seniority, spanning

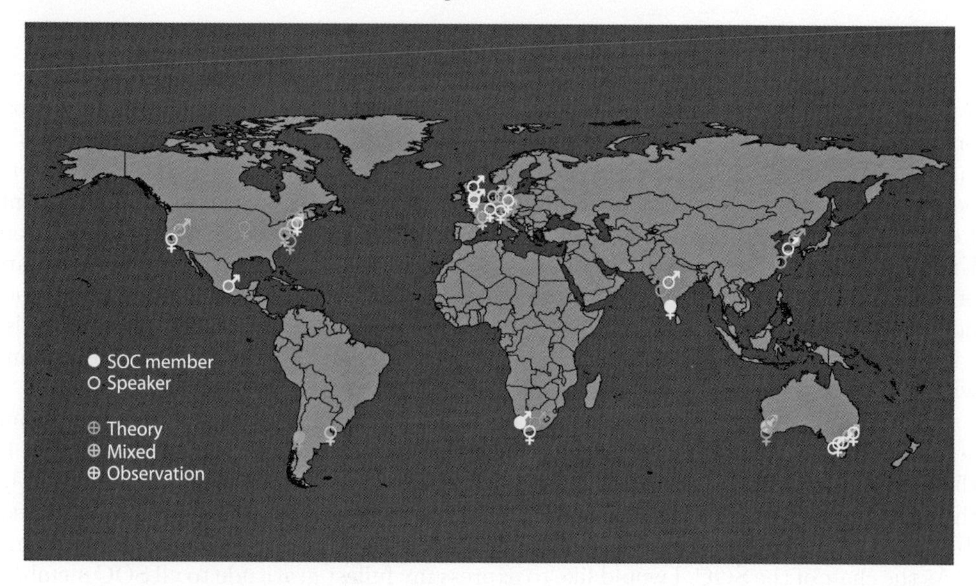

Figure 1. The geographical distribution of the SOC members and selected speakers, as well as their expertise, was fairly balanced, apart from a natural European clustering.

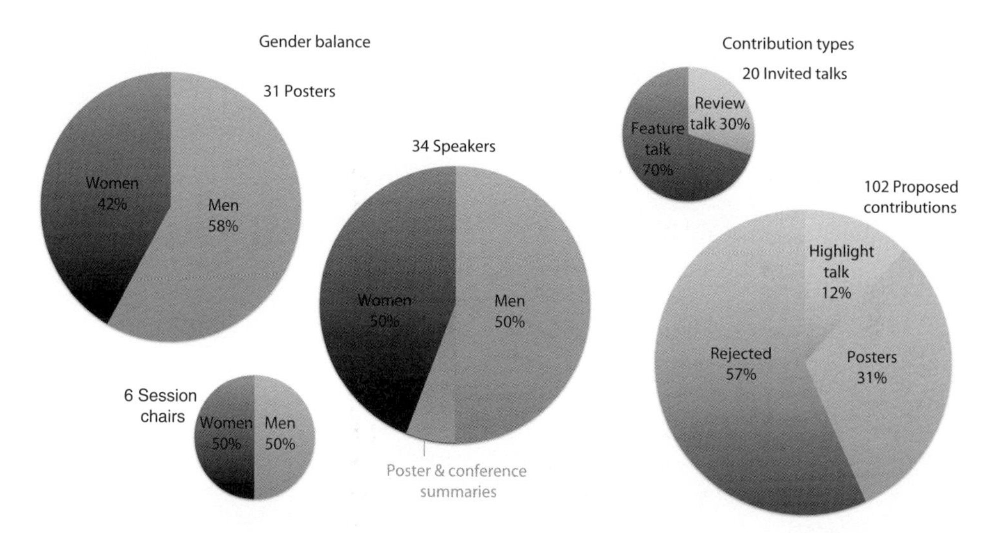

Figure 2. LEFT: An almost 50-50 gender balance was achieved for presenters and session chairs during the meeting. RIGHT: In addition to 20 invited talks, 102 proposal for additional talks and posters were received. Of those submissions, 31 were assigned a poster and 14 a short talk.

from early-career researchers, even first-time presenters, to very senior astronomers, most notably Prof James Peebles, one of the most influential theoretical cosmologists of the last 50 years and a key player in early AM science. Many PhD students and several Master students were selected for poster presentations.

In total, 374 participants of the General Assembly signed up for this meeting, but other participants were free to walk in. We estimate that 400–500 people attended parts of the meeting, with an average of 150 present at any one time.

4. Acknowledgements

On behalf of the whole SOC, I sincerely thank all our speakers and poster presenters for the high standard of their contributions. Many invited speakers, especially those presenting a review, made a respectable effort to paint a holistic picture of recent results and persisting questions. These presentations, which extended well beyond the presenters' personal research, were particularly important and deserve a special acknowledgement. I extend this gratitude to all those, whose proposal was not selected for a presentation. Undoubtedly, most rejected submissions would have been on par with the high standard of the meeting and would have enriched the scientific discussion. Those submissions helped us gauge the collective interest of our attendees and optimise the program. I also wish to thank the estimated 400–500 people who attended (parts of) this meeting and who contributed through challenging and interesting questions.

On behalf of all attendees and SOC members, I would like to send a warm thank you to the local organisers in Vienna, especially to Gerhard Hensler and his team, to the excellent technical crew at the convention centre, and to the overall coordination by the IAU, in particular to the then General Secretary Piero Benvenuti and his successor Teresa Lago, who is the chief editor of the proceedings published in Focus in Astronomy.

As the chair of the SOC, I would like to express my fullest gratitude to all SOC members for their enormous help from the early vision (mid 2016) to successful completion (late 2018) of this meeting – a two-year long process, which I would walk again with this driven team of brilliant scientists and reliable organisers.

Astronomy in Focus - XXX
Proceedings IAU Symposium No. XXX, 2018
M. T. Lago, ed.

doi:10.1017/S1743921319004071

Angular Momentum – Conference Summary

Francoise Combes

Observatoire de Paris, LERMA, College de France, CNRS, PSL Univ., Sorbonne Univ., F-75014, Paris, France
email: francoise.combes@obspm.fr

Abstract. Angular momentum (AM) is a key parameter to understand galaxy formation and evolution. AM originates in tidal torques between proto-structures at turn around, and from this the specific AM is expected to scale as a power-law of slope 2/3 with mass. However, subsequent evolution re-shuffles this through matter accretion from filaments, mergers, star formation and feedback, secular evolution and AM exchange between baryons and dark matter. Outer parts of galaxies are essential to study since they retain most of the AM and the diagnostics of the evolution. Galaxy IFU surveys have recently provided a wealth of kinematical information in the local universe. In the future, we can expect more statistics in the outer parts, and evolution at high z, including atomic gas with SKA.

Keywords. galaxies: bulges – galaxies: dwarf – galaxies: evolution – galaxies: formation – galaxies: general – galaxies: halos

---

This focus meeting has emphazised the high importance of angular momentum, to understand the formation and evolution of galaxies. It is now used extensively, given the progress of IFU instruments and large galaxy surveys.

Given these recent developments, it is difficult to imagine the debate that was occurring only 60 years ago, about the origin of the angular momentum of galaxies. The theory was first proposed by C. von Weizsäcker that galaxies were originating in large eddies of cosmic turbulence. This theory was followed by many people like G. Gamow, V. Rubin, his student or J. Oort.

Jim Peebles convinced Jan Oort that turbulence was irrelevant, that gravity and tidal torques could create the right amount of angular momentum (AM). For that he computed the torques with N-body simulations (N=90) and showed that the un-dimensional value of the AM $\lambda = \frac{J|E|^{1/2}}{GM^{5/2}} \sim 0.1$, in agreement from analytical estimations.

Since then, dark matter has been introduced, the problem is more complex, since we observe only the angular momentum of the baryons, which has to be related to the dark matter one. How are these acquired, how do they exchange?

The first cosmological simulations with baryons and dark matter, pointed out a serious problem, called the AM catastrophy: the baryons were losing their angular momentum through dynamical friction in mergers in favor of the dark matter, and were accumulating in very small disks at the bottom of the potential wells. Thanks to the feedback, and also the increase in spatial resolution of the simulations (lowering the effects of friction), the AM catastrophy is now limited (e.g. Obreja, Pedrosa and others, this meeting).

1. The "Fall" relation

In their pioneering study, Fall & Efstathiou (1980) take into account baryons and dark matter, which was only made of old stars at this epoch. Fall (1983) considers several scenarios of AM, mass or energy conservation, and concludes that the best scenario fitting the observations is that of baryonic mass M and AM conserved, while energy is dissipated. In this case, the specific angular momentum, i.e. j= J/M is a power-law function of mass, with slope 2/3. Several parallel lines can be traced, with the same slope in the logj-logM diagram, the highest one is for very late disk galaxies (Sc), while the early-type galaxies (ETG) fall below, due to their high velocity dispersion and low rotation (low V/σ). When only dark matter halos are concerned, the Virial relation combined with the hypothesis that all halos at any mass are formed out of a constant volumic density, leads to the power-law relation with slope 2/3.

Thirty years later Romanowsky & Fall (2012), and Fall & Romanowsky (2013) follow up using the much better determined AM and the much larger statistics provided by modern galaxy surveys. They show that the specific j can be used as a new classification scheme for galaxies, since all the Hubble sequence can be retrieved through parallel lines of 2/3 slopes in the logj-logM baryonic diagram. Many other versions of this diagram and classification were published (Obreschkow & Glazebrook 2014; Cortese *et al.*, 2016; Posti *et al.*, 2018; Sweet *et al.*, 2018).

All these studies led to consider a third parameter in the AM scaling relation: the relation can be viewed in a 3-dimension space, where the third axis is the bulge to total mass ratio B/T (Fall & Romanowsky 2018, also Obreschkow & Glazebrook 2014). The scaling relation M-j-B/T can then be retrieved from the well known Tully-Fisher relation for spirals, and fundamental plane for early-type galaxies. together with a structure relation (for instance the Freeman's relation $M \propto R^2$ for high-surface brightness spirals).

2. ΛCDM hydro numerical simulations

In the recent years, there has been a burst of simulation papers, interested in following angular momentum, as described by Susana Pedrosa in her review (Pedrosa & Tissera 2015; Genel *et al.* 2015; Teklu *et al.* 2015; Obreja *et al.* 2016, 2018; Lagos *et al.* 2018). Although the most realistic simulations, including star formation and feedback, have solved the AM catastrophy (through the effect of feedback and higher spatial resolution), they have revealed that the scaling relations of specific AM (j) versus baryonic mass are flatter than those observed. The various galaxies follow parallel lines in the logj-logM baryonic diagram, with the B/T parameter increasing towards the bottom right, but the slope of the lines are nearly 1/3.

Although the stellar feedback helps to solve the AM catastrophy, it also excessively thickens galaxy disks. Simulations still predict too massive bulges, and feedback is not sufficient to produce the large number of observed bulgeless galaxies.

James Bullock remarked that very different results (especially in density and temperature) can be obtained in general in cosmological simulations when using different codes, different algorithms (Eulerian, Lagrangian), different resolutions, different recipes for star formation and feedback. However, the results on angular momentum, either of stars (j_*) or gas (j_{gas}) are converging!

Due to dissipation, gaseous filaments are much thinner than dark matter filaments. This means that even before matter enters into galaxies, the specific AM of baryons is 3 times higher than the specific AM of dark matter. This changes the initial conditions in general adopted in semi-analytical models, where baryons and dark matter are assumed to have gained the same specific j through tidal torques. The virial radius R_V changes a lot with time, it increases by a factor $\sim$ 3 from z=1 to z=0. Since $j \propto \lambda R_V$, it is still possible that the size of baryonic disks are the same at the end. The final j will depend on the AM of the gas accreted in the mean time.

The size ratio between the stellar and dark matter components decreases with time for low M, this was not reproduced before by the semi-analytical models. Now abundance matching is considering sizes, as Rachel Somerville showed in her talk.

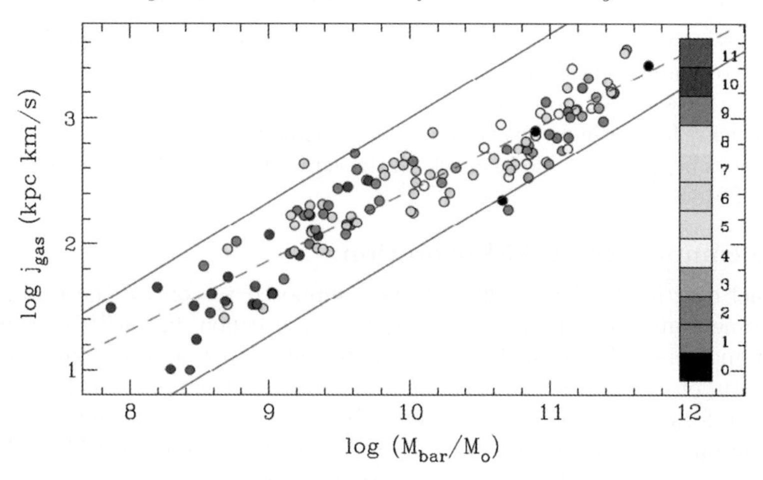

Figure 1. The specific gas angular momentum $j_{gas} \propto R_d\ V_[flat$, versus the baryonic mass $M_{bar}= M_* + M(HI)$, from the 175 spiral galaxies of the SPARC sample of Lelli *et al.* (2016). The atomic gas is rotating maximally (negligible velocity dispersion), and the diagram should follow the upper envelope with a slope 2/3. In fact, the best fit has a slope of 0.55. The colour indicates the galaxy type, 0 being a lenticular, then Sa, Sab .. 9 is Sm, 10 Im and 11 BCD.

3. Why such a scaling relation?

The observation of the $\log j_*$ - $\log M_*$ scaling relations in parallel lines with a slope 2/3 is not straightforward to interpret. The first predictions were done with the total matter, and can be applied essentialy to the dark matter, but it is not obvious why the stars would follow the same relation.

Posti *et al.* (2018a,b) have proposed some biased collapse scenario, to explain why the baryons do not retain all their initial angular momentum. However, the scenario must be rather contrived. Indeed, to derive from the dark matter relation $j_{DM} = J_{DM}/\ M_{DM} \propto M_{DM}^{2/3}$, the equivalent relation for stars, $j_* \propto f_j\ f_*^{-2/3}\ M_*^{2/3}$, we must assume that the product $f_j\ f_*^{-2/3}$ = cst, with $f_j = j_*/j_{DM}$ and $f_* = M_*/M_{DM}$. This last ratio is the well known fraction of stellar mass in a galaxy, which is much below the universal baryon fraction f_b=17%. From abundance matching, this function peaks for halos of the Milky Way mass, and then falls steeply on each side by 2 orders of magnitude (e.g. Behroozi *et al.* 2010). To interpret the AM observations, we should explain why the f_j ratio has the same behaviour, more exactly $f_j \propto f_*^{2/3}$. the biased collapse scenario proposed by Posti *et al.* (2018b) requires that the outer parts of halos, rich in AM, fail to accrete on the galaxy to form stars. This requirement looks like a conspiracy!

May be the specific AM of baryons does not always follow the scaling relation with slope 2/3. When dwarfs dominated by dark matter and gas are considered, the slope is more near 0.5, as shown in Figure 1.

4. Exchanges of AM – Secular evolution

During galaxy evolution, angular momentum is not frozen either in the baryons or dark matter, but their fraction may vary. AM can be exchanged through spiral arms within the disk, which produces radial migration. Some density breaks in the radial distribution of stars can be attributed to these processes (Athanassoula 2014, Peschken *et al.* 2017). Bars exchange AM with the dark halo, enhancing the formation of bars, which are waves of negative angular momentum. Bars can also be destroyed through torquing the gas, which is driven to the center.

It is interesting to follow AM along cosmic filaments. Galaxies have special orientations with respect to filaments: spirals have their spin parallel to them, while ellipticals, coming from mergers of spirals, have their spin perpendicular to them. The fraction of fast rotators (at least faster than the average) is increasing with the distance to the filaments. Galaxy surveys begin to be able to check all these predictions. (Welker *et al.* 2014, Xiaohu Yang *et al.* 2018)

5. Large complexity in AM evolution

Shy Genel described a long long equation, supposed to control the evolution of the angular momentum, and follow its evolution along a galaxy life, with matter accretion and major mergers. All parameters have to be taken into account, such as the stars formed in situ, or ex-situ, the gas forming stars, and what happens during the feedback, the new star formation from the gas lost, the gas accretion, the minor mergers, the radial migration, the AM exchange with DM. All this is far from the AM prediction from torques at turn-around, and the scaling relation of $j \propto M^{2/3}$.

How can we explain this miracle?

First the envelope at high j applies to pure disks, with 100% efficiency to retain AM. This is relatively obvious if material is almost in circular orbits: this plays the role of an attractor (see the talk from Francesca Rizzo, and Rizzo *et al.* 2018). Then you depart progressively from this attractor, as soon as you form bulges, spheroids, heating the stellar component, without the possibility of gas cooling.

6. Apparent contradictions

AM is a proxy for morphological types, as Fall & Romanowsky (2013) proposed. It is also well known that morphological types are segregated by the densiy of environment (Dressler *et al.*, 1980). Spirals are dominating in the field, while their abundance decreases at high galaxy density in favor of lenticulars and ellipticals. Michele Capellari (2016) in his review article proposes to apply this segregation with density to fast and slow rotators, to replace the spiral/elliptical classification. And indeed, slow rotators are found at density peaks in clusters and groups.

But in her talk, Jenny Greene claimed that there is no evidence of environment effect on the AM of early-type galaxies (Greene *et al.*, 2018). This is obtained from many surveys (MASSIVE, SAMI, MANGA), and the AM depends only on mass.

Another issue when considering AM, is to know whether studies are extending enough in radius. As described beautifully by Matthew Colless, we are witnessing a golden age for kinematical studies of galaxies, with integral field units (IFU) large surveys (Atlas3D, SAMI, CALIFA, MANGA etc.). However, large numbers (thousands) of galaxies are observed only to Re, and hundreds to 2Re. In general you need HI surveys to reach the flat portion of rotation curves, richer in AM.

In the optical, the kinematics of Globular Clusters (GC) show that the spin and ellipticity increase in S0, while they drop in Ellipticals with radius, as described with the SLUGGS survey by Jean Brodie (Brodie & Romanowsky 2016). With Planetary Nebulae (PNe) Pulsoni *et al.* (2018) go much further in radius, to 15-20 Re, where all the AM and signatures of the galaxy formation subsist. There is a large diversity of situations for ETG. Some slow rotators begin to rotate in the outer parts, and among fast rotators, 70% slowly rotate in the outer parts.

The transition radius between in-situ and ex-situ material is $\propto 1/M_*$: i.e. there is more ex-situ material in massive galaxies, formed through mergers. This is perfectly compatible with Illustris simulations (Rodriguez-Gomez *et al.* 2016).

Lagos *et al.* (2018) have measured in detail through simulations how galaxies gain and lose AM by matter accretion and mergers. Dry mergers reduce specific j by 30%, while wet mergers inscrease j by 10%.

7. Atomic gas and dwarfs

As shown in Obreschkow *et al.* (2016) and in Murugeshan's talk, the angular momentum has a large influence in the stability of spiral galaxies and their HI gas fraction. The stability criterion can be written as q = j σ_v/GM $\propto M^{-1/3}$, and the HI gas fraction f_{atm} is AM-regulated and also $\propto M^{-1/3}$. A related study by Romeo & Mogotsi (2018) on stability and AM regulation includes the thickness of the stellar disk T_*, i.e. $Q_* \sim \sigma_v\ T_*$.

In Chengalur's talk, another discrepancy between simulations and observations was revealed for dwarf galaxies: the specific AM of baryons j_b increases below a baryonic mass of $10^{9.1} M_\odot$, with respect to the $M^{2/3}$ expected scaling relation (Kurapati *et al.* 2018). For these dwarfs, disks become thicker due to star formation feedback, and to the shallow potential well. There is no dependency on large-scale environment, so this is not due to possible accretion. Another explanation is that such dwarfs are dominated by dark matter, therefore their observed rotational velocity is much higher with respect to their visible mass (M_{bar}) than for spiral of larger masses.

In FIRE simulations, dwarfs have very low rotational support: the large SF feedback gives them a rounder shape (El-Badry *et al.* 2018), and their specific j falls below the $M^{2/3}$ scaling relation.

8. Perspectives

May be all diagnostics of galaxy evolution are retained in the outer parts: accretion, ex-situ star formation, etc. In that case PNe are the best tracers of AM and evolution. It is of prime importance to acquire more statistics, for instance in the Hector IFS survey, 10^5 galaxies will be obtained. Also other parameters must be followed, metallicity, stellar populations (see Kassin's talk).

With ELT and JWST, it will be possible to track the evolution with redshift. We know already that galaxies become clumpy at $z > 2$ and have much lower j_*. While it is predicted that $j_* \sim (1+z)^{-1/2}$ (Obreschkow *et al.* 2015), F. Fraternali in his talk found no evolution with z. [*Remark by the co-editor, Danail Obreschkow: It seems important to bear in mind that the approximation* $j \propto (1+z)^{-1/2}$ *for dark haloes of fixed mass only applies to the matter-dominated era* ($z \gtrsim 1$). *Explicitly, neglecting the weak evolution of* j *expected from the cosmic evolution of the virial over-density* $\Delta_{\rm vir}$, *we find* $\langle j/M^{2/3}\rangle \propto E^{-1/3}(z) = (\Omega_m(1+z)^3+\Omega_\Lambda)^{-1/6}$, *which asymptotes to* $\Omega_m^{-1/6}(1+z)^{-1/2}$ *for* $z\to\infty$, *but Taylor-expands to* $(1+z)^{-\Omega_m/2} \approx (1+z)^{-0.15}$ *around* $z=0$.]

It is also paramount to study external accretion of gas, which contains a lot of AM, is at the origin of warps, etc. HI maps are badly needed at intermediate and high z; in the future SKA will provide a large number of these gas maps.

References

Athanassoula, E.: 2014, *MNRAS*, 438, L81

Behroozi, P. S., Conroy, C., Wechsler, R. H.: 2010, *ApJ*, 717, 379

Brodie, J., Romanowsky, A. J.: 2016, *IAUS*, 317, 190

Bullock, J. S., Boylan-Kolchin, M.: 2017, *ARA&A*, 55, 343

Capellari, M.: 2016, *ARA&A*, 54, 597

Cortese, L., Fogarty, L. M. R., Bekki, K. *et al.*: 2016 *MNRAS*, 463, 170

Dressler, A.: 1980 *ApJ*, 236, 351

El-Badry, K., Quataert, E., Wetzel, A. *et al.*: 2018, *MNRAS*, 473, 1930

Fall, S. M., Efstathiou, G.: 1980, *MNRAS*, 193, 189

Fall, S. M.: 1983, *IAUS*, 100, 391

Fall, S. M., Romanowsky, A. J.: 2013, *ApJ*, 769, L26

Fall, S. M., Romanowsky, A. J.: 2018, *ApJ*, in press, arXiv180802525

Genel, S., Fall, S. M., Hernquist, L. *et al.*: 2015, *ApJ*, 804, 40L

Greene, J. E., Leauthaud, A., Emsellem, E. *et al.*: 2018, *ApJ*, 852, 36
Kurapati, S., Chengalur, J. N., Pustilnik, S., Kamphuis, P.: 2018, *MNRAS*, 479, 228
Lagos, C., Stevens, A. R. H., Bower, R. G. *et al.*: 2018, *MNRAS*, 473, 4956
Lelli, F., McGaugh, S. S., Schombert, J. M.: 2016, *AJ*, 152, 157
Obreja, A., Stinson, G. S., Dutton, A. A. *et al.*: 2016, *MNRAS*, 459, 467
Obreja, A., Dutton, A. A., Maccio, A. V. *et al.*: 2018, *MNRAS*, in press, arXiv180406635
Obreschkow, D., Glazebrook, K.: 2014, *ApJ*, 784, 26
Obreschkow, D., Glazebrook, K., Bassett, R. *et al.*: 2015, *ApJ*, 815, 97
Obreschkow, D., Glazebrook, K., Kilborn, V., Lutz, K.: 2016 *ApJ*, 824, L26
Pedrosa, S. E., Tissera, P. B.: 2015, *A&A*, 584, A43
Peschken, N., Athanassoula, E., Rodionov, S. A.: 2017, *MNRAS*, 468, 994
Posti, L., Fraternali, F., Di Teodoro, E. M., Pezzulli, G.: 2018a, *A&A*, 612, L6
Posti, L., Pezzulli, G., Fraternali, F., Di Teodoro, E. M.: 2018b, *MNRAS*, 475, 232
Pulsoni, C., Gerhard, O., Arnaboldi, M. *et al.*: 2018, *A&A*, in press, arXiv171205833
Rizzo, F., Fraternali, F., Iorio, G.: 2018, *MNRAS*, 476, 2137
Rodriguez-Gomez, V., Pillepich, A., Sales, L. V. *et al.*: 2016, *MNRAS*, 458, 2371
Romanowsky, A. J., Fall, S.M: 2012, *ApJS*, 203, 17
Romeo, A. B., Mogotsi, K. M.: 2018, *MNRAS*, 480, L23
Somerville, R. S., Behroozi, P., Pandya, V. *et al.*: 2018, *MNRAS*, 473, 2714
Sweet, S. M., Fisher, D., Glazebrook, K. *et al.*: 2018, *ApJ*, 860, 37
Teklu, A. F., Remus, R.-S., Dolag, K. *et al.*: 2015, *ApJ*, 812, 29
Welker, C., Devriendt, J., Dubois, Y.; Pichon, C., Peirani, S.: 2014, *MNRAS*, 445, L46
Yang, X., Zhang, Y., Wang, H. *et al.*: 2018, *ApJ*, 860, 30

Astronomy in Focus - XXX
Proceedings IAU Symposium No. XXX, 2018
M. T. Lago, ed.

doi:10.1017/S1743921319004083

On the History and Present Situation

P. J. E. Peebles

Princeton University,
Princeton New Jersey, 08544, USA
email: pjep@princeton.edu

Abstract. A common thought in the 1950s was that galaxies rotate because they are remnants of primeval currents, as in turbulence. But this idea is quite unacceptable in an expanding universe described by general relativity theory. Since we are no smarter now than in the 1950s the lesson I draw is that we do well on occasion to pause to consider whether we might be missing something. An example is the pure disk galaxies that are so common nearby and so rare in simulations. We have something to learn from this.

Keywords. galaxies: bulges, galaxies: formation, cosmology: theory.

Carl Friedrich von Weizsäcker (1951) argued for the likely importance of turbulence in protoplanetary disks, and he proposed that turbulence may also play a role in galaxy formation: "I do not propose a theory of the origin of this initial turbulence" but "it is a consistent theory to think of the galaxies (or perhaps the clusters of galaxies) as the largest eddies of a cosmic turbulence that existed a couple of billion years ago." Von Weizsäcker's student Sebastian von Hoerner had a distinguished career in radio astronomy and SETI. In a paper while still at the University of Gottingen, von Hoerner (1953) considered how matter might be distributed in a galaxy that grew out of a turbulent eddy. He concluded that (in my translation) "Since we have obtained qualitatively consistent statements on the surface density run in spiral nebulae in three completely different ways, we will consider this result as an argument for the applicability of the assumed turbulence theory."

George Gamow was an intuitive genius but not always careful with details. He felt the present mean mass density is far too low for the gravitational assembly of galaxies. He and Teller had proposed that this meant galaxies formed at high redshift, when the density would have been more important. But Gamow (1954) argued that "The only escape from this difficulty is to assume the existence of very large original density fluctuations in the primordial gas ... in a turbulent state. Besides, in order to permit large variations of density, this turbulence must have been supersonic." Gamow's student Vera Rubin had a distinguished career in astronomy, with particular attention to the informative role of rotation curves of galaxies. In her PhD thesis she introduced and applied statistical measures of the galaxy distribution; she pioneered an important line of research. Rubin (1954) concluded that her statistical measures are "physically reasonable if the galaxies have condensed from a turbulent gaseous medium."

The relation between student and teacher can be complicated. We can only wonder how enthusiastic these two students were about their evidence for primeval turbulence.

In an important review Jan Henrik Oort (1958) showed that the empirical support for cosmology was scant but by no means trivial. Oort argued that "Most galaxies deviate greatly from the spherical shape and have a considerable angular momentum. The total angular momentum must have been present in the primeval clump of material from which the galaxy has contracted." Turbulence is not mentioned, but I count Oort's picture as a variant of the concept of primeval currents.

Leonid Ozernoi was a leading figure in exploration of the primeval turbulence picture in the early 1970s. We agreed that primeval turbulence calls for irrotational flow, $\nabla \times \vec{v} = 0$. This is because currents with nonzero divergence would produce density fluctuations that (with a reasonable mass density) would grow by gravitational attraction, the usual gravitational instability picture. Suppose the primeval irrotational flow has comoving coherence length y, or physical length $a(t)y$ at expansion time t, where $a(t)$ is the expansion parameter. How the flow behaves depends on the ratio

$$R(t) = \frac{v(t)t}{a(t)v}, \text{ where } R(t) \propto a(t) \text{ if } p = p/3, \; R(t) \propto a(t)^{-1/2} \, ifp = 0. \tag{1}$$

At redshift $z > z_{\rm eq} = 5000$ the universe was dominated by radiation and $R(t)$ would have been growing. If $R(t)$ approached unity flows moving in different directions would have encountered each other and been forced to change direction. That is, turbulence would have formed and decayed to viscosity. But this is far too early for galaxy formation. At $z < z_{\rm eq}$ the ratio $R(t)$ decreases, so if turbulence had not developed by then it would not develop. I concluded (Peebles 1971a) that the primeval turbulence theory is not viable. Ozernoi (1972) disagreed (by long distance, he was in the USSR). Interest in primeval turbulence continued through the 1970s, but attention was turning to the role of gravity.

At the conference where von Weizsäcker (1951) spoke about primeval turbulence Fred Hoyle (1951) proposed that gravitational transfer of angular momentum caused galaxies to rotate. He evidently was unaware that Gustaf Strömberg (1934) had expressed similar thoughts some two decades earlier. (I am grateful to Matthias Steinmetz for alerting me to Strömberg directly after my IAU lecture.) As Hoyle put it, a young protogalaxy would have been an irregular blob that could be torqued by the gravitational field gradients of neighboring blobs, transferring angular momentum. His estimate of the effect suggested this is a credible explanation of why galaxies rotate.

The NASA archive ADS lists no citations to Hoyle's paper for the next two decades, and no citations to Gustaf Strömberg (1934) until 1995. The rate of research in cosmology through most of the 20th century was modest. Another example is the exchange with Werner Heisenberg after Hoyle's talk:

Heisenberg: "How can an irregular thing like a cloud have originated otherwise than as a consequence of turbulent motion?"

Hoyle: "A cloud can form in a more or less uniform medium through gravitational instability."

Heisenberg: "This possibility exists, but wouldn't you say that if we believe in the expanding universe (I know that some of us do not but that is another matter), then we should also assume that there is an enormous energy in this primary cosmic gas which expands? Now, if there is this enormous kinetic energy of the gaseous masses, I suppose there must be turbulence, because the turbulent motion is the normal motion of the gas, whereas laminar flow is extremely exceptional."

In the late 1960s I was taken by the idea that galaxies, and their clumpy spatial distribution, grew by the gravitational instability of the expanding universe. I had to explain why galaxies rotate. I did not know Hoyle's 1951 argument then, but worked along similar lines in my computation of the gravitational angular momentum transfer. The analytic estimate in Peebles (1969) amounts to

$$\lambda \equiv \frac{L|E|^{1/2}}{GM^{5/2}} \sim 0.08. \tag{2}$$

Here L is the angular momentum of the newly assembled protogalaxy, M is its mass, and E is its gravitational binding energy. (This combines eqs. [35] and [36] in Peebles 1969. I introduced λ in Peebles 1971b).

Oort (1970) argued that I had seriously overestimated the gravitational transfer of angular momentum, and concluded that galaxies "must have been endowed with their angular momentum from the beginning." That led me to compute the angular momentum transfer in numerical N-body simulations. They indicated $\lambda = 0.07^{0+1}_{0.03}$. These simulations had $N = 90$ to 150 particles. This is ludicrous by today's standard, but that was a different age. In this paper I ventured to add that since λ is a pure number set by the scale-invariant physics of gravity and a pressureless ideal gas one might expect λ to be of order unity. What other value might it have?

I don't know which of my arguments was the more persuasive, but after publication of this paper Oort sought me out to explain, not at length but quite clearly, that he withdrew his objection to my result. It was an edifying example for this callow youth.

Efstathiou and Jones (1979) found $\lambda = 0.07 \pm 0.03$ in simulations with $N = 1000$ particles. That number also is tiny by today's standards, but far better than I did, and I suppose large enough to make the case: gravity in an expanding Einstein-de Sitter universe produces angular momentum in the neighborhood of $\lambda \sim 0.1$

The next advance was the proposal, independently by White and Rees (1978) and Gunn, Lee, Lerche, *et al.* (1978), that the luminous parts of galaxies formed by dissipative settling of baryonic gas and plasma in subluminal massive halos. (Gunn *et al.* had in mind halos of nonbaryonic matter, later known as WIMPS. White and Rees felt that remnants of early stellar generations are more likely forms of subluminal matter, but that nonbaryonic dark matter would do.) Neither paper mentions rotation of galaxies.

Gustaf Strömberg (1934) and Fall (1979) pointed out that dissipative settling could spin up a young galaxy. Fall could be more explicit: it could bring $\lambda \sim 0.1$ up to the value $\lambda \sim 1$ suitable for the disk of a spiral galaxy. Fall and Efstathiou (1980) elaborated on this point. Let v_c be the speed of rotation of a newly gravitationally assembled protogalaxy, and let $\mathrm{v}_r \sim (GM/R)^{1/2}$ be its internal speed of support, mainly random. The angular momentum parameter is, roughly, $\lambda \sim v_c/v_r$. Suppose that after assembly the bulk of the mass dissipatively settled by the factor α. The rotation speed would scale up as $v_c \propto \alpha$, and the speed of pressure support would increase as $v_r \propto \alpha^{1/2}$. So we see that the protogalaxy would have to have collapsed by a factor $\alpha \sim \lambda^{-2} \sim 100$ to get to rotational support. That seems excessive. Surely the protogalaxy would instead fragment into something like an elliptical galaxy. But if instead the diffuse baryons settled in a subluminal massive halo with density run $\rho \propto r^{-2}$, and the diffuse baryon mass were subdominant, then spin-up would require settling by the factor $\alpha \sim \lambda^{-1} \sim 10$. This factor of ten is about what Eggen, Lynden-Bell, and Sandage (1962) found could account for the metal-poor high-velocity stars in the Milky Way. It's a valuable sanity check.

By 1980 gravitational transfer of angular momentum had become the standard and accepted explanation of why galaxies rotate. That was followed by the development of increasingly detailed numerical simulations of how baryons and dark matter gather by gravity and non-gravitational stresses in all their complexities to produce what are now impressively good approximations to real galaxies. I don't imagine much attention is given to λ anymore; the simulations take care of it.

If primeval turbulence is so manifestly wrong, as I argue, why was the idea so commonly accepted in the 1950s and 1960s? There was phenomenology: spiral galaxies call to mind turbulent eddies. The expanding universe was familiar, but not so carefully considered, or so well trusted, as to make the idea of primeval turbulence seem suspicious. Recall Heisenberg's remarks. And we must bear in mind that ideas can be self-reenforcing: people analyzed primeval turbulence because others had.

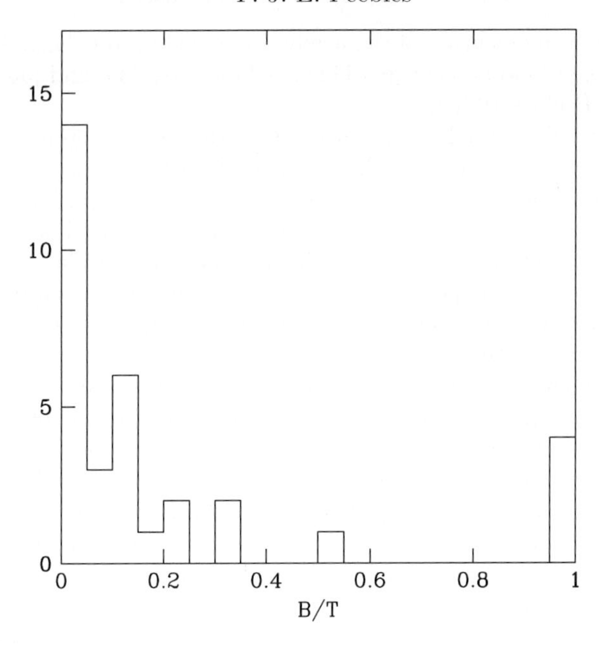

Figure 1. Distribution of bulge to total luminosities of the largest galaxies within 10 Mpc.

I began with the thought of pausing on occasion to consider whether, as for primeval turbulence, we are overlooking something. I offer the paper by Kormendy, Drory, Bender, and Cornell (2010), with the title *Bulgeless Giant Galaxies Challenge Our Picture of Galaxy Formation by Hierarchical Clustering.* I suggest that those who do not often admire the wonderfully detailed images of nearby galaxies to be seen on the web contemplate NGC 1300. (The reader is referred to the high-resolution HST image.) I don't see a classical bulge; the central region looks like a whirlpool. Would an HST image reveal its funnel? or maybe a star cluster? Also worth contemplating is the image of M 101 in Figure 3 in Peebles (2014), adapted from Kormendy *et al.* (2010). It has a central star cluster with luminosity about a part in 10^5 of the galaxy. The galaxy spiral arms are seen all the way in to this relatively tiny star cluster. I suppose gravity in the disk produces the spiral arms, and so expect that the central part of this galaxy cannot be dominated by mass in a classical stellar bulge.†

The variants of the pure disk phenomenon seen in NGC 1300 and M 101 seem to be common among the nearest large galaxies that can be examined in closest detail. Brent Tully's Local Universe catalog‡ lists 38 galaxies closer than 10 Mpc with K-band luminosity greater than 10^{10}. (I can't find K-band luminosities for a few, but they have large optical luminosities.) The fractions B/T of galaxy luminosities that are in classical bulges are given by Kormendy Kormendy *et al.* (2010) and Fisher and Drory (2011) for 33 of these 38 galaxies. Figure 1 shows the distribution. Three ellipticals are at the far right. Near the center is the Sombrero Galaxy, also well worth a visit to the web. The

† I might state my understanding that a classical bulge is supported by near isotropic stellar velocities that cause the stars to move in a near axisymmetric distribution that may rise above the disk. Stars rise above the disk in the peanut-shaped bar in the Milky Way, but I suppose this would not be termed a classical bulge. I am cautioned that a bulge luminosity derived from the excess above a pure exponental fit to the surface brightness run may be in a classical bulge, or it may be a departure from an exponential distribution of the stars moving in the disk.

‡ Available at the Extragalactic Distance Database, `http://edd.ifa.hawaii.edu`, as the catalog "Local Universe (LU)"

galaxies M 31 and M 81 are next to the left of center. The rest of these large galaxies are still further to the left, most judged to have little or no light in classical bulges.

My impression is that distributions of B/T in recent large-scale simulations peak at B/T close to 50%, quite unlike Figure 1. This is not a criticism: the authors are reporting results from painstakingly careful work. I have not detected much concern in the galaxy formation community about the failure to match Figure 1. This is sensible: galaxy model building has encountered and resolved many other problems; maybe this is just one more. But I offer the cautionary reminder of earlier thinking about primeval turbulence. We are much better informed now, but we are reaching much further, on still quite modest empirical grounds. Star formation is observed in some detail, but it still must be schematically modeled in galaxy formation simulations. Dark matter and Einstein's cosmological constant are not even observed, apart from their gravitational effect. So although the ΛCDM theory passes demanding tests it may need improving. Thinking in cosmology has been redirected, sometimes by closer consultation of the theory, as for primeval turbulence, sometimes by observation, as in the falsification of the 1948 steady state cosmology. You can think of other examples. The pure disk phenomenon is a case of déjà vu all over again; it is certain to teach us something of value. That may be about how pure disks can be understood within the present paradigm. Or it may serve to redirect our thinking once again, toward a still better cosmology.

References

Efstathiou, G., and Jones, B. J. T. 1979, *MNRAS*, 186, 133
Eggen, O. J., Lynden-Bell, D., and Sandage, A. R. 1962, *ApJ*, 136, 748
Fall, S. M. 1979, *Nature*, 281, 200
Fall, S. M., and Efstathiou, G. 1980, *MNRAS*, 193, 189
Fisher, D. B., and Drory, N. 2011, *ApJ Lett.*, 733, L47
Gamow, G. 1954, *Proc. NAS*, 40, 480
Gunn, J. E., Lee, B. W., Lerche, I., Schramm, D. N., and Steigman, G. 1978, *ApJ*, 223, 1015
Hoyle, F. 1951, in Proceedings of a Symposium on the Motion of Gaseous Masses of Cosmical Dimensions held at Paris, August 16-19, 1949, 195
Kormendy, J., Drory, N., Bender, R., and Cornell, M. E. 2010, *ApJ*, 723, 54
Oort, J. H. 1958, Eleventh Solvay Conference. Editions Stoops, Brussels, 21 pp.
Oort, J. H. 1970, *A & Ap*, 7, 381
Ozernoi, L. M. 1972, *Soviet Astron.-AJ*, 15, 923
Peebles, P. J. E. 1969, *ApJ*, 155, 393
Peebles, P. J. E. 1971a, *A&SS*, 11, 443
Peebles, P. J. E. 1971b, *A&Ap*, 11, 377
Peebles, P. J. E. 2014, in Proceedings of the 7th International Conference on Gravitation and Cosmology, Goa, 2011, *J. Phys.: Conference Series*, 484, 012001
Rubin, V. C. 1954, *Proc. NAS*, 40, 54
Strömberg, G. 1934, *ApJ*, 79, 460
von Hoerner, S. 1953, *Zeitschrift für Astrophysik*, 32, 51
von Weizsäcker, C. F. 1951, in Proceedings of the Symposium on the Motions of Gaseous Masses of Cosmical Dimensions, Paris, August 16-19, 1949, pages 158 and 200
White, S. D. M., and Rees, M. J. 1978, *MNRAS*, 183, 341

Astronomy in Focus - XXX
Proceedings IAU Symposium No. XXX, 2018
M. T. Lago, ed.

doi:10.1017/S1743921319004095

Angular Momentum Evolution of Galaxies: the Perspective of Hydrodynamical Simulations

Claudia del P. Lagos[1,2,3]

[1]International Centre for Radio Astronomy Research (ICRAR), M468, University of Western Australia, 35 Stirling Hwy, Crawley, WA 6009, Australia

[2]ARC Centre of Excellence for All Sky Astrophysics in 3 Dimensions (ASTRO 3D)

[3]Cosmic Dawn Center (DAWN), Niels Bohr Institute, University of Copenhagen, Copenhagen, Denmark

email: claudia.lagos@icrar.org

Abstract. Until a decade ago, galaxy formation simulations were unable to simultaneously reproduce the observed angular momentum (AM) of galaxy disks and bulges. Improvements in the interstellar medium and stellar feedback modelling, together with advances in computational capabilities, have allowed the current generation of cosmological galaxy formation simulations to reproduce the diversity of AM and morphology that is observed in local galaxies. In this review I discuss where we currently stand in this area from the perspective of hydrodynamical simulations, specifically how galaxies gain their AM, and the effect galaxy mergers and gas accretion have on this process. I discuss results which suggest that a revision of the classical theory of disk formation is needed, and by discussing what the current challenges are.

Keywords. Galaxy: evolution - Galaxy: formation - Galaxy: fundamental parameters

---

1. Introduction

The formation of galaxies can be a highly non-linear process, with many physical mechanisms interacting simultaneously. Despite all that potential complexity, early studies of galaxy formation stressed the importance of three quantities to describe galaxies: mass, M, energy, E, and angular momentum (AM), J (e.g. Peebles 1969); one can alternatively define the specific AM, $j \equiv J/M$, which contains information on the scale length and rotational velocity of a system. It is therefore intuitive to expect the relation between j and M to contain fundamental information.

In recent years, Integral field spectroscopy (IFS) is opening a new window to explore galaxy kinematics and its connection to galaxy formation and evolution, with IFS based measurements of the $j_{\rm stars} - M_{\rm stars}$ relations being reported in the local (e.g. Cortese *et al.* 2016) and high-z Universe (Burkert *et al.* 2016; Swinbank *et al.* 2017; Harrison *et al.* 2017). The last decade has also been a golden one for cosmological hydrodynamical simulations, with the first large cosmological volumes, with high enough resolution to study the internal structure of galaxies being possible (Schaye *et al.* 2015; Dubois *et al.* 2016; Vogelsberger *et al.* 2014; Pillepich *et al.* 2018). These simulations have been able to overcome the catastrophic loss of AM, which refers to the problem of galaxies being too low j compared to observations (Steinmetz & Navarro 1999; Navarro & Steinmetz 2000) and over-cooling problem. This problem was solved by improving the spatial resolution, adopting j conservation numerical schemes, and including efficient feedback (e.g. Kaufmann *et al.* 2007; Zavala *et al.* 2008; Governato *et al.* 2010; Guedes *et al.* 2011; DeFelippis *et al.* 2017). Fig. 1 shows examples of several cosmological

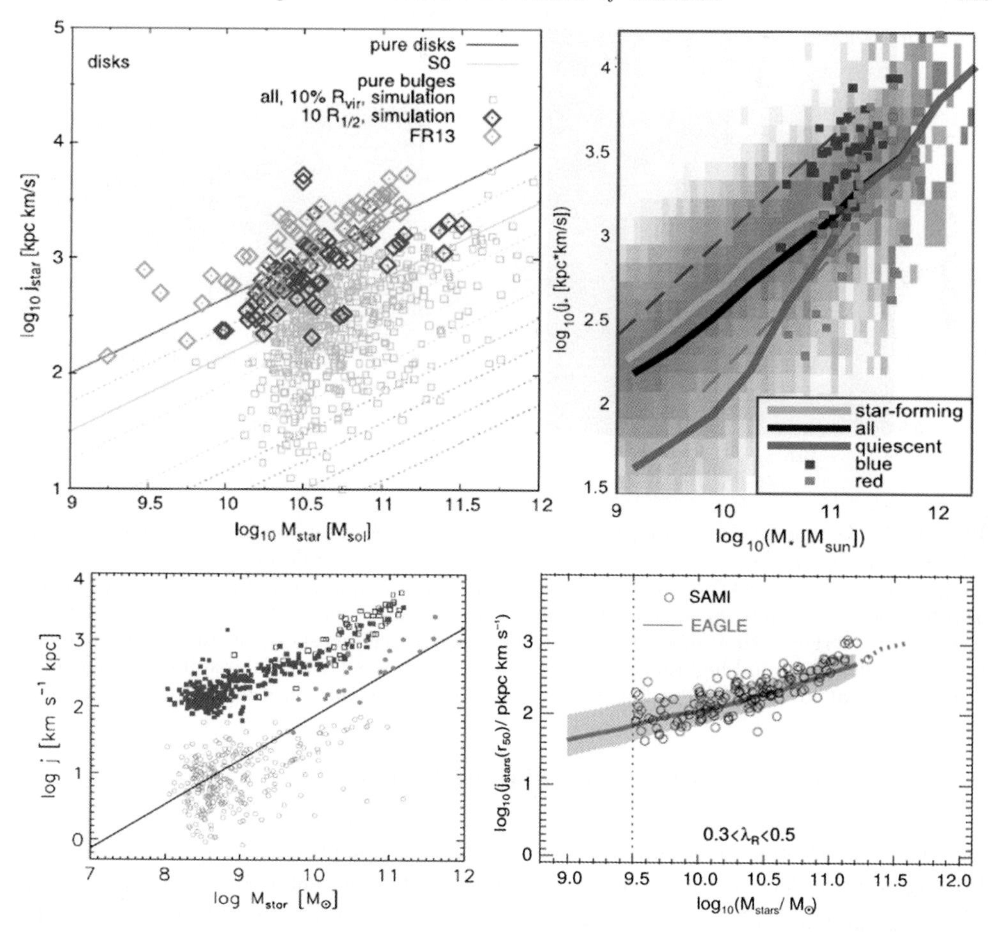

Figure 1. The $z = 0$ $j_{\rm stars} - M_{\rm stars}$ relation for several simulations: Magneticum (Teklu *et al.* 2015), Illustris (Genel *et al.* 2015), Fenix (Pedrosa & Tissera 2015) and EAGLE (Lagos *et al.* 2017). Panels (a), (b) and (c) show the total $j_{\rm stars}$ compared to the observational measurements of Fall & Romanowsky (2013), while panel (d) shows $j_{\rm stars}$ measured within an effective radius and compares with SAMI observations (Cortese *et al.* 2016).

simulations which have been shown to reproduce the observed $j_{\rm stars} - M_{\rm stars}$ relation for $z = 0$ galaxies. In addition to those above, there are several cosmological zoom-in simulations that have shown the same level of success (e.g. Wang *et al.* 2018; El-Badry *et al.* 2018).

The level of agreement of Fig. 1 gives us assurance that we can use these simulations to study how the AM of galaxies is gained/lost throughout the process of galaxy formation and evolution. The left panel of Fig. 2 shows an example of the morphologies of simulated galaxies in the $j_{\rm stars} - M_{\rm stars}$ plane at $z = 0$ in the EAGLE simulations. There is a clear correlation between a galaxy's morphology and its kinematics, as seen in observations (e.g. Cortese *et al.* 2016, Fall & Romanowski 2018). An important caveat, however, is that most, if not all, the simulations of Fig. 1 (and those with similar specifications) are currently unable to form very thin disks (with ellipticities $\gtrsim 0.7 - 0.8$) due to the insufficient resolution and simplistic interstellar medium (ISM) models. The latter prevent us from obtaining a realistic vertical structure of disks (see discussion in Lagos *et al.* 2018a).

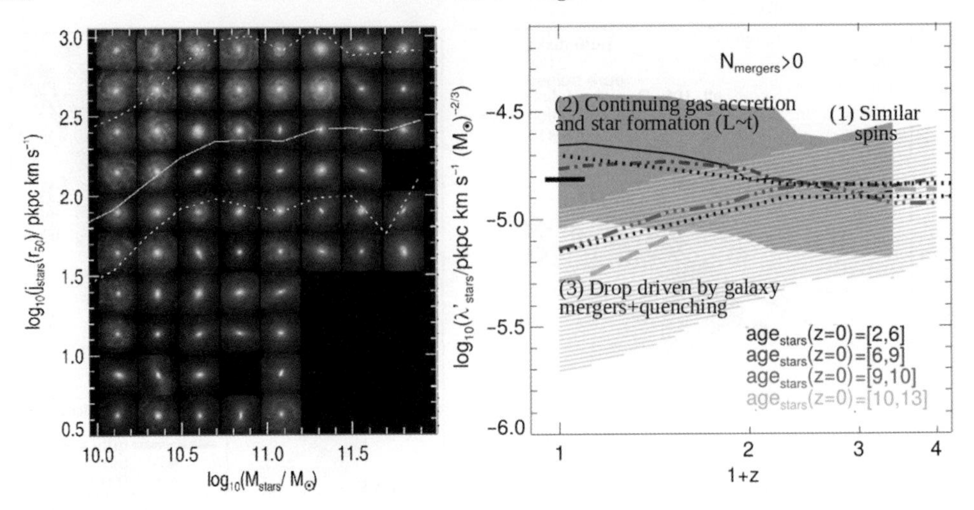

Figure 2. *Left panel:* From Lagos *et al.* (2018b). Synthetic gri optical images of randomly selected galaxies in the $j_{stars}(r_{50}) - M_{stars}$ plane at $z = 0$. The solid and dashed lines show the median and the $16^{th} - 84^{th}$ percentile range. *Right panel:* Adapted from Lagos *et al.* (2017). λ'_{stars} as a function of redshift for galaxies that by $z = 0$ have different stellar ages and that have had at least one merger with a mass ratio ≥ 0.1. The shaded regions show the $16^{th} - 84^{th}$ percentile ranges for the lowest and highest age bins. At $z \gtrsim 1.2$, galaxies have similar spins, while diverging dramatically at lower redshifts. Lagos *et al.* (2017) identify three critical features described in the figure.

2. How galaxies gain their angular momentum?

Lagos *et al.* (2017) used the EAGLE simulations (Schaye *et al.* 2015) to study how the stellar spin of galaxies at $z = 0$ evolved depending on their stellar population age and merger history. The stellar spin in this case was defined as a pseudo spin $\lambda'_{stars} = j_{stars}(r_{50})/M_{stars}^{2/3}$, with $j_{stars}(r_{50})$ being the stellar specific AM within one effective radius. The power-law index 2/3 comes from the predicted j-mass relation of dark matter halos (Fall 1983). The right panel of Fig. 2 shows the evolution of λ'_{stars} of $z = 0$ galaxies with different stellar ages that have had at least 1 galaxy merger. Progenitors display indistinguishable kinematics at $z \gtrsim 1$ despite their $z = 0$ descendants being radically different ((1) in the right panel of Fig. 2). Similarly, Penoyre *et al.* (2017) found using Illustris that the progenitors of $z = 0$ early-type galaxies that are slow and fast rotators, had very similar properties at $z \gtrsim 1$ (see also Choi & Yi 2017 for an example using the Horizon-AGN simulation). The evolution of λ'_{stars} diverges dramatically at $z \lesssim 1$, in which galaxies that by $z = 0$ have young stellar populations, grow their disks efficiently due to the continuing gas accretion and star formation ((2) in Fig. 2); galaxies that by $z = 0$ have old stellar populations went through active spinning down, due to the effects of dry galaxy mergers and quenching ((3) in Fig. 2; discussed in more detail in § 3).

The reason why continuing gas accretion drives spinning up is because the AM of the material falling into halos is expected to increase linearly with time (as predicted by tidal torque theory; Catelan & Theuns 1996). El-Badry *et al.* (2018) explicitly demonstrated this using the FIRE simulations. Garrison-Kimmel *et al.* (2018) also using FIRE, in fact argued that the most important predictor of whether a disk will be formed by $z = 0$ is the halo gas j by the time the galaxy has formed half of its stars. These simulations thus suggest that *the later the accretion the more efficient the spinning up.*

Simulations suggest the critical transition at $z \approx 1$ is driven by a change in the main mode of gas accretion onto galaxies, from filamentary accretion to gas cooling from a hydrostatic halo (e.g. Garrison-Kimmel *et al.* 2018). The latter seems to be key in

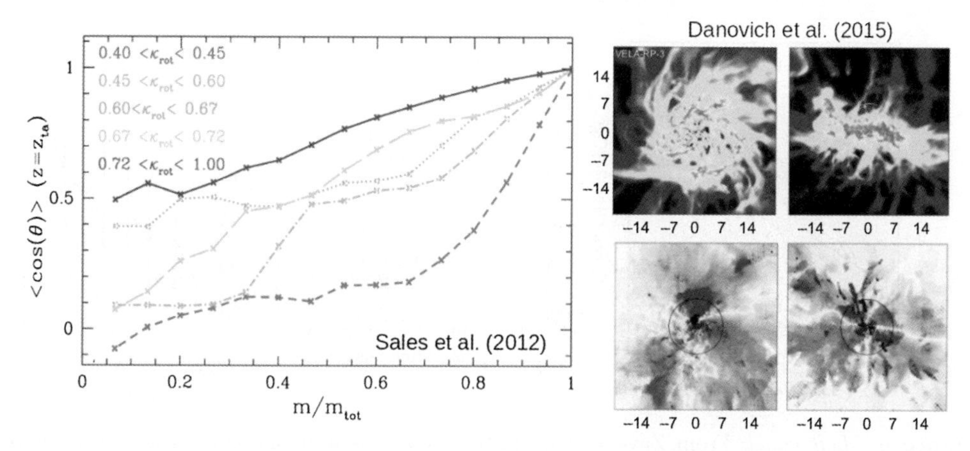

Figure 3. *Left panel:* From Sales *et al.* (2012). The angle between the AM vector enclosed within a given mass fraction (x-axis) and the total spin of the system, measured at the time of maximum expansion of the halo, for galaxies that by $z = 0$ have different fractions of kinetic energy invested in rotation ($\kappa_{\rm rot}$; the higher the $\kappa_{\rm rot}$ the more rotation dominates). This figure shows that alignments between the halo and galaxy are key to facilitate the formation of disks. *Right panel:* From Danovich *et al.* (2015). Face-on and edge-on density (top) and normalised torque (bottom) maps of a simulated $z \approx 2$ galaxy. The figure shows that the quadrupole torque pattern expected in the idealized case of the formation of a thin disk is seen in both orientations, meaning that the galaxy is being torqued in all three axes.

facilitating alignments between the accreting gas and the galaxy, while the former is by nature more stochastic. Sales *et al.* (2012) showed that galaxies that by $z = 0$ are more rotation-dominated formed in halos that had the inner/outer parts better aligned (see left panel in Fig. 3). Similarly, Stevens *et al.* (2016) showed that significant AM losses of the cooling gas to the hot halo are seen in cases where the hot halo is more misaligned with the galaxy. On the other hand, filamentary accretion at $z \gtrsim 1$ is not as efficient in spinning up galaxies mostly because gas filaments arrive from different directions (e.g. Welker *et al.* 2017), typically causing torques to act in all three axes of a galaxy (Danovich *et al.* 2015; see left panel in Fig. 3). The latter is intimately connected to high-z disks being turbulent and highly disturbed.

An important result that is robust to the details of the simulation being used, is that the circum-galactic medium (CGM) seems to have a specific AM in excess to that of the halo by factors of $3 - 5$ (Stewart *et al.* 2017; Stevens *et al.* 2017). Stevens *et al.* (2017) also showed that about $50 - 90\%$ of that excess j can be lost to the hot halo in the process of gas cooling and accretion onto the galaxy.

3. How galaxies loose their angular momentum?

The latest generation of hydrodynamical simulations has been able to approximately reproduce the morphological diversity of galaxies (e.g. Vogelsberger *et al.* 2014; Dubois *et al.* 2016; see left panel in Fig. 2). A long-standing question is therefore how do galaxies become elliptical with j significantly below disk galaxies of the same mass?

Zavala *et al.* (2016) used the EAGLE simulations to study the AM of galaxies and found a very strong correlation between the kinematic stellar bulge-to-total ratio and the net loss in AM of the stars that end up in galaxies at $z = 0$ (left panel Fig. 4), suggesting galaxy mergers to be good candidates for the physical process behind this correlation. More recently, several authors have shown (e.g. Penoyre *et al.* 2017; Lagos *et al.* 2018a,b) that the gas fraction of the merger is one of the key parameters indicating whether the merger will lead to the primary galaxy spinning up or down. Middle and

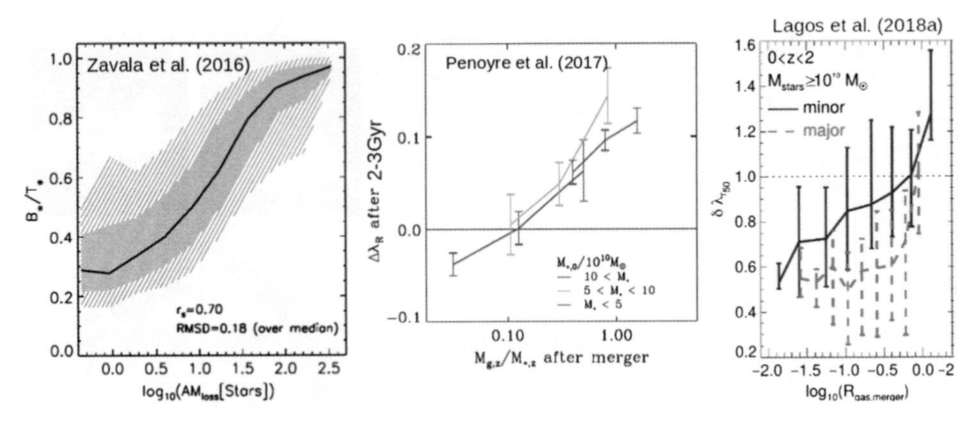

Figure 4. *Left panel:* From Zavala *et al.* (2016). There is a strong correlation between the bulge-to-total ratio of $z=0$ galaxies (x-axis) and the net AM loss suffered by the galaxy stars (i.e. the difference between the maximum and $z=0$ $j_{\rm stars}$; y-axis). *Middle panel:* From Penoyre *et al.* (2017). Change in stellar spin parameter in galaxies (as defined in Emsellem *et al.* 2007) approximately $2-3$ Gyr after a merger, as a function of the gas fraction of the secondary galaxy which merged. Galaxies can become more rotation- ($\Delta(\lambda_{\rm R})>0$) or dispersion-dominated ($\Delta(\lambda_{\rm R})<0$), depending of the gas brought up by the secondary galaxy. *Right panel:* From Lagos *et al.* (2018a). Fractional change in $\lambda_{\rm R}$ as a function of the total gas-to-stellar mass ratio of the merging system, for minor (mass ratios between $0.1-0.3$) and major (mass ratios ≥ 0.3) mergers. Gas-poor mergers are required to significantly spin galaxies down.

right panels of Fig. 4 show the clear effect of galaxy mergers on the stellar spin parameter of galaxies in Illustris and EAGLE, respectively.

Lagos *et al.* (2018b) showed that other merger parameters can have a significant effect on the $j_{\rm stars}$ structure, with high orbital j and/or co-rotating mergers driving more efficient spinning up. However, gas fraction is the single strongest parameter that determines whether a galaxy spins up or down as a result of the merger, with the mass ratio modulating the effect. Active Galactic Nuclei feedback is key to prevent further gas accretion and the regrowth of galaxy disks in elliptical galaxies (e.g. Dubois *et al.* 2016). Early works on dry mergers (e.g. Navarro & White 1994) show that dynamical friction redistribuites $j_{\rm stars}$ in a way such that most of it ends up at very large radii. However, when integrating over a large enough baseline, one should find $j_{\rm stars}$ converging to $j_{\rm halo}$. Using EAGLE, Lagos *et al.* (2018b) confirmed that to be the case: gas-poor mergers do not significantly change the *total* j of the system, but significantly re-arrange it so that the inner parts of galaxies ($r/r_{50}<5$) become highly deficient in j compared to galaxies of the same stellar mass that went through gas-rich mergers or not mergers at all. This was also seen by Teklu *et al.* (2015) as very deficient $j_{\rm stars}$ profiles in early-types compared to late-types at $r/R_{\rm vir}<0.2$. An important prediction of that process is that the cumulative $j_{\rm stars}$ radial profiles of high j galaxies are much more self-similar than those of galaxies with low j. In other words: there are few ways in which a galaxy by $z=0$ can end up with high j, but many pathways that lead to low j (e.g. Naab *et al.* 2014, Garrison-Kimmel *et al.* 2018).

4. Discussion and future prospects

The picture that has emerged from simulations in how galaxies gain their AM is significantly more complex than the classical picture of galaxy disks forming inside out observing j conservation (e.g. Mo, Mao & White 1998). This complexity, however, is driven by processes that act in different directions and that tend to compensate quite efficiently, so that galaxies follow the classical disk formation model to within 50% (Zavala *et al.* 2016; Lagos *et al.* 2017, 2018b).

This inevitably opens the following question: to what extent are we forcing $j_{\rm stars} \sim j_{\rm halo}$ in disk galaxies through the process of tuning free-parameters in simulations? State-of-the-art simulations tend to carefully tune their parameters to reproduce some broad statistics of galaxies, such as the stellar-halo mass relation, stellar mass function, and in some cases the size-mass relation (e.g. Crain *et al.* 2015). A consequence of such tuning may well result in this conspiracy: the CGM's j being largely in excess of the halo's j, but then losing significant amounts of it while falling onto the galaxy, so that by $z=0$ disk galaxies have $j_{\rm stars} \sim j_{\rm halo}$. A possible solution for this conundrum is to perform detailed, high-resolution simulations of individual disk galaxies in a cosmological context and test widely different feedback mechanisms with the aim of understanding which conditions lead to $j_{\rm stars} \sim j_{\rm halo}$ and how independent the tuning of parameters is of the evolution of specific AM.

Another important area of research will be the improvement of the description of the vertical structure of disks, as current large cosmological hydrodynamical simulations struggle to form thin disks $\epsilon \gtrsim 0.7-0.8$. This is most likely due to the ISM and cooling modelling and resolution in these simulations being insufficient. Currently, simulations tend to force the gas to not cool down below $\approx 10^4$ K, which corresponds to a Jeans length of ≈ 1 kpc, much larger than the scaleheights of disks in the local Universe. This issue could be solved by including the formation of the cold ISM, which necessarily means improving the resolution of the simulations significantly.

References

Burkert A., Förster Schreiber N. M., Genzel R. *et al.*, 2016, *ApJ*, 826, 214
Catelan P., & Theuns T., 1996, *MNRAS*, 282, 436
Choi H., & Yi S. K., 2017, *ApJ*, 837, 68
Cortese L., Fogarty L. M. R., Bekki K. *et al.*, 2016, arXiv:1608.00291
Crain R. A., Schaye J., Bower R. G. *et al.*, 2015, *MNRAS*, 450, 1937
Danovich M., Dekel A., Hahn O. *et al.*, 2015, *MNRAS*, 449, 2087
DeFelippis D., Genel S., Bryan G. L. *et al.*, 2017, *ApJ*, 841, 16
Dubois Y., Peirani S., Pichon C. *et al.*, 2016, *MNRAS*, 463, 3948
El-Badry K., Quataert E., Wetzel A. *et al.*, 2018, *MNRAS*, 473, 1930
Emsellem E., Cappellari M., Krajnović D. *et al.*, 2007, *MNRAS*, 379, 401
Garrison-Kimmel S., Hopkins P. F., Wetzel A. *et al.*, 2018, *MNRAS*
Genel S., Fall S. M., Hernquist L. *et al.*, 2015, *ApJ*, 804, L40
Governato F., Brook C., Mayer L. *et al.*, 2010, *Nature*, 463, 203
Guedes J., Callegari S., Madau P. *et al.*, 2011, *ApJ*, 742, 76
Harrison C. M., Johnson H. L., Swinbank A. M. *et al.*, 2017, *MNRAS*, 467, 1965
Kaufmann T., Mayer L., Wadsley J. *et al.*, 2007, *MNRAS*, 375, 53
Lagos C. d. P., Schaye J., Bahé Y. *et al.*, 2018a, *MNRAS*, 476, 4327
Lagos C. d. P., Stevens A. R. H., Bower R. G. *et al.*, 2018b, *MNRAS*, 473, 4956
Lagos C. d. P., Theuns T., Stevens A. R. H. *et al.*, 2017, *MNRAS*, 464, 3850
Mo H. J., Mao S., & White S. D. M., 1998, *MNRAS*, 295, 319
Naab T., Oser L., Emsellem E. *et al.*, 2014, *MNRAS*, 444, 3357
Navarro J. F., & Steinmetz M., 2000, *ApJ*, 538, 477
Navarro J. F., & White S. D. M., 1994, *MNRAS*, 267, 401
Pedrosa S. E., & Tissera P. B., 2015, *A&A*, 584, A43
Peebles P. J. E., 1969, *ApJ*, 155, 393
Penoyre Z., Moster B. P., Sijacki D. *et al.*, 2017, *MNRAS*, 468, 3883
Pillepich A., Springel V., Nelson D. *et al.*, 2018, *MNRAS*, 473, 4077
Sales L. V., Navarro J. F., Theuns T. *et al.*, 2012, *MNRAS*, 423, 1544
Schaye J., Crain R. A., Bower R. G. *et al.*, 2015, *MNRAS*, 446, 521
Steinmetz M., & Navarro J. F., 1999, *ApJ*, 513, 555
Stevens A. R. H., Lagos C. d. P., Contreras S. *et al.*, 2016, arXiv:1608.04389

Stewart K. R., Maller A. H., Oñorbe J. *et al.*, 2017, *ApJ*, 843, 47
Swinbank A. M., Harrison C. M., Trayford J. *et al.*, 2017, *MNRAS*
Teklu A. F., Remus R.-S., Dolag K. *et al.*, 2015, *ApJ*, 812, 29
Vogelsberger M., Genel S., Springel V. *et al.*, 2014, *Nature*, 509, 177
Wang L., Obreschkow D., Lagos C. D. P. *et al.*, 2018, arXiv:1808.05564
Welker C., Dubois Y., Devriendt J. *et al.*, 2017, *MNRAS*, 465, 1241
Zavala J., Okamoto T., & Frenk C. S., 2008, *MNRAS*, 387, 364

Astronomy in Focus - XXX
Proceedings IAU Symposium No. XXX, 2018
M. T. Lago, ed.

doi:10.1017/S1743921319004101

Emerging Angular Momentum Physics from Kinematic Surveys

Matthew Colless[1]

[1]Research School of Astronomy and Astrophysics, Australian National University
Canberra, ACT 2611, Australia
email: `matthew.colless@anu.edu.au`

Abstract. I review the insights emerging from recent large kinematic surveys of galaxies at low redshift, with particular reference to the SAMI, CALIFA and MaNGA surveys. These new observations provide a more comprehensive picture of the angular momentum properties of galaxies over wide ranges in mass, morphology and environment in the present-day universe. I focus on the distribution of angular momentum within galaxies of various types and the relationship between mass, morphology and specific angular momentum. I discuss the implications of the new results for models of galaxy assembly.

Keywords. galaxies: kinematics and dynamics, galaxies: structure, galaxies: evolution galaxies: formation, galaxies: stellar content

---

1. Introduction

This brief review focusses on recent integral field spectroscopy surveys of the stellar kinematics in large samples of galaxies at low redshifts. It does not cover radio HI surveys of the neutral gas in low-redshift galaxies (which are important for understanding the kinematics at large radius) nor does it extend to surveys at high redshifts (which explore the origin and evolution of galaxy kinematics). What *local* surveys of stellar kinematics can tell us about angular momentum in galaxies is its dependence on mass, morphology and other properties (if sample selection is understood) and its dependence on environment (if the sample is embedded in a fairly complete redshift survey); such dependencies can provide *indirect* evidence for the origin and evolution of angular momentum.

It is immediately apparent that all current kinematic surveys have weaknesses relating to the trade-offs demanded by instrumental constraints: firstly, between spatial resolution and spatial coverage (also between spectral resolution and spectral coverage) and, secondly, between this per-galaxy information and sample size (also sample volume and completeness). The lack of radial coverage is a serious problem for late-type disk galaxies having exponential mass profiles (i.e. having Sersic index $n \approx 1$), for which $M/M_{\rm tot} = 0.5,0.8$ at $R/R_e \approx 1.0,1.8$ and $j/j_{\rm tot} = 0.5,0.8$ at $R/R_e \approx 1.0,2.2$. But it is a much worse problem for early-type spheroidal galaxies with deVaucouleurs profiles ($n \approx 4$), for which $M/M_{\rm tot} = 0.5,0.8$ at $R/R_e \approx 1.0,3.2$ and $j/j_{\rm tot} = 0.5,0.8$ at $R/R_e \approx 4.4,>9$ (see Figure 1a). This problem is compounded by the necessary instrumental trade-off between radial coverage (field of view) and spatial resolution (spaxel scale) of integral field units (IFUs) due to constraints imposed by the limited available detector area. For example, in the SAMI sample the median major axis is $R_e = 4.4$ arcsec (10-90% range spans 1.8-9.4 arcsec) which means that the SAMI IFUs only sample out to a median radius of $1.7R_e$ (see Figure 1b).

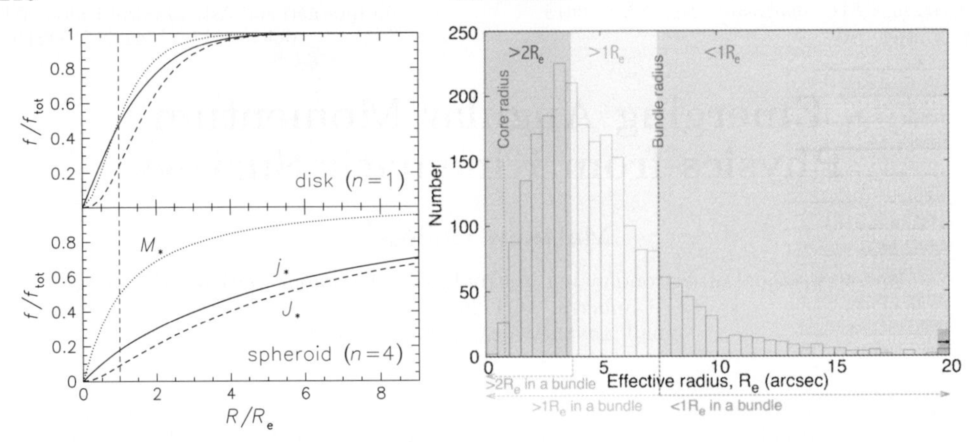

Figure 1. (a) Left panel: the fraction of mass (M), angular momentum (J), and specific angular momentum ($j = J/M$) as functions of radius (in units of the effective radius, R_e) for both an exponential disk profile (Sersic index $n = 1$; top panel) and a deVaucouleurs spheroid profile (Sersic index $n = 4$; bottom panel) [https://iopscience.iop.org/article/10.1088/0067-0049/203/2/17, Fig. 3]. (b) Right panel: the distribution of effective radius R_e (in arcsec) for the SAMI galaxy sample, showing those parts of the sample for which the integral field unit covers $<1R_e$, $>1R_e$ and $>2R_e$ [based on https://academic.oup.com/mnras/article/446/2/1567/2891878, Fig. 2].

2. Surveys

2.1. *SAMI*

SAMI is the Sydney-AAO Multi-IFU instrument on the 3.9m Anglo-Australian Telescope (AAT). It has 13 IFUs that can be positioned over a 1 degree field at the telescope's prime focus. Each hexabundle IFU has 61×1.6 arcsec fibres covering a 15 arcsec diameter field of view. SAMI feeds the AAOmega spectrograph, which gives spectra over 375–575nm at $R \approx 1800$ ($70\,\mathrm{km\,s^{-1}}$) and 630–740nm at $R \approx 4300$ ($30\,\mathrm{km\,s^{-1}}$). The SAMI Second Data Release (DR2) includes 1559 galaxies (about half the full sample) covering $0.004 < z < 0.113$ and $7.5 < \log(M_*/M_\odot) < 11.6$. The core data products for each galaxy are two primary spectral cubes (blue and red), three spatially binned spectral cubes, and a set of standardised aperture spectra. For each core data product there are a set of value-added data products, including aperture and resolved stellar kinematics, aperture emission line properties, and Lick indices and stellar population parameters. The data release is available online through AAO Data Central (`datacentral.org.au`).

2.2. *CALIFA*

CALIFA is the Calar Alto Legacy Integral Field survey, consisting of integral field spectroscopy for 667 galaxies obtained with PMAS/PPak on the Calar Alto 3.5m telescope. There are three different spectral setups: 375–750 nm at 0.6 nm FWHM resolution for 646 galaxies, 365–484 nm at 0.23 nm FWHM resolution for 484 galaxies, and a combination of these over 370–750 nm at 0.6 nm FWHM resolution for 446 galaxies. The CALIFA Main Sample spans $0.005 < z < 0.03$ and the colour-magnitude diagram, with a wide range of stellar masses, ionization conditions and morphological types; the CALIFA Extension Sample includes rare types of galaxies that are scarce or absent in the Main Sample.

2.3. *MaNGA*

MaNGA is the Mapping Nearby Galaxies at Apache Point Observatory survey (part of SDSS-IV). It is studying the internal kinematic structure and composition of gas and

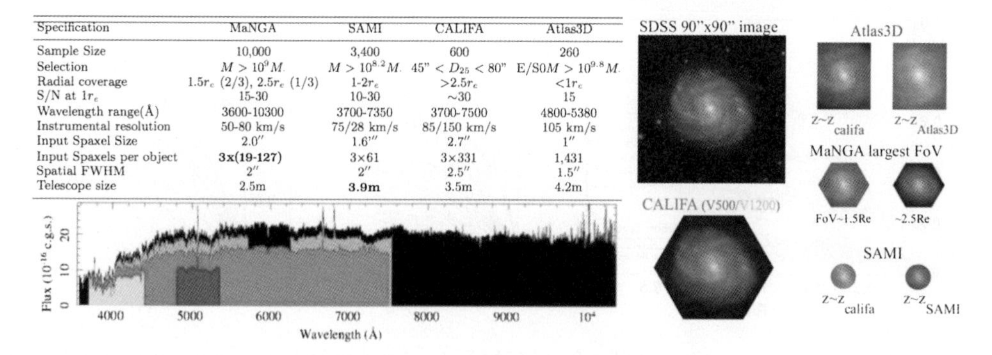

Specification	MaNGA	SAMI	CALIFA	Atlas3D
Sample Size	10,000	3,400	600	260
Selection	$M > 10^9 M$	$M > 10^{8.2} M$	45" $< D_{25} <$ 80"	E/S0$M > 10^{9.8} M$
Radial coverage	$1.5r_e$ (2/3), $2.5r_e$ (1/3)	1-$2r_e$	$>2.5r_e$	$<1r_e$
S/N at $1r_e$	15-30	10-30	~30	15
Wavelength range(Å)	3600-10300	3700-7350	3700-7500	4800-5380
Instrumental resolution	50-80 km/s	75/28 km/s	85/150 km/s	105 km/s
Input Spaxel Size	2.0″	1.6″	2.7″	1″
Input Spaxels per object	**3x(19-127)**	3×61	3×331	1,431
Spatial FWHM	2″	2″	2.5″	1.5″
Telescope size	2.5m	**3.9m**	3.5m	4.2m

Figure 2. Upper left: table of key parameters of the MaNGA, SAMI, CALIFA and Atlas3D surveys. Lower left: the wavelength coverage of the MaNGA (black), SAMI (grey), CALIFA (red & green) and Atlas3D (blue) surveys. Right: illustration of the relative fields of view covered by the IFUs used in each survey. [Based on Sanchez *et al.* (2015), Table 1 & Figure 1.]

stars in 10,000 nearby galaxies. It employs 17 fibre-bundle IFUs varying in diameter from 12 arcsec (19 fibres) to 32 arcsec (127 fibers) that feed two dual-channel spectrographs covering 360–1030 nm at $R \approx 2000$. The targets have $M_* > 10^9\,M_\odot$ based on SDSS-I redshifts and i-band luminosities. The MaNGA sample is designed to approximate uniform radial coverage in terms of R_e, a flat stellar mass distribution, and a wide range of environments. SDSS Data Release 14 (DR14) includes MaNGA data cubes for 2812 galaxies.

2.4. *Comparison*

Figure 2 provides a tabular and graphical summary of the parameters of these three surveys (and also the earlier Atlas3D survey), which helps to understand their various relative strengths and weaknesses, and consequently their complementarities. A few kinematic surveys of small samples offer greater radial coverage and higher velocity resolution: SLUGGS surveyed kinematics of 25 early-type galaxies to $\sim 3R_e$ from stars and to $\sim 10R_e$ using globular clusters (Bellstedt *et al.* 2018); PN.S surveyed the kinematics of 33 early-type galaxies to $\sim 10R_e$ using planetary nebulae (Pulsoni *et al.* 2018).

3. Results

3.1. *Role of angular momentum*

After mass, angular momentum is the most important driver of galaxy properties, with a key role in the formation of structure and morphology. For regular oblate rotators, angular momentum can be derived from dynamical models as well as direct estimates of projected angular momentum. Surveys can determine population variations in the total angular momentum and its distribution with radius, exploring dependencies on mass, morphology, ellipticity and other properties. These relations can provide insights on the assembly histories of galaxies for comparison with simulations.

3.2. *Angular momentum & spin profiles*

SAMI, CALIFA and MaNGA together now provide angular momentum profiles (or, alternatively, spin proxy, λ_R, as a function of R/R_e) for thousands of galaxies to $R/R_e \sim 1$ and for hundreds of galaxies to $R/R_e \sim 2$. These samples are large enough to be useful when split by mass, morphology or environment. Figure 3 shows spin profiles for galaxies

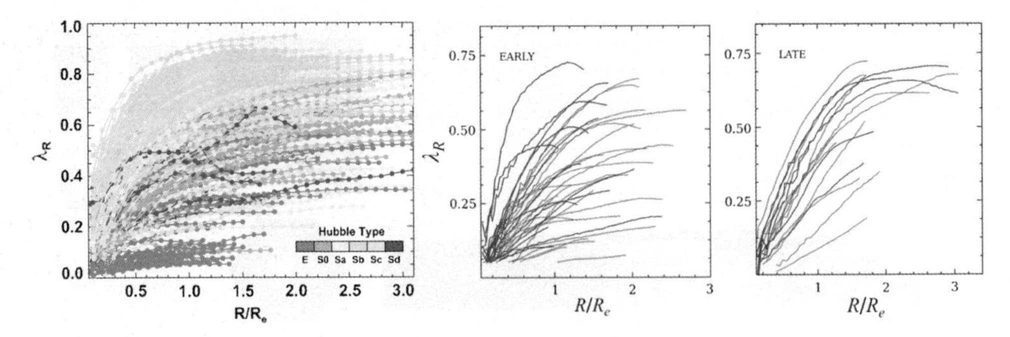

Figure 3. Left: Galaxy spin profiles from CALIFA, showing the variation with Hubble type [Falcón-Barroso (2016), Fig. 2]. Right: Galaxy spin profiles from MaNGA, showing the variation with mass for early and late-type galaxies [Greene *et al.* (2018), Fig. 3].

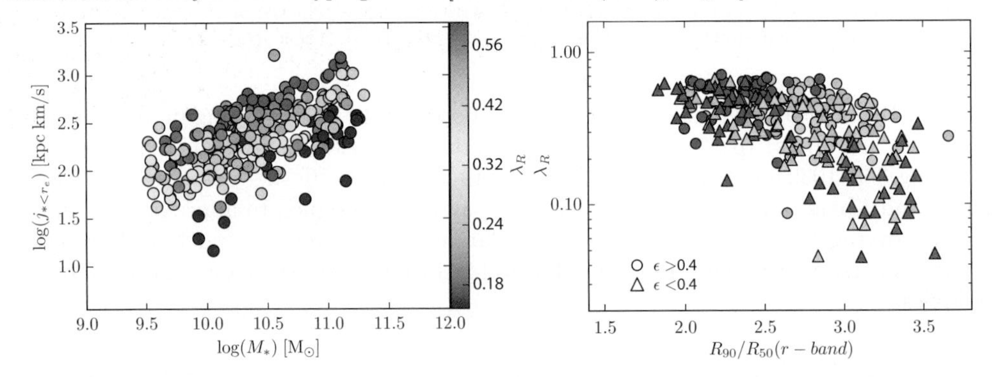

Figure 4. Left: Specific angular momentum versus stellar mass for SAMI galaxies, colour-coded by spin parameter λ_R. Right: Galaxy spin parameter versus r-band concentration for SAMI galaxies, colour-coded by morphology. [Based on Cortese *et al.* (2016), Figs. 6 & 7].

from the CALIFA survey (Falcón-Barroso 2016) and the MaNGA survey (Greene *et al.* 2018); Foster *et al.* (2018) give similar results from the SAMI survey.

3.3. *Spin, morphology & ellipticity*

Typical galaxies lie on a plane relating mass M, j and stellar distribution (quantified by, e.g., Sersic index n or photometric concentration index), with overall morphologies regulated by their mass and dynamical state (see, e.g., Cortese *et al.* 2016). The correlation shown in the left panel of Figure 4 between the offset from the mass–angular momentum (M-j) relation and spin parameter λ_R shows that at fixed M the contribution of ordered motions to dynamical support varies by more than a factor of three. The right panel of Figure 4 shows that λ_R correlates strongly with morphology and concentration index (especially if slow-rotators are removed), suggesting that late-type galaxies and early-type fast-rotators form a continuous class in terms of their kinematic properties.

The spin–ellipticity (λ_R–ϵ) diagram is a particularly revealing frame for understanding relations between kinematic and morphological properties of galaxies. This is illustrated in Figure 5, from the work of Graham *et al.* (2018) using the MaNGA survey. The left panel shows the strong correlation between the mass of a galaxy and its position in this diagram, with more massive galaxies tending to have lower spin and ellipticity. The central panel shows the areas of the diagram occupied by various morphological types: elliptical galaxies occupy the low-λ_R, low-ϵ region, while lenticular and spiral galaxies largely overlap, covering the full range of ϵ at $\lambda_R > 0.5$. The right panel shows how galaxies

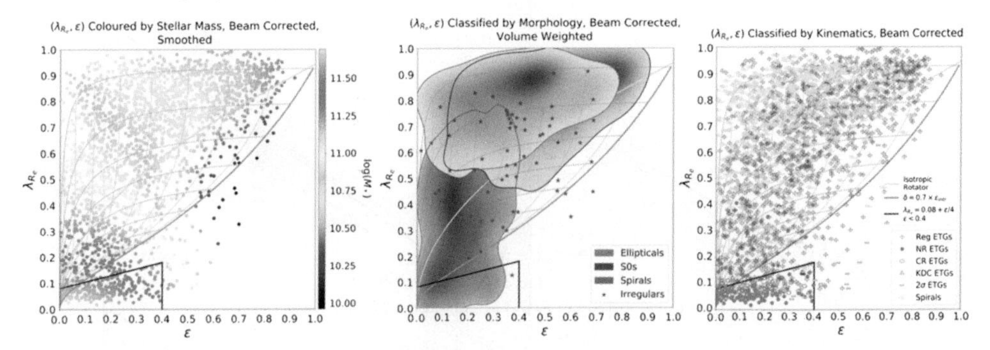

Figure 5. Distributions of galaxy properties in the spin–ellipticity (λ_R–ϵ) diagram: left—stellar mass; centre—visual morphology; right—kinematic class. [Graham *et al.* (2018), Figs. 5, 8 & 9.]

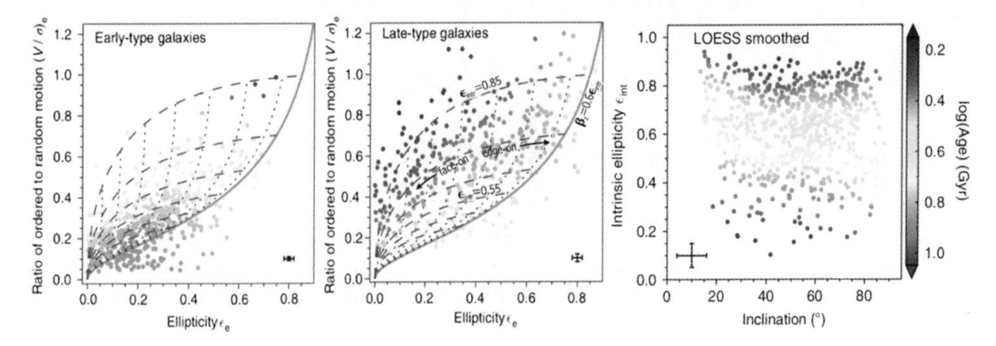

Figure 6. Left/middle panels: The ratio of ordered to random motions (V/σ) versus the apparent ellipticity for early/late-type galaxies. Right panel: assuming galaxies are oblate rotators, the derifed distribution of intrinsic ellipticity as a function of apparent inclination. [van de Sande *et al.* (2018), Figs. 3 & 4.]

belonging to different kinematic classes are distributed: spirals generally lie in the region consistent with rotationally-dominated kinematics, while regular (fast-rotating) early-type galaxies occupy a wider range of λ_R at given ϵ, with lower λ_R corresponding to systems with more pressure-support; slowly-rotating ('non-rotating') early-type galaxies mainly occupy the region with $\lambda_R < 0.15$ and $0 < \epsilon < 0.4$.

There is also an strong correlation between a galaxy's spin parameter and its intrinsic ellipticity, as demonstrated using the SAMI survey by van de Sande *et al.* (2018). Figure 6 shows the distribution of the ratio of rotation velocity to velocity dispersion (V/σ) with apparent ellipticity (ϵ) for early-type and late-type galaxies, together with the inferred distribution of intrinsic ellipticity ($\epsilon_{\rm int}$). This is derived using the theoretical model predictions for rotating, oblate, axisymmetric spheroids with varying intrinsic shape and anisotropy shown by the dashed and dotted lines in the left two panels. The galaxies are colour-coded by the luminosity-weighted age of their stellar populations, and the righthand panel shows the clear trend of age with intrinsic ellipticity. As van de Sande *et al.* (2018) discuss in detail, this newly discovered relation extends beyond the general notion that disks are young and bulges are old.

3.4. *The mass–angular momentum relation*

The mass–angular momentum (M-j) relation is discussed in detail elsewhere in these proceedings. However, it is worth noting the opportunties for studying this key relation

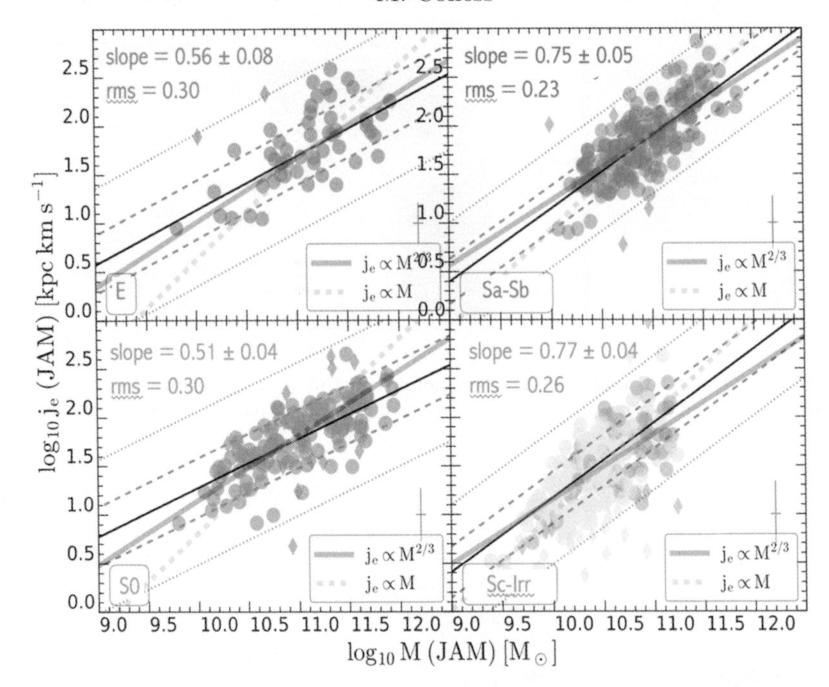

Figure 7. The relation between mass and specific angular momentum, both derived from Jeans anisotropic mass (JAM) models fitted to the SAMI kinematic data, for elliptical (E), lenticular (S0), early-spiral (Sa-Sb) and late-spiral/irregular (Sc-Irr) galaxies. [D'Eugenio *et al.*, in prep.]

that follow from large surveys providing kinematics for many galaxies. Some prelminary results from the SAMI survey are shown in Figure 7 (D'Eugenio *et al.*, in prep.), using hundreds of galaxies with masses and angular momenta derived from self-consistent dynamical models—in this case, Jeans anisotropic mass (JAM) models. This permits the study of the M-j relation for subsets of the population, such as different morphological types. While the results shown here are too preliminary to allow conclusions to be drawn, the opportunities are clear.

4. Summary

This is a golden age for studying galaxy angular momentum. Large kinematic surveys using integral field spectrographs are vastly increasing the amount and richness of the available information Sample sizes are now beginning to allow studies of the dependence on multiple simultaneous influences (mass/morphology/environment...) The main limitations remain instrumental trade-offs between spatial resolution and radial coverage, and challenges in spatial resolution and surface brightness at higher redshift.

References

Bellstedt, S., Forbes, D.A., Romanowsky, A. J., *et al.*, 2018, *MNRAS*, 476, 4543 [DOI: 10.1093/mnras/sty456]

Cortese, L., Fogarty, L.M.R., Bekki, K., *et al.*, 2016, *MNRAS*, 463, 170 [DOI:10.1093/mnras/stw1891]

Falcón-Barroso, J., 2016, *Astronomical Surveys and Big Data*, ASP Conf. Series, 505, 133 [https://ui.adsabs.harvard.edu/#abs/2016ASPC..505..133F]

Foster, C., van de Sande, J., Cortese, L., *et al.*, 2018, *MNRAS*, 480, 3105 [DOI: 10.1093/mnras/sty2059]

Graham, M.T., Cappellari, M., Li, H., *et al.*, 2018, *MNRAS*, 477, 4711 [DOI: 10.1093/mnras/sty504]

Greene, J.E., Leauthaud, A., Emsellem, E., *et al.*, 2018, *ApJ*, 852, 36 [DOI: 10.3847/1538-4357/aa9bde]

Pulsoni, C., Gerhard, O., Arnaboldi, *et al.*, 2018, *A&A*, 618, 94 [DOI: 10.1051/0004-6361/201732473]

Sánchez, Sebastián F., & The CALIFA Collaboration, 2015, *Galaxies in 3D across the Universe*, IAU Symposium 309, pp85-92 [DOI: 10.1017/S1743921314009375]

van de Sande, J., Scott, N., Bland-Hawthorn, J., *et al.*, 2018, *Nature Astronomy*, 2, 483 [DOI: 10.1038/s41550-018-0436-x]

Astronomy in Focus - XXX
Proceedings IAU Symposium No. XXX, 2018
M. T. Lago, ed.

doi:10.1017/S1743921319004113

The Fundamental Physics of Angular Momentum Evolution in a ΛCDM Scenario

Susana Pedrosa[1,2]

[1]Institute for Astronomy and Space Science, CONICET - UBA,
Ciudad Universitaria, Buenos Aires, Argentina
email: `supe@iafe.uba.ar`

[2]Dept. de Física Teórica, Univ. Autónoma de Madrid
Cantoblanco, Madrid, Espana

Abstract. Galaxy formation is a very complex process in which many different physical mechanisms intervene. Within the LCDM paradigm processes such as gas inflows and outflows, mergers and interactions contribute to the redistribution of the angular momentum content of the structures. Recent observational results have brought new insights and also triggered several theoretical studies. Some of these new contributions will be analysed here.

Keywords. galaxies: formation, galaxies: evolution

---

1. Introduction

Angular momentum exchange is ubiquitous in the structure assembly process. Every galaxy formation scenario assumes the exchange of angular momentum. Encrypted inside the today galactic morphology is stored the assembly history not only of the galaxy itself but also of the different regions of the DM halo. Estimations of the angular momentum content of the dark matter haloes require a connection between it and the galactic one.

In a hierarchical clustering universe the angular momentum budget of galaxies originates in primordial torques that act upon baryons and the dark matter. Fall (1983) proposed, in a seminal paper, the fundamental correlation between the angular moment (AM) content of the galactic components and the stellar mass of the galaxy. He found that both components follow a power law correlation with the stellar mass, with an exponent of ~ 0.6. The spheroidal component, although following a parallel sequence, presents an offset a factor of about 5 lower due to the lost of AM during the galaxy assembly. Theoretical models of galaxy formation (Fall & Efstathiou 1980, Mo *et al.* 1998) predict a linear correlation between the dark matter specific angular momentum and $M_{virial}^{2/3}$.

Within the hierarchical scenario galaxies are shaped and reshaped by several processes. For instance, supernova feedback that redistributes energy and mass through mass loaded winds is the key ingredient in the galaxy formation recipe that allows theoretical models to overcome the angular momentum catastrophe. With its inclusion more realistic galaxies could be obtained in cosmological simulations. Other processes that may ultimately determine the resulting galactic morphology are mergers, interactions and disc instabilities. For instance, mergers, in all their types, are the most accepted mechanism responsible for the formation of spheroidal galaxies.

2. Observational studies

In 2012 and 2013, Fall & Romanowsky (2013) (FR13) and Romanowsky & Fall (2012)) (FR12), revisited Fall (1983) using an improved and extended observational sample. They confirmed previous findings: all morphological types of galaxies lie along a parallel sequence with exponent $\alpha \sim 0.6$ in the stellar j - M plane. They proposed then that the j - M diagram constitutes a more physically motivated description of the galactic morphology than the typical disk to bulge classification.

Obreschkow & Glazebrook (2014), using high precision measurements of the stellar and baryonic specific AM of a THINGS sample of 16 spiral galaxies (Leroy *et al.* 2008), includes the β parameter (bulge fraction) in the AM description. They found a strong correlation for the plane fitting the 3D space of β, logM and log j. For a fixed β, the projection results in an exponent $\alpha \sim 1$, larger than the one found by Romanowsky & Fall (2012). They proposed that the contributions to the j-M plane of bulge and disks components were not independent.

After the FR12 and FR13 works, it followed a most interesting burst of numerical studies analysing to what extent their models fitted these observational constrains. Interestingly, despite the fact that this numerical experiments were perform with different numerical codes, different prescriptions for the subgrid physics, different resolutions, different feedback implementations, they all results in very good agreement with the new observational data. And through this process, many interesting contributions to the knowledge about the origin and evolution of the galactic AM during the assembly of the galaxy were developed. Numerical simulations constitute ideal tools for filling the gap between the observed and the inferred knowledge.

3. Numerical analysis

Genel *et al.* (2015) analysed the AM distribution of galaxies from the Illustris Simulation (Vogelsberger et al. 2014, Genel *et al.* 2014). They discriminate the $z = 0$ population based on their specific star formation rate, the flatness and the concentration, obtaining two parallel relations corresponding to early-and-late-type galaxies, in agreement with FR12 observations. They find that galactic winds with high mass-loading factors are crucial for getting the late-type galaxies relation that results from full conservation of the specific AM generated by cosmological tidal torques.

Using intermediate resolution cosmological simulations, Pedrosa & Tissera (2015) found the specific angular momentum of spheroidal and disk component to determine relations with the same slopes, regardless of the virial mass of their host galaxy. They found no evolution of this relation with redshift, indicating that spheroidal and disk component conserve similar relative amount of AM as they evolve, independently of virial masses. As shown in Fig.1, there is a clear correlation between the morphological type of the galaxy and its total specific AM content: higher D/T ratios are related with higher contents of specific AM. The AM of stellar bulges is consistent with elliptical galaxies indicating that bulges might be considered as mini-ellipticals, in agreement with FR12.

Teklu *et al.* (2015), using galaxies from the Magneticum Pathfinder Project (Dolag et al. 2015, in preparation), find that disk galaxies populate haloes with larger spin than those that host spheroidal galaxies. And disk galaxies live preferentially in haloes with central AM aligned with the AM of the whole halo. They also verify that their galaxies are located on the j-M plane in agreement with observations. Their stellar disk AM is lower than the cold gas one, in agreement with Pedrosa & Tissera (2015) and Obreschkow & Glazebrook (2014). They attribute this excess to the recent accretion of gas with high AM from the outer parts of the halo. This is shown in the fact that young stars (formed from this freshly accreted gas) present higher content of AM than older ones, Fig. 2.

Zavala *et al.* (2016), using a sample of over 2000 central galaxies extracted from the EAGLE simulation (Schaye *et al.* 2015, Crain *et al.* 2015) follow through time selected particles at z = 0 using a Lagrangian method. They find a correlation between the specific AM of $z = 0$ stars and that of the inner part of the DM halo. They find this link to be specially strong for stars formed before the turnaround. Spheroids, typically

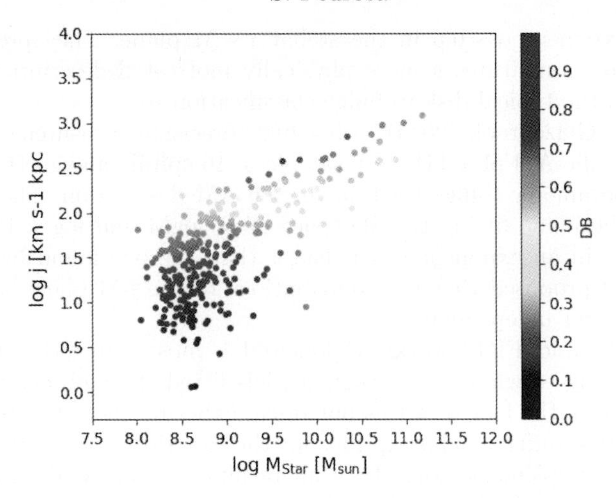

Figure 1. Relation between j and the total stellar mass for simulated galaxies with different disc-to-bulge ratios

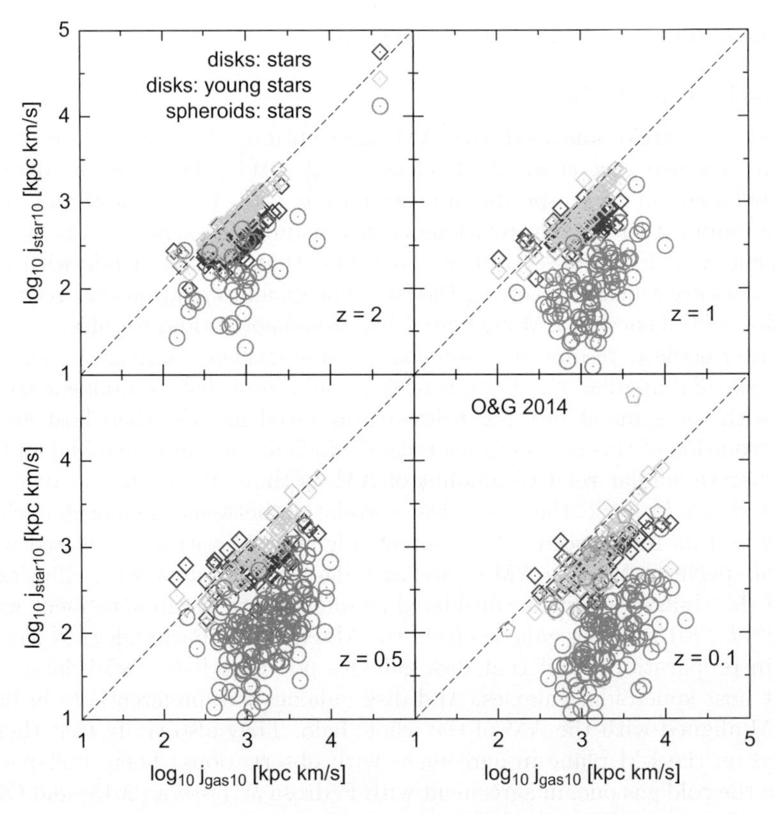

Figure 2. [Reprint from Teklu *et al.* (2015)] Specific angular momentum of the gas against the specific angular momentum of stars, both within 10% of the virial radius for galaxies that are classified as disks (blue diamonds) at four redshifts. For young stars only (turquoise diamonds). At z = 0.1 comparison with data from Obreschkow & Glazebrook (2014) (purple pentagons).

assembled at this epochs will then suffer loss of AM due to the merging activity of the inner halo assembly process. The cold gas, that mostly preserves the high specific AM acquired from the primordial tidal torques, forms stars after the turnaround and then

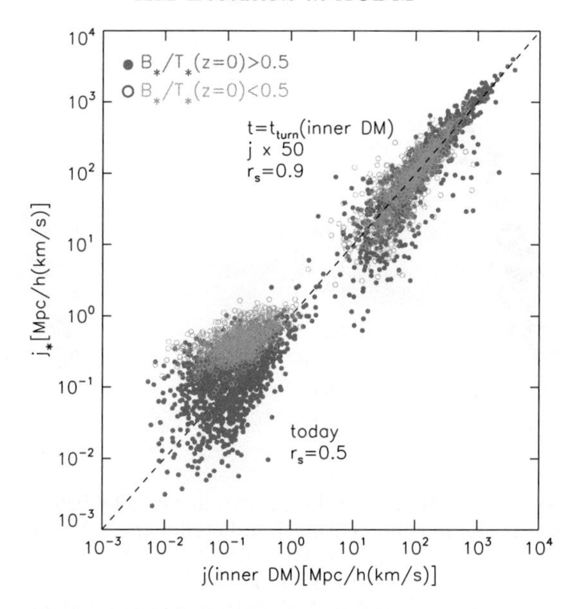

Figure 3. [Reprint from Zavala *et al.* (2016)] Correlation between the stellar specific angular momentum and that of the inner dark matter halo at $z = 0$ and at the time of turnaround of the inner dark matter halo (for the latter epoch, j is multiplied by 50). The sample of galaxies is divided into present-day bulge-dominated (solid red), and disc-dominated (open blue).

build the stellar disk component. They find that the inner DM halo loses 90% of its specific AM after turnaround through transfer to dark matter clumps. While bulge dominated objects tracks the inner halo, the disk dominated ones follow the whole DM halo closely, Fig. 3. They claim that most of the stars belonging to the today ellipticals were already formed at turnaround and then got locked inside the DM clumps that will form the inner halo.

Also using a Lagrangian method, Obreja *et al.* (2018) trace back in time structure progenitors of 25 zoom-ins simulations from the NIHAO Project (Wang *et al.* 2015), in order to dissect the AM budget evolution of the eight morphological components they identify (see also Dominguez Tenreiro *et al.* 2015 . They find that thin disks typically retain 70% of its AM while thick disks only a 40%. They also find that 90% of their velocity dispersion dominated objects in the sample retain less than 10% of the central AM. Regarding the rotation dominated structures, most of the thin disks has a retention factor greater than 50% while thick disks might loose as much as 85% of its AM.

Most recently, Fall & Romanowsky (2018) presented new observational evidence that reinforce their previous assumptions. They propose a simple model, valid for all morphological type, in which the stellar AM is the linear superposition of independent contributions from disks and bulges. They obey a power scaling relation with essentially the same coefficient but differs in their normalisation. They consider that the parallel sequences in the j - M plane corresponding to different values of β (bulge fraction) means that disks and spheroids are formed via independent processes. Spheroidal galaxies never acquired enough AM or else they loose it in violent events, while disk-dominated systems are the result of a more quiescent processes that were not affected by mergers. One of the main feature of the new dataset studied in F&R18 is that both, photometric and kinematic data, extend to large radii, taking into account tha fact that much of the AM lies beyond the effective radius. This issue was already pointed out by Lagos *et al.* (2018). Using simulated galaxies from EAGLE, they find that elliptical galaxies with the higher Sersic index have most of their stellar AM budget inhabiting beyond five half mass

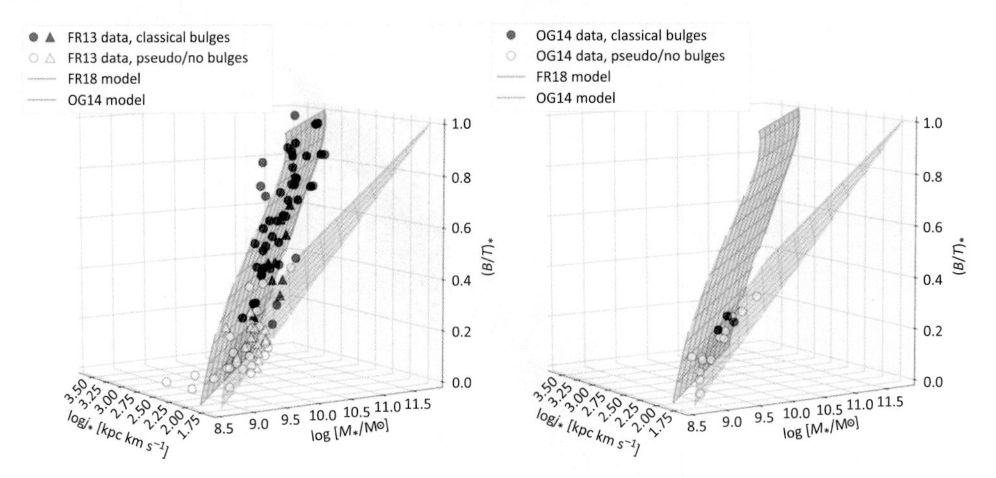

Figure 4. [Reprint from Fall & Romanowsky (2018), arXiv:1808.02525] Stellar bulge fraction beta against stellar specific angular momentum j and stellar mass M from Fall & Romanowsky (2018) (left panel) and from Obreschkow & Glazebrook (2014) (right panel). Filled symbols: galaxies with classical bulges, open symbols:pseudo bulges or no bulges. Orange surface:relation for independent disks and bulges derived in Fall & Romanowsky (2018), while the blue plane: linear regression derived by Obreschkow & Glazebrook (2014).

radius. Fall & Romanowski (2018) claim that the β - j - M 3D diagram is well fitted by a plane modeled as a linear superposition of independent contribution from disks and bulges, Fig. 4.

4. Conclusions

In the last few year substantial progress has been made in our understanding of the physical processes involved in the evolution of the galactic angular momentum throughout the assembly process of the galaxy. A bigger picture has been build that allows us to grasp the morphological classification of galaxies in terms of their positions in the angular momentum - Mass plane and the processes that determine how this position will evolve. But still important questions remains to be solved. For instance, why both, disk dominated and spheroidal galaxies, follow parallel relations independently of the mass. Higher resolution simulation and improved methods for mimic observations would probably bring some answers and, of course, new questions.

References

Crain, R. *et al.* 2015, *MNRAS*, 450, 1937
Dominguez Tenreiro, R. *et al.* 2015, *ApJL*, 800, 30
Fall, S. M. in IAU Symp. 100, Internal Kinematics and Dynamics of Galaxies, ed. E. Athanassoula 1983, *IAU*, 391,
Fall, S. M., & Efstathiou, G. 1980, *MNRAS*, 193, 189
Fall, S. M., & Romanowsky, A. J. 2013, *ApJL*, 769, L26
Genel, S., Vogelsberger, M., Springel, V., *et al.* 2014, *MNRAS*, 445, 175
Genel, S., Fall, S. M., Hernquist, L., *et al.* 2015, *ApJL*, 804, L40
Lagos, C. del P., Stevens, A. R. H., Bower, R. G., *et al.* 2018, *MNRAS*, 473, 4956
Leroy, A. K., Walter, F., Brinks, E., *et al.* 2008, *AJ*, 136, 2782
Mo, H. J., Mao, S., & White, S. D. M. 1998, *MNRAS*, 295, 319
Obreja, A., Dutton, A. A., Maccio, A. V., et al. submitted arXiv:1804.06635 2018, *MNRAS*,
Obreschkow, D., & Glazebrook, K. 2014, *ApJ*, 784, 26
Romanowsky, A. J., & Fall, S. M. 2012, *ApJS*, 203, 17
Schaye J., *et al.* 2015, *MNRAS*, 446, 521

Teklu, A. F., Remus, R.S., Dolag, K., *et al.* 2015, *ApJ*, 812, 29
Vogelsberger, M., Genel, S., Springel, V., *et al.* 2014, *Nature*, 509, 177
Wang L., Dutton A. A., Stinson G. S., Maccio A. V., *et al.* 2015, *MNRAS*, 454, 83
Zavala, J., Frenk, C. S., Bower, R., *et al.* 2016, *MNRAS*, 460, 4466

Astronomy in Focus - XXX
Proceedings IAU Symposium No. XXX, 2018
M. T. Lago, ed.

doi:10.1017/S1743921319004125

Angular Momentum Accretion onto Disc Galaxies

Filippo Fraternali[1] **and Gabriele Pezzulli**[2]

[1]Kapteyn Astronomical Institute, University of Groningen,
P.O. Box 800, 9700AV Groningen, The Netherlands,
email: `fraternali@astro.rug.nl`

[2]Department of Physics, ETH Zurich,
Wolfgang-Pauli-Strasse 27, 8093 Zurich, Switzerland
email: `gabriele.pezzulli@phys.ethz.ch`

Abstract. Throughout the Hubble time, gas makes its way from the intergalactic medium into galaxies fuelling their star formation and promoting their growth. One of the key properties of the accreting gas is its angular momentum, which has profound implications for the evolution of, in particular, disc galaxies. Here, we discuss how to infer the angular momentum of the accreting gas using observations of present-day galaxy discs. We first summarize evidence for *ongoing* inside-out growth of star forming discs. We then focus on the chemistry of the discs and show how the observed metallicity gradients can be explained if gas accretes onto a disc rotating with a velocity $20-30\%$ lower than the local circular speed. We also show that these gradients are incompatible with accretion occurring at the edge of the discs and flowing radially inward. Finally, we investigate gas accretion from a hot corona with a cosmological angular momentum distribution and describe how simple models of rotating coronae guarantee the inside-out growth of disc galaxies.

Keywords. Angular momentum, gas accretion, hot halo, metallicity gradient, corona

---

1. Introduction

At variance with massive quiescent ellipticals, which assembled most of their mass a long time ago and experienced, at some points in the past, an abrupt decline of their star formation rate, the majority of presently star forming galaxies have been undergoing, for most of the cosmic time, a rather constant or gently declining star formation history (e.g. Pacifici *et al.* 2016). This could in principle be explained either by a gradual consumption of a very large initial amount of cold gas, or by continuous accretion of new gas from the intergalactic medium. Both theory and observations strongly argue in favour of the second option, as i) gradual accretion is expected from the cosmological theory of structure formation (e.g. van den Bosch *et al.* 2014); ii) star forming galaxies have relatively short depletions times (Saintonge *et al.* 2011) and iii) preventing a huge initial reservoir of gas from very rapid exhaustion requires an implausibly low star formation efficiency, in stark contrast with observations (Kennicutt & Evans 2012; Fraternali & Tomassetti 2012).

Observing gas accretion into galaxies directly has proven challenging (Sancisi *et al.* 2008; Rubin *et al.* 2012). In a galaxy like the present-day Milky Way gas accretion does not seem to take place in the form of cold gas clouds at high column densities, like the classical high-velocity clouds (Wakker & van Woerden 1997) as their estimated accretion rate is too low (Putman *et al.* 2012) and their origin may be, at least partially, from a galactic fountain rather than a genuine accretion (Fraternali *et al.* 2015, Fox *et al.* 2016). Lower column densities have been probed in absorption but the accretion rates are more uncertain (Lehner & Howk 2011; Tumlinson *et al.* 2017). A possibility is that gas accretion takes place from the cooling of the hot gas, the galactic coronae that surround the Milky Way and similar galaxies (Miller & Bregman 2015). The cooling can be stimulated by the mixing with the disc gas through fountain condensation (Armillotta *et al.* 2016; Fraternali 2017). Alternatively, it may come from cold cosmological filaments directly reaching the discs (e.g. Kereš *et al.* 2009). To distinguish between these scenarios it is crucial to estimate the properties of the accreting gas.

A powerful tool to investigate the properties of the accreting gas is to infer them indirectly (backward approach) from observations of galaxy discs today. A simple example of this backward approach consists in the estimate of the accretion rates from the star formation histories (e.g. Fraternali & Tomassetti 2012). More elaborate estimates allow us to derive the angular momentum of the gas, the location where the accretion should take place and the properties of the medium from which the accretion originates. These topics are the focus of this proceeding.

2. Accretion of angular momentum on star forming galaxies

A very important observational fact about the evolution of currently star forming spiral galaxies, is that they have been increasing in size while increasing in mass (*inside-out growth*, e.g. Larson 1976; Dale *et al.* 2016). This is most likely due to the fact that the gas that has been accreted most recently is more rich in angular momentum with the respect to the one which was accreted at earlier epochs, a very well established prediction of the cosmological theory of tidal torques (Peebles 1969).

Crucially, observations indicate the radial growth – and therefore angular momentum accretion – is also a gentle process, which has been proceeding at a regular rate throughout galaxy evolution and is still ongoing today, as shown by studies of spatially resolved stellar populations (e.g. Williams *et al.* 2009; Gogarten *et al.* 2010) or recent star formation (e.g. Muñoz-Mateos *et al.* 2007). Pezzulli *et al.* (2015) have proposed a quantitative analysis of the phenomenon. They have shown that, *relative to* the well known exponential profile of the stellar mass surface density of spiral galaxies, the radial profile of the current *star formation rate surface density* shows a mild depletion in the inner regions and a slight enhancement in the outer ones, which agree both qualitatively and quantitatively with ongoing radial growth of stellar discs at a low but measurable rate. Figure 1 shows the distribution of the measured specific radial growth rate $\nu_R \equiv \dot{R}_\star / R_\star$ of the stellar scale-length $R_\star$ of the sample of nearby spiral galaxies from that study. The vast majority of objects are currently growing, at a rate about equal to one third of their *specific star formation rate* (sSFR, or $\nu_M \equiv \dot{M}_\star / M_\star \simeq 0.1\,\mathrm{Gyr}^{-1}$ at $z = 0$, e.g. Speagle *et al.* 2014). Furthermore, the results were shown to agree quantitatively with expectations for gradual angular momentum assembly of galaxies evolving *along* the specific angular momentum versus stellar mass (Fall) relation (Fall & Romanowsky 2013).

3. Disentangling models of accretion

We have seen that most spiral galaxies must have been (and probably are) gradually accreting angular momentum rich gas from the surrounding medium. The compelling question arises of what is the exact physical mechanism by which this happens. Two competing scenarios exist: *cold mode* accretion and *hot mode* accretion (e.g. Birnboim & Dekel 2003; Binney 2004) and different modes can dominate at different masses and redshift. In the former case, cold and angular momentum rich gas from intergalactic filaments joins the disc at large radii and then somehow drifts inwards to sustain star formation with the observed radial profile throughout the disc. We can call this *purely radial* accretion. In the second scenario, instead, the gas accreting onto the halo does not join the main body of the galaxy *directly*, but it is rather stored (together with its

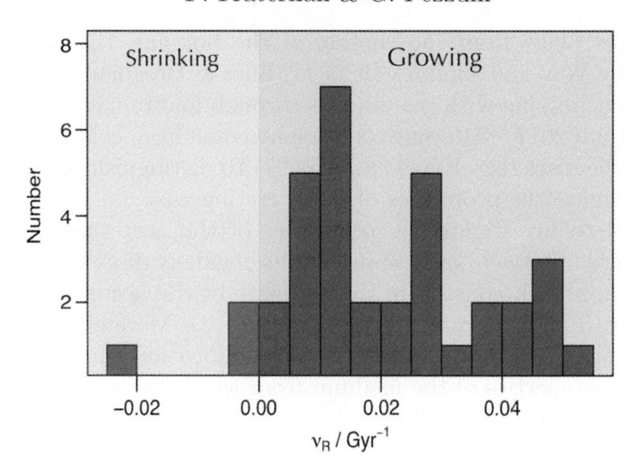

Figure 1. Distribution of the observed specific radial growth rates (ν_R) of a sample of nearby galaxies (Pezzulli *et al.* 2015). Note how the vast majority are growing inside out still at $z = 0$.

angular momentum) into a hot CGM (*corona*) and then only gradually condenses on to the disc as a gentle 'rain', which may be modelled, at least at first order, as mostly vertical accretion (perpendicular to the galaxy disc).

With appropriate choices of the parameters, both scenarios can give rise to the same *structural* evolution of the disc, as constrained by observations of the star and gas content of galaxies as a function of galactocentric radius and time (Pezzulli & Fraternali 2016a). The two models however differ enormously (and can thus be distinguished) in terms of *chemical* evolution. This is because vertical accretion of relatively metal-poor gas has a *metal-dilution* effect, which goes in the direction of counter-acting metal enrichment by local star formation, whereas radial accretion implies that, before arriving at the position where it is finally locked in to stars, each gas element will have already traversed other regions of the galaxy, where it will have been chemically enriched by the stars being formed there. We emphasize that i) this observational test is better performed on *gas-phase* abundances of α elements (as this choice minimizes uncertainties due to stellar radial migration and time delays in chemical enrichment) and ii) this kind of comparison between models is only meaningful *at fixed structural evolution*, as, otherwise, differences in other leading order effects (gas fraction, star formation efficiency and so on) dominate over those due to different geometry of accretion. With these specifications clarified, the discriminating power of the method is remarkable. This is illustrated in Figure 2, where the predictions are shown, for the abundance gradient of α-elements in the ISM of the Milky Way, for models with purely vertical, purely radial and mixed accretion. Details can be found in e.g. Pitts & Tayler (1989), Schönrich & Binney (2009) and Pezzulli & Fraternali (2016a). The latter work proposes an analytic and general approach to the problem, which can be readily applied to any galaxy or structural evolution model.

The clear result is that a combination of vertical and radial accretion is required to match the observed gradient (e.g. Genovali *et al.* 2015, marked here as a dashed line). This is actually not surprising, when angular momentum conservation is taken into account. A purely radial accretion, in fact, requires the angular momentum of the accreting gas to be transferred to some other not very well identified phase. On the other hand, purely vertical accretion is only possible if the material is accreted, at any radius, with exactly the angular momentum needed for local centrifugal balance, as any discrepancy would force the condensed gas to move radially within the disc after accretion (see also Mayor & Vigroux 1981; Bilitewski & Schönrich 2012). We now discuss whether a coherent physical picture can naturally give account of the findings discussed so far.

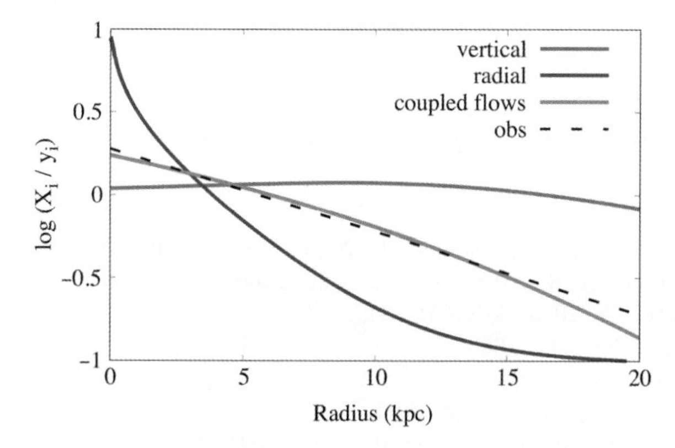

Figure 2. Abundance profiles of gas phase α-elements, as a function of the galactocentric radius, for the Milky Way at $z = 0$ predicted by disc evolution models in which the gas accretes purely vertically, purely radially or through mixed (coupled) flows. The observed profile is shown by the dashed curve. The models are calculated as described in Pezzulli & Fraternali (2016b).

4. A consistent picture

The model which better reproduces the chemical evolution of the Milky Way (§3) requires gas accretion occurring with a specific angular momentum that is $75 \pm 5\%$ of that of the disc at each radius. This is very naturally expected for a hot mode accretion scenario (i.e. from the hot corona). Virtually every hydrodynamically consistent model of the hot CGM requires in fact the hot gas to be *not* in local centrifugal equilibrium, as the high temperatures will generally imply a significant contribution of pressure support (in addition to rotation) against gravity. Note that the same model also predicts the presence of moderate radial flows (a few $\mathrm{km\,s^{-1}}$ or less) within the disc: crucially, however, this radial flow is not due to equatorial accretion of cold flows, but it is rather the natural consequence of the rotation lag of the accreting (hot) gas and angular momentum conservation.

Two main questions arise to further test whether the model is viable in a cosmological context. First, if the disc accretes material with a local deficit of angular momentum, can the accretion still *globally* provide enough angular momentum, as required to sustain the global radial growth of the disc (§2)? Second, is the implied rotation of the corona consistent with cosmological expectation from tidal torque theory?

The answer to these questions requires building self-consistent models of the hydrodynamical equilibrium of a hot rotating corona in a galaxy scale gravitational potential and with a given angular momentum distribution. Pezzulli, Fraternali & Binney (2017) described the solution to this problem and showed that a corona with a cosmologically motivated angular momentum distribution can naturally develop, *in the proximity of the disc*, rotation velocities close to the value required to match the chemical constraints. The model also predicts that the rotation velocity of the hot gas should drop significantly when approaching the virial radius.The first prediction is in excellent agreement with the recent observations by Hodges-Kluck *et al.* (2016); the second will require next-generation X-ray observations to be confirmed or discarded.

Pezzulli, Fraternali & Binney (2017) also found that the specific angular momentum of the inner corona increases rather steeply with radius and becomes larger than the *average* angular momentum of the disc at a radius $R_{\rm crit}$ slightly larger than the disc scale-length, but well within the range of direct contact between the galaxy and the hot halo. This is is sufficient to make the corona a plausible source of angular momentum growth, provided that the accretion of coronal gas is particularly efficient at relatively large radii, as predicted for instance by models of fountain-driven condensation (e.g. Marasco *et al.* 2012; Fraternali *et al.* 2013), and/or that the accretion is inhibited or counter-acted in the very central regions by star formation or AGN feedback (as suggested for instance for the Milky Way by the discovery of the Fermi bubbles; Su *et al.* 2010).

References

Armillotta, L., Fraternali, F. & Marinacci, F. 2016, MNRAS, 462, 4157
Bilitewski, T. & Schönrich, R. 2012, MNRAS, 426, 2266
Birnboim, Y. & Dekel, A. 2003, MNRAS, 345, 349
Binney, J. 2004, MNRAS, 347, 1093
Dale, D. A. 2016, AJ, 151, 4
Fall, S. M. & Romanowsky, A. J. 2013, ApJ, 769, 26
Fox, A. J., Lehner, N., Lockman, F. J., *et al.* 2016, ApJL, 816L, 11
Fraternali, F. & Tomassetti, M. 2012, MNRAS, 426, 2166
Fraternali, F., Marasco, A., Marinacci, F. & Binney, J. 2013, ApJL, 764L, 21
Fraternali, F., Marasco, A., Armillotta, L. & Marinacci, F. 2015, MNRAS, 447, L70
Fraternali, F. 2017, Gas Accretion onto Galaxies, ASSL, 430, 323
Gogarten, S. M. *et al.* 2010, ApJ, 712, 858
Genovali, K. *et al.* 2015, A&A, 580A, 17
Hodges-Kluck, E. J., Miller, M. J. & Bregman, J. N. 2016, ApJ, 822, 21
Kennicutt, R. C. & Evans, N. J. 2012 ARA&A, 50, 531
Kereš, D., Katz, N., Fardal, M., Davé, R., & Weinberg, D. H. 2009, MNRAS, 395, 160
Larson, R. B. 1976, MNRAS, 176, 31
Lehner, N., & Howk, J. C. 2011, Science, 334, 955
Marasco, A., Fraternali, F., & Binney, J. J. 2012, MNRAS, 419, 1107
Mayor, M. & Vigroux, L. 1981, A&A, 98, 1
Miller, M. J., & Bregman, J. N. 2015, ApJ, 800, 14
Muñoz-Mateos, J. C. *et al.* 2007, ApJ, 658, 1006
Pacifici, C., Oh, S., Oh, K., Lee, J. & Yi, S. K. 2016, ApJ, 824, 45
Peebles, P. J. E. 1969, ApJ, 155, 393
Pezzulli, G., Fraternali, F., Boissier, S. & Muñoz-Mateos, J. C. 2015, MNRAS, 451, 2324
Pezzulli G. & Fraternali F. 2016a MNRAS, 455, 2308
Pezzulli G. & Fraternali F. 2016b AN, 337, 913
Pezzulli, G., Fraternali, F. & Binney, J. 2017, MNRAS, 467, 311
Pitts, E. & Tayler, R. J. 1989, MNRAS, 240, 373
Putman, M. E., Peek, J. E. G., & Joung, M. R. 2012, ARA&A, 50, 491
Rubin, K. H. R., Prochaska, J. X., Koo, D. C., & Phillips, A. C. 2012, ApJL, 747L, 26
Saintonge A. *et al.* 2011, MNRAS, 415, 61
Sancisi, R., Fraternali, F., Oosterloo, T. & van der Hulst, T. 2008, A&ARv, 15, 189
Schönrich, R. & Binney, J. 2009, MNRAS, 396, 203
Speagle, J. S. *et al.* 2014, ApJS, 214, 15
Su, M. *et al.* 2010, ApJ, 724, 1044
Tumlinson, J., Peeples, M. S., & Werk, J. K. 2017, ARA&A, 55, 389
van den Bosch, F. C., Jiang, F., Hearin, A., *et al.* 2014, MNRAS, 445, 1713
Wakker, B. P., & van Woerden, H. 1997, ARA&A, 35, 217
Williams, B. F. *et al.* 2009, ApJ, 695, 15

FM7
Radial Metallicity Gradients in Star Forming Galaxies

Astronomy in Focus - XXX
Proceedings IAU Symposium No. XXX, 2018
M. T. Lago, ed.

doi:10.1017/S1743921319004137

FM 7: Radial metallicity gradients in star forming galaxies

Laura Magrini[1], **Letizia Stanghellini**[2] **and Katia Cunha**[3,4]

[1]INAF-Osservatorio Astrofisico di Arcetri- Firenze, Italy
email: laura.magrini@inaf.it

[2]National Optical Astronomy Observatory, Tucson, USA
email: letizia@noao.edu

[3]University of Arizona, Tucson, USA

[4]Observatrio Nacional, So Cristvo, Rio de Janeiro, Brazil
email: kcunha@on.br

For the XXXth IAU General Assembly in Vienna, Austria, in August 2018, we had the great opportunity to propose the Focus Meeting 7 (FM7) on *Radial metallicity gradients in star-forming galaxies* that was accepted with the endorsement and support of Division J (Galaxies & Cosmology) and of Division H (Interstellar Matter and Local Universe).

During this Focus meeting, we gathered different astrophysical communities and experts of different fields: from scientists interested in the resolved populations of our own Galaxy, to those focussed on extreme high-redshift galaxies, passing through large spectroscopic surveys in the local, intermediate- and high-redshift Universe. All these scientists have in common a strong interest in the study of radial metallicity gradients in star-forming galaxies, and each of them brings a different point of view, which is essential for a better understanding of all the facets of the topic. The aim of FM7 was to produce global view of the state-of-art of our knowledge of radial metallicity distributions in galaxies.

We had 11 invited reviews, 19 contributed talks, which are published in this Volume, and about 30 posters, which are published in the on-line version. In what follows, we summarise the content of these proceedings, highlighting the major results.

The Milky Way (MW) is an ideal model testbed for our understanding of galaxy formation and evolution. We have learned from Laura Inno how Cepheids can trace the Galactic metallicity gradient even in the extremely obscured regions of the disk, until the borders of the disk and the bulge, while Jorge García-Rojas has described the latest determinations of radial abundance gradients obtained from the analysis of HII regions and planetary nebulae in our Galaxy point towards an interesting flattening of the gradient in the innermost regions.

In recent years we have witnessed a flourishing of large spectroscopy surveys of the stellar components of the MW. The results of some of the most important ones have been presented during FM7. Maosheng Xiang has shown us results from the LAMOST Galactic spectroscopic survey characterising the stellar metallicity distributions at different positions and for different ages were presented. We have learnt that *age evolution* is not a synonym of *time evolution* and that many processes are in between what we measure and what models predict. Sofia Randich has presented the results from the Gaia-ESO survey, with particular emphasis on the Galactic radial distribution of metals in open clusters, and how results from very young star clusters present new and challenging observational constraints to be considered by theoretical models. Jo Bovy has introduced new results on the global chemical and spatial structure of the disc from the APOGEE survey were presented, highlighting the importance of mergers during the Galactic evolution.

Also, we have seen a massive increment of high-quality observational spectroscopic data for galaxies of the local and far Universe. Results from the CHAOS project, presented by Danielle Berg, have increased, by more than an order-of-magnitude, the number of H II regions with high-quality spectrophotometry available to facilitate the first detailed measurements of the chemical abundances of a statistically significant sample of nearby disc galaxies to both understand their chemical evolution and to calibrate high-redshift observations. The low-redshift Universe as seen from SDSS-IV MaNGA, the largest integral field spectroscopic survey of nearby galaxies to date, has been presented by Francesco Belfiore, builds on the understanding of the interplay between inflows, star formation and feedback processes in galaxy evolution.

Simone Bianchi and Lisa Kewley have discussed different aspects of measuring the metal content in galaxies, including alternative ways to measure the metal content through the dust-to-gas mass ratio, with an extensive overview of the application of emission-line diagnostics for measuring metallicity gradients in galaxies, summarising the current state-of-art in metallicity determinations from empirical, theoretical and Bayesian statistical methods.

Finally, we had the opportunity to hear about results from theoretical models and comparisons with the observations. Ivan Minchev showed the results on the abundance gradient evolution of groups of stars with similar ages, the so-called mono-age populations, in galaxy formation simulations, stressing the importance radial mixing, disc flaring, and inside-out disc formation to understand both models and observations. Patricia Tissera has presented a summary of the current state-of-knowledge from a numerical point of view and she has discussed the main results from the analysis of the EAGLE simulations.

Results of the large spectroscopic surveys and of several large programs were presented by many speakers (Friedrich Anders, Enrique Pérez-Montero, I-Ting Ho, Laura Sanchéz-Menguano, Jorge Barrera-Ballesteros, David Carton, Mirko Curt) together with results from less conventional tracers of the radial gradients, such as planetary nebulae (Sheila Flores Duran), blue and red supergiants (Lee Patrick) and DLA systems (Lise Christensen). Advances in theoretical models were also presented. (Mercedes Mollá, Fiorenzo Vincenzo, Francesca Fragkoudi, Lia Athanassoula). We also had exciting contributed talks on several topics, such as, isotopic ratios (Laura Colzi and Jiangshui Zhang) and the relation between dust and metals (Viviana Casasola).

We had a very stimulating and interesting Focus Meeting in Vienna, with more than 200 participants (we expected to have about 50 participants!). The interest on the topic was very strong and we even had several proposals for ensuing workshops on similar topics. It is a great pleasure to acknowledge the IAU Divisions J and H that supported us, the SOC for the help in selecting topics and speakers, and the IAU organisation for hosting us.

Laura Magrini, Letizia Stanghellini, Katia Cunha (co-Chairs and Editors of this Chapter)

Astronomy in Focus - XXX
Proceedings IAU Symposium No. XXX, 2018
M. T. Lago, ed.

doi:10.1017/S1743921319004149

The present-time Milky Way stellar Metallicity Gradient

Laura Inno

Max-Planck Institute for Astronomy, D-69117, Heidelberg, Germany
email: inno@mpia.de

Abstract. The present-time Milky Way (MW) radial metallicity gradient is a prime observable for galaxy evolution studies. Yet, a large diversity of measured gradients can be found in the literature, with values ranging from -0.01 to -0.09 dex kpc^{-1}, depending on the tracers used. In order to understand if this diversity comes from Galactic evolution processes or observational biases, stellar probes uniformly distributed across the disc and with accurately known ages and distances are needed. Classical Cepheids fulfil all these requirements and have been used to measure accurate abundance gradients in the MW. Here, I summarise some of the recent results based on Cepheids and on other stellar probes of similar age, and briefly discuss their implication for Galactic evolution.

Keywords. stars: variables: Cepheids – Galaxy: abundances

1. Cepheids as proxies of the gas-phase chemical composition

Classical Cepheids (Cepheids) have a number of properties that make them ideal proxies of the gas-phase chemical composition of the Galactic disc. In fact, they are luminous stars that can be seen even through severe dust extinction, and their individual ages and distances can be precisely determined on the basis of their periods (Bono *et al.* 2005, Inno *et al.* 2015). They are young (10-200 Myr), but compared to other young stars, they have lower temperature (T_{eff} ~5,500 K) and hence their spectra, rich in metal absorption lines, allow for precise abundance determinations of many chemical elements (e.g. da Silva *et al.* 2016 and references therein). Unfortunately, the use of Cepheids as chemical probes is currently limited by the small number of known Cepheids in the Galaxy. With respect to external galaxies (as, e.g., the Magellanic Clouds), Cepheids are poorly mapped in the MW: only ~600 have been identified until very recently, and, for the majority of them, accurate abundance measurements are available in the literature (Genovali *et al.* 2014). Even if limited in size, though, this sample allows us to probe a much larger volume of the MW disc with respect to the one accessible only through stars with geometric distances currently determined at comparable accuracy (~5%, see Panel *a*) in Fig. 1).

2. The shape of the MW metallicity gradient

The metallicity of the Cepheids in this sample is shown Panel *b*) of Fig. 1 and compared to the gradients based on stellar tracers with similarly young age, such as Open Clusters (OCs), OB stars, and Red Clump (RC) stars with ages <1 Gyr (references in the caption). The agreement is remarkably good if we consider only the range of Galactocentric radii between 5 and 10 kpc, but it vanishes when the entire range of distances is taken into account. The metallicity distribution of OCs shows a break in the slope at about 10–12 kpc, with a steepening in the inner and a flattening in the outer part, while RC stars show an opposite trend, with a break at 6 kpc, a flat gradient in the inner

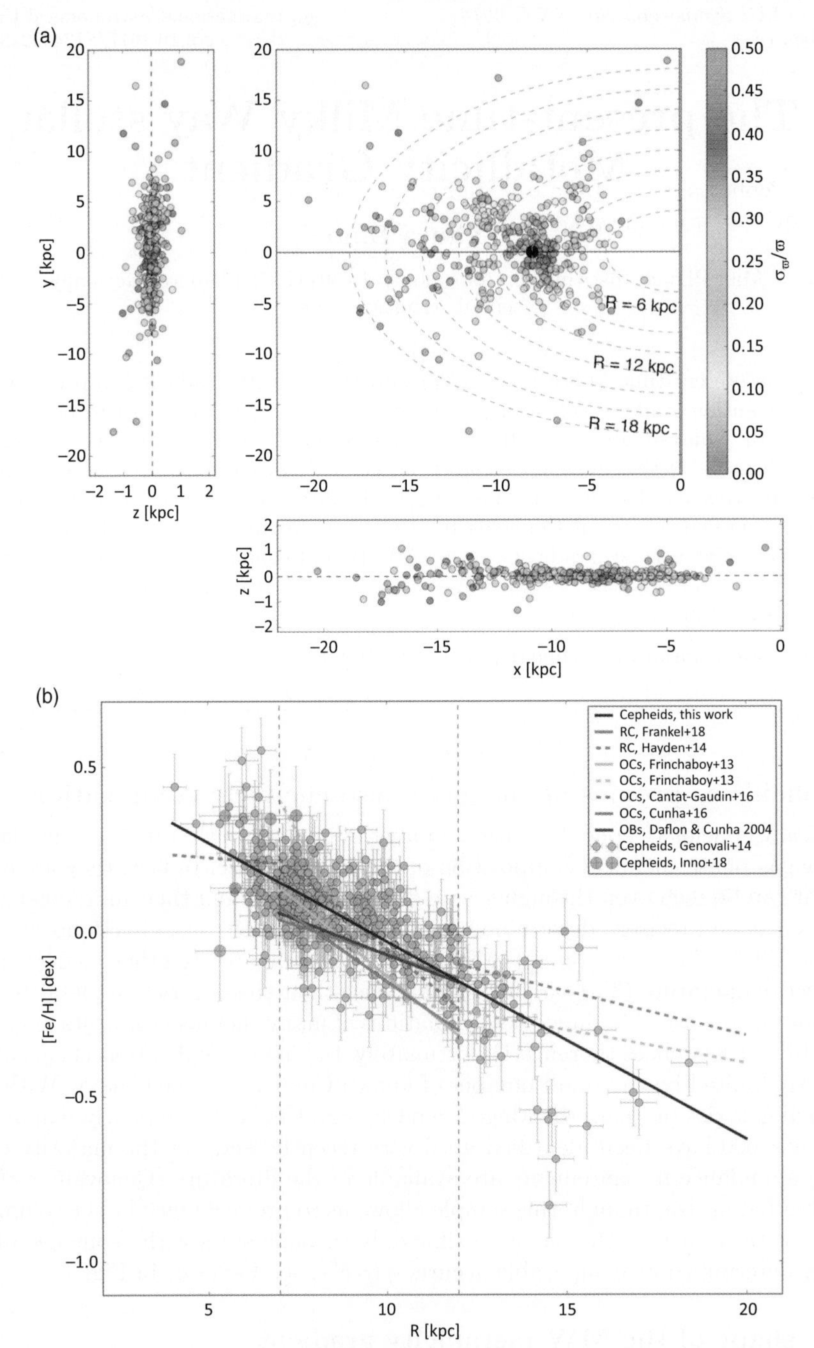

Figure 1. *Panel a)*: Distribution of the MW Cepheids projected into the Galactic plane and colour coded by the relative parallax error from Gaia DR2 (sample by Genovali *et al.* 2014). *Panel b)*: Metallicity distribution as a function of the Galactocentric radius R of the Cepheids in Panel *a*). Possible contaminants, such as Type-II (e.g. HQ Car, Lemasle *et al.* 2015) and Anomalous Cepheids (e.g. HK Cas), have been removed on the basis of their geometric parallax. A gradient of -0.057±0.002 dex kpc^{-1} is found (blue line) for the cleaned sample. The slopes of the metallicity gradient measured on the basis of OCs (Frinchaboy *et al.* 2013; Cunha *et al.* 2016; Cantat-Gaudin *et al.* 2016), OB stars (Daflon & Cunha 2004) and young RC stars (Hayden *et al.* 2014; Frankel *et al.* 2018) are also shown. The agreement is good for 6 kpc$\lesssim$R$\lesssim$12 kpc (vertical dashed lines) while different trends are found in the inner and outer regions.

region and a steep gradient in the central Galaxy. However, recent results based on new Cepheids identified in the inner disc (Inno *et al.* 2018) do not support the steepening in the inner part, but suggest that there is instead an increase of the metallicity spread around R∼5 kpc, due to the dynamical interaction with the central Bar.

The possible flattening in the outer region remains an open question, as the current sample of Cepheids at R>15 kpc is limited and potentially also biased. At R∼10 kpc, the disc deviates from the planar geometry and warps. Therefore, it is unclear whether the Cepheids are only found at larger height above the plane because of selection effects (i.e., the heavy extinction towards the middle plane), or because they trace the warp. Such selection effects will be soon better understood and eventually removed, thanks to all-sky, time-domain surveys (e.g., Gaia, OGLE-IV, ASAS-SN, PanSTARRs), which have just recently identified >2,000 new Galactic Cepheids (Jayasinghe *et al.* 2018). To take full advantage of this paradigm shift, we have designed an ongoing program with APOGEE-2 to spectroscopically observe hundreds of the newly identified Cepheids and at least threefold the current sample. This final spectroscopic sample of MW Cepheids will thus be a gold-mine for future studies on the MW structure and recent history.

References

Bono, G., Marconi, M., Cassisi, S., *et al.* 2005, *ApJ*, 621, 966
Cantat-Gaudin, T., Donati, P., Vallenari, A., et al. 2016, *A&A*, 588, A120
Cunha, K.; Frinchaboy, P. M.; Souto, D. *et al.* 2016, *AN*, 337, 922
Daflon, Simone; Cunha, Katia 2004, *ApJ*, 617, 1115
da Silva, R., Lemasle, B., Bono, G., *et al.* 2016, *A&A*, 586, A125
Frankel, N., Rix, H.-W., Ting, Y.-S., *et al.* 2018, *ApJ*, 865, 96
Frinchaboy, P. M., Thompson, B., Jackson, K. M., *et al.* 2013, *ApJL*, 777, L1
Genovali, K., Lemasle, B., Bono, G., *et al.* 2014, *A&A*, 566, A37
Hayden, M. R., Holtzman, J. A., Bovy, J., *et al.* 2014, *AJ*, 147, 116
Jayasinghe, T., Stanek, K. Z., Kochanek, C. S., *et al.* 2019, *MNRAS*, 486, 1907
Inno, L., Matsunaga, N., Romaniello, M., *et al.* 2015, *A&A*, 576, A30
Inno, L., Urbaneja, M. A., Matsunaga, N., *et al.* 2019, *MNRAS*, 482, 83
Lemasle, B., Kovtyukh, V., Bono, G., *et al.* 2015, *A&A*, 579, A47

Astronomy in Focus - XXX
Proceedings IAU Symposium No. XXX, 2018
M. T. Lago, ed.

doi:10.1017/S1743921319004150

Radial metallicity gradients with Galactic nebular probes

Jorge García-Rojas[1,2]

[1]Instituto de Astrofísica de Canarias, E-38200 La Laguna, Tenerife, Spain
email: jogarcia@iac.es

[2]Dept. de Astrofísica, Universidad de La Laguna, E-38206, La Laguna, Tenerife, Spain

Abstract. The study of radial metallicity gradients in the disc of the Milky Way is a powerful tool to understand the mechamisms that have been acting in the formation and evolution of the Galactic disc. In this proceeding, I will put the eye on some problems that should be carefully addressed to obtain precise determinations of the metallicity gradients.

Keywords. ISM: abundances, Galaxy: abundances, Galaxy: disk, H II regions, planetary nebulae: general

---

1. Introduction

It is well known that the abundance gradients in the disc of our Galaxy are more difficult to determine than in external galaxies mainly owing to distance uncertainties (especially for planetary nebulae) but also owing to the lack of objects in the inner and outer parts of the Galactic disc which, is heavily obscured by the presence of dust close to the Galactic plane. Owing to Fe is strongly depleted onto dust grains in the interstellar medium (ISM), other metallicity tracers should be used, mainly O, but also α-elements (Ne, Ar, Cl, S). In the case of H II regions, O is a probe to trace the present-day chemical composition of the ISM, and in PNe, O and α-elements are an archive of abundances in the past because in principle, these elements are not synthesized in the progenitor stars of PNe.

2. Radial metallicity gradients

There are many open problems with the abundance gradients of the Milky Way such as its possible temporal evolution, the existence or not of a flattening of the gradient in the outer (or in the inner) disc of the Galaxy, or the applicability of O as a reliable element to trace the metallicity in PNe. In the following I will briefly discuss some of the major sources of uncertainties in the determination of abundance gradients using spectrophotometric data of H II regions and PNe, that should be addressed to try to answer some of these open questions.

When trying to compute precise abundances in photoionised nebulae, and homogeneous analysis determining physical conditions and chemical abundances from the same set of spectra is mandatory (Perinotto & Morbidelli 2006). Additionally, the use of appropriate lines to compute abundances is very important; as an example, computing O^+/H^+ ratios from the trans-auroral [O II] 7320+30 lines could introduce undesired uncertainties owing to these lines could be strongly affected by telluric emission. Using physical conditions from radio recombination lines and optical or infrared lines to compute abundances can introduce systematic uncertainties owing to the different areas of the nebula covered in the different wavelength ranges. The use of appropriate atomic data is also very important; Juan de Dios & Rodríguez (2017) showed that atomic data variations could introduce differences of 0.1-0.2 dex in the derived abundances for low-density objects, but can reach or even surpass 0.6-0.8 dex at densities above 10^4 cm^{-3}, which could be very important for young and compact PNe. Finally, in spite of the big efforts made by several groups in the last years (see Stanghellini & Haywood 2010; Frew *et al.* 2016), distance determinations uncertainties are still one of the most important sources of uncertainty on the determination of the gradient, particularly for PNe. Unfortunately, precise parallaxes for PNe by the recent Gaia DR2 have only been provided for relatively nearby objects and, therefore do not significantly improve the scenario (Kimeswenger & Barría 2018).

Planetary nebulae. PNe are useful tools to constrain the chemical evolution of the Milky Way as they can probe the O abundance of the ISM over a range of epochs. There are several detailed studies in the literature that have found evidences of an evolution of the gradient with time (see e. g. Stanghellini & Haywood 2018, who find evidence of a steepening of the gradient with time). However, other studies (Henry *et al.* 2010, Maciel & Costa 2013) claim that owing to the observed scatter, and the uncertainties introduced by distance, age of the progenitor stars and other effects such as radial migration (Magrini *et al.* 2016), it is not clear that there has been an evolution of the gradient with time. Finally, an important point that should be taken into account is the recent discovery of the production of O in C-rich PNe at near-solar metallicities made by Delgado-Inglada *et al.*(2015). This overproduction of O (up to 0.3 dex) makes this tracer a doubtful archive of metallicities in the past if it is not taken into account.

H II *regions.* The radial metallicity gradient using H II regions has been widely studied (see Esteban & García-Rojas 2018 and references therein). However, there are several open problems, as the possible flattening of the gradient in the outskirts of the Galaxy, as it has been observed in other spiral galaxies (Bresolin *et al.* 2012). A recent study of the metallicity gradient covering a large range of Galactocentric distances and making use of very deep high-quality spectrophotometric VLT and GTC data has shown that there is not a flattening of the gradient in the Galactic anticentre (Esteban *et al.* 2017). Moreover, these data have also shown that the scatter at a given Galactocentric distance is of the order of the computed uncertainties, indicating that the ISM medium is well mixed at a given distance along the Galactic disc. Finally, the analysis of additional data have also shown a possible flattening or drop of the gradient in the inner disc (Esteban & García-Rojas 2018). This behaviour has been also previously found from metallicity distributions using cepheids and red giants, and in other spiral galaxies (see Esteban & García-Rojas 2018, and references therein).

References

Bresolin, F., Kennicutt, R. C., & Ryan-Weber, E. 2012, *ApJ*, 750, 122
Delgado-Inglada, G., Rodríguez, M., Peimbert, M. *et al.* 2015, *MNRAS*, 449, 1797
Esteban, C., Fang, X., García-Rojas, J., & Toribio San Cipriano, L. 2017, *MNRAS*, 471, 987
Esteban, C., & García-Rojas, J. 2018, *MNRAS*, 478, 2315
Frew, D., Parker, Q. A., & Bojičić, I. S. 2016, *MNRAS*, 455, 1459
Juan de Dios, L., & Rodríguez, M. 2017, *MNRAS*, 469, 1036
Henry, R. B. C., Kwitter, K. B., Jaskot, A. E. *et al.* 2010, *ApJ*, 724, 748
Kimeswenger, S., & Barría, D. 2018, *A&A*, 616, L2
Maciel, W. .J., & Costa, R. D. D. 2013, *RevMexAA*, 49, 333
Magrini, L., Coccato, L., Stanghellini, L. *et al.* 2016, *A&A*, 588, A91
Perinotto, M., & Morbidelli, L. 2006, *MNRAS*, 372, 45
Stanghellini, L., & Haywood, M. 2010, *ApJ*, 714, 1096
Stanghellini, L., & Haywood, M. 2018, *ApJ*, 862, 45

Astronomy in Focus - XXX
Proceedings IAU Symposium No. XXX, 2018
M. T. Lago, ed.

doi:10.1017/S1743921319004162

Stellar metallicity gradients of the Milky Way disc from LAMOST

Maosheng Xiang

Max-Planck Institute for Astronomy, Heidelberg, Germany
email: mxiang@mpia.de

Abstract. Stellar metallicity gradients set important constraints on the formation and evolution history of the Milky Way. We present radial and vertical metallicity gradients of the Galactic disc for mono-age stellar populations from the LAMOST Galactic Surveys, and discuss their constraints on the disc assemblage history.

Keywords. Galaxy: disc, Galaxy: formation, Galaxy: evolution

---

1. Introduction

The stellar metallicity gradients of the Galactic disc are consequences of a series of fundamental astrophysical processes, such as gas infall (and inflow), star formation and element enrichment, galaxy merger and secular dynamic evolution (stellar migration), they thus serve as a very useful tool for Galactic archaeology (e.g. Grisoni *et al.* 2018).

Both the radial and the vertical stellar metallicity gradients have been derived with various tracers (see a summary in Xiang *et al.* 2015). It is widely found that both the radial and the vertical metallicity gradients exhibit significant spatial variations (e.g. Cheng *et al.* 2012, Hayden *et al.* 2014, Huang *et al.* 2015). On the other hand, although there are some efforts to characterise the temporal evolution of the disc stellar metallicity gradients (Maciel *et al.* 2003, Nordström *et al.* 2004, Magrini *et al.* 2009, Stanghellini 2010, Casagrande *et al.* 2011), the results are still not conclusive due to the lack of large stellar samples with reliable age estimates. The LAMOST Spectroscopic Survey of the Galactic Anti-center (Liu *et al.* 2014, Yuan *et al.* 2015, Xiang *et al.* 2017) delivers robust stellar parameters (including metallicity and age) for millions of disc stars with simple yet statistically non-trivial target selection function, thus provides opportunities for systematic studies on the disc metallicity gradients, especially their time evolution.

2. Results and discussion

Fig. 1 plots the disc stellar metallicity gradients derived from a sample of 0.3 million LAMOST main-sequence turn-off (MSTO) stars with different ages (Xiang *et al.* 2015). The left panel shows that the radial gradients vary significantly with both $|Z|$ and age. The oldest populations (age > 11 Gyr) have almost *zero* radial gradients at all heights above the disc mid-plane. The younger populations exhibit negative radial gradients, which flatten with $|Z|$. From the population with age > 11 Gyr to 8–11 Gyr, the radial gradient shows an abrupt change from zero gradient to a strong negative gradient. The 6–8 Gyr population exhibits the steepest gradient at all heights, which means that the trend of radial gradient with age is not a monotonous one but exhibits a reverse behaviour: from steepening to flattening with time (i.e., with decreasing age). Such a reverse trend has been further confirmed by a recent work using a million stars from LAMOST with improved stellar parameters (Wang et al., MNRAS, in press). The right panel shows that

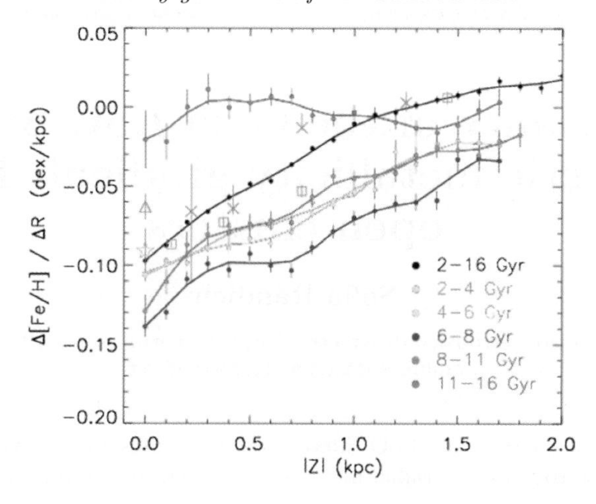

Figure 1. *Left*: Radial stellar metallicity gradient as a function of $|Z|$. *Right*: Vertical stellar metallicity gradient as a function R. Solid lines are smoothed results of the individual measurements (dots with error bars). Symbols in grey are measurements in literatures with different tracers (see Xiang *et al.* 2015 for details). The figures are adapted from Xiang *et al.* (2015).

the vertical metallicity gradients flatten significantly with R, and also vary with age. All populations, including the oldest one, exhibit negative vertical gradients.

Our results suggest the formation mechanism of the older (thick) disc is different to that of the younger (thin) disc. The old disc may have formed from gas with homogeneous metallicity distribution, which is probably consequence of a radially homogeneous star formation history or a fast and violent formation process. However, the negative vertical gradients suggest that the old (thick) disc was unlikely formed too fast and violent, and the upside-down mechanism may have played a significant role. The flattening trend of metallicity gradients with R and Z for the younger, mono-age populations are probably consequences of stellar migration and disc flaring (Minchev *et al.* 2014, Kawata *et al.* 2017). A flaring phenomenon for almost all mono-age disc populations have been predicted by simulations (Minchev 2016), and also been observed recently (Xiang *et al.* 2018). Nevertheless, mechanisms cause the reverse trend of metallicity gradient as a function of age need to be further understood.

References

Casagrande, L. *et al.* 2011, *A&A*, 530, 138
Cheng, J.Y. *et al.* 2012, *ApJ*, 746, 149
Grisoni, V., Spitoni, E. & Matteucci, F. 2018, *MNRAS*, 481, 2570
Hayden, M.R. *et al.* 2014, *AJ*, 147, 116
Huang, Y. *et al.* 2015, *RAA*, 15, 1240
Kawata, D. *et al.* 2017, *MNRAS*, 464, 702
Liu, X.-W. *et al.* 2014, *IAUS*, 298, 310
Maciel, W.J., Costa, R.D.D., & Uchida, M.M.M. 2003, *A&A*, 397, 667
Magrini, L., Sestito, P., Randich, S., & Galli, D. 2009, *A&A*, 494, 95
Minchev, I., Chiappini, C. & Martig, M. 2014, *A&A*, 572, 92
Minchev, I. 2016, *AN*, 337, 703
Nordström, B. *et al.* 2004, *A&A*, 418, 989
Stanghellini, L. & Haywood, M. 2010, *ApJ*, 714, 1096
Xiang, M.-S. *et al.* 2015, *RAA*, 15, 1209
Xiang, M.-S. *et al.* 2017, *MNRAS*, 467, 1890
Xiang, M.-S. *et al.* 2018, *ApJS*, 237, 33
Yuan, H.-B. *et al.* 2015, *MNRAS*, 448, 855

Astronomy in Focus - XXX
Proceedings IAU Symposium No. XXX, 2018
M. T. Lago, ed.

doi:10.1017/S1743921319004174

The relevance of the Gaia-ESO Survey on the Galactic metallicity gradient: Focus on open clusters

Sofia Randich

INAF-Osservatorio Astrofisico di Arcetri, Largo E. Fermi 5, I-50125, Firenze, Italy
email: sofia.randich@inaf.it

Abstract. An overview of the Gaia-ESO Survey project is presented, with focus on open star clusters and their use to trace the radial metallicity gradient in the thin disc.

---

1. The Gaia-ESO Survey

The Gaia-ESO Survey (GES- Gilmore *et al.* 2012; Randich *et al.* 2013) is a large public spectroscopic survey that has completed 340 observing nights on the ESO-VLT. GES used FLAMES to target 10^5 field stars in the bulge, the thick and the thin discs, and the halo, as well as a significant sample of open star clusters (OCs), providing a homogeneous overview of the distributions of kinematic, metallicity, and elemental abundances. Observations have been carried out using both Giraffe and UVES in parallel (Pancino *et al.* 2017). The analysis of the spectra yields radial and rotational velocities, stellar parameters and properties, metallicity and a few abundances for Giraffe targets; for the stars observed at high resolution with UVES detailed abundances for up to 32 chemical elements are also measured. The parameters and abundances of all the sample stars are homogenized to the same scale using a variety of calibrators.

GES has several advantages for the investigation of the radial metallicity gradient. Among them we mention that it is the largest stellar survey performed on a 8m class telescope, hence reaching fainter magnitudes and larger distances than other surveys. It is also the only one that has targeted all stellar populations, from pre-main sequence stars to evolved giants, from O-type stars to cool M dwarfs, from very young star forming regions (SFRs) to very old populations.

2. Open Clusters in the Gaia-ESO Survey

As mentioned above, GES has put a special emphasis on OCs, in order to address a variety of science goals, including cluster formation and dynamical evolution, stellar physics, and thin disc formation and evolution. The cluster sample is not only important to derive the gradient, but also to calibrate stellar ages (Randich *et al.* 2018), which are in turn fundamental to investigate the evolution of the gradient. GES observations have covered 65 and eight science and calibration OCs, respectively, while data for another 20 clusters have been retrieved from the ESO archive and homogeneously analyzed with the GES sample. The OC sample well covers the parameter space, distance from the Sun, age, and Galactocentric distance, in particular. Within each cluster several secure members are observed with UVES, to provide robust median metallicities and abundances (e.g., Spina *et al.* 2017 –S17; Magrini *et al.* 2017).

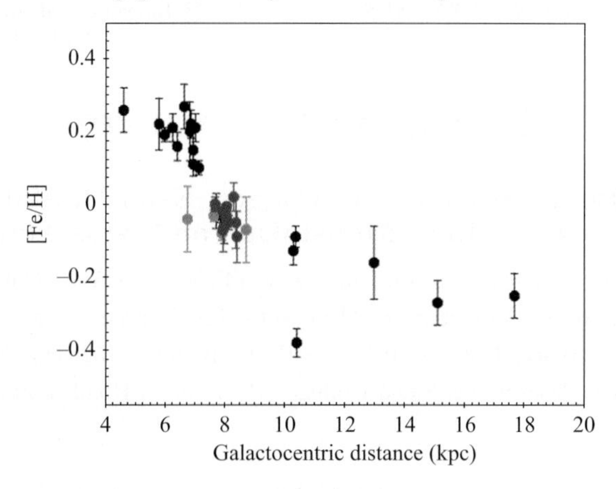

Figure 1. Update of Figure 5 in S17. [Fe/H] vs. Galactocentric distance for 38 OCs included in GES analysis cycle 5. Red/blue/black symbols: SFRs; age 10–100 Myr; age > 100 Myr.

3. A few highlights

Science analysis based on the first five cycles has confirmed the huge potential of GES OC sample to address the issue of the radial metallicity gradient (e.g., Jacobson *et al.* 2016). In particular, GES crucially allows the determination of the present-day metallicity distribution: S17 derived that based on the young clusters and SFRs analyzed in the fourth GES cycle, a few of which are located in the inner parts of the disc. Most surprisingly they found that all the young regions have close-to-solar or slightly sub-solar metallicities. At variance with the older OCs, the distribution traced by the young ones is almost flat, with the innermost SFRs having [Fe/H] values $0.10 - 0.15$ dex below the older counterparts located at similar Galactocentric radii (see also Figure 1). These findings favour models that predict a flattening of the radial gradient with time. On the other hand, the decrease of the average [Fe/H] at young ages is not easily explained by the models, and reveal a complex interplay of the several process that controlled the recent evolution of the thin disc.

4. Gaia-ESO and *Gaia*

The final analysis of GES will be completed soon, yielding homogeneous metallicities for the full sample of OCs (science, calibration, and archival) for a total of more than 90 objects. At the same time *Gaia* DR2 will allow the homogeneous determination of distances and ages, as well as to integrate the cluster orbits using proper motions. The joint dataset will allow further insights on the gradient, its time evolution, and dependence on the azimuth and height on the plane.

References

Gilmore, G., *et al.* 2012, *Msngr*, 147 25
Jacobson, H.R., *et al.* 2017, *A&A*, 591 37
Magrini, L., *et al.* 2017, *A&A*, 603 2
Pancino, E., *et al.* 2017, *A&A*, 598 5
Spina, L., *et al.* 2017, *A&A*, 601 70
Randich, S., Gilmore, G., and the Gaia-ESO Consortium 2013, *Mesngr*, 154 47
Randich, S., Tognelli, E., Kackson, R., and the Gaia-ESO collaboration 2018, *A&A*, 612 99

Astronomy in Focus - XXX
Proceedings IAU Symposium No. XXX, 2018
M. T. Lago, ed.

doi:10.1017/S1743921319004186

The CHAOS Survey

Danielle A. Berg[1], Richard W. Pogge[1], Evan D. Skillman[2], Kevin V. Croxall[3], John Moustakas[4] and Ness Mayker[1]

[1]Dept. of Astronomy, The Ohio State University, 140 W 18th Ave., Columbus, OH, 43210
[2]Institute for Astrophysics, University of Minnesota, 116 Church St., Minneapolis, MN 55455
[3]Illumination Works LLC, 5650 Blazer Parkway, Suite 152, Dublin OH 43017
[4]Dept. of Physics & Astronomy, Siena College, 515 Loudon Road, Loudonville, NY 12211

Keywords. galaxies: abundances, galaxies: evolution, galaxies: ISM, galaxies: spira l, ISM: HII regions, ISM: abundances, ISM: lines and bands

---

1. Introduction

Our very best understanding of the abundances in spiral galaxies comes from detailed spectroscopic studies and measurements of their faint, sensitive emission lines using either temperature-sensitive auroral collisionally-excited lines (CELs) or recombination lines (RLs). However, the RL method is known to produce systematically higher oxygen abundance measurements than the CEL-method (e.g., García-rojas & Esteban 2007), and which method is more accurate is still debated in the astronomical community. Because of the inherent faintness of both the auroral CELs ($\sim 10^2\times$ fainter than Hβ) and the RLs ($\sim 10^3\times$ fainter than Hβ), extragalactic abundance measurements are commonly based on empirical calibrations to stronger emission lines or photoionization modeling. Unfortunately, these so-called strong-line calibrations, which are the typical method of determining galactic abundances in lower S/N spectra, including IFU surveys and high redshift targets, demonstrate large systematic discrepancies amongst different methods (Kewley & Ellison 2008). Thus, all ISM abundances derived from emission-line spectra, especially those based on strong-line calibrations, are fundamentally limited by the current lack of an accurate absolute abundance scale.

The CHemical Abundances of Spirals (CHAOS) survey leverages the combined power of the dual 8.4m mirrors on the Large Binocular Telescope (LBT) with the broad spectral range (3400Å $< \lambda < 1\mu$m) and sensitivity of the Multi Object Double Spectrograph (MODS; Pogge *et al.* 2010) to measure the physical conditions and abundance gradients of nearby spiral galaxies. With a field of view comparable to the angular diameters of the nearest spiral galaxies, we have used laser-machined slit masks to obtain multi-object spectroscopy of $\sim$650 H II regions across 11 face-on spiral galaxies. To date, 5 galaxies have been analyzed with an unprecedented number of temperature-sensitive auroral CEL detections: NGC 628 (46 regions; Berg *et al.* 2015), NGC 5194 (30 regions; Croxall *et al.* 2015), NGC 5457 (75 regions; Croxall *et al.* 2016), NGC 3184 (32 regions; in prep), and M 33 (in prep).

2. First Results from CHAOS

CHAOS has produced the richest data set to date of direct elemental abundances of HII regions in nearby bright spiral galaxies derived from electron temperature (T_e) measurements. In Berg *et al.* (2015) we observed significant auroral line detections from one or more ions in 47 HII regions across the disk of our first CHAOS target, NGC 628, amassing an unprecedented total of 126 individual auroral line measurements within a single galaxy. Comparing derived temperatures from these auroral line measurements revealed some surprising results. While the [OIII] $\lambda4363/(\lambda\lambda4959{,}5007)$ auroral-to-strong-line ratio has long been the principle nebular temperature diagnostic, we found large discrepancies for temperatures based on [OIII] $\lambda4363$. Yet, temperatures based on [SIII] $\lambda6312$ and [NII] $\lambda5755$ showed a very tight relationship, indicating that these temperatures are reliable. Further, the dispersion in the O/H gradients in our sample is minimized when prioritizing abundances based on T_e[SIII] and T_e[NII] over T_e[OIII].

The CHAOS $T_e - T_e$ results have been confirmed in additional targets: NGC 5194 (Croxall *et al.* 2015), NGC 5457 (Croxall *et al.* 2016), and NGC 3184 (Berg *et al.*, in prep). This CHAOS dataset reveals offsets from the commonly used $T_e - T_e$ relationships predicted by photoionization modeling (e.g., Garnett et al. 1992). In Croxall *et al.* (2016) we provide updated $T_e - T_e$ relationships that are empirically calibrated using T_e detections from 74 HII regions in NGC 5457. These data also allowed us to robustly measure the physical conditions in a significant number of low-ionization HII regions, and propose new empirical ionization correction factors for S and Ar.

3. Next Steps: True Chemical Abundances of Spiral Galaxies

While the unprecedented number of auroral line detections in the CHAOS dataset has led us to breakthroughs in our understanding of HII region abundance determinations, further progress requires additional avenues of investigation. First, we have been able to measure the *Balmer Jump* in the nebular continuum near $\lambda3650$ in a significant number of CHAOS spectra (CHAOS team, in prep.). This provides an estimate of the mean T_e in an HII region derived independently from the CEL temperatures, allowing us to evaluate the importance of fluctuations within the regions (c.f., Guseva *et al.* 2006, 2007). A second measure of temperature inhomogeneities comes from the handful of CII RL measurements observed in each of two CHAOS galaxies to date (M101 and M33; Skillman et al., in prep.). Because RLs are far less sensitive to the T_e (linear dependence) than CELs (exponential dependence), temperature fluctuations have been proposed as the source of the discrepancies in abundances derived from the two methods (e.g., Peimbert 1969). Follow-up UV spectra will be obtained for these regions in M101 using the Cosmic Origins Spectrograph (COS) on the Hubble Space Telescope (*HST*) in Cycle 26 (HST-GO-15126; PI: Berg). The OIII] $\lambda\lambda1661{,}1666$ and CIII] $\lambda\lambda1907{,}1909$ CELs will be measured from these spectra, allowing the most accurately measured carbon abundance gradient in a spiral galaxy to date. Many of the HII regions in our sample have also been observed in the mid-IR with Spitzer or Herschel. These spectra allow us to compare abundances from the temperature-insensitive fine-structure IR emission lines to our CHAOS abundance for a third measure of temperature fluctuations (e.g., Garnett *et al.* 2004; Croxall *et al.* 2013).

Finally, we can measure present-day abundances from blue super giant (BSG) stars, which recently formed from and have the same chemical composition as the gas in their surrounding HII regions, as standards for the absolute abundance scale. The extreme brightness of BSGs makes it possible to obtain resolved spectra of individual BSGs out to 10 Mpc with current instrumentation. From the CHAOS sample, 10 galaxies are within this 10 Mpc distance necessary to accurately measure abundance from BSGs. BSG abundance gradients have been measured for several nearby spiral galaxies, including NGC 300 (Kudritzki *et al.* 2008), NGC 3031 (Kudritzki *et al.* 2012), and M83 (Bresolin *et al.* 2016). Spectroscopic observations of BCGs and/or HII regions are planned for the CHAOS sample, from which we will compare CHAOS nebular abundances with the BSG abundances as a function of radius from each galaxy's center. This will allow us to further resolve the CEL/RL abundance discrepancy problem and better anchor the strong-line calibrations.

4. Summary

The CHAOS survey has measured high-quality optical spectra in 100s of HII regions across 11 nearby spiral galaxies. This dataset represents the most accurate and precise study of direct CEL abundances across a broad range of physical conditions in extragalactic HII regions to date, and so provides a rich database from which an absolute nebular abundance scale can be calibrated.

References

Berg, D. A., *et al.* 2015, *ApJ*, 806, 16
Bresolin, F., *et al.* 2016, *ApJ*, 830, 64
Croxall, K. V., *et al.* 2013, *ApJ*, 777, 96
Croxall, K. V., *et al.* 2015, *ApJ*, 808, 42
Croxall, K. V., *et al.* 2016, *ApJ*, 830, 4
Kewley, L. J., & Ellison, S. L., 2008, *ApJ*, 681, 1183
Kudritzki, R.-P., *et al.* 2012, *ApJ*, 747, 15
Kudritzki, R.-P., *et al.* 2008,*ApJ*, 681, 269
García-Rojas & Esteban 2007, *ApJ*, 670, 457
Garnett, D. R. 1992, *AJ*, 103, 1330
Garnett, D. R., *et al.* 2004, *AJ*, 128, 2772
Guseva, N. G., *et al.* 2006, *ApJ*, 644, 890
Guseva, N. G., *et al.* 2007, *ApJ*, 464, 885
Pogge, R. W., *et al.* 2010, *Proc. SPIE*, 7735, 77350A

Astronomy in Focus - XXX
Proceedings IAU Symposium No. XXX, 2018
M. T. Lago, ed.

doi:10.1017/S1743921319004198

Metallicity gradients in nearby star forming galaxies

Francesco Belfiore

University of California Observatories - Lick Observatory, University of California Santa Cruz, 1156 High St., Santa Cruz, CA 95064, USA
email: fbelfior@ucsc.edu

Abstract. I study the gas phase metallicity (O/H) radial profiles in a representative sample of 550 nearby star forming galaxies with resolved spectroscopic data from the SDSS-IV MaNGA survey. Using strong-line ratio diagnostics (R23 and O3N2) and referencing to the effective (half-light) radius (R_e), I find that the metallicity gradient steepens with stellar mass going from $\log(M_\star/M_\odot) = 9.0$ to $\log(M_\star/M_\odot) = 10.5$. At higher masses a flattening of the metallicity radial profile is observed in the central regions ($R < 1\ R_e$). These findings are in agreement with recent independent analysis of other large samples of nearby galaxies.

Keywords. galaxies: abundances, galaxies: ISM, galaxies: evolution

1. Introduction

Metals are direct products of stellar nucleosynthesis, making chemical abundances ideal tracers of the integrated history of star formation and gas flows in and out of galaxies. In particular, metallicity radial gradients in galaxies have long been a source of considerable interest, as they are directly related to the process of disc assembly (Larson 1976; Matteucci & François 1989; Pezzulli & Fraternali 2016)

Observationally, it is well known that in the local Universe disc galaxies present a negative metallicity gradient (e.g. Vila-Costas & Edmunds 1992; Van Zee *et al.* 1998). Only recently, however, integral field spectroscopy (IFS) surveys of nearby galaxies have provided large data sets suitable for systematically studying the shape of the gas phase metallicity gradients in the nearby Universe. In particular the recent CALIFA (Sánchez *et al.* 2012), SAMI (Croom *et al.* 2012) and SDSS IV MaNGA (Bundy *et al.* 2015) surveys aim to observe $\sim$ 700, 3000 and 10000 galaxies respectively at $\sim$kpc resolution.

2. Metallicity gradients in the local Universe

In Belfiore *et al.* (2017) I make use of data from the MaNGA survey to derive gas-phase metallicity radial profiles for a sample of 550 local star forming galaxies. The uniformity of the dataset over the stellar mass range $9.0 < \log(M_\star/M_\odot) < 11.0$, where the MaNGA sample is fully representative of the local galaxy population, makes this study the most comprehensive analysis of metallicity radial profiles in galaxies to date. Star forming regions are selected from the [SII]-BPT diagram (Baldwin, Phillips & Terlevich 1981, also Belfiore *et al.* 2016), since at the resolution of the MaNGA survey ($\sim$2 kpc) one cannot identify individual HII regions. Gas-phase metallicity is derived with two complementary strong line calibrations, that of Maiolino *et al.* (2008) based on the $R23 = ([OIII]\lambda\lambda4959, 5007 + [OII]\lambda\lambda3727, 29)/H\beta$ index, and that of Pettini & Pagel (2004) based on the $O3N2 = ([OIII]\lambda5007/H\beta)/([NII]\lambda6584/H\alpha)$ index. Metallicity radial profiles for different mass bins are shown in Fig. 1. Using either metallicity calibration,

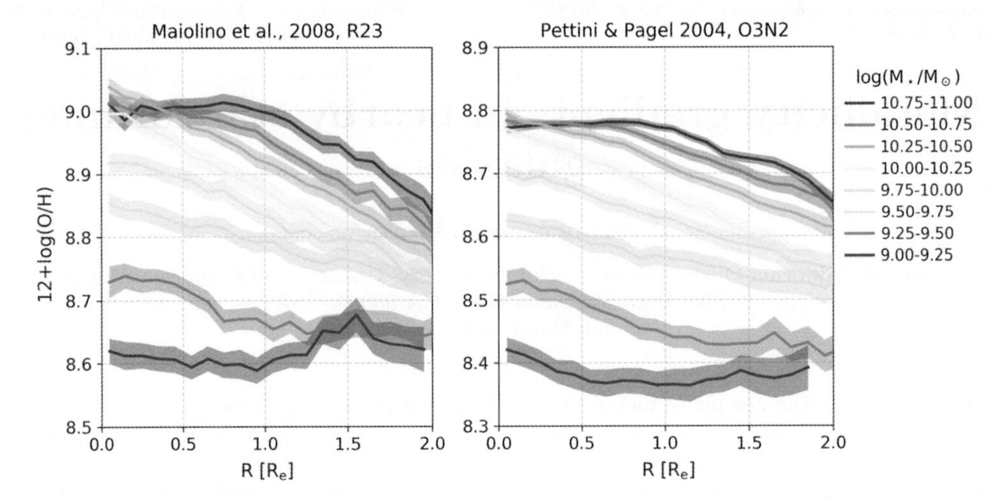

Figure 1. *Adapted from Belfiore et al. (2017).* Gas-phase metallicity radial profiles gradients for 550 star forming galaxies from the MaNGA survey. Radial metallicity profiles are shown in 0.25 dex mass bins and for two different strong line metallicity calibrators (the Maiolino *et al.* 2008 calibration on the R23 on the left and the Pettini & Pagel 2004 metallicity calibration based on O3N2 on the right).

the shape of the metallicity gradient changes as a function of stellar mass, lying roughly flat among galaxies with $\log(M_\star/M_\odot) = 9.0$ but exhibiting slopes as steep as -0.14 dex R_e^{-1} at $\log(M_\star/M_\odot) = 10.5$ (using R23, but equivalent results are obtained using O3N2). Galaxies of higher stellar mass, on the other hand, show a flattening in the regions $<1.0\ R_e$. Interestingly, these features have also been described in the most recent analysis of the CALIFA data (Sánchez-Menguiano *et al.* 2016) and in a study based on higher-resolution MUSE data (see these proceedings and Sánchez-Menguiano *et al.* 2017). In light of these recent results it appears that the change in shape of the metallicity radial profiles as a function of stellar mass is now a robust observational result, borne out in independent analysis of different datasets. Future theoretical work will need to address the origin of these trends.

References

Baldwin J. A., Phillips M. M., Terlevich R. 1981, *PASP*, 93, 5
Belfiore F. *et al.* 2016, *MNRAS*, 461, 3111
Belfiore F. *et al.* 2017, *MNRAS*, 469, 151
Bundy K. *et al.* 2015, *ApJ*, 798, 7
Croom S. *et al.* 2012, *MNRAS*, 421, 872
Larson R. B. 1976, *MNRAS*, 176, 31
Maiolino R. *et al.* 2008, *A&A*, 488, 463
Matteucci F. & François P. 1993, *MNRAS*, 239, 885
Pettini M. & B. E. J. Pagel 2004, *MNRAS* 348, L59
Pezzulli G. & Fraternali F. 2015, *MRNAS*, 455, 2308
Van Zee L., Salzer J. J., Haynes M P. 1998, *ApJ*, 497, 10
Vila-Costas N. B. & Edmunds M. 1992, *MNRAS*, 259, 121
Sánchez S. F. *et al.* 2012, *A&A*, 538, A8
Sánchez-Menguiano L. *et al.* 2016, *A&A*, 587, A70
Sánchez-Menguiano L. *et al.* 2017, *A&A*, 609, A119

Astronomy in Focus - XXX
Proceedings IAU Symposium No. XXX, 2018
M. T. Lago, ed.

doi:10.1017/S1743921319004204

Dust-to-gas ratio and metallicity gradients in DustPedia galaxies

Simone Bianchi[1] **and the DustPedia consortium**

[1]INAF-Osservatorio Astrofisico di Arcetri, Largo E. Fermi, 5, 50125, Firenze, Italy
email: simone.bianchi@inaf.it

Abstract. Far-infrared photometric observations from the *Herschel* Space Observatory offer the opportunity to study the dust-to-gas ratio at a resolved scale in nearby galaxies. The amount, and gradient, of solid-phase metals can thus be compared with metallicity measurements in the gas phase. We describe our preliminary work on the topic with data from the DustPedia project.

Keywords. dust, extinction, galaxies: ISM, galaxies: abundances, galaxies: structure

---

1. Introduction

A considerable fraction of the heavy elements in the ISM is locked-up in dust: for the local medium, the observed depletions show that half of the total metal mass is in the gas, half in grains (with a dust-to-metal ratio, DTM$\approx$0.5; see, e.g., Clark et al. 2016). The dust-to-gas mass ratio (DGR) can thus be used as a tracer of the solid-phase metallicity. For galaxies, it is tempting to use the DGR also as a proxy of the gas-phase metallicity, assuming the constant DTM derived in the local MW ISM. A direct association of DGR and metallicity (typically the O/H ratio) is however prevented by several uncertainties. For instance, the estimate of the dust mass depends on the opacity cross sections: the mass can vary by a factor of three or more, depending on the assumed dust-grain model (Casasola *et al.* 2017); also, the derivation of the DGR requires an estimate of the gas molecular component, which is usually derived indirectly from CO. Thus, the uncertainties in the DGR are not significantly different from those on metallicity, i.e. the usually-quoted 0.7-dex spread due to the use of different metallicity calibrators. The possible dependence of some of the factors (e.g. DTM and the CO-to-H_2 conversion factor, α_{CO}) on metallicity itself also affects the comparison between DGR and metallicity gradients. With these caveats, we nevertheless attempted a few study on the subject for some of the galaxies in the DustPedia sample.

2. The DustPedia database

DustPedia - A definitive study of Cosmic Dust in the local Universe (Davies *et al.* 2017; http://www.dustpedia.com) is a collaborative-focused project supported by the EU FP7 grant 606847. Its goal is to perform a complete characterisation of dust in local galaxies including: its composition and physical properties in different galactic environments; its origin and evolution; its relation to other ISM components; its effects on stellar radiation; the connection between SF indicators and dust emission; the contribution of local galaxies to the cosmic far-IR background. The research is carried out on a sample of 875 objects, i.e. almost all the large (D25>1') and nearby (z<0.01) galaxies observed by the *Herschel* Space Observatory. A legacy database was constructed from *Herschel* and other UV-to-microwave observations, containing multi-wavelength imagery and aperture-matched photometry in up to 42 bands (Clark *et al.* 2018; data are publicly available at http://dustpedia.astro.noa.gr/). Eventually, the database will include data on the atomic and molecular gas from the literature (Casasola, see this volume) and on metallicity from literature and public MUSE observations (De Vis *et al.* 2019). Homogeneously-derived metallicities are available for over 10000 position across about 500 galaxies, for seven different calibrators. Characteristic (global) metallicities at $0.4R_{25}$ have been derived using Bayesian gradient estimates. They are currently being analysed in the framework of dust/metallicity evolutionary models (see, e.g. De Vis *et al.* 2017) About 70 galaxies have low-uncertainty estimates of the metallicity gradient.

3. The DGR as a tracer of metallicity

In Magrini *et al.* (2011) we studied the DGR and O/H gradients in four galaxies of the *Herschel* Virgo Cluster Survey (Davies *et al.* 2012), a precursor to the DustPedia project. Maps of the dust surface density were obtained by pixel-to-pixel fitting of the far-infrared/submm spectral energy distribution (SED); the DGR was derived with different choices of α_{CO}; it was converted into the same scale as O/H observations, assuming a MW DTM; and its gradient compared to that of the metallicity. We found that an α_{CO} lower than the value used for the MW provide a better match between the gradients, supporting a dependence of the factor on metallicity. We plan to extend the work to the largest DustPedia galaxies, for which Casasola *et al.* (2017) found that radial gradients in the dust surface density are flatter than the stellar, and intermediate between those of the molecular and gas components.

Conversely, for a given α_{CO} and constant DTM, one can derive the dust mass from the metallicity, and compare it with the value obtained using standard SED fitting methods. The comparison yields an estimate for the dust absorption cross-section, independently of a dust-grain model. Such measurements have been made at a global scale, using integrated galactic properties (see, e.g. Clark *et al.* 2016) and are now being attempted at a resolved scale for a few large DustPedia galaxies (Clark *et al.* 2019). Preliminary results for M83 show an increase of the cross section at 500μm with galactocentric distance, with values encompassing those of current state-of-the-art dust models and ranging over more than 1-dex. The results could be due to real changes in the dust properties with radius, or depend on the assumption of a constant DTM ratio. In fact, a DGR steeper than the metallicity gradient, and thus a variation of DTM with radius, is predicted when most of the dust mass is due to accretion of atoms onto seed-grains in the ISM; this has been verified on observations of M31 (Mattsson *et al.* 2014). Also, a super-linear dependence of DGR on the metallicity has been found also for M101 (Chiang *et al.* 2018) and for the MW (Giannetti *et al.* 2017). We are currently investigating the impact of DTM gradients in our analysis of DustPedia galaxies.

References

Casasola, V., Cassarà, L., Bianchi, S., *et al.* 2017, *A&A*, 605, A18
Chiang, I., Sandstrom, K. M., Chastenet, J., *et al.* 2018, *ApJ*, 865, 117
Clark, C. J. R., Schofield, S. P., Gomez, H. L., & Davies, J. I. 2016, *MNRAS*, 459, 1646
Clark, C. J. R., Verstocken, S., Bianchi, S., *et al.* 2018, *A&A*, 609, A37
Clark, C. J. R., De Vis, P., Baes, M., *et al.* 2019, *MNRAS*, in press
Davies, J. I., Bianchi, S., Cortese, L. *et al.* 2012, *MNRAS*, 419, 3505
Davies, J. I., Baes, M., Bianchi, S., *et al.* 2017, *PASP*, 129, 044102
De Vis, P., Gomez, H. L., Schofield, S. P., *et al.* 2017, *MNRAS*, 471, 1743
De Vis, P., Jones, A., Viaene, S., *et al.* 2019, *A&A*, 623, A5
Giannetti, A., Leurini, S., König, C., *et al.* 2017, *A&A*, 606, L12
Magrini, L., Bianchi, S., Corbelli, E., *et al.* 2011, *A&A*, 535, A13
Mattsson, L., Gomez, H. L., Andersen, A. C., *et al.* 2014, *MNRAS*, 444, 797

Astronomy in Focus - XXX
Proceedings IAU Symposium No. XXX, 2018
M. T. Lago, ed.

doi:10.1017/S1743921319004216

Galactic archaeology: Understanding the metallicity gradients with chemo-dynamical models

I. Minchev, F. Anders and C. Chiappini

Leibniz-Institut für Astrophysik Potsdam (AIP), An der Sternwarte 16,
14482 Potsdam, Germany
email: iminchev@aip.de

Abstract. Radial metallicity gradients measured today in the interstellar medium (ISM) and stellar components of disk galaxies are the result of chemo-dynamical evolution since the beginning of disk formation. This makes it difficult to infer the disk past without knowledge of the ISM metallicity gradient evolution with cosmic time. We show that abundance gradients are meaningful only if stellar age information is available. The observed gradient inversion with distance from the disk mid-plane seen in the Milky Way can be explained as the effect of inside-out disk formation and disk flaring of mono-age populations. A novel recent method is presented for constraining the evolution of the Galactic ISM metallicity with radius and time directly from the observations, while at the same time recovering the birth radii of any stellar sample with precise metallicity and age measurements.

Keywords. stars: abundances, ISM: abundances, Galaxy: abundances, Galaxy: evolution, Galaxy: disk, (Galaxy:) solar neighbourhood, Galaxy: kinematics and dynamics

---

1. Inversion in radial abundance gradients with height above the disk mid-plane

An inversion of the radial metallicity gradient from negative close to the disk plane, to positive at higher distance above it, $|z|$, has been found in SEGUE, RAVE, APOGEE and LAMOST surveys. Similar inversion, but from positive to negative, exists for $[\alpha/\mathrm{Fe}]$. This phenomenon was also seen in the Milky Way chemo-dynamical model of Minchev, Chiappini & Martig 2013 (MCM13) and was interpreted by Minchev, Chiappini & Martig (2014) as the result of inside-out disk formation and disk flaring of mono-age stellar populations. While older, kinematically hot stars dominate in the inner disk at high $|z|$, due to disk flaring, younger stars populate the outer disk high $|z|$ regions; this also predicts a negative age gradient at high $|z|$ (Minchev et al. 2015), indeed found by Martig *et al.* (2016) using APOGEE data. This interplay among the density of stars with different ages as a function of r and $|z|$ can be seen in Fig. 1, where the vertical rectangles represent the stellar density.

2. Estimating stellar birth radii and the time evolution of the ISM metallicity gradient

One of the most important goals in Galactic Archaeology is finding the birth places of the stars we currently observe in the Milky Way. The recent work by Minchev *et al.* (2018) (see also Frankel *et al.* 2018) described a way to do that solely based on metallicity and age measurements (see Fig. 2). This also allowed to constrain the ISM metallicity evolution with Galactic radius for the first time, finding that at the onset of disk formation it was twice as steep as it is currently found (~ -0.15 vs ~ -0.07 dex/kpc). It should be

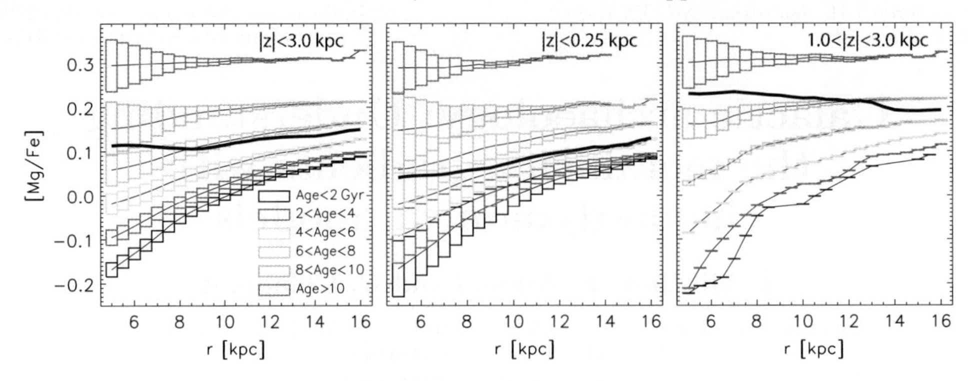

Figure 1. Inversion of the [Mg/Fe] gradient with distance from the disk mid-plane, $|z|$ in the MCM13 model. Thick black curves show the total sample for three different $|z|$ slices, as indicated. The height of rectangular symbols reflects the stellar density of each bin. The positive gradient seen at small $|z|$ (middle) is reversed at high distance above the disk mid-plane (right).

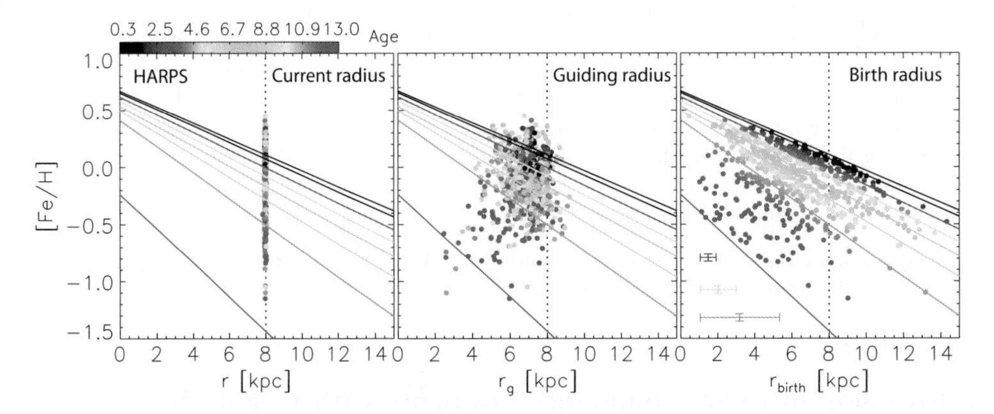

Figure 2. [Fe/H] of HARPS data (Adibekyan et al. 2011) coloured by age, versus current Galactic disc radius (left), guiding radius (middle), and estimated birth radius (right), as described in Minchev *et al.* (2018). The recovered ISM metallicity gradient evolution with time is shown by the coloured lines.

kept in mind that the age dependence of the abundance radial profile measured today (Anders *et al.* 2017) is significantly different from the ISM gradient at the time of stellar birth, more so for older populations, due to the effect of radial migration resulting from the angular momentum exchange in the disk.

Ages for turn-off stars from the Gaia (Gaia Collaboration *et al.* 2018) and asteroseismic missions, combined with ground-based spectroscopic surveys (e.g., SDSS-V, WEAVE, MOONS, and especially 4MOST) will much improve the determination of the Galactic chemical evolution and migration history needed to constrain the Milky Way formation.

References

Adibekyan, V. Z., Santos, N. C., Sousa, S. G., & Israelian, G. 2011, *A&A*, 535, L11
Anders, F., Chiappini, C., Minchev, I., et al. 2017, *A&A*, 600, A70
Frankel, N., Rix, H.-W., Ting, Y.-S., Ness, M. K., & Hogg, D. W. 2018, arXiv:1805.09198
Gaia Collaboration, Brown, A. G. A., Vallenari, A., et al. 2018, *A&A*, 616, A1
Minchev, I., Chiappini, C., & Martig, M. 2014, *A&A*, 572, A92
Minchev, I., Anders, F., Recio-Blanco, A., et al. 2018, *MNRAS*, 481, 1645

Astronomy in Focus - XXX
Proceedings IAU Symposium No. XXX, 2018
M. T. Lago, ed.

doi:10.1017/S1743921319004228

Galaxy Evolution in the context of radial metallicity gradients

Patricia B. Tissera

Physics Department, Universidad Andres Bello, 700 Fernandez Concha, Santiago, Chile.
email: patricia.tissera@unab.cl

Abstract. The chemical abundances of the gas-phase and stellar components of disc galaxies are relevant to understand their formation and evolution. It has been shown that an inside-out disc formation yields negative chemical profiles. However, a large spread in metallicity gradients, including positive ones, has been reported by recent and more precise observations, suggesting the action of other physics processes such as gas outflows and inflows, radial migration, and mergers and interactions. Cosmological simulations that includes chemical models provide a tools to tackle the origin of the metallicity profiles and the action of those processes that might affect them as a function of time. I present a summary of the current state-of-knowledge from a numerical point of view and discuss the main results from the analysis of the EAGLE simulations.

Keywords. galaxies:formation, galaxies:evolution

1. Introduction

The distribution of chemical elements in the interstellar medium (ISM) and the stellar populations (SPs) of galaxies is the result of action of different physical processes such as star formation, stellar nucleosynthesis, gas accretion, galaxy interactions and mergers, secular evolution, migration. In spite of the complexity posed by the variety of involved temporal and spatial scales, chemical radial gradients are observed. In the Local Universe, galaxies with well-defined discs show a correlation between metallicity gradients of HII regions and stellar mass. These correlations store information on the processes that took place in the formation history of galaxies. In the Local Universe, IFU surveys, such as CALIFA, MaNGA and SAMI, have collected a large database of metallicity gradients in disc galaxies of different stellar masses. These new estimates report a larger variety of metallicity gradients at a given stellar mass. Radial oxygen gradients in the star-forming galaxies have been also detected across cosmic times. These data contribute key information to constraint galaxy formation models.

2. Summary

Disc metallicity gradients: inside out formation. Cosmological hydrodynamical simulations are powerful tool to study the chemical evolution of galaxies (Taylor & Kobayashi 2017). Tissera *et al.* (2016, 2017) analysed the radial oxygen gradients in cosmological simulations, finding that if the discs formed inside out then, negative metallicity gradients are detected in the gas-phase, in agreement with previous results (Pilkington *et al.* 2012).

Disc metallicity gradients: feedback and mergers. Observations report the existence of inverted (positive) metallicity gradients in galaxies generally associated to interactions and mergers. Numerical simulations are able to reproduce this inversion of the metallicity gradients as a results of both mergers and interactions and strong feedback that removes enriched material preferentially from the inner regions (Tissera *et al.* 2019).

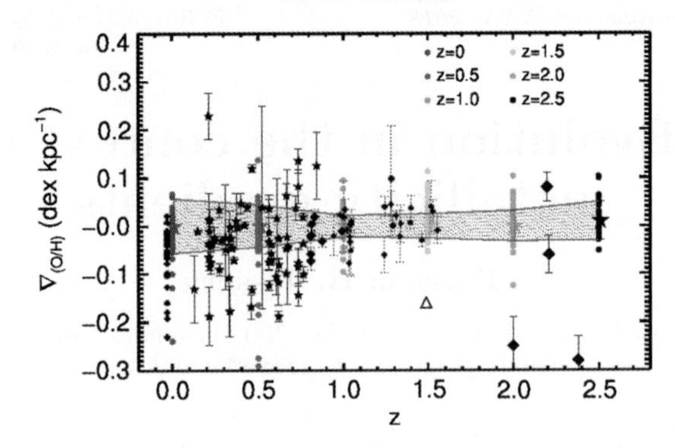

Figure 1. Oxygen gas-phase abundance gradients in discs selected from the EAGLE project (L25R756RECL) as a function of redshift. The shaded area shows the 25 and 75 percentiles. A compilation of available observations at different redshifts are shown for comparison (black symbols).

Both mechanisms can act together, contributing to regulate the transformation of gas into stars, which is a crucial process to shape metallicity gradients.

Disc metallicity gradients: redshift evolution. The available observational data of metallicity gradients across time show a large dispersion with no clear trend. Results from hydrodynamical simulations are consistent with no evolution when enhanced stellar feedback is adopted (e.g. Gibson *et al.* 2013 and Ma *et al.* 2018) while weaker stellar feedback shows a slight trend with redshift (e.g. Tissera *et al.* 2017). As shown in Figure 1, the metallicity gradients obtained from the EAGLE project (Schaye *et al.* 2015) are consistent with the observed large dispersion and show no clear evolution in time (Tissera *et al.* in prep).

The distribution of chemical abundances in the ISM and SPs reflect the complex history of assembly and star formation history of galaxies and provides a challenging opportunity to understand galaxy formation.

Acknowledgments

This work has been partially funded by Fondecyt Regular 1150334 and LACEGAL Network of Horizon 2020 EC program.

References

Gibson, B.K:, Pilkington, K., Brook C. B., Stinson G. S., Bailing J., 2013, *A&A*, 554, 47
Ma, X., *et al.* 2018, *MNRAS*, 477, 219
Pilkington, K. *et al.* 2012, *MNRAS*, 452, 969
Schaye J., *et al.*, 2015, *MNRAS*, 446, 521
Taylor, R, & Kobayashi,C. 2017, *MNRAS*, 471, 3856
Tissera, P.B., Pedrosa, S. E., Sillero, E., Vilchez, J.M. 2016, *MNRAS*, 456, 2982
Tissera, P., Machado, R.E. G., Vilchez, J.M., *et al.* 2017, *A&A*, 604,118
Tissera, P., Rosas-Guevara, Y., *et al.* 2019, *MNRAS*, 482, 2208–2221

Astronomy in Focus - XXX
Proceedings IAU Symposium No. XXX, 2018
M. T. Lago, ed.

doi:10.1017/S174392131900423X

The evolution of the Milky Way's radial metallicity gradient as seen by APOGEE, CoRoT, and Gaia

Friedrich Anders, Ivan Minchev and Cristina Chiappini

Leibniz-Institut für Astrophysik Potsdam (AIP),
An der Sternwarte 16, 14482 Potsdam, Germany

Abstract. The time evolution of the radial metallicity gradient is one of the most important constraints for Milky Way chemical and chemo-dynamical models. In this talk we reviewed the status of the observational debate and presented a new measurement of the age dependence of the radial abundance gradients, using combined asteroseismic and spectroscopic observations of red giant stars. We compared our results to state-of-the-art chemo-dynamical Milky Way models and recent literature results obtained with open clusters and planetary nebulae, and propose a new method to infer the past history of the Galactic radial abundance profile.

Keywords. stars: asteroseismology, stars: late-type, stars: abundances, stars: distances, stars: ages, Galaxy: stellar content, Galaxy: evolution

---

In Anders *et al.* (2017), we used combined asteroseismic CoRoT and spectroscopic APOGEE observations of 418 red giants located close to the Galactic disc plane (6 kpc $< R_{\rm Gal} <$ 13 kpc) to derive the age dependence of the Milky Way's radial metallicity gradient. The radial iron gradient traced by the youngest red-giant population ($-0.058 \pm 0.008 \pm 0.003$ dex/kpc) reproduces the results obtained with young Cepheids and HII regions, while for the 1-4 Gyr population we obtain a slightly steeper gradient ($-0.066 \pm 0.007 \pm 0.002$ dex/kpc). For older ages, the gradient flattens again to reach values compatible with zero at around 10 Gyr.

While these measurements are in good agreement with other nebular and stellar tracers, the debate about the evolution of the Milky Way's radial metallicity gradient is still ongoing. It should be kept in mind, however, that the age dependence of the metallicity gradient is not the same as its evolution. The age dependence of the abundance profile is seriously affected by secular processes, in particular stellar radial migration.

As shown independently by Minchev *et al.* (2018) and Frankel *et al.* (2018), it is possible – and necessary – to infer both the evolution of the radial abundance profile and the stellar migration history at the same time. These measurements are crucially dependent on the availability of precise and accurate stellar ages for large samples of stars. We expect that ages for turn-off stars with precise *Gaia* DR2 parallaxes (Gaia Collaboration *et al.* 2018) and precise stellar parameters from spectroscopic surveys (and eventually the PLATO mission; see Miglio *et al.* 2017) will enable a much more detailed archaeological determination of the evolution of the Milky Way's radial abundance profile.

References

Anders, F., Chiappini, C., Minchev, I., *et al.* 2017, *A&A*, 600, A70
Frankel, N., Rix, H.-W., Ting, Y.-S., *et al.* 2018, *ApJ*, 865, 96
Gaia Collaboration, Brown, A. G. A., Vallenari, A., *et al.* 2018, *A&A*, 616, A1
Miglio, A., Chiappini, C., Mosser, B., *et al.* 2017, *Astron. Nachr.*, 338, 644
Minchev, I., Anders, F., Recio-Blanco, A., *et al.* 2018, *MNRAS*, 481, 1645

Astronomy in Focus - XXX
Proceedings IAU Symposium No. XXX, 2018
M. T. Lago, ed.

doi:10.1017/S1743921319004241

The radial distribution in nearby galaxies of the ionizing field of radiation of HII regions using photoionization models

Enrique Pérez-Montero, Rubén García-Benito and José M. Vílchez

Instituto de Astrofísica de Andalucía - CSIC, Apdo 3004, E-18080, Granada, Spain

Abstract. HII regions in galaxy disks can be used as a powerful tool to trace the radial distribution of several of their properties and shed some light on the different relevant processes on galaxy formation and evolution. Among the properties that can be extracted from the study of the ionized gas are the metallicity, the excitation and the hardness of the ionizing field of radiation. In this contribution we focus on the determination of both the ionization parameter (U) and the effective temperature of the ionizing clusters (T_*) by means of a bayesian-like comparison between the observed relative fluxes of several emission-lines with the predictions from a set of photoionization models. We also show the implications that the use of our method has for the study of the radial variation of both U and T_* in some very well-studied disk galaxies of the Local Universe.

Keywords. methods: data analysis – ISM: abundances – galaxies: abundances

---

The equivalent effective temperature (T_*) of the ionizing field of radiation is one of the functional parameters that dominates the relative flux of the most prominent emission lines in HII regions. In unresolved objects Vilchez & Pagel (1988) propose the so-called softness parameter, defined as log $\eta\prime$ = log([OII] 3727 Å/[OIII] 4959,5007 Å)/([SII] 6717,31 Å/[SIII] 9069,9532 Å) in the optical range, that can be an estimator for T_*. According to Pérez-Montero & Vílchez (2009) most nearby disk galaxies present negative radial slopes of log $\eta\prime$ what could be interpreted as a radial hardening of the ionizing stellar radiation in these objects. To confirm this trend we built a large grid of photoionization models covering a wide range in T_*, metallicity, and ionization parameter and we designed a code, called HII-CHI-MISTRY-TEFF with similar features as those described in Pérez-Montero (2014), which finds by means a bayesian-like approach, the best solution to fit the relative observed intensities of the four emission lines defined in log $\eta\prime$. We applied our code to a large sample of HII regions in the galaxies of the CHAOS project, described in Berg *et al.* (2015), with good-quality spectroscopical optical data of the four required emission lines and a determination of the metallicity based on the direct method. We verified that the analyzed galaxies present radial increases of T_* that correlate with the negative slopes of their metallicity gradients. On the contrary, no radial variations of the ionization parameter are found for the same objects.

References

Berg, D. A., Skillman, E. D., Croxall, K. V., *et al.* 2015, *ApJ*, 806, 16
Pérez-Montero, E. 2014, *MNRAS*, 441, 2663
Pérez-Montero, E., & Vílchez, J. M. 2009, *MNRAS*, 400, 1721
Vilchez, J. M., & Pagel, B. E. J. 1988, *MNRAS*, 231, 257

Astronomy in Focus - XXX
Proceedings IAU Symposium No. XXX, 2018
M. T. Lago, ed.

doi:10.1017/S1743921319004253

Go beyond radial gradient: azimuthal variations of ISM abundance in 3D

I-Ting Ho

Max Planck Institute for Astronomy, Knigstuhl 17, 69117 Heidelberg, Germany

Abstract. Using 3D spectroscopy data from the TYPHOON Project (PI: B. Madore), I show convincing observational evidence that the ISM oxygen abundance traced by HII regions presents systematic azimuthal variations in NGC 1365 and NGC 2997. I discuss a possible physical origin and on-going efforts to explore the prevalence and cause of such variations.

Keywords. galaxies: ISM, galaxies: abundances, galaxies: spiral, galaxies: individual (NGC 1365, NGC 2997)

While radial abundance gradients shed light on the inside-out formation history of galactic disks, the azimuthal variations of ISM abundances can place critical constraints on gas mixing in galaxies (Roy & Kunth 1995). The degree of the azimuthal variations depends on how efficiently metals synthesised inside massive stars can be mixed with the surrounding medium when gas and stars orbit around the gravitational potential. Although radial gradients are now routinely measured by observations, azimuthal variations are still poorly constrained.

Recently, the increasing number of high quality integral field spectroscopy observations has begun to reveal the presence of azimuthal variations in the nearby Universe (Sanchez-Menguiano *et al.* 2016; Vogt *et al.* 2017). Robust mapping of metallicity in the warm ionised medium is facilitated by 3D spectroscopy in nearby galaxies achieving spatial resolutions approaching the typical scale of HII regions (< 100 pc). Using IFU data from the TYPHOON Project (PI: B. Madore), I have recently reported that in two nearby galaxies, NGC 1365 (Ho *et al.* 2017) and NGC 2997 (Ho *et al.* 2018), the HII region (strong line) oxygen abundances present clear, systematic azimuthal variations.

In NGC 1365, the variations are particularly pronounced, 0.2 dex. The oxygen abundances peak on the two $m = 2$ spiral arms and are lower in the inter-arm regions. Similar signatures are also seen in NGC 2997 but less pronounced, only about 0.05 dex. In Ho *et al.* (2017), we show that the azimuthal variations in NGC 1365 can be explained by two physical processes: gas undergoes localised, sub- kiloparsec-scale self-enrichment when orbiting in the inter-arm region, and experiences efficient, kiloparsec-scale mixing-induced dilution when spiral density waves pass through.

On-going IFU observations of nearby galaxies (e.g. MUSE large program by the PHANGS collaboration) and future surveys (e.g. SDSS IV Local Volume Mapper) will soon begin to address the prevalence and degree of azimuthal variations in nearby galaxies.

References

Ho, I.-T., Seibert, M., Meidt, S. E., *et al.* 2017, *ApJ*, 846, 39
Ho, I., Meidt, S. E., Kudritzki, R.-P., *et al.* 2018, arXiv:1807.02043
Roy, J.-R., & Kunth, D. 1995, *A&A*, 294, 432
Sánchez-Menguiano, L., Sánchez, S. F., Pérez, I., *et al.* 2016, *A&A*, 587, A70
Vogt, F. P. A., Pérez, E., Dopita, M. A., Verdes-Montenegro, L., & Borthakur, S. 2017, *A&A*, 601, A61

Astronomy in Focus - XXX
Proceedings IAU Symposium No. XXX, 2018
M. T. Lago, ed.

doi:10.1017/S1743921319004265

Oxygen abundance profiles with MUSE: Radial gradients and widespread deviations

Laura Sánchez-Menguiano[1,2], Sebastián F. Sánchez[3] and Isabel Pérez[4]

[1]Instituto de Astrofísica de Canarias, La Laguna, Tenerife, Spain

[2]Dpto. de Astrofísica, Universidad de La Laguna, La Laguna, Tenerife, Spain
email: lsanchez@iac.es

[3]Instituto de Astronomía, Universidad Nacional Autónoma de México, México

[4]Dpto. de Física Teórica y del Cosmos, Universidad de Granada, Granada, Spain

Abstract. This study has been published in Sánchez-Menguiano *et al.* (2018). We encourage the reader to that article for more details on the study and the results.

In this study we characterise the oxygen abundance radial distribution of a sample of 102 spiral galaxies observed with VLT/MUSE for which a total of 14345 H II regions are detected. We develop a new methodology to automatically fit the abundance radial profiles, that are derived using the calibration proposed in Marino *et al.* (2013) for the O3N2 indicator. We find that 55 galaxies of the sample exhibit a single negative gradient. In addition to this negative trend, 47 galaxies also display either an inner drop in the abundances (21), an outer flattening (10), or both (16), which suggests that these features are a common property of disc galaxies. The presence and depth of the inner drop depends on the galaxy mass, with the most massive systems presenting the deepest drops, while there is no such dependence for the outer flattening. The inner drop appears always around $0.5\,r_e$, while the position of the outer flattening varies over a wide range of galactocentric distances. Regarding the main negative gradient, we find a characteristic slope in the sample of $\alpha_{\mathrm{O/H}} = -0.10 \pm 0.03\,\mathrm{dex/r_e}$. This slope is independent of the presence of bars and the density of the environment. However, when inner drops or outer flattenings are detected, slightly steeper gradients are observed, suggesting that radial motions might play an important role in shaping the abundance profiles. We define a new normalisation scale $r_{O/H}$ for these profiles (tightly correlated with r_e) based on the characteristic abundance gradient, with which all the galaxies show a similar position for the inner drop ($0.5\,r_{O/H}$) and the outer flattening ($1.5\,r_{O/H}$). Finally, we find no significant dependence of the dispersion around the gradient with any galaxy property, with values compatible with the uncertainties associated with the derivation of the abundances.

We have reproduced the analysis using the calibration described in Dopita *et al.* (2016), and the one proposed in Marino *et al.* (2013) for the N2 index. In general, we obtain a very good agreement between the results based on the three calibrators. The most significant difference lies on the number of inner drops and outer flattenings detected in the abundance distribution, that is reduced when using these two calibrators. However, the overall distributions are rather similar in the three cases, strengthening our conclusions.

References

Dopita, M. A., Kewley, L. J., Sutherland, R. S., & Nicholls, D. C. 2016, *Ap&SS*, 361, 61
Marino, R. A., Rosales-Ortega, F. F., Sánchez, S. F., *et al.* 2013, *A&A*, 559, A114
Sánchez-Menguiano, L., Sánchez, S. F., Pérez, I., Ruiz-Lara, T., *et al.* 2018, *A&A*, 609, A119

Astronomy in Focus - XXX
Proceedings IAU Symposium No. XXX, 2018
M. T. Lago, ed.

doi:10.1017/S1743921319004277

Gas and stellar metallicity gradients in face-on disc galaxies

P. Sánchez-Blázquez

Universidad Autónoma de Madrid, 28049, Madrid, Spain

Abstract. We present an analysis of the stellar and gaseous metallicity gradients in a sample of 260 disc galaxies from the CALIFA survey. The slope of the different components are compared with the main characteristics of the galaxies, such as mass, morphology, presence of a bar, or gas fraction.

Keywords. galaxies: stellar content, galaxies: evolution, galaxies: abundances

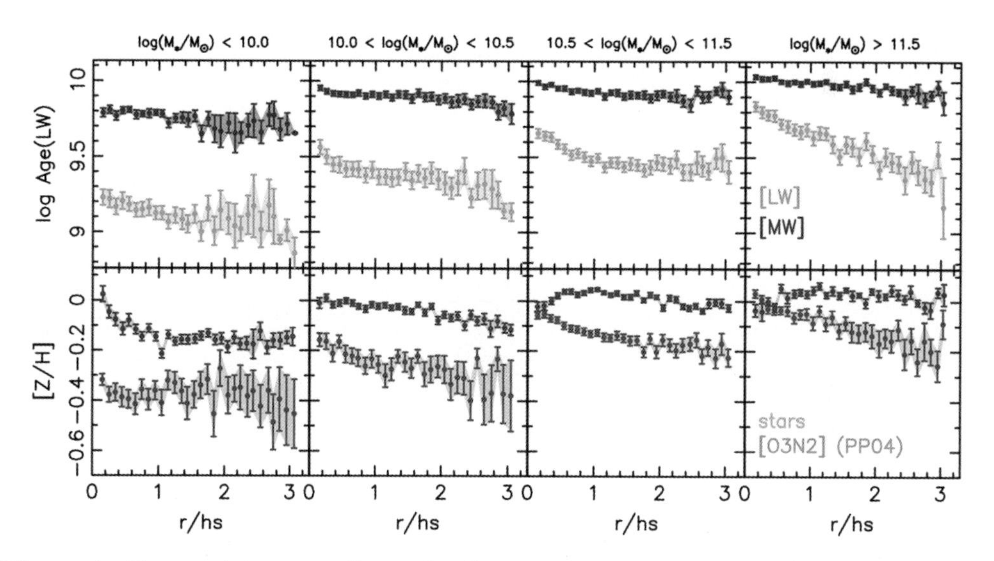

Figure 1. Top panel: mean gradients of stellar age. Bottom panel: mean metallicity gradients for the stellar (blue) and gaseous compoennts (green).

While studies of gas-phase metallicity gradients in disc galaxies are common, very little has been done in the acquisition of stellar abundance gradients. Furthermore, very rarely the gradients of both components are compared for the same galaxies (although see Lian *et al.* 2018). We present here an analysis of the stellar populations and gas-phase metallicities for a sample of 260 galaxies from the CALIFA survey Sanchez *et al.* (2012). Fig. 1 shows the mean metallicity gradient for galaxies in different mass bins. The mean stellar age and metallicity gradients in the disc are shallow and negative. If normalised to the scale-length of the disc, they do not show clear trends with the mass or the presence of a bar although there are some weak trends with the gas fraction or the morphology. On average, gas-phase metallicity gradients are flatter than those of stars. The majority of the stars in the disc (at least up to 4 scale-lengths) are old, although there is a larger percentage of young stars at larger radii, which is compatible with the inside-out formation scenario.

References

Lian, J., Thomas, D., Maraston, C., *et al.* 2018, *MNRAS*, 476, 3883
Sánchez, S. F., Kennicutt, R., Gil de Paz, A., *et al.* 2012,*A&A*, 538, 8

Astronomy in Focus - XXX
Proceedings IAU Symposium No. XXX, 2018
M. T. Lago, ed.

doi:10.1017/S1743921319004289

SDSS-IV MaNGA: Testing the Metallicity Distribution across the Merging Sequence

Jorge K. Barrera-Ballesteros[1], **Li-hwai Lin**[2], **Bu-Ching Hsieh**[2], **Hsi-An Pan**[2], **Sebastian Sánchez**[3] and **Timothy Heckman**[1]

[1]Dept. of Physics and Astronomy, Johns Hopkins University, 3400 N Charles St., Baltimore, MD 21218, USA
email: jbarrer@jhu.edu

[2]Academia Sinica, Institute of Astronomy & Astrophysics (ASIAA), P.O. Box Box 23-141, Taipei 10617, Taiwan

[3]Instituto de Astronomía, Universidad Nacional Autonoma de Mexico

Abstract. Interactions and mergers have been known as key scenarios to enhance global star formation rates and to lower the metal content of galaxies. However, little is known on how interactions affect the spatial distribution of gas metallicities. Thanks to the SDSS-IV MaNGA survey we are able to statistically constrain the impact of interactions across the optical distributions of galaxies. In this study, we compare the radial distribution of the ionized gas metallicity from a sample of 329 interacting objects – covering different interaction stages – with a statistical robust control sample. Our results suggest that galaxies close to coalesce tend to have flat, lower metallicities than non-interacting star-forming galaxies.

Keywords. Galaxy: evolution, galaxies: interactions, galaxies: abundances

---

Large surveys have shown that interactions dilute the central metallicity in comparison to normal star-forming galaxies (Ellison *et al.*, 2008). Here we use 329 interacting galaxies included in the IFU SDSS-IV MaNGA survey (>4600 observed targets, see an example in panel (a) of Fig. 1). It allows us to derived a sample of non-interacting galaxies with similar stellar mass for each of these galaxies. We measure the metallicity gradient for each of these galaxies and compare it with the average gradient from its corresponding control sample. In Fig. 1 (b) we show the residuals of this comparison. We find that pairs of galaxies tend to have similar or slightly larger metallicity gradients compare to normal galaxies. On the other hand, galaxies near to coalesce tend to have lower, flat metallicity

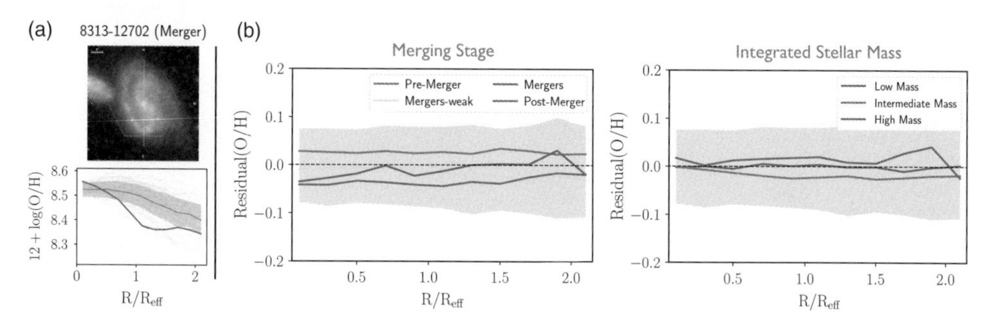

Figure 1. (a) Interacting galaxy (top) metallicity gradient (bottom, red line) with its corresponding control sample (blue line). (b) Distribution of gradients residuals (between interacting and control samples) binned in different interacting stages (left) and stellar masses (right).

gradient residuals. These results indicate that interactions affect the distribution of metals across the entire galaxy, not only at its central region.

Reference

Ellison, S., Patton D., Simmard, L., & McConnachie W. 2008, *ApJ*, 135, 1877

Astronomy in Focus - XXX
Proceedings IAU Symposium No. XXX, 2018
M. T. Lago, ed.

doi:10.1017/S1743921319004290

The evolution of the oxygen radial gradients in spiral galaxies

M. Mollá[1], O. Cavichia, B. Gibson, P. Tissera, P. Sánchez-Blázquez, A. I. Díaz, Y. Ascasibar, C. G. Few, S. F. Sánchez and W. J. Maciel

[1]Depto. de Investigación Básica, CIEMAT, Avda. Complutense 40. E-28040 Madrid. (Spain)
email: mercedes.molla@ciemat.es

Abstract. We analyse the evolution with redshift of the radial gradient of oxygen abundances in spiral disks resulting from our MULCHEM chemical evolution models, computed for galaxies of different sizes or masses, studying the relationships between the gradients and galaxy characteristics as the stellar mass, the size, the gas fraction or the star formation rate for $z < 4$.

Keywords. Galaxies: Spirals; Galaxies: Abundances; Galaxies: Formation.

In Mollá *et al.* (2019) we studied the evolution of the oxygen radial gradient. Now, with our set of MULCHEM chemical evolution models, we predict how the radial gradients were in the past for other galaxy with different sizes and masses. We compute models for galaxies with dynamical masses in the range $\mathrm{M_{vir}}$ $[10^{10}$ -$10^{13}]$ $\mathrm{M_{\odot}}$, with new stellar yields and initial mass function (Mollá *et al.* 2015), updated gas accretion rates (Mollá *et al.* 2016), and new prescriptions to form molecular cloud (Mollá *et al.* 2017). We take into account the growth of the disks, measuring the radial gradients always within the optical disk corresponding to each redshift. We analyse the correlations of the radial gradients with the stellar mass, the disk size and the star formation rate at the present time and in the past. For the present time, we find that, as expected, the effective radius, $R_{\rm eff}$, depends on the stellar mass $\mathrm{M_*}$. The radial gradient ∇, measured as $\mathrm{dex\,kpc^{-1}}$, shows a clear dependence on $R_{\rm eff}$, in agreement with data. It also varies with the fraction of gas and with the specific SFR, showing a different behaviour between low and high star formation efficiency, ϵ_s, models. The gradient ∇ depends on $\mathrm{M_*}$ for $M_* < 4 \times 10^9\ M_{\odot}$ with a high dispersion. Above this mass, gradients are very similar. The normalised radial gradient is $\nabla_{\rm Reff} \sim -0.10\,\mathrm{dex\,R_{eff}^{-1}}$ for all galaxies with differences depending on ϵ_s. These correlations at $z = 0$ may change at other redshifts: $R_{\rm eff}$ is smaller for a similar $\mathrm{M_*}$ at higher z; In the SFR–∇ correlation, points move at higher sSFR showing steeper gradients at $z = 1$ than at $z = 0$. Correlations ∇ – $\mathrm{M_*}$ or $R_{\rm eff}$ appear at all redshifts, but with more dispersion at high redshift for low ϵ_s and $\mathrm{M_*}$. A light correlation of $\nabla_{\rm Reff}$–$\mathrm{M_*}$ and $R_{\rm eff}$ appears at high redshift which disappear at $z = 0$, but a similar value is obtained for high ϵ_s models at all redshifts. The evolution of $\nabla(z)$ is smooth, except when the disk begins form, when a very steep radial gradient appears. This occurs at any time depending on the dynamical mass; thus these models explain all negative ∇ for isolated spiral galaxies. (See details in Mollá *et al.* 2019).

References

Mollá M., Cavichia O., Gavilán M., Gibson B. K. 2015, *MNRAS*, 451, 3693
Mollá M., Díaz Á. I., Gibson B. K., Cavichia O., *et al.* 2016, *MNRAS*, 462, 1329
Mollá M., Díaz Á. I., Ascasibar Y., Gibson B. K. 2017, *MNRAS*, 468, 305
Mollá, M., Díaz, A. I., Cavichia, O., Gibson, B. K., *et al.* 2019, *MNRAS*, 482, 307

Astronomy in Focus - XXX
Proceedings IAU Symposium No. XXX, 2018
M. T. Lago, ed.

doi:10.1017/S1743921319004307

Supergiant Stars as Abundance Probes

L. R. Patrick[1,2], C. J. Evans[3], B. Davies[4] and R.-P. Kudritzki[5,6]

[1]IAC
[2]ULL, Tenerife, Spain
email: lpatrick@iac.es
[3]UKATC, Edinburgh, UK
[4]LJMU, Liverpool, UK
[5]IfA, Hawaii, USA
[6]UOM, Germany

Abstract. By compiling abundances from red and blue supergiants (SGs) within the Local Universe, I present the Mass-Metallicity relation (MZR) using stellar tracers, demonstrating the excellent internal consistency. Comparing this result with nebular tracers, those empirically calibrated to direct-method studies provide the most consistent results.

Keywords. stars: abundances, (stars:) supergiants, galaxies: abundances, galaxies: dwarf

---

Massive stars are important probes of chemical evolution in star-forming galaxies. These tracers represent the youngest stellar population and provide robust, independent abundance estimates, important to constrain models of galactic chemical evolution and to anchor the more uncertain nebular estimates at larger distances. Massive stars have been used to examine the metal-content and -distribution in many Local Universe galaxies out to distances of ∼20 Mpc.

Blue and Red SG stars are the brightest components in stellar populations in the optical and near-IR, respectively. Even though these stars are evolved products, they

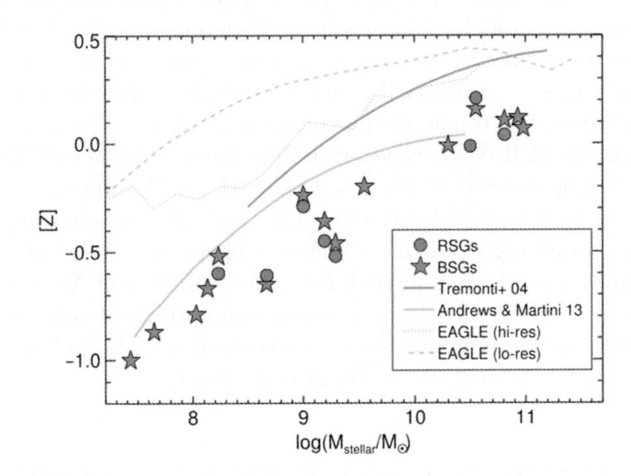

Figure 1. MZR estimated using stellar tracers from Davies *et al.* (2017), using both red and blue SG metallicities in Local Universe galaxies. The MZR from direct line measurements of H II regions shows the best agreement.

are still remarkably young objects (<50 Myr). Intermediate resolution multi-object spectroscopy combined with state-of-the-art stellar model atmospheres, are vital for realising the potential of these stars (e.g. Evans *et al.* 2011).

Recent results demonstrate the internal consistency between the stellar tracers in different systems (e.g. Gazak *et al.* 2015; Patrick *et al.* 2017). These results are compiled in Fig. 1 and shown against nebular tracers at larger distances. Using this comparison, we determine that nebular tracers empirically calibrated to direct-method studies provide the most consistent results.

References

Andrews, B. H., & Martini, P. 2013, *ApJ*, 765, 140
Davies, B., Kudritzki, R.-P., Lardo, C., *et al.* 2017, *ApJ*, 847, 112
Evans, C. J., Davies, B., Kudritzki, R.-P., *et al.* 2011, *A&A*, 527, A50
Gazak, J. Z., Kudritzki, R., Evans, C., *et al.* 2015, *ApJ*, 805, 182
Patrick, L. R., Evans, C. J., Davies, B., *et al.* 2017, *MNRAS*, 468, 492
Tremonti, C. A., Heckman, T. M., Kauffmann, G., *et al.* 2004, *ApJ*, 613, 898

Astronomy in Focus - XXX
Proceedings IAU Symposium No. XXX, 2018
M. T. Lago, ed.

doi:10.1017/S1743921319004319

Metallicity gradients in M31, M33, NGC 300, and Milky Way using Argon abundances

Sheila N. Flores-Durán and Miriam Peña

Instituto de Astronomía, Universidad Nacional Autónoma de México,
Apdo. Postal 70264, 04510 Ciudad de México, México
email: sflores, miriam@astro.unam.mx

Abstract. We studied Planetary Nebulae (PNe) metallicity gradients using Ar abundances. We compared them with H II regions in the galaxies of the local universe M 31, M 33, NGC 300 and in the Milky Way. Galactocentric radio (R_G) and chemical abundances were collected from the literature, carefully selecting an homogeneous sample for each galaxy. In these galaxies, metallicity gradients computed with PNe abundances are flatter than those of H II regions.

Keywords. ISM: planetary nebulae, galaxies: ISM

It has recently been shown that O in PNe may be enriched in low-metallicity environments (Peña *et al.* 2007; Flores-Durán *et al.* 2017) and also in Galactic PNe with carbon-rich dust (Delgado-Inglada *et al.* 2015). On one hand, metallicity gradients derived from O abundances in PNe might be affected by stellar nucleosynthesis and, thus, the O/H gradient might not be always adequate for chemical evolution studies. On the other hand, Ar abundances in PNe are not modified during the PN progenitor lifetime. Hence we propose to use Ar to trace the metallicity gradients in the galaxies of the local universe M 31, M 33, NGC 300, and the Milky Way.

The results for each galaxy are listed below. For M 31 the linear fits of the Ar gradient for PNe is $(6.12 \pm 0.04) - (0.00005 \pm 0.0006)R_G$ and for H II regions is $(6.38 \pm 0.18) - (0.021 \pm 0.013)R_G$. In the case of M 33, Ar metallicity gradient fits are $(6.20 \pm 0.06) - (0.018 \pm 0.014)R_G$ and $(6.27 \pm 0.04) - (0.045 \pm 0.016)R_G$ for PNe and H II regions respectively. We found that Ar metallicity gradients in NGC 300 are $(6.31 \pm 0.02) - (0.051 \pm 0.012)R_G$ for PNe and $(6.33 \pm 0.04) - (0.104 \pm 0.014)R_G$ for H II regions.

In the Milky Way we compare metallicity gradients of PNe using Galactocentric distances from Frew *et al.* (2016) and Stanghellini & Haywood (2010), no big differences were found. The Ar metallicity gradient fits for PNe computed with Frew *et al.* (2016) distances is $(6.48 \pm 0.01) - (0.015 \pm 0.001)R_G$ and for H II regions is $(6.83 \pm 0.14) - (0.042 \pm 0.021)R_G$.

By comparing PNe metallicity gradients with H II regions in M 31, M 33, NGC 300, and the MW we found that PNe show flatter gradients.

References

Delgado-Inglada, G., Rodríguez, M., Peimbert, M., Stasińska, G. & Morisset, C. 2015, *MNRAS*, 449, 1797
Flores-Durán, S. N., Peña, M., & Ruiz, M. T. 2017, *A&A*, 601 A147
Frew D. J., Parker Q. A., & Bojičić I. S. 2016, *MNRAS*, 455, 1459
Peña, M., Richer, M. G, & Stasińska, G. 2007a, *A&A*, 466, 75
Staghellini, L. & Haywood, M. 2010, *ApJ*, 714, 1096

Astronomy in Focus - XXX
Proceedings IAU Symposium No. XXX, 2018
M. T. Lago, ed.

doi:10.1017/S1743921319004320

The first census of precise metallicity radial gradients at cosmic noon from HST

Xin Wang[1], Tucker A. Jones[2] and Tommaso Treu[1]

[1]Dept. of Physics & Astronomy, University of California, Los Angeles, CA, USA 90095-1547
[2]University of California Davis, 1 Shields Avenue, Davis, CA 95616, USA

The chemo-structural evolution of galaxies at the peak epoch of star formation (i.e., $z\sim2$, a.k.a. cosmic noon) is a key issue in galaxy evolution physics that we do not yet fully understand. A key diagnostic of this process is the spatial distribution of gas-phase oxygen abundance (i.e.metallicity). There have been numerous attempts to investigate the evolution of metallicity radial gradients using ground-based data, yet these data are often acquired under natural seeing, whose median resolution is around $\sim0\farcs6$ at best. Multiple papers have shown that intrinsically steep radial gradients can only be recovered by imaging spectroscopy with sub-kpc spatial resolution; seeing-limited observations will result in spuriously flat, ergo systematically biased, radial gradients. This beam-smearing problem is even more aggravating at higher z, due to intrinsically smaller galaxy sizes. One solution to this problem is relying on the technique of adaptive optics. Though effective, this method is highly costly due to a number of reasons: the relatively low Strehl ratio in H/K-band necessary to probe cosmic noon sources, the constraining proximity between targets and bright tip/tilt reference stars, etc. As a consequence, the progress of obtaining unbiased (i.e.*precise*) metallicity radial gradients at sub-kpc resolution in cosmic noon galaxies has been slow, yielding only $\sim$20 measurements before our work.

We, on the other hand, take a different route. In a series of works (Jones *et al.* 2015; Wang *et al.* 2017, 2018, in prep.), we invented and improved a much more effective method to measure *precise* metallicity gradients in cosmic noon galaxies. By combining the deep Hubble Space Telescope near-infrared grism data and a novel Bayesian method inferring metallicity directly from emission line fluxes, we obtained over 80 unbiased metallicity maps at $z\sim1.2$-2.3. *This improves the sample size by one order of magnitude!* Our maps reveal diverse galaxy morphologies, indicative of various effects such as efficient radial mixing from tidal torques, rapid accretion of low-metallicity gas, and other physical processes which can effectively affect the gas and metallicity distributions in individual galaxies. In particular, we found two dwarf galaxies at $z\sim2$ displaying greatly positive (i.e.inverted) gradients, strongly suggesting that powerful galactic winds triggered by central star bursts carry the bulk of stellar nucleosynthesis yields to the outskirts (Wang *et al.* 2018). We also observe an intriguing correlation between stellar mass and metallicity gradient, consistent with the "downsizing" galaxy formation picture that more massive galaxies are more evolved into a later phase of disk growth, where they experience more coherent mass assembly at all radii and thus show shallower metallicity gradients. Furthermore, 10% of the gradients measured in our sample are inverted, which are hard to explain by currently existing hydrodynamic simulations and analytical chemical evolution models. Our data analysis techniques can also be applied to data from future space missions employing grism instruments, i.e., JWST, WFIRST, Euclid, etc.

References

Jones, T. *et al.*. 2015, *AJ*, 149, 107
Wang, X. *et al.*. 2017, *ApJ*, 837, 89
Wang, X. *et al.*. 2018, *ApJ* submitted, arXiv:1808.08800

Astronomy in Focus - XXX
Proceedings IAU Symposium No. XXX, 2018
M. T. Lago, ed.

doi:10.1017/S1743921319004332

Resolving gas-phase metallicity gradients of $0.1 \lesssim z \lesssim 0.8$ galaxies

David Carton[1,2]

[1]Univ Lyon, Univ Lyon1, Ens de Lyon, CNRS, Centre de Recherche Astrophysique de Lyon, UMR5574, 69230, Saint-Genis-Laval, France
email: `david.carton@univ-lyon1.fr`

[2]Leiden Observatory, Leiden University, PO Box 9513, 2300 RA, Leiden, The Netherlands

Abstract. We present gas-phase metallicity gradients of 84 star-forming galaxies between $0.08 < z < 0.84$. Using the galaxies with reliably determined metallicity gradients, we measure the median metallicity gradient to be negative ($-0.039^{+0.007}_{-0.009}$ dex/kpc). Underlying this, however, is significant scatter: $(8 \pm 3)\%$ [7] of galaxies have significantly positive metallicity gradients, $(38 \pm 5)\%$ [32] have significantly negative gradients, $(31 \pm 5)\%$ [26] have gradients consistent with being flat. (The remaining $(23 \pm 5)\%$ [19] have unreliable gradient estimates.)

Keywords. galaxies: evolution – galaxies: abundances – galaxies: ISM

In Carton *et al.* (2018) we present the first sample of metallicity gradients from intermediate redshift galaxies ($0.1 \lesssim z \lesssim 0.8$). Galaxies at higher redshifts are typically are found to have flat or positive metallicity gradients (e.g. Queyrel *et al.* 2012), whereas here we find the average metallicity gradient to be negative. While we do observe some galaxies with inverted metallicity gradients, we do not recover a previously identified trend between metallicity gradient and the star-formation intensity in a galaxy. The lack of trend does not preclude the ability of mergers to flatten metallicity gradients, but suggests that this flattening may only occur on a much shorter timescale than that for which we might observe an elevated star formation rate.

Instead, we identify a curious trend between the metallicity gradient and the size of the galaxy (Fig. 1); we identify no large galaxies ($r_d > 3$ kpc) with inverted metallcity

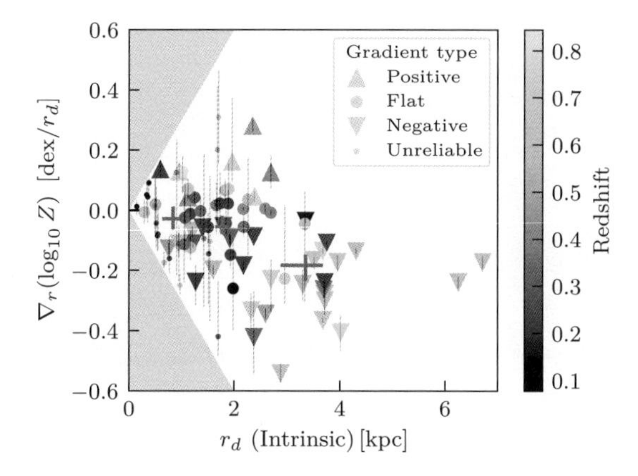

Figure 1. Metallicity gradients of galaxies as a function of their disc scale-length. Red crosses, which indicate the median trend. A grey shading denotes the region disallowed by our prior on the metallicity gradient.

gradients. We speculate that these large galaxies might be more similar to galaxies we observe at low-redshifts (where almost all have negative metallicity gradients), suggesting that a common negative metallicity gradient is only established in well-evolved systems.

References

Carton, D., Brinchmann, J., Contini, T., Epinat, B., Finley, H., Richard, J., Patrício, V., Schaye, J., Nanayakkara, T. Weilbacher, P. M., Wisotzki, L. 2018, *MNRAS* 478, 4293

Queyrel, J., Contini, T., Kissler-Patig, M., Epinat, B., Amram, P., Garilli, B., Le Fèvre, O., Moultaka, J., Paioro, L., Tasca, L., Tresse, L., Vergani, D., López-Sanjuan, C., Perez-Montero, E. 2012, *A&A* 539, A93

Astronomy in Focus - XXX
Proceedings IAU Symposium No. XXX, 2018
M. T. Lago, ed.

doi:10.1017/S1743921319004344

Metallicity gradients in intermediate-redshift absorption-selected galaxies

Lise Christensen[1] **Henrik Rhodin**[1] **and Palle Møller**[2]

[1]Dark Cosmology Centre, Niels Bohr Institute, University of Copenhagen, Juliane Maries Vej 30, 2100 Copenhagen O, Denmark
email: `lichrist@nbi.ku.dk`

[2]European Southern Observatory, Karl-Schwarzschildstrasse 2, D-85748 Garching bei München, Germany

Abstract. Metallicity gradients are most frequently investigated directly from galaxies observed in emission. We have shown that galaxies detected via strong quasar absorption lines also exhibits a metallicity gradient in the outskirts and circumgalactic medium out to ~ 40 kpc distance. We infer a metallicity gradient of -0.022 dex kpc^{-1} for absorption-selected systems at redshifts $0.1 \lesssim z \lesssim 3$. Applying this metallicity gradient and a flattening of the gradient beyond 12 kpc, we demonstrate that absorption-selected galaxies obey the same mass-metallicity relation (MZR) as observed for luminosity-selected galaxies.

Strong absorption lines in quasar spectra arise when the lines of sight to distant quasars intersect intervening galaxies. Associated metal absorption lines from the strongest hydrogen absorption lines, the damped Lyman-α absorbers (DLAs), allow us to trace accurate metallicities of galaxies back to redshifts $z > 5$. This has revealed metallicities around 0.1–100% solar values with a gradual increase in metallicity with increasing cosmic time as expected when DLAs get progressively enriched by star-formation processes. DLAs have metallicity distributions roughly similar to that of Milky Way halo stars, but with a 2–3 decade range in metallicities at all redshifts.

In order to understand the connection between the DLAs and the host galaxies and how the metallicities vary with radial distances, we first need to detect the host galaxy in emission. This search has evolved rapidly in the past decade (e.g. Fynbo *et al.* 2010), and we now have a sufficiently large sample of absorber-galaxy pairs out to redshifts $z > 2$, where we can compare galaxy metallicities in emission and absorption.

Based on the established galaxy mass-metallicity relation (MZR) and its evolution with redshift Møller *et al.* (2013) predicted a similar relation for absorption-selected galaxies by connecting the absorption metallicity to the host galaxy stellar masses. This prediction was verified in Christensen *et al.* (2014). Based on the differences between the predicted versus the measured stellar masses, we can indirectly infer a metallicity gradient. Expanding this sample to include 19 DLA and sub-DLA systems at $0.1 < z < 3$, Rhodin *et al.* (2018) derive an average metallicity gradient $\langle \Gamma \rangle = -0.022 \pm 0.001$ dex kpc^{-1}, and propose a truncation with a flattening of the gradient at larger radii ($r \gtrsim 12$ kpc). In turn, this gradient implies that absorption selected galaxies obey the MZR of luminosity selected galaxies at these redshifts.

References

Christensen, L., Møller, P., Fynbo, J. P. U., & Zafar, T. 2014, *MNRAS*, 445, 225
Fynbo, J. P. U., Laursen, P., Ledoux, C., et al. 2010, *MNRAS*, 408, 2128
Møller, P., Fynbo, J. P. U., Ledoux, C., & Nilsson, K. K. 2013, *MNRAS*, 430, 2680
Rhodin, N. H. P., Christensen, L., Møller, P., Zafar, T., & Fynbo, J. P. U. 2018, arXiv:1807.01755

Astronomy in Focus - XXX
Proceedings IAU Symposium No. XXX, 2018
M. T. Lago, ed.

doi:10.1017/S1743921319004356

Metallicity gradients in high-z galaxies: insights from the KLEVER Survey

Mirko Curti[1,2]

[1]Cavendish Laboratory
[2]Kavli Institute for Cosmology, University of Cambridge,
Madingley Road CB3 0HA, Cambridge, United Kingdom

Abstract. We present reconstructed source plane metallicity maps for a sample of ~ 30 gravitationally lensed galaxies between $1.2 < z < 2.5$, observed in the framework of the KLEVER Survey. Oxygen abundance is derived exploiting a variety of different emission line diagnostics, as provided by the full coverage of the near-infrared bands. The majority of galaxies in our sample present flat radial metallicity gradients, in agreement with galaxy evolution models predicting strong feedback mechanisms in place at these epochs. However, complex patterns as seen in some of our metallicity maps warn against the use of azimuthally-averaged radial gradients as the only observable to constrain chemical evolution models.

Keywords. galaxies: high-redshift galaxies: abundances galaxies: evolution

---

Studying the spatial distribution of heavy elements within galaxies represents a powerful tool to constrain their baryonic and chemical assembly history, including the effects of gas flows. In the local Universe, galaxies are generally characterised by negative gradients (e.g., Sanchez *et al.* 2014), which are generally interpreted as indicative of an "inside-out" scenario of galaxy formation. On the contrary, the situation at higher redshifts is instead much more complicated: the growing number of observational efforts in the last years produced sometimes conflicting results (e.g., Cresci *et al.* 2010; Wang *et al.* 2017), whose interpretation in terms of galaxy evolution proves indeed to be rather difficult.

In this work, we have analysed a sample of gravitationally lensed galaxies at $1.2 < z < 2.5$, observed with the SINFONI and KMOS near-infrared integral field spectrographs on the VLT as part of the KLEVER Survey. Our data provides spatially resolved information in the Y, J, H and K bands, enabling us to map the full suite of rest-frame optical emission lines, which can be used to robustly infer the chemical properties of the ISM in our galaxies from a combination of complementary diagnostics. Thanks to a careful de-lensing procedure, we could reconstruct source-plane metallicity maps with a typical resolution of 2-3 kpc and investigate the presence of radial gradients. Almost the entirety of our sample is characterised by flat metallicity gradients (albeit with large dispersion), similarly to what previously reported in the literature (e.g., Wuyts *et al.* 2016). Comparison with cosmological simulations that explore the effect of stellar feedback, suggests a scenario where efficient mixing processes are active in redistributing a significant amount of gas over large scales. However, despite the apparent homogeneity, many of our galaxies exhibit clumpy and irregular patterns in the metallicity maps, suggesting to move beyond radial gradients as the main constraints for galaxy evolution models. We also find tentative evidence of a spatial anti-correlation between metallicity and the star formation rate surface density (Σ_{SFR}), in particular for high Σ_{SFR}. This may be driven by the infall of pristine gas which locally dilutes the metal content whilst it is triggering new star formation.

References

Cresci, G., *et al.*, 2010, *Nature*, 467, 811
Sanchez, S., *et al.*, 2014, *A&A*, 563, A49
Wuyts, E. *et al.*, 2016, *ApJ*, 827, 74
Wang, X. *et al.*, 2017, *ApJ*, 837, 89

Astronomy in Focus - XXX
Proceedings IAU Symposium No. XXX, 2018
M. T. Lago, ed.

doi:10.1017/S1743921319004368

The dust/gas/metallicity scaling relations in the Local Universe

V. Casasola[1], S. Bianchi[1], P. De Vis[2], L. Magrini[1], E. Corbelli[1] and DustPedia collaboration

[1]INAF – Osservatorio Astrofisico di Arcetri, Largo E. Fermi 5, 50125 Firenze, Italy
email: casasola@arcetri.astro.it

[2]IAS, CNRS, Univ. Paris-Sud, Univ. Paris-Saclay, 91405, Orsay Cedex, France

Abstract. We have combined data of the DustPedia project with observations of gas components of the interstellar medium (ISM) and metallicity abundances for late-type DustPedia galaxies to definitively characterize the ISM scaling relations in the Local Universe. In particular, we have focused on the comparison of the dust-to-gas mass ratio with gas phase metallicities.

Keywords. dust, ISM, ISM: abundances

---

1. Dust-to-gas mass ratio as metallicity tracer

The project DustPedia (www.dustpedia.com, Davies *et al.* 2017) is an European collaborative-focused program aimed at performing the complete characterization of dust in local galaxies. This research is carried out on a sample of 875 objects, i.e. all the large ($D_{25} > 1'$) and nearby ($z < 0.01$) galaxies observed by *Herschel* Space Observatory. A legacy database was constructed from *Herschel* and other UV-to-microwave observations (http://dustpedia.astro.noa.gr/, Clark *et al.* 2018).

Dust constitutes an important property to understand chemical evolution of galaxies. Metals are produced mainly by the stellar nucleosynthesis and then returned to the interstellar medium (ISM), either as gas and as solid grains condensed during the later stages of stellar evolution; they can later be destroyed and incorporated into new generations of stars. We have presented main scaling relations between dust, gas (CO & HI), and metallicity for a sample of $\sim$450 DustPedia late-type galaxies. The CO and HI data have been extracted from the literature, uniformly homogenized, and reported to their values within r_{25} according to Casasola *et al.* (2017) and Wang *et al.* (2014). Using literature data, we have also determined the characteristic gas-phase metallicity for each sample galaxy (De Vis *et al.* 2019).

The main preliminary results are: *i)* the DGR correlates with metallicity, stellar mass and star formation rate, and hence the DGR can be used to estimate metallicity; *ii)* the DGR depends on the morphological type (for late-type galaxies) and the CO-to-H_2 conversion factor (X_{CO}). These findings are confirmed by using a very large local galaxy sample and various assumptions on X_{CO} (constant X_{CO} and X_{CO} depending on metallicity). The definitive results will be published in Casasola *et al. submitted.*

References

Casasola, V., Cassarà, L. P., Bianchi, S., *et al.* 2017, *A&A*, 605, A18
Clark, C. J. R., Verstocken, S., Bianchi, S., *et al.* 2018, *A&A*, 609, A37
Davies, J. I., Baes, M., Bianchi, S., *et al.* 2017, *PASP*, 129, 044102
De Vis, P., Jones, A., Viaene, S., *et al.* 2019, *A&A*, 623, A5
Wang, J., Fu, J., Aumer, M., *et al.* 2014, *MNRAS*, 441, 2159

Astronomy in Focus - XXX
Proceedings IAU Symposium No. XXX, 2018
M. T. Lago, ed.

doi:10.1017/S174392131900437X

Nitrogen isotopic ratio across the Galaxy through observations of high-mass star-forming cores

L. Colzi[1,2], F. Fontani[2], V. M. Rivilla[2], A. Sánchez-Monge[3], L. Testi[2,4], M. T. Beltrán[2] and P. Caselli[5]

[1]Università degli studi di Firenze, Italy
[2]INAF-Osservatorio Astrofisico di Arcetri, Florence, Italy
[3]I. Physikalisches Institut of the Universitt zu Kln, Cologne, Germany
[4]ESO, Garching, Germany
[5]Max-Planck-Institüt für extraterrestrische Physik, Garching, Germany

Abstract. There is a growing evidence that our Sun was born in a rich cluster that also contained massive stars. Therefore, the study of high-mass star-forming regions is key to understand our chemical heritage. In fact, molecules found in comets, in other pristine Solar System bodies and in protoplanetary disks, are enriched in ^{15}N, because they show a lower ^{14}N/^{15}N ratio (100-150) with respect to the value representative of the Proto-Solar Nebula (PSN, 441±6), but the reasons of this enrichment cannot be explained by current chemical models. Moreover, the ^{14}N/^{15}N ratio is important because from it we can learn more about the stellar nucleosynthesis processes that produces both the elements. In this sense observations of star-forming regions are useful to constrain Galactic chemical evolution (GCE) models.

Keywords. ISM: abundances, Galaxy: evolution, nucleosynthesis

---

We have derived the ^{14}N/^{15}N ratio in a sample of 87 high-mass star-forming cores, from observations of HCN(1-0) and HNC(1-0), observed with the IRAM-30m telescope (Colzi *et al.* 2018a and Colzi *et al.* 2018b) and we have found that the ratio spans the range 100-1000, with most of the sources (25%) having values in the range 310 ¡ ^{14}N/^{15}N ¡ 350 (see Fig. 2 in Colzi *et al.* 2018b), namely below the PSN value and just above the terrestrial atmosphere (TA) value (272). The abundance ratio ^{14}N/^{15}N is also considered a good indicator of stellar nucleosynthesis. Both isotopes have indeed an important secondary production in the CNO cycles. There is the cold CNO cycle that takes place in main-sequence stars and in the H-burning shells of red giants, and the hot CNO cycle, that occurs instead in novae outbursts and is the main way to produce ^{15}N. However, there is also a strong primary component of ^{14}N created in the Hot Bottom Burning (HBB) of asymptotic giant branch (AGB) stars. These differences lead to an increase of ^{14}N/^{15}N ratio with the Galactocentric distance (D_{GC}), up to 8 kpc, as predicted by models of GCE (e.g. Romano *et al.* 2017). However, the relative importance of these processes is still unclear. The only way to test this is to provide more observational contraints. The 87 dense cores that we have observed span D_{GC} in the range 212 kpc, and we have obtained a new Galactocentric trend of ^{14}N/^{15}N. The GCE model is able to reproduce this trend (see Fig. 6 in Colzi *et al.* 2018b).

References

Colzi, L., *et al.* 2018a, *A&A*, 609, A129
Colzi, L., *et al.* 2018b, *MNRAS*, 478, 3693
Romano, D., *et al.* 2017, *MNRAS*, 470, 401

Astronomy in Focus - XXX
Proceedings IAU Symposium No. XXX, 2018
M. T. Lago, ed.

doi:10.1017/S1743921319004381

Systematic observations on Galactic Interstellar isotope ratios

J. S. Zhang[1], Y. T. Yan[1], W. Liu[1], H. Z. Yu[1], J. L. Chen[1] and C. Henkel[2,3]

[1]Center for Astrophysics, Guangzhou University, Guangzhou 510006, China
email: jszhang@gzhu.edu.cn

[2]Max-Planck-Institut für Radioastronomie, Auf dem Hügel 69, D-53121 Bonn, Germany

[3]Astron. Dept., King Abdulaziz University, P.O. Box 80203, Jeddah 21589, Saudi Arabia

Abstract. We are performing systematic observation studies on the Galactic interstellar isotopic ratios, including $^{18}O/^{17}O$, $^{12}C/^{13}C$, $^{14}N/^{15}N$ and $^{32}S/^{34}S$. Our strategy focuses on combination of multi-transition observation data toward large samples with different Galactocentric distances. Our preliminary results show positive Galactic radial gradients of $^{18}O/^{17}O$ and $^{12}C/^{13}C$. In both cases, the ratio increases with the Galactocentric distance, which agrees with the inside-out scenario of our Galaxy. Observations of other isotopes such as $^{14}N/^{15}N$ and $^{32}S/^{34}S$ are on-going.

Keywords. ISM, isotopic ratios, systematic observations, gradient

1. Our works & Preliminary results

We have performed a $C^{18}O$ and $C^{17}O$ mapping of molecular clouds in Galactic center (Zhang *et al.* 2015) and single pointing of a small sample of Galactic disc molecular clouds with the Delingha 13.7m telescope (DLH) in Purple mountain observatory (Li *et al.* 2016). Now we are performing $C^{18}O$ and $C^{17}O$ multi-transition observations toward a larger sample. Through observations obtained with ARO12, SMT and IRAM 30m, JCMT and DLH13.7, We detected the J=1-0 lines of $C^{18}O$ and $C^{17}O$ in 122 out of 192 sources, and the J=2-1 lines in 270 out of 359 sources. Our preliminary results

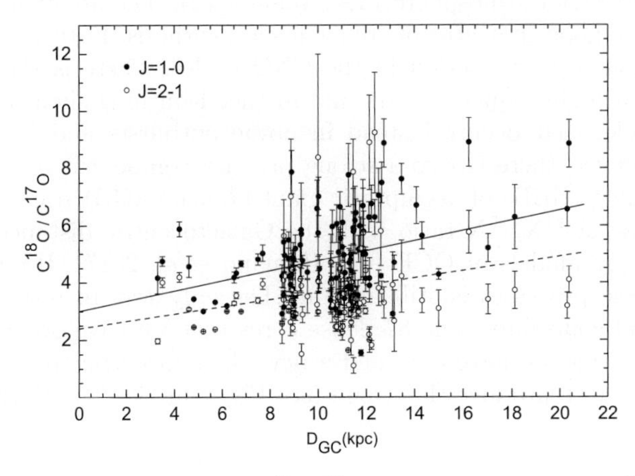

Figure 1. The Galactic radial gradient of $^{18}O/^{17}O$, from observations of $C^{18}O$ and $C^{17}O$ J=2-1 and J=1-0. The solid and dashed line show linear fits for J=1-0 and J=2-1 results, respectively.

support a positive Galactic radial gradient of $^{18}O/^{17}O$, i.e., the ratio increasing with the Galactocentric distance (Figure 1), which agrees with the inside-out formation scenario of our Galaxy. Other projects on different isotopic ratios are mainly based on observations at the Tianma radio telescope (TMRT) in Shanghai. Through TMRT observations of H_2CO and $H_2{}^{13}CO$ absorption lines and of the continuum in C and Ku bands, we determined the isotopic ratio $^{12}C/^{13}C$ for whole sample. We obtained a linear relation between the C isotopic ratio and the Galactocentric distance $^{12}C/^{13}C = (5.50{\pm}1.15)\,D_{GC}{+}4.70{\pm}6.91$. TMRT observations on $^{14}NH_3$ and $^{15}NH_3$ for isotopic ratio $^{14}N/^{15}N$, and $C^{32}S$ and $C^{34}S$ for $^{32}S/^{34}S$ will been soon available.

References

Zhang *et al.* 2015, ApJS, 219, 2, 28
Li *et al.* 2016, RAA, 16, 47

Astronomy in Focus - XXX
Proceedings IAU Symposium No. XXX, 2018
M. T. Lago, ed.

doi:10.1017/S1743921319004393

Radial elemental abundance gradients in galaxies from cosmological chemodynamical simulations

Fiorenzo Vincenzo and Chiaki Kobayashi

Centre for Astrophysics Research, University of Hertfordshire
College Lane, AL10 9AB, Hatfield, United Kingdom
emails: f.vincenzo@herts.ac.uk, c.kobayashi@herts.ac.uk

Abstract. Cosmological chemodynamical simulations are nowadays among the best tools to study how chemical elements are produced within galaxies, to reconstruct also the spatial distribution of the chemical elements as a function of time within different galaxy environments. Our simulation code includes the main stellar nucleosynthetic sources in the cosmos (core-collapse and Type Ia supernovae, hypernovae, asymptotic giant branch stars, and stellar winds from stars of all masses and metallicities). We present the predictions of our simulation for the evolution of the radial gradients of O/H, N/O and C/N in the gas-phase of a sample of ten star-forming disc galaxies, all characterised by very different star formation histories at the present time (see Figure 1). On average, our simulated disc galaxies show a clear inside-out growth of the stellar mass as a function of time, and more negative slopes of the radial gas-phase O/H versus radius at earlier epochs of the galaxy evolution; we predict negative slopes of N/O and positive slopes of C/N at almost all redshifts, because of the main secondary origin of N in stars, even though

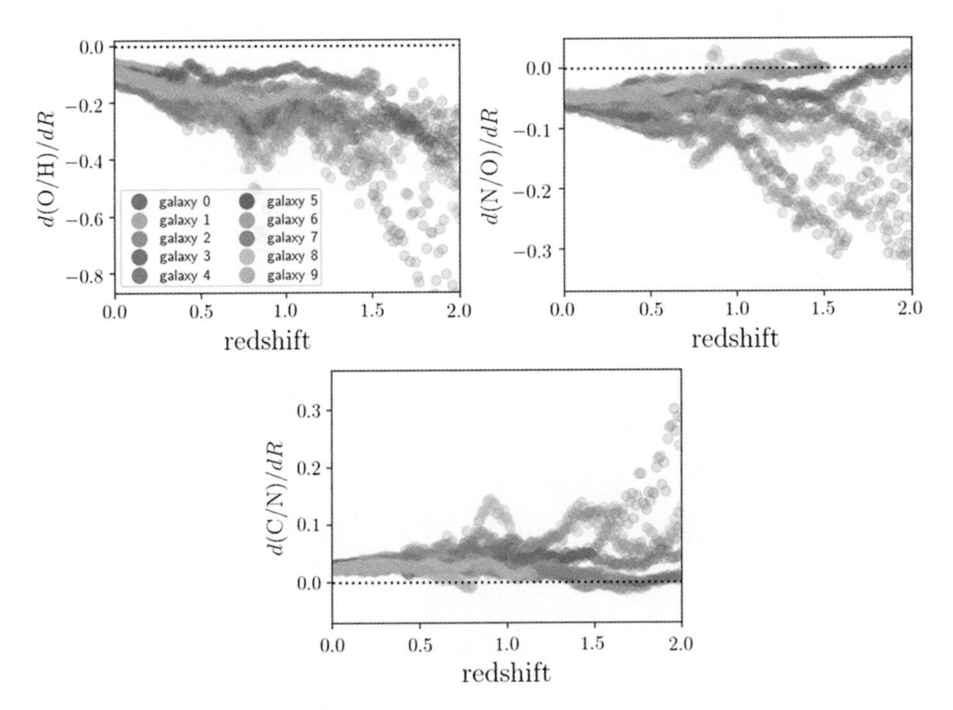

Figure 1. Redshift evolution of the slope of O/H, N/O and C/N versus radius in the gas-phase of our ten simulated disc galaxies, which are the same as in Vincenzo & Kobayashi (2018). From "galaxy 0" to "galaxy 9", the star formation history is concentrated towards later epochs. We only show the redshifts when the total galaxy stellar mass $M_{\star} > 1.0 \times 10^{9}$ $M_{\odot}$

the high-redshift simulation data are highly scattered because of the more turbulent conditions of the interstellar medium. Finally, we show that similar results are found with zoom-in simulations, where a spiral galaxy is re-simulated with a larger number of resolution elements. With zoom-in simulations, we study how stellar migrations (particularly old and metal-poor stellar populations migrating outwards) and radial gas flows are capable of influencing the galaxy chemical evolution at different galactic radii.

Keywords. galaxies: abundances, galaxies: evolution, ISM: abundances, hydrodynamics

---

Reference

Vincenzo, F., & Kobayashi, C. 2018, *MNRAS*, 478, 155

Astronomy in Focus - XXX
Proceedings IAU Symposium No. XXX, 2018
M. T. Lago, ed.

doi:10.1017/S174392131900440X

What the Milky Way bulge reveals about the initial metallicity gradients in the disc

F. Fragkoudi[1], P. Di Matteo[1], M. Haywood[1], S. Khoperskov[1], A. Gomez[1], M. Schultheis[2], F. Combes[3,4] and B. Semelin[3]

[1]GEPI, Observatoire de Paris, Place Jules Janssen, 92195, Meudon, France
[2]Laboratoire Lagrange, Observatoire de la Côte d'Azur, Bd de l'Observatoire, Nice, France
[3]Observatoire de Paris, LERMA, CNRS, PSL Univ., F-75014, Paris, France
[4]College de France, 11 Place Marcelin Berthelot, 75005, Paris, France

Abstract. We examine the metallicity trends in the Milky Way (MW) bulge – using APOGEE DR13 data – and explore their origin by comparing two N-body models of isolated galaxies which develop a bar and a boxy/peanut (b/p) bulge. Both models have been proposed as scenarios for reconciling a disc origin of the MW bulge with a negative vertical metallicity gradient. The first is a superposition of co-spatial disc populations, different scaleheights and metallicities (with flat gradients) where the thick, metal-poor populations contribute significantly to the stellar mass budget in the inner galaxy. The second model is a single disc with an initial steep radial metallicity gradient which gets mapped by the bar into the b/p bulge in such a way that the vertical metallicity gradient of the MW bulge is reproduced – as shown already in previous works in the literature. As we show here, the latter model does not reproduce the positive longitudinal metallicity gradient of the inner disc, nor the metal-poor innermost regions seen in the data. The model with co-spatial thin and thick disc populations reproduces all the aforementioned trends. We therefore see that it is possible to reconcile a (primarily) disc origin for the MW bulge with the observed trends in metallicity by mapping the inner thin and thick discs of the MW into a b/p.

Keywords. Galaxy: bulge, Galaxy: abundances, Galaxy: disk

---

1. Results

We show in Fig. 1 the metallicity map for the Milky Way (MW) bulge from APOGEE DR13 and for the two models considered. We see that while M2 (Martinez-Valpuesta *et al.* 2013) can reproduce the vertical metallicity gradient in the bulge it does not reproduce the global metallicity trends. On the other hand, we see that M1 reproduces well all the trends seen in the data, such as the vertical and longitudinal gradients, and

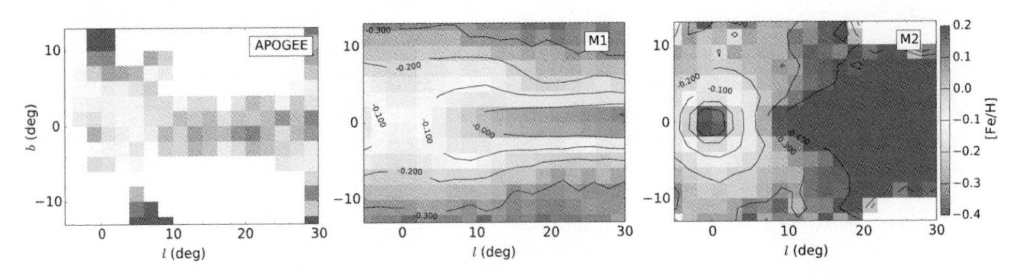

Figure 1. Mean metallicity along the line of sight as a function of galactic longitude l and galactic latitude b for the APOGEE DR13 data (left), model `M1` (middle) and model `M2` (right).

the metal-poor inner regions. We can thus reconcile the disc-origin of the MW bulge with a thin+thick disc model, without the need for an initial steep metallicity gradient.

Reference

Martinez-Valpuesta, I.,& Gerhard, O. 2013, *ApJ*, 776, L3

Astronomy in Focus - XXX
Proceedings IAU Symposium No. XXX, 2018
M. T. Lago, ed.

doi:10.1017/S1743921319004411

Structure and evolution of metallicity and age radial profiles in Milky-Way-like galaxies

E. Athanassoula

Aix Marseille Univ, CNRS, CNES, LAM, Marseille, France

Abstract. I will give here a very short summary of some of the work I have been doing lately on the chemical evolution in Milky-Way-like spiral galaxies.

Keywords. methods: n-body simulations, galaxies: abundances, galaxies: spiral, galaxies: evolution, galaxies: structure, Galaxy: abundances, Galaxy: evolution

---

In this talk, I discussed the structure and evolution in time of the radial metallicity and age profiles, using high resolution chemo-dynamical N-body simulations of Milky-Way-like galaxies. The simulations and the code used are described in Athanassoula *et al.* (2016, 2017). This sample of simulations includes both barred and non-barred disc galaxies, formed and evolved either in isolation or following a major merger.

I presented in some detail two examples: one non-barred formed from the evolution of a merger remnant, and one strongly barred formed and evolving in isolation. For the first I chose a case having a type II surface density radial profile, since observations show that this is the most common type (for a review of the dynamical aspects see Athanassoula (2017) and references therein). For the non-barred example, I obtained the radial profiles by an azimuthal averaging. This, however, was not a good solution for the strongly barred example, so I used two narrow slits, along the direction of the bar major and minor axes, respectively. For both examples I found a clear link between the metallicity or age profiles and the surface density ones.

Extending this analysis to a large number of such examples I reached the conclusion that the properties and evolution of the metallicity and age radial profiles display a considerable variety, depending on several parameters such as e.g. those determining the disc angular momentum and its morphology. In particular, the existence and strength of a bar and of spirals will strongly influence both the radial migration and the star formation and thus the radial metallicity and age profiles and their evolution.

I also argued that for radial profiles it is preferable to avoid units such as arcsec, or kpc and, instead, to normalise by a length with some physical significance. In non-barred galaxies one can use e.g. the (inner) disc scale length, or, even better, the break radius. In barred galaxies the best is to use the barlength, which, although not trivial to calculate accurately, has a physical meaning, separating regions with clearly different orbital structure. Such a normalisation will allow better comparisons between different galaxies, or galaxies and simulations.

References

Athanassoula, E. 2017, in IAU Symposium 321, Formation and Evolution of Galaxy Outskirts, ed. A. Gil de Paz, J. Knapen, J. Lee, (Cambridge: Cambridge Univ. Press), 61
Athanassoula, E., Rodionov, S., Peschken, N., Lambert, J.C. 2016, *ApJ*, 821, 90
Athanassoula, E., Rodionov, S., Prantzos, N. 2017, *MNRAS*, 467L, 46

FM8
New Insights in Extragalactic Magnetic Fields

Astronomy in Focus - XXX
Proceedings IAU Symposium No. XXX, 2018
M. T. Lago, ed.

doi:10.1017/S1743921319004423

New Insights in Extragalactic Magnetic Fields

Luigina Feretti[1], Federica Govoni[2], George Heald[3], Lawrence Rudnick[4] and Melanie Johnston-Hollitt[5,6]

[1]INAF - Istituto di Radioastronomia, Via Gobetti 101, I-40129 Bologna, Italy

[2]INAF - Osservatorio Astronomico di Cagliari, Via della Scienza 5, I-09047 Selargius (CA), Italy

[3]CSIRO Astronomy and Space Science, PO Box 1130, Bentley WA 6102, Australia

[4]Minnesota Institute for Astrophysics, University of Minnesota, 116 Church St. SE, Minneapolis, MN 55455, USA

[5]International Centre for Radio Astronomy Research, Curtin University, Bentley, WA 6102, Australia

[6]Peripety Scientific Ltd., PO Box 11355 Manners Street, Wellington 6142, New Zealand

Abstract. In this contribution we introduce the motivation and goals of IAU Focus Meeting 8, "New Insights in Extragalactic Magnetic Fields". We provide a background for the nine contributions included in these proceedings, as well as the online contributions. A recap of the meeting is provided in the form of audience feedback that was collected during the wrap-up session at the conclusion of FM8.

1. Introduction

Magnetic fields are a key ingredient of the extragalactic Universe on many different spatial scales, from individual galaxies and active galactic nuclei (AGN) to clusters of galaxies and the cosmic web. They play an important role in the process of large-scale structure formation and enrichment of the intergalactic medium, having effects on turbulence, cloud collapse, large-scale motions, heat and momentum transport, convection, viscous dissipation, etc. They are of utmost importance in the growth of radio galaxies and AGN and are crucial for the formation of spiral arms in spiral galaxies, outflows and star formation processes.

Despite their importance and ubiquity, magnetic fields remain poorly understood components of the Universe. The origin of the fields that are currently observed remains largely uncertain. A commonly accepted hypothesis is that they result from the amplification of much weaker pre-existing "seed" fields via shock/compression and/or turbulence/dynamo amplification during merger events, and different magnetic field scales survive as the result of turbulent motions. The origin of "seed" fields is unknown. They could be either "primordial", i.e. generated in the early Universe prior to recombination, or produced locally at later epochs of the Universe, in early stars and/or (proto)galaxies, and then injected in the interstellar and intergalactic medium. The cosmic origin of magnetic "seed fields" and the subsequent processes through which they are amplified give us critical information on the growth of structure in the universe. The history of these processes can be uncovered through accurate knowledge of the strength and structure of magnetic fields in clusters, in the intergalactic medium, at the boundary of galaxy clusters, in the filamentary cosmic web, and in the relation of magnetic fields to gas flows in spiral galaxies, radio galaxies and AGN.

The coming years and decades will see a burst in our capacity to collect information and to develop our understanding of extragalactic magnetic fields. This change is due to the next generation of radio astronomy facilities, especially the SKA and its pathfinders and precursors, as well as major advances in MHD numerical simulations and algorithmic improvements to extract magnetism information from the databases. The Focus Meeting brought together the scientific community to discuss the challenges and opportunities for understanding the magnetized Universe from scales of galaxies to the cosmic web and to connect information from across the spectrum, to address the fundamental questions that remain unanswered:

- How did magnetic fields form and evolve, and how are they maintained?
- How do they control the acceleration and dynamics of relativistic particles in astrophysical plasmas?
- How do magnetic fields affect the evolution of thermal plasma in galaxies and clusters?
- How do magnetic fields illuminate otherwise invisible processes in the thermal plasma?
- How do the insights from magnetic field studies contribute to the larger questions about origins and evolution of structures in the Universe, from galactic to cosmic web scales?

Most of what is known about magnetic fields in the Universe comes from sensitive observations at radio frequencies, which directly prove the existence of relativistic electrons gyrating around magnetic field lines. In addition, measurements of the Faraday rotation effect on the polarized emission of radio galaxies provide information on both the magnetic field strength and its structure. Beside studies at cm and mm wavelengths, magnetism can be investigated at a broader range of other wavelengths. Indeed, the density of the cluster thermal gas obtained from X-ray data is a crucial parameter for the interpretation of Rotation Measure data. In addition, upper bounds to the magnetic fields in voids and large scale structure can be obtained from studies of the Cosmic Microwave Background radiation anisotropies and dust polarization, and from studies of the multi-TeV gamma-ray flux of distant blazars, whose emission is deflected by the extragalactic magnetic fields.

Given the huge shift in processing capacity that will be required by the observational facilities seeking to provide new breakthroughs, the time was ripe to take stock of the state of the field and to understand our successes and opportunities. This Focus Meeting engaged the observational and theoretical communities to consider the results already in hand, to present new algorithms and numerical techniques for the interpretation of the observations, and to address underlying theoretical issues. With this approach, we hoped to accelerate our ability to explore the massive volumes of data that will be delivered by next-generation instruments, toward our ultimate quest to obtain a deeper understanding of the magnetized Universe.

2. Selected Highlights

The aim of the Focus Meeting was to develop a diverse scientific program, covering a large range of topics related to extragalactic magnetic fields. The collection of papers presented in Astronomy in Focus is representative of the diversity of the contributions that animated the meeting. The meeting started with a welcome by L. Feretti and was structured around a series of invited talks, as well as a rich collection of contributed oral and poster presentations. The meeting concluded with a final discussion lead by L. Rudnick. Readers are encouraged to peruse all the contributions, but we present here some of the highlights identified by the attendees themselves at the conclusion of the meeting. During the final discussion, audience members were invited to provide short written descriptions of what they found most interesting and exciting. They were admonished to cite work other than their own, and sometimes they even did that. Note that the citations provided here refer to these proceedings; where relevant, references to the original work are found in the individual contributions. So, with a modicum of editorializing, here's what they identified:

Early Universe, Cosmology

- Standard picture for the origin involves turbulent dynamo amplification in the interstellar or intracluster medium (*Subramanian's talk, these proceedings)*.
- Primordial magnetic fields could have a measurable effect on the $B-modes$ of the CMB at low l: good news for magnetic field afficionados, but for disentagling inflation signatures, not so much! (*Yamazaki's talk, online)*.
- MHD turbulence accompanying primordial fields might be capable of distorting the gravitational wave spectrum at levels accessible to LISA (*Kahniashvili's talk, these proceedings)*.
- During the period of reionization, magnetic fields create Faraday rotation which can be observed in the CMB. Current estimates of $10^{-8}\,\mathrm{rad\,m^{-2}}$ at $z=10$ are very encouraging for seeding the dynamos for later amplification in galaxies and clusters (*Ruiz-Granados's poster, online)*.
- In the ongoing quest to identify the origins of primordial magnetic fields, a mechanism invoking photoionization in an inhomogeneous IGM around Pop III stars, primeval galaxies and quasars might do the job (*Langer's talk, online)*.

Field amplification and large scale structures

Progress continues on a number of fronts to understand how the μG fields that we see today were amplified from much weaker seed fields. Recent insights include:

- It isn't clear that we can always generate the large scale coherence we need (*Sur's poster, online)*, whether we can align the magnetic fields post-shock in cluster "relics" (*Ryu's talk)*, and whether we can distinguish between astrophysical and primordial field origins in clusters (*Dominguez-Fernandez's poster, online)*.
- Magnetic fields can suppress turbulence in the ICM, homogenize both temperature and density fluctuations (*Shukurov's poster, online)*, and affect galaxy evolution and star formation (*talks by Tabatabaei, and Ramos-Martinez, online)*.
- Power spectra do not sufficiently capture the structure of fields, which might form as filaments, ribbons, sheets, etc. *(Shukurov's talk, see online contribution by Seta; Jones' poster)*

Fast Radio Bursts

- The combination of Dispersion Measures and Rotation Measures for Fast Radio Bursts suggests that there are at least two different types. Contributions can be either local to the source or indicators of cosmological magnetic fields at levels $<20\,\mathrm{nG}$ (*Johnston's talk)*.

Extragalactic radio sources

- Exquisite fine structure in Rotation Measures and magnetic fields is now being revealed by LOFAR *(talk by Hoeft, poster by Heesen)*.
- ALMA is in the polarization business! including the discovery of RMs $>10^5\,\mathrm{rad\,m^{-2}}$ in 3C273's jet (*talk by Nagai)*, and progress is being made in reconstruction of 3D fields in jets *(talk by Laing)*.
- The first indicators of the relation between turbulence in the thermal plasma in clusters and the resulting fields are emerging *(talk by Bonafede, these proceedings)*.

Normal Galaxies

• Progress continues on the challenge of dynamos on various scales *(talks by Chamandy, and Sokoloff, online)*.

• On the observational side, field reversal now seen outside of Milky Way in the axisymmetric spiral NGC 4666 *(poster by Stein, online)*, and even the pitch angle of the fields with respect to the spiral arms can be measured with good accuracy in M31 *(poster by Beck, online)*.

• Large scale dynamos are sometimes sub-critical and will not generate strong coherent fields; causes are under investigation *(poster by Rodrigues, online)*.

• Among many highlights from the CHANG-ES survey *(poster by Stil, online)*, the Virgo galaxy NGC4388 has magnetized outflows extending to 5 kpc above the plane *(poster by Damas-Segovia)*.

• Surprisingly, the small group of galaxies in Stephan's Quintet hosts an intergalactic field of strengths comparable to those seen in clusters *(talk by Nikiel-Wroczyński)*.

Techniques and the Future

• Signatures of the primordial field can be decoded from local structure if we're clever enough to overcome a factor of 10^{24} confusion from other fields *(talk by Enßlin)*.

• An incredible variety of new polarization information, including circular polarization, is coming in the next few to 10 years, new telescopes, surveys and techniques to exploit them: LoTSS, MWA, VLASS, POSSUM, MeerKAT, QUOCKA, SKA and even SOFIA in the far IR *(talks and posters by Mao, these proceedings; Loi, Ma, Rudnick, Heald, Horellou, Lopez, online; Gaensler, Hoeft, McKinven, Zinneker)*.

Acknowledgements

The organisers of this Focus Meeting (LF, FG, GH, MJH) would like to extend their gratitude to the organizers of the IAU General Assembly and to the LOC for their hard work and support. We would also like to thank the contributors to the Focus Meeting: those who delivered invited talks, contributed talks or posters, and all of those who participated in the discussions and provided input to the collection of highlights that directly led to this contribution. Our particular thanks go to those of the speakers and poster presenters who delivered contributions to our proceedings.

Partial support for LR comes from U.S. National Science Foundation grant AST1714205 to the University of Minnesota.

Astronomy in Focus - XXX
Proceedings IAU Symposium No. XXX, 2018
M. T. Lago, ed.

doi:10.1017/S1743921319004435

Origins of Cosmic magnetism

Kandaswamy Subramanian

IUCAA, Post Bag 4, Ganeshkhind, Pune 411007, India
email: kandu@iucaa.in

Abstract. The standard picture for the origin of magnetic fields in astrophysical systems involves turbulent dynamo amplification of a weak seed field. Dynamos convert kinetic energy of motions to magnetic energy. While it is relatively easy for magnetic energy to grow, explaining the observed degree of coherence of cosmic magnetic fields generated by turbulent dynamos, remains challenging. We outline potential resolution of these challenges. Another intriguing possibility is that magnetic fields originated at some level from the early universe.

Keywords. magnetic fields, galaxies: magnetic fields, galaxies: clusters: general, early universe

1. Introduction

The universe is magnetized from planets, stars, nearby and high redshift galaxies, the plasma in galaxy clusters and perhaps even the inter galactic medium in void regions devoid of galaxies! Understanding the coherence of magnetic fields detected in these astronomical systems presents an outstanding challenge of modern astrophysics. The general paradigm for extragalactic magnetogenesis involves dynamo amplification of a seed magnetic field due to electromagnetic induction by motions of a conducting plasma. The seed itself could be due to a cosmic battery effect or more intriguingly primordial. We briefly review the dynamo paradigm and then touch upon a possible primordial scenario.

Plasma in all astrophysical systems are turbulent. In galaxies, turbulence is driven by supernovae explosions and in galaxy clusters during its formation by collapse and mergers. Turbulence combined with large scale shearing motions in a highly conducting fluid of galaxies and clusters, generically leads to dynamo action, a process referred to as a turbulent dynamo. Turbulent dynamos are conveniently divided into the small-scale (or fluctuation) dynamos and large-scale (or mean-field) dynamos. The distinction depends respectively on whether the generated magnetic field is ordered on scales smaller or larger than the scale of the turbulent motions. The small-scale dynamos would be relevant for the magnetization of galaxy clusters and young galaxies, while the large-scale dynamo for understanding the system scale magnetic fields in disk galaxies.

2. Small-scale dynamos

In a highly conducting plasma, magnetic flux through any area moving with the fluid is conserved. Consider a flux tube containing plasma of density ρ, magnetic field strength B, area of cross section A and going through fluid parcels separated by a length l. Flux conservation implies $BA =$ constant. Mass conservation in the flux tube gives $\rho Al =$ constant, which implies $B/\rho \propto l$. In any turbulent flow, fluid parcels random walk away from each other and l increases due to random stretching, and if ρ is roughly constant, then B increases. This of course comes at the cost of $A \propto 1/\rho l \propto 1/B$ decreasing, the field being concentrated on smaller and smaller scales l_B at least in one direction, till decay rate due to resistive diffusion, η/l_B^2 is of same order as growth rate due to random stretching v/l. Here v and l are the velocity and coherence scale respectively of turbulent eddies, while η is the resistivity of the plasma. This gives $l_B \sim l R_m^{-1/2}$ where the magnetic Reynolds number $R_m = vl/\eta$ is typically very large in astrophysical systems, which also implies a resistive scale $l_B \ll l$.

Kazantsev (1967) first showed that a specialized short correlated random flow can be a dynamo and cause net magnetic field growth on the very rapid eddy turn over time l/v, provided the R_m exceeds a very modest critical value $R_c \sim 100$. This growth due to the small scale dynamo has since been verified by many direct numerical simulations where a seed field is introduced into a turbulent flow (Haugen *et al.*, (2004), Schekochihin*et al.*, (2004), Bhat & Subramanian (2013)). Thus generically turbulence in the interstellar or intra cluster medium, which have $R_m \gg R_c$, is expected to rapidly amplify magnetic fields on a timescale $l/v \sim 10^7$ yr in galaxies (with $v \sim 10$ km s^{-1} and $l \sim 100$ pc) and $l/v \sim 3 \times 10^8$ yr in galaxy clusters (with $v \sim 300$ km s^{-1} and $l \sim 100$ kpc), much smaller than their age. However as $l_B \ll l$, the field in the growing phase is extremely intermittent and concentrated in to the small resistive scales. The big challenge is then whether these fields can become coherent enough on dynamo saturation to explain for example observations of the Faraday rotation inferred in galaxy clusters and young galaxies.

We have used direct numerical simulations of fluctuation dynamos in forced compressible turbulence with various values of R_m, fluid Reynolds number R_e and up to rms Mach number of $\mathcal{M} = 2.4$, to directly measure the resulting Faraday rotation measure (RM) and the degree of coherence of the magnetic field (Subramanian *et al.* 2006; Bhat & Subramanian 2013; Sur *et al.* 2018). The measured values of a normalized RM, $\bar{\sigma}_{RM}$, normalized by that expected in a model where fields with the rms strength B_{rms} are assumed to be coherent on the forcing scale of turbulence, are shown in Fig. 1 for some of these runs. At dynamo saturation, for a range of parameters, we find $\bar{\sigma}_{RM} \sim 0.40 - 0.55$, or an rms RM contribution which is about half the value expected if the field is coherent on the turbulent forcing scale. This arises in spite of the highly intermittent nature of the field. The left panel of Fig. 1, also shows that when regions with the field above $2B_{rms}$ are excluded from the RM computation for the subsonic case, there is only a modest 20% decrease of $\bar{\sigma}_{RM}$ (Bhat & Subramanian 2013). Thus the dominant contribution to the RM in this case (and also when the flow is transonic) comes from the general sea of volume filling fields, rather than from the rarer, strong field structures. However, in the supersonic case, strong field regions as well as moderately overdense regions contribute significantly to RM. The density dependence is illustrated in the right panel of Fig. 1, where the $\bar{\sigma}_{RM}$ contribution by various overdensity ranges is shown for the cases with $\mathcal{M} = 1.1$ and 2.4 (Sur *et al.* 2018). Our results can account for the observed RMs in galaxy clusters and in young galaxies just as due to fluctuation dynamo action. We also find that the coherence scale of the generated intermittent field is about 1/3 to 1/4 of the velocity coherence scale for the parameters so far explored, and so not too small.

3. Mean-field dynamos

Magnetic fields with coherence scales larger than that of the stirring can be amplified if the turbulence is helical. In disk galaxies, supernovae drive turbulent motions which become helical due to the rotation and vertical stratification of the disk. Helical turbulent motions of the gas draw out any toroidal field in the galaxy into a loop and twists it to look like a twisted Omega (called the α-effect). These lead to the generation of poloidal magnetic fields from toroidal fields. The shear due to galactic differential rotation winds up radial component of the poloidal field to generate a toroidal component. The combination of these two effects lead to a mean-field dynamo generation of disk galaxy magnetic fields on the differential rotation time scales of about $10^8 - 10^9$ yr.

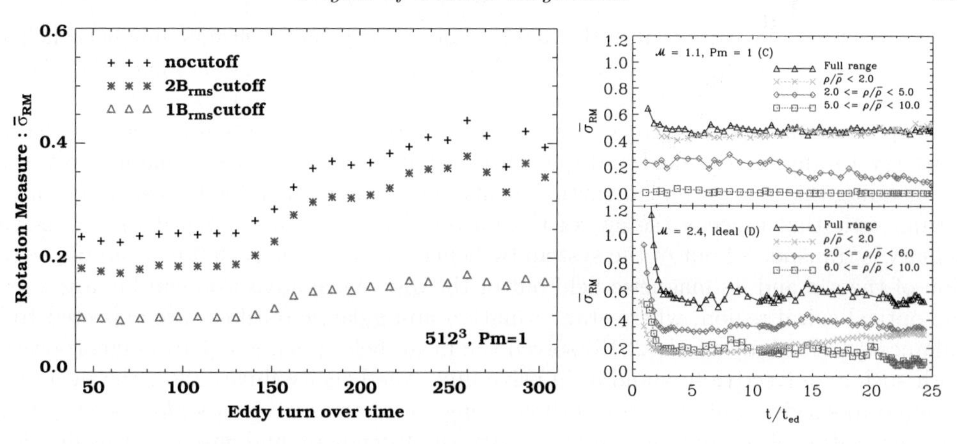

Figure 1. Left panel shows the time evolution $\bar{\sigma}_{RM}$ for a 512^3 subsonic run with $\mathcal{M}=0.14$, $R_m = R_e = 622$ from Bhat & Subramanian (2013). The stars and triangles show respectively the result of excluding regions with field above $2B_{rms}$ and B_{rms} while the crosses correspond to not imposing any cutoff. The right panel shows the corresponding results from Sur *et al.* (2018) of higher Mach number flows, with also overdensity cuts as indicated in the figure.

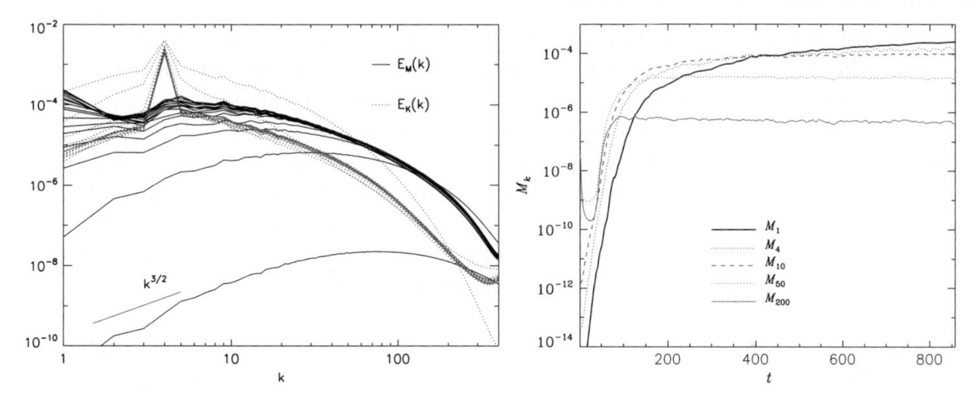

Figure 2. Left panel: time evolution of the magnetic spectra $E_M(k)$ and kinetic spectra $E_K(k)$ from Bhat *et al.* (2016) for a 1024^3 simulation of helically forced (at $k=4$), turbulence with $\mathcal{M} = 0.135$, $R_m = 10R_e = 3375$. Right panel: The time evolution of different modes of the magnetic spectra $M_k(t)$ for $k = 1, 4, 10, 50$, and 200. Initially all scales grow together, with magnetic power peaked on small scales. Lorentz forces act to order the field on larger and larger scales as the dynamo saturates.

This idea faces two important challenges. First, magnetic fluctuations due to the fluctuation dynamo in a disk galaxy grow on small scales much faster ($10^7 yr$) than the growth time of the mean field. Lorentz forces can then become important to saturate the field growth much before the mean field has grown significantly. Can the large-scale field then grow at all? Bhat *et al.* (2016) examined this issue using direct simulations of magnetic field amplification due to fully helical turbulence in a periodic box. The results of one such run where both the fluctuation and mean-field dynamos arise in a unified manner is shown in Fig. 2. We find that initially scales both larger and much smaller than the stirring scale grow together as an eigenfunction dominated by small scales. But crucially on saturation of small scales due to the Lorentz force, larger and larger scales come to dominate due to the mean-field dynamo action. Finally system scale fields (here the scale of the box) develop provided small-scale magnetic helicity can be efficiently removed (see below), which in this simulation is due to resistive dissipation.

The second challenge is that in the highly conducting galactic plasma, magnetic helicity which measures the linkages between field lines, is nearly conserved. Then when helical motions writhe the toroidal field to generate a poloidal field, an oppositely signed twist develops on smaller scales, to conserve magnetic helicity and Lorentz forces due to this twist try to unwind the field and quench the dynamo. Large-scale dynamos only work by shedding this small-scale magnetic helicity. This can happen due to resistivity, but on a time scale that exceeds the age of the universe! It can also happen if the small-scale helicity is transported out of the system by helicity fluxes. One such flux is simply advection of the gas and its magnetic field out of the disk. Such advection can be larger from the optical spiral region, where star formation and galactic outflows are expected to be enhanced. Chamandy *et al.* (2015) solved the mean-field dynamo equation incorporating both such an advective flux and a diffusive flux. The helicity fluxes allow the mean-field dynamo to survive, but stronger outflow along spiral arms led to a suppression of mean field generation there and an interlaced pattern of magnetic and gaseous arms develops, as seen in the galaxy NGC6946 (Beck & Hoernes 1996). Interestingly a wide spread magnetic spiral only results if the optical spiral is allowed to wind up and thus here we are constraining spiral structure theory using magnetic field observations!

4. Primordial magnetic fields

An intriguing possibility is that magnetic fields are an early Universe relic, arising during inflation along with density fluctuations, or being generated in QCD or electroweak phase transitions (for a review see Subramanian 2016). A number of problems have been raised about inflationary magnetogenesis. A model by Sharma *et al.* (2018) which addresses these, predicts a blue magnetic field spectrum $d\rho_B/d\ln k \propto k^4$ and requires a low energy scale of inflation and reheating. The field is also helical and so orders itself considerably as it decays. A scenario with reheating at a temperature of 100 GeV leads to present day field strengths of order $B_0 = 4 \times 10^{-11}$ G with a coherence scale of 70 kpc. Such models can be constrained by space gravitational wave detectors like LISA in the future.

In summary the origin of cosmic magnetism on galactic and extragalactic scales is still an area of active research with many interesting ideas which will continue to fascinate astronomers of the future.

Acknowledgements

I thank Pallavi Bhat, Axel Brandenburg, Luke Chamandy, T. R. Seshadri, Ramkishor Sharma, Sharanya Sur and Anvar Shukurov for the enjoyable collaborations reported here and the FM8 organisers and IAU for financial support to attend this meeting.

References

Beck, R., & Hoernes, P. 1996, *Nature*, 379, 47
Bhat, P., and Subramanian, K. 2013, *MNRAS*, 429, 2469–2481
Bhat, P., Subramanian, K., & Brandenburg, A. 2016, *MNRAS*, 461, 240
Chamandy, L., Shukurov, A., & Subramanian, K. 2015, *MNRAS*, 446, L6
Haugen, N. E., Brandenburg, A., and Dobler, W. 2004, *PRE*, 70 (1), 016308
Kazantsev, A. P. 1967, *JETP*, 53, 1807 (English translation: Sov. Phys. JETP, 26, 1031, 1968)
Schekochihin, A. A., Cowley, S. C., Taylor, S. F., Maron, J. L., and McWilliams, J. C. 2004, *ApJ*, 612, 276
Sharma, R., Subramanian, K., & Seshadri, T. R. 2018, *PRD*, 97, 083503
Subramanian, K., Shukurov, A., and Haugen, N. E. L. 2006, *MNRAS*, 366, 1437
Subramanian, K. 2016, *Reports on Progress in Physics*, 79, 076901
Sur, S., Bhat, P., & Subramanian, K. 2018, *MNRAS*, 475, L72

Astronomy in Focus - XXX
Proceedings IAU Symposium No. XXX, 2018
M. T. Lago, ed.

doi:10.1017/S1743921319004447

Magnetism in the Early Universe

Tina Kahniashvili[1,2] Axel Brandenburg[1,3,4], Arthur Kosowsky[5], Sayan Mandal[1,2] and Alberto Roper Pol[2,4,6]

[1]Department of Physics, Carnegie Mellon University, USA

[2]Abastumani Astrophysical Observatory, Ilia State University, Georgia

[3]Nordita, KTH Royal Institute of Technology & Stockholm University; Department of Astronomy, Stockholm University, Sweden; and JILA, University of Colorado at Boulder, USA

[4]Laboratory for Atmospheric and Space Physics, University of Colorado at Boulder, USA

[5]Department of Physics and Astronomy and PITT PACC, University of Pittsburgh, USA

[6]Department of Aerospace Engineering, University of Colorado at Boulder, USA

Abstract. Blazar observations point toward the possible presence of magnetic fields over intergalactic scales of the order of up to ~ 1 Mpc, with strengths of at least $\sim 10^{-16}$ G. Understanding the origin of these large-scale magnetic fields is a challenge for modern astrophysics. Here we discuss the cosmological scenario, focussing on the following questions: (i) How and when was this magnetic field generated? (ii) How does it evolve during the expansion of the universe? (iii) Are the amplitude and statistical properties of this field such that they can explain the strengths and correlation lengths of observed magnetic fields? We also discuss the possibility of observing primordial turbulence through direct detection of stochastic gravitational waves in the mHz range accessible to LISA.

Keywords. Early Universe, Cosmic Magnetic Fields, Turbulence, Gravitational Waves

---

1. Introduction

Magnetic fields of strengths of the order of $\sim 10^{-16}$ G are thought to be present in the voids between galaxy clusters; see Neronov & Vovk (2010) for the pioneering work and Durrer & Neronov (2013) for a review and references therein. These are thought to be the result of the amplification of the seed magnetic field, with two scenarios of the origin currently under discussion; see Subramanian (2016) for a review: a bottom-up (astrophysical) scenario, where the seed is typically very weak and magnetic field is transferred from local sources within galaxies to larger scales, and a top-down (cosmological) scenario where a magnetic field is generated prior to galaxy formation in the early universe on scales that are large at the present epoch. We discuss two different scenarios of primordial magnetogenesis: magnetic fields produced during inflation or during cosmological phase transitions. We address cosmic magnetohydrodynamic (MHD) turbulence, in order to understand the magnetic field evolution. Turbulent motions can also affect cosmological phase transitions. We argue that even a small total energy density in turbulence (less than 10% of the total thermal energy density) can have substantial effects because of strong nonlinearity of the relevant physical processes; see also Vazza *et al.* (2017).

2. Overview

The evolution of a primordial magnetic field is determined by various physical processes that result in amplification and damping of the field. Complexities arise in the problem due to the strong coupling between magnetic field and plasma motions (Kahniashvili *et al.* 2010), producing MHD turbulence, which then undergoes free decay after the forcing is switched off (Brandenburg *et al.* 1996; Dimopoulos & Davis 1997; Jedamzik *et al.* 1998; Subramanian & Barrow 1998); see Kahniashvili *et al.* (2016) for a recent overview. The presence of initial kinetic and/or magnetic helicity strongly affects the development of turbulence. In several models of phase transition magnetogenesis, parity (mirror symmetry) violation leads to a non-zero chirality (helicity) of the field (Cornwall 1997; Giovannini & Shaposhnikov 1998; Field & Carroll 2000; Giovannini 2000; Vachaspati 2001). We also underline the importance of possible kinetic helicity: our recent simulations have shown that through the decay of hydromagnetic turbulence with initial kinetic helicity, a weak nonhelical magnetic field eventually becomes fully helical (Brandenburg *et al.* 2017).

The anisotropic stresses of the resulting turbulent magnetic and kinetic fields are a source of gravitational waves, as already pointed out by Deryagin *et al.* (1986). The amplitude of the gravitational wave spectrum depends on the strength of the turbulence, and its characteristic wavelength depends on the energy scale at which the gravitational wave source is generated (Gogoberidze *et al.* 2007).

3. Results

Understanding the mechanisms for generating primordial turbulence is a major focus of our investigation. Turbulence may be produced during cosmological phase transitions when the latent heat of the phase transition is partially converted to kinetic energy of the plasma as the bubbles expand, collide, and source plasma turbulence (Christensson *et al.* 2001). The two phase transitions of interest in the early universe are (i) the electroweak phase transition occurring at a temperature of $T \sim 100\,\mathrm{GeV}$, and (ii) the QCD phase transition occurring at $T \sim 150\,\mathrm{MeV}$. Turbulence at the electroweak phase transition scale is more interesting for the gravitational wave detection prospects, since the characteristic frequency of the resulting stochastic gravitational wave background, set by the Hubble length at the time of the phase transition, falls in the Laser Interferometer Space Antenna (LISA) frequency band; see Kamionkowski *et al.* (1994), and Kosowsky *et al.* (2002) for pioneering studies, and Caprini & Figueroa (2018) for a recent review.

Since the electroweak phase transition is probably a smooth crossover in the Standard Model of particle physics, it would not proceed through bubble collisions and follow up turbulence. However, our knowledge of electroweak scale physics is incomplete; at least two lines of reasoning point toward a first-order phase transition in the very early universe. First, such a transition can provide the out-of-equilibrium environment necessary for successful baryogenesis; see, e.g., Morrissey & Ramsey-Musolf (2012). Secondly, as discussed above, turbulence induced in a first-order transition naturally amplifies the seed magnetic fields which can explain the magnetic fields that might be present in cosmic voids; see Fig. 1 and Brandenburg *et al.* (2017). Arguments in favor of a primordial origin of such fields were also given by Dolag *et al.* (2011).

If significant magnetic fields exist after the phase transitions, they can source turbulence for long durations, extending even until recombination. For these sources, the damping due to the expansion of the universe cannot be neglected. Numerical simulations show only a slow decay of turbulent energy, especially at the large-scale end of the spectrum, along with the generation of significant energy density in velocity fields; see Fig. 4 of Kahniashvili *et al.* (2010), and Brandenburg & Kahniashvili (2017). Turbulence in the early universe can also be generated during inflation, whereby the magnetic field energy is injected into primordial plasma ensuring a strong coupling between the magnetic field and fluid motions. The correlation scale of induced turbulent motions is limited by the Hubble scale, as required by causality; see Kahniashvili *et al.* (2012) for the non-helical case and Kahniashvili *et al.* (2017) for the helical case, while the magnetic field stays frozen-in at superhorizon scales. The strength of the turbulent motions is determined by the total energy density of the magnetic field; a sufficiently strong field can lead to a detectable gravitational wave signal (Kahniashvili *et al.* 2008).

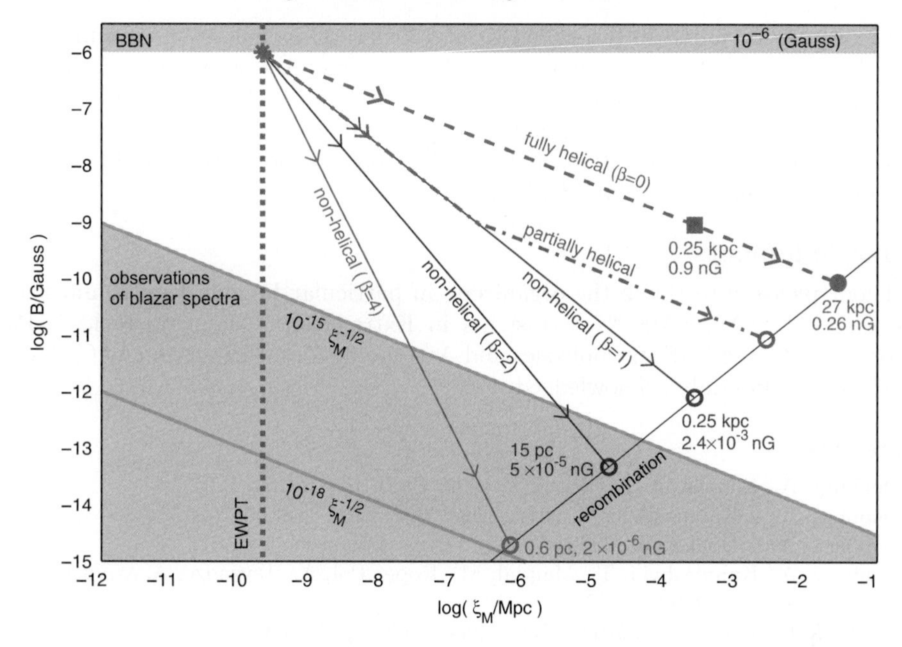

Figure 1. Turbulent evolution of the strength $B_{\rm rms}$ and correlation length ξ_M of the magnetic field starting from their upper limits given by the Big Bang Nucleosynthesis (BBN) bound and the horizon scale at the electroweak phase transitions (from Brandenburg *et al.* 2017, Fig. 11).

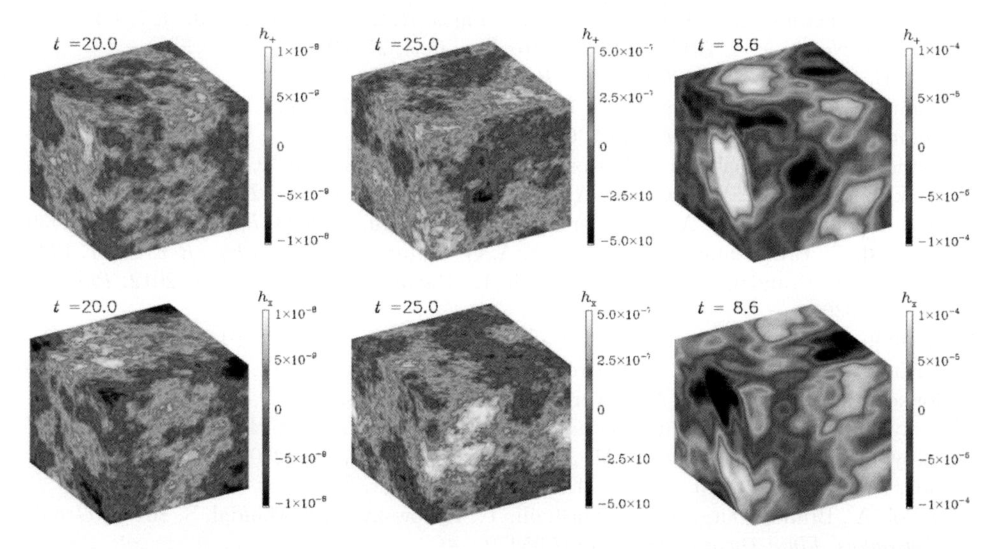

Figure 2. Visualizations of h_+ (top) and $h_\times$ (bottom) on the periphery of the computational domain for different positions of the initial turbulent spectrum peak frequency $k_f/k_H = 300$, 60, 2 from left to right respectively. (in press)

The Pencil Code (Brandenburg & Dobler 2002) is a general public domain tool box to solve sets of partial differential equations on large, massively parallel platforms. It has recently been applied to early universe simulations of mesh size up to 2304^3 (Brandenburg & Kahniashvili 2017), which was necessary for modeling turbulence at the phase transitions (Brandenburg *et al.* 2017) and the inflationary stage (Kahniashvili *et al.* 2017). We have recently added a module to evolve the gravitational waves in the simulation domain from the dynamically evolving MHD stresses. Details of the numerical simulations can be found in Roper Pol *et al.* (2019a). Our first results are presented in Roper Pol *et al.* (2019b) and in Fig. 2, where we plot the gravitational wave strain components h_+ and $h_\times$ sourced by fully helical hydromagnetic turbulence. It must be highlighted that the presence of initial magnetic helicity significantly affects the detection prospects. However, the detection of the circular polarization degree by LISA seems to be problematic (Smith & Caldwell 2017).

Acknowledgments

It is our pleasure to thank the organizers, in particular Luigina Feretti and Federica Govoni of IAU XXX FM8 "New Insights in Extragalactic Magnetic Fields". Partial support through the NSF Astrophysics and Astronomy Grant Program (AAG) (1615940 & 1615100) is gratefully acknowledged.

References

Brandenburg, A., Enqvist, K. & Olesen, P. 1996, *Phys. Rev. D*, 54, 1291
Brandenburg, A. & Kahniashvili, T. 2017, *Phys. Rev. Lett.*, 118, 055102
Brandenburg, A. & Dobler, W. 2002, *Comput. Phys. Commun.*, 147, 471
Brandenburg, A., Kahniashvili, T., Mandal, M., Roper Pol, A., Tevzadze, A. & Vachaspati, T. 2017, *Phys. Rev. D*, 96, 123528
Caprini, C. & Figueroa, D. G. 2018, *Class. Quant. Grav.*, 35, 163001
Cornwall, J. M. 1997, *Phys. Rev. D*, 56, 6146
Christensson, M., Hindmarsh, M. & Brandenburg, A. 2001, *Phys. Rev. E*, 64, 056405
Deryagin, D., Grigoriev, D., Rubakov, V. & Sazhin, M. 1986, *Mod. Phys. Lett. A*, 1, 593
Dimopoulos, K. & Davis, A. C. 1997, *Phys. Lett. B*, 390, 87
Dolag, K., Kachelriess, M., Ostapchenko, S. & Tomas, R. 2011, *Astrophys. J.*, 727, L4
Durrer, R. & Neronov, A. 2013, *Astron. Astrophys. Rev.*, 21, 62
Field, G. B. & Carroll, S. M. 2000, *Phys. Rev. D*, 62, 103008
Giovannini, M. & Shaposhnikov, M. E. 1998, *Phys. Rev. D*, 57, 2186
Giovannini, M. 2000, *Phys. Rev. D*, 61, 063004
Gogoberidze, G., Kahniashvili, T. & Kosowsky, A. 2007, *Phys. Rev. D*, 76, 083002
Jedamzik, K., Katalinic, V. & Olinto, A. V. 1998, *Phys. Rev. D*, 57, 3264
Kahniashvili, T., Kosowsky, A., Gogoberidze, G. & Maravin, Y. 2008, *Phys. Rev. D*, 78, 043003
Kahniashvili, T., Brandenburg, A., Tevzadze, A. G. & Ratra, B. 2010, *Phys. Rev. D*, 81, 123002
Kahniashvili, T. Brandenburg, A., Campanelli, L., Ratra, B. & Tevzadze, A. 2012, *Phys. Rev. D*, 86, 103005
Kahniashvili, T., Brandenburg, A. & Tevzadze, A. 2016, *Phys. Scripta*, 91, 104008
Kahniashvili, T. Brandenburg, A., Durrer, R., Tevzadze, A. & Yin, W. 2017, *JCAP*, 1712, 002
Kamionkowski, M., Kosowsky, A. & Turner, M. S. 1994, *Phys. Rev. D*, 49, 2837
Kosowsky, A., Mack, A. & Kahniashvili, T. 2002, *Phys. Rev. D* 66, 024030
Morrissey, D. E. & Ramsey-Musolf, M. J. 2012, *New J. Phys.*, 14, 125003
Neronov, A. & Vovk, I. 2010, *Science*, 328, 73
Roper Pol, A., Brandenburg, A., Kahniashvili, T., Kosowsky, A. & Mandal, S. 2019a, *Geophys. Astrophys. Fluid Dynam.*, arXiv:1807.05479
Roper Pol, A., Mandal, S., Brandenburg, A., Kahniashvili, T., & Kosowsky, A., 2019b, Phys. Rev. Lett. Submitted, arXiv:1903.08585
Smith, T. L. & Caldwell, R. 2017, *Phys. Rev. D*, 95, 044036
Subramanian, K. & Barrow, J. D. 1998, *Phys. Rev. Lett.*, 81, 3575
Subramanian, K. 2016, *Rept. Prog. Phys.*, 79, 076901
Vachaspati, T. 2001. *Phys. Rev. Lett.*, 87, 251302
Vazza, F., Brggen, M., Gheller, C., Hackstein, S., Wittor, D. & Hinz, P. M. 2017, *Class. Quant. Grav.*, 34, 234001

Astronomy in Focus - XXX
Proceedings IAU Symposium No. XXX, 2018
M. T. Lago, ed.

doi:10.1017/S1743921319004459

Constraining magnetic fields in galaxy clusters

Annalisa Bonafede[1,2,3] Chiara Stuardi[1,2], Federica Savini[3], Franco Vazza[1,2,3] and Marcus Brüggen[3]

[1]Dipartimento di Fisica e Astronomia, Università di Bologna, via P. Gobetti 93/2, 40129, Bologna, Italy.

[2]INAF - Istituto di Radioastronomia, Via Gobetti 101, I-40129 Bologna, Italy.

[3]Hamburger Sternwarte, Universität Hamburg, Gojenbergsweg 112, 21029, Hamburg, Germany.
email: annalisa.bonafede@unibo.it

Abstract. Magnetic fields originate small-scale instabilities in the plasma of the intra-cluster medium, and may have a key role to understand particle acceleration mechanisms. Recent observations at low radio frequencies have revealed that synchrotron emission from galaxy clusters is more various and complicated than previously thought, and new types of radio sources have been observed. In the last decade, big steps forward have been done to constrain the magnetic field properties in clusters thanks to a combined approach of polarisation observations and numerical simulations that aim to reproduce Faraday Rotation measures of sources observed through the intra-cluster medium. In this contribution, I will review the results on magnetic fields reached in the last years, and I will discuss the assumptions that have been done so far in light of new results obtained from cosmological simulations. I will also discuss how the next generation of radio instruments, as the SKA, will help improving our knowledge of the magnetic field in the intra-cluster medium.

Keywords. Galaxy clusters, magnetic field, non-thermal phenomena

---

1. Introduction

Galaxy clusters host magnetic fields that are responsible for a variety of phenomena on a large range of spatial scales. The intra-cluster medium (ICM) is an almost perfect plasma, and the magnetic field originates small-scale instabilities that can amplify the field itself and modify the ICM microphysics. On Mpc scale, the magnetic field interacts with Cosmic Ray electrons (CRe) and produce diffuse emission such as radio halos, mini halos, and radio relics. Recent observations with the LOw Frequency ARray (LOFAR, van Haarlem *et al.* 2013) have shown that the radio emission in the ICM is more complex than initially thought. In particular, Savini *et al.* (2018) have discovered that steep-spectrum emission on Mpc scale is observed also in clusters that do not show signs of major merger. In Fig. 1 (right panel), the radio emission from the galaxy cluster RXCJ1720.1+2638 is shown. This cluster is known to host a mini halo confined between two cold fronts, i.e. discontinuities in the X-ray surface brightness profile. It has been proposed that the cold fronts are formed during a minor merger, where the dense core of the cluster is perturbed and starts a sloshing motion in the dark matter potential well. LOFAR reveals that the radio emission extends well beyond the cold fronts. The radio emission inside and outside the core can be well separated both in brightness and in spectral index. The core emission (mini halo) is relatively flat ($\alpha \sim -1$) and the spectral index distribution is uniform. The emission SW and NE of the core emission, beyond

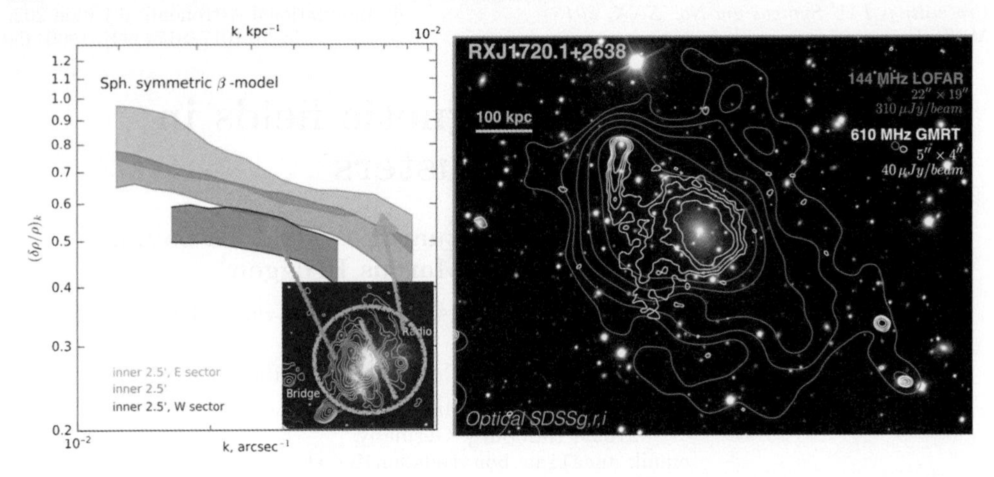

Figure 1. Left: power spectrum of the gas density fluctuations in the cluster MACSJ0717 in the regions with and without radio emission, as shown in the inset (Bonafede *et al.* 2018). Right: The cluster RXCJ1720.1+2638: optical emission and radio emission from GMRT at 610 MHz (white) and LOFAR at 150 MHz (red), from Savini *et al.* (2018, in press).

the cold fronts, is steeper ($\alpha \sim -2$), suggesting either a different acceleration process and/or a different magnetic field strength in the two regions.
New emission has also been observed by LOFAR in the cluster MACSJ0717.0+3745 (Bonafede *et al.* 2018). This massive ($M_{500} \sim 1.1 \times 10^{15} M_{\odot}$) high-z (z=0.546) cluster was known to host diffuse radio emission visible at GHz frequencies, which is not co-spatial with the X-ray emission from the gas. The radio emission is confined in the Eastern part of the cluster. LOFAR observations at 150 MHz have confirmed this asymmetry in the radio emission (see Fig. 1, left panel), and a new radio arc with a linear size of $\sim$ 1.7 Mpc has been discovered at the NW of the cluster centre. The analysis of the power spectrum of the gas density fluctuations - performed following the method by Zhuravleva *et al.* (2014) - has shown that there is a correlation between the amplitude of the power spectrum and the presence of radio emission. This might indicate a higher ratio of kinetic versus thermal energy in the region with radio emission than in the region without.
Recently, theoretical studies have focussed on the role of magnetic fields to accelerate particles in the cluster outskirts. Radio relics are linked to low Mach number shocks, that have a low acceleration efficiency (e.g. Brüggen *et al.* 2011). Shock waves should also amplify the magnetic field, but only the magnetic field amplification in the relic of the Coma cluster has been analysed so far (Bonafede *et al.* 2013) and no net amplification in the shocked region has been found.

2. Constraints on magnetic fields and assumptions

The analysis of the Faraday effect from sources inside and in the background of clusters is the most used technique to constrain the magnetic field. Using this method, Bonafede *et al.* (2010) used the Rotation Measure (RM) of 7 sources in the background of the Coma cluster and the Faraday code (Murgia *et al.* 2004), and derived that the magnetic field strength in the cluster centre is $B_0 \sim 5\,\mu$G, with a radial profile that scales with the square-root of the gas density profile (i.e. $B(r) \propto n^{0.5}$). To derive these values, assumptions need to be made on the magnetic field properties. In particular, the magnetic field is assumed to be a Gaussian random field, characterised by a single power law power spectrum. Thanks to the increasing resolution reached by recent cosmological

simulations, it is now possible to compare observed and simulated RM. Using an adaptive mesh refinement method, Vazza *et al.* (2018) have performed MHD cosmological simulations reaching a resolution of $\sim$ 4 kpc in the cluster centres. A primordial magnetic field $B_i = 0.1$ nG at z= 30 is evolved, and its amplification is observed to be above the one predicted from a pure adiabatic compression. As the resolution increases, the magnetic field amplification increases as well, reaching $B \sim \mu$G in the cluster centres. The magnetic field components show a departure from a Gaussian distribution, with a tail towards high values of the magnetic field. The amplitude of this tail depends on time, cluster dynamics, and resolution of the simulation. The net effect is that the RM profile for a Coma-like cluster can be reproduced with slightly lower values of B_0 than those derived by Bonafede *et al.* (2010). In the specific case analysed by Vazza *et al.* (2018), the average value of the magnetic field in the core is $B_{core} \sim 1.5\mu$G.
Cosmological simulations also predict that magnetic field power spectrum deviates from a simple power law. The relevant scales to understand the role of magnetic field in accelerating particles would be the injection scale ($L_{inj} \sim$ few 100 kpc) and the viscous scale ($L_{visc} \sim$ few kpc, or even lower if the Spitzer viscosity is suppressed in the ICM). The power spectrum can be approximated by a power law between L_{inj} and L_{visc} (see e.g. Ryu, this meeting, Xu *et al.* 2009).
However, it must be noted that the number of free parameters would increase if non-Gaussian magnetic fields and non power law power spectra were considered. The constraints on the magnetic fields that have been obtained so far are based on few sources per cluster (from 1 to 7), and the modelling done by the authors to reproduce mock RM images already have a total of 5 free-parameters. Hence, we conclude that - despite more complicated models for the magnetic field should be ideally considered - the limiting factor is the low number of sources per cluster. Considering more complicated models of magnetic field and increasing the number of free parameters will make sense only when future observations will be able to sample the RM through several tens of lines on sight per cluster.

3. New results with present instruments and future perspectives

The advent of spectro-polarimetric and wide-band receivers enables us to make an important step forward in the study of cluster magnetic fields. We started a project to constrain the magnetic field amplification by low Mach number shocks in clusters. To overcome the limit of few sources observed per cluster, we built up a sample of cluster with double relics, and obtained $\sim$ 80 h observing time at the Jansky Very Large Array. In Fig. 2, we show the cluster RXCJ1314-2515 (Stuardi et al in prep). We analysed the Faraday spectrum of the emission using the Rotation Measure synthesis technique (Brentjens & de Bruyn 2005). So far, the regions that we have analysed seem consistent with the Galactic RM, suggesting little/no amplification of the magnetic field in the shocked region.
The next generation of radio instruments, i.e. the Square Kilometer Array (SKA) will permit a study of the magnetic field in individual clusters. The expected number density of polarised sources that SKA will observe is $\sim$ 300 per square degree (Rudnick & Owen 2014). This means that we will be able to study the magnetic field in a Coma-like cluster using $\sim$ 50 radio sources, i.e. a factor 7 higher than present-day studies. Bonafede *et al.* (2015) have analysed the RM grid that SKA will provide for clusters of different mass, finding that we will have enough sources to constrain the magnetic field in the background of low-mass galaxy clusters and galaxy groups. Having such a high number of RM samples through the line of sight, one may be able to investigate different configurations for the magnetic field, i.e. non-Gaussian distributions of the components and different functions for the power spectrum.

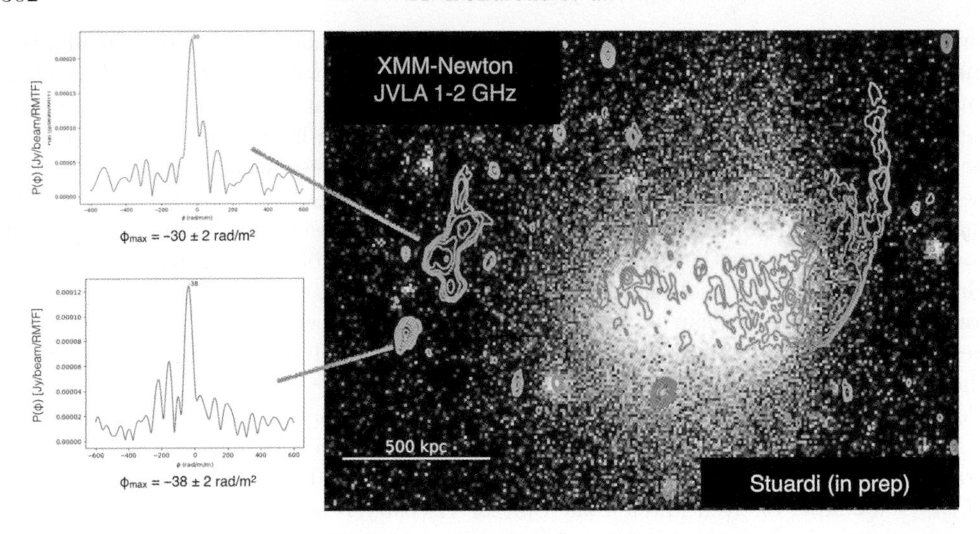

Figure 2. Radio emission from the cluster RXCJ1314 at 1-2 GHz (Stuardi *et al.*, in prep.). The left panels show the Faraday spectra through one source in the relic (top) and in the pre-shock region (bottom). The Galactic foreground RM is $\phi \sim -30 rad/m^2$, consistent with the values found for the two sources.

4. Conclusions

Obtaining constraints on the magnetic field in the ICM is crucial to understand particle acceleration mechanisms and the ICM microphysics. One must bear in mind that magnetic fields are not directly observable, and the constraints derived through observations rely on the assumptions that have been made. State-of the art instruments, such as the JVLA are very promising to obtain statistical constraints on the magnetic field using samples of clusters. Using this approach, Stuardi *et al.* (in prep) are already investigating for the first time the magnetic field amplification by low Mach number shocks in radio relics. In the next years, the advent of the SKA will permit a detailed study on individual clusters, and will in principle allow us to investigate different model assumptions for the magnetic field structure.

Acknowledgments

AB and CS acknowledge financial support from the ERC-Stg DRANOEL, no 714245. FV acknowledges financial support from the ERC-Stg MAGCOW, no.714196.

References

Bonafede A., *et al.*, 2018, *MNRAS*, 478, 2927
Bonafede A., *et al.*, 2015, aska.conf, 95
Bonafede A., Vazza F., Brüggen *et al.* 2013, *MNRAS*, 433, 3208
Bonafede A., Feretti L., Murgia M., Govoni *et al.* 2010, *A&A*, 513, A30
Brentjens M. A., de Bruyn A. G., 2005, *A&A*, 441, 1217
Brüggen M., Bykov A., Ryu D., Röttgering H., 2011, *SSR*, p. 138
Murgia M., Govoni F., Feretti L., *et al.* 2004, *A&A*, 424, 429
Rudnick L., Owen F. N., 2014, *ApJ*, 785, 45
van Haarlem M. P., Wise M. W., Gunst A. W., Heald G., *et al.* 2013, *A&A*, 556, A2
Savini F., *et al.* 2018, *MNRAS*, 478, 2234
Vazza F., Brunetti G., Brüggen M., Bonafede A., 2018, *MNRAS*, 474, 1672
Xu H., Li H., Collins D. C., Li S., Norman M. L., 2009, *ApJ*, 698, L14
Zhuravleva I., Churazov E. M., Schekochihin A. A., *et al.* 2014, *ApJ*, 788, L13

Astronomy in Focus - XXX
Proceedings IAU Symposium No. XXX, 2018
M. T. Lago, ed.

doi:10.1017/S1743921319004460

Magnetic fields in the intergalactic medium and in the cosmic web

Marcus Brüggen[1], Shane O'Sullivan[1], Annalisa Bonafede[1,2,3] and Franco Vazza[1,2,3]

[1]University of Hamburg, Gojenbergsweg 112, 21029 Hamburg, Germany
email: mbrueggen@hs.uni-hamburg.de

[2]Dipartimento di Fisica e Astronomia, Universita di Bologna, via P. Gobetti 93/2, 40129, Bologna, Italy

[3]INAF - Istituto di Radioastronomia, Bologna Via Gobetti 101, I-40129 Bologna, Italy

Abstract. In these proceedings we discuss advances in the theory and observation of magnetic fields in the intergalactic medium and in the cosmic web. We make the point that, despite perhaps unsurmountable obstacles in simulating a small-scale dynamo, currently most cosmological magnetohydrodynamical simulations paint a similar picture of magnetic field amplification in the cosmos. However, observations of magnetic fields in the intergalactic medium turn out to be very difficult. As a case in point, we present recent work on Faraday rotation measurement in the direction of a giant galaxy with the Low Frequency Array (LOFAR). These observations demonstrate the currently unique capability of LOFAR to measure Faraday rotation at the high accuracy and angular resolution required to investigate the magnetisation of large-scale structure filaments of the cosmic web.

Keywords. (galaxies:) intergalactic medium, (cosmology:) large-scale structure of universe, magnetic fields, methods: numerical, (magnetohydrodynamics:) MHD

---

1. Introduction

Much has been written about the origin of cosmic magnetism, the processes that can seed fields in the early universe and the various mechanisms to amplify magnetic fields. Here we can refer to the review by Donnert *et al.* (2018). For want of clear indications to the contrary, it is almost always assumed that magnetohydrodynamics (MHD) is a good theory that describes the evolution of fields in the cosmos. Still one should mention some brave attempts to go beyonds the simplest descriptions, e.g. by Schekochihin *et al.* (2004).

Under the set of equations of MHD, it appears that in the presence of turbulence a few percent of the kinetic energy is transferred to magnetic energy in a fast operating small-scale fluctuation dynamo (Miniati & Beresnyak 2015). Such a turbulent dynamo has been shown to operate in galaxy clusters (e.g. Jaffe 1980; Roland 1981; Ruzmaikin *et al.* 1989; De Young 1992; Goldshmidt & Rephaeli 1993; Kulsrud *et al.* 1997; Sánchez-Salcedo *et al.* 1998; Subramanian *et al.* 2006). In such dynamos, magnetic fields are amplified though an inverse cascade up until a scale, where the field starts to act back onto the fluid flow. Simulations seem to agree that field amplification is primarily caused by compression in cosmological filaments, whereas at higher overdensities such as in galaxy clusters, turbulence is increasingly solenoidal and there are a sufficient number of eddy turn-overs for a dynamo to cause a fast amplification beyond what you would get via compression. The initial exponential increase in magnetic field strength is followed by a non-linear growth phase. The timescale of exponential growth is set by the magnetic Prandtl number and is determined by the spatial resolution and the algorithm. The real growth rate thus may never be determined by direct numerical simulations. Moreover, on smaller scales magnetic fields may get injected by galactic outflows and active galaxies. Galactic winds can transport magnetic fields into the circum-galactic medium where it can be stripped and enter the ICM (Donnert *et al.* 2009; Xu *et al.* 2009). The modelling of this is still in its infancy.

A recent example of such simulation work is presented in Vazza *et al.* (2018) where it was demonstrated how a small-scale dynamo develops (see Fig. 1). Interestingly but not unexpectedly, a significant non-Gaussian distribution of field components is found which results from the superposition of plasma that has gone through different amplification histories. Evidence for the presence of a dynamo is the anti-correlation of magnetic field strength and its curvature $\vec{K}$,

$$\vec{K} = \frac{\left(\vec{B}\cdot\nabla\right)\vec{B}}{\vec{B}^2}, \tag{1.1}$$

so that $\vec{B}\vec{K}^{\frac{1}{2}} = \text{const}$ (Schekochihin *et al.* 2004) which can be tested in simulations of small-scale dynamos.

2. Observations of Faraday rotation from giant radio galaxy

On the observational side, progress has been fairly slow which is largely due to the fact that measuring extragalactic fields is unreasonably difficult. On scales of galaxies and beyond, magnetic fields are best traced via radio observations. Cluster magnetic fields were first inferred from upper limits on the diffuse synchrotron emission by Burbidge (1958). Later estimates based on the Rotation Measure (RM) of background sources to the Coma cluster obtain central magnetic fields of $3-7\mu$G. Consequently, the ICM is a high $\beta = 8\pi n_{\rm th} k_{\rm B} T/B^2 \approx 100$ plasma, meaning that thermal pressure is much larger than magnetic pressure. There are attempts to measure the magnetic fields in cluster outskirts using Faraday rotation of the polarised emission from radio relics (e.g. Kierdorf *et al.* 2017). Radio relics or cluster radio shocks trace shock waves in merging galaxy clusters and sometime show large degrees of polarisation and μG magnetic fields.

In these proceedings, we would like to bring attention to a recent attempt by O'Sullivan *et al.* (2019) to analyse the FRII radio galaxy (J1235+5317) at redshift $z = 0.34$ with a linear size of 3.4 Mpc whose polarised emission may have been rotated by intergalactic magnetic fields. This work was conducted with the Low Frequency Array (LOFAR) whose broad bandwidth provides excellent precision to measure Faraday rotation and, at the same time, is sensitive to emission on large angular scales. On the downside, LOFAR observations suffer from Faraday depolarisation, which renders many sources undetectable in polarisation (Farnsworth *et al.* 2011). Figure 2 shows the RM distributions of this giant radio galaxy. The mean and standard deviations of the RM are +7.42 rad m^{-2} and 0.07 rad m^{-2} for the North-Western radio lobe, and +9.92 rad m^{-2} and 0.11 rad m^{-2} for the South-Eastern radio lobe.† The mean RM difference between the two lobes is 2.5 ± 0.1 rad m^{-2}. Next, we relied on dynamical modelling of the radio lobes to infer the density of the ambient gas, which came out to be $n_{\rm e} \sim 10^{-7}$cm^{-3}. This suggests that the radio galaxy is expanding into a very underdense region. However, the observed Faraday depolarisation of ~ 0.1 rad m^{-2} which is most likely caused by plasma local to the source, requires $n_{\rm e} \sim 10^{-5}$cm^{-3} with a turbulent magnetic field of strength ~ 0.09 μG at a distance of about 1.5 Mpc from the host galaxy. Hence, we are either underestimating the density of the external medium or the depolarisation does not occur

† The RM errors are 0.04 rad m^{-2} and 0.06 rad m^{-2}, for the NW and SE lobes, respectively.

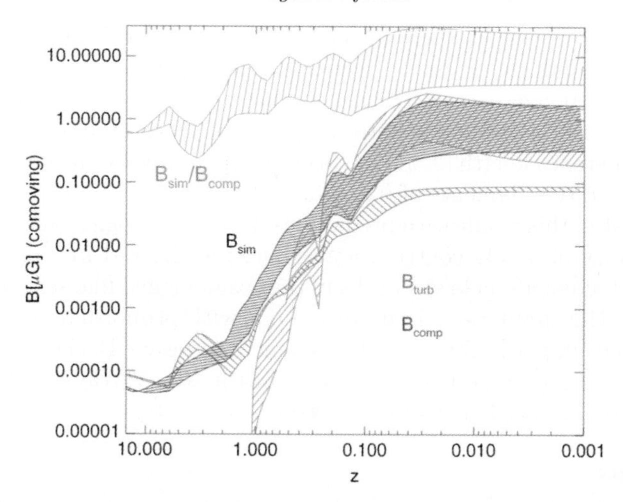

Figure 1. Magnetic field growth in a cosmological MHD simulation performed with the ENZO code. The black curve shows the magnetic field strength in innermost comoving Mpc^3 as a function of redshift. In comparison, there is the prediction from compression alone (blue) and from dynamo amplification (red) Beresnyak & Miniati (2016), assuming a 4% amplification efficiency. The grey curve shows the ratio between the simulated field and the expectation from compression alone. The dashed areas show the scatter. From Vazza *et al.* (2018).

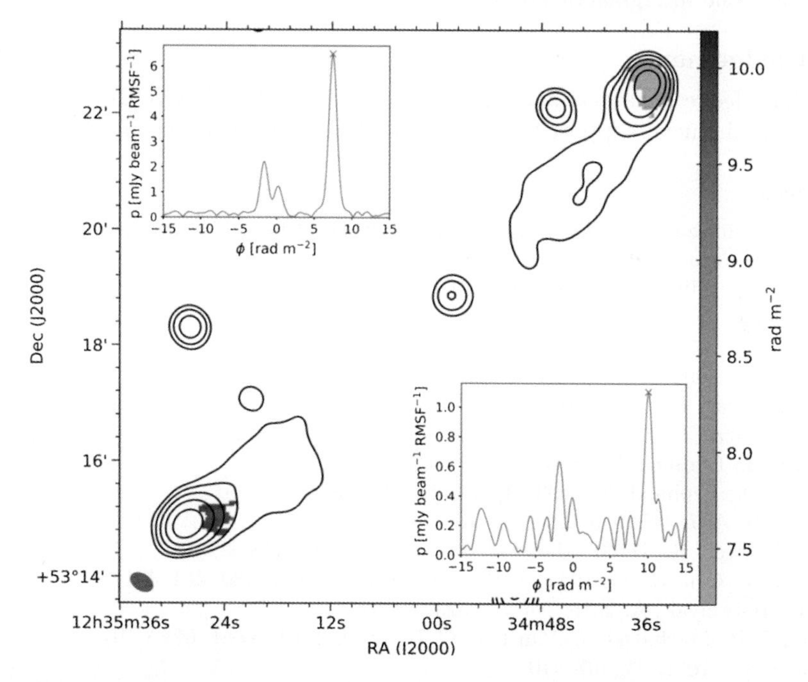

Figure 2. Faraday RM distribution of the NW and SE lobes, overlaid by the total intensity contours starting at 5 mJy/beam and increasing in factors of two. Insets: The absolute value of the Faraday dispersion function for the NW lobe (top) and SE lobe (bottom). From O'Sullivan *et al.* (2019).

in the environment close to the source. Better models for the evolution of radio jets within a realistic cosmological environment may help refine these estimates in the future. With the current value, the estimated magnetic field strength is unable to account for the observed difference of the mean RM of 2.5 rad m^{-2} between the two lobes. We then searched a catalogue of cosmological filaments that is derived from optical spectroscopic observations and found an excess of filaments that intersect lines-of-sight towards the Northwestern lobe. If magnetised plasma in these filaments causes the RM difference between the lobes and assuming a path length through each filament of 3 Mpc, and a magnetic field coherence length of 300 kpc, this would imply a density-weighted magnetic field strength inside the filaments of 0.3 μG.

We then compared this result with predictions from cosmological simulations and found that the probability of a RM contribution as large as 2.5 rad m^{-2} is only $\sim$ 5%. This estimate assumed magnetic field strengths in the cosmological filaments of 10 to 50 nG, as suggested from MHD simulations that started out with primordial magnetic fields of $\sim$1 nG, close to current upper limits from the Cosmic Microwave Background. Alternatively, the RM difference could come from variations in the Milky Way and finer observations of Milky Way RMs are needed to obtain better constraints.

3. Conclusions

There remains the hope that with the advent of the Square Kilometre Array (SKA) and the LOFAR upgrade this field will gain fresh momentum (e.g. Bull *et al.* 2018). Large samples of RMs from radio galaxies with known redshifts will permit more advanced statistical analyses, such as RM structure functions (e.g. Akahori *et al.* 2014). Only then will we be able to disentangle the effect of the Milky Way on RM measurements for filaments and the intergalactic space.

Acknowledgements

AB acknowledges financial support from the ERC-StG DRANOEL, no. 714245. FV acknowledges financial support from the ERC-StG MAGCOW, no. 714196.

References

Akahori, T., Kumazaki, K., Takahashi, K., & Ryu, D. 2014, *Publications of the Astronomical Society of Japan*, 66, 65
Beresnyak, A., & Miniati, F. 2016, *ApJ*, 817, 127
Bull, P., Camera, S., Kelley, K., *et al.* 2018, arXiv:1810.02680
Burbidge, G. R. 1958, *ApJ*, 128, 1
De Young, D. S. 1992, *ApJ*, 386, 464–472
Donnert, J., Dolag, K., Lesch, H., Müller, E. 2009, *MNRAS*, 392, 1008–1021
Donnert, J., Vazza, F., Brueggen, M. ZuHone, J. 2018, *SSRv*, 214, 122
Farnsworth, D., Rudnick, L., & Brown, S. 2011, *ApJ*, 141, 191
Goldshmidt, O., Rephaeli, Y. 1993, *ApJ*, 411, 518–528
Jaffe, W. 1980, *ApJ*, 241, 925–927
Kierdorf, M., Beck, R., Hoeft, M., *et al.* 2017, *A&A*, 600, A18
Kulsrud, R. M., Cen, R., Ostriker, J. P., Ryu, D. 1997, *ApJ*, 480, 481–491
Miniati, F., Beresnyak, A. 2015, *Nature*, 523, 59–62
O'Sullivan, S. P., Machalski, J., Van Eck, C. L., *et al.* 2019, *A&A*, 622A, 16
Roland, J. 1981, *A&A*, 93, 407–410
Ruzmaikin, A., Sokolov, D., Shukurov, A. 1989, *MNRAS*, 241, 1–14
Sánchez-Salcedo, F. J., Brandenburg, A., Shukurov, A. 1998, *APSS*, 263, 87–90
Schekochihin, A. A., Cowley, S. C., Taylor, S. F., Maron, J. L., McWilliams, J. C. 2004, *ApJ*, 612, 276–307
Subramanian, K., Shukurov, A., Haugen, N. E. L. 2006, *MNRAS*, 366, 1437–1454
Vazza, F., Brunetti, G., Brüggen, M., Bonafede, A. 2018, *MNRAS*, 474, 1672–1687
Xu, H., Li, H., Collins, D. C., Li, S., & Norman, M. L. 2009, *ApJL*, 698, L14

Astronomy in Focus - XXX
Proceedings IAU Symposium No. XXX, 2018
M. T. Lago, ed.

doi:10.1017/S1743921319004472

Magnetism in the Square Kilometre Array Era

S. A. Mao

Max Planck Institute for Radio Astronomy, Auf dem Hügel 69, D-53121 Bonn, Germany
email: mao@mpifr-bonn.mpg.de

Abstract. The unprecedented sensitivity, angular resolution and broad bandwidth coverage of Square Kilometre Array (SKA) radio polarimetric observations will allow us to address many long-standing mysteries in cosmic magnetism science. I will highlight the unique capabilities of the SKA to map the warm hot intergalactic medium, reveal detailed 3-dimensional structures of magnetic fields in local galaxies and trace the redshift evolution of galactic magnetic fields.

Keywords. polarization, cosmology: large-scale structure of universe, galaxies: magnetic fields

---

1. Introduction

The Square Kilometre Array (SKA) will be the most powerful radio telescope in the world and it is currently in pre-construction phase. It will be hosted at two separate sites: SKA1-MID (350 MHz−24 GHz) with 133 15-m SKA dishes and 64 13.5-m MeerKAT dishes over a maximum baseline of 150 km will be located in the Karoo site in South Africa, while SKA1-LOW (50−350 MHz) with 130,000 antennas over a maximum baseline of 65 km will be located in the Boolardy site in Western Australia. The full SKA1-MID will have $\sim$ 5 times better sensitivity and 4 times higher angular resolution than the Karl G. Jansky Very Large Array, while the full SKA1-LOW will be a factor of $\sim$8 more sensitive than the Low Frequency Array. Science commissioning of SKA1 will commence in 2022 and full operation will begin in 2025[†]. The origin and evolution of cosmic magnetism is one of the five original SKA Key Science Projects (Gaensler, Beck & Feretti 2004). In the updated SKA science book *Advancing Astrophysics with the SKA*, the magnetism community continues to have a strong presence, contributing a total of 19 chapters. This proceedings will highlight selected topics covered in these chapters, as well as latest developments in the field which were not included in the science book.

Radio polarization observations encode rich information on particle densities and magnetic fields in the Universe on different scales: from Mpc down to sub-pc scales. Besides mapping the polarized synchrotron emission from the astrophysical object of interest, a key measurement of cosmic magnetism is the Faraday rotation towards polarized background extragalactic radio sources. These measurements, forming a so-called rotation measure (RM) grid, enable us to directly probe the magnetic field strength and direction, as well as the gas density in the foreground intervening medium.

Our knowledge of the rotation measure sky has improved significantly in the past 20 years. In the early 2000s, only $\sim 10^3$ extragalactic sources have Faraday rotation measurements (Johnston-Hollitt 2003). At present, the NRAO VLA Sky Survey rotation measure catalog of Taylor *et al.* (2009) (1 source deg^{-2} at DEC$>-40°$) along with the

† For the anticipated SKA1 science performance, see document number: SKA-TEL-SKO-0000818 (Braun, Bonaldi, Bourke, Keane & Wagg 2017). For the current timeline, see document number: SKA-TEL-SKO-00000822 (SKAO Science & Op Teams 2017). Both documents are available on https://astronomers.skatelescope.org/documents/.

S-PASS/ATCA catalog of Schnitzeler *et al.* (2018) (0.2 source deg^{-2} at DEC<0°) provide us with ~40,000 extragalactic RMs across the entire sky, enabling magnetic field measurements in a range of different foreground astrophysical objects. In addition to the ever increasing density of the all-sky RM grid, the advent of broadband polarimetry — a more than 10-fold increase in the instantaneous observing bandwidth in frequency — have brought revolution to our field in the recent years as well. Precise and nπ-ambiguity free Faraday rotation (Ma *et al.* 2018) along with other properties of the magnetized gas can be derived using newly-developed broadband polarization analysis tools.

An all-sky full Stokes survey at 2" resolution with SKA1-MID (Band 2) down to 4 μJy beam^{-1} will provide ~ 7 to 14 million extragalactic radio sources with Faraday rotations (Johnston-Hollitt *et al.* 2015). This dense RM grid will facilitate the characterization of astrophysical magnetic fields, gas densities and turbulence in unprecedented details.

2. Extragalactic Magnetism Science with the SKA

2.1. *Revealing the elusive missing baryons and the magnetic fields in the cosmic web*

Only about 50% of the expected baryons in the Universe can be accounted for, while the rest – missing baryons – are thought to reside in the warm-hot intergalactic medium (WHIM) in the form of shock-heated gas at 10^5–10^7K. The WHIM can be traced by absorption lines in X-ray and UV spectra, but extremely long integration time is required to produce high significance detection along a single sight line (Nicastro *et al.* 2018). Radio observations offer a highly complementary approach to characterize the particle density and magnetic fields in the cosmic web. Shocks produced by the accretion of baryonic matter as large-scale structures of the Universe form are sufficient to accelerate particles to relativistic energies, illuminating the cosmic web in synchrotron emission in the presence of intergalactic magnetic fields. The WHIM should also produce imprints on the RM of background sources if magnetic fields permeate the cosmic web.

Currently, limits on the surface brightness and the magnetic fields of the cosmic web can be placed by cross-correlating tracers of large-scale structures and diffuse synchrotron emission (e.g., Brown *et al.* 2017, Vernstrom *et al.* 2017). Intergalactic magnetic field strength can also be estimated by determining whether a difference in the number of large scale structure filaments intercepting the line-of-sight corresponds to a difference in Faraday rotation between two nearby sight lines (O'Sullivan *et al.* 2018). A potentially new population of radio sources in environments connecting galaxy clusters was discovered recently (Vacca *et al.* 2018). These sources have properties similar to those expected from the brightest patches of the diffuse emission associated with the WHIM.

With the advent of the SKA comes new methods to reveal the cosmic web. Direct imaging of the brightest filaments of the cosmic web will be possible with deep (>1000 hours) SKA1-LOW observations (Vazza *et al.* 2015), provided that confusion from Galactic foreground synchrotron emission and extragalactic point source emission are minimized and accurately removed. A dense (>10^3 sources deg^{-2}) and precise rotation measure grid (error~1 rad m^{-2}) towards filaments of the cosmic web and its structure function can probe the turbulent scale, gas densities and magnetic fields in the intergalactic medium (Akahori & Ryu 2011, Taylor *et al.* 2015). Performing a joint Faraday rotation and dispersion measure analysis of the ~10^4 localized fast radio bursts anticipated by the SKA (Macquart *et al.* 2015) can put tight limits on properties of the magnetized intergalactic medium if the *in situ* and host galaxy contributions can be reliably subtracted (Akahori *et al.* 2016, Johnston *et al.* this volume). The all-sky RM grid with redshift information will enable the employment of new Bayesian algorithms (Vacca *et al.* 2016) to statistically isolate Faraday rotation produced by the cosmic web. These different approaches will yield densities and magnetic fields of the WHIM, addressing the missing baryon problem and ultimately distinguishing between different magnetogenesis scenarios.

2.2. *Mapping the 3-dimensional magnetic fields in nearby galaxies*

The leading theories of the amplification of magnetic fields in galaxies are the fluctuation dynamo and the large-scale α-Ω dynamo, but details of these processes remain poorly constrained observationally. Rigorous comparisons between theoretical predictions and observations of 3-dimensional magnetic fields in nearby galaxies are necessary to fully understand these dynamos. Magnetic fields in approximately 100 galaxies have been studied by measuring their diffuse polarized synchrotron emission at limited angular resolution with mostly narrowband data (Beck & Wielebinski 2013). Only 3 nearby large-angular extent galaxies, the Large and the Small Magellanic Clouds and M31, have had their magnetic fields probed via the RM-grid approach (Gaensler *et al.* 2005, Mao *et al.* 2008, Han *et al.* 1998). These studies have established some basic properties of galactic magnetic fields, such as the typical field strength and the dominant disk field symmetry, but a 3-dimensional picture of galactic magnetic fields is still lacking. Broadband radio polarimetry in combination with models of the magnetized interstellar medium (ISM) can be used to conduct tomography to characterize both large and small-scale galactic magnetic fields as a function of the line-of-sight depth, thus yielding 3-D pictures of galactic magnetic fields (Fletcher *et al.* 2011, Kierdorf *et al.* this volume).

While until now only a few galaxies have broadband polarization data that allow for tomography studies, the SKA will revolutionize this area. With the SKA, sensitive diffuse polarized synchrotron emission can be measured with excellent λ^2 coverage, specifically with SKA1-MID band 2, 4 and possibly 3 (Beck *et al.* 2015, Heald *et al.* 2015), which is ideal for magnetic field tomography. The SKA will have enough surface brightness sensitivity to map the polarized emission at extremely high angular resolution (1 kpc at $z{\sim}0.04$). Moreover, at least 200 nearby galaxies will have enough polarized background sources for us to conduct RM-grid experiments on. A joint analysis of the broadband diffuse polarized emission and the RM grid of nearby galaxies will provide a complete 3-D view of their magnetic fields. The strength and structure of the disk and halo magnetic fields and their radial and vertical dependencies can be determined. The nature (isotropic vs. anisotropic) and the power spectrum of random magnetic fields can be derived. With much improved angular resolution, an extensive search for extragalactic large-scale magnetic field reversals will also be feasible. A clear link between galaxy properties and their magnetic fields will emerge from these SKA data and will allow us to constrain the field generation processes.

2.3. *Tracing the redshift evolution of galactic magnetic fields*

Since galactic magnetic fields play important roles in processes that are closely linked to galaxy evolution, it is crucial to understand how galaxies and their magnetic fields have co-evolved over cosmic time. Directly tracing the redshift evolution of galactic magnetic fields is a challenging task: polarized synchrotron emission from cosmologically distant galaxies is faint and Faraday rotation produced by these distant galaxies when seen against background polarized sources is subjected to redshift dilution and is difficult to isolate from other sources of Faraday rotation along the line of sight. As a result, measurements of magnetic fields in galaxies beyond the local Universe are scarce. Recently, Mao *et al.* (2017) have demonstrated that strong gravitational lensing of polarized background quasars by galaxies offers a clean and effective probe of the *in situ* magnetic fields in individual cosmologically distant galaxies. Using differential polarization properties (Faraday rotation and fractional polarization) derived from broadband observations of a lensing system at z=0.44, the authors have derived both the magnetic field strength and geometry in the lensing galaxy as seen 4.6 billion years ago, making it the most distant galaxy with such a measurement.

At present, the number of systems for which this technique can be applied to is limited (Mao *et al.* in prep). With an expected discovery of $>10^4$ new radio-bright gravitational lenses (a factor of $>$ 100 more than the currently known systems, McKean *et al.* 2015), the SKA will provide significantly more lensing systems that are well-suited for magnetism studies, extending measurements of magnetic fields in cosmologically distant galaxies to a much wider range in redshift and in mass. Along with other tracers of galactic magnetic fields at high redshifts (e.g., Basu *et al.* 2018), the SKA will enable one to firmly establish the observational trend of galactic magnetic fields as a function of cosmic time.

3. Summary

The SKA will transform our understanding of the origin and evolution of cosmic magnetic fields. The dense, broadband all-sky RM-grid together with additional pointed observations of selected targets will greatly advance our knowledge on the magnetized WHIM, 3-D magnetic fields in galaxies and their redshift evolution.

References

Akahori, T. & Ryu, D. 2011, *ApJ*, 738, 134

Akahori, T., Ryu, D., & Gaensler, B. M. 2016, *ApJ*, 824, 105

Basu, A., Mao, S. A., Fletcher, A., *et al.* 2018, *MNRAS*, 477, 2528

Beck, R., & Wielebinski, R. 2013, in Planets, Stars and Stellar Systems, Vol. 5, ed. T. D. Oswlat & G. Gilmore (Dordrecht: Springer), 641, updated in 2018 (arXiV: 1302.5663)

Beck, R., Bomans, D., Colafrancesco, S., *et al.* 2015, *PoS, AASKA14*, 94

Brown, S., Vernstrom, T., Carretti, E., *et al.* 2017, *MNRAS*, 458, 4246

Fletcher, A., Beck, R., Shukurov, A., Berkhuijsen, E. M., & Horellou, C. 2011, *MNRAS*, 412, 2396

Gaensler, B. M., Beck, R., & Feretti, L. 2004, *New Astron. Revs*, 48, 1003

Gaensler, B. M., Haverkorn, M., Staveley-Smith, L., *et al.* 2005, *Science*, 307, 1610

Han, J. L., Beck, R., & Berkhuijsen, E. M. 1998, *A&A*, 335, 1117

Heald, G., Beck, R., de Blok, W. J. G., *et al.* 2015, *PoS, AASKA14*, 106

Johnston-Hollitt, M. 2003, *Phd Thesis*, University of Adelaide

Johnston-Hollitt, M., Govoni, F., Beck, R., *et al.* 2015, *PoS, AASKA14*, 92

Macquart, J. P., Keane, E., Grainge, K., *et al.* 2015, *PoS AASKA14*, 55

Ma, Y. K., Mao, S. A., Stil, J., *et al.* 2018, *MNRAS*, submitted

Mao, S. A., Gaensler, B. M., Stanimirović, S., *et al.* 2008, *ApJ*, 688, 1029

Mao, S. A., Carilli, C., Gaensler, B. M., *et al.* 2017, *Nature Astronomy*, 1, 621

McKean, J., Jackson, N., Vegetti, S., *et al.* 2015, *PoS, AASKA14*, 92

Nicastro, F., Kaastra, J., Krongold, Y., *et al.* 2018, *Nature*, 558, 406

O'Sullivan, S. P., Machalski, J., Van Eck, C. L., *et al.* 2018, *MNRAS*, submitted

Schnitzeler, D. H. F. M., Carretti, E., Wieringa, M. H., *et al.* 2018, *MNRAS*, submitted

Taylor, A. R., Stil, J. M., & Sunstrum, C. 2009, *ApJ*, 702, 1230

Taylor, A. R., Agudo, I., Akahori, T., *et al.* 2015, *PoS AASKA14*, 113

Vacca, V., Oppermann, N., Ensslin, T., *et al.* 2016, *A&A*, 519, A13

Vacca, V., Murgia, M., Govoni, F., *et al.* 2018, *MNRAS*, 479, 776

Vazza, F., Ferrari, C., Bonafede, A., *et al.* 2015, *PoS AASKA14*, 97

Vernstrom, T., Gaensler, B. M., Brown, S., Lenc, E., & Norris, R. P. 2017, *MNRAS*, 467, 4914

Astronomy in Focus - XXX
Proceedings IAU Symposium No. XXX, 2018
M. T. Lago, ed.

doi:10.1017/S1743921319004484

Capabilities of next generation telescopes for cosmic magnetism

Jeroen M. Stil†

Department of Physics and Astronomy, The University of Calgary,
2500 University Drive NW, Calgary, AB, T2N 1N4, Canada
email: jstil@ucalgary.ca

Abstract. The next generation of radio telescopes offer significant improvement in bandwidth and survey speed. We examine the ability to resolve Faraday thick objects in Faraday space as a function of survey parameters. The necessary combination of $\lambda_{\rm max}$ and $\lambda_{\rm min}$ to resolve objects with modest Faraday thick components requires one or two surveys with instantaneous bandwidth 300 MHz to 750 MHz offered by next generation telescopes. For spiral galaxies, bandwidths in excess of 1.5 GHz are required. Correction for Galactic Faraday rotation must account for common gradients of order 10 rad m^{-2} per degree. How effective a new rotation measure grid is in probing the foreground depends on off-axis polarization calibration.

Keywords. magnetic fields, polarization, techniques: polarimeters, techniques: polarimetric

---

1. Introduction

Research in cosmic magnetism has entered a transformational period with the commissioning of new and upgraded radio telescopes that have much larger fractional bandwidth and larger field of view than the previous generation of telescopes. Among the new facilities are meter wave telescopes Low Frequency Array (LOFAR) and the Murchison Widefield Array (MWA), and centimeter wave telescopes Australian SKA Pathfinder (ASKAP) and MeerKAT, and large filled-aperture FAST. Existing telescopes that have made major upgrades for broad-band, wide-field imaging are the Jansky Very Large Array (JVLA), the Westerbork Synthesis Radio Telescope (WSRT/Apertif), the upgraded GMRT, eMERLIN, and the Arecibo radio telescope. On the horizon are the Square Kilometre Array (SKA) and the next generation Very Large Array (ngVLA).

Most of these facilities have polarization surveys planned or well underway. Naturally, these surveys are optimized for the main science drivers of the survey, given the capabilities of the observatory with which they are made. Factors that impact the effectiveness of an observatory for cosmic magnetism include the instantaneous bandwidth, operating frequency range, and the spatial frequencies probed over the observed frequency range. Important secondary factors are survey speed and available time for PI-driven follow-up observations. After all, one of the main drivers of a survey is discovery, to be followed by more detailed targeted observations.

Most extragalactic radio emission is synchrotron radiation. Its intensity and polarization contain information on magnetic field strength and structure. High angular resolution is a major factor as polarization angle structure within the beam causes depolarization. On the other hand, as we explore more of the low-surface-brightness universe, short spacing information can become a higher priority for extragalactic magnetism than it

†Present address: Department of Physics and Astronomy, The University of Calgary, 2500 University Drive NW, Calgary, AB, T2N 1N4, Canada.

has been in the past. Interpretation of broad-band polarization observations of well-resolved sources requires careful consideration of the spatial frequencies sampled by an interferometer across the observed frequency range.

In this paper, I focus on two aspects of extragalactic cosmic magnetism enabled by the next generation radio telescopes: the ability to resolve a source in Faraday depth, and correction for Galactic Faraday rotation to the extent that it does not dominate the uncertainty in extragalactic Faraday rotation.

2. Resolving structure in Faraday depth

Structure in Faraday depth arises from differential Faraday rotation across the beam and from differential Faraday rotation along the line of sight (see R. Laing, this meeting, for a discussion how higher angular resolution reduces complexity, up to a point). The ability of a survey with continuous data between a minimum wavelength $\lambda_{\rm min}$ and a maximum wavelength $\lambda_{\rm max}$ to resolve an object in Faraday depth depends on Faraday depth resolution (Brentjens & De Bruyn 2005),

$$\Delta\phi = \frac{2\sqrt{3}}{\lambda_{\rm max}^2 - \lambda_{\rm min}^2}, \tag{2.1}$$

and the largest observable continuous Faraday depth range in the source,

$$\delta\phi = \frac{\pi}{\lambda_{\rm min}^2}. \tag{2.2}$$

Figure 1 shows a diagram of Faraday depth resolution and largest observable Faraday depth scale. A survey occupies a particular locus in this diagram through Equations 2.1 and 2.2. We can now choose a bandwidth, and vary the lowest frequency $\nu_{\rm min}$ (or $\lambda_{\rm max}$), to define a curve that represents all surveys with a set bandwidth. Figure 1 shows four curves representing surveys with bandwidths of 50 MHz (red), 100 MHz (blue), 750 MHz (green), and 1.5 GHz (magenta). Dots on the curves mark $\nu_{\rm min} = 100$, 600, 1000, and 2000 MHz (from left to right). The black diagonal marks the locations where Faraday depth resolution is equal to the largest detectable Faraday depth scale. We can only resolve objects in a survey with a particular bandwidth if their internal Faraday depth range is between the curve and the black diagonal. Deconvolution of Faraday depth structure below the resolution limit is difficult because both the phase and the amplitude of blended Faraday depth components matter (Kumazaki *et al.* 2014, Sun *et al.* 2015).

Different classes of objects can now be painted into the diagram, with the horizontal range limited by depolarization on the low-frequency side and the black diagonal on the high frequency side. Disks of spiral galaxies for example depolarize strongly below 1 GHz, but polarization has been detected as low as 350 MHz (Giessübel *et al.* 2013). The green and yellow regions in the left panel of Figure 1 mark approximate loci for diffuse polarized emission from spiral galaxy disks. Lobes of radio galaxies may have little internal Faraday rotation, but a range of Faraday depth may arise from structure in the surrounding medium. Steep spectrum sources such as relics in clusters may be better detectable at lower frequencies, as indicated by the purple region in Figure 1.

Figure 1 shows that in practice, it is easier to resolve objects with a modest Faraday depth range than objects with a larger intrinsic Faraday depth range. The latter require multiple surveys with telescopes with instantaneous bandwidth in the range 300 MHz to 750 MHz. Transformation of wavelength from the observer's frame to the source frame at redshift z amounts to translation along the black diagonal line in the sense that the same source at high z would be less resolved in Faraday depth.

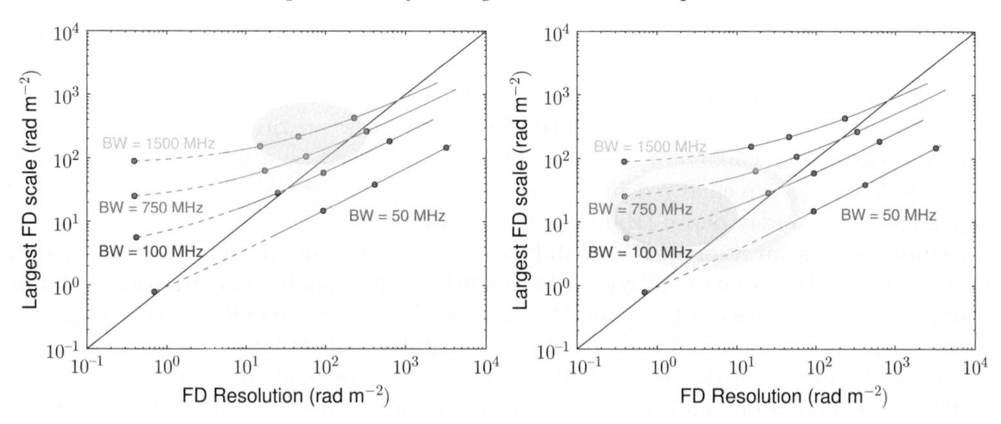

Figure 1. Diagram of Faraday depth resolution and largest observable Faraday depth scale for surveys with total bandwidth 50 (red), 100 (blue), 750 (green), and 1500 MHz (magenta). Dots mark surveys with $\nu_{\min} =$ 100 MHz, 600 MHz, 1 GHz and 2 GHz (left to right). Dashed curves (frequency less than 350 MHz) are within the realm of aperture plane arrays, while the continuous curves have frequencies accessible with traditional arrays. The black diagonal line indicates where resolution is equal to the largest detectable Faraday depth scale. In order to resolve an object in Faraday depth, the Faraday depth range of the object must be between the curve and the black diagonal line. In the left panel, the green region marks the approximate parameter space for edge-on galaxies, and the yellow region the same for face-on galaxies. In the right panel, the blue region marks the approximate parameter space for lobes of radio galaxies, and the purple region a fiducial range for relics in galaxy clusters.

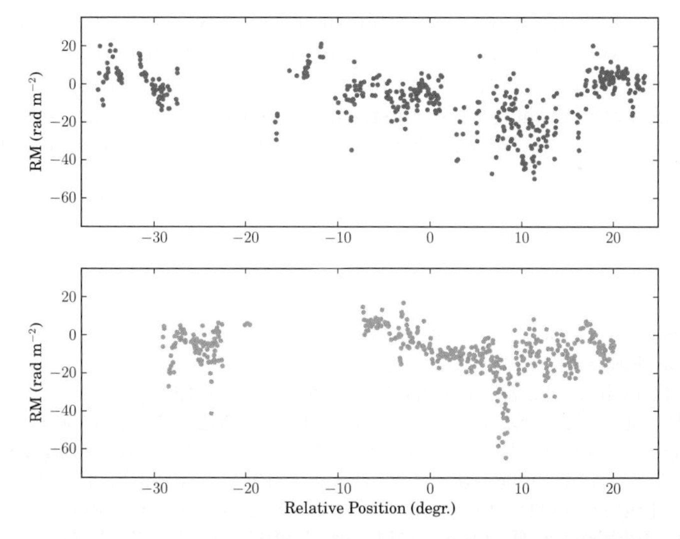

Figure 2. Variation of RM with position from high-latitude diffuse emission in the GALFACTS survey. The panels show two 0.3 degree wide strips, separated by 3 degrees in Galactic latitude in the range 65° to 70°. The horizontal axis is relative separation in the direction of Galactic longitude, corrected for latitude. Only data with strong polarized signal and RM error less than 5 rad m^{-2} are shown. Gaps in the data arise from regions with weaker signal.

3. Importance of off-axis polarization calibration

Arguably the foremost data set that next generation surveys will produce is a dense grid of rotation measures of polarized extragalactic sources across the sky (Beck & Gaensler 2004, Johnston-Hollitt *et al.* 2015). Correcting for Faraday rotation by the Galactic interstellar medium will be the most common application of the RM grid for extragalactic cosmic magnetism. The density of the grid and the quality of the RM data will determine the accuracy of this correction. Far from the Galactic plane, the errors in Oppermann *et al.* (2012, 2015) are of the order of 10 rad m^{-2}, which constitutes 22° rotation at 20 cm. An error in the Galactic Faraday rotation is magnified by a factor $(1+z)^2$ for the Faraday depth of a screen at redshift z, introducing significant uncertainty in the redshift dependence of Faraday depth (Hammond *et al.* 2012).

Figure 2 shows variation of RM of diffuse Galactic emission at high Galactic latitude from the GALFACTS survey (Taylor & Salter 2010). Although Faraday rotation of diffuse emission does not measure the same Faraday depth as extragalactic sources, we get an impression of the foreground RM structure on angular scales that are not yet accessible otherwise. Gradients of the order of 10 rad m^{-2} per degree and changes in the variance of RM are common on small scales. These small-scale structures cannot be recognized with the current sampling density of $\sim$1 polarized source per square degree. The average density of RMs in Figure 2 is $\sim$30 per square degree, comparable to the RM grid expected from the POSSUM survey (Gaensler *et al.* 2010, Rudnick & Owen 2014).

Constructing an RM grid that allows for consistent subtraction of Galactic Faraday rotation to the level of 1 rad m^{-2} is within reach of next generation sky surveys. The RM grid relies on sources measured across the field of view, far from the field centre where traditional polarization calibration solutions apply. In the absence of direction-dependent polarization calibration, we can expect residual leakage at the level of a few percent of total intensity at Faraday depth zero that blends with the astrophysical signal, unless the Faraday depth resolution of the survey is much better than 10 rad m^{-2} (Figure 2). This requires frequencies below 600 MHz (see Figure 1). The significance of blending of instrumental polarization with astrophysical Faraday rotation was discussed at this meeting by Y. K. Ma *et al.* Off-axis polarization calibration requires an investment by the observatory in terms of commissioning observations and software (e.g. Jagannathan *et al.* 2017). The performance of next generation telescopes for extragalactic magnetism depends in no small way on their ability to calibrate instrumental polarization across the field of view.

References

Beck, R. and Gaensler, B. M., 2004, *New Astronomy Reviews*, 48, 1289
Brentjens, M. A., & De Bruyn, A. G. 2005, *A&A*, 441, 1217
Giessübel, R., Heald, G., Beck, R. & Arshakian, T. G. 2013, *A&A*, 559, A27
Gaensler, B., Landecker, T. L., & Taylor, A. R. 2010, *BAAS*, 42, 515
Hammond, A., Robishaw, T., & Gaensler B. M. 2012, arXiv:1209.1438
Jagannathan, P., Bhatnagar, S., Rau, U., & Taylor A.R. 2017, *AJ*, 154, 56
Johnston-Hollitt, M., Govoni, F., Beck, R., Dehghan, S., *et al.*., 2015, PoS (AASKA14) 092
Kumazaki, K., Akahori, T., Ideguchi, S., Kurayama, T., & Takahashi, K., 2014, *PASJ*, 66, 61
Oppermann, N. Junklewitz, H., Robbers, G., *et al.* 2012, *A&A*, 542, A93
Oppermann, N., Junklewitz, H., Greiner, M., *et al.* 2015, *A&A*, 575, A118
Rudnick, L. & Owen, F. 2014, *ApJ*, 785, 45
Sun, X., Rudnick, L., Akahori, T., Anderson, C. S., Bell, M. R., *et al.* 2015, *AJ*, 149, 60
Taylor, A. R. & Salter, C. J. 2010, *ASP Conf. Ser.* 438, 402

Astronomy in Focus - XXX
Proceedings IAU Symposium No. XXX, 2018
M. T. Lago, ed.

doi:10.1017/S1743921319004496

A fresh view of magnetic fields and cosmic ray electrons in halos of spiral galaxies

Ralf-Jürgen Dettmar[1], Volker Heesen[2] and the CHANG-ES Team[3]

[1]Ruhr-Universität Bochum, Universitätsstrasse 150, 44801 Bochum, Fakultät für Physik und Astronomie, Astronomisches Institut (AIRUB), Germany
email: dettmar@astro.rub.de

[2]Universität Hamburg, Hamburger Sternwarte, Gojenbergsweg 112, 21029 Hamburg, Germany
email: volker.heesen@hs.uni-hamburg.de

[3]Judith Irwin (PI), Dept. of Physics, Engeneering Physics, & Astronomy, Queen's University, Kingston, Ontario, Canada, K7L 3N6
email: irwinja@queensu.ca

Abstract. Recent numerical models of the multiphase ISM underline the importance of cosmic rays and magnetic fields for the physics of the ISM in disc galaxies. Observations of properties of the ISM in galactic halos constrain models of the expected exchange of matter between the star-forming disc and the environment (circumgalactic medium, CGM). We present new observational evidence from radio-continuum polarization studies of edge-on galaxies on magnetic field strength and structure as well as cosmic ray electron transport in galactic halos. The findings are discussed in the context of the disk-halo interaction of the interstellar medium. In addition, it is also briefly demonstrated how recent LOFAR observations of edge-on galaxies further constrain the extent of magnetic fields in galactic halos.

Keywords. galaxies: magnetic fields, galaxies: spiral, galaxies: halos, radio continuum: galaxies

1. Introduction

In the context of galaxy formation scenarios various feedback mechanisms (e.g., Silk 2013) have been discussed to explain the observed dependence of the global and long-term star formation efficiency in dark halos of different mass expressed as the ratio of stellar mass to the mass of the dark matter halo (e.g., Behroozi *et al.* 2013; Moster *et al.* 2013). For disc galaxies many different processes related to star formation are proposed: besides supernova explosions, stellar winds, and radiation pressure a number of recent papers discuss the possible importance of magnetic fields and cosmic ray pressure as an additional factor for the global dynamics of the interstellar medium (ISM) as well as for the launching of galactic winds (e.g., Girichidis *et al.* 2016; Pakmor *et al.* 2016). In the following we discuss the possible evidence for such cosmic ray driven galactic winds as observed by cosmic ray electrons (CREs) propagating in the magnetic fields in galactic halos. Recent progress in instrumentation such as the introduction of broad-band multi-channel receivers at all large radio facilities allow for a fresh look at this long standing problem. Progress has also been made with regard to the wavelength coverage: with LOFAR (van Haarlem *et al.* 2013) it is now possible to observe the emission from CREs at lower energies which are less affected by energy losses. They represent an old population that has traveled furthest from the site of origin in the star forming regions of the mid-plane.

2. Transport models for cosmic ray electrons

The polarized and frequency dependent radio-synchrotron radiation of CREs allows us to constrain the magnetic field strength and structure in galaxies. The frequency dependence of the synchrotron intensity also contains information on the transport processes for the CREs (e.g., Beck 2015). However, the observed emission has to be corrected for the frequency dependent contribution of the thermal radio-continuum which is also strongly correlated with star-formation. Here considerable progress has been made due to the availability of infrared data from satellite observations such as *Spitzer* and *WISE*. The dust emission in the ISM is a very good proxy for the star formation rate and can thus be used to correct for the thermal emission (e.g., Vargas *et al.* 2018). The clean synchrotron emission at various frequencies can then be used to study the emission by CREs with regard to magnetic field strengths and cosmic ray propagation processes such as diffusion or advection.

In a case study of the southern edge-on galaxies NGC 7090 and NGC 7462 based on Australia Telescope Compact Array (ATCA) data, Heesen *et al.* (2016) demonstrated that advective and diffusive transport of cosmic ray electrons into the halo can be distinguished by fitting a 1-D cosmic ray transport model to the intensity and spectral profiles perpendicular to the galaxy disc. Based on the analysis of archival data from the Very Large Array (VLA) and the Westerbork Synthesis Radio Telescope (WSRT), Heesen *et al.* (2018) discuss results of a larger sample (Fig. 1) with the conclusion that the radio halos of most galaxies studied so far are best described by advective transport models. For these galaxies the bulk velocity of the cosmic ray electrons correlates with the escape velocity of the galaxies as expected for galactic winds.

3. The CHANG-ES survey: first results

This analysis method is now to be applied to observations from the Karl G. Jansky Very Large Array (VLA) in the context of the "Continuum HAlos in Nearby Galaxies – an Evla Survey" (CHANG-ES) project. The CHANG-ES sample consists of 35 edge-on spiral galaxies in the local universe selected by angular size and total flux. The targets were observed in the B-, C- and D-array-configurations in C- and L-band (Irwin *et al.* 2012). First results for the complete sample have been presented by Wiegert *et al.* (2015) in combination with a first data release of the D-array data products. The study provides an image of an averaged radio-continuum halo or thick disk for the sample. The conclusion that the average edge-on galaxy exhibits an extended radio-continuum halo is different from the finding by Singal *et al.* (2015) claiming that edge-on galaxies do not possess radio-continuum thick disks or halos that would make the emission look similar to that observed in the Milky Way. Further analyses study general properties of the CHANG-ES sample such as typical scale heights of the radio-continuum disks (Krause *et al.* 2018) as presented in another contribution to this meeting (Krause 2018). The case study of NGC 4666 based on CHANG-ES data with an indication of a possible reversal of the magnetic field direction in the plane of the disk is also presented elsewhere in these proceedings by Stein *et al.* (2018). A more detailed description of the CHANG-ES project and of its first results is presented by Stil *et al.* (2018) at this conference.

4. Magnetic fields in galactic halos

Radio-continuum halos of star-forming disc galaxies frequently exhibit a large scale X-shaped magnetic field as, e.g., described for the prototypical case of NGC 5775 in Tüllmann *et al.* (2000). The extent of the synchrotron emission itself is proof for the presence of a magnetic field reaching far into the halo. Since the low energy cosmic rays are expected to be transported furthest, studies of the corresponding low frequency

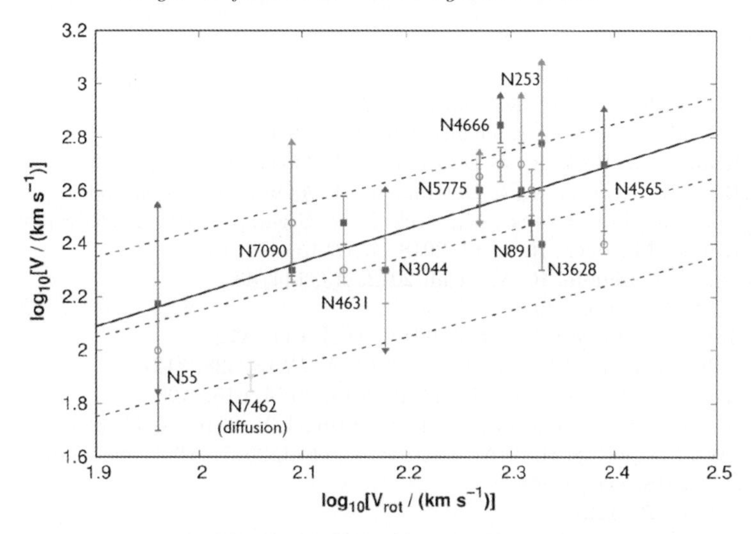

Figure 1. The correlation of the advection speed of cosmic ray electrons with the maximum rotation speed, i.e. total mass, of a sample of galaxies studied by Heesen *et al.* (2018). Most galaxies in this sample are best described by advective models for the cosmic ray electron propagation.

Figure 2. The edge-on galaxy NGC 4217 with a dumbbell-shaped radio-continuum halo is observed by LOFAR at 145 MHz as part of the Data Release 1 of the LoTSS survey (Shimwell *et al.* 2017; Shimwell 2018). The reprocessing by A. Miskolczi resulted in the reproduced map with a noise level of $\sim 70\mu$Jy/beam at 6 arcsec resolution.

emission allow us to probe the magnetic field far into the halo. LOFAR now provides observations with unprecedented sensitivity and resolution at such low frequencies as, e.g., demonstrated by the study of NGC 891 (Mulcahy *et al.* 2018). The LOFAR LoTSS survey (Shimwell *et al.* 2017) will provide a very good base for in depth studies of many more edge-on galaxies. To demonstrate this potential, Fig. 2 shows a map of NGC 4217 observed as part of the LoTSS Data Release 1 (Shimwell 2018). The dumbbell-shape radio halo (or thick disk) results from the combined effect of a radial decline of the magnetic field strength and the energy losses of the CREs transported into the halo (see Heesen *et al.* 2009a,b for a more detailed discussion).

References

Beck, R. 2015, *A&A Rev.*, 24, 4
Behroozi, P. S., Wechsler, R. H., & Conroy, C. 2013, *ApJ*, 770, 57
Girichidis, P., Naab, T., Walch, S., et al. 2016, *ApJL*, 816, L19
Heesen, V., Krause, M., Beck, R., & Dettmar, R.-J. 2009a, *A&A*, 506, 1123
Heesen, V., Beck, R., Krause, M., & Dettmar, R.-J. 2009b, *A&A*, 494, 563
Heesen, V., Dettmar, R.-J., Krause, M., Beck, R., & Stein, Y. 2016, *MNRAS*, 458, 332
Heesen, V., Krause, M., Beck, R., et al. 2018, *MNRAS*, 476, 158
Irwin, J., Beck, R., Benjamin, R. A., et al. 2012, *AJ*, 144, 43
Krause, M. 2018, *these proceedings*
Krause, M., Irwin, J., Wiegert, T., et al. 2018, *A&A*, 611, A72
Moster, B. P., Naab, T., & White, S. D. M. 2013, *MNRAS*, 428, 3121
Mulcahy, D. D., Horneffer, A., Beck, R., et al. 2018, *A&A*, 615, A98
Pakmor, R., Pfrommer, C., Simpson, C. M., & Springel, V. 2016, *ApJL*, 824, L30
Shimwell, T. W., Röttgering, H. J. A., Best, P. N., et al. 2017, *A&A*, 598, A104
Shimwell, T. W. 2018, *priv. comm.*
Silk, J. 2013, *ApJ*, 772, 112
Singal, J., Kogut, A., Jones, E., & Dunlap, H. 2015, *ApJL*, 799, L10
Stein, Y., Dettmar, R.-J., Irwin, J., et al. 2018, *these proceedings*
Stil, J. 2018, *these proceedings*
Tüllmann, R., Dettmar, R.-J., Soida, M., Urbanik, M., & Rossa, J. 2000, *A&A*, 364, L36
van Haarlem, M. P., Wise, M. W., Gunst, A. W., et al. 2013, *A&A*, 556, A2
Vargas, C. J., Mora-Partiarroyo, S. C., Schmidt, P., et al. 2018, *ApJ*, 853, 128
Wiegert, T., Irwin, J., Miskolczi, A., et al. 2015, *AJ*, 150, 81

Astronomy in Focus - XXX
Proceedings IAU Symposium No. XXX, 2018
M. T. Lago, ed.

doi:10.1017/S1743921319004502

The Magnetized Disk-Halo Transition Region of M51

M. Kierdorf[1], S. A. Mao[1], A. Fletcher[2], R. Beck[1], M. Haverkorn[3], A. Basu[4], F. Tabatabaei[5] and J. Ott[6]

[1]Max-Planck-Institut für Radioastronomie, Auf dem Hügel 69, 53121, Bonn, Germany
email: kierdorf@mpifr-bonn.mpg.de

[2]School of Mathematics and Statistics, Herschel Building, Newcastle University, NE1 7RU U.K.

[3]Department of Astrophysics/IMAPP, Radboud University Nijmegen; P.O. Box 9010, 6500 GL Nijmegen, Netherlands

[4]Fakultät für Physik, Universität Bielefeld, Universitätsstr. 25, 33615 Bielefeld

[5]Instituto de Astrofísica de Canarias, San Cristóbal de La Laguna Santa Cruz de Tenerife, Spain

[6]National Radio Astronomy Observatory, 1003 Lopezville Road, Socorro, NM 87801, USA

Abstract. An excellent laboratory for studying large scale magnetic fields is the grand design face-on spiral galaxy M51. Due to wavelength-dependent Faraday depolarization, linearly polarized synchrotron emission at different radio frequencies gives a picture of the galaxy at different depths: Observations at L-band (1 – 2 GHz) probe the halo region while at C- and X-band (4 – 8 GHz) the linearly polarized emission probe the disk region of M51. We present new observations of M51 using the Karl G. Jansky Very Large Array (VLA) at S-band (2 – 4 GHz), where previously no polarization observations existed, to shed new light on the transition region between the disk and the halo. We discuss a model of the depolarization of synchrotron radiation in a multilayer magneto-ionic medium and compare the model predictions to the multi-frequency polarization data of M51 between 1 – 8 GHz. The new S-band data are essential to distinguish between different models. Our study shows that the initial model parameters, i.e. the total regular and turbulent magnetic field strengths in the disk and halo of M51, need to be adjusted to successfully fit the models to the data.

Keywords. polarization, Galaxies: spiral, individual (M51), magnetic fields

---

1. Introduction

M51 is a nearby face-on grand design spiral galaxy. Modern radio interferometers with high spatial resolution allow us to probe detailed structures of the galaxy in total intensity and linear polarization. Investigating depolarization effects of linearly polarized synchrotron emission at different wavelengths is a powerful tool to put constraints on the magneto-ionic properties of the interstellar medium (ISM) in galaxies. One effect which can cause depolarization is Faraday rotation in magnetized thermal plasma which rotates the plane of linear polarization of an electro-magnetic wave by an angle proportional to λ^2. The proportionality constant is called the rotation measure (RM) and is measured in units of rad m^{-2}.

By comparing the observed degree of polarization as a function of wavelength with models of depolarization, one can investigate the underlying magnetic field properties (e.g. the regular and turbulent magnetic field strengths in the ISM). In the ISM of spiral galaxies, cosmic rays (CRs) as well as thermal electrons are mixed with magnetic fields in the same spatial volume. This causes emission of synchrotron radiation and

Faraday rotation at the same locations. In such a case the polarization plane experiences *differential Faraday rotation* which results in a sinc-function variation of the fractional polarization with λ^2 (Burn 1966). Therefore, at different frequencies one can probe the linearly polarized emission of a face-on galaxy at different physical depths: at high radio frequencies the polarized signal from the disk of the galaxy experiences low Faraday depolarization whereas at low radio frequencies, the polarized signal from the disk is almost completely depolarized. Fig. 1 shows the observed degree of polarization of M51 at frequencies between 1 – 8 GHz at the same angular resolution and the same color scale. One can see that the degree of polarization decreases with increasing wavelength (from bottom to top). Especially at L-band (top panel of Fig. 1) the central region of M51 is strongly depolarized. A detailed description of the data reduction and analysis and the full result of this work will be included in a forthcoming paper (Kierdorf *et al.*, in prep.).

2. M51's "unknown" polarization layer

Polarization studies of M51 shows that different configurations of the regular magnetic field exist in the disk and in the halo (e.g. Fletcher *et al.* 2011). According to Fletcher *et al.* (2011), the regular field in the disk is best described by a superposition of two azimuthal modes (axisymmetric plus quadrisymmetric), whereas the halo field has a strong bisymmetric azimuthal mode. The clear difference in the magnetic field configuration between the disk and the halo of M51 is still poorly understood. A better understanding will come from observations of the transition region between the disk and the halo. To investigate the "unknown" polarized layer between the disk and the halo we observed M51 in S-band (2 – 4 GHz, 7.5 - 15 cm) where no polarization data existed previously. We used the Karl G. Jansky Very large Array (VLA) in Soccoro, New Mexico operated by the National Radio Astronomy Observatory (NRAO) which provides large antenna separation and wideband receivers resulting in high spatial resolution and wide frequency coverage with high sensitivity. Our new broadband S-band polarization data fill the gap between data observed with the VLA at L-band (1-2 GHz) by Mao *et al.* (2015), and C-band (4.85 GHz) and X-band (8.35 GHz) by Fletcher *et al.* (2011). With this combined high quality and broad frequency coverage data set we are able to investigate the magneto-ionic properties in different layers of M51.

3. Faraday depolarization in a multi-layer magneto-ionic medium

Shneider *et al.* (2014) developed a model of the depolarization of synchrotron radiation in a multilayer magneto-ionic medium. They developed model predictions for the degree of polarization as a function of wavelength for a two-layer system with a disk and a halo and a three-layer system with a far-side halo, a disk and a near-side halo. The model includes differential Faraday rotation caused by regular magnetic fields and internal Faraday dispersion due to random magnetic fields. In the case of a three-layer system, the near and far-side halo have identical properties. Fig. 2 shows the model predictions of the normalized degree of polarization (p/p_0) for a two-layer (left panel) and three-layer (right panel) system, respectively. p_0 is the intrinsic degree of polarization which is assumed to be 70 % corresponding to the theoretical injection spectrum for electrons accelerated in supernova-remnants with a synchrotron spectral index of $\alpha_{\rm syn} = -0.5$ (Shneider *et al.* 2014). The nomenclature of the different models is as follows: 'D' and 'H' stands for regular fields in the disk and halo, respectively. 'I' and 'A' denotes isotropic and anisotropic turbulent fields where the first one is for the disk and the second for the halo. The observed degrees of polarization at X-band, C-band, S-band and L-band are also shown. For the purpose of this proceeding, we consider the total and polarized intensity integrated in a sector with an azimuthal angle centered at 100° and an opening angle

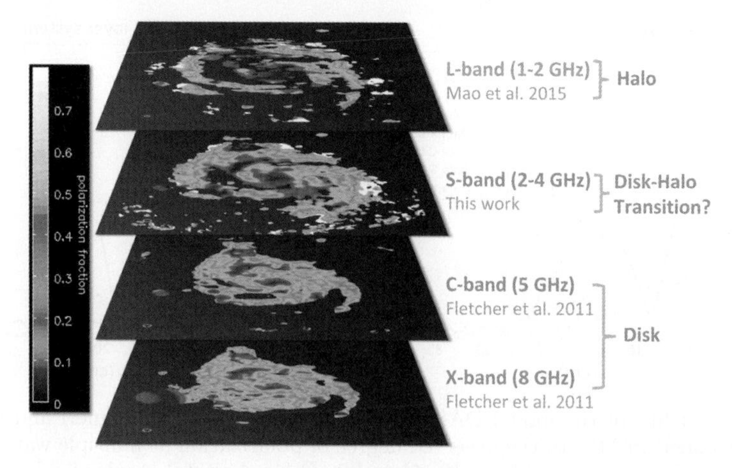

Figure 1. Observed degree of polarization of M51 at different frequencies. All images have the same color scale and are smoothed to the same resolution of 15 arcsec (which corresponds to about 550 pc at the distance of M51). Note that the total intensity images used to calculate the degree of polarization were not corrected for thermal emission.

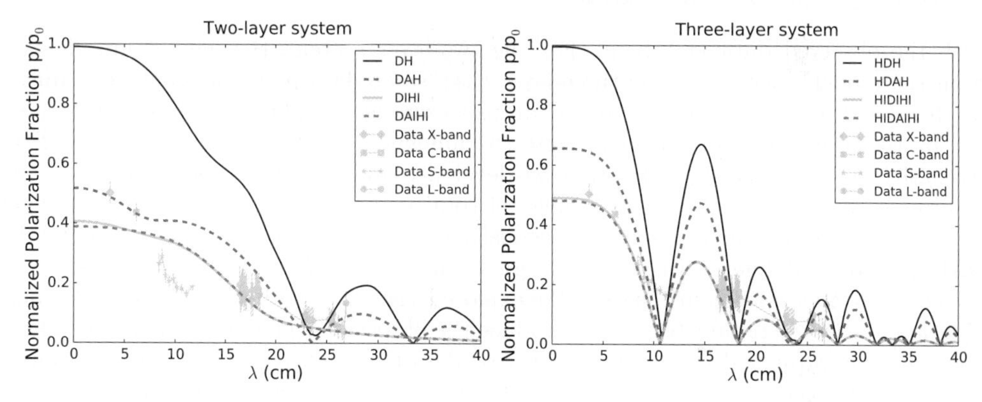

Figure 2. Depolarization models from Shneider *et al.* (2014) for a two-layer system (left panel) and a three-layer system (right panel) in M51 plotted together with the observed degree of polarization at multiple wavelengths. All model profiles featured have been constructed from a set of the following parameters: a total regular magnetic field strength of 5 μG in the disk and halo, a disk turbulent random field of 14 μG, and a halo turbulent random field of 4 μG and a thermal electron density of 0.11 cm^{-3} and 0.01 cm^{-3} in the disk and halo, respectively. For nomenclature and description of the model types appearing in the legend we refer to the text.

of 20° and radial boundaries 2.4 – 3.6 kpc†. With only the data points at 3.6 cm, 6.2 cm, and L-band (15 – 30 cm) it is not possible to assess if a two-layer or three-layer system is more likely for M51. Since the model predictions strongly differ within the wavelength range of S-band (7.5–15 cm), our new S-band data are essential to distinguish between the different systems.

By comparing the observed degree of polarization to the models, one can directly rule out models with only regular magnetic fields in the disk and halo (DH) since the observed data deviate most from those model predictions. However, it appears that none of the model predictions with the parameters given in Shneider *et al.* (2014) are in agreement

† To obtain the non-thermal total flux density at this location, we assumed a thermal fraction $f^{\rm th}_{\nu_0}$ of 9 % at $\nu_0 = 3$ GHz (Tabatabaei *et al.* 2017) and extrapolated the non-thermal flux densities at frequency ν via $S_\nu = f^{\rm th}_{\nu_0} \left(\frac{\nu}{\nu_0}\right)^{-0.1} S_{\nu_0}$.

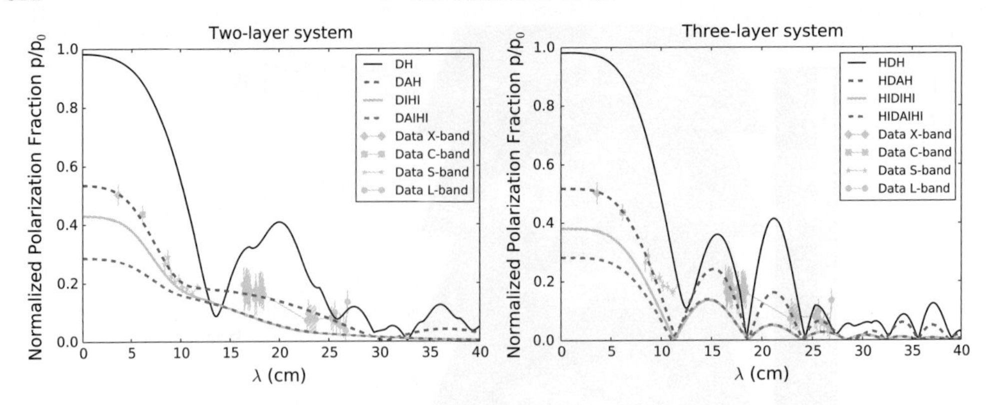

Figure 3. "Best-fit" of the model DAH for a two-layer system (left panel) and a three-layer system (right panel) in M51 to the observed degree of polarization at multiple wavelengths. The "best fit" magnetic field strengths are $10\,\mu$G and $3\,\mu$G for the total regular field in the disk and in the halo for the two-layer and three-layer system, respectively. The total random magnetic field strengths in the disk amounts to $14\,\mu$G and $16\,\mu$G for the two - and three-layer system, respectively. For the electron density a value of $0.07\,\mathrm{cm}^{-3}$ and $0.01\,\mathrm{cm}^{-3}$ in the disk and halo fits the data best in both layer systems.

with the observed data at S-band. For the two-layer system, the data points deviate from the model whereas for the three-layer case, some data points are well reproduced by the model predictions but the model drops to zero at $\lambda \approx 11$ cm which is clearly ruled out by the observed data. To investigate the influence of different total magnetic field strengths on the degree of polarization, we developed an interactive tool which allows one to produce model predictions for a range of total regular and turbulent magnetic field strengths in the disk and halo simultaneously. Fig. 3 shows the "best fit" of the model DAH (red dashed line) with regular magnetic fields in disk and halo and anisotropic turbulent magnetic fields in the disk. We explored visually whether any reasonable combination of the free parameters can reproduce the observed degree of polarization. These "best fit" magnetic field strengths and electron densities are listed in the caption of Fig. 3 and are all physically plausible values. For the three-layer system it is not possible to lift up the zero points in the model by changing any parameter. Therefore, this three-layer model can be ruled out. In other words, we do not detect any polarized emission from the far side halo.

4. Future Work

We show that the comparison of the observed degree of polarization to the wavelength-dependent depolarization models is a powerful tool to put constraints on the magnetic field strengths and thermal electron density in different regions of the galaxy. As a next step, we will apply the same method to other sectors with different azimuthal angles and radii in M51. Consistency between fits to different sectors is a strong indication that the model is physically meaningful. Further, since the different models distinguish between isotropic and anisotropic turbulent magnetic fields, by comparing the observed degree of polarization in different regions of the galaxy with the model predictions, we can investigate turbulent magnetic field configurations in different locations in M51.

References

Burn B. J., 1966, *MNRAS*, 133, 67
Fletcher A., Beck R., Shukurov A., Berkhuijsen E. M., Horellou C., 2011, *MNRAS*, 412, 2396
Mao S. A., Zweibel E., Fletcher A., Ott J., Tabatabaei F., 2015, *ApJ*, 800, 92
Shneider C., Haverkorn M., Fletcher A., Shukurov A., 2014, *A&A*, 567, A82
Tabatabaei F. S., *et al.*, 2017, *ApJ*, 836, 185

Astronomy in Focus - XXX
Proceedings IAU Symposium No. XXX, 2018
M. T. Lago, ed.

doi:10.1017/S1743921319004514

Techinques and algorithmic advances in the SKA era

V. Vacca[1], F. Govoni[1], M. Murgia[1], T. Enßlin[2], N. Oppermann[3], L. Feretti[4], G. Giovannini[4,5], J. Jasche[6], H. Junklewitz[7] and F. Loi[5]

[1]INAF - Osservatorio Astronomico di Cagliari, Via della Scienza 5, Selargius, Italy
email: valentina.vacca@inaf.it

[2]MPA, Karl-Schwarzschild-Str 1, Garching, Germany

[3]CITA, University of Toronto, 60 St. George Street, Toronto, Canada

[4]INAF - Istituto di Radioastronomia, Via Gobetti 101, Bologna, Italy

[5]DIFA, Università degli Studi di Bologna, Viale Berti Pichat 6/2, Bologna, Italy

[6]Excellence Cluster Universe, TUM, Boltzmannstrasse 2, Garching, Germany

[7]Argelander-Institut für Astronomie, Auf dem Hügel 71, Bonn, Germany

Abstract. The new generation of radio interferometers will deliver an unprecedented amount of deep and high resolution observations. In this proceedings, we present recent algorithmic advances in the context of the study of cosmic magnetism in order to extract all the information contained in these data.

Keywords. magnetic fields, polarization, methods: numerical methods: statistical

---

1. Introduction

Magnetism in the cosmic web is widely unknown. Magnetic fields have revealed themselves in galaxy clusters in the form of diffuse synchrotron sources and via the Faraday effect on background radio galaxies that indicate strengths of a few μG and fluctuation scales up to a few hundreds of kpc (e.g., Feretti *et al.* 2012). Beyond galaxy clusters, along filaments and in the voids of the cosmic web, some indication of their presence has been found so far but needs to be confirmed. To shed light on cosmological magnetic field origin and evolution, the analysis of magnetic properties in these environments is crucial. Nowadays, magnetic fields are thought to originate from a seed magnetic field whose strength has been amplified and geometry modified during processes of structure formation. The investigation of non-thermal components in galaxy clusters and in the low density environments between them has the potential to shed light on their evolution and formation. In the following, we give an overview of the effort done by the scientific community in the last years in this direction. In §2 and §3, we respectively describe the progress concerning the study via diffuse synchrotron emission and Faraday effect and in §4 we present the conclusions.

2. Diffuse emission

Diffuse synchrotron sources called radio halos have been observed to permeate the central volume of about 60 galaxy clusters where they reveal μG magnetic fields and ultra-relativistic particles ($\gamma \gtrsim 10^4$). Their emission permits to directly probe the intracluster magnetic field. Magnetic fields fluctuating on scales larger than the resolution of the observations are expected to generate radio halos with disturbed morphology and high degrees of polarization. Magnetic fields with fluctuation scales smaller than the beam could be responsible of regular morphology and no polarized signal (e.g., Vacca *et al.* 2010). To date only in three systems a polarized signal likely associated with the diffuse emission of the halo has been detected (Girardi *et al.* 2016). Numerical three-dimensional simulations by Govoni *et al.* (2013) indicate that, at 1.4 GHz, radio halos show intrinsic fractional polarization levels of 15-35% at the cluster center that increase in the cluster outskirts. This level of polarization corresponds to a polarized signal of about 2-0.5 μJy/beam for strong and intermediate luminosity radio halos at $3''$ of resolution but, due to instrumental limitations (resolution and sensitivity), it is hard to detect. The JVLA has the potential to detect already polarized emission in high-luminosity radio halos, while for intermediate-luminosity sources only SKA1 can succeed. The detection of faint-luminosity radio halos instead will be very difficult even with the SKA1.

Properly imaging these sources is very important to investigate the magnetization of the medium. However, standard imaging tools as CLEAN (e.g., Högbom 1974) are unsatisfactory since they rely on the assumption of a completely uncorrelated point source sky, while radio halos are diffuse and extended. In the last years new imaging algorithms have been developed based on either compressed sensing (e.g., MORESANE Dabbech *et al.* 2015) or Bayesian statistics (e.g., RESOLVE, Junklewitz *et al.* 2016), capable of exploiting the capabilities of the new generation of radio telescopes to accurately reproduce the complexity of the radio sky in total intensity as well as in polarization. This progress is particularly important in the context of the investigation of faint diffuse emission beyond galaxy clusters, from the filaments of the cosmic web. Recently, we observed a region of the sky of $8^\circ \times 8^\circ$ containing several galaxy clusters of which about ten at $z \approx 0.1$, with the Sardinia Radio Telescope (SRT) at 1.4 GHz (Vacca *et al.* 2018). The data revealed a field very bright in radio especially crowded at the location of galaxy clusters and between them (Fig. 1). To overcome the limited spatial resolution of $\sim 13'$ and separate possible diffuse emission from embedded discrete radio sources, we combined these data with higher resolution data from the NRAO VLA Sky Survey (NVSS, $45''$, Condon *et al.* 1998) and subtracted point-sources. In the resulting image, we identified 28 new diffuse sources with radio emissivity and X-ray emission 10-100 times lower than cluster radio sources. The comparison with magneto-hydro-dynamical simulations suggests that they could represent the tip of the iceberg of the emission associated with the warm-hot intergalactic medium and correspond to magnetic field strengths of $\sim$20-50 nG.

3. Radio galaxies

A complementary and alternative approach to investigate magnetic fields in the cosmic web is given by the analysis of polarimetric properties of background radio galaxies. While crossing the magneto-ionic media in between the source and the observer, their signal suffers a rotation $\Delta\Psi$ of the polarization angle

$$\Delta\Psi = \Psi_{\rm obs} - \Psi_{\rm int} = \phi\lambda^2 \tag{3.1}$$

where $\Psi_{\rm int}$ and $\Psi_{\rm obs}$ are respectively the intrinsic and observed polarization angle, λ the wavelength of observation, and ϕ the Faraday depth, related to the magnetic field along the line of sight $B_{||}$ and to the thermal gas density $n_{\rm e}$ of the magneto-ionic medium

$$\phi = 812 \int_0^{l[\rm kpc]} n_{\rm e}[{\rm cm}^{-3}]{\rm B}_{||}[\mu{\rm G}]{\rm d}l \qquad {\rm rad/m}^2. \tag{3.2}$$

In the case of a screen external to the radio source, the Faraday depth does not change as a function of λ^2 and is defined as Rotation Measure (RM, Burn 1966). If the radio and X-ray plasma are mixed, the Faraday depth is no more constant and the observed

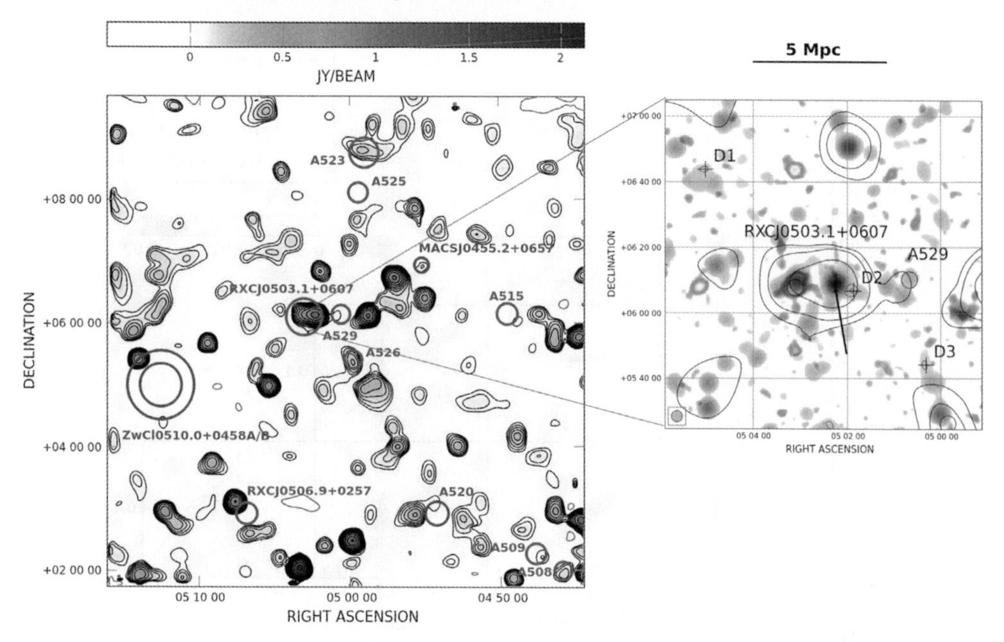

Figure 1. Left: SRT image (angular resolution $13.9' \times 12.4'$) in colors and contours at 1.4 GHz. The circles and labels identify the position of the galaxy clusters in the field of view with known redshift: in red clusters with $0.08 < z < 0.15$, in green clusters with redshift outside this range. Right: Zoom of the central region of the full field of view. SRT contours (negative in magenta and positive in black) and SRT+NVSS after compact-source subtraction contours (in blue, resolution $3.5' \times 3.5'$) overlaid on X-ray emission from the RASS in red colors and radio emission from the SRT+NVSS in grey colors. Images from Vacca *et al.* (2018).

polarized intensity may be associated with a range of Faraday depths. In this case more advanced approaches are necessary to properly recover polarimetric properties of the radio sources as QU-fitting, RM Synthesis, and Faraday Synthesis (e.g., O'Sullivan *et al.* 2012; Brentjens & de Bruyn 2005; Bell & Enßlin 2012).

In case of radio galaxies in the background of galaxy clusters, deep and high resolution polarimetric data at multiple frequencies permit to obtain detailed rotation measure images. When the Galactic contribution is negligible, these images represent a two-dimensional picture of the intracluster magnetic field and can be used to derive its power spectrum via the comparison with simulations (e.g., Vacca *et al.* 2012). Unfortunately, to date detailed Faraday depth images are available typically for one source per clusters, apart from a few exceptions (e.g., Govoni *et al.* 2006). With the SKA1, we expect a number of several tens of sources for nearby galaxy clusters ($z < 0.1$) allowing us to study the magnetic field over the complete cluster volume (Bonafede *et al.* 2015).

Thanks to the high sensitivity and resolution of the polarization surveys planned with the SKA precursors and path-finders, the upcoming Faraday depth catalogs will contain several thousands of sources and up to 7-14 million with the SKA1 (Johnston-Hollitt *et al.* 2015). These data will simultaneously carry out the information of all the structures between the source and the observer (our Galaxy, intervening sources, galaxy clusters, filaments, voids, etc). To isolate the Faraday effect due to the cosmic web different strategies have been developed as, e.g., filtering out unwanted contributions by Akahori *et al.* (2014). In this context, we recently developed statistical Bayesian approaches to disentangle the Galactic and extragalactic Faraday rotation while properly taking into account the noise (Oppermann *et al.* 2015) and to further decompose the extragalactic Faraday

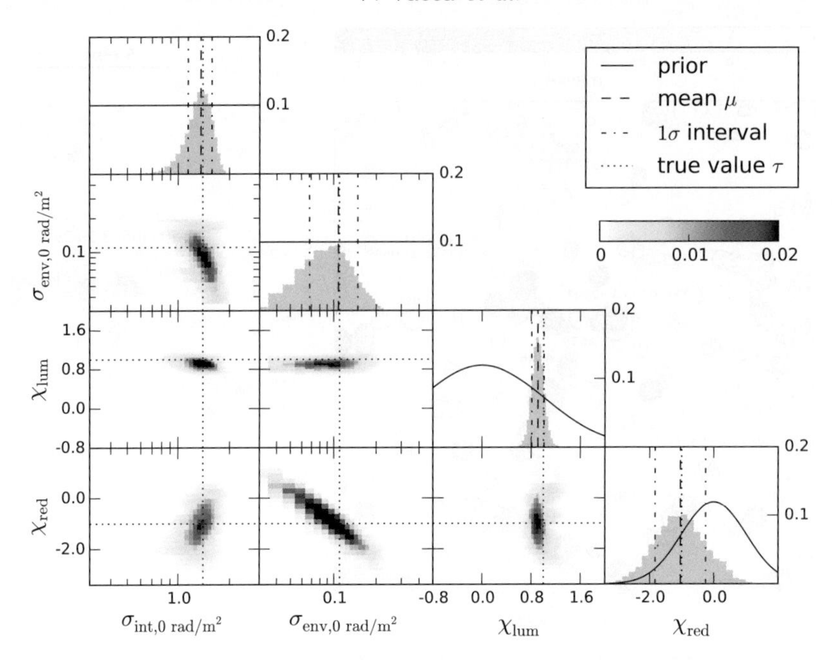

Figure 2. Example of the posterior from the algorithm presented in Vacca *et al.* (2016) for a mock catalog of Faraday depth values for 3500 sources in the frequency range covered by SKA1-LOW and an overall extragalactic contribution of $0.7\,\mathrm{rad/m^2}$.

rotation in the contributions intrinsic to the emitting source, due to any intervening galaxy and associated with different regions of the large scale structure (Vacca *et al.* 2016). We showed that high-quality low-frequency data for a few thousand of sources have the potential to investigate magnetic fields with strengths of $\approx$0.2-2 nG associated with the large scale structure (Fig. 2).

4. Conclusions

The SKA, its path-finders and precursors are expected to permit a breakthrough in the study of cosmic magnetism, thanks to new data of unprecedented high-quality. To exploit these data and finally uncover the signature of the faint magnetic fields of the cosmic web, advanced and sophisticated techniques of analysis as those developed in the last decades are essential.

References

Akahori, T., Gaensler, B. M., & Ryu, D. 2014, *ApJ*, 790, 123

Bell, M. R., & Enßlin, T. A. 2012, *A&A*, 540, A80

Bonafede, A., Vazza, F., Brüggen, M., *et al.* 2015, *AASKA14*, 95

Brentjens, M. A., & de Bruyn, A. G. 2005, *A&A*, 441, 1217

Burn, B. J. 1966, *MNRAS*, 133, 67

Condon, J. J., Cotton, W. D., Greisen, E. W., *et al.* 1998, *AJ*, 115, 1693

Dabbech, A., Ferrari, C., Mary, D., *et al.* 2015, *A&A*, 576, A7

Feretti, L., Giovannini, G., Govoni, F., & Murgia, M. 2012, *A&ARv*, 20, 54

Girardi, M., Boschin, W., Gastaldello, F., *et al.* 2016, *MNRAS*, 456, 2829

Govoni, F., Murgia, M., Xu, H., *et al.* 2013, *A&A*, 554, A102

Govoni, F., Murgia, M., Feretti, L., *et al.* 2006, *A&A*, 460, 425

Högbom, J. A. 1974, *A&AS*, 15, 417

Johnston-Hollitt, M., Govoni, F., Beck, R., *et al.* 2015, *AASKA14*, 92

Junklewitz, H., Bell, M. R., Selig, M., & Enßlin, T. A. 2016, *A&A*, 586, A76
Oppermann, N., Junklewitz, H., Greiner, M., *et al.* 2015, *A&A*, 575, A118
O'Sullivan, S. P., Brown, S., Robishaw, T., *et al.* 2012, *MNRAS*, 421, 3300
Vacca, V., Murgia, M., Govoni, F., *et al.* 2018, *MNRAS*, 479, 776
Vacca, V., Oppermann, N., Enßlin, T., *et al.* 2016, *A&A*, 591, A13
Vacca, V., Murgia, M., Govoni, F., *et al.* 2012, *A&A*, 540, A38
Vacca, V., Murgia, M., Govoni, F., *et al.* 2010, *A&A*, 514, A71

FM9
Solar Irradiance: Physics-Based Advances

Astronomy in Focus - XXX
Proceedings IAU Symposium No. XXX, 2018
M. T. Lago, ed.

doi:10.1017/S1743921319004526

FM9 - Solar Irradiance: Physics-Based Advances

Greg Kopp[1,2] and Alexander Shapiro[2]

[1]University of Colorado / Laboratory for Atmospheric & Space Physics
Boulder, CO 80303 U.S.A.
email: `Greg.Kopp@LASP.Colorado.edu`

[2]Max Planck Institut für Sonnensystemforschung, Göttingen, Germany
email: `ShapiroA@MPS.MPG.DE`

Keywords. Sun: irradiance, TSI, variability, MHD, radiative transfer; Stars: variability; instrumentation: radiometry, photometry

---

1. The Importance of Understanding Solar Variability

Solar irradiance varies on all timescales at which it has ever been observed and possibly also on multi-decadal to century timescales. Knowledge of the magnitudes and timescales of this variability extends to many fields other than solar, including Earth-climate modeling, stellar-variability studies, and exoplanet detection. The terrestrial atmospheric and climate systems respond to variations in solar radiative output on timescales from days to decades, and there is also evidence for solar influences on climate over longer timescales. Stellar astronomers have been comparing variability of the Sun with that of other lower main sequence stars to determine how typical the Sun's solar activity cycle is. The magnitude and timescales of stellar variability can limit the detectability of exoplanets via transit methods relying on precision photometry. Understanding the physics behind solar variability helps assess the resulting effects on the Earth as well as the causes of similar stellar brightness variations and thus even the habitability of exoplanets.

2. Recent Improvements in Solar-Irradiance Understandings

Currently available empirical and semi-empirical models of solar-irradiance variability and some of the measurements of this variability have discrepancies that limit knowledge particularly of the long-term changes in the Sun's radiative output. Fortunately, recent advances in modeling and observing the solar atmosphere make it possible to create a new generation of significantly more realistic physics-based irradiance models. Benefitting from the enormous recent progress in solar observations and models, it is now possible to develop a new generation of irradiance models based on the current state-of-the-art in solar physics. In particular, these advances in understanding solar variability include:

• *3D magneto-hydrodynamic* (MHD) simulations of flows and magnetic fields in the near-surface layers of the Sun and stars have reached a high level of realism, and can now reproduce many sensitive observational tests. These simulations make it possible to replace 1D representations of the solar atmosphere with realistic 3D simulations and also enable assessment of the contributions of granulation to short-term solar-irradiance variability.

• New time-efficient *radiative transfer codes* and approaches have been developed. These allow calculated emergent spectra from 3D MHD cubes to account for effects from millions of atomic and molecular lines as well as deviations from local thermodynamic equilibrium, giving more accurate estimates of outgoing radiation as a function of position on the solar disk.

• *New atomic and molecular data* allow more reliable computation of the opacities in the solar atmosphere. The irradiance variability in the UV, violet, blue, and green spectral domains is fully controlled by millions of the Fraunhofer lines. Recent advances in laboratory astrophysics and in collecting the data (e.g. a major upgrade of the Vienna atomic line database, which now also includes molecular data) make possible significantly more accurate calculations of solar-irradiance variability.

• *Surface flux transport models* (SFTMs) now more realistically simulate the evolution of the large-scale surface magnetic field over the solar cycle. This allows reconstructing the evolution of the solar surface magnetic field and irradiance over long timescales, which is crucial to understanding the pre-anthropogenic solar contributions to climate change, from which natural sensitivities of climate can be estimated.

• *Irradiance-monitoring instrument improvements* are providing more stable long-term measurements, helping constrain the range of possible secular variations in the Sun's radiative output, as well as better short-term sensitivity, which can help refine solar-irradiance models using other indicators of solar activity.

• *High-resolution imagery of magnetic features* on the solar surface, which are the main driver of solar irradiance variability, from recent solar missions such as the Solar Dynamics Observatory (SDO), STEREO, SUNRISE, HINODE, etc. SDO in particular provides frequent space-based magnetograms, which are needed inputs to the newest physics-based solar-irradiance models.

With these recent advances in understanding solar-irradiance variations, we organized a dedicated focus meeting to summarize such physics-based advances and enable discussions between the many research areas involved in these improved understandings.

3. Focus Meeting 9

Focus Meeting 9, titled "Solar Irradiance: Physics-Based Advances," brought together researchers from around the world to discuss new insights, measurements, and models related to solar-irradiance variability. After an introduction session highlighting the prominent questions in the fields of Earth climate's sensitivity to solar variability, solar-surface magnetic features causing solar-irradiance variability, and the brightness variability of Sun-like stars, the meeting included the following four primary sessions:

• Session 1: Available Solar-Irradiance Data Sets and Models
• Session 2: Brightness Contrasts of Solar-Surface Magnetic Features
• Session 3. Structure and Evolution of Solar-Surface Magnetic Fields
• Session 4. Extrapolating Solar Models to Sun-like Stars

This chapter in the IAU General Assembly XXX proceedings summarizes but a few of the presentations and posters from this focus meeting to provide a sample of the many diverse presentations at the meeting itself.

Astronomy in Focus - XXX
Proceedings IAU Symposium No. XXX, 2018
M. T. Lago, ed.

doi:10.1017/S1743921319004538

Recent progresses in the use of 3D MHD simulations for solar irradiance reconstructions

Serena Criscuoli

National Solar Observatory
3665 Discovery Drive, 80303 Boulder, CO, USA
email: scriscuo@nso.edu

Abstract. The use of 3D magneto-hydrodynamic simulations of the solar atmosphere in modeling irradiance variations seems a natural evolution of the current irradiance reconstruction techniques making use of one-dimensional, static, atmosphere models. Nevertheless, the development of such new models poses serious computational challenges. This contribution focuses on recent progresses made in the development of novel irradiance reconstruction models making use of 3D MHD simulations and discusses current and future challenges.

1. Introduction

Variations of solar and stellar irradiance affect the atmosphere and climate of the Earth and exo-planets. Moreover, understanding solar variability provides useful insight in modeling stellar magnetic variability, which in turn, is of paramount importance for improving the detection of exo-planets and for characterizing the habitability zones of stars (Fabbian *et al.* 2017). Measurements of solar irradiance variations are intermittent, dis-homogeneous, and often affected by instrumental calibration and degradation effects. Moreover, systematic measurements started only in the late seventies. Time series of irradiance variations necessary to assess the influence of irradiance variations on the Earth atmosphere are therefore complemented and extended in time with irradiance estimates obtained by reconstructions. Unfortunately, even reconstructions present disagreements that do not allow to correctly assess the influence of solar irradiance variability on the Earth atmosphere (Ermolli *et al.* 2013). This contribution focuses on recent progresses in developing novel irradiance reconstruction techniques based on the use of three dimensional magneto-hydrodynamic (3D-MHD) simulations of the solar atmosphere, which we expect to greatly improve reconstruction techniques based on semi-empirical approach.

Semi-empirical irradiance reconstructions are performed by combining measures of the variation of surface magnetism, as derived by the analysis of full-disk data, with estimates of the radiative emission synthesized with atmosphere models. These reconstruction techniques have been proven successful in reproducing more than 90% of the variability of the measured Total Solar Irradiance (TSI, i.e. the irradiance integrated over the whole spectrum), but the agreement is less good when studying timescales longer than one decade and/or restricting the analysis to finite spectral ranges. Among the different aspects that may contribute to explain such uncertainties, the employed synthetic spectra are a point of concern. Synthesis of stellar spectra strongly depend on the assumptions adopted for the radiative transfer calculations as well as on the employed atmosphere models. Concerning this last aspect, irradiance reconstructions have relied so far on the use of semi-empirical one-dimensional static atmosphere models, which have been most typically derived to reproduce measured disk-integrated spectra (e.g. Fontenla, Avrett & Loeser 1993). By their nature, such atmosphere models cannot capture the fine spatial and temporal scales of features observed with modern high spatial resolution instrumentation, and more in general the complex three-dimensional nature of the processes that describe the propagation of radiation through convective magnetized plasma. More specifically, the use of one-dimensional static atmosphere models present the following drawbacks: 1) May fail in reproducing irradiance variability at spectral and temporal scales other than those employed to derive the models; 2) May fail in reproducing observed properties of quiet and magnetic structures; 3) Radiative transfer codes other than the ones used to derive the atmosphere models may produce different spectra; 4) Can be hardly validated with independent measurements of properties of quiet and magnetic plasma (as for instance doppler or spectro-polarimetric measurements).

2. Use of 3D-MHD models for solar irradiance reconstructions

Uitenbroek & Criscuoli (2011) employed 3D MHD simulations of the solar photosphere to show that semi-empirical one-dimensional models do not describe the physical average properties of the solar atmosphere, thus explaining the issues 1) and 2) listed above. The use of 3D MHD simulations of the solar photosphere and chromosphere for irradiance reconstructions is in this respect a huge step forward, as they are derived solving basic magneto-hydrodynamic equations with a few observational constraints (Stein 2012, Nurdlund, Stein & Asplund 2009). These simulations, which typically represent small areas of the solar atmosphere (a few square-arcsec horizontally, and vertically a few Mm above and below the optical depth unity surface; *box-in-a-star* regime) with high spatial resolution (a few tens of kilometers or better), have been proven to reproduce the observed properties of the solar spectrum with a higher degree of accuracy than one-dimensional models (e.g. Asplund *et al.* 2009; Pereira *et al.* 2013).

The use of these simulations for irradiance reconstruction purposes has been for long hampered by the highly demanding resources required for both the atmosphere modeling and the radiative output computations. First applications have been therefore limited to investigations of the radiative properties of magnetic structures (e.g. Tritschler & Uitenbroek 2006; Afram *et al.* 2011, Criscuoli 2013; Thaler & Spruit 2014; Criscuoli & Uitenbroek 2014a) or to qualitative estimates of irradiance variability (Criscuoli & Uitenbroek 2014b). The crescent availability of supercomputers has recently allowed to overcome computational difficulties, at least for what concerns simulations of the solar photosphere and spectral synthesis performed under Local Thermodynamic Equilibrium (LTE). Nowadays, time series of solar and stellar photospheres obtained at different level of magnetization produced with different MHD codes are available under request or at specific databases (e.g. Beeck *et al.* 2013; Rempel 2014; Kitiashvili *et al.* 2015; Salhab *et al.* 2018). Computations of the whole spectrum at the spectral resolution necessary to model the observed variability (1 nm or better) is possible making use of pre-computed tables of opacities (e.g. Norris *et al.* 2017). Thanks to such advancements, Shapiro *et al.* 2017 recently estimated the contribution of granulation to high frequency (minutes to hours) TSI variations using 3D hydrodynamic simulations. The first reconstruction of TSI irradiance variations making use of 3D-MHD simulations was presented in Yeo *et al.* 2017. These reconstructions combine photospheric magnetograms acquired with the Helioseismic and Magnetic Imager (HMI) with spectra synthetized with photospheric simulations obtained with the Max Planck Institute for Solar System Research/University of Chicago Radiation Magneto-hydrodynamics (MURaM, Vögler *et al.* 2005) code. The agreement between the synthetic and observed TSI variations is at 95% level, which is comparable with the agreement obtained with other models.

3. Future challenges

The major challenge we are currently facing is the use of MHD simulations to model and reconstruct the irradiance in the UV. Most of the radiation at these spectral ranges originate in the chromosphere, which is a layer of the atmosphere that is extremely difficult to model for both what concerns the correct description of the physical processes involved and the computational resources required (e.g. Freytag *et al.* 2012). At the moment, Bifrost (Gudiksen *et al.* 2011) is the most advanced code for reproducing the properties of the solar chromosphere. Nevertheless, recent studies indicate that synthetic spectra obtained with Bifrost snapshots fail in fully capturing observed properties of the chromosphere (e.g. Bastian *et al.* 2017), thus suggesting that the Bifrost code still does not properly take into account physical processes occurring in the higher layers of the solar atmosphere. Furthermore, the production of statistically significant time series of snapshots to use for irradiance reconstruction as well as non-LTE radiative transfer computations are still prohibitive. Given the great interest shown by the solar community in recent years for both modeling the chromosphere and improving multi-dimensional non-LTE radiative transfer codes (e. g. Sukhorukov & Leenaarts 2017), together with the rapid increase in the computational power, these issues will be likely overcome in the next future.

References

Afram, N., Unruh, Y. C., Solanki, S. K. *et al.* 2011, *A&A*, 526, 120

Asplund, M., Grevesse, N., Sauval, A. J., Scott, P. 2009, *ARA&A*, 47, 481

Bastian, T. S., Chintzoglou, G., De Pontieu, B. *et al.* 2017, *ApJ*, 845, 19

Beeck, B., Cameron, R. H., Reiners, A., Schüssler, M. 2013, *A&A*, 558, 48

Criscuoli, S. 2013, *ApJ*, 778, 27

Criscuoli, S. & Uitenbroek, H. 2014a, *A&J*, 562, 1

Criscuoli, S. & Uitenbroek, H. 2014b, *ApJ*, 788, 151

Ermolli, I., Matthes, K., Dudok de Wit, T. *et al.* 2013, *ACP*, 13.3945

Fabbian, D., Simoniello, R., Collet, R., Criscuoli, S., Korhonen, H., Krivova, N. A. *et al.* 2017, *AN*, 338, 753

Freytag, B., Steffen, M., Ludwig, H.-G. *et al.* 2012, *JCoPh*, 231, 919

Fontenla, J. M., Avrett, E. H. & Loeser, R. 1993, *ApJ*, 406, 319

Gudiksen, B. V., Carlsson, M., Hansteen, V. H. *et al.* 2011, *A&A*, 531, 154

Kitiashvili, I.N., Kosovichev, A.G., Mansour, N.N., Wray, A.A. 2015, *ApJ*, 809, 84

Norris, C. M., Beeck, B., Unruh, Y. C. *et al.* 2017, *A&A*, 605, 45

Nordlund, Å, Stein, R. F.& Asplund, M. 2009, *LRSP*, 6, 2

Pereira, T. M. D., Asplund, M., Collet, R. *et al.* 2013, *A&A*, 554, 118

Rempel, M. 2014, *ApJ*, 789, 132

Salhab, R. G., Steiner, O., Berdyugina, S. V. *et al.* 2018, *A&A*, 614, 78

Shapiro, A. I., Solanki, S. K., Krivova, N. A. *et al.* 2018, *NatAs*, 1, 612

Sukhorukov, A. V. & Leenaarts, J. 2017, *a&A*, 597, 46

Stein, R., F. 2012, *LRSP*, 9, 4

Thaler, I. & Spruit, H. C. 2014, *A&A*, 566, 11

Uitenbroek, H. & Criscuoli, S. 2011, *ApJ*, 736, 69

Tritschler, A. & Uitenbrock, H. 2006, *ApJ*, 648, 741

Vögler, A., Shelyag, S., Schüssler, M., *et al.* 2005, *A&A*, 429, 335

Astronomy in Focus - XXX
Proceedings IAU Symposium No. XXX, 2018
M. T. Lago, ed.

doi:10.1017/S174392131900454X

Solar irradiance: from multiple observations to a single composite

Thierry Dudok de Wit[1] **and Greg Kopp**[2]

[1]LPC2E, CNRS and University of Orléans,
3A avenue de la Recherche Scientifique, 45071 Orléans, France
email: ddwit@cnrs-orleans.fr

[2]University of Colorado, Laboratory for Atmospheric and Space Physics,
3665 Discovery Drive, Boulder 80303 CO, USA
email: greg.kopp@lasp.colorado.edu

Abstract. We review recent developments in combining solar irradiance datasets from different instruments to obtain one single composite, which is the key to understanding how irradiance varies on decadal timescales and beyond.

Keywords. Solar irradiance, Solar variability

1. Context

The solar electromagnetic spectrum and its evolution in time are paramount for understanding solar variability and for quantifying its impact on Earth's climate. Several instruments have been monitoring the total radiative output of the Sun (TSI, or total solar irradiance) and the spectrally-resolved solar irradiance (SSI, or spectral solar irradiance) since the 1970's. These observations have deeply impacted our perception of how the Sun varies in time (Ermolli *et al.* 2013, Solanki *et al.* 2013). Today there is a growing demand for understanding solar variability on timescales of decades and beyond. This is considerably longer than the lifetime of an individual instrument. Such an objective can be met only by doing data fusion, i.e. by carefully stitching together the different data sets to make one single and homogeneous composite.

2. Fragmented observations

Both the SSI and the TSI have been monitored by multiple missions. However, as Figure 1 shows, the observations are highly fragmented in time and in spectral coverage, which considerably complicates making a composite record. We are facing two practical problems here. First, as shown by Figure 1, there are frequent periods during which specific spectral bands are not observed. Such voids can be filled only by having recourse to SSI models. Conversely, there are also periods when several instruments are observing simultaneously. One then has to make the best use of all available datasets, which often differ in spectral resolution, temporal resolution, stability, etc.

Different approaches have been developed for merging simultaneous observations. There is a long history of applying such methods to combine observations of the sunspot number. This problem has recently received considerable interest because of the impact the choice of the method may have on estimates of the long-term solar variability (Clette *et al.* 2014).

One common composite-creation method is the backbone method, which relies on a small subset of overlapping time series ("backbones") that offer the longest duration of observations. Daisy-chaining is a method that consists of stitching together records

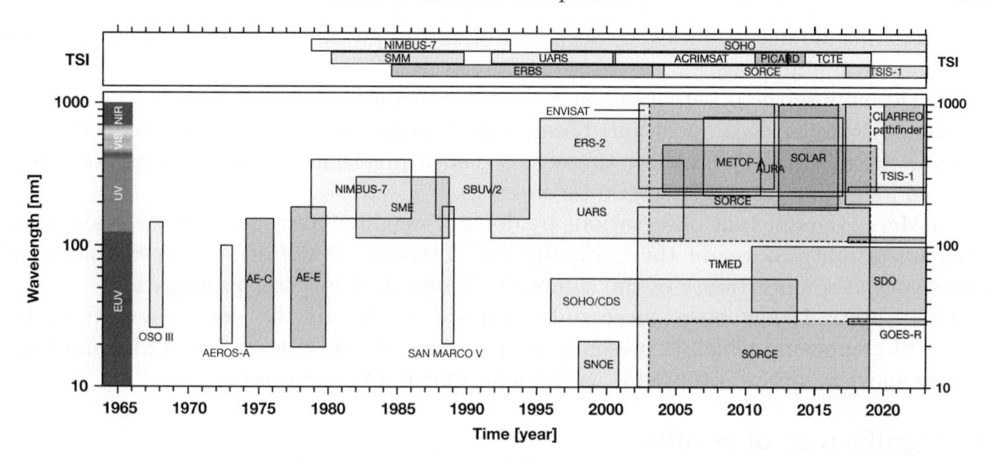

Figure 1. Overview of the main satellite missions that have been making SSI or TSI observations. The quality and resolution are highly mission-dependent.

by comparing them during an overlap period when there are at least two observers. Daisy-chaining has also been used for making solar irradiance composites, such as the intensity of the Lyman-α line (Woods *et al.* 2000), the MgII index (Viereck *et al.* 2004) and the TSI (Fröhlich 2006). In all these reconstructions, at each time step the data from one single instrument are selected, generally the "best" one. DeLand & Cebula (2008) were the first to use daisy-chaining with SSI data and came up with a composite of the UV flux between 120 and 400 nm. Because relative errors tend to be considerably larger in the SSI than in the TSI for the MgII index, the transition from one instrument to another tends to produce discontinuities that hamper the analysis of long-term variations.

A different attempt for combining SSI records was used by Haberreiter *et al.* (2017), who decided instead to combine all existing observations by weighted averaging. In addition they bypassed the daisy-chaining by doing a more global merging. The same approach has been applied to the TSI by Dudok de Wit *et al.* (2017).

3. Making the composite

Although the backbone and the daisy-chain methods are appreciated for their conceptual simplicity, they fail to meet some of the important standards of transparent and reproducible science. First, by selecting one instrument only out of several (and discarding the others) one throws away precious information. Second, the selection of the backbone or the preferred instrument is often based on subjective evaluation. In the absence of quantitative criteria for determining which observations are of the best quality, it becomes difficult to assess the uncertainty of the end product and update the latter when new data become available.

To overcome these shortcomings we developed a new statistical framework that exploits all available information and explicitly states what the assumptions are. Its main ingredient is the notion of uncertainty: each observation comes with a data-driven uncertainty that quantifies its quality. This uncertainty then allows us to assign different weights to the observations when combining them for building a composite. The Bayesian framework is ideally suited for this (Gelman *et al.* 2013). However, while we are still working on a fully Bayesian method, at this stage we consider a simpler maximum likelihood approach.

The main steps of the method are (see Dudok de Wit *et al.* 2017 for details):

(a) Determine the uncertainty of the observations by applying the same estimator to all records. This is crucial for allowing them to be meaningfully compared. Although some data sets include uncertainties, these values often cannot be compared because they rely on different assumptions.

(b) Decompose the observations and their uncertainties into different timescales (e.g. by using a wavelet decomposition) because their properties are timescale dependent. In general the uncertainty on short timescales (called precision) is better understood than that on long timescales (called stability).

(c) Merge the different observations by doing a weighted average, scale-by-scale, using their uncertainty to weight them. Finally, the composite is obtained by recombining all timescales. The uncertainty of the composite is obtained by error propagation.

This approach has been successfully applied so far to the determination of the new TSI composite, which is presently undergoing validation tests and TSI-community endorsement before we propose it as the new official TSI composite.

4. Significance of results

Our new approach for making composites by data fusion raises several methodological questions, such as the importance of distinguishing the statistical problem (What is the best way of assembling the observations to obtain a composite?) from the scientific one (What prior information may I use to correct the original data sets?).

Interestingly, we often find the uncertainties of SSI and TSI records to behave as pink noise, with a $1/f$ scaling of their power spectral density. Such a scaling is a hallmark of non-stationarity. It also means that the uncertainty of the difference between the SSI (or TSI) taken at two different dates depends on the time interval, unlike what would happen with white noise. Consequently, the popular notion of stability as a time-invariant rate is not appropriate and should be replaced by a scale-dependent uncertainty.

Finally, thanks to the possibility of estimating uncertainties at different timescales and propagating them to the end product, for the first time we are able to obtain realistic and time-dependent values of uncertainty of the composite. For the TSI, the standard deviation of the uncertainty ranges from 0.4 W/m^2 in the 1980's to less than 0.1 W/m^2 after 2000, with the improvement largely being due to newer instruments having lower noise. (Note that these are *relative* uncertainties, not *absolute* scale uncertainties.) These values may seem small, but they imply that differences between TSI levels observed during different solar minima are barely or not statistically significant. In particular, they warn us against the risk of over-interpreting weak trends.

References

Clette, F., Svalgaard, L., Vaquero, J. M., & Cliver, E. W. 2014, *SSR*, 186, 35
DeLand, M. T., & Cebula, R. P. 2008, *JGR*, 113, 11103
Dudok de Wit, T., Kopp, G., Fröhlich, C., & Schöll, M. 2017, *GRL*, 44, 1196
Ermolli, I., Matthes, K., Dudok de Wit, T. *et al.* 2013, *ACP*, 13, 3945
Fröhlich, C. 2006, *SSR*, 125, 53
Gelman, A., Carlin, J. B., Stern, H. S., Rubin, D. B., & Dunson, D. B. 2013, *Bayesian Data Analysis* (London: Chapman & Hall)
Haberreiter, M., Schöll, M., Dudok de Wit, T., *et al.* 2017, *JGR*, 122, 5910
Solanki, S. K., Krivova, N. A., & Haigh, J. D. 2013, *ARAA*, 51, 311
Viereck, R. A., Floyd, L. E., Crane, P. C. *et al.* 2004, *Space Weather*, 2, S10005
Woods, T. N., Tobiska, W. K., Rottman, G. J., & Worden, J. R. 2000, *JGR*, 105, 27195

Astronomy in Focus - XXX
Proceedings IAU Symposium No. XXX, 2018
M. T. Lago, ed.

doi:10.1017/S1743921319004551

Synoptic maps in three wavelengths of the Chromospheric Telescope

Andrea Diercke[1,2] and Carsten Denker[1]

[1]Leibniz-Institut für Astrophysik Potsdam, An der Sternwarte 16, 14482 Potsdam, Germany
email: adiercke@aip.de

[2]Universität Potsdam, Institut für Physik und Astronomie, 14476 Potsdam, Germany

Abstract. The Chromospheric Telescope (ChroTel) observes the entire solar disk since 2011 in three different chromospheric wavelengths: Hα, Ca II K, and He I. The instrument records full-disk images of the Sun every three minutes in these different spectral ranges. The ChroTel observations cover the rising and decaying phase of solar cycle 24. We started analyzing the ChroTel time-series and created synoptic maps of the entire observational period in all three wavelength bands. The maps will be used to analyze the poleward migration of quiet-Sun filaments in solar cycle 24.

Keywords. Sun: chromosphere, Sun: filaments, methods: data analysis, techniques: image processing

1. Introduction

Synoptic maps allow us to see a more global view of the Sun and enable studying large-scale relations over a longer period of time. The first synoptic maps of the Sun were created for sunspot observations (Carrington 1858). Later physical relations were determined from such synoptic maps such as Spörer's Law (Cliver 2014). Nowadays, synoptic maps are available for many physical parameters. One prominent example are the hand-drawn McIntosh synoptic maps (Gibson *et al.* 2017), which facilitate determining a relation between open magnetic structures (coronal holes) and closed magnetic structures (filaments or active regions). Another example are long-term studies of filaments with full-disk images from the Kodaikanal Observatory from 1914–2007 (Chatterjee *et al.* 2017). In the following sections, we will present synoptic maps in three different chromospheric wavelengths, which will be used to study the polarward migration of high-latitude filaments during solar cycle 24.

2. Observations

The Chromospheric Telescope (ChroTel, Kentischer *et al.* 2008; Bethge *et al.* 2011) is a full-disk imager mounted on the telecope building of the Vaccum Tower Telescope (VTT, von der Lühe 1998) at the Observatorio del Teide in Tenerife, Spain. ChroTel is a robotic 10-cm aperture telescope observing the solar chromosphere in three different wavelengths since 2011. The telescope obtains full-disk images with 2048 × 2048 pixels in Hα λ6562.8 Å, Ca II K λ3933.7 Å, and He I λ10 830 Å using narrow-band Lyot filters. At the last spectral region, the instrument observes at seven filter positions, ±3 Å around the line core of the He I red component. In principle, ChroTel contributes to a variety of scientific topics (Kentischer *et al.* 2008), e.g, the dynamic response of the chromosphere to photospheric driving or the chromospheric source of the fast solar wind. For both topics, the calculation of Doppler velocities from the spectroscopic He I data is required.

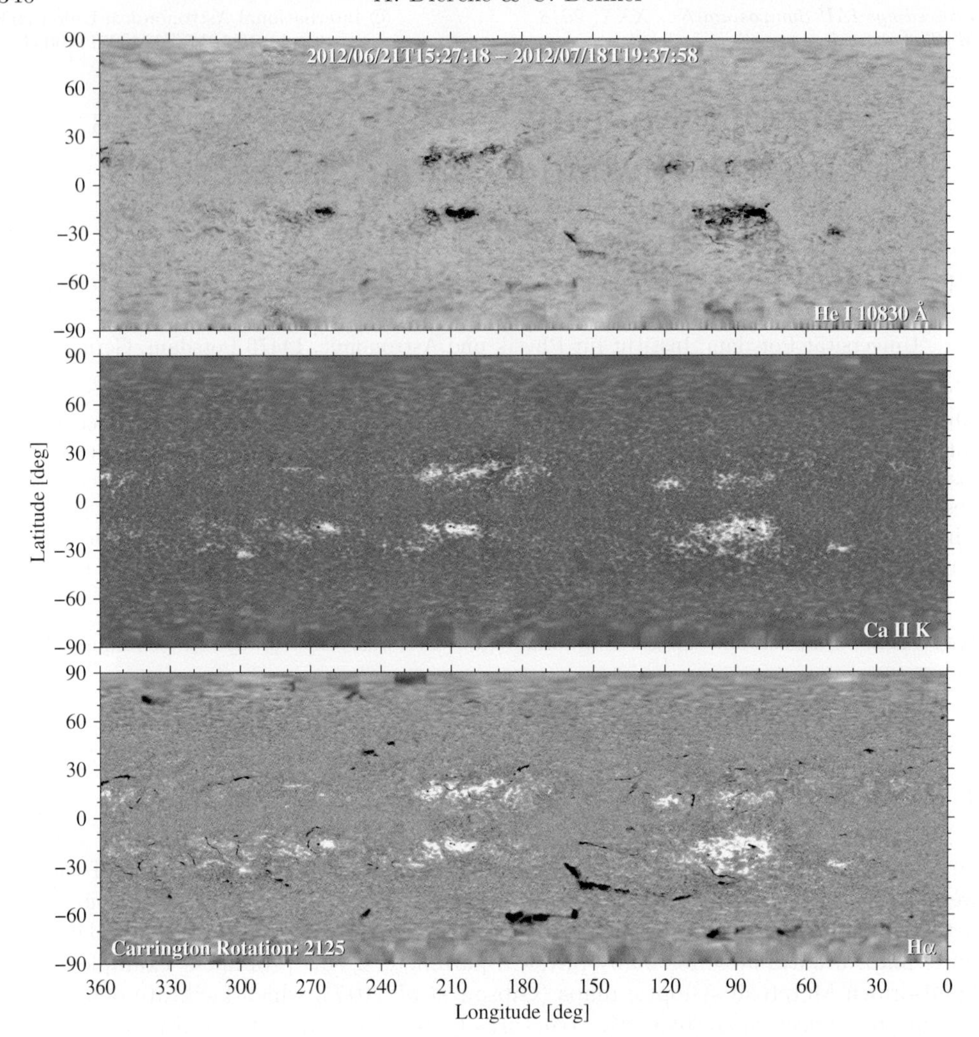

Figure 1. Synoptic map for Carrington rotation 2125 in different wavelengths: He I λ10 830 Å (line core), Ca II K, and Hα. The full-disk images were corrected for limb darkening and intensity variations, derotated to the corresponding time in the Carrington grid, and merged together to create a Carrington map with a sampling of 0.1°.

In addition, ChroTel full-disk images assist in finding large-scale structures and track the chromospheric plasma during and after eruptive events. The cadence between the images of the same filter is three minutes in the standard observing mode. Higher cadences of up to 10 s are possible, for example, during periods with enhanced flare activity.

The synoptic maps from ChroTel contain, among others, information about the number, location, area, and orientation of the filaments. The location of filaments follows the solar activity cycle. We focus in our study on large-scale filaments, polar crown filaments, and high-latitude filaments and their propagation towards the pole, which is known as "dash-to-the-pole" (Cliver 2014). Furthermore, other properties of filaments can be determined from these data such as length, width, and lifetime, which can be statistically analyzed throughout the cycle.

3. Methods

Between 2012 and October 2018, ChroTel observed the Sun in Hα on 950 days. For each day we downloaded all images in Hα and selected the best image of the day by calculating the Median Filter-Gradient Similarity (MFGS, Deng *et al.* 2015; Denker *et al.* 2018). In addition, we downloaded the closest images in time for Ca II K and He I. In the pre-processing, we rotated and rescaled the images, so that the radius corresponds to $r = 1000$ pixels, which yields an image scale of about $0.96''$ pixel^{-1}. Furthermore, all images are corrected for limb-darkening as described in Diercke *et al.* (2018). Due to intensity variation introduced by filter transmission, we had to correct all the images with a newly developed method using Zernike polynomials to compute an even background (Shen *et al.* 2018, submitted). To create the Carrington maps, one image a day was derotated to the corresponding longitudinal positions on a Carrington grid. These image slices were merged together to create a Carrington map for each rotation with a sampling of 0.1°. The process is repeated for all three wavelength bands (Fig. 1). In the Hα data we clearly recognize the filaments as elongated black structures, whereby the active regions appear bright. The map of Carrington Rotation 2125 also contains polar crown filaments (PCFs). In the He I data, the active regions and filament regions appear dark. The Ca II K line is sensitive to magnetic fields and appears bright at these locations. The filaments are not recognizable at this wavelength.

4. Future Plans

The final goal is to create a super-synoptic map with the location of the filaments to study the "dash-to-the-pole" of the PCFs in cycle 24, not only in Hα, but also in the He I triplet, as well as with full-disk Doppler maps. To extract the filament information, we will compare two methods: morphological image processing and neural networks. In addition, the statistical properties of the filaments will be scrutinized. Xu *et al.* (2018) describe solar cycle 24 as "abnormal" because of a faster poleward migration of the southern PCFs compared to other cycles. This poleward movement will be validated with the ChroTel data set.

References

Bethge, C., Peter, H., Kentischer, T. J., *et al.* 2011, Astron. Astrophys., 534, A105

Carrington, R. C. 1858, Mon. Not. R. Astron. Soc., 19, 1

Chatterjee, S., Hegde, M., Banerjee, D., & Ravindra, B. 2017, Astrophys. J., 849, 44

Cliver, E. W. 2014, Space Sci. Rev., 186, 169

Deng, H., Zhang, D., Wang, T., et al. 2015, Sol. Phys., 290, 1479

Denker, C., Dineva, E., Balthasar, H., et al. 2018, Sol. Phys., 293, 44

Diercke, A., Kuckein, C., Verma, M., & Denker, C. 2018, Astron. Astrophys., 611, A64

Gibson, S. E., Webb, D., Hewins, I. M., et al. 2017, in IAU Symposium, Vol. 328, Living Around Active Stars, ed. D. Nandy, A. Valio, & P. Petit, 93–100

Kentischer, T. J., Bethge, C., Elmore, D. F., et al. 2008, in Proc. SPIE, Vol. 7014, Ground-Based and Airborne Instrumentation for Astronomy II, ed. C. M. M. McLean, I. S., 701413

Shen, Z., Diercke, A., & Denker, C. 2018, Astron. Astrophys., submitted

von der Lühe, O. 1998, New Astron. Rev., 42, 493

Xu, Y., Pötzi, W., Zhang, H., et al. 2018, Astrophys. J., 862, L23

Astronomy in Focus - XXX
Proceedings IAU Symposium No. XXX, 2018
M. T. Lago, ed.

doi:10.1017/S1743921319004563

Cycle-dependent and cycle-independent surface tracers of solar magnetic activity

D. D. Sokoloff[1,2], V. N. Obridko[2], I. M. Livshits[2,3] and A. S. Shibalova[1]

[1]Department of Physics, Moscow State University,
119991, Moscow, Russia
email: sokoloff.dd@gmail.com; as.shibalova@physics.msu.ru

[2]IZMIRAN, 4 Kaluzhskoe Shosse, Troitsk, Moscow, 142190, Russia
email: obridko@izmiran.ru; ilivsh@gmail.com

[3]Sternberg State Astronomical Institute,
Moscow State University, 119991, Moscow, Russia

Abstract. We consider several tracers of magnetic activity that separate cycle-dependent contributions to the background solar magnetic field from those that are independent of the cycle. The main message is that background fields include two relative separate populations. The background fields with a strength up to 100 Mx cm^{-2} are very poorly correlated with the sunspot numbers and vary little with the phase of the cycle. In contrast, stronger magnetic fields demonstrate pronounced cyclic behaviour. Small-scale solar magnetic fields demonstrate features of fractal intermittent behaviour, which requires quantification. We investigate how the observational estimate of the solar magnetic flux density B depends on resolution D in order to obtain the scaling $\ln B_D = -k \ln D + a$ in a reasonably wide range. The quantity k demonstrates cyclic variations typical of a solar activity cycle. k depends on the magnetic flux density, i.e. the ratio of the magnetic flux to the area over which the flux is calculated, at a given instant. The quantity a demonstrates some cyclic variation, but it is much weaker than in the case of k. The scaling is typical of fractal structures. The results obtained trace small-scale action in the solar convective zone and its coexistence with the conventional large-scale solar dynamo based on differential rotation and mirror-asymmetric convection. Here we discuss the message for solar dynamo studies hidden in the above results.

Keywords. solar activity, solar dynamo, solar magnetic field

---

1. Introduction

Solar magnetic field underlying the famous 11-year solar activity cycle is generated by solar dynamo driven by differential rotation and mirror-asymmetric convection. This dynamo, known as the large-scale one, produces large-scale magnetic field $\mathbf{B}$ as well as small-scale one $\mathbf{b}$. Both $\mathbf{B}$ and $\mathbf{b}$ are expected to demonstrate 11-year cyclic behaviour. The point is that one more type of dynamo, so-called small-scale or turbulent dynamo, can act in the solar interior to produce cycle-independent small-scale magnetic field. The question is to what extent we can separate contributions of both type of dynamos, i.e. cycle-dependent and cycle-independent components in surface small-scale solar magnetic field.

2. Fractal properties of solar magnetic field as tracers for cyclic behaviour of small-scale field

The problem to be resolved addressing small-scale magnetic field behaviour is that smallest structures of surface solar small-scale magnetic fields cannot be resolved even by modern observations. A possible approach here is to separate small-scale magnetic fields according to its strength measured as magnetic flux per a resolved area. In the framework of this approach we demonstrate (Obridko *et al.* 2017) that the background fields with a strength up to 100 Mx cm^{-2} are very poorly correlated with the sunspot numbers and vary little with the phase of the cycle. In contrast, stronger magnetic fields demonstrate pronounced cyclic behaviour.

We identify this distinction as that one between small-scale magnetic field generated by small-scale and large-scale solar dynamo correspondingly.

Another approach is to use concepts of fractal geometry to quantify intermittent magnetograms where very weak as well as very strong magnetic fields are present (Shibalova *et al.* 2017). For this purpose we emulate magnetograms with lower resolution D rather the actual one smoothing a given magnetogram with a kernel with a size D. We investigate how the observational estimate of the solar magnetic flux density B_D at a magnetogram smoothed up to resolution D depends on D in order to obtain the fractal scaling $\ln B_D = -k \ln D + a$ in a reasonably wide range. The quantity k demonstrates cyclic variations typical of a solar activity cycle. k depends on the magnetic flux density, i.e. the ratio of the magnetic flux to the area over which the flux is calculated at a given instant. The quantity a demonstrates some cyclic variation, but it is much weaker than in the case of k.

Our interpretation of the result obtained is that we see again a coexistence of cycle-dependent and cycle-independent small-scale magnetic field components.

3. Conclusion and Discussion

We conclude that surface solar small-scale magnetic field contains cycle-dependent and cycle-independent components. The first one can be identified as a result of a large-scale dynamo action while the second one appears to be a result of a more or less independent small-scale dynamo.

We note however that this separation is far from perfect and perhaps it is more adequate to say that the above two kinds of solar dynamo are in practice quite interacting and have to be considered as two theoretical extrema of a joint process.

The research is supported by RFBR projects 18-52-06002, 18-02-00085, 17-02-00300 as well as by the Foundation for the Advancement of Theoretical Physics and Mathematics “BASIS” under grant 18-1-1-77-1.

References

Obridko, V.N., Livshits, I.M., & Sokoloff, D.D. 2017, *MNRAS*, 472, 2575

Shibalova, A.S., Obridko, V.N., & Sokoloff, D.D. 2017, *Solar Phys.*, 292, 44

Astronomy in Focus - XXX
Proceedings IAU Symposium No. XXX, 2018
M. T. Lago, ed.

doi:10.1017/S1743921319004575

General Features of Solar Cycle 24

Fulin Gursoy and Remziye Canbay

Istanbul University, Department of Astronomy and Space Sciences, Science Faculty
email: fulingursoy@gmail.com and rmzycnby@gmail.com

Abstract. In this study, using the data of Istanbul University Observatory, general features of Solar Cycle 24 are presented.

Keywords. Sun, Sunspot areas, Solar cycle

---

1. Introduction

The length of a solar cycle is about 11 years. The number of sunspot and sunspot groups on the Sun increases and decreases over time through this cycle. Solar Cycle 24 started in January 2008 after Solar Cycle 23 with an extraordinary length. Because the minimum of Solar Cycle 23 lasted a long time, the first spots of Solar Cycle 24 were at lower latitudes than normal. It is noted that the first half of the cycle has the lowest activity of the last 100 years.

2. Method-Result

In this study, Istanbul University Observatory data were used. Telescope Properties; Astrograph D=30 cm, f=150 cm and Sun telescope "Photosphere D=13 cm f=200 cm and Chromosphere D=12 cm f= 232 cm (H-alpha- Lyot Filter)" The maximum of Solar Cycle 24 was expected to be in 2013; but there was not much activity at that time. In January of 2014, a huge sunspot group was observed on the Sun. This group, which is the largest group of spots observed in Solar Cycle 24, is called "AR1944". The largest spot in AR1944 is about 2 times larger than Earth, and the entire sunspot group is about 7 times larger than Earth. This sunspot group was also observed at the Istanbul University Observatory.

One of the biggest flares of Solar Cycle 24 was observed at the Istanbul University Observatory on 25 June, 2015. It was observed that the flare started at 10:55 and ended at 13:55, with the maximum amount of flare coverage being an area of about 20 Earths wide on the surface of the Sun.

Solar Cycle 24 has been a very weak cycle. It reached its maximum in April 2014. The extensions of Solar Cycle 24 in the northern and southern hemispheres seem to be compatible because they are seen at similar latitudes and times. Our observational results were affected because of some technical disruptions in the telescope of the Istanbul University Observatory.

References

This research was made help of Istanbul University Observatory Sunspot Observations.
https://solarscience.msfc.nasa.gov/images/Zurichal Color Small.jpg
https://science.nasa.gov/science-news/science-at-nasa/2008/10jan solarcycle24
http://astronomi.istanbul.edu.tr/images/news/parlama-20150625.jpg
http://astronomi.istanbul.edu.tr/gozlemarsiv/arsiv/2014/20140109.JPG
https://solarscience.msfc.nasa.gov/predict.shtml

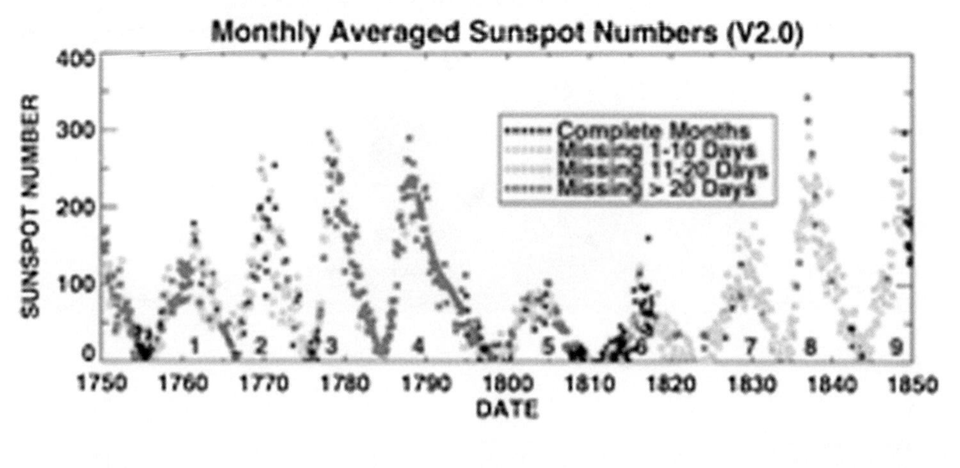
Monthly Averaged Sunspot Numbers (V2.0)
Complete Months
Missing 1-10 Days
Missing 11-20 Days
Missing > 20 Days
SUNSPOT NUMBER
DATE

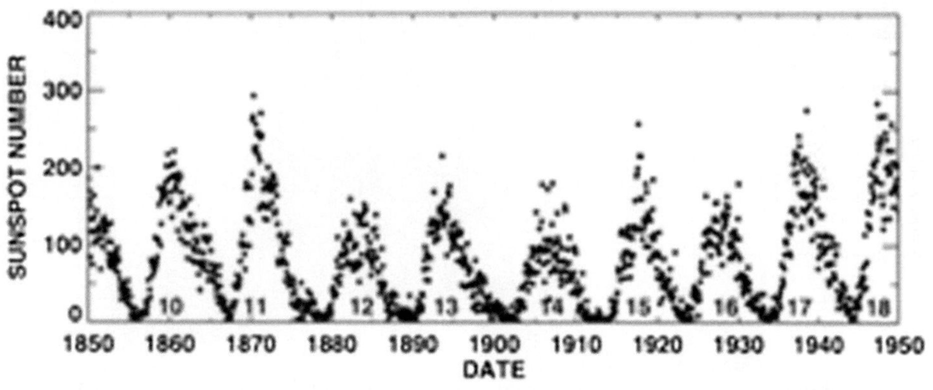
SUNSPOT NUMBER
DATE

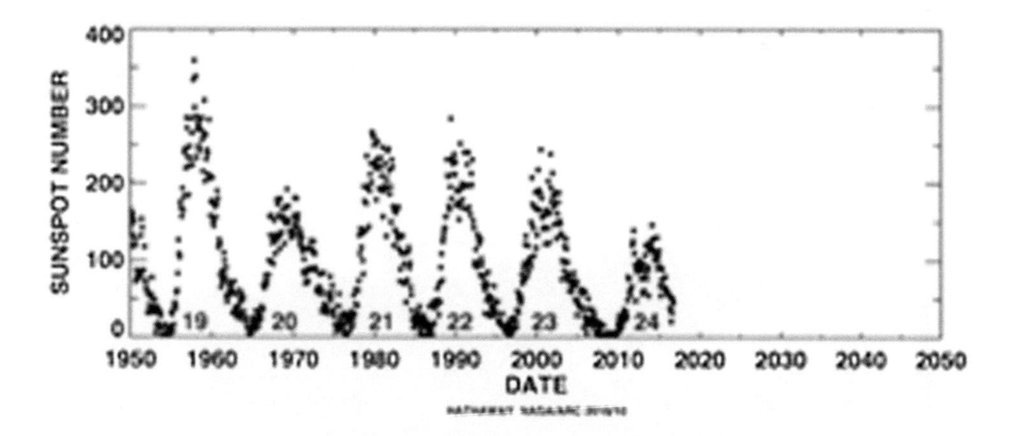
SUNSPOT NUMBER
DATE

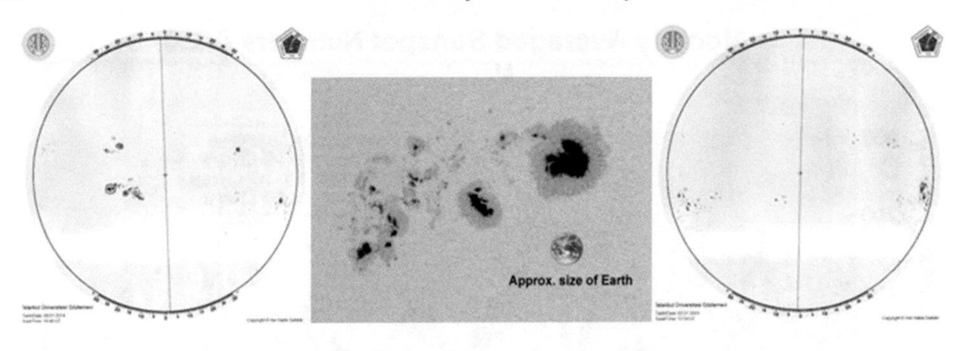
Approx. size of Earth

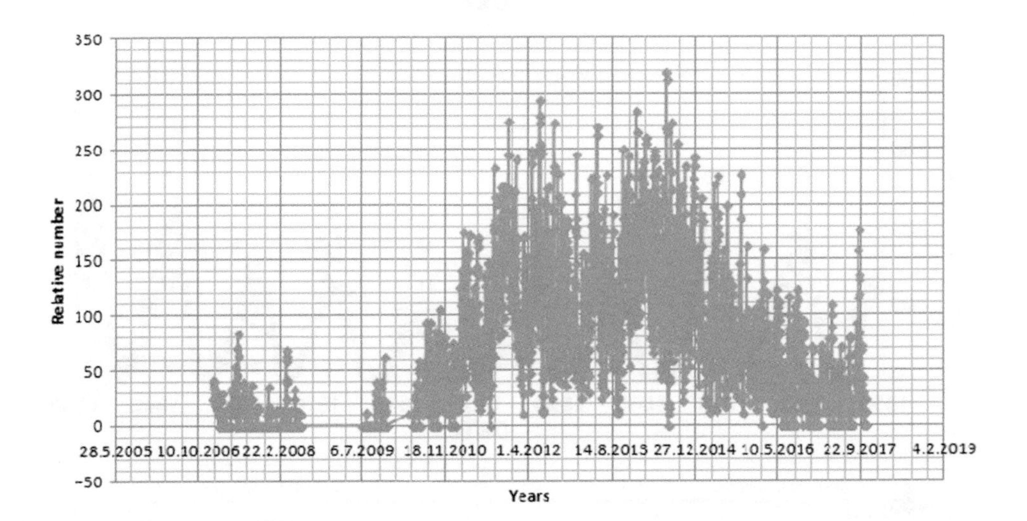
350
300
250
200
150
100
50
0
-50
Relative number
28.5.2005 10.10.2006 22.2.2008 6.7.2009 18.11.2010 1.4.2012 14.8.2013 27.12.2014 10.5.2016 22.9.2017 4.2.2019
Years

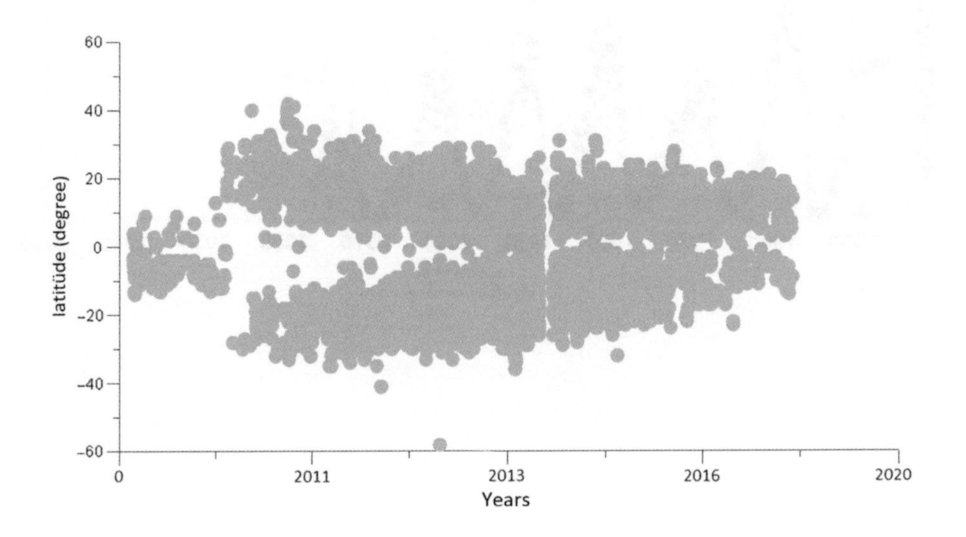
60
40
20
0
-20
-40
-60
latitüde (degree)
0
2011
2013
2016
2020
Years

Astronomy in Focus - XXX
Proceedings IAU Symposium No. XXX, 2018
M. T. Lago, ed.

doi:10.1017/S1743921319004587

Large-scale transport of solar and stellar magnetic flux

Emre Işık[1,2]

[1]Max-Planck-Institut für Sonnensystemforschung, 37077, Göttingen, Germany
[2]Feza Gürsey Center for Physics and Mathematics, Boğaziçi University, 34684 Istanbul, Turkey
email: isik@mps.mpg.de

Abstract. Surface flux transport (SFT) models have been successful in reproducing how magnetic flux at the solar photosphere evolves on large scales. SFT modelling proved to be useful in reconstructing secular irradiance variations of the Sun, and it can be potentially used in forward modelling of brightness variations of Sun-like stars. We outline our current understanding of solar and stellar SFT processes, and suggest that nesting of activity can play an important role in shaping large-scale patterns of magnetic fields and brightness variability.

Keywords. Sun: activity, Sun: magnetic fields, stars: magnetic fields

---

1. Introduction

The transport of magnetic flux on the Sun is one of the few astrophysical phenomena that occurs in front of our eyes, thanks to synoptic observations of the line-of-sight magnetic field. At a given solar cycle, bipolar magnetic regions (BMRs) emerge with their dipole-moments showing a mean tilt angle with respect to the east-west direction, such that the leading polarities are closer to the equator than the trailing ones. The tilt angle exhibits a large scatter around its mean value of a few degrees. Following their emergence, BMRs are subject to differential rotation, meridional flow, and turbulent convective motions that disperse magnetic flux elements in a random-walk fashion, which is generally modelled as a two-dimensional diffusion problem (Leighton 1964, but see also Schrijver 2001). The diffusion occurs at the convective length scales of supergranulation, at a rate of about 250 km^2 s^{-1}. For extensive reviews of SFT we refer the reader to Mackay & Yeates (2012) and Jiang *et al.* (2014a).

2. Flux transport on the Sun

The main motivation for SFT modelling has been to understand the surface physics relevant to the magnetic butterfly diagram of azimuthally averaged radial magnetic field. SFT has also been useful in constraining the boundary conditions of flux-transport dynamo models (Muñoz-Jaramillo *et al.* 2010, Cameron *et al.* 2012), as well as generating synthetic irradiance variations for the past solar activity (Dasi-Espuig *et al.* 2016), coronal field extrapolations (e.g. Nandy *et al.* 2018), and the evolution of the open magnetic flux shaping the heliosphere (Jiang *et al.* 2011).

When data-driven SFT models are compared with observations, several observed features are reproduced, such as the poleward plumes of signed magnetic flux, which eventually reverse the polar fields. Some missing pieces of the model have been incorporated recently, such as the observed tilt angle scattter (Jiang *et al.* 2014b) and inflows around active regions (Jiang *et al.* 2010, Cameron & Schüssler 2012). In rare occasions, BMRs can emerge across the equator, or with a negative or abnormally large tilt angle

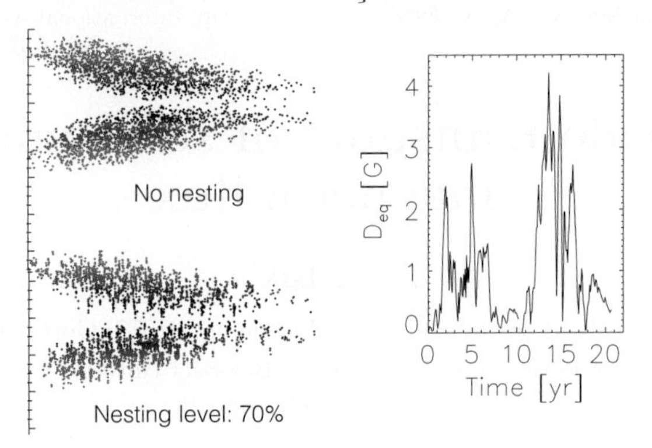

Figure 1. Cycle variation of the equatorial dipole moment for random BMR longitudes (first decade, blue; right panel), and for two-dimensional nesting (second decade, red) with a probability of 70%, each corresponding to the relevant butterfly diagram on the left panel.

nearby the equator. As Cameron *et al.* (2013) demonstrated, one such event can change the hemispheric magnetic flux by 60% (see also Jiang *et al.* 2015).

In an SFT model based on random-walk flux dispersal, Schrijver (2001) considered nesting of emerging BMRs for individual flux emergence events, with a probability of 40%. To demonstrate the importance of nesting in shaping the large-scale field geometry, we show in Fig. 1 a comparison of the global equatorial dipole moment, $D_{\rm eq}$ for two SFT simulations, over an activity cycle. As opposed to random BMR longitudes, a strong degree of nesting (70%) in both longitude and latitude amplifies the fluctuations of $D_{\rm eq}$, reaching levels comparable to observations (Wang 2014). This is because the probability for a nonzero equatorial dipole-moment contribution from ensembles of nested BMRs from both hemispheres becomes higher when BMRs tend to emerge into active nests.

3. Flux transport on late-type stars

Application of the SFT model to other cool stars is useful to better assess possible drivers of stellar brightness and spectral variability. As an example, a long-standing question concerning the morphology of starspots is whether the observed starspots are monolithic or conglomerates. As demonstrated by Işık *et al.* (2007), the corresponding spot lifetimes can be very different in these two cases, which should be considered when interpreting observational relationships between spot sizes and lifetimes. Another example is the effect of meridional flow speeds that are much faster than solar values, leading to strong polar fields with intermingled polarities on rapidly rotating active stars (Holzwarth *et al.* 2006).

Later, Işık *et al.* (2011) developed a model combining a deep-seated $\alpha\Omega$ dynamo, flux-tube rise and SFT to demonstrate how surface fields evolve over cycle timescales. For stars of type G2V, K0V, and K1IV, they showed that the observed surface variations can be very different from the internal dynamo, owing to a combination of rotational effects on rising flux, convection-zone geometry and SFT. Scaling an SFT model with coronal feedback to higher levels of activity, Lehmann *et al.* (2018) found a good match between their simulations and the observed 3D geometry of large-scale magnetic fields on fast-rotating Sun-like stars.

The effect of nesting considered in Sect. 2 can be very important for stellar variability in rotational timescales. A comprehensive simulation framework including this effect has been developed by Işık *et al.* (2018), who presented SFT models driven by a Sun-like butterfly diagram at the base of the convection zone, and the emergence latitudes and tilt angles calculated using flux-tube simulations. Figure 2 shows pole-on snapshots of radial

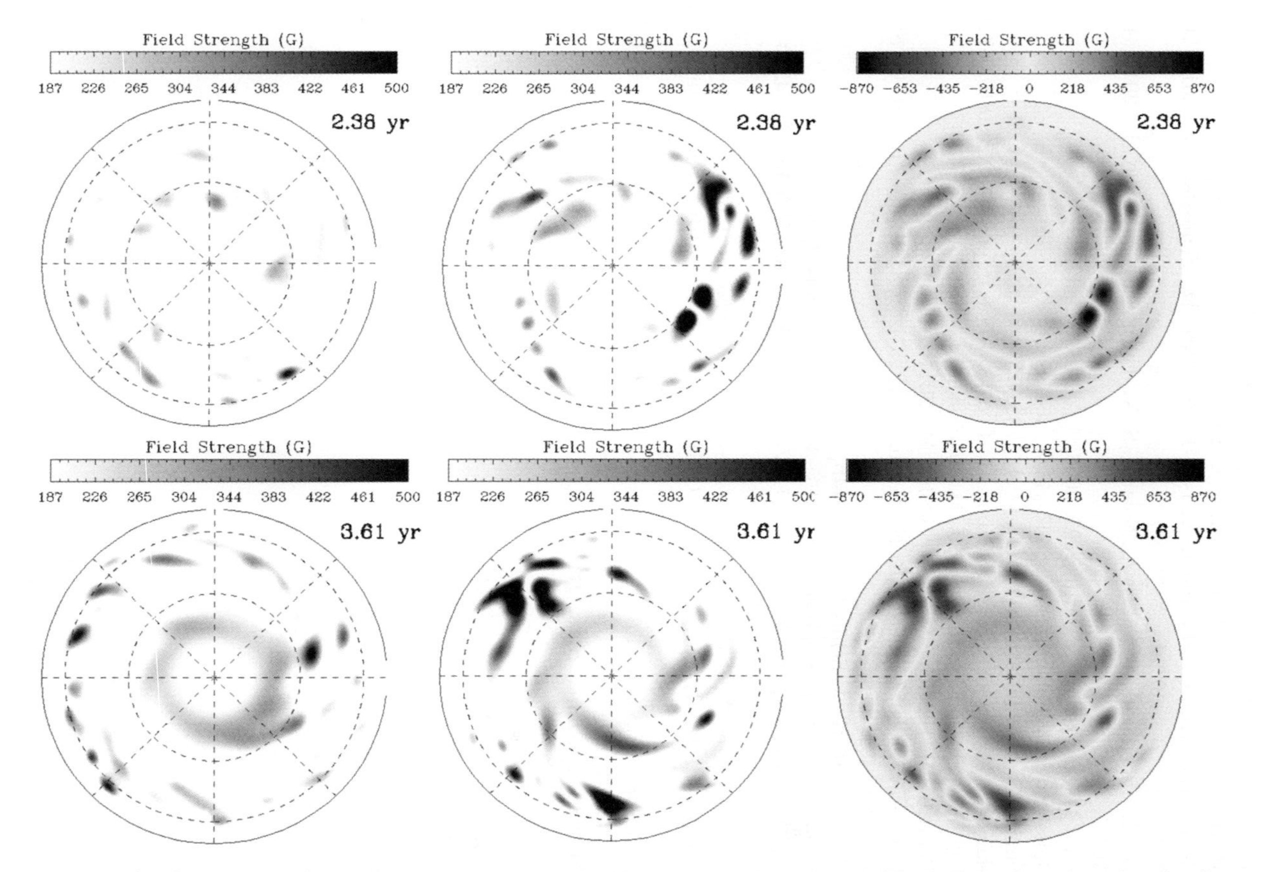

Figure 2. Pole-on views of a Sun-like model with a rotation rate and BMR emergence rate eight times higher than for the Sun, at two different phases of the activity cycle. First and second columns show unsigned field above a threshold to represent starspots, without and with nesting as in Fig. 1, respectively. The last column shows the signed-field snapshots corresponding to those on the second column.

magnetic field for a Sun-like star rotating eight times faster and more active than the Sun. With nesting, the spot distributions become highly non-axisymmetric as compared to random longitudes. This can affect the rotational modulation of activity indicators.

References

Cameron, R. H., Schmitt, D., Jiang, J., Işık, E. 2012, *A&A* 542, A127

Cameron, R. H. & Schüssler, M. 2012, *A&A*, 548, A57

Cameron, R. H., Dasi-Espuig, M., Jiang, J., Işık, E., Schmitt, D., Schüssler, M. 2013, *A&A* 557, A141

Dasi-Espuig, M., Jiang, J., Krivova, N. A., Solanki, S. K., Unruh, Y. C., Yeo, K. L. 2016, *A&A* 590, A63

Holzwarth, V., Mackay, D. H., & Jardine, M. 2006, *MNRAS*, 369, 1703

Işık, E., Schüssler, M., & Solanki, S. K. 2007, *A&A*, 464, 1049

Işık, E., Schmitt, D., & Schüssler, M. 2011, *A&A*, 528, A135

Işık, E., Solanki, S. K., Krivova, N. A., Shapiro, A. I. 2018, *A&A*, in press, arXiv:1810.06728

Jiang, J., Işık, E., Cameron, R. H., Schmitt, D., Schüssler, M. 2010, *ApJ*, 717, 597

Jiang, J., Cameron, R. H., Schmitt, D., Schüssler, M. 2011, *A&A*, 528, A83

Jiang, J., Hathaway, D. H., Cameron, R. H., Solanki, S.K., Gizon, L., Upton, L. 2014a, *Space Sci. Revs.* 186, 491

Jiang, J., Cameron, R. H., & Schüssler, M. 2014b, *ApJ* 791, 5

Jiang, J., Cameron, R. H., & Schüssler, M. 2015, *ApJ* (Letters), 808, L28

Lehmann, L. T., Jardine, M. M., Mackay, D. H., Vidotto, A. A. 2018, *MNRAS*, 478, 4390

Leighton, R. B. 1964, *ApJ* 140, 1547

Mackay, D. H., & Yeates, A. R. 2012, *Living Reviews in Solar Physics* 9, 6

Muñoz-Jaramillo, A., Nandy, D., Martens, P. C. H., Yeates, A. R. 2010, *ApJ* (Letters), 720, L20

Nandy, D., Bhowmik, P., Yeates, A. R., Panda, S., Tarafder, R., Dash, S. 2018, *ApJ*, 853, 72

Schrijver, C. J. 2001, *ApJ*, 547, 475

Wang, Y.-M. 2014, *Space Sci. Revs.*, 186, 387

Astronomy in Focus - XXX
Proceedings IAU Symposium No. XXX, 2018
M. T. Lago, ed.

doi:10.1017/S1743921319004599

Monitoring solar activity with PEPSI

Ekaterina Dineva[1,2], Carsten Denker[1], Klaus G. Strassmeier[1], Ilya Ilyin[1] and Alexei A. Pevtsov[3]

[1]Leibniz Institute for Astrophysics Potsdam (AIP), An der Sternwarte 16, 14882 Potsdam, Germany

[2]Institute for Physics and Astronomy, University of Potsdam, Karl-Liebknecht-Str. 24/25, 14476 Potsdam-Golm, Germany

[3]National Solar Observatory, 3665 Discovery Drive, Boulder, CO 80303, U.S.A.

Abstract. Synoptic Sun-as-a-star observations are carried out with the Potsdam Echelle Polarimetric and Spectroscopic Instrument (PEPSI), which receives light from the Solar Disk-Integration (SDI) telescope. Daily spectra are produced with a high signal-to-noise ratio, providing access to unprecedented quasi-continuous, long-term, disk-integrated spectra of the Sun with high spectral and temporal resolution. We developed tools to monitor and study solar activity on different time-scales ranging from daily changes, over periods related to solar rotation, to annual and decadal trends. Strong chromospheric absorption lines, such as the Ca II H & K λ3934 & 3968 Å lines, are powerful diagnostic tools for solar activity studies, since they trace the variations of the solar magnetic field. Other lines, such as Hα λ6563 Å line and the near-infrared (NIR) Ca II λ8542 Å line, provide additional information on the physical properties in this highly complex and dynamic atmospheric layer. Currently, we work on a data pipeline for extraction, calibration, and analysis of the PEPSI/SDI data. We compare the SDI data with daily spectra from the Integrated Sunlight Spectrometer (ISS), which is part of the Synoptic Long-Term Investigation of the Sun (SOLIS) facility operated by the U.S. National Solar Observatory (NSO). This facilitates cross-calibration and validation of the SDI data.

Keywords. Sun: chromosphere, Sun: activity, instrumentation: spectrographs, techniques: spectroscopic, methods: data analysis

---

1. Introduction

For decades scientists monitor one important aspect of our Sun, *i.e.*, the Sun's activity and magnetic cycle, by measuring the magnetic field strength or by building time-series of various indices serving as indications of the solar activity and magnetism (Keil *et al.* 1998; Livingston *et al.* 2007; Bertello *et al.* 2010). There is a well established 11-year activity cycle: part of the 22-year Hale cycle, *i.e.*, the Sun's magnetic cycle. Its existence was first recognized from observations of sunspot appearance, number, and migration. Active regions mark location on the solar surface with strong magnetic fields. These fields are rooted in the solar interior, where they are shaped and guided by dynamo processes. The magnetic fields facilitate the energy transport through solar atmosphere, provide energy and serve as a trigger for powerful solar flares and eruptions. Thus, solar magnetism is the force shaping many aspects of the solar system environment, *i.e.*, the heliosphere.

Solar activity monitoring provides the much-needed background for theoretical models explaining the solar and stellar dynamo and as input for planning and operating space missions. Many diagnostic tools were developed to support this effort, *e.g.*, magnetic field extrapolations based on synoptic full-disk vector magnetograms, precise measurements and predictions of the solar radio, X-ray, and particle fluxes, and classification of the active regions, sunspots, and eruptive or energetic events (Ermolli *et al.* 2014). Using spectroscopic proxies to study the variations of the global solar magnetic field is an integral part of this effort to characterize the solar cycle. In this work, we introduce preliminary results of calibration and evaluation procedures, which later will ensure the precise derivation of the activity indices based on PEPSI/SDI Sun-as-a-star observations.

The Ca II H & K lines are powerful tools to examine the solar atmosphere's stratification, using their formation height and magnetic sensitivity. Magnetic heating introduced in plage and network regions cause emissions in the cores of these strong chromospheric absorption lines. An important feature is the sensitivity of the line-core intensity to variations of the magnetic fields. This aspect was used by Pevtsov *et al.* (2016) to reconstruct long-term magnetic variations. Other chromospheric lines such as the He I NIR triplet, the Ca II NIR line, and the Hα line show a similar behavior but less pronounced. Hα is a good tracer of the development and evolution of, for example, filaments and prominences.

2. Observations and standard data reduction

PEPSI is a state-of-the-art, thermally stabilized, fiber-fed, high-resolution spectrograph for the 11.8-meter (light-gathering ability) Large Binocular Telescope (LBT) at Mt. Graham, Arizona. Typically the LBT with its large light-gathering power feeds starlight to PEPSI. However, the spectrograph can also receive sunlight from the SDI telescope. The observed spectra contain a multitude of photospheric and chromospheric spectral lines in the wavelength range of 380–910 nm. The spectral resolution is $\mathcal{R} = 250\,000$, and an average exposure time 0.3 s. The signal-to-noise ratio varies between 2 000:1 and 8 000:1 depending on the wavelength. The standard data reduction steps of the spectra includes dark and flat-field corrections, scattered light subtraction, wavelength and flux calibration, etc. A detailed description of the PEPSI data reduction pipeline is given in Strassmeier *et al.* (2018). In 2018, PEPSI/SDI started the routine daily observations. However, the data for a comparative study discussed in this paper were obtained during two separate campaigns in 2015 and 2016.

3. Software development and data reduction specific to SDI

Line profiles in the desired wavelength region are retrieved from PEPSI FITS files using the IDL software repository SolarSoft (Freeland & Handy 2012). The Optical Solar Physics research group at AIP actively develops an IDL software library 'sTools' (Kuckein *et al.* 2017) for high-resolution imaging and spectropolarimetry. Initially intended for the data reduction of the post-focus instruments at the 1.5-meter GREGOR solar telescope (Schmidt *et al.* 2012), the scope of sTools was expanded to include other instruments such as PEPSI/SDI.

In order to calculate reliable activity indices, we need a precise wavelength and intensity calibration. Thus, we developed a method for fine tuning the continuum calibration of the PEPSI/SDI spectra. In the first step, we align and add the all spectra taken over the course of a single day. On average PEPSI/SDI takes approximately 200 single-exposure spectra per day. Comparing this daily average profile to the NSO Fourier Transform Spectral (FTS, Kurucz *et al.* 1984) flux atlas, we determine accurate continuum levels. We are using the ratio between PEPSI/SDI daily average profile and FTS atlas to correct the continuum if necessary. This task include the following steps: (1) we compute the ratio of the SDI and FTS spectra, which is then decomposed into its Fourier components; (2) the ratio is restored for a chosen (smaller) number of Fourier coefficients, which strongly depend on line profile shape; (3) the restored ratio is applied to the SDI spectrum, thus, providing the proper continuum correction. For validation, single and average spectra are compared to spectra from the same day but obtained with SOLIS/ISS (Keller *et al.*

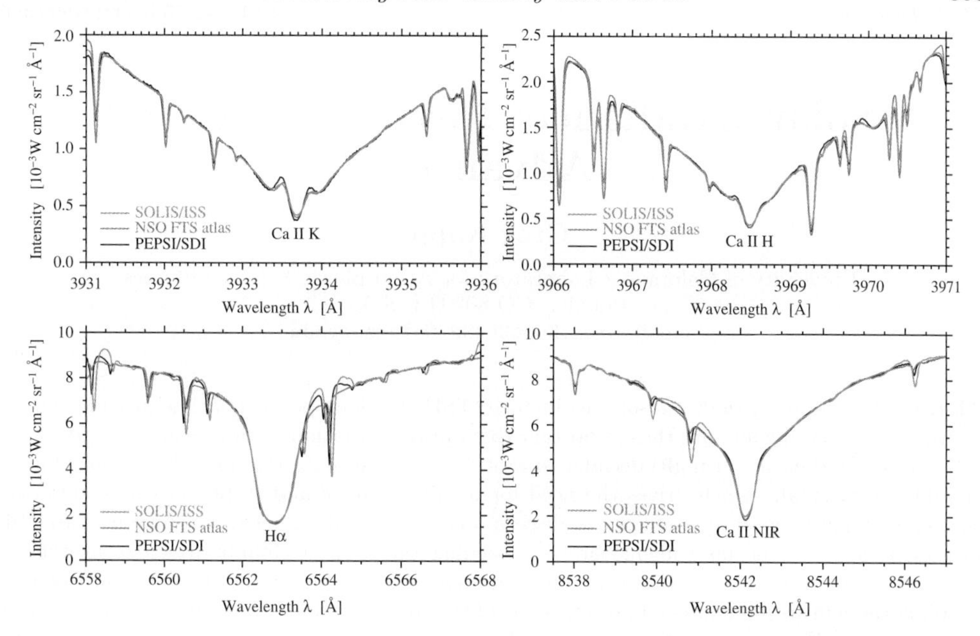

Figure 1. Sun-as-a-star spectra of the Ca II H & K lines (*top*) and the Hα & Ca II NIR lines, commonly referred to as the infrared triplet (*bottom*). The PEPSI/SDI (*black*) and SOLIS/ISS (*blue*) spectra on 2016 November 17 are compared to the NSO FTS spectral atlas (*red*).

2003). The results are presented in Fig. 1. In general, we see a relatively good agreement between PEPSI/SDI and SOLIS/ISS spectra in wings of selected spectral lines. Some differences are present in the cores of strong lines. These processed PEPSI/SDI spectral profiles are ready for further analysis, including comparing Sun-as-a-star spectra with spatially resolved solar images linking spectral features to various solar phenomena.

References

Bertello, L., Ulrich, R. K., & Boyden, J. E. 2010, *Sol. Phys.*, 264, 31

Ermolli, I., Shibasaki, K., Tlatov, A., & van Driel-Gesztelyi, L. 2014, *Space Sci. Rev.*, 186, 105

Freeland, S. L. & Handy, B. N. 2012, SolarSoft: Programming and Data Analysis Environment for Solar Physics, Astrophysics Source Code Library

Keil, S. L., Henry, T. W., & Fleck, B. 1998, in *ASP Conf. Ser.*, Vol. 140, Synoptic Solar Physics, eds. K. S. Balasubramaniam, J. Harvey, & D. Rabin, 301

Keller, C. U., Harvey, J. W., & Giampapa, M. S. 2003, in *Proc. SPIE*, Vol. 4853, Innovative Telescopes and Instrumentation for Solar Astrophysics, eds. S. L. Keil & S. V. Avakyan, 194–204

Kuckein, C., Denker, C., Verma, M., *et al.* 2017, in *IAU Symp.*, Vol. 327, Fine Structure and Dynamics of the Solar Atmosphere, eds. S. Vargas Domínguez, A. G. Kosovichev, L. Harra, & P. Antolin, 20–24

Kurucz, R. L., Furenlid, I., Brault, J., & Testerman, L. 1984, Solar Flux Atlas from 296 to 1300 nm (Sunspot, New Mexico: National Solar Observatory)

Livingston, W., Wallace, L., White, O. R., & Giampapa, M. S. 2007, *Astrophys. J.*, 657, 1137

Pevtsov, A. A., Virtanen, I., Mursula, K., Tlatov, A., & Bertello, L. 2016, *Astron. Astrophys.*, 585, A40

Schmidt, W., von der Lühe, O., Volkmer, R., *et al.* 2012, *Astron. Nachr.*, 333, 796

Strassmeier, K. G., Ilyin, I., & Steffen, M. 2018, *Astron. Astrophys.*, 612, A44

Astronomy in Focus - XXX
Proceedings IAU Symposium No. XXX, 2018
M. T. Lago, ed.

doi:10.1017/S1743921319004605

Solar Irradiance: Instrument-Based Advances

Greg Kopp

University of Colorado / Laboratory for Atmospheric & Space Physics
Boulder, CO 80303 U.S.A.
email: Greg.Kopp@LASP.Colorado.edu

Abstract. Variations of the total solar irradiance (TSI) over long periods of time provide natural Earth-climate forcing and are thus important to monitor. Variations over a solar cycle are at the 0.1 % level. Variations on multi-decadal to century timescales are (fortunately for our climate stability) very small, which drives the need for highly-accurate and stable measurements over correspondingly long periods of time to discern any such irradiance changes. Advances to TSI-measuring space-borne instruments are approaching the desired climate-driven measurement accuracies and on-orbit stabilities. I present a summary of the modern-instrument improvements enabling these measurements and present some of the solar-variability measurement results from recent space-borne instruments, including TSI variations on timescales from solar flares and large-scale convection to solar cycles.

Keywords. Sun: irradiance, TSI, variability; instrumentation: radiometry

---

1. Climate-Driven Requirements of Solar-Variability Measurements

Solar forcing is one of the natural influences of climate change on the Earth. While the climate system cannot respond quickly to short-term solar variability, changes over the 11-year solar cycle are evident in climate records, causing radiative forcings of $\sim 0.1°$C (Lean 2017), or a climate sensitivity of $\sim 0.6\ C\ W^{-1}\ m^2$. Longer-term changes, which allow the Earth's climate system more time to equilibrate, should have at least as great a radiative-forcing sensitivity. Detecting small, potential solar variations over multi-decadal to century timescales drives stringent total solar irradiance (TSI) accuracy and stability requirements. Measuring such variability requires the uncertainties shown in Table 1.

2. Instrument Improvements Enabling Required Solar-Variability Measurements

All TSI instruments prior to the SORCE Total Irradiance Monitor (TIM), launched in 2003, measured erroneously high values largely because of internal instrument scatter due to aperture placement (Kopp & Lean 2011). Those instruments had a large view-limiting aperture at the front of each radiometer, which allowed two to three times the amount of sunlight intended to be collected into the interior on all non-TIM instruments. Ground-based estimates have now been applied to correct most of the afflicted instruments.

Including the scatter-eliminating aperture layout, the TIM introduced several innovations to improve TSI-measurement accuracy:

- Precision apertures located at front of instrument
- Nickel Phosphorus black absorptive cavity interiors demonstrate the best on-orbit stability of any TSI instrument
- Phase sensitive detection reduces sensitivity to out-of-band noise in the instrument's servo system and in ground-based data processing

Table 1. Climate-Driven TSI Measurement Requirements

Parameter	Requirement
Radiometric Uncertainty (Accuracy)	10^{-4}
Stability (Long-term Precision)	10^{-5} yr^{-1}
Noise (Short-term Precision)	$<10^{-5}$

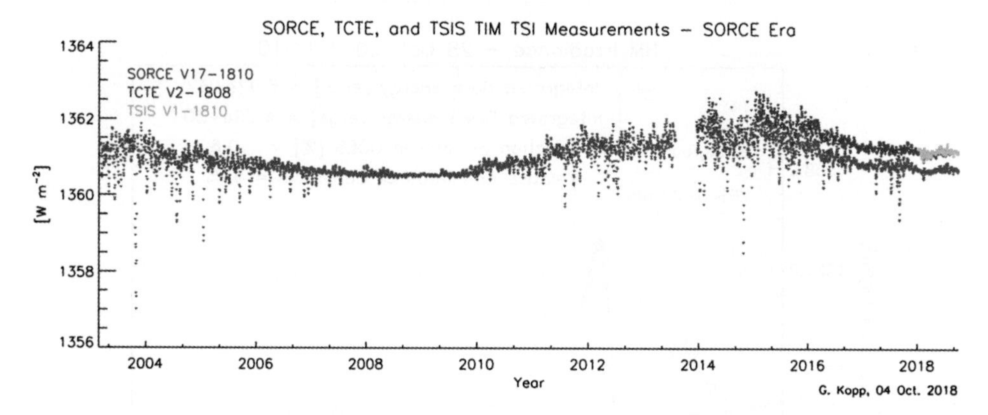

Figure 1. TIM TSI instruments show calibration consistency over many years

- Feedforward maintains cavity thermal stability and servo balance
- The instrument's thermal background is both measured on-orbit and modeled
- Pulse-width modulation of precision DC constant-voltage references applies power linearly (rather than as V^2, as is the case in varying-voltage references)
- Digital servo system with proportional-integral-derivative control allows servo tuning

Three on-orbit TIMs calibrated independently over a 15-year span show good consistency (see Fig. 1). Many of these TIM-introduced improvements are now being incorporated into other new instruments.

3. Solar-Variability Magnitudes and Timescales

The instrument improvements described above enabled the first solar-flare detection by a TSI instrument thanks to a combination of the SORCE/TIM's low noise and a large flare (Woods *et al.* 2006). While flares are easily detected at ultraviolet and x-ray wavelengths, where the solar disk is very dark, they are nearly insignificant in terms of net contribution to the total radiative solar energy, making them difficult to discern in TSI above the ever-present variations of 0.005 % to 0.01 % due to globally-averaged solar convection and oscillations. The value of a flare detection in the TSI is that the time-integrated TSI enables an estimate of the net radiant flare energy, as shown in Fig. 2.

Low instrument noise also enables observations of Venus and Mercury transits, as shown in Fig. 3, where the effects of limb darkening during the transits are evident. Observations (red) show intrinsic short-term solar variability on 3- to 5-minute timescales due to convection and oscillations. These transit observations are representative of exoplanet transits of Sun-like stars and the potential obscuring effects of stellar variability.

The TSI varies on all timescales over which it has been observed. Kopp (2014) discusses these timescales and variabilities, summarized here in Table 2. Figs. 2 and 3 show solar variability on short timescales and Fig. 1 shows it on daily to solar-cycle timescales.

Table 2. Solar-Variability Magnitudes and Timescales

Timescale	Magnitude
minutes	0.01%
days	<0.3%
11-yr solar cycle	0.1%
multi-decadal to centuries	0.05-0.3% (unknown)
stellar evolution	10^{-10} yr^{-1}

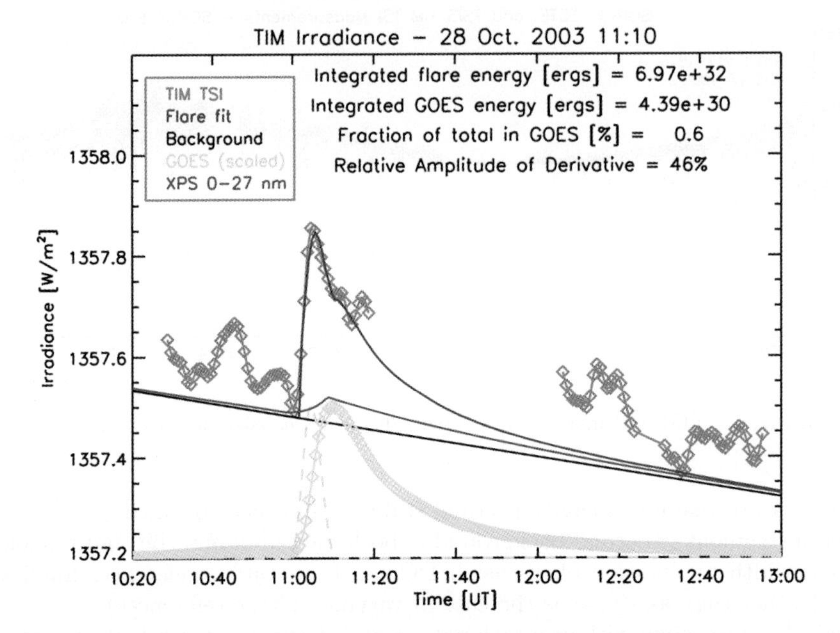

Figure 2. The first solar flare ever detected in TSI (red) enabled flare-energy estimates.

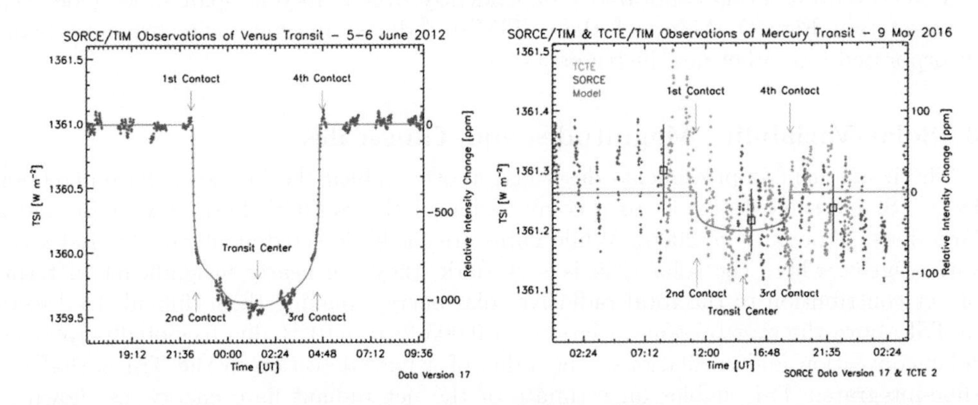

Figure 3. Venus (left) and Mercury (right) transits observed by the SORCE/TIM and TCTE/TIM. Observations (red and green) closely match predictions including limb-darkening effects (grey). The Mercury transit was not significant enough to be considered a "detection."

4. Conclusions

I describe recent TSI-instrument improvements. These have enabled more accurate and stable measurements of the solar variability on short (solar flare and convection/oscillation) and long (daily to solar-cycle) timescales. I summarize the magnitudes and timescales of measured solar variability benefitting from these improved TSI instruments.

References

Kopp, G. & Lean, J.L. 2011, *Geophys. Res. Letters*, **38**, L01706, doi:10.1029/2010GL045777
Kopp, G. 2014, *J. Sp. Weather & Sp. Climate*, **4**, A14, doi:10.1051/swsc/2014012
Kopp, G. 2016, *J. Sp. Weather & Sp. Climate*, **6**, A30, doi:10.1051/swsc/2016025
Lean, J.L. 2017, *Oxford Res. Encycl. Climate Sci.*, doi:10.1093/acrefore/9780190228620.013.9
Woods, T.N., Kopp, G., & Chamberlin, P.C. 2006, *J. Geophys. Res.*, **111**, A10S14, doi:10.1029/2005JA011507

Astronomy in Focus - XXX
Proceedings IAU Symposium No. XXX, 2018
M. T. Lago, ed.

doi:10.1017/S1743921319004617

First TSI results and status report of the CLARA/NorSat-1 solar absolute radiometer

Benjamin Walter[1], Bo Andersen[2], Alexander Beattie[3] Wolfgang Finsterle[1], Greg Kopp[4], Daniel Pfiffner[1] and Werner Schmutz[1]

[1]Physikalisch Meteorologisches Observatorium Davos and World Radiation Center (PMOD/WRC), Davos, Switzerland
email: benjamin.walter@pmodwrc.ch

[2]Norwegian Space Center, Oslo, Norway

[3]Space Flight Laboratory, Institute for Aerospace Studies, University of Toronto, Canada

[4]Laboratory for Atmospheric and Space Physics, Boulder, Colorado, USA

Abstract. The Compact Lightweight Absolute Radiometer (CLARA) is orbiting Earth on-board the Norwegian NorSat-1 micro-satellite since 14th of July 2017. The first light total solar irradiance (TSI) measurement result of CLARA is 1360.18 W m^{-2} for the so far single reliable Channel B. Channel A and C measured significantly lower (higher) TSI values and were found being sensitive to satellite pointing instabilities. These channels most likely suffer from electrical interference between satellite components and CLARA, an effect that is currently under investigation. Problems with the satellite attitude control currently inhibit stable pointing of CLARA to the Sun.

Keywords. Absolute radiometer, Sun, Total Solar Irradiance

---

1. Introduction

The Compact Lightweight Absolute Radiometer (CLARA), built by the Physikalisch-Meteorologisches Observatorium Davos and World Radiation Center (PMOD/WRC) in Davos, Switzerland, is an electrical substitution radiometer (ESR, e.g. Brusa & Fröhlich 1986 or Schmutz *et al.* 2013) based on a new three-cavity design. It is small and lightweight, thus suitable for flying on low-cost micro satellites. CLARA is one of three payloads on the Norwegian micro satellite NorSat-1, which was launched on the 14th of July 2017 from Baikonour, Kasachstan.

CLARA includes several innovations compared to the previous generation of PMO6-type (Brusa & Fröhlich 1986) radiometers built at PMOD/WRC: i) Three-cavity design for degradation tracking and redundancy; ii) Digital control-loop with feed-forward system allowing for measurement cadences of 30s; iii) Aperture arrangement to reduce internal scattered light; iv) Cavity and heat-sink design to minimize non-equivalence, size, and weight of the instrument (Walter *et al.* 2017 and Suter 2014). CLARA was end-to-end calibrated against the SI-traceable cryogenic radiometer of the TSI Radiometer Facility (TRF, Kopp *et al.* 2007) at the Laboratory for Atmospheric and Space Physics (LASP) in Boulder (Colorado). Details about the CLARA instrument design, characterization and calibration can be found in Walter *et al.* (2017).

During the four weeks of the outgassing and commissioning phase, performance tests of basic electrical signals and temperatures of CLARA were executed. The passive thermal control of the radiometer head via the front shield (Walter *et al.* 2017) was found to work very well, with temperature difference < 1K between the solar and eclipsed portions of the low-Earth-orbiting spacecraft.

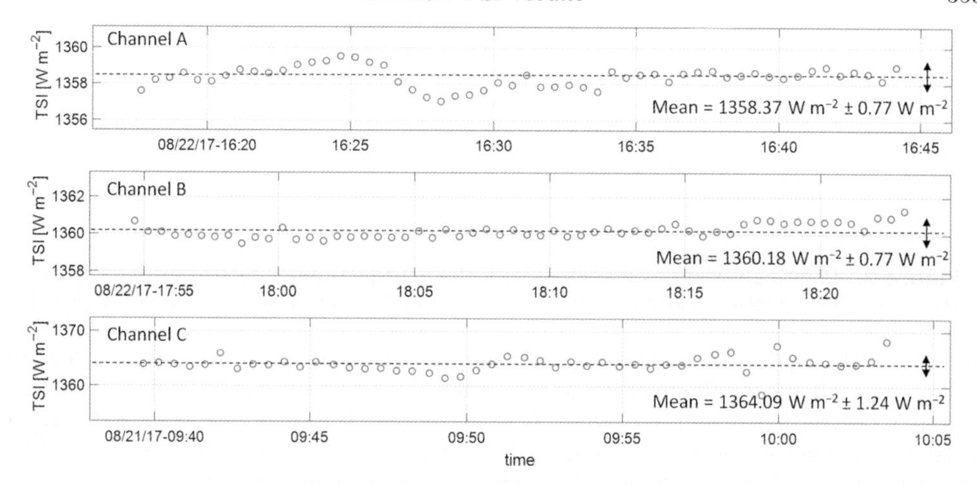

Figure 1. First light TSI measurement results of CLARA Channel A, B and C with measurement uncertainties (k=1). Channel B results are considered being the only reliable measurements so far. Channel A and C results are affected by satellite pointing instabilities (Section 3).

2. First Light TSI Results

The CLARA shutters were opened for the first TSI measurements on the 21st and 22nd of August 2017 (Fig. 1). Each time-series in Fig. 1 corresponds to TSI measurements during the second half of a solar portion of one orbit. Only the second half has been chosen because the satellite attitude control resulted in pointing instabilities during the first half where CLARA was not well enough pointed at the Sun to perform reliable TSI measurements. The averages of the three CLARA channels were 1358.37 W m^{-2} for Channel A, 1360.18 W m^{-2} for Channel B and 1364.09 W m^{-2} for Channel C. Channel B measured stable TSI values and was found being very little affected by the pointing instabilities or electromagnetic interference of satellite components (see Section 3) compared to Channel A and C. Therefore, CLARA Channel B TSI values are considered being reliable and correct (unlike Channel A and C) within the stated measurement uncertainty of ± 0.77 W m^{-2} ($k = 1$, Fig. 1), which is proven by the good agreement with the VIRGO measurements. The daily average of the VIRGO radiometer was 1360.14 W m^{-2} for the 21st and 1360.15 W m^{-2} for the 22nd. The rather high measurement uncertainties of 0.77-1.24 W m^{-2} are mainly a result of the limitations of the end-to-end calibration at the TRF (Walter *et al.* 2017).

3. CLARA Status

3.1. *Pointing sensitivity*

The large TSI variations on the 22nd of August 2017 around 16:25 from Channel A in Fig. 1 are correlated with deviations of $\approx 0.3°$ and $\approx 0.4°$ for X- and Y- angles of the satellite pointing attitude (Fig. 2). Generally, the optical design of CLARA should be insensitive to pointing variations of up to $\pm$ 1°. Detailed analysis of the basic CLARA signals preclude optical and thermal effects being responsible for the TSI measurement disturbance. That basic current and voltage signals are affected during the pointing instabilities suggests that electromagnetic disturbances affect the TSI measurements. This assumption is consistent with the fact that the two strongly affected Channels A and C are oriented towards the satellite electrical components like the magnetorquers, reaction wheels or the on-board computer, whereas Channel B is oriented towards space. Further analysis of this effect is ongoing. The satellite pointing instabilities were a result of the NorSat-1 attitude control algorithm, which was improved by the satellite manufacturer

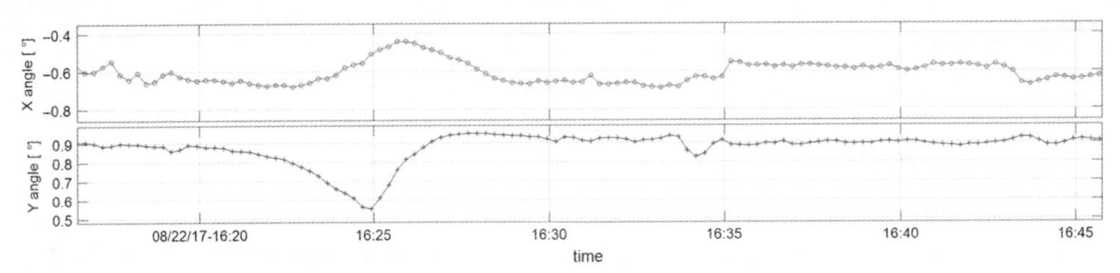

Figure 2. Pointing variations measured with the four quadrant (4Q) sensor of CLARA. A strong satellite pointing instability at 16:25 results in TSI variations of Channel A (Fig. 1).

in May 2018, significantly reducing the pointing deviations to $< 0.1°$. Merely few TSI values are available for Channel B and C for the improved pointing, showing a clear trend towards a better agreement of Channel C with Channel B measurements. The differences between Channel A and C and Channel B are thus expected being largely the result of the pointing sensitivity or the electromagnetic disturbances, respectively.

3.2. *NorSat-1 attitude control issues*

Issues with increased friction on one of the three reaction wheels providing attitude control were detected in early May 2018 by the satellite manufacturer (UTIAS-SFL). As a result, NorSat-1 tumbled several times, resulting in critically low (around -30°C) temperatures of CLARA at the end of May. The satellite manufacturer is working on improving the attitude control to be able to point CLARA back to the Sun. Because of the critically low temperatures and for safety reasons, CLARA was switched off from the end of May until mid August 2018, when CLARA was switched on for performance tests. All basic signals looked good and CLARA is ready for further TSI measurements once the attitude control issue is resolved and stable Sun pointing is obtained.

4. Conclusions

Generally, the CLARA/NorSat-1 experiment shows that flying small and lightweight TSI radiometers on low-cost micro-satellites provides new challenges. The experiment suggests that low-cost satellite components may affect TSI radiometers, at least CLARA-type radiometers. Therefore, we recommend performing more sophisticated electromagnetic susceptibility tests for CLARA-type radiometers in the future. Nevertheless, the experiment is considered successful and lessons have been learned for future missions (e.g. DARA-JTSIM/FY-3E and DARA/PROBA-3) of this new radiometer type.

References

Brusa, R.W., Fröhlich, C. 1986, *Appl. Opt. 25*, p. 173

Kopp, G., Heuerman, K., Harber, D. & Drake, G. 2007, *Society of Photo-Optical Instrumentation Engineers (SPIE) Conference Series, volume 6677*

Schmutz, W., Fehlmann, A., Finsterle, W., Kopp, G. & Thuillier, G. 2013, *AIP Conf. Proc. 1531*, p. 624-7

Suter, M. 2014, *PhD thesis, University of Zürich*

Walter, B., Levesque, P.-L., Kopp, G., *et al.* 2017, *Metrologia 54*, p. 5

Astronomy in Focus - XXX
Proceedings IAU Symposium No. XXX, 2018
M. T. Lago, ed.

doi:10.1017/S1743921319004629

Solar disk radius measured by Solar occultation by the Moon using bolometric and photometric instruments on board the PICARD satellite

G. Thuillier[1], P. Zhu[2], A. I. Shapiro[3], S. Sofia[4], R. Tagirov[5], M. van Ruymbeke[2], J.-M. Perrin[6], T. Sukhodolov[1], and W. Schmutz[1]

[1]Physikalisch-Meteorologisches Observatorium Davos, 7260 Davos Dorf, Switzerland, email: gerard.thuillier91@gmail.com

[2]Royal Observatory of Belgium, 3 avenue circulaire, 1180 Bruxelles,

[3]Max Planck Institute for Solar System Research, Gottingen, Germany

[4]Astronomy Department, Yale University, PO Box 208101, New Haven, CT 06520-8101, USA

[5]Imperial College London, Blackett Laboratory, Prince Consort Road SW7 2AZ, London, UK

[6]Observatoire de Haute-Provence, F-04870 St Michel l'Observatoire, France

---

1. Scientific rational

The solar disk radius is a basic metrological quantity as the planet's radius of the solar system. Despite the importance of having an accurate value of the solar disk radius, the existing measurements show differences likely due to large uncertainties due to the use of different measurement techniques, instrument calibration (if any), wavelength domain of measurements, and atmospheric effects for instruments running from the ground. Furthermore, is the solar radius constant or changing with time in particular with solar activity? The solar radius value depends on the solar atmosphere opacity, which allows solar model validation by comparing model predictions with the observations.

2. PICARD solar radius measurements and results

PICARD is a spacecraft developed by the Centre National d'Etudes Spatiales (CNES). It was launched on 15 June 2010. PICARD mission is described by Thuillier *et al.* (2006). Several instruments are carried, among them, the Bolometric Oscillations Sensor (BOS, B), the PREcision MOnitoring Sensor (PREMOS, CH), and a solar sensor (SES, F). They are used to derive the solar disk radius using the light curves produced when the Sun is occulted by the Moon. 17 occultations occurred from 2010 to 2013. One of them was not usable due to lack of spacecraft stability. Our approach is the use of the lunar radius as the reference. The solar disk radius for each occultation was simultaneously obtained for five spectral domains in the visible and IR solar continuum (535, 607, 750, 782, 784 nm) and two in UV (210, 263 nm). The calculation of the solar disk radius uses a simulation of the light curve taking into account the center to limb variation provided by the Non-local thermodynamic Equilibrium Spectral SYnthesis (NESSY, Tagirov *et al.* 2017). The positions of the Sun and Moon are provided by IMCCE (Institut de mécanique céleste et de calcul des éphémérides), and the spacecraft position by CNES. For a set of solar to Moon ratios, the quadratic difference between the measured and simulated light curve is calculated for each ratio. Among these differences, a minimum is found from which the solar radius is derived from knowing the lunar radius. The solar radius determined as explained above is referred to the lunar radius given by the Kaguya mission (Kato *et al.* 2010). At one astronomical unit, the solar radius is 959.78 arcseconds at 782 nm; 959.79 arcseconds at 750 nm; 959.76 arcseconds at 535 nm. We found 960.07 arcseconds at 210nm, which is a larger value than the others given the photons at this wavelength originate from the upper photosphere and lower chromosphere. The minimum solar disk radius is found around 600 nm. This is shown in Figure 1. The decrease from UV to visible wavelengths and increase in the near IR was expected by the solar models (Thuillier *et al.* 2011). Figure 1 also displays results from different techniques from the ground, planetary transits, imaging telescopes, and the Solar Disk Sextant (#3) on board a stratospheric balloon, which is in close agreement with our results. For the period of observations within solar cycle 24, no relation with the solar activity was found. However, this cycle was weak and the F10.7 only varies from 90 to 130 units during the period of observations. The center to limb variation is manifested at the beginning and end of the occultation and at maximum occultation. It varies with wavelength from 35 milliarcseconds (mas) in the UV to 2 mas at 784 nm. Furthermore, Figure 1 shows the recent solar radius obtained by other techniques obtained at different wavelengths. No consistent organization is revealed by these measurements as a function of wavelength as models suggest. Likely, each technique has its own scale and no direct relationship between these scales exists up to now. We note that the solar radius measured by the Venus and Mercury transits are systematically greater by 100 to 300 mas than our results calibrated on the Moon radius as well as the measurements using the total solar eclipses and Sun transits from the ground. The best agreement between our results is found with the Solar Disk Sextant (Sofia *et al.* 2013), which operates on board a stratospheric balloon, and uses an angular reference incorporated in the instrument. The details of this work are given in Thuillier *et al.* (2017).

Haberreiter *et al.* (2008) uses the seismic radius obtained by helioseismology from Schou *et al.* (1997), who provided a solar radius equal to 959.2 arcseconds. Calculating the angular distance between the region where waves are reflected and the inflection point of the limb at 500 nm is estimated by the COSI model (Shapiro *et al.* 2010) to be 0.46 arcsecond. By adding this value to the seismic radius, the optical solar radius is obtained to be 959.66 arcseconds at 500 nm. The B3 IAU solar radius reference (Prša *et al.* 2016) recommends the solar radius provided by Haberreiter *et al.* (2008).

3. Uncertainties

The uncertainty affecting the measured solar radius is calculated by taking into account several sources:

— the noise of the measurement,
— the uncertainty of the Moon radius,
— the uncertainty of the three bodies' positions (Earth, Moon, spacecraft) with respect to the Sun,
— the uncertainty on the pre and post solar irradiance to determine the light curve.

These uncertainties have been quadratically combined, and their mean for the seven wavelengths is 26 mas.

4. Conclusions

Sixteen solar occultations by the Moon observed from orbit by the PICARD spacecraft from June 2011 to April 2014 have allowed us to simultaneously measure the solar radius in seven spectral domains from the UV to the near IR calibrated on the lunar radius. For each occultation, the light curve is modeled taking into account the position of the

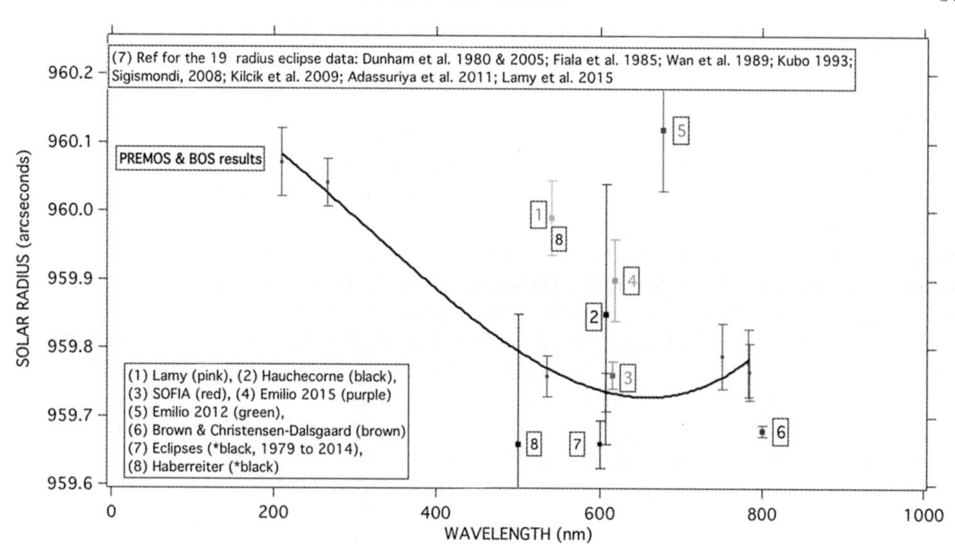

Figure 1. Display of all available solar disk values. 1 to 8 references are given in Thuillier *et al.* (2017).

Sun, Moon, and spacecraft, the center to limb darkening provided by the wavelength-dependent NESSY model, and a given set of solar radius to Moon ratios. For these ratios, the quadratic differences between the measured light curve and the modeled light curve are calculated. The minimum is found, the corresponding Sun to Moon ratio is determined from which the solar radius is extracted using the lunar radius measured by the Kaguya mission. Normalization to one AU is made afterward. The results are:

— the center to limb variation decreases with wavelength from the UV (35 mas at 210 nm) to the near IR (2 mas at 784 nm)

— With the 16 occultations, no significant radius variation related to solar activity was detected meaning that, if extant, it would be smaller than 26 mas. However, during that period, the F10.7 solar index varied only from 90 to 130 units.

— Based on the absence of variation with solar activity, the results were averaged for each spectral domain. The results show a variation with wavelength consisting of a decrease of 130 mas from 210 nm to 600 nm and followed by an increase of 50 mas at 784 nm. Such a variation was expected by the solar models. At one AU and 535 nm, the solar radius is 969.759 arcseconds.

— Solar models have, in general, the capability of producing the radius variation with wavelength using the inflection point position (IPP). In order to avoid a possible scale bias between two different techniques (IPP and lunar radius), the radii were referred to the radius at 535 nm. Comparing our results with NESSY calculations, it appears that NESSY has provided consistent results in the visible and near IR. In the UV, the agreement is at the three σ limit, which is interpreted as due to the present knowledge of the Fraunhofer lines structure.

— We point out that the Moon provides a stable reference for instruments in space, allowing long-term studies by using simple photometers and bolometers, which are robust instruments able to survive without significant aging in the harsh orbital environment.

— Our results referred to the Moon radius show the dependence of the solar radius with wavelength. This is why it is suggested that future solar radius measurements should be made in several wavelength domains in the solar continuum if possible, to ease the comparison between different data sets.

References

Haberreiter, M., Schmutz, W. & Kosovichev, A.G. 2008, *The Astrophysical Journal Letters*, 675, L53

Kato, M., S. Sasaki, Y. Takizawa, *et al.*. 2010, *Space Sci. ref.*, 3, 154

Prša, A., Harmanec, P., Torres, G., *et al.*. 2016, *AJ*, 152, 41

Schou, J., Kosovichev, A. G., Goode, P. R., & Dziembowski, W. A. 1997, *Astrophysical Journal Letters*, 498, L197

Sofia, S., Girard, T. M., Sofia, U. J., *et al.* 2013, *MNRAS*, 436, 2151

Shapiro, A., Schmutz, W., Schoell, M., Haberreiter, M., & Rozanov, E. 20010, *A&A*

Tagirov, R., Shapiro, A. I., & W. Schmutz 2017, *A&A*, 581, A116

Thuillier, G., Dewitte, S., Schmutz, W., & Picard Team 2006, *Adv. Space Res.*, 38, 1792

Thuillier, G., Claudel, J., Djafer, D., Haberreiter, M., Mein, N., Melo, S., Schmutz, W., Shapiro, A., Short, C. I., & Sofia, S. 2011, *Sol. Phys.*, 268, 125

Thuillier, G., Zhu, P., Shapiro, A. I., Sofia, S., Tagirov, R., van Ruymbeke, M., Perrin, J.-M., Sukhodolov, T. and Schmutz, W., 2017, *A & A*, 0.1051/0004-6361/

Astronomy in Focus - XXX
Proceedings IAU Symposium No. XXX, 2018
M. T. Lago, ed.

doi:10.1017/S1743921319004630

The Solar-Stellar Dynamo-Irradiance Connection

Ricky Egeland

High Altitude Observatory, National Center for Atmospheric Research, Boulder, CO, USA

Nearly everything that is known observationally about distant stars comes from their electromagnetic radiation. These observations are in limited bandpasses that form only part of the total solar irradiance that is observed for the Sun using space-based bolometers. Like the Sun, stellar spectral irradiance varies on multiple timescales, many of which are driven by surface magnetism. These time scales range from minutes (e.g. acoustic p-modes and surface granulation "flicker," (Cranmer *et al.* 2014) to a decade or more, analogous to the ~11 year solar sunspot cycle. Long-term stellar variability is often not as regular or well-behaved as in our nearest star, for example the erratic "Var" class and multiple-period cycles of Baliunas *et al.* (1995).

The Sun's regular magnetic cycle is beautifully demonstrated in "butterfly diagrams" (e.g. Hathaway 2015) that reveal a remarkable level of order in a cycle that lasts ~22 years when considering polarity. In these diagrams we observe that magnetic bipole regions have an opposite order of leading and following polarity across the equator, and this order changes in the same hemisphere from one sunspot cycle to the next (Hale's Law). The higher latitude following polarity (Joy's Law) is seen to be advected to the poles, which in turn flip their polarity some time after hemispheric sunspot maximum. There are three overarching questions when presented with this view of solar magnetism:

- What are the processes which generate and change the solar magnetic field?
- How do these processes depend on the solar structural and kinematic properties?
- How do these processes change on stellar evolutionary timescales?

Analogous questions may be asked of spectral irradiance. The first of these might be fruitfully explored with solar observations alone, but the second and third demand either observations of other stars or an unassailable dynamo theory, which nonetheless cannot be obtained without confirmation by stellar observations. Hence the interest in long-term synoptic observations of proxies for stellar magnetism that underlies the sub-field known as the "solar-stellar connection" and extends to studies of stellar spectral irradiance variability.

Figure 1 shows the bandpasses of six synoptic stellar surveys that capture variability of interest for studying stellar magnetism or spectral irradiance. The bandpasses are shown using the Kurucz *et al.* (1984) solar spectrum degraded to 1 Å resolution. The rough order of the number of target stars is indicated, which ranges from 10^2 to 10^9. Three of these are ground-based (HK Surveys, Fairborn APT Survey, and LSST) and three are space-based (*Kepler*, *TESS*, *Gaia*).

Four important parameters determine the science that can be obtained from a magnetic proxy or spectral irradiance variability program: duration, cadence, precision, and number of targets. In Figure 2 I plot each of the synoptic surveys with the duration of the program on the x-axis and the mean observation frequency on the y-axis. The marker size is scaled to the number of stars observed in the program. The Nyquist frequency

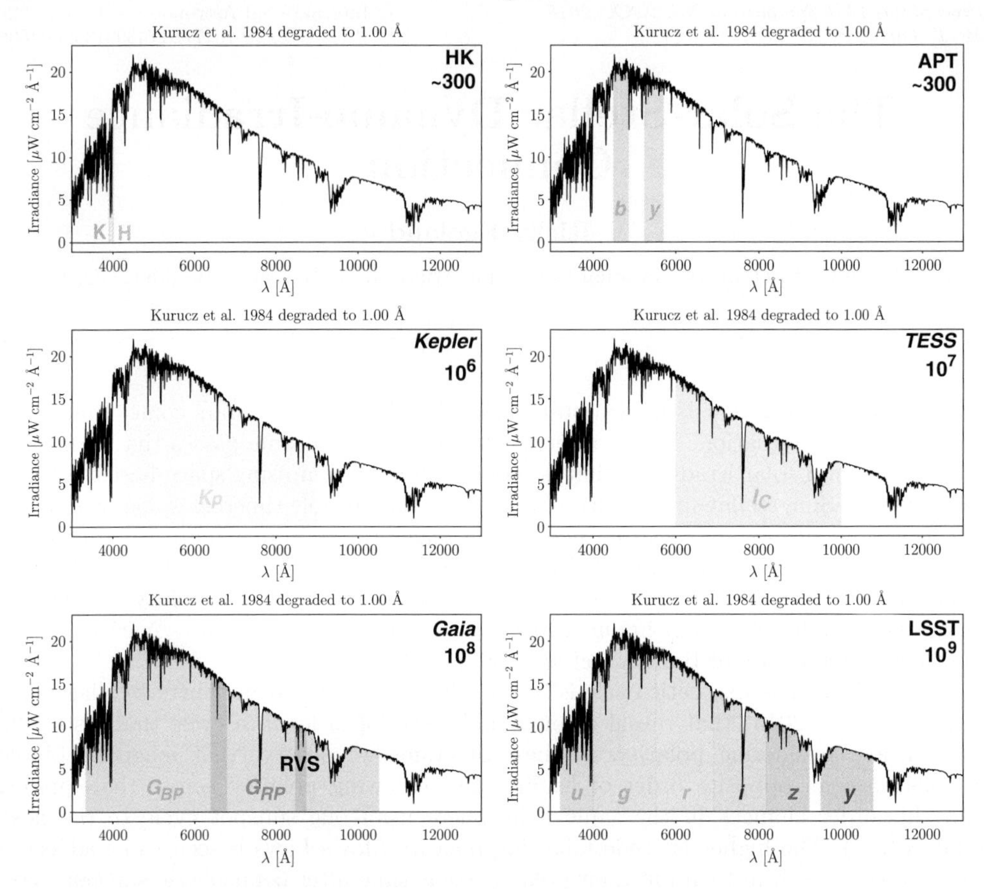

Figure 1. Synoptic stellar survey bandpasses.

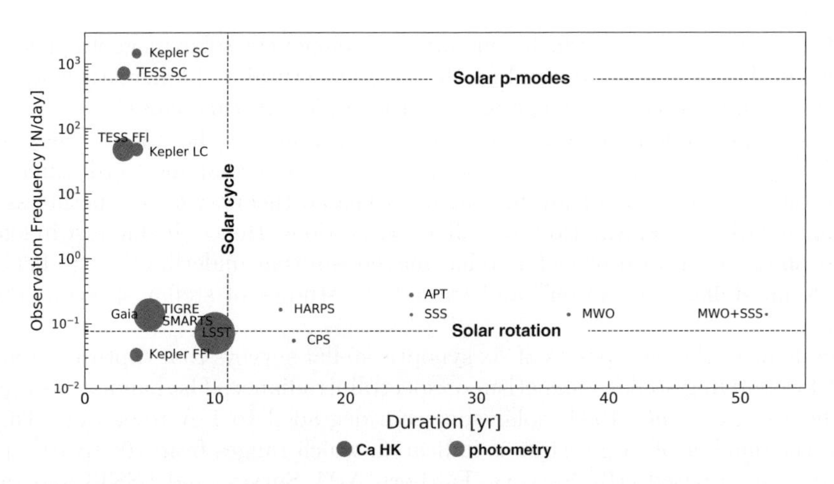

Figure 2. Synoptic stellar survey duration and mean cadence.

required to capture solar *p*-modes and rotation is indicated, as well as the solar cycle duration.

The ground-based Ca II HK surveys are the current champions of duration, and will remain so for quite some time. Prominent among these are the Mount Wilson Observatory

HK Project (MWO; 1966–2003; Baliunas *et al.* 1995) and the Lowell Observatory Solar-Stellar Spectrograph (SSS; 1994–present; Hall *et al.* 2007). Only these have sufficient duration to reveal Sun-like stellar activity cycles, and when combined they produce time series exceeding 50 years (Egeland 2017). Other stellar HK surveys include SMARTS and TIGRE, as well as radial velocity exoplanet surveys CPS and HARPS with irregular cadence. The Fairborn Observatory Automated Photometric Telescope program (APT; 1993–present; Henry *et al.* 1995) is the longest running photometric survey sensitive enough to detect the $\sim$1 mmag cycle-scale variation of the Sun. Recently, Radick *et al.* (2018) compares SSS HK activity to APT photometry to reveal the patterns of their correlation as well as the trend for more active stars to have larger amplitude variability.

The space-based exoplanet missions *Kepler* (2009–2018) and *TESS* (Ricker *et al.* (2014); launched April 2018) lead in cadence and precision. Lightcurves from the short-cadence (SC) targets of these missions can detect the p-mode oscillations and characterize stellar structure using asteroseismology. The 30 minute cadence of the long cadence (LC) targets is more than sufficient to detect stellar rotation, with over 34,000 rotation periods reported in McQuillan *et al.* (2014) and revealing a curious upper limit that may be a signal that magnetic braking stops operating when the dynamo becomes sufficiently weak (van Saders *et al.* 2018). The remarkable technique of exoplanet spot-transit photometry has revealed the location and size distribution of spots on another star (Morris *et al.* 2017). *Kepler* full-frame images (FFI) have recently been calibrated to remove instrument systematics allowing them to be used to study medium-term (4 year) photometric variability at a monthly cadence (Montet *et al.* 2017). *TESS* will provide similar FFIs at a 30 minute cadence and 27 day duration for the majority of the sky over its nominal 2 year mission. Overlapping observation sectors will have coverage for up to 1 year duration, and mission extension combined with plentiful fuel reserves could potentially extend *TESS* time series to the decade scale (G. Ricker; this meeting). The *Gaia* stellar astrometry mission obtains photometry with sufficient precision and cadence to detect variability and rotation. Lanzafame *et al.* (2018) reports on over 8×10^8 sources with > 20 observations, $\sim 5 \times 10^5$ classified as "variable", and over 15,000 with rotation period detected.

The Large Synoptic Survey Telescope (LSST; LSST Science Collaboration 2009) will regularly observe an unprecedented $\sim 2 \times 10^{10}$ targets over a nominal 10 year period starting in 2021. Each target will be "visited" approximately 80 times per year, with observations spread among its 6 bandpasses. The target 5 mmag zeropoint precision, combined with bandpass and seasonal averaging, should yield seasonal means at $\sim$1 mmag precision. While this is just at the threshold of detecting a weak Sun-like photometric cycle, many millions of more variable active stars will be detected by LSST (Hawley *et al.* 2016). Furthermore, actual zeropoint precision may reach the 2 mmag level (Ž. Ivezić, this meeting), and even more precise *relative* photometry may be obtained with additional processing, enabline LSST to detect Sun-like photometric cycles.

References

Baliunas, S. L., Donahue, R. A., Soon, W. H., *et al.* 1995, *ApJ*, 438, 269
Cranmer, S. R., Bastien, F. A., Stassun, K. G., & Saar, S. H. 2014, *ApJ*, 781, 124
Egeland, R. 2017, PhD thesis, Montana State University, Bozeman, Montana, USA
Hall, J. C., Lockwood, G. W., & Skiff, B. A. 2007, *AJ*, 133, 862
Hathaway, D. H. 2015, Living Reviews in Solar Physics, 12, arXiv:1502.07020
Hawley, S. L., Angus, R., Buzasi, D., *et al.* 2016, ArXiv e-prints, arXiv:1607.04302
Henry, G. W., Fekel, F. C., & Hall, D. S. 1995, *AJ*, 110, 2926
Kurucz, R. L., Furenlid, I., Brault, J., & Testerman, L. 1984, Solar flux atlas from 296 to 1300 nm
Lanzafame, A. C., Distefano, E., Messina, S., *et al.* 2018, *A&A*, 616, A16

LSST Science Collaboration. 2009, ArXiv e-prints, arXiv:0912.0201
McQuillan, A., Mazeh, T., & Aigrain, S. 2014, *ApJS*, 211, 24
Montet, B. T., Tovar, G., & Foreman-Mackey, D. 2017, *ApJ*, 851, 116
Morris, B. M., Hebb, L., Davenport, J. R. A., Rohn, G., & Hawley, S. L. 2017, *ApJ*, 846, 99
Radick, R. R., Lockwood, G. W., Henry, G. W., Hall, J. C., & Pevtsov, A. A. 2018, *ApJ*, 855, 75
Ricker, G. R., Winn, J. N., Vanderspek, R., *et al.* 2014, in *Proc.* SPIE, Vol. 9143, Space Telescopes and Instrumentation 2014: Optical, Infrared, and Millimeter Wave, 914320
van Saders, J. L., Pinsonneault, M. H., & Barbieri, M. 2018, ArXiv e-prints, arXiv:1803.04971

Astronomy in Focus - XXX
Proceedings IAU Symposium No. XXX, 2018
M. T. Lago, ed.

doi:10.1017/S1743921319004642

Statistical properties of starspots on solar-type stars and their correlation with flare activity

Hiroyuki Maehara

Okayama Observatory, Kyoto University,
3037-5 Honjo, Kamogata, Asakuchi, Okayama, Japan
email: maehara@kwasan.kyoto-u.ac.jp

Abstract. We analyzed the statistical properties of starspots on solar-type stars and the correlation between properties of starspots and flare activity using observations from the *Kepler* mission. We found the size distribution of starspots on solar-type stars shows the power-law distribution and both size distributions of starspots on slowly-rotating solar-type stars and of relatively large sunspots are roughly lie on the same power-law line. We also found that the frequency-energy distributions for superflares and solar flares from spots with different sizes are the same for solar-type stars and the Sun. These results suggest that the magnetic activity on solar-type stars and that on the Sun are caused by the same physical processes

Keywords. stars: flare, stars: spots

1. Introduction

Recent high-precision photometry by the Kepler mission found a large number of "superflares" on solar-type stars (e.g., Maehara *et al.* 2012). Most superflare stars show quasi-periodic brightness modulations caused by the rotation of the star with starspots (e.g., Notsu *et al.* 2013). According to Shibata *et al.* (2013), the energy of the largest flares on the solar-type stars increases as the area of starspots estimated from the amplitude of the rotational modulations increases. They also found that the energy of the largest flares is comparable to ~ 10 % of the magnetic energy which can be stored near the spots. These results suggest that the existence of large starspots is a key factor to produce superflares. We analyzed the statistical properties of starspots on solar-type stars and their correlation with the flare activity by using the data from the *Kepler* (rotational variations: McQuillan *et al.* 2014; flares: Shibayama *et al.* 2013) combined with the stellar parameters updated by *Gaia* DR2 (Berger *et al.* 2018).

2. Starspots on solar-type stars

Assuming the empirical relation between temperatures of spot and photosphere (Berdyugina 2005), we estimated the area of starspots on the stars from the amplitude of brightness variations caused by the rotation (McQuillan *et al.* 2014) and radius of the stars (Berger *et al.* 2018). Figure 1 (a) shows the scatter plot of the area of starspots on solar-type stars (main sequence stars with $5600 < T_{\rm eff} < 6000$K) in units of micro solar hemispheres (MSH; 1 MSH$= 3 \times 10^{16}$ cm^2) as a function of the rotation period. Small dots indicate solar-type stars in the Kepler field and open squares indicate solar-type stars showing superflares with the energy of 10^{33} erg. In the case of solar-type stars with $P_{\rm rot} > 13$ days, the area of the largest starspots rapidly decreases as the rotation period increases. On the other hand, the largest starspots area among the stars with $P_{\rm rot} < 13$

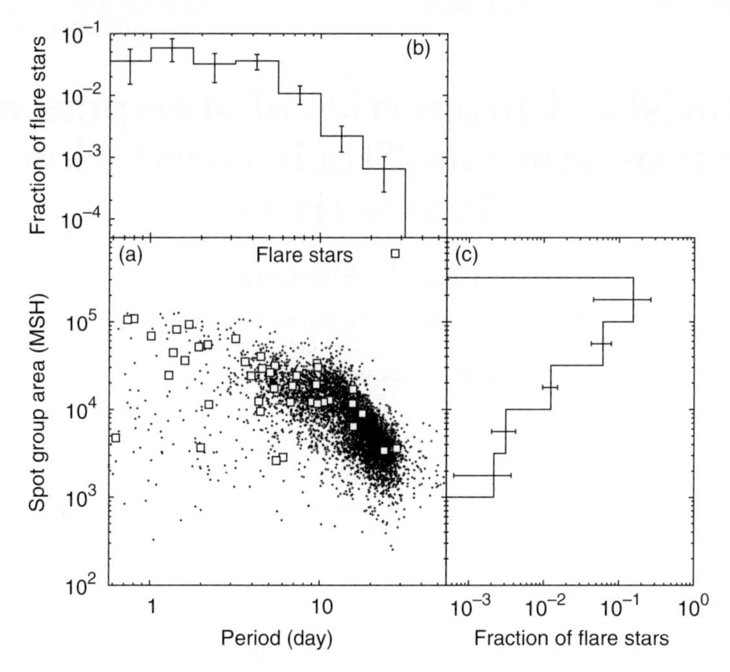

Figure 1. (a) Scatter plot of the area of starspots as a function of rotation period. (b) Fraction of flare stars as a function of rotation period. (c) Fraction of flare stars as a function of the area of starspots.

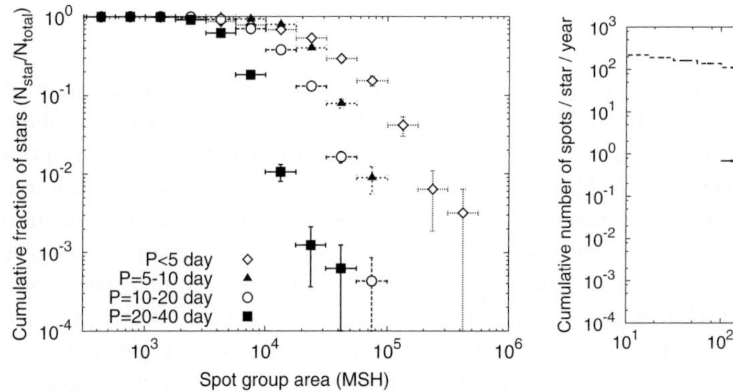

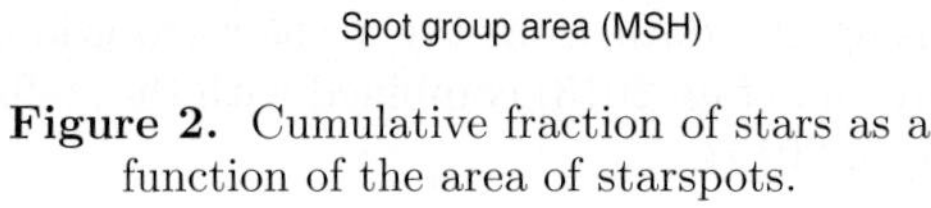

Figure 2. Cumulative fraction of stars as a function of the area of starspots.

Figure 3. Size distributions of starspots and sunspots.

days is roughly constant ($\sim 5 \times 10^4$-1×10^5 MSH)). Figure 1 (b) and (c) show the fraction of flare stars as a function of rotation period and starspot area respectively. The fraction of flare stars decreases as the rotation period increases ($P_{\rm rot} > 3$ days) and as the area of starspots decreases.

Figure 2 shows the cumulative fraction of stars as a function of the starspot area for the stars with different rotation periods. The fraction of the stars with a given starspot area increases as the rotation period decreases. This suggests that rapidly-rotating stars can produce large starspots more frequently than slowly-rotating stars like our Sun.

We compared the size distribution of large starspot groups on slowly-rotating ($P_{\rm rot} > 20$ days) solar-type stars with that of sunspot groups observed during recent 140 years. The size distribution of starspots shows the power-law distribution and the size distribution

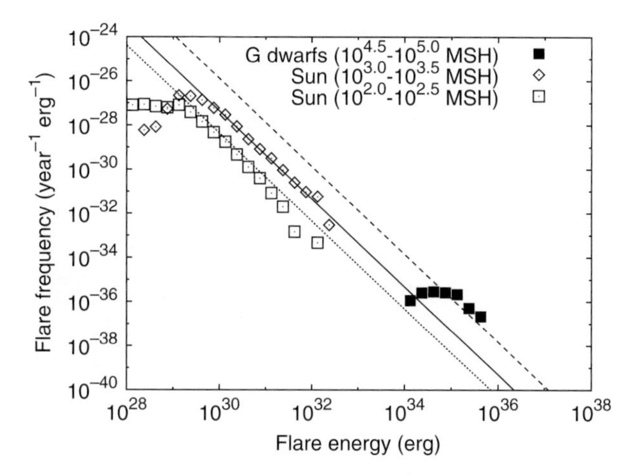

Figure 4. Comparison between occurrence frequency distribution of superflares from the solar-type stars with the starspot area of $10^{4.5}$-$10^{5.5}$ MSH and those of solar flares from the sunspot with the area of $10^{2.0}$-$10^{2.5}$ and $10^{3.0}$-$10^{3.5}$ MSH, respectively.

of relatively large sunspots lies on this power-law line (Figure 3). This result implies that the large starspots with the area of $\sim 1 \times 10^4$ MSH could appear once in a few hundred years on the slowly-rotating solar-type stars like our Sun.

3. Correlation between the area of spots and flare activity

Figure 4 shows the comparison between the occurrence frequency distribution of superflares and solar flares originating from the different spot sizes. The solid line is the power-law fit to the occurrence frequency distribution of solar flares from the sunspots with the area of $10^{3.0}$-$10^{3.5}$ MSH (power-law index: -1.99 ± 0.05). Dashed and dotted lines indicate power-law distribution with the same power-law index but 30 times and 1/10 of the occurrence frequency of the solid line. The frequency distributions of superflares from the stars with the spot area of $10^{4.5}$-$10^{5.5}$ MSH and solar flares from the sunspot with the area of $10^{2.0}$-$10^{2.5}$ MSH are roughly on the dotted and dashed lines respectively. This suggests that the frequency of flare with a given energy is roughly proportional to the spot area.

4. Summary

We analyzed the statistical properties of starspots and their correlation with the activity level of flares. Our analysis suggest that the area of the largest starspots on solar-type stars rapidly decreases as the rotation period increases for slowly-rotating solar-type stars ($P_{\rm rot} > 13$ days). We found the size distribution of starspots on solar-type stars with $P_{\rm rot} = 20$-40 days and that of relatively large sunspots lie on the same power-law line. We also found that the occurrence frequency distributions for flares originating from spots with different sizes are roughly the same for solar-type stars and the Sun. These results suggest that the magnetic activity on solar-type stars with superflares and that on the Sun are caused by the same physical processes (for more details, refer to Maehara *et al.* 2017).

References

Berdyugina, S. V. 2005, *Living Reviews in Solar Physics*, 2, 8
Berger, T. A., Huber, D., Gaidos, E., & van Saders, J. L. 2018, *ApJ*, 866, 99
Maehara, H., Shibayama, T., Notsu, S., *et al.* 2012, *Nature*, 485, 478

Maehara, H., Notsu, Y.., Notsu, S., *et al.* 2017, *PASJ*, 69, id.41
McQuillan, A., Mazeh, T., & Aigrain, S. 2014, *ApJS*, 211, 24
Notsu, Y., Shibayama, T., Maehara, H., *et al.* 2013, *ApJ*, 771, 127
Shibata, K., Isobe, H., Hillier, A., *et al.* 2013, *PASJ*, 65, 49
Shibayama, T., Maehara, H., Notsu, S., *et al.* 2013, *ApJS*, 209, 5

Astronomy in Focus - XXX
Proceedings IAU Symposium No. XXX, 2018
M. T. Lago, ed.

doi:10.1017/S1743921319004654

On long-duration 3D simulations of stellar convection using *ANTARES*

F. Kupka[1], D. Fabbian[1], D. Krüger[1,2,3], N. Kostogryz[2] and L. Gizon[2,1]

[1]Institut für Astrophysik Göttingen, Georg-August-Universität Göttingen,
Friedrich-Hund-Platz 1, 37077 Göttingen, Germany
email: `friedrich.kupka@uni-goettingen.de`

[2]Max-Planck-Institut für Sonnensystemforschung,
Justus-von-Liebig-Weg 3, 37077 Göttingen, Germany

[3]Fakultät für Mathematik, Universität Wien,
Oskar-Morgenstern-Platz 1, 1090 Wien, Österreich

Abstract. We present initial results from three-dimensional (3-D) radiation hydrodynamical simulations for the Sun and targeted Sun-like stars. We plan to extend these simulations up to several stellar days to study p-mode excitation and damping processes. The level of variation of irradiance on the time scales spanned by our 3-D simulations will be studied too. Here we show results from a first analysis of the computational data we produced so far.

Keywords. Convection, hydrodynamics, radiative transfer; Sun: oscillations; stars: oscillations

---

1. Tools and simulation setup

We performed 3-D simulations of solar and stellar convection in a Cartesian box ranging ≈ 3.61 Mm vertically and ≈ 5.88 Mm horizontally with the *ANTARES ("Advanced Numerical Tool for Astrophysical RESearch")* Fortran90 code (Muthsam *et al.* 2010). Most post-processing has been done with our own Fortran90 analysis tool *statistics*. The output of *ANTARES* and *statistics* has been visualized with *Paraview* and *gnuplot*, respectively. Additionally, we used these tools to produce figures.

2. Results

2.1. *Snapshots of solar convection*

Starting from a mean stratification given by a 1-D stellar structure model with a small perturbation to initiate a flow, the 3D simulation with *ANTARES* rapidly recovers the usual granulation pattern of solar convection, as we show in Fig. 1 with a snapshot from our latest solar simulation ("SLOPMD1", covering $\sim$ 1 solar day of relaxed state).

2.2. *Power spectrum*

We aim at a careful study of power spectra and line asymmetries for helio- and asteroseismology applications (see, e.g., Benomar *et al.* 2018). In Fig. 2 we show the power spectrum of vertical velocity some 600 km from the top of the simulation box, i.e., where $< T > \approx T_{\rm eff}$, as obtained from two of our solar granulation simulations. Two power peaks are clearly visible, the first around 2.7 mHz and the second around 3.5 mHz, corresponding respectively to oscillations with periods of ~ 6.2 and ~ 4.8 min. The first peak has become better defined in our recent, longer-duration simulation, since we are getting closer to resolve it in frequency space. The second peak is already resolved in the shorter

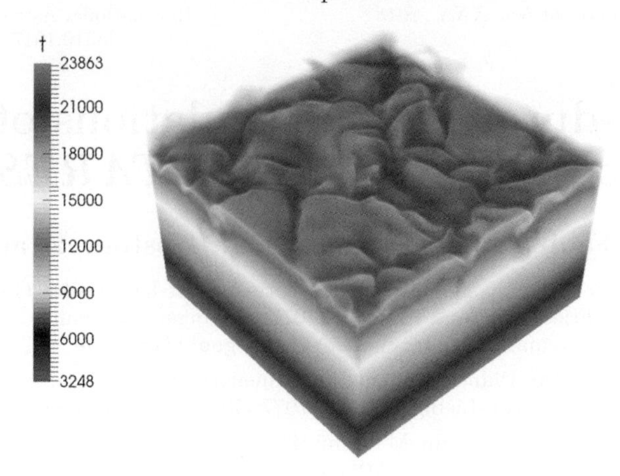

Figure 1. The imprint of granules on temperature (in units of K) at around 6 solar hours of time evolution within the relaxed part of the simulation.

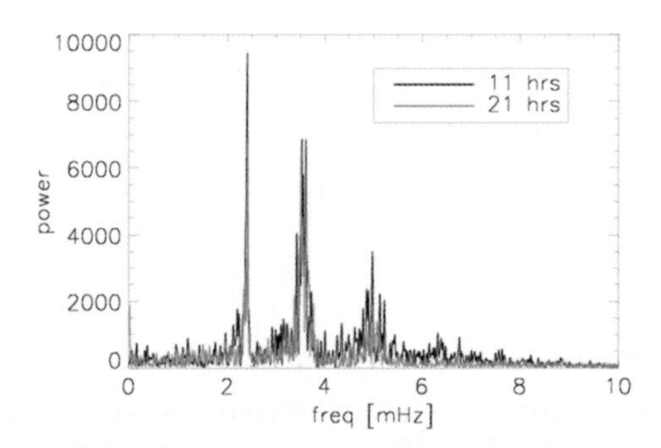

Figure 2. The vertical velocity power spectrum as function of frequency for a height where $< T > \approx T_{\rm eff}$, obtained from two of our solar granulation simulations: a run of 11 solar hours length (Kupka *et al.* 2017) and a new one of above 21 solar hours of relaxed state (black and blue line, respectively). The power spectrum shows, in both cases, three radial p-modes, of which the first two are clearly above the convective background.

simulation. A third peak close to 5.0 mHz (equivalent to a period of ~ 3.3 min) turns out to have significantly-reduced power in our longer-duration simulation, to a level not clearly distinguishable from the convective background, a further confirmation of the importance of performing simulations with sufficiently long duration (ideally, several solar/stellar days).

3. *ANTARES* code improvements

3.1. *Input data*

We developed the possibility of using 1-D models from YREC (the *Yale Rotating Stellar Evolution Code*, see Spada *et al.* 2013) as input for *ANTARES*. The input models from the stellar structure code can then be patched with 1-D atmospheric models available from the literature to provide better initial conditions for the outer layers. For compatibility with the setup used to produce the 1-D models, we are also updating the equation of state and opacity input tables for arbitrary metal abundances, based on the solar chemical mixture in Grevesse & Sauval (1998).

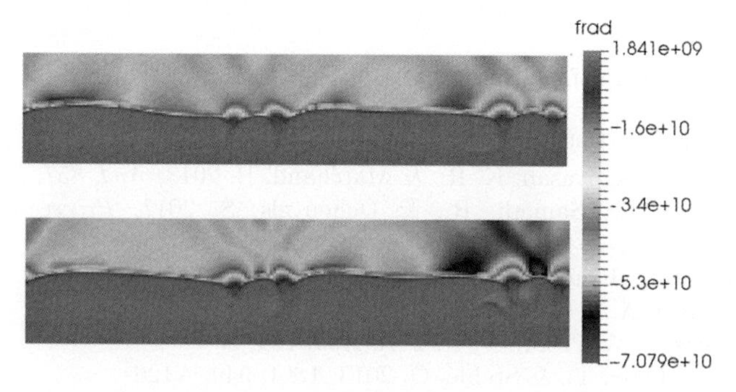

Figure 3. Radiative flux (in units of erg $\mathrm{cm}^{-2}\,\mathrm{s}^{-1}$) in the outer layers of the simulation box, for our new radiative transfer scheme (upper panel) and for the short-characteristics method (lower panel). Due to the coordinate system, outwards directed flux is negative and $F_{\mathrm{rad}} \approx -6.3 \cdot 10^{10}$ for solar $\mathrm{T}_{\mathrm{eff}}$ at the very top after relaxation.

3.2. *Improved radiative transfer*

So far ANTARES uses the short characteristics method for the calculation of intensity as the default radiative transfer (RT) method, the trapezoidal rule for calculating optical depth, and weighted parabolas (or the monotonic method of Steffen 1990) for interpolation. But overshooting of parabolas can lead to unphysical results which required introducing limiters and which can cause artifacts along the calculated rays. A new scheme, which prevents overshooting, is implemented. It is based on quadratic, monotonic Bezier-like splines (see de la Cruz Rodríguez & Piskunov 2013). The 3-D implementation of this scheme is done as in Ibgui *et al.* (2013). Fig. 3 shows how artifacts are reduced by the new RT implementation. Due to improved RT boundary conditions in the vertical direction the new method achieves a significantly-improved stability, too.

3.3. *Possibility of using the Eddington approximation*

To achieve faster calculations with only a limited increase in error we are implementing the non-grey Eddington approximation avoiding the expensive angular integration for radiative transfer. We calculate the mean intensity J_ν from

$$-\nabla \cdot \left(\frac{1}{3\kappa_\nu \rho} \nabla J_\nu \right) + \kappa_\nu \rho J_\nu = \kappa_\nu \rho B_\nu.$$

This can speed up calculations by a factor of 10−100.

4. Outlook

We are developing long-duration 3D simulations for the study of the interaction between convection and oscillations for several stellar targets of interest, including the Sun and planet-hosting stars (e.g., Kepler-409), with a focus on helio- and asteroseismology.

5. Acknowledgements

FK, DF, and DK gratefully acknowledge support by the Austrian Science Fund (FWF), project P29172. Computations were performed on the Vienna Scientific Cluster VSC-3, project 70950. We thank Dr. Hannah Schunker for creating the power spectrum plot, Figure 2. Post-processing and analyses used computing resources at the Faculty of Mathematics, Univ. Wien. NASA's Astrophysics Data System and the online tool Overleaf have been used to source relevant scientific literature and to prepare this contribution.

References

Muthsam, H. J., Kupka, F., Löw-Baselli, B., Obertscheider, C., Langer, M., & Lenz, P. 2010, *New Astron.*, 15, 460

Benomar, O., Goupil, M., Belkacem, K., Appourchaux, T., Nielsen, M. B., Bazot, M., Gizon, L., Hanasoge, S., Sreenivasan, K. R., & Marchand, B. 2018, *ApJ*, 857, 119

Kupka, F., Belkacem, K., Samadi, R., & Deheuvels, S. 2017, *Proceedings of the Polish Astronomical Society*, 5, 222

Spada, F., Demarque, P., Kim, Y.-C., & Sills, A. 2013 *ApJ*, 776, 87

Grevesse, N., & Sauval, A. J. 1998 *SSRev.*, 85, 161

de la Cruz Rodríguez, J., & Piskunov, N. 2013 *ApJ*, 764, 33

Ibgui, L., Hubeny, I., Lanz, T., & Stehlé, C. 2013 *A&A*, 549, A126

Steffen, M. 1990 *A&A*, 239, 443

FM10
Nano Dust in Space and Astrophysics

Astronomy in Focus - XXX
Proceedings IAU Symposium No. XXX, 2018
M. T. Lago, ed.

doi:10.1017/S1743921319004666

Nano dust in space and astrophysics

Ingrid Mann[1], Aigen Li[2] and Kyoko K. Tanaka[3]

[1]UiT The Arctic University of Norway, Tromsø, Norway
email: ingrid.b.mann@uit.no

[2]University of Missouri, Columbia, Missouri, USA
email: lia@missouri.edu

[3]Tohoku University, Sendai, Japan
email: kktanaka@astr.tohoku.ac.jp

Abstract. The theme of this focus meeting is related to the detection, characterization and modeling of nano particles — cosmic dust of sizes of roughly 1 to 100 nm — in space environments like the interstellar medium, planetary debris disks, the heliosphere, the vicinity of the Sun and planetary atmospheres, and the space near Earth. Discussions focus on nano dust that forms from condensations and collisions and from planetary objects, as well as its interactions with space plasmas like the solar and stellar winds, atmospheres and magnetospheres. A particular goal is to bring together space scientists, astronomers, astrophysicists, and laboratory experimentalists and combine their knowledge to reach cross fertilization of different disciplines.

Nano dust particles are intermediate between molecules and bulk matter. Because of their finite small size and large surface-to-volume ratio, the physical properties of nano grains are often peculiar, being qualitatively different from those of bulk materials. Different behavior is found, e.g., in heat capacity, melting temperature, surface energy, diffusion coefficient, and optical properties. Especially, clusters of ~ 1–10 nm are expected to reveal strongly variable size-dependent properties such as electronic structure, binding energy and dielectric function which determine how they interact with gas particles and the electromagnetic radiation. Larger clusters, with many thousands of atoms and diameters in the range of 10 nm and more, have a behavior smoothly varying with size and approaching bulk properties as size increases.

While the exact role of nano dust is not fully understood yet, those nanoclusters should play an important role, since, because of their large surface area (relative to their small mass), they interact more efficiently with particles and fields. Interstellar nano dust grains dominate the far ultraviolet extinction as well as the near- and mid-infrared emission of the interstellar medium (ISM) of the Milky Way and external galaxies. The heating of the interstellar gas and the surface layers of protoplanetary disks are dominated by nano grains through the photoelectrons provided by them. The presence of charged nano dust likewise influences other space plasma, also leading to dusty plasma effects, like waves and instabilities. Nano-sized (or smaller) polycyclic aromatic hydrocarbon (PAH) molecules, C60, diamonds reveal their presence in astrophysical regions through their characteristic vibrational spectral features. Nanodiamonds and nano TiC crystals have also been identified as presolar grains in primitive meteorites through their isotope anomalies. Their path from formation in the late stages of stellar evolution to identification in the laboratory is sketched in Figure 1.

For many years nano dust has been detected with in-situ instruments from spacecraft in different regions of the solar system. In the inner heliosphere of our solar system, nano dust forms from the dust-dust collisions in the zodiacal cloud and from sun-grazing

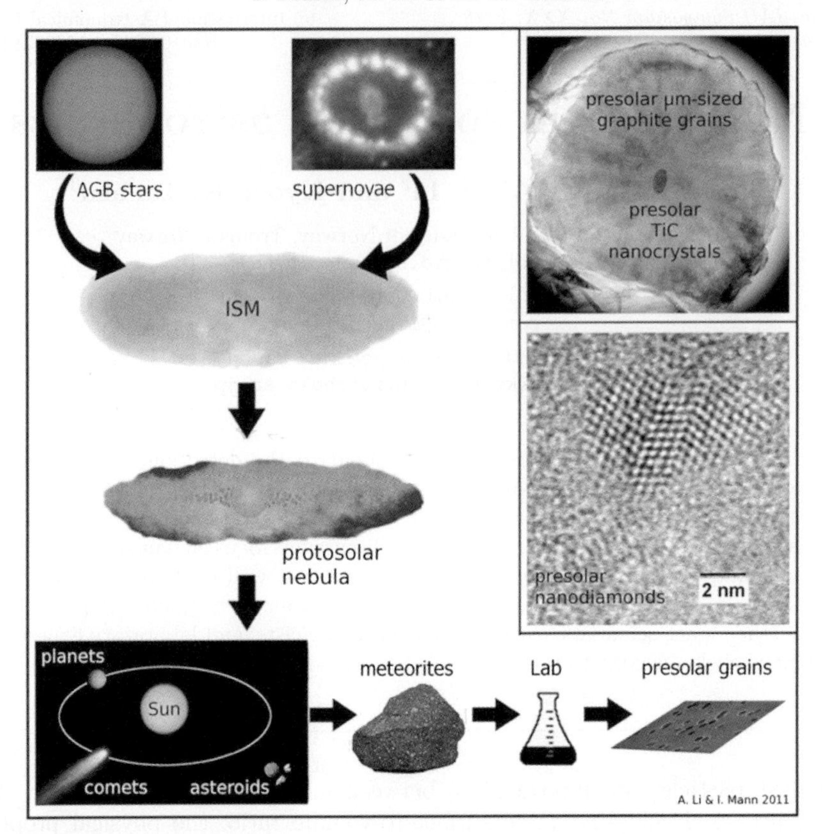

Figure 1. Schematic illustration of the history of cosmic dust grains, from their condensation in stellar winds of asymptotic giant branch (AGB) stars or in supernova ejecta, to their injection into the ISM, and subsequent incorporation into the dense molecular cloud from which our solar system formed (i.e., protosolar nebula). These grains survived all the violent processes occurring in the ISM (e.g., sputtering by shock waves) and in the early stages of solar system formation and were incorporated into meteorite parent bodies. Finally, they were collisionally liberated from their parent bodies and entered the Earth atmosphere, making them available for experimental studies in terrestrial laboratories, and therefore allowing one to separate them from the meteorite or interplanetary dust material in which they are embedded. Inserted are the TEM (*Transmission Electron Microscopy*) images of presolar nanodiamond grains and a presolar TiC nanocrystal within a micrometer-sized presolar graphite spherule. Taken from A. Li & I. Mann (2012, in Astrophys. Space Sci. Library, Vol. 385, *Nanodust in the Solar System: Discoveries and Interpretations*, ed. I. Mann, N. Meyer-Vernet, & A. Czechowski, Berlin, Springer-Verlag, 5).

comets. A notable recent finding is that nano dust in the heliosphere is deflected and accelerated in the solar wind. Astronomical observations suggest the presence of nano dust also in circum-stellar debris disks around main-sequence stars under conditions similar to the inner heliosphere. In-situ measurements from sounding rockets detect nano dust in the Earth's upper atmosphere (mesosphere), where it forms from the re-condensation of metallic compounds produced from ablating meteoroids. This dust — termed meteoric smoke — provides condensation nuclei for noctilucent clouds (first reported in 1886, and almost certainly a harbinger of climate change in the upper atmosphere). Meteoric smoke is also implicated in the formation and freezing of stratospheric clouds (which cause polar ozone depletion) and in the chemistry of clouds and atmospheres of, e.g., Mars, Venus, and Titan. Nano dust is probably also observed in comets. Planetary vulcanic plumes and impacts on planetary objects are sources of interplanetary nano dust as, e.g., observed near the surface of the Moon.

Nanodust interacts efficiently with particles and fields and in plasmas. The vast majority of our universe is plasma in which the heavy chemical elements are often contained in small solid dust particles that carry electric surface charge. A large fraction of the plasma is therefore dusty plasma, where dust participates in and gives rise to charge collective effects. Examples for dusty plasma are the ISM, the Earth's ionosphere, the ring systems of planets as well as the surface layers of moons and in general of solar system objects that are not surrounded by an atmosphere. Although dusty plasma is extensively studied, only a few observations in space are fully described with existing theory.

The research on the dynamics of nano dust in the heliosphere at present progresses motivated by the detection with space instruments on several different spacecraft. We expect a wealth of new observations in the near future. NASA has launched the *Parker Solar Probe* in August 2018 and ESA will soon launch the *Solar Orbiter*. Both spacecraft explore the most inner heliosphere and plasma in the vicinity of the Sun, including a region where we expect nanodust is being formed. Observing the near- and mid-infrared emission of the ISM which is dominated by nano dust is one focus of *JWST* which will be launched in 2021. Laboratory astrophysics is today a well-established field and in recent years further progress on dust studies was made with sample returns, from which the knowledge, often referring to larger dust, still provides information on nano dust.

During the General Assembly in Vienna, the Focus Meeting "*Nano Dust in Space and Astrophysics*" (FM10) on 28–29 August 2018 has brought together space physicists who study nano dust in the heliosphere and specialists from physics, astrophysics, as well as atmospheric research to make progress in understanding nano dust particles by combining their knowledge on dust under a wide range of space conditions. Knowledge on nano dust is also gained from studies of larger particles that progressed through laboratory astrophysics and analysis of returned samples.

We thank all our colleagues who participated in the meeting and the members of the science committee who participated in preparing the programme: Alexander G.G.M. Tielens (The Netherlands), Anja C. Andersen (Denmark), Anny-Chantal Levasseur-Regourd (France), Biwei Jiang (China), Chris M. Wright (Australia), Farid Salama (USA), John Plane (United Kingdom), Joseph A. Nuth (USA), Khare Avinash (India), Sun Kwok (Hong Kong, China), Thomas Pino (France) and Veronica Motta (Chile).

Astronomy in Focus - XXX
Proceedings IAU Symposium No. XXX, 2018
M. T. Lago, ed.

doi:10.1017/S1743921319004678

Nano dust and the far ultraviolet extinction

Biwei Jiang

Department of Astronomy, Beijing Normal University, Beijing 100875, China
email: bjiang@bnu.edu.cn

The ultraviolet (UV) extinction is determined by comparing the color or spectrum of reddened stars with un-reddened stars. Fitzpatrick & Massa (1998) obtained the UV extinction of 45 OB stars by using the IUE spectra and found that the UV extinction curves unanimously have a prominent feature around 2175 Å and keep rising in the far-UV ($\lambda \lesssim 1800$ Å) until ~ 115 nm — the short-wavelength limit of the IUE spectrophotometry. This law is confirmed by other studies, including those based on the GALEX UV photometry (e.g. Sun *et al.* 2018). Meanwhile, the UV extinction curve is steeper towards diffuse sightlines than dense environment. The sightline to HD 210121 has an extremely steep far-UV (FUV) extinction at $\lambda^{-1} \gtrsim 6\,\mu\mathrm{m}^{-1}$ with a remarkably small $R_V = 2.1$. M31, the other giant galaxy in the local group, displays similar UV extinction curves as the Milky Way galaxy. In the neighbouring Magellanic clouds that are metal-poor and have a much lower dust-to-gas ratio, the FUV extinction curve is steeper than our Galaxy, and the 2175 Å bump becomes weaker. Distant AGNs show almost no bump at 2175 Å and the FUV extinction becomes flat.

Nano dust grains are inferred from the rising FUV extinction because a dust grain is most effective in extinction when its size is comparable to the wavelength. Several species of nano dust grains are suggested to explain the observed UV extinction curve, such as nano carbon grains and nano silicate grains (Li & Mann 2012). The size distribution of nano dust is derived from dust infrared emission other than the FUV extinction that constrains the dust volume rather than the dust size (Wang *et al.* 2015). Weingartner & Draine (2001) explained the lack of 2175 Å bump by models lacking carbonaceous grains (graphite and PAH) with radii $< 0.02\,\mu$m and the rise in FUV by models with nano silicate grains.

The questions remain. Because the UV extinction is much more severe than in the optical wavebands ($A_{2000\,\mathrm{Å}}$ is about $3 \times A_V$), the UV extinction law probes generally only diffuse environment or the shallow surface area of dense clouds. Deep UV survey is needed to investigate the UV extinction law in dense medium. The identification of nano dust species is another open question.

References

Fitzpatrick, E., & Massa, D. 1998, *ApJ*, 328, 734
Li, A., & Mann, I. 2012, *ASSL*, 385, 5
Sun, M.X., Jiang, B.W., Zhao, H. *et al.* 2018, *ApJ*, 861, 153
Wang, S., Li, A., & Jiang, B. W. 2015, *ApJ*, 811, 38
Weingartner J. C., & Draine B. T. 2001, *ApJ*, 548, 296

Astronomy in Focus - XXX
Proceedings IAU Symposium No. XXX, 2018
M. T. Lago, ed.

doi:10.1017/S174392131900468X

Lower-temperature formation of silicate and oxide nano dust

Yuki Kimura

Hokkaido University
email: ykimura@lowtem.hokudai.ac.jp

Nucleation from vapor to solid has a large hindrance because of disadvantage for creation of a new surface. To overcome the large barrier for phase transition, larger supersaturation is required. However, it is not easy to predict how large supersaturation is required in a system, such as gas ejecta of evolved stars, although understanding of nucleation processes of dust are essential to understand whole processes of material evolution in a history of the universe.

Surface free energy and sticking probability are critical parameters to establish dust formation model based on nucleation theories. Nevertheless, those physical quantities have large uncertainty, because, I believe, nucleation processes are always passes through the size of nano-scale during agglomeration of atoms or molecules to bulk materials. Therefore, those physical quantities of nanoparticles have to be determined.

Especially, in a gas outflow of late-type stars, dust is only able to form at significantly lower-temperature compared with thermal equilibrium because of rareness of solid materials for heterogeneous nucleation. Recent years, we have tackled to know how dust forms in such extreme condition; how large supersaturation is required, how different physical properties they have in an environment far from thermal equilibrium, and whether dust formation follows classical nucleation theory or multistep nucleation.

Our laboratory and microgravity experiments using sounding rockets gave us following results. For homogeneous nucleation is required very large supersaturation (10^5 to 10^{14}). The surface energy is sometime 30% larger than that of corresponding bulk (Kimura *et al.* 2012) and sticking probability of Fe to be solid from supersaturated gas is as low as 0.002% against 100% as conservatively thought (Kimura *et al.* 2017). Formation of alumina dust around oxygen-rich late-type stars and its 13 μm feature was successfully duplicated by a specially designed experimental system (Ishizuka *et al.* 2018). Before formation of crystalline alumina, supercooled alumina particles formed, indicating two step nucleation processes. Our laboratory experiments using an in-situ IR measurement system of dust analogues during nucleation and growth succeeded reproduction of the spectrum of astronomical silicate with Mg-bearing silicate particles and found two step crystallization process that a liquid droplet form from a supersaturated gas at first and, then, forsterite nucleates and grows from the supercooled droplet (Ishizuka *et al.* 2015).

These homogeneous nucleation experiments suggest that dust seems to be able to form when the size of critical nuclei becomes single atoms or molecules because of largest hindrance for formation of dimer, which requires to release their binding energy. For instance, water cannot be cooler than the temperature of supercooled limit (around -42 degree C), because the size of critical nuclei becomes one molecule and, then the nucleation is unavoidable. The case of microgravity experiment of Fe, actually, the size of critical nuclei was a single atom. To expect formation condition of first dust, which formed via homogeneous nucleation around evolved stars, supersaturation limit should be considered rather than thermal equilibrium.

References

Ishizuka, S., Kimura Y., & Sakon, I. 1993, *The Astrophysical Journal*, 803, 88

Ishizuka, S., Kimura, Y., Sakon, I., Kimura, H., Yamazaki, T., Takeuchi, S., & Inatomi, Y. 2018, *Nature Communications*, in press

Kimura, Y., Tanaka, K. K., Miura, H., Tsukamoto, K. 2012, *Crystal Growth & Design*, 12, 3278

Kimura, Y., Tanaka, K. K., Nozawa, T., Takeuchi, S., Inatomi, Y. 2017, *Science Advances*, 3, e1601992

Astronomy in Focus - XXX
Proceedings IAU Symposium No. XXX, 2018
M. T. Lago, ed.

doi:10.1017/S1743921319004691

Interstellar and Circumstellar Fullerenes

Jan Cami[1,2]

[1]Department of Physics & Astronomy and Centre for Planetary Science and Exploration (CPSX), The University of Western Ontario, London N6A 3K7, Canada
email: jcami@uwo.ca

[2]SETI Institute, 189 Bernardo Ave, Suite 100, Mountain View, CA, USA

In recent years, it has become clear that fullerenes (and in particular C_{60}) are widespread and abundant in space, following their detection in a variety of astrophysical environments (see e.g. Cami *et al.* 2010, and many others) and the identification of several diffuse interstellar bands (DIBs) as due to C_{60}^+ (Campbell *et al.* 2015). However, many aspects about their formation and excitation are not clear yet.

Much research has focused on understanding fullerene formation in these environments. Laboratory experiments have shown that the temperature determines carbon condensation in bottom-up routes (Jäger *et al.* 2009): high temperatures ($T \geqslant 3500$ K) result in fullerenes (and fullerenic soot) while polycylic aromatic hydrocarbon (PAH) molecules (and graphitic soot) form at lower temperatures ($T \leqslant 1700$ K). A H-poor environment also results in fullerenes (even at those low temperatures; see Wang *et al.* 1995). Fullerenes also form from UV irradiation of large PAHs in a top-down process (Zhen *et al.* 2014); this is probably at work in the reflection nebula NGC 7023 (Berné & Tielens 2012).

By far the majority of all infrared C_{60} detections corresponds to young, low-excitation planetary nebulae (PNe), where the fullerenes are typically located far away from the central star (Bernard-Salas *et al.* 2012). This excludes an in-situ bottom-up formation process. However, PAH photo-processing is not a likely formation route either, given that there are many mature PNe that display copious amounts of PAH emission: if PAH photoprocessing would result in fullerenes far away from the central star in low-excitation objects, then the PAHs in these more mature objects should all have been converted to fullerenes as well. The key to resolving the formation of C_{60} in PNe may be in the dust. The C_{60}-PNe may represent objects where – for some reason – dust condensation (in an earlier phase, presumably on the asymptotic giant branch) happened at higher temperatures (or in a H-poor environment), producing fullerenic rather than graphitic dust. When the object becomes a planetary nebula, a fast wind overtakes a slow wind, and an ionization front develops. These processes could perhaps destroy much of the dust, and only the fullerenes (as the most stable species) survive. This would explain why we only see the fullerenes in young PNe. Further research about the nature and evolutionary status of these objects will help to pin down the C_{60} formation route.

References

Bernard-Salas, J., Cami, J., Peeters, E., *et al.* 2012, *ApJ*, 757, 41
Berné, O., & Tielens, A. G. G. M. 2012, *PNAS*, 109, 401
Cami, J., Bernard-Salas, J., Peeters, E., & Malek, S. E. 2010, *Science*, 329, 1180
Campbell, E. K., Holz, M., Gerlich, D., & Maier, J. P. 2015, *Nature*, 523, 322
Jäger, C., Huisken, F., Mutschke, H., Jansa, I. L., & Henning, T. 2009, *ApJ*, 696, 706
Wang, X. K., Lin, X. W., Mesleh, M., *et al.* 1995, *J. Mat. Res.*, 10, 1977
Zhen, J., Castellanos, P., Paardekooper, D. M., Linnartz, H., & Tielens, A. G. G. M. 2014, *ApJL*, 797, L30

Astronomy in Focus - XXX
Proceedings IAU Symposium No. XXX, 2018
M. T. Lago, ed.

doi:10.1017/S1743921319004708

Nano dust in stellar atmospheres and winds

Susanne Höfner

Department of Physics & Astronomy, Uppsala University, Sweden
email: susanne.hoefner@physics.uu.se

The extended dynamical atmospheres of cool, luminous asymptotic giant branch stars (AGB stars), are places where solid particles condense out of the gas. This stardust leaves its marks on the observable stellar spectra, and also on the structure and dynamics of the stellar atmospheres, since radiation pressure on the newly-formed grains is a key factor for driving the massive winds of these evolved low- and intermediate mass stars. The mass loss of AGB stars is discussed in detail in a review by Höfner & Olofsson (2018). In recent years, considerable progress has been made in understanding the underlying physical processes, and in characterizing the properties of the dust particles. In particular, improvements in high-angular-resolution techniques have led to spatially resolved observations of the dust-forming atmospheric layers of close-by cool giant stars, making detailed comparisons with predictive models possible. Mg-Fe silicates and Al_2O_3 are prominent dust species in mid-IR spectra of M-type AGB stars; interferometric measurements indicate that alumina form closer to the star than silicates (e.g., Karovicova *et al.* 2013). Large dust grains (0.1–0.5 μm) are found at distances of about 2 stellar radii, e.g. Norris *et al.* (2012); Ohnaka *et al.* (2017), as required for driving winds by photon scattering on near-transparent silicate grains with a low Fe/Mg ratio. RHD models of winds driven by photon scattering on Mg_2SiO_4 grains with sizes of $0.1 - 1\,\mu$m show realistic mass loss rates and wind velocities, as well as visual and near-IR photometry in good agreement with observations, see Höfner (2008); Bladh *et al.* (2013). The formation of composite grains with an Al_2O_3 core and a silicate mantle can give grain growth a head start, leading to higher mass loss rates and wind velocities, and an even better agreement with observed variations in visual and near-IR colors, cf. Höfner *et al.* (2016). High-angular-resolution imaging of scattered visual and near-IR light shows clumpy dust clouds surrounding AGB stars; cloud morphologies and grain sizes change on time scales of weeks to months, e.g. Khouri *et al.* (2016), Ohnaka *et al.* (2017). Clumpy dust clouds emerge naturally in 3D RHD star-in-a-box models, as a consequence of atmospheric shock waves induced by giant convection cells, see Freytag & Höfner (2008), Freytag *et al.* (2017), Höfner & Freytag (in prep.).

References

Bladh, S., Höfner, S., Aringer, B., & Eriksson, K. 2015, *A&A*, 575, A105
Bladh, S., Höfner, S., Nowotny, W., Aringer, B., & Eriksson, K. 2013, *A&A*, 553, A20
Freytag, B., & Höfner, S. 2008, *A&A*, 483, 571
Freytag, B., Liljegren, S., & Höfner, S. 2017, *A&A*, 600, A137
Höfner, S. 2008, *A&A*, 491, L1
Höfner, S., Bladh, S., Aringer, B., & Ahuja, R. 2016, *A&A*, 594, A108
Höfner, S. & Olofsson, H. 2018, *A&ARv*, 26:1
Karovicova, I., Wittkowski, M., Ohnaka, K., *et al.* 2013, *A&A*, 560, A75
Khouri, T., Maercker, M., Waters, L. B. F. M., *et al.* 2016, *A&A*, 591, A70
Norris, B. R. M., Tuthill, P. G., Ireland, M. J., *et al.* 2012, *Nature*, 484, 220
Ohnaka, K., Weigelt, G., & Hofmann, K.-H. 2017, *A&A*, 597, A20

Astronomy in Focus - XXX
Proceedings IAU Symposium No. XXX, 2018
M. T. Lago, ed.

doi:10.1017/S1743921319004848

Nano Dust as a Possible Cause of Hot Emission in Planetary Debris Disks

Kate Y. L. Su

Steward Observatory, University of Arizona, USA

Planetary debris disks are tenuous disks consisting of dust replenished by collisions of leftover planetesimals and cometary activity, events that are driven through gravitational shepherding and stirring by planets. The majority of the disks show warm and cold dust emission in a structure analogous to that of minor body belts in the solar system with asteroid- and/or Kuiper-belt components. Roughly 20–30% of main sequence stars observed interferometrically in the near-infrared (NIR) show extended excess emission that has been attributed to very small ($\lesssim$200 nm) hot dust in the vicinity of the stars. These NIR hot excesses have no obvious correlation with the presence of cold dust measured by far-infrared (Ertel *et al.* 2014). Detailed models on two dozens of such NIR excesses (e.g., Kirchschlager *et al.* 2017) suggest that the NIR excesses are (1) dominated by thermal dust emission from refractory material (amorphous carbon, graphite, or Mg/Fe oxides), (2) with grain sizes less than 200 nm, (3) located within $\sim$0.01–1 au from the star, depending on its luminosity, and (4) a total dust mass in the range of $(0.2\text{–}3.5)\times10^{-9}$ $M_{\oplus}$. Various mechanisms have been proposed to explain the origins of these nanograins, particularly with regard to how they are retained in the presence of strong radiative force around early-type stars. It is unlikely that these NIR excesses arise from in-situ dust emission produced by collosional cascades of parent bodies or transient events as discussed by Kral *et al.* (2017). These nanograins most likely come from the outer part of the planetary system either transported under the influence of Poynting-Robertson drag (Kobayashi *et al.* 2008) and/or scattered by multiple low-mass planets (Bonsor *et al.* 2014) or exterior eccentric planet(s) (Faramaz *et al.* 2017). Due to the radiation pressure, the lifetime of the nanograins is very short (less than days to weeks), requiring a high replenished rate that is difficult to sustain by the transport and delivery hypotheses. Possible mechanisms to prolong the lifetime of nanograins include either interaction with gas (Lebreton *et al.* 2013) or magnetic trapping with a weak stellar magnetic field (Su *et al.* 2013; Rieke *et al.* 2016) similar to the nanograins detected and modeled in the solar system (Czechowski & Mann 2010). Howevre, further refinements on theories and future observations are needed to advance our understanding of this phenomenon.

References

Bonsor, A., Raymond, S. N., Augereau, J.-C., & Ormel, C. W. 2014, *MNRAS*, 441, 2380
Czechowski, A., & Mann, I. 2010, *ApJ*, 714, 89
Ertel, S., Absil, O., Defrère, D., *et al.* 2014, *A&A*, 570, A128
Faramaz, V., Ertel, S., Booth, M., Cuadra, J., & Simmonds, C. 2017, *MNRAS*, 465, 2352
Kirchschlager, F., Wolf, S., Krivov, A. V., *et al.* 2017, *MNRAS*, 467, 1614
Kobayashi, H., Watanabe, S.-I., Kimura, H., & Yamamoto, T. 2008, *Icarus*, 195, 871
Kral, Q., Krivov, A. V., Defrère, D., *et al.* 2017, The Astronomical Review, 13, 69
Lebreton, J., van Lieshout, R., Augereau, J.-C., *et al.* 2013, *A&A*, 555, A146
Rieke, G. H., Gáspár, A., & Ballering, N. P. 2016, *ApJ*, 816, 50
Su, K. Y. L., Rieke, G. H., Malhotra, R., *et al.* 2013, *ApJ*, 763, 118

Astronomy in Focus - XXX
Proceedings IAU Symposium No. XXX, 2018
M. T. Lago, ed.

doi:10.1017/S174392131900485X

Heterogeneous chemistry on nano dust in the terrestrial and planetary atmospheres (including Titan)

John Plane

School of Chemistry, University of Leeds
email: j.m.c.plane@leeds.ac.uk

Cosmic dust particles are produced in the solar system from the sublimation of comets as they orbit close to the sun, and also from asteroidal collisions between Mars and Jupiter. Recent advances in interplanetary dust modelling provide much improved estimates of the fluxes of cosmic dust particles into planetary (and lunar) atmospheres throughout the solar system (Plane *et al.* 2018). Combining the dust particle size and velocity distributions with a new chemical ablation model enables the injection rates of individual elements to be predicted as a function of location and time (Carrillo-Sánchez *et al.* 2016). This information is essential for understanding a variety of atmospheric impacts, including the formation of layers of metal atoms and ions, the subsequent production of meteoric smoke particles, and the role of these particles in ice cloud nucleation and heterogeneous chemistry (Plane *et al.*, 2015). Specific examples that will be discussed are: in the terrestrial atmosphere, the formation of mesospheric and stratospheric ice clouds, and polar vortex chemistry (James *et al.* 2018); for Venus, the oxidation of CO and removal of O_2 on meteoric smoke particles in the hot troposphere (Frankland *et al.* 2017); for Mars, production of an Mg^+ layer which has recently been observed by the MAVEN spacecraft (Crismani *et al.* 2017), and the formation of metal carbonate-rich ice particles which nucleate CO_2 clouds in the Martian mesosphere (Plane *et al.* 2018); and for Titan, the production of benzene in the troposphere by the cyclo-trimerization of acetylene on dust particles (Frankland *et al.* 2016).

References

Carrillo-Sánchez, J. D., Nesvorný, D., Pokorný, P., Janches, D. & Plane, J. M. C. 2016, *Geophys. Res. Lett.*, 43, 11,979–11, 986

Crismani, M. M. J., Schneider, N. M., Plane, J. M. C., *et al.* 2017, *Nat. Geosci.*, 10, 401–404

Frankland, V. L., James, A. D., Carrillo-Sánchez, J. D., Mangan, T. P., Willacy, K., Poppe, A. R. & Plane, J. M. C. 2016, *Icarus,* 278, 88–99

Frankland, V. L., James, A. D., Carrillo-Sanchez, J. D., Nesvorný, D., Pokorn, P. & Plane, J. M. C. 2017, *Icarus*, 296, 150–162

James A. D., Brooke, J. S. A., Mangan, T. P., Whale, T. F., Plane, J. M. C. & Murray B. J. 2018, *Atmos. Chem. Phys.*, 18, 4519–4531

Plane, J. M. C., Flynn, G. J., Määttänen, A., Moores, J. E., Poppe, A. R., Carrillo-Sánchez, J. D., & Listowski, C. 2018, *Space Sci. Rev.*, 214, 23

Plane, J. M. C., Cárrillo-Sanchez, J. D., Mangan, T. P., Crismani, M. M. J., Schneider, N. M. & Määttänen, A. 2018, *J. Geophys. Res.-Planets*, 123, 695–707

Astronomy in Focus - XXX
Proceedings IAU Symposium No. XXX, 2018
M. T. Lago, ed.

doi:10.1017/S1743921319004861

Dusty plasma interactions near the Moon and in the system of Mars

Sergey I. Popel and Lev M. Zelenyi

Space Research Institute, Russian Academy of Sciences, Moscow, 117997 Russia
email: popel@iki.rssi.ru

We present results of recent self-consistent studies Popel *et al.* (2017), Popel *et al.* (2018a), and Popel *et al.* (2018b) which consider dust and dusty plasmas at the Moon and in the system of Mars. These studies are associated with the future space missions Luna-25 and Luna-27 as well as Phobos-Grunt 2 and ExoMars 2020. The dusty plasma system over the Moon includes charged dust, photoelectrons, and electrons and ions of the solar wind and Earth's magnetosphere (see Figure 1). The electrostatically ejected dust population can exist in the near-surface layer over the Moon while the dust appearing in the lunar exosphere owing to impacts of meteoroids present everywhere. Dusty plasmas are shown to be formed in the surface layer over the illuminated part of Mars' satellites Phobos and Deimos owing to photoelectric and electrostatic processes. In view of a weak gravitational field, dust particles rising over the surfaces of Phobos and Deimos are larger than those over the surface of the Moon. In this case, the role of adhesion, which is a significant process preventing the separation of dust particles from the lunar surface, is much smaller on Phobos and Deimos. We discuss also dusty plasmas in Martian atmosphere. This work was supported by the Russian Foundation for Basic Research (project no. 18-02-00341).

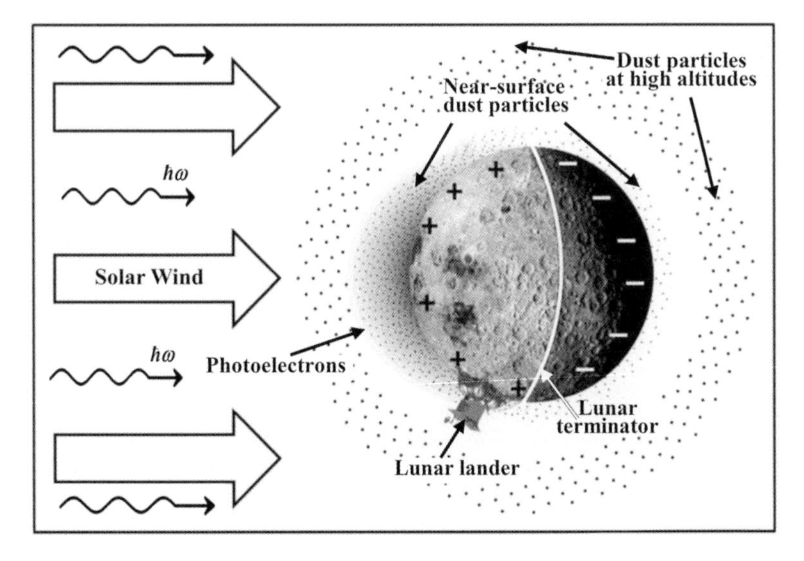

Figure 1. The main elements characterizing the dusty plasma system over the Moon (the terminator, the photoelectrons, the near-surface dust particles, dust particles at high altitudes, photons of solar radiation ($\hbar\omega$), and the solar wind) as well as the lunar lander at a high lunar latitude in the South Hemisphere.

References

Popel, S. I., Golub', A. P., Zakharov, A. V., & Zelenyi, L. M. 2017, *JETP Letters*, 106, 485
Popel, S. I., Golub', A. P., Zelenyi, L. M., & Dubinskii, A. Yu. 2018a, *Planetary and Space Science*, 156, 71
Popel, S. I., Golub', A. P., & Zelenyi, L. M. 2018b, *Plasma Physics Reports*, 44, 723

Astronomy in Focus - XXX
Proceedings IAU Symposium No. XXX, 2018
M. T. Lago, ed.

doi:10.1017/S1743921319004873

Processing of nano dust particles in galaxies†

T. Onaka[1], T. Nakamura[1,‡], I. Sakon[1], R. Ohsawa[1], R. Wu[2], H. Kaneda[3], V. Lebouteille[4] and T. L. Roellig[5]

[1]University of Tokyo,
[2]Observatoire de Paris,
[3]Nagoya University,
[4]Laboratory AIM - CEA Saclay,
[5]NASA Ames Research Center
[‡](present address): Recruit Communications
email: onaka@astron.s.u-tokyo.ac.jp

A family of emission bands observed in the near- to mid-infrared are attributed to the emission from nano-sized dust containing polycyclic aromatic hydrocarbons (PAHs) or PAH-like atomic groups. Investigations of variations of the emission bands (hereafter PAH emission) in violent conditions are thus significant for the study of the processing of nano-sized dust particles. Infrared observations of the Infrared Camera (IRC) on board *AKARI* (Onaka *et al.* 2007) have clearly shown that the PAH emission is detected in an Hα filament produced by winds from super star clusters in the nearby starburst dwarf galaxy NGC 1569 (Onaka *et al.* 2010). Recent analysis of IRC observations of two merger galaxies, NGC 2782 and NGC 7727, also shows that the PAH emission is prominently seen in extended structures produced by merger events (Fig. 1a, Onaka *et al.* 2018). The mid-infrared spectral energy distribution (SED) of the extended structures in both galaxies shows a sharp decline at 24 μm, which cannot be accounted for even if the contribution from very small grains (VSGs) is removed (Fig. 1b). These results suggest nano-sized

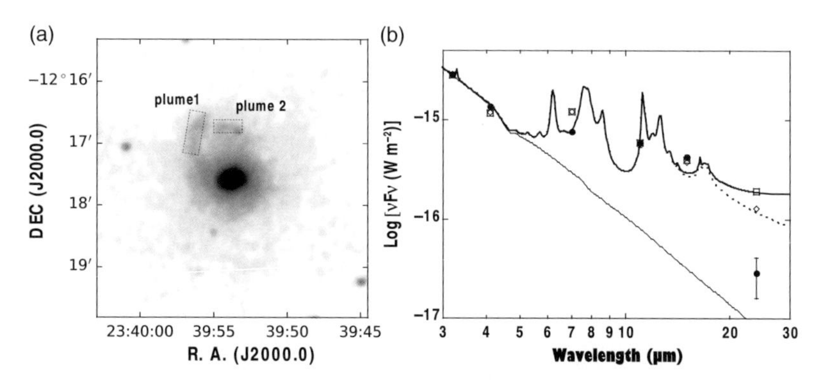

Figure 1. (a) IRC 7 μm band image of NGC 7727 and (b) the SED of Plume 2. The thick line and dotted line indicates the model SEDs with VSGs and without VSGs, respectively, using the DUSTEM model (Compiégne *et al.* 2011). The thin solid line indicates the assumed stellar component. See Onaka *et al.* (2018) for details.

†This work is based on observations with *AKARI*, a JAXA project with the participation of ESA.

dust particles may be produced by fragmentation of larger carbonaceous grains (e.g., VSGs) in violent conditions (see Onaka *et al.* (2018) for detailed discussion).

References

Compiégne, M., Verstraete, L., Jones, A., *et al.* 2011, *A&A*, 525, A103
Onaka, T., Matsuhara, H., Wada, T., *et al.* 2007, *PASJ*, 59, S401
Onaka, T., Matsumoto, H., Sakon, I., & Kaneda, H. 2010, *A&A*, 514, A15
Onaka, T., Nakamura, T., Sakon, I., *et al.* 2018, *ApJ*, 853, 31

Astronomy in Focus - XXX
Proceedings IAU Symposium No. XXX, 2018
M. T. Lago, ed.

doi:10.1017/S1743921319004885

Iron dust growth in the Galactic interstellar medium: clues from element depletions

Svitlana Zhukovska[1], Thomas Henning[2] and Clare Dobbs[3]

[1]Max Planck Institute for Astrophysics,
[2]Max Planck Institute for Astronomy,
[3]Exeter University
email: szhukovska@mpa-garching.mpg.de

Iron is severely depleted from the interstellar gas, but the long-standing question "Where is the missing interstellar iron?" remains unclear. We address it using a model of dust evolution in homogeneous interstellar medium based on three-dimensional hydrodynamic simulations of the Galactic disk (Zhukovskka *et al.* 2016, Zhukovska *et al.* 2018). The model includes dependence of dust destruction in SN shocks and growth by accretion of gas-phase metals on local physical conditions. Dust destruction process efficiently releases Fe back to the gas phase. This results in the lower depletions compared to the observed value, if all Fe is placed in nanoparticles or silicate grains. In order to reproduce the observed trend of interstellar Fe depletion with gas density, our model requires that solid iron resides in two dust components: (i) metallic iron nanoparticles with sizes in the range of 1–10 nm and (ii) small inclusions in silicate grains (Fig. 1).

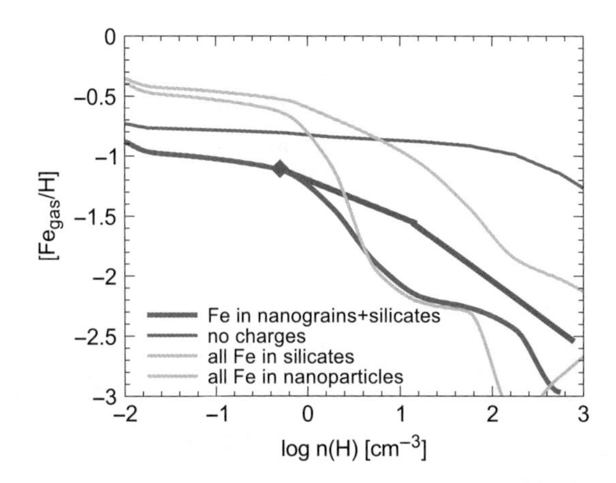

Figure 1. Red lines show the relation between Fe gas-phase abundance and gas density derived from observational data. It is best reproduced by the model in which 70% of Fe is locked in silicates and the rest resides in nanoparticles (blue solid line). The dashed line shows the same model without account for the grain charges. Green solid line and dash-dotted line show the models in which solid iron is in silicates and nanoparticles, respectively. Adapted from Zhukovska *et al.* (2018).

References

Zhukovska, S. and Henning, Th. and Dobbs, C. 2018, *ApJ*, 857, 94
Zhukovska, S., Dobbs, C., Jenkins, E. B., & Klessen, R. S. 2016, *ApJ*, 831, 147

Astronomy in Focus - XXX
Proceedings IAU Symposium No. XXX, 2018
M. T. Lago, ed.

doi:10.1017/S1743921319004897

Small-scale clustering of nano-dust grains in turbulent interstellar molecular clouds

Lars Mattsson

Nordita, KTH Royal Institute of Technology & Stockholm University, Sweden
email: lars.mattsson@nordita.org

It is well established that large grains will decouple from a turbulent gas flow, while small grains will tend to trace the motion of the gas. Small grains may still cluster on scales smaller than those typical for a turbulent flow due to centrifuging of particles away from vortex cores and accumulation of particles in convergence zones (a.k.a. preferential concentration). However, this has not yet been demonstrated for *compressible* flows.

We have studied clustering and dynamics of nano-dust in simulations of high-resolution (1024^3) simulations of forced homogeneous isothermal turbulence, mimicking the conditions in centres of molecular clouds. Fig. 1 shows the correlation dimension D_2 and the average relative increase of the dust density $\langle F_{\rm incr}\rangle$ as a function of the grain-size parameter α for simulations with different assumptions regarding kinetic drag and forcing of the flow. The clustering has a maximum for around a certain α, which lies in the nano-dust range (shaded area) for a typical mass-scaling of the simulations. Combined with the fact that nano dust may be abundant, and the increased interaction rate due to turbulent motions, the values of $\langle F_{\rm incr}\rangle$ suggest aggregation of nano-dust may be quite efficient. Comparing coagulation models based on the MOMIC code by Mattsson (2016), with and without corrections for turbulent clustering and relative motion of nano dust, we can see an order-of-magnitude increase of the coagulation rate.

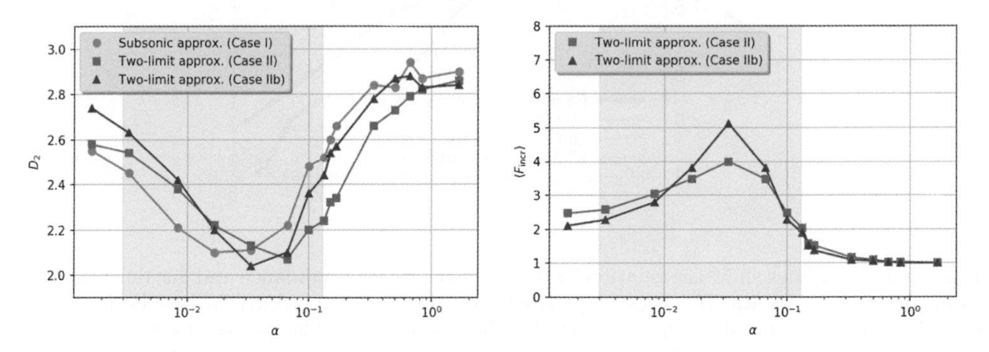

Figure 1. Correlation dimension D_2 (left)and the average relative increase of the dust density $\langle F_{\rm incr}\rangle$ (right) as a function of the grain-size parameter α for three cases: compressive forcing and a subsonic approximation for the kinetic drag (Case I); a transsonic approximation for the drag (Case II); solenoidal forcing and the transsonic approximation (Case IIb).

Finally, we note that charged nano-dust grains will have a different behaviour compared to the passive-scalar type dust in the present simulations. Whether this will further accentuate the clustering or lead to dispersion, counteracting the small-scale clustering, is unclear. Determining this will be the goal of future simulations.

Reference

Mattsson L. 2016, *Planetary & Space Science*, 133, 107

Astronomy in Focus - XXX
Proceedings IAU Symposium No. XXX, 2018
M. T. Lago, ed.

doi:10.1017/S1743921319004903

Dust Formation from Vapor through Multistep Nucleation in Astrophysical Environments

Kyoko K. Tanaka

Tohoku University
email: kktanaka@astr.tohoku.ac.jp

Cosmic dust grains are believed to form in outflows in the late stages of evolution of stars such as AGB stars and supernovae. The condensation and crystallization processes are important for understanding the origin of cosmic dusts and have seen by various observations. For instance, the silicate dusts condense in outflows with amorphous structure, as evidenced by the broad and smooth appearance of around 9.7 μm spectrum of silicate. Some observations suggest an increase in the fraction of crystalline as it cools from an intrinsic change in optical properties of the dust (Waters *et al.* 1996). Despite the transition from vapor to solid is a familiar process, the process is not fully understood yet. One reason is that size of nuclei is usually very small ($<$ nm) and the properties of nuclei are poorly understood.

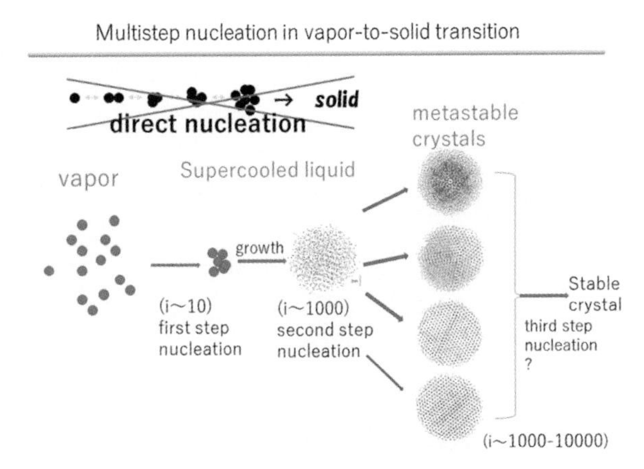

Figure 1. Overview of vapor-to-solid transition obtained by our molecular dynamics simulation (Tanaka *et al.* 2017).

In the study, we present molecular dynamics (MD) simulations of vapor-to-solid phase transition with a simple potential model (Lennard-Jones type) and discuss the transition process. In the simulations, the nuclei of supercooled liquid appear and growth. After the growth of nuclei, the crystallizations of supercooled nano-clusters are observed and the crystallized nano clusters have various structures of metastable phase (Tanaka *et al.* 2017). Our simulations indicate that the vapor-to-solid transition occurs through multistep nucleation which is vapor-to-liquid nucleation (first step nucleation) and crystallization in the supercooled liquid droplets (second step nucleation), even though the temperature is much lower than the triple temperature (Fig. 1). Recent experimental studies support the multiple processes of nucleation for various substances including silicate materials (Kimura *et al.* 2012, Ishizuka, Kimura & Sakon 2015). Our results with the experiments indicate that the multistep nucleation is a common phenomenon in the first stage of condensation from vapor to solid in the astrophysical environments.

This work was supported by JSPS KAKENHI Grants No. 18K03689 and No. 15H05731.

References

Waters, L.B. F. M., *et al.* 1996 *A&A*, 315, L361
Tanaka, K. K., Diemand J., Tanaka, H. & Angelil, R. 2017, *Physical Rev. E*, 96, 022804
Kimura, Y., Tanaka, K. K., Miura, H., Tsukamoto, K. 2012, *Crystal Growth & Design*, 12, 3278

Ishizuka, S., Kimura Y., & Sakon, I. 1993, *ApJ*, 803, 88

Astronomy in Focus - XXX
Proceedings IAU Symposium No. XXX, 2018
M. T. Lago, ed.

doi:10.1017/S1743921319004915

Spatially Resolved Studies of DIBs in Galaxies outside the Local Group

Ana Monreal-Ibero[1,2], Peter M. Weilbacher[3] and Martin Wendt[3,4]

[1]Instituto de Astrofísica de Canarias (IAC), E-38205 La Laguna, Spain,
[2]Universidad de La Laguna, Dpto. Astrofísica, E-38206 La Laguna, Spain,
[3]Leibniz-Institut für Astrophysik Potsdam, An der Sternwarte 16, D-14482 Potsdam, Germany,
[4]Institut für Physik und Astronomie, Universität Potsdam, Karl-Liebknecht-Str. 24/25, 14476 Golm, Germany
email: amonreal@iac.es

Diffuse interstellar bands (DIBs) are faint spectral absorption features of unknown origin associated to the interstellar medium (see Herbig 1995). Research on DIBs beyond the Local Group will surely blossom in the era of the ELTs but we can already now start paving the way. In Monreal-Ibero *et al.* (2015), we proposed the use of high-sensitivity IFSs as tools to detect and map DIBs. We used MUSE commissioning data, obtaining the first determination of a DIB radial profile in a galaxy outside the Local Group. Next, we derived the first maps for the DIBs at $\lambda5780$ and $\lambda5797$ in galaxies outside the Local Group using GTO MUSE data of the Antennae Galaxy (Monreal-Ibero *et al.* 2018). The strongest of the two DIBs (at $\lambda5780$) was detected in an area of $\sim 0.6\square'$, corresponding to a linear scale of ~ 25 kpc^2 (see Fig. 1). This region was sampled using >200 out of ~1200 independent lines of sight. The DIB $\lambda5797$ was detected in >100 independent lines of sight. These maps where compared with the 2D distribution of the extinction, atomic and molecular gas, and emission in the mid-infrared. The derived results illustrate the enormous potential of integral field spectrographs for extragalactic DIB research.

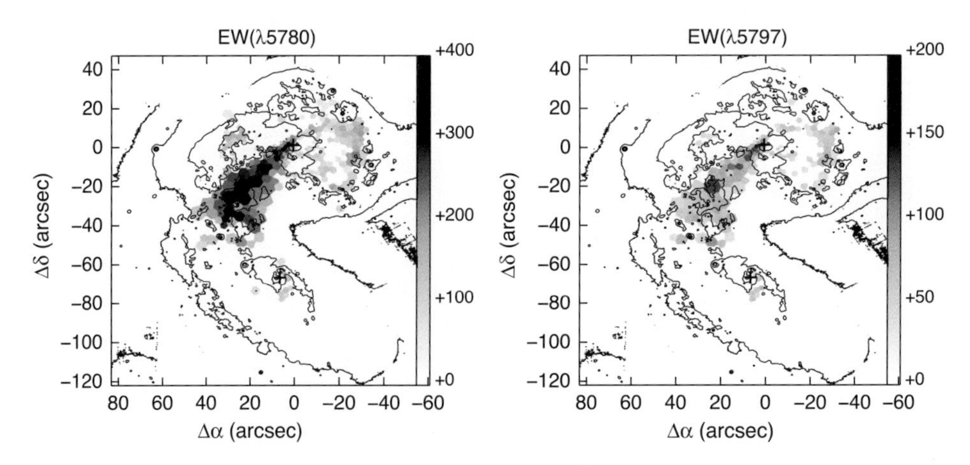

Figure 1. Maps of the derived equivalent width in mÅ for DIBs at $\lambda5780$ (*left*) and $\lambda5797$ (*right*) in the Antennae Galaxy. The reconstructed white-light image is overplotted as reference with contours in logarithmic stretching in steps of 0.5 dex.

Acknowledgements

AMI acknowledges support from the Spanish MINECO through project AYA2015-68217-P. PMW received support through BMBF Verbundforschung (projects MUSE-AO, grant 05A14BAC, and MUSE-NFM, grant 05A17BAA).

References

Herbig, G. H. 1995, *ARA&A*, 33, 19
Monreal-Ibero, A., *et al.* 2015, *A&A*, 576L, 3
Monreal-Ibero, A., Weilbacher, P. M., & Wendt, M. 2018, *A&A*, 615A, 33

Astronomy in Focus - XXX
Proceedings IAU Symposium No. XXX, 2018
M. T. Lago, ed.

doi:10.1017/S1743921319004927

Orion Bar as a window to evolution of small carbonaceous dust grains

Maria Murga, Dmitri Wiebe and Maria Kirsanova

Institute of Astronomy of the Russian Academy of Sciences
email: murga@inasan.ru

The Orion Bar is one of the most well-known photodissociation regions. An enormous volume of observational data in various spectral ranges makes it a versatile tool for checking theoretical ideas. Specifically, it allows studying small carbonaceous grains, which reveal themselves through mid-infrared (IR) emission bands. Their lifecycle strongly depends on external conditions that vary dramatically within this object. Thus, it is possible to trace the evolutionary changes of dust at different conditions within a single object.

We use the archival photometric and spectroscopic observations in the mid-IR range from several telescopes: UKIRT, Spitzer, ISO, SOFIA. Fluxes in bands at 3.3, 3.4, 6.4, 6.6, 7.7, 11.2 μm are measured. These IR-bands are usually associated with vibrations of polycyclic aromatic hydrocarbons (PAHs). Ratios between band fluxes indicate variations in dust size ($F_{3.3}/F_{11.2}$), ionization stage ($F_{6.4}/F_{11.2}$, $F_{7.7}/F_{3.3}$, $F_{7.7}/F_{11.2}$), and fraction of grains with aliphatic bonds ($F_{3.4}/F_{3.3}$). We have found that in the Orion Bar the band ratios change with the distance from the Trapezium Cluster (TC). The ratio $F_{3.3}/F_{11.2}$ increases with the distance from the TC indicating growing abundance of small PAHs relative to larger ones. The ratios $F_{6.4}/F_{11.2}$, $F_{7.7}/F_{3.3}$, $F_{7.7}/F_{11.2}$ decrease toward the molecular cloud, implying that abundance of positively charged PAHs drops relative to abundance of neutral PAHs. The ratio $F_{3.4}/F_{3.3}$ has a minimum approximately at the ionisation front and increases toward the molecular cloud, which likely characterises the increasing contribution of hydrogenated grains whether they are PAHs or small amorphous carbons.

We have fitted the spectrum of the Orion Bar by PAHs' spectra from the NASA Ames PAH IR database presented by Bauschlicher *et al.* (2018). The spectrum is well fitted by the PAH-mixture of ≈ 50 species. Most of them are large (number of carbon atoms $N_{\rm C} > 80$) pericondensed neutral species and cations, medium ($40 < N_{\rm C} < 80$), and small ($N_{\rm C} < 40$) dehydrogenated pericondensed neutral species and cations, and also small hydrogenated PAHs like hydrogenated coronene.

We have modelled the evolution of PAHs at the conditions of the Orion Bar using our dust evolution model described in the work of Murga *et al.* (2016). We have obtained the results which are generally consistent with observations but with some exceptions. Specifically, it was found that there should be large cations and neutral PAHs at the ionization front, and the small fraction of medium dehydrogenated PAHs can exist also, but the model cannot predict the the presence of small PAHs, neither hydrogenated nor dehydrogenated. We assume that such small PAHs can be result of destruction of PAH clusters or restructuring of amorphous carbonaceous grains from H-rich to H-poor state.

References

Bauschlicher, C. W., & Ricca, Jr. A., & Boersma, C., Allamandola, L. J. 2018, *ApJS*, 234, 32
Murga, M. S., & Khoperskov, S. A., & Wiebe, D. S 2016, *Astron. Rep.*, 60, 2, 333

Astronomy in Focus - XXX
Proceedings IAU Symposium No. XXX, 2018
M. T. Lago, ed.

doi:10.1017/S1743921319004939

Mixed Aromatic Aliphatic Organic Nanoparticles (MAON) as Carriers of Unidentified Infrared Emission Bands

Sun Kwok[1,2], SeyedAbdolreza Sadjadi[2] and Yong Zhang[2]

[1]Dept. of Earth, Ocean and Atmospheric Sciences, University of British Columbia, Canada
[2]Laboratory for Space Research, The University of Hong Kong, Hong Kong, China
email: skwok@eoas.ubc.ca, sunkwok@hku.hk

The unidentified infrared emission (UIE) phenomenon consists of a family of emission bands and broad emission plateaus superimposed on an underlying continuum. The most popular explanation for the UIE bands is the polycyclic aromatic hydrocarbon (PAH) hypothesis, but this model has a number of problems (Kwok & Zhang 2013; Zhang & Kwok 2015). While the UIE bands are likely to arise from stretching and bending modes of aromatic and aliphatic groups in a carbonaceous compound, the exact vibrational modes creating these bands and the exact chemical structure of the compound are not known.

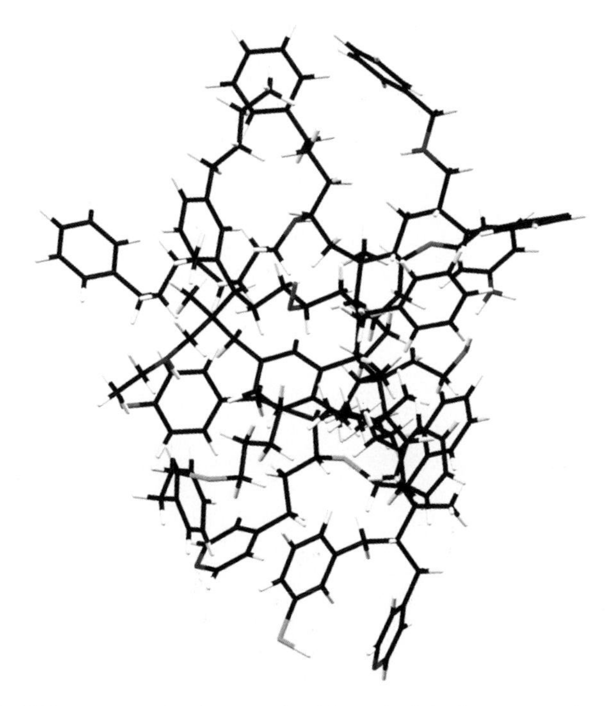

Figure 1. The MAON structure is characterized by a highly disorganized arrangement of small units of aromatic rings linked by aliphatic chains. This structure contains 169 C atoms (in black) and 225 H atoms (in white). Impurities such as O (in red), N (in blue), and S (in yellow) are also present. A typical MAON particle may consist of multiple structures similar to this one.

Some of the UIE bands are likely to be due to coupled vibrational modes and the identification of their exact nature is not trivial (Sadjadi, Zhang & Kwok 2015, 2017). We report results of quantum chemistry calculations of large (>100 carbon atoms) molecules with mixed aromatic/aliphatic structures (MAON, Kwok & Zhang 2011, Figure 1) with the goal of identifying the origin of the UIE bands and exploring various possibilities of the chemical nature of the UIE carrier.

References

Kwok, S., & Zhang, Y. 2011, *Nature*, 479, 80
Kwok, S., & Zhang, Y. 2013, *ApJ*, 771, 5
Sadjadi, S., Zhang, Y., & Kwok, S. 2015, *ApJ*, 801, 34
Sadjadi, S., Zhang, Y., & Kwok, S. 2017, *ApJ*, 845, 123
Zhang, Y., & Kwok, S. 2015, *ApJ*, 798, 37

Astronomy in Focus - XXX
Proceedings IAU Symposium No. XXX, 2018
M. T. Lago, ed.

doi:10.1017/S1743921319004940

Graphene and Carbon Nanotubes in Space

Xiuhui Chen[1], Zichun Xiao[2], Aigen Li[3] and Jianxin Zhong[1]

[1]Xiangtan University, China
[2]Ravenscroft School, Raleigh, NC 27615, USA
[3]University of Missouri, USA
email: lia@missouri.edu

As the fourth most abundant element in the universe, carbon plays an important role in the physical and chemical evolution of the interstellar medium (ISM). Due to its unique property to form three different types of chemical bonds through sp^1, sp^2, and sp^3 hybridizations, carbon can be stabilized in various allotropes, including amorphous carbon, graphite, diamond, polycyclic aromatic hydrocarbon (PAH), fullerenes, graphene, and carbon nanotubes (CNTs).

Many allotropes of carbon are known to be present in the ISM (Henning & Salama 1998). Presolar graphite grains and nanodiamonds have been identified in primitive meteorites based on their isotopically anomalous composition (see Nittler 2018). While hydrogenated amorphous carbon grains reveal their presence in the diffuse ISM through the ubiquitous $3.4\,\mu$m aliphatic C–H absorption feature (Pendleton & Allamandola 2002), the aromatic C–H and C–C emission features at 3.3, 6.2, 7.7, 8.6 and $11.3\,\mu$m infer the widespread presence of PAHs in a wide variety of interstellar regions (see Hudgins & Allamandola 2005). The detections of interstellar C_{60} and C_{70} (Cami *et al.* 2010; Sellgren *et al.* 2010) and their cations (Berné *et al.* 2013; Strelnikov *et al.* 2015) have also been reported based on their characteristic infrared (IR) emission features.

Graphene was first experimentally synthesized in 2004 by A.K. Geim and K.S. Novoselov for which they received the 2010 Nobel Prize in physics. More recently, García-Hernández *et al.* (2011; 2012) reported for the first time the presence of unusual IR emission features at ~ 6.6, 9.8, and $20\,\mu$m in several planetary nebulae, both in the Milky Way and the Magellanic Clouds, which are coincident with the strongest transitions of planar C_{24}, a piece of graphene. In principle, graphene could be present in the ISM as it could be formed from the photochemical processing of PAHs, which are abundant in the ISM, through a complete loss of their H atoms (e.g., see Berné & Tielens 2012). On the other hand, as illustrated in Figure 1a, both quantum-chemical computations and laboratory experiments have shown that the exciton-dominated π–π^* electronic transitions in graphene cause a strong absorption band near 2755 Å or 4.5 eV (Yang *et al.* 2009; Nelson *et al.* 2010) which is not seen in the ISM. This allows us to place an upper limit of ~ 5 ppm of C/H on the abundance of graphene in the diffuse ISM (see Chen *et al.* 2017). Moreover, the nondetection of the 6.6, 9.8, and $20\,\mu$m emission features of graphene C_{24} in the observed IR emission spectra of the diffuse ISM is also consistent with an upper limit of ~ 5 ppm of C/H in the C_{24} graphene sheet (see Chen *et al.* 2017).

CNTs can be envisioned as a layer of graphene sheet rolled up into a cylinder. They are novel 1D materials made of an sp^2-bonded wall one atom thick. Strong confinement (~ 1 nm) of charge carriers results in their unique optical properties being dominated by strongly bound excitons as revealed by the sharp optical absorption features. It would be interesting to explore whether CNTs could be responsible for some of the mysterious diffuse interstellar bands (e.g., see Zhou *et al.* 2006).

As illustrated in Figure 1b, CNTs also exhibit a broad and intense absorption feature at ~ 4.5 eV that is typically attributed to a π-plasmon excitation (e.g., see Kataura *et al.*

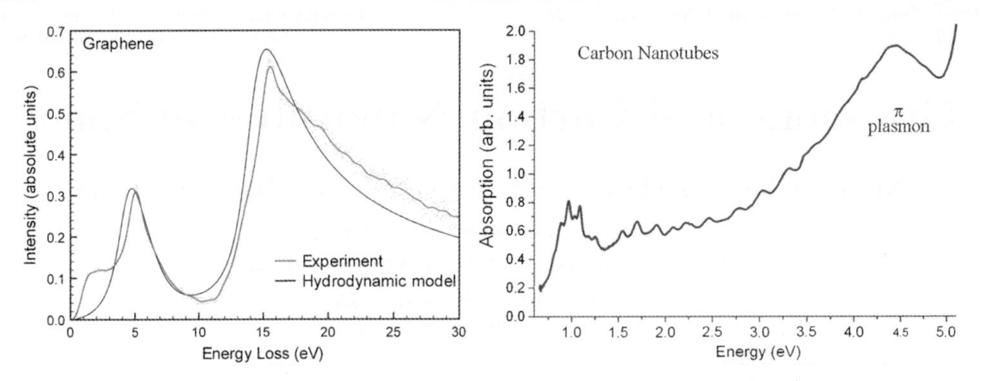

Figure 1. *Left* (a): The energy-loss function (ELF) of graphene, obtained experimentally and calculated using a two-dimensional hydrodynamic model by Nelson*et al.* (2014), is dominated by two peaks at $\sim$4.5 and $\sim$15 eV, known as π and $\pi+\sigma$ plasmons, respectively. *Right* (b): The optical absorption spectrum from dispersed single-wall CNTs measured by Kataura *et al.* (1999) exhibits a number of electronic transitons, with the π plasmon also peaking at $\sim$4.5 eV being the most prominent. The nondetection in the Galactic interstellar extinction curve of the $\sim$4.5 eV absorption peak of graphene and CNTs would allow one to place an upper limit on the amounts of graphene and CNTs in the ISM.

1999), although the exact position of this feature appears to vary with nanotube diameter (see Rance *et al.* 2010). Similar to graphene, the absence of the $\sim$4.5 eV absorption in the ISM would allow us to place an upper limit on the abundance of interstellar CNTs.

Like graphene, CNTs would emit in the IR through their C–C vibrational modes. CNTs are actually more IR-active than graphene due to their cylindrical boundary condition. Many vibrational modes in the $\sim$680–1730 cm^{-1} range have been experimentally detected for CNTs (e.g., see Kim *et al.* 2005). *JWST*'s unique high sensitivity and high resolution IR capabilities will open up an IR window unexplored by *Spitzer* and unmatched by *ISO* observations and thus will allow us to explore the possible presence of graphene and CNTs in the ISM in greater detail.

Acknowledgements

We are supported in part by NSF AST-1816411 and an AAS international travel grant.

References

Berné, O., & Tielens, A.G.G.M. 2012, *PNAS*, 109, 401
Berné, O., Mulas, G., & Joblin, C. 2013, *A&A*, 550, L4
Cami, J., Bernard-Salas, J., Peeters, E., & Malek, S. E. 2010, *Science*, 329, 1180
Chen, X.H., Li, A., & Zhang, K. 2017, *ApJ*, 850, 104
García-Hernández, D. A., Kameswara Rao, N., & Lambert, D. L. 2011, *ApJ*, 729, 126
García-Hernández, D. A., Villaver, E., García-Lario, P., et al. 2012, *ApJ*, 760, 107
Henning, Th., & Salama, F. 1998, *Science*, 282, 2204
Hudgins, D. M., & Allamandola, L. J. 2005, *IAU Symp.* 231, 443
Kataura, H., Kumazawa, Y., Maniwa, Y., et al. 1999, *Synth. Metals*, 103, 2555
Kim, U.J., Liu, X.M., Furtado, C.A., et al. 2005, *Phys. Rev. Lett.*, 95, 157402
Nelson, F. J., Kamineni, V. K., Zhang, T., et al. 2010, *Appl. Phys. Lett.*, 97, 253110
Nelson, F. J., Idrobo, J.-C., Fite, J.D., et al. 2014, *Nano Lett.*, 14, 3827
Nittler, L.R. 2018, *Geochim. Cosmochim. Acta*, 221, 1
Pendleton, Y. J., & Allamandola, L. J. 2002, *ApJS*, 138, 75
Rance, G.A., Marsh, D.H., Nicholas, R.J., et al. 2010, *Chem. Phys. Lett.*, 493, 19
Sellgren, K., Werner, M. W., Ingalls, J. G., et al. 2010, *ApJL*, 722, L54
Strelnikov, D., Kern, B., & Kappes, M. M. 2015, *A&A*, 584, A55
Yang, L., Deslippe, J., Park, C.-H., et al. 2009, *Phys. Rev. Lett.*, 103, 186802
Zhou, Z., Sfeir, M. Y., Zhang, L., et al. 2006, *ApJL*, 638, L105

Astronomy in Focus - XXX
Proceedings IAU Symposium No. XXX, 2018
M. T. Lago, ed.

doi:10.1017/S1743921319004952

Constraining dust properties in circumstellar envelopes of C-stars in the Magellanic Clouds: Optical constants and grain size of carbon dust

Ambra Nanni[1], Paola Marigo[1], Martin A. T. Groenewegen[2], Bernhard Aringer[1], Stefano Rubele[1], Alessandro Bressan[4], Léo Girardi[3], Giada Pastorelli[1] and Sara Bladh[5]

[1]Dipartimento di Fisica e Astronomia Galileo Galilei, Università di Padova, Vicolo dell'Osservatorio 3, I-35122 Padova, Italy

[2]Royal Observatory of Belgium, Ringlaan 3, B-1180 Brussel, Belgium

[3]Osservatorio Astronomico di Padova, Vicolo dell'Osservatorio 5, I-35122 Padova, Italy

[4]SISSA, via Bonomea 265, I-34136 Trieste, Italy

[5]Department of Physics and Astronomy, Uppsala University, 75120 Uppsala, Sweden
email: ambra.nanni@unipd.it

In galaxies with sub-solar metallicity, such as the Magellanic Clouds (MCs), a large fraction of the thermally pulsing asymptotic giant branch (TP-AGB) stars evolve through the carbon-rich phase (C-stars), shaping the near and mid-infrared colours of the resolved stellar populations. The spectra of C-stars are largely affected by the presence of carbon dust that condenses in their circumstellar envelopes (CSEs). The study of dust growth and radiative transfer in the CSEs of these stars allows us to investigate the properties of carbon dust. In particular, the main uncertainties in the input physics of radiative transfer models are related to the choice of the grain size distribution and of the optical constants for carbon dust. The former cannot be directly derived from observations, while for the latter several sets of lab measurements, very different from each other, are available. The results obtained by using different combinations of those inputs can be tested against the observations of thousands of C-stars in the MCs, providing constraints on the carbon dust properties. By requiring our models to simultaneously reproduce several observed infrared colour-colour diagrams (CCDs), we found the best agreement with the observations of the MCs for nano dust particles, with sizes between 0.035–0.1 μm, rather than by larger grains, of 0.2–0.7 μm (Nanni *et al.* 2016). The inability of large grains to reproduce the infrared colours is independent of the adopted optical data set and the deviations between models and observations tend to increase for increasing grain sizes. In addition to that, some sets of optical constants are always unable to reproduce the infrared colours. This investigation also allows us to identify a possible trend between the grain size or grain structure (more graphite-like or diamond like) and the mass-loss rate. The optical constants that satisfactorily reproduce the observed CCDs are adopted to compute grids of spectra as a function of the input stellar quantities. By employing these grids, we fit the spectral energy distribution of the C-stars in the MCs and we estimate their mass-loss rates and dust production rates (Nanni *et al.* 2018).

References

Nanni, A., Marigo, P., Groenewegen, M. A. T., *et al.* 2016, *MNRAS*, 462, 1215
Nanni, A., Marigo, P., Girardi, L., et al. 2018, *MNRAS*, 473, 5492

Astronomy in Focus - XXX
Proceedings IAU Symposium No. XXX, 2018
M. T. Lago, ed.

doi:10.1017/S1743921319004964

The role of alumina in triggering stellar outflows

David Gobrecht[1], John Plane[2], Stefan Bromley[3,4] and Leen Decin[1]

[1]KU Leuven,
[2]University of Leeds,
[3]Universitat de Barcelona
[4]ICREA
email: david.gobrecht@kuleuven.be

An important dust component in Asymptotic Giant Branch (AGB) stars is aluminum oxide or alumina (stoichiometric formula Al_2O_3) showing a spectral emission feature around $\sim$13 μm attributed to Al$-$O streching and bending modes (Posch *et al.* 1999; Sloan *et al.* 2003). Alumina presolar grains are also found in pristine meteorites with typical sizes of a few tens of nm to μms (Stroud *et al.* 2004). Owing to their refractory nature (thermal stability) and the large abundances of Al- and O-bearing compounds, alumina grains are thought to represent the first condensates to emerge in the atmospheres of oxygen-rich AGB stars. In the bulk phase, alumina exists predominantly in two crystalline forms (corundum and clay). The properties of nanoparticles with sizes below $\sim$50 nm, however, differ significantly from bulk properties. Quantum and surface effects of these small particles lead to non- crystalline structures, whose characteristics (geometry, coordination, density, energy) may differ by orders of magnitude, compared to the bulk material. A top-down approach, like classical nucleation theory, is thus not applicable. Therefore, we follow a bottom-up approach, starting with molecular precursors (AlO, AlOH) and the smallest stoichiometric clusters (Al_2O_3, Al_4O_6). Then, we successively build up larger-sized clusters.

We present the results of the quantum-mechanical structure calculations of $(Al_2O_3)_n$ clusters with $n = 1-10$, including potential energies, rotational constants, charge distributions and structure-specific infrared spectra (vibrational frequencies and intensities). We find new global minima candidates for cluster sizes $n = 8$, 9, and 10, that are partly reported in Gobrecht *et al.* (2018). A homogeneous nucleation, where Al_2O_3 monomers are successively added, is energetically viable in circumstellar conditions ($p = 10^{-5} - 10$ Pa, $T = 500-6000$ K). However, the formation of the monomer itself represents an energetic bottleneck. Moreover, the most stable monomer structure is a triplet state and has a spin barrier. A potential loophole are formation routes towards the dimer (Al_4O_6) without requiring the monomer as intermediary and pathways involving SiO as a catalyst.

The most intense vibrations of the small alumina clusters occur around 10-11 μm. Around 13 μm the overall IR intensity is rather low. Also other characteristics (energy, coordination, bond lengths) indicate that the bulk limit is not reached for $(Al_2O_3)_n$ clusters, $n = 1-10$, with a size range of d$\leq$1 nm.

Acknowledgements for support by the ERC consolidator grant 646758 "AEROSOL".

References

Posch, T., Kerschbaum, F., Mutschke, H., Fabian, D., Dorschner, J., & Hron, J. 1999, *A&A*, 352, 609

Sloan, G. C., Kraemer, K. E., Goebel, J. H., & Price, S. D. 2003, *ApJ*, 594, 483

Stroud, R. M., Nittler, L. R., & Alexander, C. M. O'D. 2004, *Science*, 305, 1455

Gobrecht, D., Decin, L., Cristallo, S., & Bromley, S. T. 2018, *Chem. Phys. Lett.*, in press

Astronomy in Focus - XXX
Proceedings IAU Symposium No. XXX, 2018
M. T. Lago, ed.

doi:10.1017/S1743921319004976

Polycyclic Aromatic Hydrocarbons in Protoplanetary Disks:

The 6.2/7.7 and 11.3/7.7 Band Ratios as a Diagnostic Tool

Ji Yeon Seok[1] and Aigen Li[2]

[1]National Astronomical Observatories of China,
[2]University of Missouri
email: jiseok@nao.cas.cn

Protoplanetary disks (PPDs) frequently emanate the so-called unidentified infrared emission (UIE) features in their infrared (IR) spectra (e.g., Seok & Li 2017). Major UIE features appear at 3.3, 6.2, 7.7, 8.6, 11.3, and 12.7 μm, commonly attributed to polycyclic aromatic hydrocarbon (PAH) molecules. PAHs play crucial roles in the evolution of PPDs physically and chemically. Exposed to ultraviolet photons from the central star, PAHs re-emit the absorbed energy through their vibrational relaxation via available internal modes at IR wavelengths. The relative strengths of the PAH emission bands prominently vary depending on the physical properties of PAHs such as their size (N_C) and charge state (ϕ_{ion}), which sensitively reflect the local conditions of PPDs.

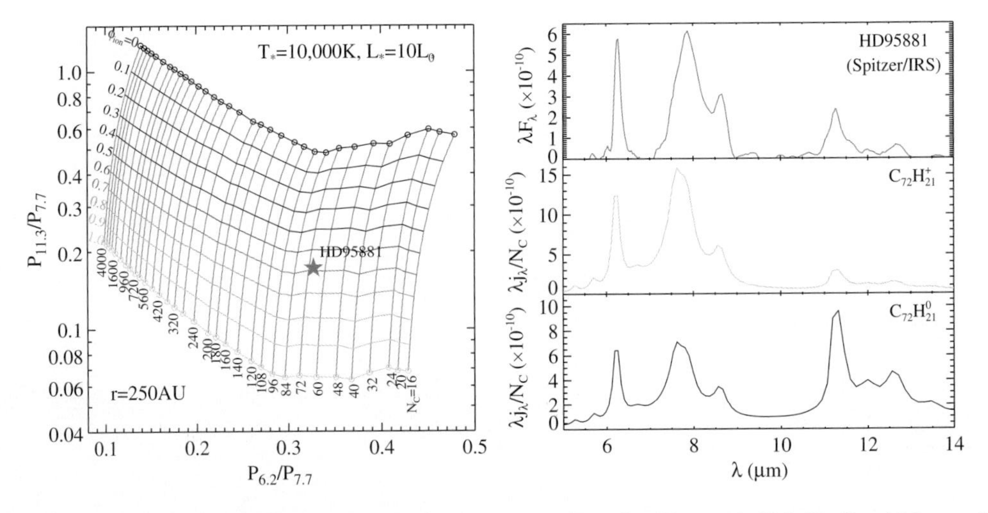

Figure 1. *Left*: A model grid of $I_{6.2}/I_{7.7}$ vs. $I_{11.3}/I_{7.7}$ for $T_{eff} = 10,000$ K, $L = 10L_\odot$ and $r = 250$ au with band ratios of HD 95881 (star symbol). *Right*: The Spitzer/IRS spectrum (top) is compared with model spectra of ionized (middle) and neutral (bottom) PAHs with $N_C = 72$.

Adopting the Astro-PAH model of Draine & Li (2007) and Li & Draine (2001), we carried out model calculations to quantitatively interpret the PAH features observed in various PPDs, taking a wide range of stellar properties such as effective temperature (T_{eff}) and luminosity (L_*) as well as PAH properties such as N_C, ϕ_{ion}, and radial distance from the central star (r) into account. For a given T_{eff}, the model emission spectra of ionized and neutral PAHs with N_C at various L_* and r are calculated. Grid diagrams of the 6.2/7.7 band ratio versus the 11.3/7.7 band ratio as a diagnostic tool allow us to directly compare observed band ratios with the models and to easily infer N_C, ϕ_{ion}, and the local physical conditions where PAHs reside. As an example, a model grid for HD 95881 is shown in Fig. 1, from which we infer that PAHs in HD 95881 at $r = 250$ au are dominated by a mixture of ionized and neutral PAHs ($\phi_{ion} \approx 0.6$) with a moderate size of $N_C = 72$.

References

Draine, B. T., & Li, A. 2007, *ApJ*, 657, 810
Li, A., & Draine, B. T. 2001, *ApJ*, 554, 778
Seok, J. Y., & Li, A. 2017, *ApJ*, 835, 291

Astronomy in Focus - XXX
Proceedings IAU Symposium No. XXX, 2018
M. T. Lago, ed.

doi:10.1017/S1743921319004988

Dust dynamics on adaptive-mesh-refinement grids: application to protostellar collapse

Ugo Lebreuilly, Benoît Commerçon and Guillaume Laibe

École normale supérieure de Lyon, CRAL, UMR CNRS 5574, Université de Lyon,
46 Allée d'Italie, 69364 Lyon Cedex 07, France
email: ugo.lebreuilly@ens-lyon.fr

We propose a method to follow the dynamics of gas and dust mixtures. We implement an algorithm (Lebreuilly *et al.* Submitted) in the adaptive-mesh-refinement code RAMSES (Teyssier 2002) that solves the monofluid equation of gas and dust mixtures in the diffusion approximation (Laibe & Price 2014). This algorithm allows an efficient simultaneous treatment of multiple dust species and is tested against canonical tests, e.g., the Dustywave, the Dustyshock and the Dustydiffuse.

A Boss and Bodenheimer (Boss & Bodenheimer 1979) protostellar collapse test of a rotating $1M_\odot$ core is performed to study the dust dynamics during the formation of the first Larson core (Larson 1969). Eight dust species are simultaneously considered. The grains, with sizes ranging from 5nm to 0.5mm, are distributed as a power-law consistent with the MRN distribution (Mathis *et al.* 1977).

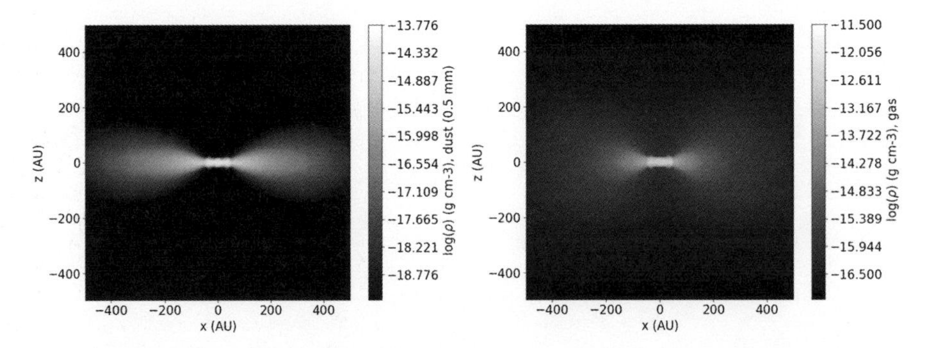

Figure 1. Edge-on cut of the collapse at $t = 120$ kyr. Dust density of the 0.5 mm grains (left) and gas density (right) for the Boss and Bodenheimer test. The color range is set so that, if the dust ratio was constant, the two maps would be the same.

Figure 1 shows the dust (0.5 mm grains) and the gas densities at $t = 120$ kyr. Dust decouples from the gas for grains larger than 100μm which is consistent with previous works (Bate & Lorén Aguilar 2017). Nano-dust remains very well coupled to the gas.

References

Bate, M. R . & Lorén, P. 2017, *Monthly Notices of the Royal Astronomical Society*, 465,1089
Boss, A. P. & Bodenheimer, P. 1979, *The Astrophysical Journal*, 234, 289
Laibe, G. & Price, D. 2014, *Monthly Notices of the Royal Astronomical Society*, 440, 2136
Larson, R. B. 1969, *Monthly Notices of the Royal Astronomical Society*, 145, 271
Lebreuilly, U. & Commerçon, B. & Laibe, G., Submitted, *Astronomy and Astrophysics*
Mathis, J. S., Rumpl, W. & Nordsieck, K. H. 1977, *The Astrophysical Journal*, 217, 425
Teyssier, R. 2002, *Astronomy and Astrophysics*, 385, 337

Astronomy in Focus - XXX
Proceedings IAU Symposium No. XXX, 2018
M. T. Lago, ed.

doi:10.1017/S174392131900499X

Dusty plasma effects in the atmosphere of Mars and near the Martian Surface

Yulia N. Izvekova and Sergey I. Popel

Space Research Institute, Russian Academy of Sciences, Moscow, 117997 Russia
email: popel@iki.rssi.ru

Dusty plasma effects in the Martian atmosphere [Izvekova & Popel (2017)] are discussed. A specific feature of the Martian atmosphere is the presence of dust grains in a wide range of altitudes (see Figure 1). Taking into account the presence of the Martian ionosphere and the high conductivity of the medium at lower altitudes, the appearance

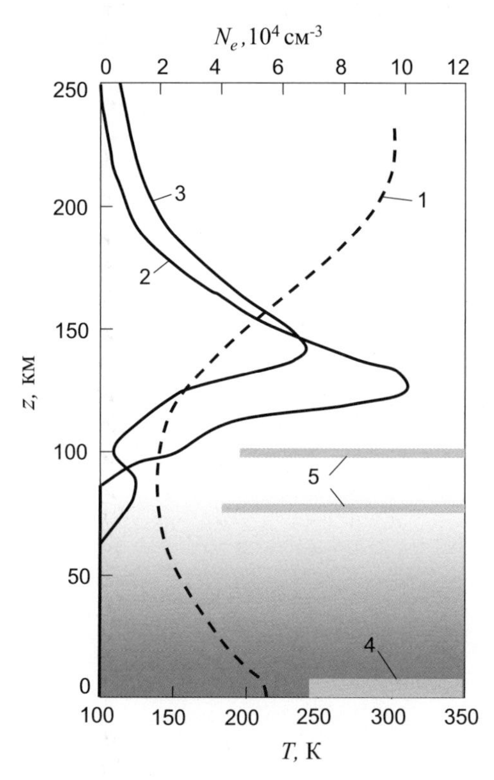

Figure 1. Altitude profiles of the (1) temperature, (2) electron number density in the daytime ionosphere (below 80 km, the number density of charged grains drops to ~ 1 cm^{-3}), and (3) number density of ionospheric electrons at the time of sporadic layer formation at altitudes of 65–100 km. Stripes 4 and 5 show the positions of the layers of clouds consisting of frozen carbon dioxide and water ice. According to observations, dust grains with number densities depending substantially on the latitude and time can be present at altitudes of up to 100 km.

of plasma systems in the Martian atmosphere can be considered quite a common phenomenon. Special attention is paid to Dust Devils that frequently form in the Martian atmosphere and can efficiently lift dust grains. The processes of dust grain charging as a result of triboelectric effect and generation of electric fields in a Dust Devil are discussed.

The dynamics of dust grains in such a vortex is simulated with allowance for their charging and the generated electric field. This work was supported by the Russian Foundation for Basic Research (project no. 18-02-00341).

Reference

Izvekova, Yu. N., & Popel, S. I. 2017, *Plasma Physics Reports*, 43, 1172

Astronomy in Focus - XXX
Proceedings IAU Symposium No. XXX, 2018
M. T. Lago, ed.

doi:10.1017/S1743921319005003

Fragmentation and molecular growth of polycyclic aromatic hydrocarbons in the interstellar medium

Tao Chen[1,2]

[1]Leiden University, Leiden Observatory, Niels Bohrweg 2, NL-2333 CA Leiden, Netherlands

[2]School of Engineering Sciences in Chemistry, Biotechnology and Health, Department of Theoretical Chemistry & Biology, Royal Institute of Technology, 10691, Stockholm, Sweden

email: chen@strw.leidenuniv.nl

The interstellar medium (ISM) can be considered a molecular factory, in which a rich organic inventory has been identified through radio and infrared observations (Tielens 2008). This work review the recent experimental and theoretical studies on fragmentation and molecular growth of polycyclic aromatic hydrocarbon molecules in the ISM.

The gas-phase molecules can be ionized, isomerized, dissociated following the impacts of ions, atoms, or photons (Chen *et al.* 2014; Stockett *et al.* 2014, 2015; Zhen *et al.* 2016). Two types of fragmentation processes are found: statistical and non-statistical fragmentations. In statistical fragmentation processes, the losses of $C_{2n}H_x$ dominate the mass spectra (Chen *et al.* 2015). Non-statistical fragmentations represent direct knock-out of atoms from a molecule, in which single C-losses are commonly observed on the mass spectra (Stockett *et al.* 2014). The fragmentation of molecules lead to the formation of small molecules or isomerize to more stable structure, such as C_{60} or C_{70} (Zhen *et al.* 2014). On the other hand, small molecules may conglomerate a weakly bonded cluster, and form large molecules in the cluster following statistical or non-statistical fragmentations (Zettergren *et al.* 2013; Delaunay *et al.* 2015; Zhen *et al.* 2018). New molecules can be formed in such cycle, e.g., a bowl-shape molecule can be formed in the photodissociation experiement of bisanthenequinone cations (Chen *et al.* 2017), dumbbell C_{118} and C_{119} are formed inside clusters of C_{60} by collision with α particles (Zettergren *et al.* 2013), etc. These reactions enrich the molecular inventory in the ISM and may act as the first step towards life.

References

Chen, T., Zhen, J., Wang, Y., Linnartz, H., & Tielens, A. G. 2017, *Chem. Phys. Lett.*, 692, 298
Chen, T., Gatchell, M., Stockett, M. H., *et al.* 2014, *J. Chem. Phys.*, 140, 224306
Chen, T., Gatchell, M., Stockett, M. H., *et al.* 2015, *J. Chem. Phys.*, 142, 144305
Delaunay, R., Gatchell, M., Rousseau, P., *et al.* 2015, *J. Phys. Chem. Lett.*, 6, 1536
Stockett, M. H., Zettergren, H., Adoui, L., *et al.* 2014, *Phys. Rev.* A, 89, 032701
Stockett, M. H., Gatchell, M., de Ruette, N., *et al.* 2015, *Intl. J. Mass Spec.*, 392, 58
Tielens, A.G.G.M. 2008, *Annu. Rev. Astron. Astrophys.*, 46, 289
Zettergren, H., Rousseau, P., Wang, Y., *et al.* 2013, *Phys. Rev. Lett.*, 110, 185501
Zhen, J., Castellanos, P., Paardekooper, D. M., Linnartz, H., & Tielens, A. G. 2014, *ApJL*, 797, L30
Zhen, J., Chen, T., & Tielens, A. G. 2018, *ApJ*, 863, 128
Zhen, J., Castillo, S. R., Joblin, C., *et al.* 2016, *ApJ*, 822, 113

Astronomy in Focus - XXX
Proceedings IAU Symposium No. XXX, 2018
M. T. Lago, ed.

doi:10.1017/S1743921319005015

Molecular dynamics simulations and anharmonic spectra of large PAHs

Tao Chen[1,2]

[1]Leiden University, Leiden Observatory, Niels Bohrweg 2, NL-2333 CA Leiden, Netherlands
[2]School of Engineering Sciences in Chemistry, Biotechnology and Health, Department of Theoretical Chemistry & Biology, Royal Institute of Technology, 10691, Stockholm, Sweden
email: `chen@strw.leidenuniv.nl`

Due to the difficulties in obtaining high-resolution infrared (IR) spectra of polycyclic aromatic hydrocarbon molecules (PAHs) from experiments, current study of PAHs have led to an ever-increasing reliance on computational quantum chemistry. Our recent results show that the second-order vibrational perturbations theory (VPT2) produce accurate anharmonic spectra, which are consistent well with the high-resolution low-temperature gas-phase experimental spectra of PAHs (Chen 2018). However, such method suffers from low efficiency of calculation, it only works for small molecules (less than 24 C-atoms). Moreover, high symmetric (D6h) molecules (e.g. coronene and circumcoronene) can not be calculated with such method (Mackie *et al.* 2016), but these species are actually expected to be highly abundant in space given their remarkable stability (Bauschlicher *et al.*2008).

Recently, we apply molecular dynamics (MD) simulations for producing anharmonic IR spectra of PAHs. In order to reduce the computational cost, the semi-empirical methods are utilized, which produce the potential energy surface (PES) efficiently at each step. The results are validated against the experimental spectra. A rather low value of the mean absolute error can be achieved with certain semi-empirical method and appropriate settings of the MD simulations, see our recent article for details (Chen *et al.* 2018).

As no assumptions about the shape of the PES is made, MD intrinsically accounts for anharmonicity, ro-vibrational couplings and temperature effects. In addition, MD is a time-dependent method, which has no restriction on the symmetry of the molecules. Therefore anharmonic IR spectra of molecules with D6h symmetry can also be produced by MD simulations. Using MD simulations, we manage to produce high-temperature anharmonic IR spectra of D6h PAHs, e.g. coronene and circumcoronene (Chen *et al.* 2018).

References

Bauschlicher Jr, C. W., Peeters, E., & Allamandola, L. J. 2008, *ApJ*, 678, 316
Chen, T. 2018, *ApJS*, in press
Chen, T., Luo, Y., Duan, S., *et al.* 2018, in press
Mackie, C. J., Candian, A., Huang, X., *et al.* 2016, *J. Chem. Phys.*, 145, 084313

Astronomy in Focus - XXX
Proceedings IAU Symposium No. XXX, 2018
M. T. Lago, ed.

doi:10.1017/S1743921319005027

The Cassini RPWS/LP Observations of Dusty Plasma in the Kronian System

M. W. Morooka[1], J.-E. Wahlund[1], L. Hadid[1], A. Eriksson[1], E. Vigren[1], N. Edberg[1], D. Andrews[1], A. M. Persoon[2], W. S. Kurth[2], S.-Y. Ye[2], G. Hospodarsky[2], D. A. Gurnett[2], W. Farrell[3], J. H. Waite[4], R. S. Perryman[4], M. Perry[5] and O. Shebanits[6]

[1]Swedish Institute of Space Physics, Uppsala, Sweden
[2]University of Iowa, Department of Physics and Astronomy, Iowa City, IA, USA
[3]NASA/Goddard Space Flight Center, Greenbelt, MD, USA
[4]Southwest Research Institute, San Antonio, TX, USA
[5]Johns Hopkins University, Applied Physics Laboratory, Laurel, MD, USA
[6]Imperial College London, London, UK
email: morooka@irfu.se

Saturn's icy moons and ring system are ideal places to study the nm-dust. Nearly thirteen years of its orbital period around Saturn Cassini revealed the presence and the size/charge state of nm to μm dust in many places. Observations in different places shows the different type of generation mechanisms (coagulation and fragmentation) of grains depending on the different environments. The electrical dust-plasma interaction was first identified in the E ring (Wahlund *et al.* 2009). The E ring dust consists of mostly 0.01-100 μm ice grains, and the ring has a structure that the dusty plasma of smaller nm-grains surrounds the ring core consisting of larger (μm) sized dust. Similar dusty plasma structure was found near the F ring as well (Morooka *et al.* 2011). The Water jets out of the Enceladus Tiger strips create a plume of dusty plasma. Multi-instruments study confirmed a size distribution of the dust in the plume (e.g., Morooka *et al.* 2011) and that magnetic field in the plume is affected by the charged dust (Engelhardt *et al.* 2015). In Titans atmosphere and ionosphere Multi-instruments observations confirmed series of hydrocarbon-nitrile compounds (Waite *et al.* 2007) as well as positively and negatively charged heavy ions of up to 50,000 amu (Coates *et al.* 2007). The observation is a signature of aerosol formation. Statistical study revealed the EUV control of this formation (Shebanits *et al.* 2017). Aerosols grows as a result of the complex chemistry and the charge balance. During the last phase of the mission, called the Grand Finale, Cassini made 22 Saturn flybys drove in between the D ring and atmosphere of Saturn. The results of different dust and plasma characteristics depending on the altitude of Saturn indicates that the ring materials are falling down to the planet, accumulated in the ionosphere to create a dust layer, and create a dusty negative ion ionosphere of Saturn (Morooka *et al.* 2018b).

References

Coates, A., *et al.*, *Geophys. Res. Lett.*, 34, 22, 2007
Engelhardt, I.A.D., *et al.*, *Planet. Space Sci.*, 117, 453, 2015
Morooka, M.W., *et al.*, *J. Geophys. Res.* 116, A12, 2011
Morooka, M.W., *et al.*, *J. Geophys. Res.*, 123, 4668, 2018a

Morooka, M.W., *et al.*, *J. Geophys. Res.*, submitted, 2018b
Shebanits, O., *et al.*, *J. Geophys. Res.*, 122, 7491, 2017
Waite, J.H., *et al.*, *Science*, 316, 870, 2007
Wahlund, J.-E., M. André, A. Eriksson, *et al.*, *Planet. Space Sci.*, 57, 1795, 2009
Wahlund J.-E., M. W. Morooka, L. Hadid, *et al.*, *Science, eaao4134*, 2017

Astronomy in Focus - XXX
Proceedings IAU Symposium No. XXX, 2018
M. T. Lago, ed.

doi:10.1017/S1743921319005039

Formation and interaction of nano dust in planetary debris discs

Ingrid Mann[1], Johann Stamm[1], Margareta Myrvang[1], Carsten Baumann[1], Saliha Eren[1], Andrzej Czechowski[2] and Aigen Li[3]

[1]UiT The Arctic University of Norway, Tromsø, Norway
[2]Space Research Center, Polish Academy of Sciences, Warsaw, Poland
[3]University of Missouri, Columbia, Missouri, USA
email: ingrid.b.mann@uit.no

The circum-stellar planetary debris discs are typically observed by thermal emission in mid-infrared and located at large distance from the star, comparable to the solar system's Kuiper belt. They are produced by fragmentation of small objects and dust-dust collisions, but with higher rates and the dust clouds are denser than in the solar system. Most of the small dust particles are pushed away from the vicinity of the star by radiation pressure. Some stars, however, reveal thermal emission spectra that suggest the existence of dust relatively close to the star, denoted as hot debris discs (Absil *et al.* 2013; Su *et al.* 2013, 2016). To explain the observations, some authors consider trapping of nanodust by the stellar magnetic field (Su *et al.* 2016), a process that we previously found for the dust in the vicinity of the Sun (Czechowski & Mann 2010).

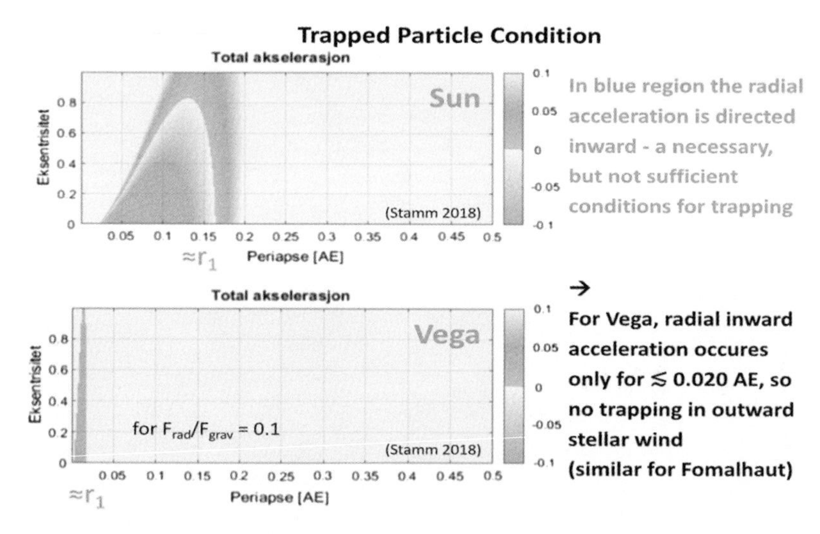

Figure 1. Comparison of the nanodust trapping for Sun and Vega. The blue area in the eccentricity-periapsis plot indicates orbital parameters where trapping is possible (from Stamm 2018).

We here discuss the cases of Vega and Fomalhaut. From trajectory calculations we find that the trapping conditions are different from the case of the Sun and that a hypothetical trapping zone would only occur for a small range of orbital parameters in the very close vicinity of the star (see Fig. 1). Our estimate is based on applying a Parker-type magnetic field model (Stamm 2018) to the conditions of Vega and Fomalhaut, as well as radiation pressure force which for both stars is larger than gravity. Other groups find trapping based on a different magnetic field model. We also consider the thermal emission brightness that was brought up as another argument to support the existence of nm-sized dust. We find that the brightness observations can be explained with a dust component within 1 AU around the stars and with dust sizes below micrometer (Myrvang 2018).

This research is funded by the Research Council of Norway (grant number 262941). Results include master thesis project work of J.S. and M.M.

References

Absil, O., Defrére, D., Coudé du Foresto, V., *et al.* 2013, *Astron. Astrophys.* 555, A104
Czechowski, A., Mann, I. 2010, *Astrophys. J.* 714, 89
Lebreton, J., van Lieshout, R., Augereau, J.-C. *et al.* 2013, *Astron. Astrophys.* 555, 2013, A146
Myrvang, M. 2018, *Master Thesis*, UiT The Arctic University of Norway, Tromsø
Stamm, J., 2018, *Master Thesis*, UiT The Arctic University of Norway, Tromsø
Su, K.Y.L., Rieke, G.H., Malhotra, R. *et al.* 2013, *Astrophys. J.* 763, A118
Su, K.Y.L., Rieke, G.H., Defrére, D., *et al.* 2016, *Astrophys. J.* 818, A45

Astronomy in Focus - XXX
Proceedings IAU Symposium No. XXX, 2018
M. T. Lago, ed.

doi:10.1017/S1743921319005040

Flow behind an exponential cylindrical shock in a rotational axisymmetric mixture of small solid particles of micro size and non-ideal gas with conductive and radiative heat fluxes

G. Nath

Motilal Nehru National Institute of Technology Allahabad, Allahabad,
Uttar Pradesh, India, 211004
email: gnath@mnnit.ac.in

Similarity solutions for the propagation of exponential cylindrical shock wave in a rotational axisymmetric mixture of small solid particles of micro size and non-ideal gas, with conductive and radiative heat fluxes are discussed. The shock wave is driven out by a piston moving with time according to exponential law and the gas is taken to be non-ideal like Wu & Roberts (1993) and Roberts & Wu (1996). The axial and azimuthal components of the fluid velocity in the undisturbed medium are assumed to be varying and obeying exponential laws. The dusty gas is taken to be a mixture of small solid particles of micro size and non-ideal gas, in which solid particles are continuously distributed in the mixture like Pai *et al.* (1980), Miura & Glass (1983). The radiation is considered to be of diffusion type for an optically thick grey gas model and heat conduction is expressed in terms of Fouriers law like Vishwakarma & Nath (2010). The effects of the variation of the mass concentration of solid particles in the mixture, the ratio of the density of solid particles to the initial density of the gas, the heat conductive and radiative heat transfer parameters, and the parameter of the non-idealness of the gas are worked out in detail. It is found that an increase in the ratio of the density of solid particles to the initial density of the gas or the conductive heat transfer parameter or the radiative transfer parameter increases the compressibility of the mixture in the flow field behind the shock front, and hence there is an increase in the shock strength. Also, it is shown that an increase in the parameter of non-idealness of the gas has decaying effect on the shock wave. The potential applications of this study include analysis of data from exploding wire experiments in dusty medium, and cylindrically symmetric hypersonic flow problems associated with meteors or reentry vehicles (c.f. Hutchens (1995)).

Acknowledgement

Readers are referred to Nath, G., 2018, *Acta Astronatica*, 148, 355.

References

Hutchens, G. J. 1995, *J. Appl. Phys.*, 77, 2912
Miura, H., & Glass, I. I. 1983, *Proc. Roy. Soc. London A*, 385, 85
Pai, S.I., Menon, S, & Fan, Z. Q. 1980, *Int. J. Eng. Sci.*, 18, 1365
Roberts, P.H., & Wu, C.C. 1996, *Phys. Lett. A*, 213, 59
Vishwakarma, J. P., & Nath, G. 2010, *Phys. Scri.* 81, 045401
Wu, C.C., & Roberts, P.H. 1993, *Phys. Rev. Lett.*, 70, 3424

FM11
JWST: Launch, Commissioning, and Cycle 1 Science

Astronomy in Focus - XXX
Proceedings IAU Symposium No. XXX, 2018
M. T. Lago, ed.

doi:10.1017/S1743921319005052

JWST: Launch, Commissioning, and Cycle 1 Science

Bonnie Meinke[1] & Stefanie Milam[2], eds

[1]Space Telescope Science Institute,
3700 San Martin Drive, Baltimore, MD, 21218, USA
email: meinke@stsci.edu

[2]NASA Goddard Space Flight Center,
8800 Greenbelt Road, Greenbelt, MD, 20771, USA
email: stefanie.n.milam@nasa.gov

Introduction

The James Webb Space Telescope (JWST) is expected to revolutionize our understanding of the near- to mid-infrared sky by enabling observations with an unprecedented combination of superb angular resolution and sensitivity. Since JWST is a general-purpose observatory, its scientific success is dependent on the broader scientific community to make new discoveries. The 2018 IAU General Assembly served as an important international platform and offered an ideal opportunity to inform the broader community about current JWST status, plans for commissioning and Cycle 1 science, and what to expect in the near-future for JWST. This was accomplished as part of a JWST focus meeting, held at the beginning of the General Assembly, August 20-22, 2018. During the first day of the meeting, we opened our session with a Key Note from Dr. Ewine van Dishoeck, the new IAU president, who provided a brief overview of the mission and the revolutionary science anticipated. This was then followed by members of the design team for the four JWST science instruments (MIRI, NIRCam, NIRSpec, and NIRISS) discussed the science potential of their instruments. Presentations set the context for the many technologies developed and challenges overcome along the way to a space telescope, alongside the anticipated science returns. The focus meeting also highlighted the science to be enabled by JWST early in its life cycle and touched on commissioning, Early Release Science (ERS), Guaranteed Time Observer (GTO), and General Observer (GO) programs slated for cycle 1. Over the second two days of the meeting, talks and discussion sessions centered around the broad science topics enabled by JWST, from our solar system to the edge of the universe. Speakers represented topics across the entire astronomical community, including: interstellar matter, the local universe, galaxies, cosmology, stars and stellar physics, planetary systems, and bioastronomy. Given the cross-disciplinary and international nature of JWST's mission, this focus meeting was the ideal opportunity to discuss the science that will be enabled with JWST in the near-term. In addition, the focus meeting welcomed officials from NASA and ESA to update delegates on the current status of JWST and what to expect before science operations commence in 2021. Meanwhile, poster presentations offered delegates an opportunity to explore certain science topics more in depth, learn about proposal tools, and to explore possible new observation techniques. Details, including presentation slides and posters can be found on the JWST Observers website: https://jwst.stsci.edu/news-events/events/events-area/stsci-events-listing-container/jwst-launch-commissioning-and-cycle-1-science?mwc=4

A taster of JWST science

Ewine F. van Dishoeck Leiden Observatory, Leiden University, P.O. Box 9513, NL-2300 RA, Leiden, the Netherlands
email: ewine@strw.leidenuniv.nl

Abstract. JWST promises to examine every phase of cosmic history: from the epoch of re-ionization after the Big Bang to the formation of galaxies, stars and planets, the atmospheres of exoplanets and the evolution of our own solar system. Its leap in sensitivity, angular resolution and broad wavelength coverage from optical to mid-infrared compared with other missions will ensure major steps forward in many areas. At the GA, a brief 'taste' of JWST science has been presented, and synergies with other major facilities have been emphasized.

The four JWST instruments – NIRCam, NIRSpec, MIRI and FGS/NIRISS – cover the 0.6 – 28.8 μm range with imaging and spectroscopy up to R=3000 and with improved sensitivities up to two orders of magnitude. Examples of key science areas are:

• *High-redshift galaxies:* Because JWST sees much sharper and deeper it can identify better the youngest galaxies and star clusters. With many more filters, photometric redshifts are more reliable, and thus also the luminosity function at the faint end $z > 8$. Spectra provide key diagnostics of star formation rates, metallicity and hardness of the UV, as well as the photon escape fraction around the time of reionization.

• *Galaxy assembly:* The bulk of the stars in the Universe were formed at $1 < z < 6$. JWST can study the structure (e.g., gas accretion, mergers, disks) and physics of galaxies (e.g., ionization, outflows, AGN feedback) up to $z \approx 7$ with the same detail as is now done at $z \approx 2$. Spatially resolved stellar populations in nearby galaxies allow their star formation histories to be reconstructed ('archaeology'). JWST will be most powerful in the outskirts of galaxies (out to the Virgo cluster) whereas ELTs can sample the more crowded galaxy centers.

• *Galactic protostars and disks.* JWST can probe the physical processes by which stars and disks form and evolve through continuum imaging and spatially resolved infrared diagnostic lines that probe accretion, shocks, PDRs and high energy photons, e.g., H I recombination, atomic fine structure and H_2, OH and H_2O pure rotational lines, PAH features. Also, the onset of chemical complexity in ices that may become part of new solar systems can be observed in the 5–10 μm fingerprint region. In mature disks, JWST can probe the hot chemistry in the inner few au where terrestrial planets are forming.

• *Exoplanets.* JWST can perform deep searches for young planets in transitional and debris disks for which ALMA or scattered light images have suggested their presence. Direct imaging spectroscopy of mature exoplanets at large distances from their parent star will provide unprecedented information on their atmospheres. The broad wavelength coverage of JWST will allow much more accurate retrieval of atmospheric abundances (i.p., C/O) through transit spectroscopy. Earth-size ocean planets around M stars may be just reachable.

NIRCam: New Science Near and Far

Marcia Rieke
Steward Observatory, University of Arizona,
933 N. Cherry Ave., Tucson, AZ 85721, United States
email: mrieke@as.arizona.edu

Abstract. The near-infrared camera, NIRCam, for JWST is a versatile instrument capable of obsevations ranging from direct imaging of exoplanets to searching for distant galaxies.

NIRCam provides two types of data for the JWST mission: it is the 0.6 to 5 um imager, and it is also the wavefront sensor used in maintaining the shape and phasing of the primary mirror. NIRCam is fully redundant with two benches mounted back-to-back, each with a full optical train. Either module can acquire the needed data for wavefront sensing with a substantial benefit to survey work as both modules can be run simulataneously which doubles the field of view to ~9.7 square arc min. The weak lenses used in wavefront sensing may also be useful for science imaging to spread starlight over a large number of pixels. Because NIRCam covers such a large wavelength range, each module is comprised of short and long wavelength arms with a dichroic beamsplitter dividing the light. The short wavelength arm covers 0.6 to 2.3μm and has a pixel scale of ~0.32 arc sec per pixel while the long wavelegnth arm covers 2.45 to 5μm with a pixel scale of ~0.64 arc sec per pixel. Both arms can be used at the same time to improve survey efficiency with two wavelengths observed at once. Each arm is equipped with filter and pupil wheels with a selection of broad (R~4), medium (R~10) and narrow (R~100) filters. The long wavelength arms also includes R~1500 grisms for slitless spectroscopy and transit observations. See Greene *et al.* (2017) for more details and Beichman *et al.* (2014) for transit use. Each module also includes a selection of coronagraphic masks and each arm includes Lyot stops for coronagraphy. Use of NIRCam's coronagraphs for direct imaging of planets is explored in Beichman *et al.* (2010). A prime program for the NIRCam instrument team will be carried out in collaboration with the NIRSpec instrument team to find and characterize the most distant galaxies. NIRCam will be used to observe ~46 square arc min to 1-σ =0.4 nJy at 2μm and ~190 square arc min to 1-σ =0.9 nJy at 2μm. These areas will be observed using seven wide filters covering 0.9 to 5μm. Photometric redshifts will be estimated from the NIRCam data, and then a selection of targets for detailed NIRSpec spectroscopy at both R~100 and R~1000 will be made. These observations should allow the collaboration to examine galaxy evolution through z~10 and beyond.
For more information about NIRCam, see http://ircamera.as.arizona.edu/nircam/ and https://jwst-docs.stsci.edu/display/JTI/NIRCam+Observing+Modes

The JWST near-infrared spectrograph NIRSpec

Catarina Alves de Oliveira[1]
This presentation is made on behalf of the NIRspec instrument team, and is based on the work done by a large number of people involved in the project.
European Space Agency, c/o STScI, 3700 San Martin Drive, Baltimore, USA
email: `catarina.alves@esa.int`

Abstract. The near-infrared spectrograph NIRSpec is one of four instruments aboard the James Webb Space Telescope (JWST). NIRSpec is developed by ESA with AIRBUS Defence & Space as prime contractor. It offers seven dispersers covering the wavelength range from 0.6 to 5.3 micron with resolutions from R~100 to R~2700. Using an array of micro-shutters, NIRSpec will be capable of obtaining spectra for over 100 objects simultaneously. It also features an integral field unit with a 3 by 3 arcseconds field of view, and various slits for high contrast spectroscopy of individual objects and time series observations, including those of transiting exoplanets. We will provide an overview of the capabilities and performances of the three observing modes of NIRSpec, and how these are linked to the four main JWST scientific themes.

Keywords. instrumentation: spectrographs

Overview of NIRSpec's capabilities

NIRSpec is a versatile near-IR spectrograph equipped with a micro-shutter array (MSA), an integral field unit for 3D spectroscopy, and five fixed slits for high-contrast spectroscopy, including a wider slit for bright object times series. It will be the first MOS in space, with a quarter of a million micro-shutters that can be individually addressed to be open or closed. Some aspects of planning NIRSpec/MOS observations are somewhat different than what is done on the ground. The MSA is fixed grid structure with several factors that impact the degree of multiplexing: it is a strong function of the density of targets in the input catalogue; in a typical use, the small shutters are combined to make up a *slitlet*, which together with the dithering strategy constrains the number of observable objects; not all shutters are viable (avoid spectral overlap, truncation, operability); constrains on centering are relevant for spectrophotometric accuracy but strongly impact multiplexing. Simulations show that with the current state of the MSA, for the densest target fields, one expects to observe in a single exposure about ~200 targets for the PRISM (R~100) and ~60 targets at R~1000. An important aspect to highlight is that NIRCam imaging and NIRSpec MOS follow-up observations will be allowed within the same observing cycle, resulting in a fast turn-around of scientific results. Another opportunity to maximize science return with JWST is the possibility to acquire NIRCam imaging in parallel to NIRSpec MOS observations. The impressive versatility of the NIRSpec instrument and its modes makes it a powerful and attractive instrument from the scientific point of view, but turns it into a challenge when it comes to calibration. Data from ground-test campaigns and simulations are being extensively used to develop and test the NIRSpec calibration concept, which will be verified and revisited once JWST is in orbit.

Science with NIRISS: The Near-Infrared Imager and Slitless Spectrograph

Chris J. Willott[1] *and the NIRISS Instrument Science Team*
[1]*NRC Herzberg,*
5071 West Saanich Rd, Victoria, BC V9E 2E7, Canada
email: chris.willott@nrc.ca

Abstract. The NIRISS instrument on JWST provides four imaging and spectroscopic modes at wavelengths from 0.6 to 5 microns. These observing modes enable diverse science applications, from characterizing exoplanets that could host life to understanding the formation of galaxies prior to cosmic reionization.

The four observing modes of NIRISS are selected by choosing the relevant filters, dispersers or mask in the Filter Wheel and Pupil Wheel. The detector is a HAWAII-2RG HgCdTe device with 5.2μm cutoff, covering 2.2×2.2 arcmin^2 in full-frame.

1. Wide-Field Slitless Spectroscopy (WFSS)

NIRISS contains two orthogonally-oriented, low-resolution ($R \approx 150$) grisms that disperse spectra of all the sources in the field across the detector. One of six band-limiting filters from 0.8 to 2.3μm is used to limit the spectra length, mitigating source overlap and reducing background. WFSS enables highly multiplexed spectroscopy with several thousand spectra per field in deep observations. Galaxy emission line maps can be generated to measure resolved metallicity and ionization of nebular gas, to investigate how gas cycles in and out of galaxies. Other WFSS applications include searches for blind emission line galaxies prior to cosmic reionization and cool substellar objects.

2. Single-Object Slitless Spectroscopy (SOSS)

The SOSS mode uses a cross-dispersing grism to provide a spectrum from 0.6 to 2.8μm at $R \approx 700$. The primary use case is time-series observations of bright, variable sources. By observing exoplanet host stars during transit, the planet atmospheric composition and physical properties can be measured. NIRISS SOSS covers transitions of important molecules such as water, carbon monoxide, hydrogen cyanide, methane, and ammonia.

3. Aperture Masking Interferometry (AMI)

The non-redundant mask in NIRISS enables high-contrast imaging at resolution $0.5\lambda/D$, the smallest inner working angle of any JWST imaging mode. This enables the detection of close-in planets and disks and high-resolution study of some extended objects such as galactic nuclei.

4. Imaging

The 12 filters in NIRISS can be used for imaging. Although NIRCam is the primary near-IR imager for JWST, NIRISS can be used in parallel with NIRCam for a 50% increase in imaging area. NIRISS imaging is also critical in WFSS operations to provide the source astrometry and wavelength solution, and to model spectral overlap.

More information on NIRISS and science applications at https://jwst-docs.stsci.edu.

Observing the evolving Interstellar Medium in galaxies with the James Webb Space Telescope

Francisca Kemper[1,2]
[1]*Academia Sinica, Institute of Astronomy and Astrophysics, 10617 Taipei, Taiwan*
email: ciska@asiaa.sinica.edu.tw
Present address: [2]*European Southern Observatory, Karl-Schwarzschild-Str. 2, 85748 Garching bei München, Germany*

Abstract. The James Webb Space Telescope (JWST) will excel at studying the Interstellar Medium (ISM) of galaxies at unprecedented levels of detail owing to its combination of sensitivity, the availability of integral field units, and high spectral resolution in the MIRI instrument. It is well-placed to enhance our understanding of the chemical and physical changes of the ISM in the context of galaxy evolution.

Keywords. infrared: galaxies, infrared: ISM, ISM: lines and bands, ISM: evolution, galaxies: ISM

Previous space-based infrared observatories, like *ISO* and *Spitzer*, have revealed a plethora of spectroscopic features with diagnostic properties to observe the ISM. These features include amorphous and crystalline silicates; carbonaceous features such as polycyclic aromatic hydrocarbons (6.2, 7.7, 8.6, 11.3 and 12.7 μm) and fullerenes; various ices (methanol, CO, CO_2 and water ice); and molecular hydrogen (H_2).

The infrared molecular hydrogen lines provide a way to study the molecular content of low-metallicity galaxies, as in low-metallicity environments the CO dissociation occurs much further into the molecular clouds due to the low abundance of CO and the lower level of self-shielding. Using CO as a proxy for the molecular mass in low-metallicity galaxies overlooks the CO-dark gas, which may be a substantial fraction of the total molecular gas reservoir.

Another important diagnostic is provided by the Polycyclic Aromatic Hydrocarbons (PAHs), which show several resonances in the mid-infrared, due to specific bending and stretching modes in these molecules. For instance, the band ratio between the 11.3 and 12.7 μm features is a measure for the compactness of the PAHs, with less compact molecules presumed to be eroded by UV irradiation. The 11.3/7.7 μm versus 6.2/7.7 μm ratio provides a diagnostic that is sensitive to both grain size and ionization fraction at the same time. In spectral maps, changes in these band ratios reveal gradients in the conditions in the ISM.

Using mid-infrared spectroscopy it is also possible to determine the crystalline fraction of silicates. Crystalline silicates are only formed above $\sim$1000 K, while the silicates retain their lattice structure upon cooling. Amorphization occurs due to prolongued exposure to cosmic ray hits. Thus, the crystalline fraction reveals information about the formation history of silicates in stellar ejecta, and their residence time in the ISM. Crystalline silicates are absent in normal galaxies, but starburst galaxies show detectable amounts of crystallinity, and spectrally mapping such galaxies with MIRI may reveal hotspots in the star formation activity of these galaxies.

Galactic Nuclei Studies with JWST

Nora Luetzgendorf[1]

This presentation is made on behalf of the NIRspec/MIRI GTO team, and is based on the work done by a large number of people involved in the project.

[1] *ESA/Space Telescope Science Institute,*
3700 San Martin Dr, Baltimore, MD 21218, United States
email: nluetzge@cosmos.esa.int

Abstract. The nuclei of galaxies and their immediate vicinity are unique laboratories for a number of complex physical processes. Observing a small number of selected nearby galactic nuclei with the IFUs of both NIRSpec and MIRI in the framework of the JWST GTO program will allow us to study them in unprecedented detail over the entire near- and mid-infrared spectral range ($0.7 - 28\mu$m). The unrivaled sensitivity and continuous wavelength coverage offered by the combination of JWST and NIRSpec/MIRI will provide access to a multitude of diagnostic spectral features, and mapping these over the central few hundred pc with the unique spatial resolution of JWST ($\sim$0.1, i.e. a few pc for nearby galaxies) will break new ground even for these relatively well-studied objects.

Keywords. galaxies: active, galaxies: nuclei, stars: kinematics

---

Table 1. Overview of the target sample.

Name	Distance	Comment
Merkarian 231	175 kpc	nearest IR-bright quasar
Centaurus A	4 Mpc	closest radio galaxy
NGC 6240	97 Mpc	binary black hole
Arp 220	77 Mpc	closest ULIRG
Galactic Center	8 kpc	closest SMBH
NGC 4654	16 Mpc	double nuclear star cluster

Introduction

For the foreseeable future, JWST will be the most sensitive (and, in fact, only) facility to provide continuous spectra over the entire near- and mid-infrared spectral range (0.7 – 28μm). The NIRSpec and MIRI IFU data will provide 2-d maps for across this entire range with a resolution of $R \sim 2700 - 3000$, and thus will enable a full characterization the nuclear environment of the target galaxies on scales of a few pc.

The key objective of this joint GTO program between the NIRSpec and MIRI GTO teams is to obtain deep IFU data cubes of selected galaxy nuclei over the entire JWST wavelength range, and to employ the versatile tool kit of IR spectral diagnostics to better understand their physics. Because many of the diagnostic lines are rather faint, and ground-based IR instruments cannot reach the required S/N ratios due to the enormous thermal background from the Earth's atmosphere (or are absorbed by it), the use of JWST is mandatory for this purpose.

Scientific goals

The galaxies to be observed are summarized in Table 1. In what follows, we describe some of the specific questions that will be addressed by this program.

Highly Obscured Nuclei All nuclei in our sample exhibit strong dust extinction, especially the Galactic Center and Cen A. At optical wavelengths, it is therefore difficult to derive the true emission line fluxes and - especially - ratios. In the NIR/MIR regime, in contrast, these problems are much more manageable because the extinction is drastically reduced, and can be more easily quantified. The hydrogen recombination line ratios in the NIRSpec spectral range, in particular, will provide maps of the patchy obscuration towards the ionized gas on scales of a few pc.

Kinematics of Gas and Stars The combined integral-field spectra from NIRSpec and MIRI provide a unique data set to investigate the kinematics of stars as well as of the different phases of the ISM, from warm molecular gas to hot coronal gas. For the stars, the strong CO absorption bands at 2.4m will provide the stellar velocity field on the finest spatial scales, and thus establish accurate measurements of the enclosed dynamical mass. A similar analysis can be done for the various gas phases, and comparison of gaseous and stellar rotation curves yields additional insight into the energetics in the various nuclei.

Quantitative comparison between AGN and star formation One of the outputs of the proposed IFU spectra will be a quantitative estimate of the fraction of the bolometric luminosity attributed to the AGN on the one hand, and the star formation activity on the other. Infrared spectra obtained with the sensitivity and spatial resolution of JWST will allow us to measure the relative strength of various emission lines (many of them too weak to be detected from the ground) that trace either type of activity. In order to further characterize the AGN in our sample, we will employ state-of-the-art photoionisation models in order to infer the continuum shape of the (hidden) ionizing SED, and hence derive the Eddington ratios.

Icy Dust in Molecular Cores: A JWST/NIRCam GTO Project

Klaus W. Hodapp[1], *Adwin Boogert, Jacqueline Keane, Don Hall, Laurie Chu, Tom Greene, Michael Meyer, Doug Johnstone, Karl Misselt, Yancy Shirley, Roberta Paladini, and Marcia Rieke*
[1]*Institute for Astronomy, U. of Hawaii,*
640 N. Aohoku Place, Hilo, HI 96720, USA
email: hodapp@ifa.hawaii.edu

Abstract. This GTO project will study the spatial distribution of continuum extinction and ice absorption features in three nearby molecular cores using JWST/NIRCam slitless spectroscopy.

Keywords. space vehicles: instruments, (ISM:) dust, extinction, infrared: ISM

Our project aims at studying the early phases of ice mantle deposition on grains in molecular clouds. The onset of ice formation is observationally identified by the strong H_2O and CO_2 features that indicate the adsorption and subsequent processing of H, O and CO. At higher densities, molecular cloud CO is adsorbed so rapidly as to overwhelm the surface chemical reactions, so that a CO ice feature becomes observable, followed by the reaction product CH_3OH, the first step in reaction chains leading to more complex organic molecules, as recently reviewed by Boogert *et al.* (2015).

We will study the ice features in absorption against background stars, which, after spectral classification, serve as a known illumination source. Our target objects are nearby, isolated small molecular clouds (globules): B68, a quiescent core, L694-2, a collapsing core, and B335, a star-forming core. They are all situated in front of a dense field of background stars so that numerous lines of sight are available for absorption spectroscopy.

JWST offers two instruments capable of multi-object medium resolution spectroscopy over the wavelength range of 2.5 - 5 μm for our studies: MSA spectroscopy with NIRSpec and slitless spectroscopy with NIRCam (Greene *et al.* 2017). Despite its lower sensitivity due to higher sky backgrounds, we decided to use the slitless grism spectroscopy capabilities of NIRCam since this method does not require prior precise knowledge of the star positions. We will obtain slitless spectra of all stars in the field in the six wide and medium bandwidth NIRCam filters over the wavelength range from 2.5 - 5.0 μm. Our observing program uses multiple dither pointings to completely cover the dense cores of the globules. For each of our three objects, we expect about 100 usable background star spectra. These will give maps of the continuum and ice feature absorption depth with unprecedented spatial resolution and will allow us to study the formation and early chemical processing of ice mantles in detail.

Lensing-corrected 1.1mm number counts in the ALMA Frontier Fields Survey: A science case for JWST

Alejandra M. Muñoz Arancibia[1]
and ALMA Frontier Fields Team,
Present address: Fluid Mech Inc., 24 The Street, Lagos, Nigeria.
[1]*Instituto de Física y Astronomía, Universidad de Valparaíso,*
Av. Gran Bretaña 1111, Valparaíso, Chile
email: alejandra.munozar@uv.cl

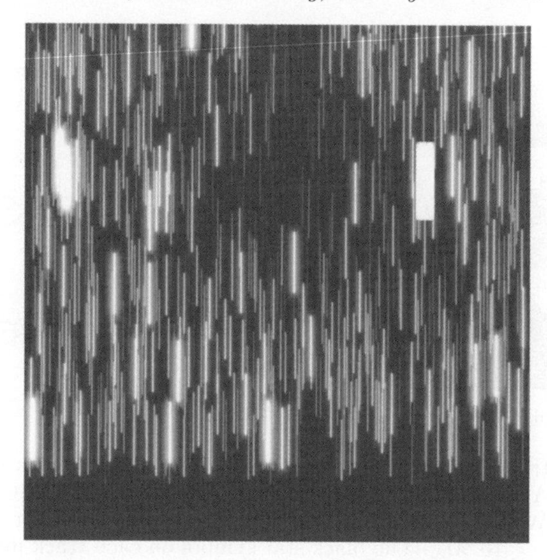

Figure 1. AXESIM simulation of NIRCam slitless spectroscopy in the F430M filter of B335 using UKIRT K-band data for the bright stars and a model of the faint star population based on the TRILEGAL galaxy model. It shows that the problem of spectrum overlap is quite manageable, and that at least 50% of all spectra can be extracted in a straightforward way.

Abstract. We present galaxy number counts around five strong-lensing galaxy clusters as part of the ALMA Frontier Fields Survey. This aims to characterize the population of faint, dusty star-forming galaxies at high redshift, benefiting from the magnification power of the clusters. Our study combines the analysis of deep (rms ~55-71 μJy/beam) ALMA 1.1 mm continuum data over ~23 square arcmin (lens plane) from this survey, with gravitational lensing models produced by different groups. Our estimates for the lensing-corrected number counts consider source detections down to S/N=4.5. Most of these detections lack spectroscopic redshifts, and from those having NIR counterparts, the majority are quite red. Moreover, some detections lack counterparts at other wavelengths, despite the extremely deep Hubble and Spitzer data available for the Frontier Fields clusters. Our ALMA detections thus comprise an interesting population for follow-up observations with JWST, as a robust determination of the missing spectroscopic redshifts will provide better constraints on the source properties, as well as more accurate estimates for the derived number counts.

Keywords. gravitational lensing, galaxies: high-redshift, submillimeter

---

Characterizing the sub-mm number counts of faint dusty star-forming galaxies (especially at sub-100 μJy; Casey *et al.* 2014) is currently a major challenge even for deep, high-resolution, observations with recent sensitive interferometers. These sources are predicted to account for approximately half of the total extragalactic background light at those sub-mm wavelengths. We exploit ALMA's unique capabilities to search for sources behind five well-studied galaxy clusters, which are part of the Hubble Frontier Fields survey (Lotz *et al.* 2017). In González-López *et al.* (2017) we present the ALMA observations for three of these galaxy clusters: Abell 2744, MACSJ0416.1-2403 and MACSJ1149.5+2223. Using these data, in combination with recent publicly available lensing models, we derive the faint end of the 1.1 mm number counts in these fields (Muñoz Arancibia *et al.* 2018). Our counts are consistent with previous estimates from

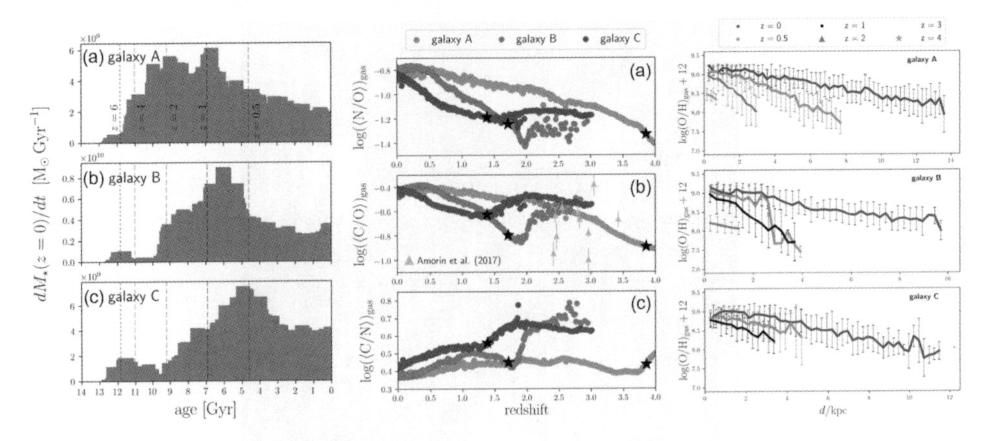

Figure 2. *Left*: Distribution of the present-day total stellar mass as a function of the star particle age. *Middle*: Average SFR-weighted gas-phase log (N/O), log (C/O) and log (C/N) as functions of redshift. We only show the predictions of our simulation for the redshifts when the galaxy stellar mass $M_* \geq 10^8 M_\odot$. The black star symbol on each track marks the redshift when $M_* \simeq 10^9 M_\odot$. See VK18a for the observational data source (grey triangles with the error bars). *Right*: Redshift evolution of radial gradients of elemental abundance ratios at $z = 0$ (blue), 0.5 (red), 1 (black), 2 (green), 3 (yellow), and 4 (magenta lines).

deep ALMA observations (e.g. Fujimoto *et al.* 2016, Aravena *et al.* 2016) at a 3σ level, nevertheless, below ≈ 0.1 mJy, our cumulative counts are lower by ≈ 1 dex, suggesting a flattening in counts at faint fluxes. These results are supported by a preliminary analysis of the remaining two galaxy clusters in our survey (Abell 370 and Abell S1063).

Elemental Abundances across Cosmic Time

Chiaki Kobayashi[1], *Fiorenzo Vincenzo*[1], *and Philip Taylor*[2]

[1] *Centre for Astrophysics Research, University of Hertfordshire, College Lane, Hatfield, UK*
email: c.kobayashi@herts.ac.uk

[2] *Research School of Astronomy and Astrophysics, Australian National University, Canberra; ARC Centre of Excellence for All Sky Astrophysics in 3 Dimensions (ASTRO 3D), Australia*

Stars in a galaxy are fossils that retain the information on star formation and chemical enrichment histories in the galaxy. This approach is called the galactic archaeology, and can be applied not only to our Milky Way Galaxy but also to other galaxies (e.g., Kobayashi 2016). Thanks to the collaboration between nuclear physics and astrophysics, we now have good understanding of the origin of elements in the Universe. The observed trends and scatters of elemental abundances from O to Zn in the solar neighborhood can be well reproduced with a chemodynamical simulation of a Milky Way-type galaxy (Kobayashi & Nakasato 2011). We show with our cosmological simulations that elemental abundances can provide strong constraints on the formation and evolutionary histories of galaxies, which can be studied with the James Webb Space Telescope.

Our simulation code is based on Gadget-3 and includes all relevant baryon physics such as UV background radiation, radiative cooling, star formation, supernova feedback, and chemical enrichment from asymptotic giant branch (AGB) stars, core-collapse supernovae (SNe II and hypernovae), and Type Ia supernovae (Kobayashi *et al.* 2007). Taylor & Kobayashi (2014) introduced a new model of AGN feedback in the code, where seed blackholes (BHs) originate from the formation of the first stars, which is different from the 'standard' model by other simulation groups such as EAGLE and Illustris. The modelling of the growth of BHs and feedback is the same as in other simulations. Our model parameters are determined in order to match the observational constraints, i.e., cosmic star formation rates, size-mas relation of galaxies, and $M_{\rm BH}$–σ relation.

On a cosmological scale, chemical enrichment takes place even more dramatically (see a movie by Philip Taylor, https://www.youtube.com/watch?v=jk5bLrVI8Tw). There is gas accretion along the cosmological filaments, where star formation and chemical enrichment are already occurring. This results in strong supernova-driven winds because of the shallow potential in the filaments. As the central galaxy grows through the accretion, a super-massive BH also grows (following the $M_{\rm BH}$–σ relation), which eventually causes even stronger winds driven by the active galactic nuclei. Metallicity is very spatially inhomogeneous; the center of massive galaxies can reach super-solar metallicity at high redshifts, while the accreted component has only one-hundredth of solar metallicity as it is mainly fed from the intergalactic medium. The wind component has about one-tenth of solar metallicity as it is a mixture of the inflow gas and supernova ejecta.

As a result, we obtain good agreement with the observed stellar mass function and mass–metallicity relations (MZRs) of galaxies for both gas-phase and stellar populations (Taylor & Kobayashi 2015). These relations evolve as a function of time; the stellar MZR does not change its shape, but the metallicity significantly increases from $z \sim 2$ to ~ 1, while the gas-phase MZR does change its shape, having a steeper slope at higher redshifts ($z \lesssim 3$, Taylor & Kobayashi 2016). Within galaxies, metallicity radial gradients are produced. We find a weak correlation between the gradients and galaxy mass, which is consistent with available observations (Taylor & Kobayashi 2017).

In Vincenzo & Kobayashi (2018a, hereafter VK18a), from another cosmological simulation (with side 10 h^{-1} Mpc) we create a catalogue of 33 stellar systems at redshift $z = 0$, all embedded within dark matter (DM) halos with virial masses in the range $10^{11} \leq M_{\rm DM} \leq 10^{13}\ M_\odot$. The mass and spacial resolutions of gas are $6.09 \times 10^6\ h^{-1} M_\odot$ and $0.84\,h^{-1}$ kpc, respectively. We first focus on three disc galaxies (Galaxy A, B, and C) with different star formation histories (SFHs, left panels of Fig. 2). We then predict how the C, N, and O abundances within the interstellar medium (ISM) of galaxies evolve as functions of the galaxy SFH. At the beginning of galaxy formation, CNO are produced by core-collapse supernovae, N is enhanced by intermediate-mass AGB stars ($\gtrsim 4 M_\odot$), then C is enhanced by low-mass AGB stars ($\lesssim 4 M_\odot$). Therefore, we predict that the average N/O and C/O steadily increase as functions of time, while the average C/N decreases, due to the mass and metallicity dependence of the yields of AGB stars; such variations are more marked during more intense star formation episodes.

Within a galaxy, the distributions of elements are not uniform either, and the central parts of the galaxies are more metal-rich than the outskirts of the galaxies. This radial gradient of elemental abundances also evolves as a function of time, which are shown in the right panels of Figure 2. In disk galaxies, the metallicity gradients become steeper at higher redshifts because of inside-out formation of discs. In Vincenzo & Kobayashi (2018), using all disk galaxy sample in the catalog, we succeed in reproducing the observed N/O–O/H relations, both for individual ISM abundances within single spatially-resolved galaxies and for average abundances in the whole ISM of many unresolved galaxies.

AGN demography with *JWST*

Hugo Messias[1] *and Mark Lacy*[2] *and CAST team*[3]

[1]*Joint ALMA Observatory, Alonso de Córdova 3107, Vitacura , Santiago, Chile*
email: `hugo.messias@alma.cl`

[2]*National Radio Astronomy Observatory, 520 Edgemont Road, Charlottesville, VA 22903*
email: mlacy@nrao.edu

[3]*CAST stands for Chasing dusty-AGN up to redShift Two and is a team comprised by ∼40 members who contributed to the current status of the project.*

Abstract. With a 5 to 10 year life-span and being a 6 m-class telescope in space, the *James Webb Space Telescope* (*JWST*) will be a highly competitive relatively short-lived tool toward knowledge revolution, with considerable operation overheads. As a result, it is both of interest to the community and facility to conduct observations as efficiently as possible. This brief manuscript highlights a colour criterion to select active galactic nuclei (AGN) from the local Universe as far back as the end of the epoch of reionization ($0 < z \lesssim 6$). Depending on the targetted Universe cosmic time, one is able to conduct a demographic study of dusty AGN with only up to four broad-band filters required (F200W, F440W, F770W, F1800W), three of which can be observed at the same time. Such observations will also allow for the community to assess stellar assembly in galaxies or to identify very high-redshift sources.

Keywords. galaxies: active, galaxies: photometry, galaxies: structure, infrared: galaxies

---

Infrared broad-band selection of AGN

When dust in the vicinity of AGN is heated by the radiation emanating from the accretion disc, it will radiate in the near- to mid-IR (∼2–60 μm) as a continuum continuously rising with wavelength. This appears distinct from a host-like spectral energy distribution (SED) characterised by a stellar continuum decaying long-wards from the 1.6 μm-bump together with Polycyclic Aromatic Hydrocarbon (PAH) emission-band features (mostly dominant at $\gtrsim 6\,\mu$m) and a colder dust continuum (dominating at $> 10\,\mu$m). This yields the $1 - 7\,\mu$m rest-frame spectral range as the one to pinpoint dusty AGN.

Ideally, one would use spectroscopy to assess this spectral range, but it is time consuming and likely limited to one galaxy at a time with the MIRI IFU. The alternative is to do a contiguous multi-band imaging survey, providing a very low spectral-resolution IFU. However, the time to achieve such survey is also time consuming and eventually difficult to schedule. This will be attempted by the JADES GTO and the CEERS ERS teams, which combined will only provide 40 arcmin^2 worth of multi-band MIRI imaging. For the community at large wishing to target different fields and larger ones, especially important for the $z < 2$ cosmic epoch, a less telescope-time consuming approach is needed.

In Messias *et al.* (2014) we proposed two colour criteria using respectively the F200W, F440W, and F770W filters for the $0 < z < 2.5$ range, adding F1800W for higher-redshifts. The fine spatial resolution enabled by JWST will allow one to deblend host and AGN light, hence selecting less-luminous AGN, a phase where AGN pass most of their life-cycle.

Finding Embedded AGN with MIRI

G. H. Rieke,
Present address: Fluid Mech Inc., 24 The Street, Lagos, Nigeria.

Jianwei Lyu, & Jane Morrison

Steward Observatory, The University of Arizona, Tucson, AZ, 85721, USA
email: grieke@as.arizona.edu

Abstract. The many photometric bands available in the JWST instruments allow construction of multi-color diagrams that are much more diagnostic than similar diagrams constructed from *Spitzer* or *WISE* data. This capability is illustrated with an example that successfully separates galaxies with subtle indications of embedded AGN from purely star forming galaxies.

Keywords. galaxies: active, quasars: general

Infrared photometric surveys are a powerful means to identify Active Galactic Nuclei (AGN), and can complement other approaches to discover forms of AGN that are not apparent by other selection methods (e.g., Donley *et al.* (2005), Del Moro *et al.* (2016)). A number of powerful methods for finding AGN have been developed using color-color diagrams of *Spitzer* IRAC and *WISE* photometry (e.g., Lacy *et al.* (2004), Stern *et al.* (2005), Donley *et al.* (2012), Stern *et al.* (2012)). Nonetheless, these methods are primarily useful for AGN with power law infrared spectra, e.g., type-1 objects of large enough luminosity to dominate the stellar output (Donley *et al.* (2012)); in addition, star forming galaxies invade the AGN identification criteria at redshifts above $\sim$ 1.5 (Donley *et al.* (2008)).

Color-color diagrams and related methods complement more sophisticated modeling because they can be less affected by initial assumptions; model fitting is efficient at finding objects that resemble the input models but may miss ones that do not. Fortunately, the many bands available with the Mid-Infrared Instrument (MIRI) on JWST allow, particularly in combination with the NIRCam bands, construction of more complex and more diagnostic diagrams. An example is shown in Figure 1, which combines 11 different photometric bands. This diagram is tuned to differentiate objects where the SED minimum near 4.5 μm in a stellar photospheric dominated SED is filled in by the emission of an AGN. The diagram has been tested using approximately 300 galaxies with high-quality *Spitzer* IRS spectra (not just with templates). As shown in the figure, it is effective in isolating spectra where this filling in is very subtle. The success of this diagram should encourage development of other similar approaches using alternative combinations of the multiple photometric bands provided by the JWST instruments.

AGB Stellar Populations in Resolved Galaxies with JWST

Paola Marigo

Present address: Fluid Mech Inc., 24 The Street, Lagos, Nigeria.,

On Behalf of the STARKEY Team and the JWST Resolved Stellar Populations Early Release Science Program

Department of Physics and Astronomy, University of Padova,
Vicolo dell'Osservatorio 3, 35122, Padova, Italy
email: paola.marigo@unipd.it

Abstract. Thanks to its spatial resolution and infrared filters, JWST is expected to greatly expand the volume accessible for studies of resolved AGB star populations, hence potentially impacting the calibration of theoretical models for this critical evolutionary phase. In this talk, I will present the predicted appearance of evolved stars in nearby galaxies using the JWST NIRCam and MIRI filters, investigating, in particular, which filter combinations allow for a better separation of the different types (M and C) of AGB stars, and their expected numbers in SMC-like galaxies located at 4 Mpc. Finally, I will discuss the expectations from The Resolved Stellar Populations Early Release Science Program (ID 1334, PI Dan Weisz), which includes the nearby star-forming dwarf WLM.

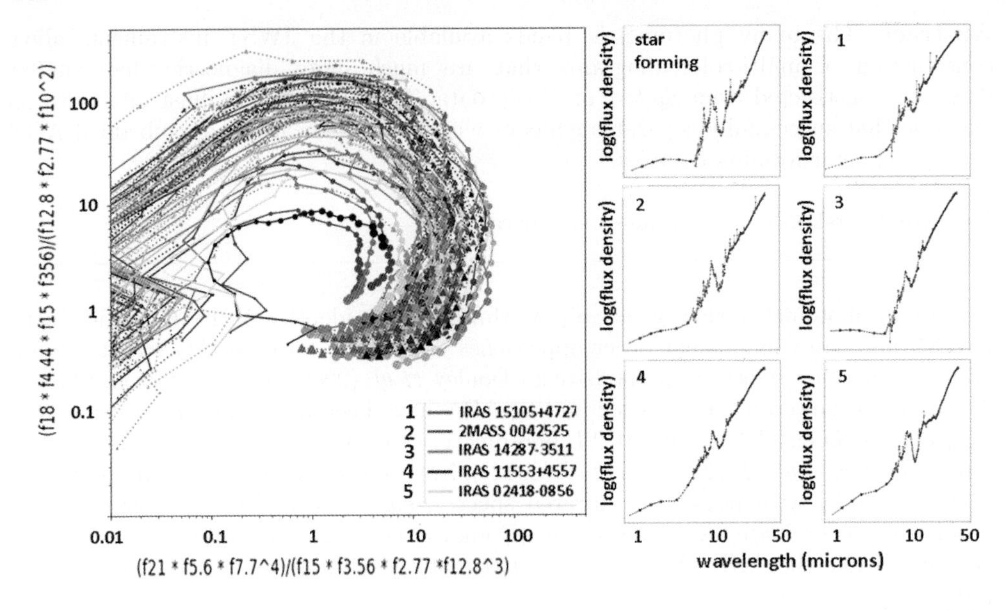

Figure 3. Performance of a MIRI/NIRCam multi-color diagram in identifying embedded AGN. The left panel shows the tracks for star forming galaxies (gray-scale) running from z = 0.5 to z = 2.4 with symbols growing in size with increasing z. It superimposes similar tracks for five galaxies selected for very subtle AGN characteristics in the 4 to 6 μm range. The panels to the right show a typical star forming galaxy SED and the SEDs of the five galaxies with embedded AGN. The SEDs for both parts of the figure combine spectra from *Spitzer* IRS with a fit to JHK, W1, and W2 photometry to represent the stellar continua. The individual SEDs show how the characteristic SED peak from stars in the 1.5 − 3 μm range is subtly modified by the emission of an embedded AGN, which also dilutes the aromatic features at 6 - 8 and 11 μm. The multi-color diagram successfully identifies these differences independent of redshift.

Keywords. stars: AGB and post-AGB, evolution, mass loss, galaxies: stellar content, infrared: stars

---

Overview

The Thermally Pulsing Asymptotic Giant Branch (TP-AGB) phase, experienced by low- and intermediate-mass stars at the end of their lives, plays a critical role across astrophysics, affecting the interpretation of astronomical data from various sources, e.g. from the chemical composition of presolar meteoric grains to the integrated light of distant galaxies. Despite its relevance, the modelling of the TP-AGB phase suffers from severe uncertainties due to the complexity of the physics involved. It follows that a proper calibration of the uncertain parameters, mainly related to the efficiencies of mass loss and convective mixing (the third dredge-up), needs to be carried out with the help of high-quality observations of resolved stellar populations. JWST will provide and excellent tool to investigate AGB stars. The use of suitable combinations of NIRCam and MIRI filters will allow to classify resolved AGB stars (M or C types; see e.g. Boyer *et al.* 2013). Moreover, with JWST we will greatly expand the statistics of AGB stars up to large distances, enabling a detailed analysis of AGB stellar populations in resolved galaxies up to few Mpc far from us. In this contribution I show a few examples of colour-colour diagrams where the evolutionary tracks of O-rich (with surface C/O$<$1) and C-rich (with surface C/O$>$1) stars form separate sequences, and

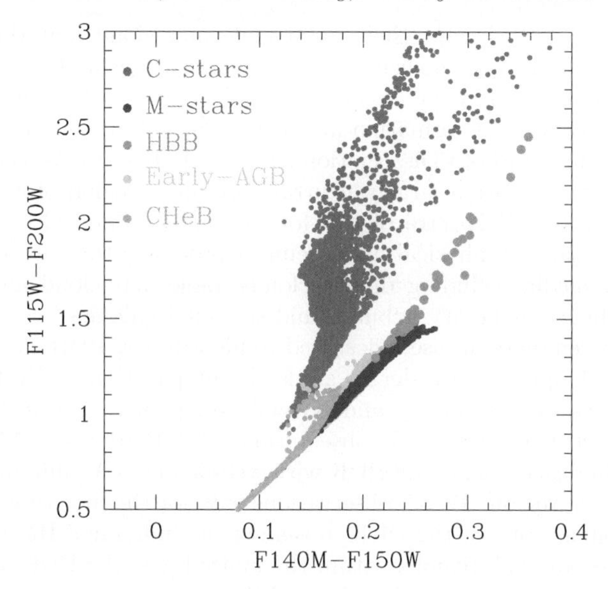

Figure 4. Synthetic stellar populations for an SMC-like galaxy as predicted in a JWST colour-colour diagram. Stars in various post-main sequence evolutionary stages are marked in colour. Note the clear separation of the C-star sequence (red) from the other stars (blue: M-type TP-AGB stars, light blue: Early AGB stars; green: Core-helium burning stars), and the location of AGB stars with hot-bottom burning (magenta). The effect of circumstellar dust forming in the winds of AGB stars is included.

others in which the location of AGB stars that undergo hot-bottom burning (with initial masses $> 3 - 4\,M_{\odot}$) stands out clearly (see Fig. 4). I discuss the main expected properties of resolved stellar populations in typical galaxies (irregulars, spirals, and ellipitcals) as seen in appropriate JWST infrared colour-magnitude diagrams, taking into account the effect of circumstellar dust from mass-losing AGB stars. The STARKEY team (ERC project, PI Marigo) is ready to provide the community with all theoretical tools (tracks, isochrones, spectra, bolometric corrections; Marigo *et al.* 2017) necessary for analysing the data of resolved AGB stars that JWST will release in the future.

Radiative Feedback from Massive Stars

The JWST-ERS PDR team[1,2,3]
J. Cami[1], *E. Habart*[2], *E. Peeters*[1], *O. Berné*[3]
and Radiative feedback from massive stars ERS team

email: `olivier.berne@irap.omp.eu`, `emilie.habart@ias.u-psud.fr`,`epeeters@uwo.ca`, [1]*Physics and Astronomy Department, University of Western Ontario, 1150 Richmond Street, ON N3A 6K7, London, Canada*
email: `jcami@uwo.ca`, `epeeters@uwo.ca`

[2]*Institut dAstrophysique Spatiale, UMR 8617-CNRS Universite Paris Sud, 91405 Orsay, France*
email: `emilie.habart@ias.u-psud.fr`

[3]*Institut de Recherche en Astrophysique et Planétologie, CNRS, CNES and Université Paul Sabatier*
9 avenue du Colonel Roche, Toulouse, France
email: `olivier.berne@irap.omp.eu`

Abstract. Massive stars disrupt their natal molecular cloud material by dissociating molecules, ionizing atoms and molecules, and heating the gas and dust. These processes drive the evolution of interstellar matter in our Galaxy and throughout the Universe from the era of vigorous star formation at redshifts of 1-3, to the present day. Much of this interaction occurs in Photo-Dissociation Regions (PDRs, Fig.5) where far-ultraviolet photons of these stars create a largely neutral, but warm region of gas and dust. PDR emission dominates the IR spectra of star-forming galaxies and also provides a unique tool to study in detail the physical and chemical processes that are relevant for inter- and circumstellar media including diffuse clouds, molecular cloud and protoplanetary disk surfaces, globules, planetary nebulae, and starburst galaxies.

We will provide template datasets designed to identify key PDR characteristics in the full 1-28 μm JWST spectra in order to guide the preparation of Cycle 2 proposals on star-forming regions in our Galaxy and beyond. We plan to obtain the first spatially resolved, high spectral resolution IR observations of a PDR using NIRCam, NIRSpec and MIRI. We will observe a nearby PDR with well-defined UV illumination in a typical massive star-forming region. JWST observations will, for the first time, spatially resolve and perform a tomography of the PDR, revealing the individual IR spectral signatures from the key zones and sub-regions within the ionized gas, the PDR and the molecular cloud (Fig. 1). These data will test widely used theoretical models and extend them into the JWST era. We will assist the community interested in JWST observations of PDRs through several science-enabling products (maps of spectral features, template spectra, calibration of narrow/broad band filters in gas lines and PAH bands, data-interpretation tools e.g. to infer gas physical conditions or PAH and dust characteristics). This project is supported by a large international team of 140 scientists.

Keywords. (ISM:) HII regions, ISM: lines and bands, ISM: atoms, ISM: molecules, (ISM:) dust, extinction, ISM: globules, infrared: ISM

Establishing Extreme Dynamic Range with JWST: Decoding Smoke Signals in the Glare of a Wolf-Rayet Binary

Thomas Madura[1] *and Ryan Lau*[2]

[1]*San José State University, San José, CA 95192-0106, USA*
email: thomas.madura@sjsu.edu

[2]*California Institute of Technology, Pasadena, CA 91125, USA*

Abstract. Dust is a key ingredient in the formation of stars and planets. However, the dominant channels of dust production throughout cosmic time are still unclear. With its unprecedented sensitivity and spatial resolution in the mid-IR, the James Webb Space Telescope (JWST) is the ideal platform to address this issue by investigating the dust abundance, composition, and production rates of various dusty sources. In particular, colliding-wind Wolf-Rayet (WR) binaries are efficient dust producers in the local Universe, and likely existed in the earliest galaxies. Our planned JWST observations of the archetypal colliding-wind binary WR 140 will study the dust composition, abundance, and formation mechanisms. We will utilize two key JWST observing modes with the medium-resolution spectrometer (MRS) on the Mid-Infrared Instrument (MIRI), and the Aperture Masking Interferometry (AMI) mode with the Near Infrared Imager and Slitless Spectrograph (NIRISS). Our observations will investigate the dust forming properties of WR binaries and establish a benchmark for key observing modes for imaging bright sources with faint extended emission. This will be valuable in various

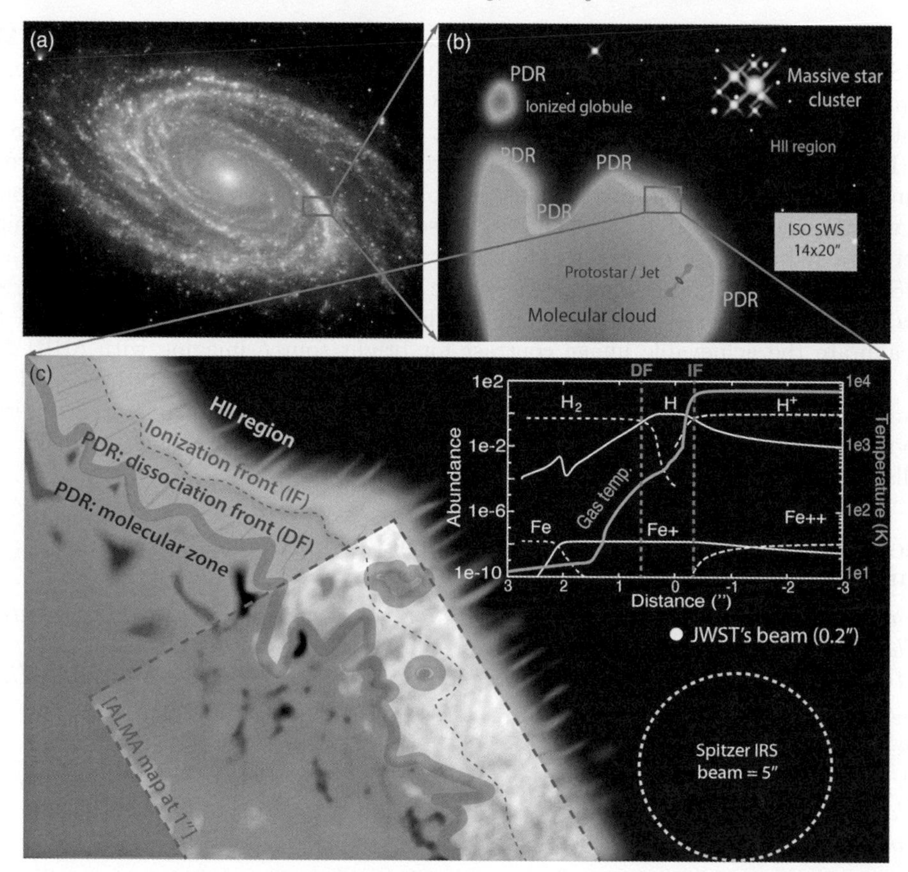

Figure 5. Zooming into a PDR. **a)** Multi-wavelength view of a Galaxy (M81): UV tracing massive stars (blue), optical light tracing HII regions (green), and PAH emission tracing PDRs (red). **b)** Sketch of a typical massive star-forming region (at a distance of 2 kpc). **c)** Zoom in on one of the numerous PDRs, showing the complex transition from the molecular cloud to the PDR dissociation front, the ionization front and the gas flow into the ionized region. Inserted is the ALMA molecular gas data of the Orion Bar, at a resolution of 1" (dashed lines; Goicoechea et al. 2016). The inset shows a model of the structure of the PDR. The scale length for FUV photon penetration corresponds to a few arcsec. The beam sizes of ISO-SWS, Spitzer-IRS and JWST-MIRI are indicated. JWST will resolve the 4 key regions.

astrophysical contexts, including mass-loss from evolved stars, dusty tori around active galactic nuclei, and protoplanetary disks. We are committed to designing and delivering science-enabling products for the JWST community that address technical issues such as bright source artifacts in addition to testing optimal image reconstruction algorithms for observing extended structures with NIRISS/AMI.

Keywords. infrared:stars, stars: Wolf-Rayet, stars: individual: WR 140, stars: winds, outflows

---

The goals of this JWST Early Release Science (ERS) program are to investigate how dust forms and evolves around massive stars, and to provide community resources and tools regarding key modes of JWST to observe faint IR emission near bright central sources with high spatial resolution and imaging contrast. We plan to use the MIRI MRS and NIRISS/AMI to perform high-resolution mid-IR imaging and spectroscopy of a prototypical dusty WR binary system, WR 140, which produces shells of dust every eight years due to colliding-wind interactions near periastron passage. MIRI MRS and NIRISS/AMI observations will probe the detailed dust morphology and measure the abundance, energetics, and mass of dust surrounding the bright, central heating source. We will provide the JWST community with science-enabling products including higher level data reduction software for addressing the bright source artifacts for the MIRI MRS, documentation describing bright source artifacts in MIRI MRS and NIRISS/AMI data, and observing strategies to mitigate these effects. Our data products will help to calibrate the achievable image contrast for bright source imaging with MIRI MRS and NIRISS/AMI. Deriving the maximum achievable image contrasts from real data is important to help the JWST community assess the viability of potential observing programs to measure faint emission from structures such as protoplanetary disks, dust shells, and accretion disks around bright sources.

The JWST ERS Program for the Direct Imaging of Extrasolar Planetary Systems

Sasha Hinkley[1], *Andrew Skemer*[2], *Beth Biller*[3], *Aarynn Carter*[1] *and ~120 Additional Collaborators*

[1] *University of Exeter, Physics Building, Stocker Rd. Exeter EX4 4QL, United Kingdom*
email: S.Hinkley@exeter.ac.uk

[2] *Department of Astronomy & Astrophysics, University of California, Santa Cruz, 1156 High St., Santa Cruz, CA 95064*
email: askemer@ucsc.edu

[3] *Institute for Astronomy, The University of Edinburgh, Royal Observatory, Blackford Hill, Edinburgh EH9 3HJ, United Kingdom*
email: bb@roe.ac.uk

Abstract. We describe our accepted JWST Early Release Science program, which will perform: a) NIRCAM & MIRI coronagraphy of a newly discovered exoplanet and a well-studied circumstellar debris disk; b) NIRSPEC & MIRI spectroscopy of a wide separation planetary mass companion; and c) deep NIRISS aperture masking interferometry. These observations have been tailored to generate representative datasets in common modes and deliver science enabling products to empower a broad user base to develop successful future investigations. Along with the approved GTO programs, these will be among the first observations to characterize exoplanets for the first time over their full spectral range from 2-28 μm, and debris disk out to 15 μm. We present a summary of these observations and our planned science enabling products in order to inform the community ahead of the launch of JWST.

Keywords. instrumentation: high angular resolution, techniques: high angular resolution, telescopes, techniques: image processing

---

Science Background & Rationale for an ERS Program

Exoplanet Direct Imaging (Bowler *et al.* 2016) remains an extraordinarily powerful technique to constrain the overall frequencies of extrasolar planets at very wide separations (~10-1000 AU), and to obtain *direct* spectroscopy of these objects. Indeed, aside from some Spitzer/AKARI observations free-floating brown dwarfs (Cushing *et al.* 2006), observations of exoplanets redward of $\sim 5\,\mu$m remain completely out of reach. Thus JWST is expected to be transformative for understanding many of the physical characteristics of exoplanets (e.g. mass, gravity, and atmospheric composition), and allow us to measure the abundances of dominant molecules in the atmosphere (e.g. CH_4, CO, CO_2, H2O, NH_3). At the same time, JWST should have the sensitivity to discover completely new classes of directly imaged planets: Saturn analogues in many cases, and down to Neptune mass planets in the most favorable cases. However, JWST can only achieve these tasks if the user base can rapidly develop a deep understanding of the optimal strategies for observations, calibrations and data post-processing. Specifically, the community dedicated to exoplanet imaging will need an exquisite understanding of the instrument response, PSF stability, and the most effective strategies for PSF subtraction. Indeed, the performance of HST required several cycles to fully understand, and the techniques are still being perfected nearly 25 years later (e.g. Schneider *et al.* 2016). Disseminating an understanding of the performance of JWST to the community as rapidly as possible after launch will be essential for preparing for Cycle 2 and beyond.

Program Description & Expected Science Enabling Products

Our 52-hour program can be divided into three categories which we describe in more detail below: Exoplanet and debris disk coronagraphy (39 hours); Spectroscopy of planetary mass companions (6 hours); and Aperture Masking Interferometry (7 hours).
1) Coronagraphy of Exoplanets & Debris Disks: We will perform coronagraphy of the directly imaged exoplanet HIP6526b (Chauvin *et al.* 2017) using NIRCAM from 2-5 μm and MIRI at 11 and 15 μm. We will also observe the HR 4796 debris disk at 3.0 and 3.6 μm using NIRCAM, and at 15 μm with MIRI. We will rapidly disseminate information about the performance of JWST in these bands (e.g. contrast curves), identify the optimal PSF subtraction techniques, and release a Python-based high-contrast imaging pipeline based on the existing pyKLIP package.
2) Spectroscopy of Planetary Mass Companions: Using the NIRSpec IFU from 1.7-5.3 μm and the MIRI Medium Resolution Spectrograph from 5-28 μm, we will gather spectroscopy of VHS 1256b (Gauza *et al.* 2016), a substellar companion with an angular separation of $\sim 8''$, which greatly reduces any contaminating host starlight.
3) Aperture Masking Interferometry (AMI): We will use NIRISS operating in the AMI mode on the HIP65426 system to characterize the residual phase errors that set the overall contrast floor. We will test the expected sensitivity of 8-9 magnitudes at $\sim \lambda/D$ angular separations, providing sensitivity to young planets at orbital separations of $\sim$15 AU. We will make public a Python-based pipeline for processing AMI data enabling rapid sensitivity estimates for Cycle 2 proposers, and fast analysis of Cycle 2 data.

The Transiting Exoplanet Community Early Release Science Program with *JWST*

Kevin B. Stevenson[1], *Jacob L. Bean*[2], *Natalie M. Batalha*[3] *and The Transiting Exoplanet Community ERS Team*
[1]*Space Telescope Science Institute,*
3700 San Martin Drive, Baltimore, MD 21218, USA
email: kbs@stsci.edu

[2]*Department of Astronomy & Astrophysics, University of Chicago,*
5640 S. Ellis Avenue, Chicago, IL 60637, USA
email: jbean@astro.uchicago.edu

[3]*Department of Astronomy & Astrophysics, University of California, 1156 High Street, Santa Cruz, CA 95064, USA*
email: natalie.batalha@ucsc.edu

Abstract. The *James Webb Space Telescope (JWST)* presents the opportunity to transform our understanding of planets and the origins of life by revealing the atmospheric compositions, structures, and dynamics of transiting exoplanets in unprecedented detail. However, the high-precision, time-series observations required for such investigations have unique technical challenges, and prior experience with *Hubble*, *Spitzer*, and other facilities indicates that there will be a steep learning curve when *JWST* becomes operational. Here, we briefly describe the science objectives, observations, and community engagement plans of the recently-approved Transiting Exoplanet Community Early Release Science (ERS) Program. Bean *et al.* (2018) provide a more detailed description, including scientific and technical motivations, for this program.
Keywords. planets and satellites: general

Science Objectives

The goal of this project is to accelerate the acquisition and diffusion of technical expertise for transiting exoplanet observations with *JWST*, while also providing a compelling set of representative datasets that will enable immediate scientific breakthroughs. To reach this overarching goal, we have three strategic objectives: (1) determine the spectrophotometric time-series performance of the key instrument modes on timescales relevant to transits for a representative range of target star brightnesses, (2) jump-start the process of developing remediation strategies for instrument-specific systematic noise, and (3) provide the community with a comprehensive suite of transiting exoplanet data to fully demonstrate *JWST*'s scientific capabilities in this area.

Observations

The Transiting Exoplanet Community ERS Program will exercise the time-series modes of all four *JWST* instruments that have been identified as the consensus highest priorities, observe the full suite of transiting exoplanet characterization geometries (transits, eclipses, and phase curves), and target planets with host stars that span an illustrative range of brightnesses (see Figure 6).

The Panchromatic Transmission Program will obtain a near-infrared (NIR, 0.6 – 5.2 m) transmission spectrum of a single planet to demonstrate *JWST*'s ability to obtain precise atmospheric composition measurements and to exercise the instrument modes that will likely be the workhorses for observations of planets ranging from hot giants to temperate terrestrials. The program has been designed to include the necessary wavelength coverage to cross-compare and validate the three NIR instruments, and thus establish the best strategy for obtaining transit spectroscopy measurements in future cycles.

The MIRI Phase Curve Program will test the hour-to-hour stability of *JWST* and MIRI/LRS. Phase-curve observations pose unique challenges that will not be tested with shorter transit- or eclipse-only observations (e.g., high-gain-antenna moves occur every 10,000 s and may disrupt the pointing). We will evaluate potential instrumental noise sources, including long-term flux variations caused by the thermal background and/or the detectors themselves using the science light curves, as prior experience with *Hubble* and *Spitzer* has shown that standard spacecraft calibration data do not characterize such effects at the required precision. We will also investigate other potential sources of

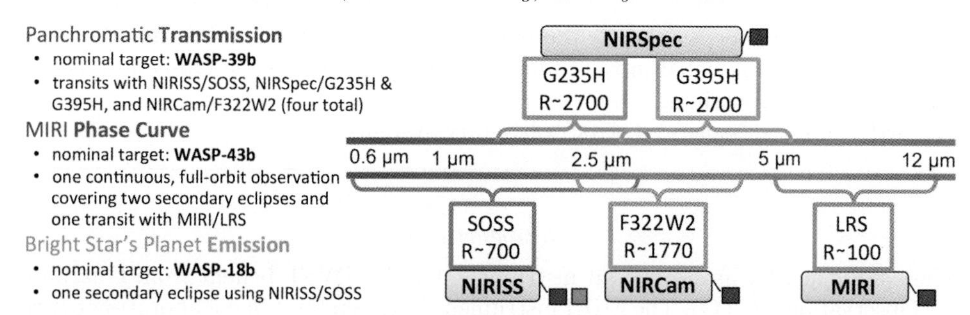

Figure 6. Summary of the three *JWST* observing programs that comprise the Transiting Exoplanet Community ERS program. The schematic on the right indicates the wavelength coverage of the instrument modes that will be utilized. Note that the color coding on the text to the left corresponds to the instrument mode labels on the right. Figure from Bean *et al.* (2018).

systematics including persistence, pointing drifts combined with intra-pixel sensitivity variations, flat-field errors, cosmic ray latency, and jitter.

The Bright Star Program will observe a single secondary eclipse of a hot Jupiter orbiting a bright host star using NIRISS/SOSS. This observation will not only demonstrate the utility of *JWST* data for revealing the atmospheric thermal structures and energy budgets of transiting exoplanets, it will also enable us to determine how precisely *JWST*'s instruments can measure transit spectra in the limit of low photon noise (i.e., a high number of recorded photoelectrons). By pushing the expected noise to very low levels, we will test *JWST*'s behavior at the limit of its achievable precision, in preparation for the compelling transiting exoplanets that TESS will find around bright stars. The performance of *JWST* in this regard is unknown, as there are no design requirements, yet it is a key metric that will ultimately determine if terrestrial exoplanet atmospheres are accessible.

The observations in this program were defined through an inclusive and transparent process that had participation from *JWST* instrument experts and international leaders in transiting exoplanet studies. The targets have been vetted with previous measurements, will be observable early in the mission, and have exceptional scientific merit.

Community Engagement

A core goal of this ERS program is to catalyze broad engagement in *JWST* and to train a community of capable *JWST* exoplanet observers. To address this goal, we will host a multi-phase Data Challenge to spark world-wide collaboration and focus the exoplanet community's creativity on analyzing *JWST* data. This Challenge will comprise online interaction and two face-to-face meetings, bringing together instrument/telescope specialists, observers, and theorists. It will facilitate the speedy validation of our scientific results and construction of our science-enabling products. These activities are not limited to those scientists who were on the original ERS proposal; they are open to the entire community.

Exoplanet Atmosphere Characterization in the framework of the MIRI European Consortium Guaranteed Time Observations

Pierre-Olivier Lagage On behalf of MIRI European Consortium exoplanet team.
Astrophysics Department at CEA, Paris-Saclay University,
F-91191, Gif-sur-Yvette, France
email: pierre-olivier.lagage@cea.fr

Abstract. In this paper, we shortly present the program of characterization of exoplanet atmospheres to be conducted in the framework of Guaranteed Time Observations of the MIRI European Consortium.
Keywords. infrared: general, (stars:) planetary systems

Introduction

The consortia who have built an instrument for the JWST benefit from Guaranteed Time Observations (GTO). The MIRI instrument has been built in collaboration between US and Europe (Rieke *et al.* 2015, Wright *et al.* 2015) and the European Consortium has got half of the GTO that an instrument consortium can get. About 25% of that time (110 hours) is used to study exoplanets.

Sources to be observed and challenges to be faced

Out of the 110 hours of GTO, 60 hours are devoted to observe 3 transiting exoplanets (HAT-P-12 b, WASP-107 b and TRAPPIST-1 b), 40 hours to observe 10 exoplanets detected by direct imaging and 10 hours to observe 7 brown dwarfs. The program has been elaborated in coordination/collaboration with the other GTO holders. Details on the program are available at: jwst-docs.stsci.edu/. Coronagraphic observations are discussed in Danielski *et al.* (2018). Given the JWST launch delay, the list may be revised.

Given the large wavelength coverage and the large sensitivity provided by the JWST, uncertainties in the atmospheric models can become a limiting factor (for example Baudino *et al.* 2017), model simplification usually made in retrieval technics can no longer be valid (for example Rocchetto *et al.* 2016), and novel data reduction techniques have to be developed and tested on simulated data. Since January 2018, such activities are developed in the framework of the ExoplANETS-A project (http://exoplanet-atmosphere.eu), partially funded by the European Commission Grant N° 776403.

JWST Commissioning from Launch to Science Observations

Michael W. McElwain[1], *George Sonneborn*[1], *Erin C. Smith*[1], *and Scott D. Friedman*[2]

Overview

The *James Webb Space Telescope* (JWST) has an extensive ground test campaign to confirm workmanship and verify Observatory performance prior to launch (e.g., McElwain *et al.* 2018, Kimble *et al.* 2018). JWST commissioning will be used to activate, checkout, and make initial calibrations for the Observatory, which is estimated to take roughly 6 months. Commissioning will conclude when the Observatory is ready to begin the Cycle 1 science program. Following commissioning, each observing Cycle will have dedicated calibration time to build a full suite of calibration data and characterize the technical nuances of the Observatory.

These commissioning activities are separated into three major phases: 1) orbital insertion, spacecraft, and deployments, 2) the telescope, and 3) the science instruments. The commissioning timeline provides a baseline sequence of activities that will be carried out to enable Cycle 1 science observations, depicted with annotations of major events/milestones in Figure 7. In practice, some activities can be moved within the timeline and we anticipate certain scenarios encountered will require contingency activities.

The timeline begins with launch aboard the Ariane 5 rocket. Shortly thereafter, JWST separates from the launch vehicle (LV) and the solar array is deployed to be power

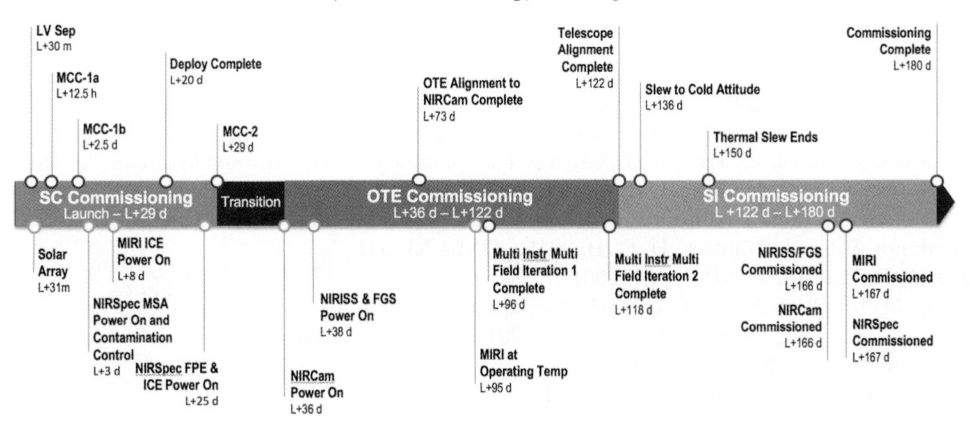

Figure 7. The JWST commissioning timeline consists of the baseline activities and sequence needed to enable Cycle 1 science. The major phases are designated by spacecraft (SC), optical telescope element (OTE), and science instrument (SI), with key events/milestones annotated. During the transition period following the deployments, the telescope and science instruments are passively cooled to their operational temperatures. Additional cooling of the Mid-Infrared Instrument (MIRI) is provided by an active cryocooler.

positive. The next critical activity is a trajectory correction burn, called a mid-course correction (MCC), to add energy required to reach L2. A second trajectory correction burn will be made after ~2.5 days, which is followed by numerous deployments that need to take place to transform from the stowed to operational Observatory. The most complex deployment is the sunshield, which enables passive cooling of the telescope and science instruments.

When cold, the segmented telescope will be aligned by identifying segments, providing wavefront control at the segment-level, co-phasing segments, and correcting the image quality over the large focal plane feeding the science instruments. It will initially be aligned and phased in the NIRCam field of view, followed by the full multi-instrument alignment. The wavefront error will be monitored every 2 days through the remainder of commissioning and throughout the life of the mission. Wavefront corrections are expected to be made approximately every 2 weeks to keep the error within specifications. Telescope commissioning is the longest duration activity in the plan, expected to take approximately 80 days.

As soon as the telescope commissioning is complete, about 4 months after launch, the science instrument commissioning activities associated with observing astronomical sources will commence. These include photometric, astrometric, flatfield, and dark current characterizations; PSF characterizations in the imagers, spectrographs, and coronagraphs; and scattered light measurements. Operational functionality will also be demonstrated including target acquisition and moving target performance, as well as the capabilities of the Operations Script Subsystem (OSS) to execute the observations specified by observers using the Astronomers Proposal Tool (APT). Some observing modes of each instrument will be commissioned and enabled for science before other, more complicated modes. We anticipate starting science observations with each mode as it is enabled in order to begin the science return of this successor to the *Hubble Space Telescope*.

With the exception of the launch segment, all of these activities will be carried out at the Science & Mission Operations Center, located at the Space Telescope Science Institute (STScI). Commissioning will be completed by the JWST mission operations team (MOT), which is comprised of the international partners from NASA, ESA, and CSA, as well as aerospace industry partners such as Observatory contractor Northrop Grumman Aerospace Systems, and STScI.

References

Aravena, M., Decarli, R., Walter, F., *et al.* 2016, *ApJ* 833, 68

Baudino, J.-L. *et al.* 2017, *ApJ*, 850, 150

Bean, J. L., Stevenson, K. B., Batalha, N. M., & 95 others, 2018. *PASP*, 130, 114402

Beichman, C., Krist, J., Trauger, J., Greene, T., Oppenheimer, B., Sivaramakrishnan, A., Doyon, R., Boccaletti, A, Barmen, T., Rieke, M. 2010, *PASP*, 122, 162

Beichman C., and 47 co-authors 2014, *PASP* 126, 1134

Gerakines, P. A., & Whittet, D. C. B. 2015, *ARAA* 53, 541

Bowler, B. P. 2016, *PASP*, 128, 102001

Boyer, M. L., Girardi, L., Marigo, P., *et al.* 2013, *ApJ*, 774, 83

Casey, C. M., Narayanan, D. & Cooray, A. 2014, *Phys. Rep.* 541, 45

Chauvin, G. *et al.* 2017, *A&A*, 605, L9

Cushing, M. C. *et al.* 2006, *ApJ*, 648, 614

Danielski, C. *et al.* 2018, *ApJ*, in press

Del Moro, A., Alexander, D. M., Bauer, F. E. *et al.* 2016, *MNRAS*, 456, 2105

Donley, J. L., Rieke, G. H., Rigby, J. R., & Pérez-González, P. G. 2005, *ApJ*, 634, 169

Donley, J. L., Rieke, G. H., Pérez-González, P. G., & Barro, G. 2008, *ApJ*, 687, 111

Donley, J. L., Koekemoer, A. M., Brusa, M. *et al.* 2012, *ApJ*, 748, 142

Fujimoto, S., Ouchi, M., Ono, Y., *et al.* 2016, *ApJS* 222, 1

Gauza, B. *et al.* 2015, *ApJ*, 804, 96

Goicoechea, J. R., Pety, J. Cuadrado, S., *et al.* 2016, *Nature*, 537, 207

González-López, J., Bauer, F. E., Romero-Cañizales, C., *et al.* 2017, *A&A* 597, A41

Greene, T., Kelly, D., Stansberry, J., Leisenring, J., Egami, E., Schlawin, E. Chu, L., Hodapp, K., Rieke, M. 2017, *JATIS* 3, 5001

Kimble, R. A., Feinberg, L. D., Voyton, M. F., Lander, J. A., Knight, J. S., Waldman, M., Whitman, T., Vila Costas, M. B., Reis, C. A., & Yang, K. 2018, *Proc. of the SPIE*, 10698, 1069805

Kobayashi, C. 2016, Nature, 540, 205

Kobayashi, C., Karakas, I. A., & Umeda, H. 2011, MNRAS, 414, 3231

Kobayashi, C., & Nakasato, N. 2011, ApJ, 729, 16

Kobayashi, C., Springel, V, & White, S. D. M. 2007, MNRAS, 376, 1465

Lacy, M., Storrie-Lombardi, L. J., Sajina, A., *et al.* 2004, *ApJS*, 154, 166

Lotz, J. M., Koekemoer, A., Coe, D., *et al.*2017, *ApJ* 837, 97

Marigo, P., Girardi, L., Bressan, A., *et al.* 2017, *ApJ*, 835, 77

McElwain, M. W., Niedner, M. B., Bowers, C. W., Kimble, R. A., Smith, E. C., & Clampin, M. 2018, *Proc. of the SPIE*, 10698, 1069802

Messias, H., Afonso, J. M., Salvato, M., Mobasher, B., & Hopkins, A. M. 2014, *A&A*, 562, A144

Muñoz Arancibia, A. M., González-López, J., Ibar, E., *et al.*2018, *A&A*, in press (arXiv:1712.03983)

Rieke, G. *et al.* 2015, *PASP*, 127, 584

Rocchetto, M. *et al.* 2016, *ApJ*, 833, 120

Schneider, G. *et al.* 2016, *AJ*, 152, 64

Stern, D., Eisenhardt, P., Gorjian, V., *et al.* 2005, *ApJ*, 631, 163

Stern, D., Assef, R. J., Benford, D. J. *et al.* 2012, *ApJ*, 753, 30

Taylor, P. & Kobayashi, C. 2014, MNRAS, 442, 2751

Taylor, P. & Kobayashi, C. 2015, MNRAS, 448, 1835

Taylor, P. & Kobayashi, C. 2016, MNRAS, 463, 2465

Taylor, P. & Kobayashi, C. 2017, MNRAS, 471, 3856

Vincenzo, F. & Kobayashi, C. 2018a, A&A, 610, L16 (VK18a)

Vincenzo, F. & Kobayashi, C. 2018b, MNRAS, 478, 155

Wright, G. *et al.* 2015, *PASP*, 127, 595

FM12
Calibration and Standardization Issues in UV-VIS-IR Astronomy

Astronomy in Focus - XXX
Proceedings IAU Symposium No. XXX, 2018
M. T. Lago, ed.

doi:10.1017/S1743921319005064

CALSPEC: HST Spectrophotometric Standards at 0.115 to 32 μm with a 1% Accuracy Goal

Ralph C. Bohlin

Space Telescope Science Institute, 3700 San Martin Drive, Baltimore, MD 21218, USA
email: bohlin@stsci.edu

Abstract. The flux distributions of spectrophotometric standard stars were initially derived from the comparison of stars to laboratory sources of known flux but are now based on calculated model atmospheres. For example, pure hydrogen white dwarf (WD) models provide the basis for the HST CALSPEC archive of flux standards. There is good evidence that relative fluxes from the visible to the near-IR wavelength of ~2.5 μm are currently accurate to ~1% for the primary reference standards.

Keywords. techniques: spectroscopic, astronomical data bases: miscellaneous, stars: fundamental parameters, (stars:) white dwarfs

---

1. Introduction

The measurement of precise absolute fluxes for stellar sources has been pursued with increased vigor since the discovery of the dark energy and the realization that its detailed understanding requires accurate spectral energy distributions (SEDs) of redshifted Ia supernovae in the rest frame. In addition, accurate absolute stellar fluxes as a function of wavelength are required for many astrophysical purposes. The fundamental parameters of stars, including mass, radius, metallicity, and age are inferred by matching accurate stellar atmosphere models to precisely calibrated UV spectroscopic data from which the effective temperature, surface gravity, composition, and interstellar reddening are determined for all types of stellar objects,

The Space Telescope Imaging Spectrograph (STIS), the Near Infrared Camera and Multi-Object Spectrograph (NICMOS), and the Wide Field Camera 3 (WFC3) low dispersion spectrometers on the Hubble Space Telescope (HST) have a calibration heritage that is tracable to the three unreddened, pure-hydrogen WDs, G191B2B, GD153, and GD71, which are in the effective temperature range 33590–59000 K. STIS covers the 0.115–1 μm range with 5 gratings at a resolution of R~500, while NICMOS has three grisms covering 0.8–2.5 μm at R~100. WFC3 has two bands G102 and G141 covering 1–1.7 μm at R=210 and R=130, respectively. The fluxes, i.e. spectral energy distributions (SEDs), of these primary standard stars are established using the Rauch Tübingen NLTE model atmosphere code (Rauch *et al.* 2013) to determine the relative flux vs. wavelength. Alternate NLTE Tlusty models for the same temperature and gravity differ in their relative flux distributions and are used to define the systematic uncertainties (Bohlin *et al.* 2014). The absolute normalization of the three models is from a weighted average of the absolute flux of Vega at 5556 Å and the MSX absolute flux measures of Sirius at 8–21 μm. The result of this reconciliation of the visible and mid-IR implies a flux for Vega at 5556 Å of 3.44 10^{-9} $erg\ cm^{-2}\ s^{-1}\ Å^{-1}$ ±0.5% (Bohlin 2014).

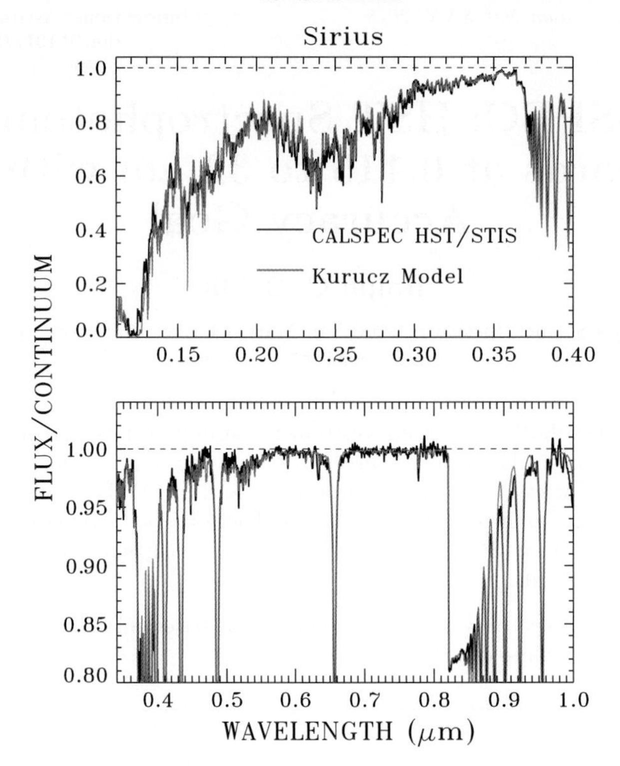

Figure 1. Comparison of STIS (black) to the Kurucz model (red) with $T_{\rm eff} = 9850\ K$, $\log g = 4.30$, and $[M/H] = +0.4$. Both the STIS SED and the model are divided by the same theoretical continuum for comparison on a magnified scale. The STIS data cover the wavelength range below 1 μm and are supplemented by IUE data below 1675 Å. The UV line-blanketing is severe, and the stellar flux goes to zero at Lα.

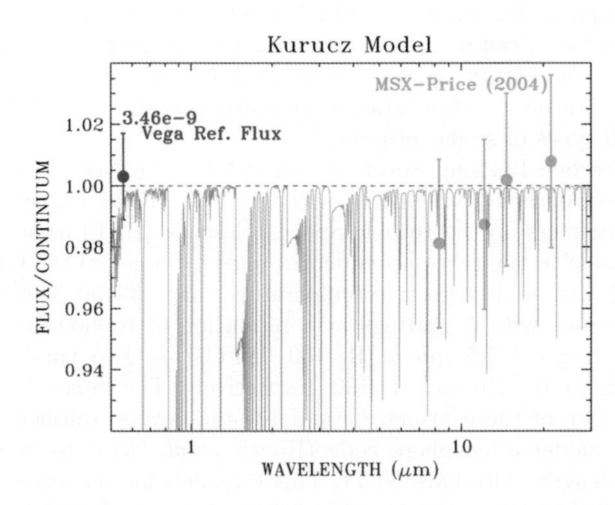

Figure 2. The green filled circles are the MSX values. The blue circle at 0.556 μm corresponds to the 3.46×10^{-9} erg cm^{-2} s^{-1} Å^{-1} value of Megessier (1995) for Vega. The adopted normalization of the model (red) is within ~0.5% of both the Megessier normalization and the average of the four MSX absolute flux measures, which implies a monochromatic best flux for Vega at 0.556 μm of 3.44×10^{-9} erg cm^{-2} s^{-1} Å^{-1}.

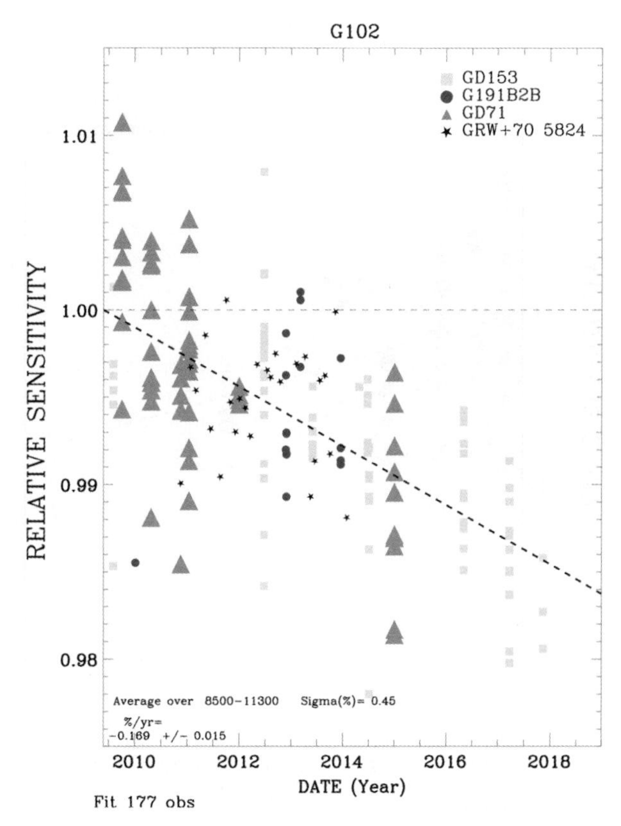

Figure 3. The change in sensitivity with time for the WFC3 G102 grism. The loss rate of 0.17 % yr^{-1} is written near the bottom of the plot.

Once calibrated using the three standard candles, the STIS and NICMOS spectrographs on HST measure the absolute flux of secondary stars from 0.115–2.5 μm with the WFC3 results replacing lower resolution NICMOS at 1–1.7 μm. The SEDs of several dozen secondaries and the three primary stars reside in the CALSPEC† database along with the covariance error matrix of uncertainties. Challenges to achieving the goal of 1% precision in the measured CALSPEC fluxes include non-linearities, changing sensitivity with time, and the high premium on HST time. However, synthetic photometry from the CALSPEC stars agrees with precision Landolt photometry to ~1% (10 mmag) in the B, V, R, and I bands (Bohlin & Landolt 2015). Model stellar atmospheres that fit these measured SEDs to ~1% extend the wavelength range to the James Webb Space Telescope (JWST) limit of ~30 μm (Bohlin *et al.* 2017).

2. Sirius as the Primary IR Reference Standard

Cohen *et al.* (1992) and, more recently, Engelke *et al.* (2010) and Rieke *et al.* (in prep.) recommend the use of Sirius as the primary IR absolute flux standard, because rapid rotation and the cool debris disk of Vega complicate the modeling of its IR flux distribution. STIS has measured the flux for Sirius from 0.17–1.01 μm on the *HST* CALSPEC scale. The measured STIS flux agrees well with predictions of a special Kurucz model atmosphere† in Fig. 1, adding confidence to the modeled IR flux predictions (Bohlin 2014). This model agrees to 0.5% with both the 5556Å zeropoint for Vega (Megessier

† http://www.stsci.edu/hst/observatory/crds/calspec.html

† http://kurucz.harvard.edu/stars/SIRIUS/ and an update (Kurucz private comm. 2013)

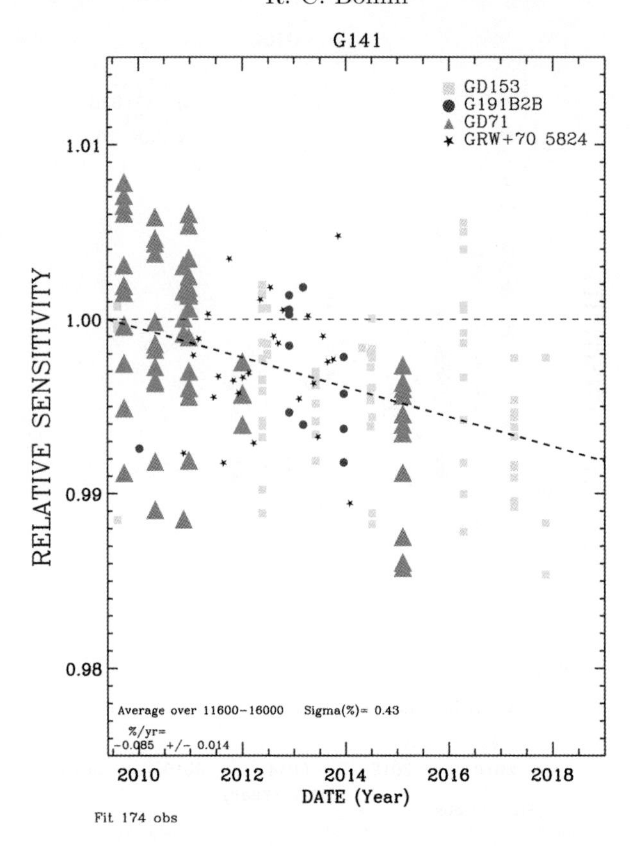

Figure 4. The change in sensitivity with time for the WFC3 G141 grism. The loss rate is 0.08 % yr^{-1}.

1995) and with the Price et al. (2004) MSX absolute flux measurements in the mid-IR at 8–21 μm, as illustrated in Fig. 2. Both the model and the STIS SED reside in CALSPEC.

3. Future CALSPEC Improvements

Old NICMOS fluxes at 1-1.7 μm will be replaced by WFC3 IR G102 and G141 IR grism SEDs, which have better wavelength accuracy, better spectral resolution, better repeatability, and, consequently, more precise flux distributions of ~1% accuracy vs. ~2% for NICMOS. As a first step in providing precision WFC3 SEDs, the sensitivity loss rates are found to be 0.17 and 0.08 % yr^{-1} for G102 and G141, respectively, as shown in Fig. 3 and Fig. 4. These rates are comparable to the STIS average loss rate since 2009 of 0.2 % yr^{-1} at 7900–9900 Å. Currently, only 19 CALSPEC stars have WFC3 grism data, but this sample should be expanded.

There is good evidence that the CALSPEC flux scale is correct to 1% from 0.2 to 7 μm, and there is agreement at the sub-percent level with the best absolute estimates in the visible at 0.5556 μm and at 8–21 μm. An external check against precision Landolt photometry also confirms CALSPEC flux distributions to better than 1% in the B, V, R, and I bandpasses. However, the CALSPEC SEDs all depend on the shape of theoretical WD flux distributions and on very few absolute flux measurements. To check and improve the precision of standard star SEDs, the theoretical WD model results should be tested by transferring modern laboratory flux standards to the stars.

References

Bohlin, R. C., Gordon, K. D., & Tremblay, P.-E. 2014, *PASP*, 126, 711

Bohlin, R. C. 2014, *AJ*, 147, 127

Bohlin, R. C., & Landolt, A. U. 2015, *AJ*, 149, 122

Bohlin, R. C., Mészáros, S., Fleming, S. W., Gordon, K. D., Koekemoer, A. M. & Kovács, J. 2017, *AJ*, 153, 234

Cohen, M., Walker, R. G., Barlow, M. J., & Deacon, J. R. 1992, *AJ*, 104, 1650

Engelke, C. W., Price, S. D., & Kraemer, K. E. 2010, AJ, 140, 1919

Megessier, C. 1995, *A&A*, 296, 771

Price, S. D., Paxson, G., Engelke, C., & Murdock, T. L. 2004, *AJ*, 128, 889

Rauch, T., Werner, K., Bohlin, R., & Kruk, J. W. 2013, *A&A*, 560, A106

Astronomy in Focus - XXX
Proceedings IAU Symposium No. XXX, 2018
M. T. Lago, ed.

doi:10.1017/S1743921319005076

Strategies for flux calibration in massive spectroscopic surveys

Carlos Allende Prieto

Instituto de Astrofísica de Canarias,
38205, La Laguna, Tenerife, Spain

Dept. de Astrofísica, Universidad de La Laguna,
38205, La Laguna, Tenerife, Spain
email: `callende@iac.es`

Abstract. Optical large-scale medium-resolution spectroscopic surveys such as SDSS, LAMOST, DESI, WEAVE or 4MOST are subject to constraints that limit the choice of flux calibrators, and the attained precision. The use of optical fibers, a large but limited field of view, the tiling strategies and tight schedules, are all factors that call for a careful evaluation of the flux calibration procedures.

The density of stars with well-known spectral energy distributions is so low that makes them unsuitable for flux calibration of large scale spectroscopic surveys. The alternative is to use stars with relatively simple spectra, which can be approximated well by synthetic spectra based on model atmospheres. One example are white dwarfs (Bohlin 1996), but their density is also too low for practical purposes: a few per square degree down to 19th magnitude. An alternative choice, exploited by the SDSS, are halo turn-off F-type stars (Stoughton *et al.* 2002). A-type stars offer another option, albeit with lower densities at high Galactic latitudes (Allende Prieto & del Burgo 2016). Ideally, one would use stars of various spectral types. The most common type, halo turn-off stars, can be used for the actual calibration, and the others for quality assessment.

The spectral typing needs to be performed before spectra are flux calibrated. Our group has explored various strategies for continuum normalization (the removal of the instrument response), finding good results using a running mean filter (Aguado *et al.* 2017; Allende Prieto *et al.* 2014). Interpolation in the models speeds up the model fitting process, but it is important to ensure that interpolations are sufficiently accurate (see, e.g. Mészáros & Allende Prieto 2013).

Fiber-fed spectrographs are particularly challenging, since errors in positioning fibers, guiding errors, or differential atmospheric refraction, add up. In our tests with data from the Baryonic Oscillations Spectroscopic Survey (BOSS; Dawson *et al.* 2016), we conclude that while the flux calibration is statistically accurate ($< 5\%$), individual spectra can exhibit much larger excursions, in excess of 20%.

Keywords. techniques: spectroscopic, surveys, stars: atmospheres, stars: fundamental parameters

---

References

Aguado, D. S., Allende Prieto, C., González Hernández, J. I., Rebolo, R., & Caffau, E. 2017, *A&A*, 604, A9
Allende Prieto, C., & del Burgo, C. 2016, *MNRAS*, 455, 3864
Allende Prieto, C., Fernández-Alvar, E., Schlesinger, K. J., *et al.* 2014, *A&A*, 568, A7
Bohlin, R. C. 1996, *AJ*, 111, 1743
Dawson, K. S., Kneib, J.-P., Percival, W. J., *et al.* 2016, *AJ*, 151, 44
Mészáros, S., & Allende Prieto, C. 2013, *MNRAS*, 430, 3285
Stoughton, C., Lupton, R. H., Bernardi, M., *et al.* 2002, *AJ*, 123, 485

Astronomy in Focus - XXX
Proceedings IAU Symposium No. XXX, 2018
M. T. Lago, ed.

doi:10.1017/S1743921319005088

Standardization in the UV with Astrosat and its issues related to star cluster studies

Priya Shah

Department of Physics, Maulana Azad National Urdu University, Gachibowli,
Hyderabad 500 032, India
email: priya.hasan@gmail.com

Abstract. The Ultra-Violet Imaging Telescope (UVIT) is one of the payloads in Astrosat, the first Indian Space Observatory. The UVIT instrument has two 375 mm telescopes: one for the far-ultraviolet (FUV) channel (1300-1800 Å), and the other for the near-ultraviolet (NUV) channel (2000-3000 Å) and the visible (VIS) channel (3200-5500 Å). We shall discuss the issues with standardization in the UV with reference to Astrosat Observations (Cycle A04). I shall discuss the problems faced in data-analysis and how these in turn lead to serious issues dealing with the color-magnitude diagarms, membership and age of the young embedded clusters studied.

Keywords. Galaxy:) halo, open clusters and associations: individual (C438, C439), ultraviolet: stars

---

1. Introduction

Astrosat is Indias first dedicated multi wavelength space observatory and was launched on Sep 28, 2015. One of the unique features of Astrosat is that it enables the simultaneous multi-wavelength observations of varied astronomical objects with a single satellite. The payloads (telescopes) observe in the visible, ultraviolet and x-ray region of the electromagnetic spectrum. They are the The Ultraviolet Imaging Telescope (UVIT), Large Area X-ray Proportional Counter (LAXPC), Soft X-ray Telescope (SXT), Cadmium Zinc Telluride Imager (CZTI) and Scanning Sky Monitor (SSM).

The UVIT consists of two 375 mm telescopes a far-ultraviolet (FUV) channel (1300-1800$\mathring{A}$), near-ultraviolet (NUV) channel (2000-3000$\mathring{A}$) and the visible (VIS) channel (3200-5500 $\mathring{A}$). It provides simultaneous imaging in the two ultraviolet channels with spatial resolution better than 1.8″, along with a provision for slit-less spectroscopy in the NUV and FUV channels (http://uvit.iiap.res.in/).

2. Target clusters

Camargo *et al.* (2015, 2016) used WISE data to identify clusters in high latitudes of the galaxy and 2MASS data to find the cluster parameters after careful decontaminations procedures were followed (Table 1) †.

Figure 1 shows the spatial distribution of the newly found clusters (green circles) compared to the earlier studies (red circles) by Carmago et al. (2015, 2016).

The discovery of these high latitude clusters are very crucial to our understanding of the galactic halo. If these young stars are formed in the halo, then it is possible that these clusters may get unbound before they reach the disc and young stars may reach the disc isolated. We also need to assess if this is an episodic event or a regular feature.

† $A_V$in the cluster central region, age, from 2MASS photometry, R_{GC} calculated using $R_\odot$ = 8.3 kpc as the distance of the Sun to the Galactic centre, x_{GC}, y_{GC}, z_{GC}: Galactocentric components

Table 1. Fundamental parameters and Galactocentric components for the ECs (Camargo *et al.* 2015)

Cluster	A_V (mag)	Age (Myr)	$d_\odot$ (kpc)	R_{GC} (kpc)	x_{GC} (kpc)	y_{GC} (kpc)	z_{GC} (kpc)
C 438	0.99 ± 0.03	2± 1	5.09 ± 0.70	8.69 ± 0.40	-07.04 ± 0.02	+0.97 ± 0.13	-4.99 ± 0.69
C 439	0.99± 0.03	2 ± 1	5.09 ± 0.47	8.70 ± 0.26	-07.05 ± 0.02	+1.06 ± 0.10	-4.97 ± 0.46
C 932	1.40± 0.03	2 ± 1	5.7 ± 0.53	10.55 ± 0.29	-9.07 ± 0.17	-0.29 ± 0.03	-5.38 ± 0.50
C 934	1.46 ± 0.06	2 ± 1	5.31 ± 0.51	10.27 ± 0.27	-8.97 ± 0.17	-0.27 ± 0.03	-5.01 ± 0.48
C 939	1.30 ± 0.06	3 ± 2	5.40 ± 0.50	10.34 ± 0.27	-9.00 ± 0.17	-0.31 ± 0.03	-5.09 ± 0.47
C 1074	0.93 ± 0.06	3 ± 1	4.14 ± 0.39	9.12 ± 0.15	-8.18 ± 0.09	-2.66 ± 0.25	3.02 ± 0.28
C 1099	0.71 ± 0.06	5 ± 1	4.32 ± 0.61	7.32 ± 0.30	-6.03 ± 0.17	-3.61 ± 0.51	2.05 ± 0.28
C 1100	0.93 ± 0.06	1 ± 1	6.87 ± 0.36	8.00 ± 0.23	-4.76 ± 0.13	-5.59 ± 0.29	3.16 ± 0.16
C 1101	0.96 ± 0.06	3 ± 1	3.91 ± 0.55	6.83 ± 0.27	-5.78 ± 0.20	-3.16 ± 0.44	1.78 ± 0.25

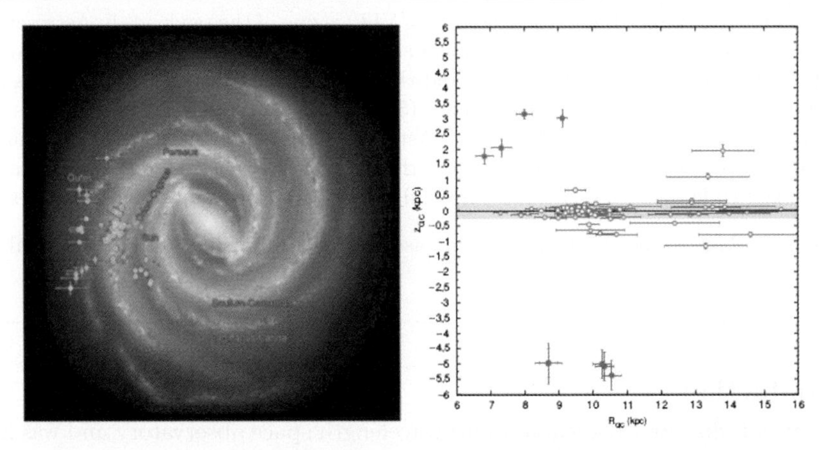

Figure 1. Spatial distribution of the ECs in this study (green circles) compared to ECs in previous works (red circles). Credit: Robert Hurt (NASA/JPL) and Camargo *et al.* (2015).

There are two possible scenarios that can explain star formation at such high galactic latitudes. One possibile scenario could be Galactic fountains or infall. The expansion of substructures powered by massive stellar winds and supernovae can trigger star formation in various shells and rings, inputting energy to the superbubble (Lee *et al.* 2009).

The other possible scenario is extragalactic in nature. The Milky Way galaxy has several satellite galaxies in its vicinity. Tidal interactions of the galaxy with its satellites is also a possible reason for star formation to take place so far from the disc of the galaxy. There are 12 known satellites of our galaxy.

3. Observations

We proposed simultaneous observations of these clusters using UVIT and the Xray telescopes on Astrosat, the Indian Astronomy Satellite. We shall concentrate only on the UVIT data. Our proposal A04-080 was granted a total observation time of 4500 secs where we observed TPhe (calibration source) and the two clusters C438, C439. We observed in the FUV (Filter: 2 - Barium Fluoride for 300 secs, Filter: 3 - Sapphire for 1200 secs) and NUV (Filter: 3 - NUV13 for 300 sec, Filter: 2 - NUV15 for 1200 secs).

Postma *et al.* (2011) describe calibration data and discuss performance of the photon-counting flight detectors for the UVIT. Tandon *et al.* (2017) reported on the performance of the (UVIT) on-board AstroSat. Murthy *et al.* (2017) wrote a software package (JUDE) to convert the Level 1 data from UVIT into scientifically useful photon lists and images. The routines are written in the GNU Data Language (GDL) and are compatible with the IDL software package. The level 1 data was analysed using the UVIT pipeline as well

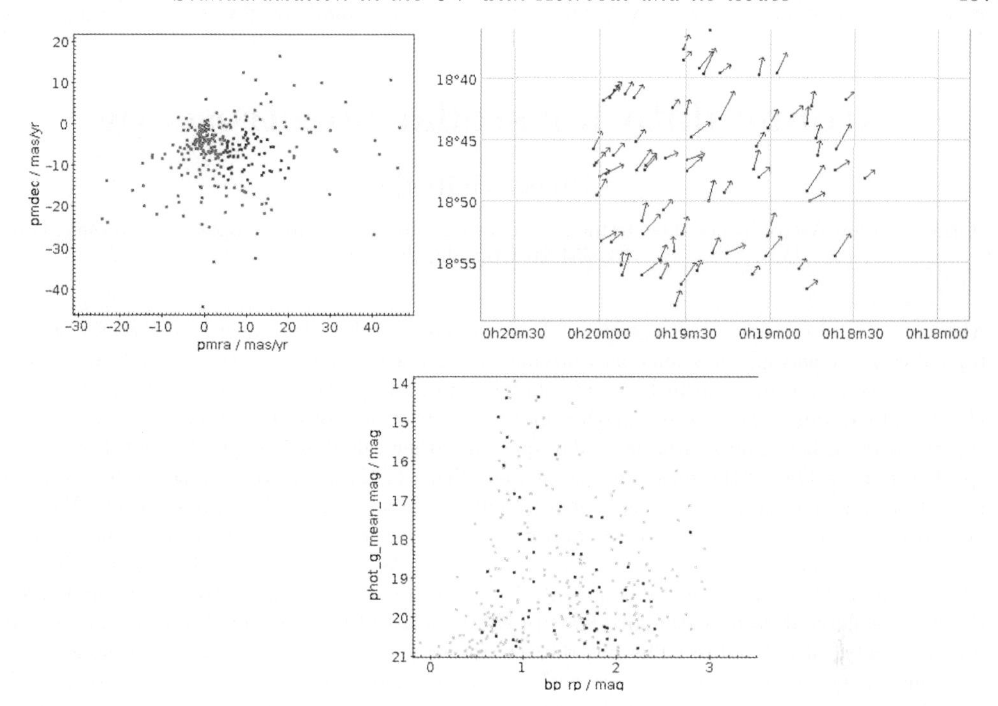

Figure 2. Spatial distribution of the ECs in this study (green circles) compared to ECs in previous works (red circles). Credit: Robert Hurt (NASA/JPL) and Camargo *et al.* (2015).

as the the JUDE pipeline, we found reasonable agreement with the two but were unable to construct good color-magnitude diagrams for the clusters.

4. Results

We used Gaia DR2 Gaia Collaboration *et al.* (2016) data to study our clusters.

Figure 2 shows the proper motion plot, probable members and the color magnitude diagram for C438 using Gaia DR2. However, the analysis is not fully conclusive because of the uncertainities, but our impression is that the groups are not real clusters (Private communication with Carme Jordi). Hence the problem of existence of these clusters still remains unsolved.

5. Acknowledgement

The author would like to thank Bhargavi S G for the help in preparation of the proposal submitted to APPS.

References

Camargo, D., Bica,E., Bonatto, C. and Salerno, G. 2015, *MNRAS*, 448, 1930
Camargo D., Bica E., Bonatto C., 2016, *A&A*, 593, 95
Gaia Collaboration *et al.*, *A& A*, Vol. 595, A1, 2016
Hasan, P., 2016, *New Advances in Physics*, 10, 1
Lee, H.-T., Chen, W. P. 2009, *ApJ*, 694, 1423
Murthy, J., Rahna, P. T., Sutaria, F., et al. 2017, *Astronomy and Computing*, 20, 120
Postma, J., Hutchings, J. B., & Leahy, D. 2011, *PASP*, 123, 833
Tandon, S. N., Hutchings, J. B., Ghosh, S. K., *et al.* 2017, *JA&A*, 38, 28
Turner, D. G., Carraro, G., Panko, E. A. 2017, MNRAS, 470, 481

Astronomy in Focus - XXX
Proceedings IAU Symposium No. XXX, 2018
M. T. Lago, ed.

doi:10.1017/S174392131900509X

Atomic data for stellar spectroscopy

Ulrike Heiter†

Observational Astrophysics, Department of Physics and Astronomy, Uppsala University, Box 516, 751 20 Uppsala, Sweden

Abstract. High-precision spectroscopy of large stellar samples plays a crucial role for several topical issues in astrophysics, such as studying the chemical evolution of the Milky Way Galaxy. Data are accumulating from instruments that obtain high-quality spectra of stars in the ultraviolet, optical and infrared wavelength regions on a routine basis. The interpretation of these spectra is often based on synthetic stellar spectra, either calculated on the fly or taken from a spectral library. One of the most important ingredients of these spectra is a set of high-quality transition data for numerous species, in particular neutral and singly ionized atoms. We rely heavily on the continuous activities of laboratory astrophysics groups that produce and improve the relevant experimental and theoretical atomic data. As an example, we briefly describe the efforts done in the context of the Gaia-ESO Public Spectroscopic Survey to compile and assess the best available data in a standard way, providing a list of recommended lines for analysis of optical spectra of FGK stars. The line data, together with specialised analysis methods, allow different surveys to obtain abundances with typical precisions of ∼0.1 dex on an industrial scale for ∼10 chemical elements. Several elements with urgent need for better atomic data have been identified.

Keywords. atomic data, stars: late-type, techniques: spectroscopic, surveys

---

1. Introduction

High-precision spectroscopy of large numbers of stars provides a good basis for studying the chemical and dynamical structure and evolution of the *Milky Way*, deriving the *origin of chemical elements*, and characterizing *planetary host stars*. In recent years, high-quality spectra have been accumulating from surveys and individual programs. The interpretation of these data using synthetic stellar spectra requires high-quality atomic transition data.

The wavelength regions that are mainly used for the above science cases span from the UV – mostly relying on Hubble Space Telescope spectra – to the optical and infrared, where spectra are obtained by ground-based 2- to 10-m telescopes around the world (e.g. ESO/Chile, France, USA, Australia, China). The relevant species are mainly neutral and singly ionized atoms, as well as diatomic and triatomic molecules, as most targets for galactic and planetary studies are F-, G-, or K-type stars.

The types of atomic and molecular data needed for a transition between two states (corresponding to a spectral line) can be broadly divided into two categories: 1) *transition probabilities* (oscillator strengths, gf-values), which can either be measured by laboratory astrophysics groups or calculated by atomic physics groups, and 2) parameters for *line-broadening by collisions* with neutral or charged particles, for which experimental data are very scarce, and the majority of which are therefore calculated by atomic physics groups (Barklem 2016).

† and the Gaia-ESO line list group (Karin Lind, Maria Bergemann, Martin Asplund, Paul S. Barklem, Šarunas Mikolaitis, Thomas Masseron, Patrick de Laverny, Laura Magrini *et al.*)

Data published by different laboratory astrophysics and atomic phsics groups have been collected and are being distributed by a number of databases, each with a different specialisation, although there is considerable overlap in data content. A few examples are the NIST Atomic Spectra Database (Kramida *et al.* 2018), the VALD database (Ryabchikova *et al.* 2015), and the STARK-B database (Sahal-Bréchot *et al.* 2017). The Virtual Atomic and Molecular Data Centre (Dubernet *et al.* 2016 and this conference, VAMDC, http://www.vamdc.eu) is an electronic infrastructure providing access to ~30 databases simultaneously, both via a web interface (the VAMDC portal) and via various Virtual Observatory tools.

2. Stellar spectroscopic surveys and their line lists

A handful of surveys are currently collecting or processing stellar spectroscopic data on an industrial scale, and several more are planned for the near future. The common goal of these surveys is to provide a homogeneous overview of the distributions of motions and chemical abundances in the Milky Way. Each of the surveys is approaching this goal in a somewhat different way. We give three examples in chronological order, all of which are obtaining spectra with a resolution of $\lambda/\Delta\lambda \gtrsim 20\,000$ and are targeting on the order of 10^5 to 10^6 stars.

The APOGEE survey (Majewski *et al.* 2017, USA) is working at infrared wavelengths (H-band), with a focus on the dust-obscured parts of Galaxy. The line list (Shetrone *et al.* 2015) comprises ~130 000 lines for 36 atoms and 6 molecules, with the "best" atomic data from the literature, and astrophysical atomic data calibrated on the Sun and Arcturus for ~20 000 lines. The Gaia-ESO Public Spectroscopic Survey (Gilmore *et al.* 2012; Randich *et al.* 2013, ESO) covers a considerable part of the optical spectral region and was designed to complement the ESA space mission Gaia by obtaining high-resolution spectra for faint stars. Also the GALAH survey (De Silva *et al.* 2015; Buder *et al.* 2018, Australia) is operating at optical wavelengths, and has its focus on chemical tagging.

Within the Gaia-ESO consortium a large effort has been put into the construction of a common line list which is being used throughout the survey (involving up to 14 abundance analysis groups). The Gaia-ESO line list has also constituted the starting point for the GALAH line list. In brief, ~1300 transitions were preselected in the relevant wavelength ranges (475 nm to 685 nm and 850 nm to 895 nm), which were presumed to allow accurate determination of stellar parameters, and of abundances for many elements for FGK-type stars. A compilation of the best atomic data for these lines defined the standard line list, comprising 44 neutral and singly ionised species (atomic numbers 3, 6, 8, 11–14, 16, 20–30, 38–42, 44, 56–60, 62–64, and 66). The preferred sources for gf-values were accurate laboratory measurements (usually from more recent publications, i.e. 1980s onwards), which were supplemented by less accurate laboratory gf-values (usually from older publications), and by theoretical data. We emphasise that no astrophysical gf-values were included or derived. A simple flag for recommended use was assigned to each line, according to the quality of the transition probabilities: *gf-flag* = **Y**es / **U**ndecided / **N**o, often (but not always) corresponding to the three levels of sources mentioned above.

The preselected lines were complemented with available data for all lines in the observed spectral range of the target stars, extracted from the VALD database in the case of atoms, and calculated and compiled by T. Masseron for 12 diatomic molecules. These data are needed to identify blends for the preselected lines, and as "background" for synthetic spectrum calculations. They were used to assign a second flag to each preselected line, according to blending properties: *syn-flag* = Y / U / N. Hence, the best lines to use in an abundance analysis would be those for which both *gf-flag* and *syn-flag* are equal to "Y". This is the case for about 15% of the preselected lines. The remainder

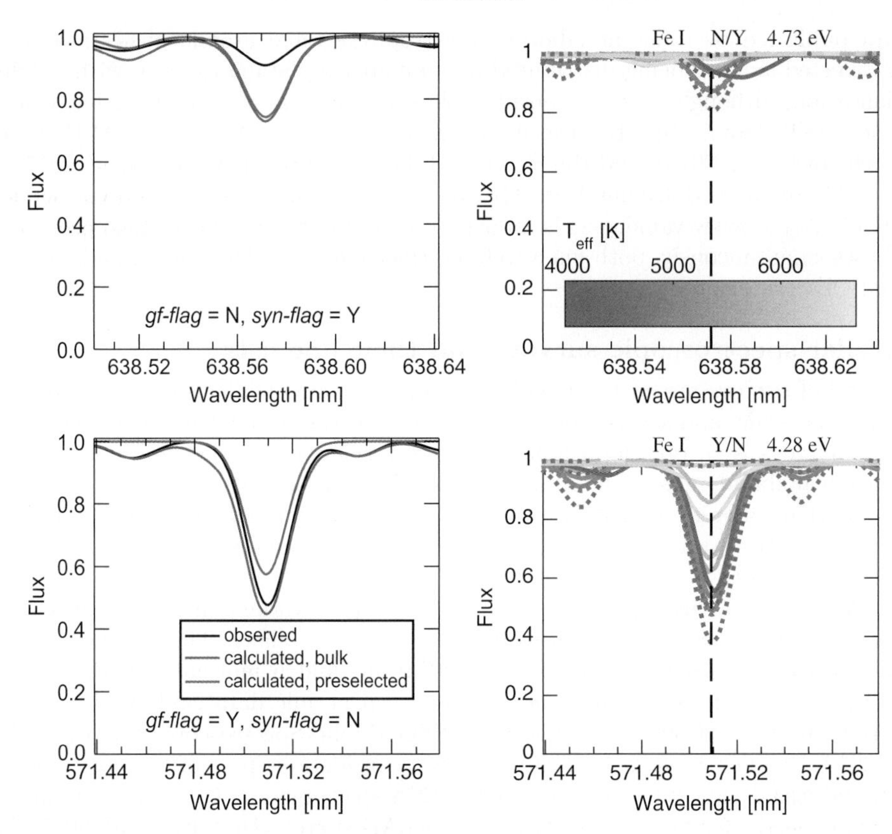

Figure 1. *Left:* Comparison of observed and calculated line profiles for Arcturus around two of the preselected Fe I lines with different combinations of *gf-flag* and *syn-flag*, convolved to a uniform spectral resolution of $R=47000$. Black lines: observations, red lines: calculations including preselected spectral lines only, blue lines: calculations including blends from background line list. *Right:* Observed line profiles for selected Gaia FGK benchmark stars (Heiter *et al.* 2015; Blanco-Cuaresma *et al.* 2014) for the same lines at the same resolution. Quality flags and lower level energy are indicated at the top of each panel. The vertical dashed line indicates central wavelength. Colour coding indicates effective temperature, solid lines are dwarfs, and dotted lines are giants.

should be regarded with care (different combinations of Y/U/N values for the two flags) or completely discarded (N/N case). An example for the giant star Arcturus is given in Fig. 1, which shows a line with inaccurate gf-value which is blend-free (upper panels) and a line with good gf-value which is blended (lower panels). The reason for this assessment becomes apparent only when observed line profiles are compared to calculated ones (left panels). The right panels show the same lines for several stars with different effective temperatures, surface gravities, and metallicities. For similar temperatures and gravities (e.g. red dotted lines representing cool giants) the variation in line strength is due to the difference in metallicity.

More examples and an extensive description and discussion of the Gaia-ESO line list is to be found in Heiter *et al.*, to be submitted to A&A.

3. Line list impact and data needs

One way to evaluate the impact of atomic data is to investigate the abundance precisions achieved by the different surveys. However, we caution that the significance of this approach is limited, since the abundance precisions also depend on stellar parameters, analysis methods, and the definition of "precision". For example, in the Gaia-ESO survey the method-to-method dispersion or line-to-line scatter is used, while the APOGEE survey refers to the star-to-star scatter within clusters. With this caveat the general picture emerging within the Gaia-ESO survey (Smiljanic *et al.* 2014; Mikolaitis *et al.* 2014; Lanzafame *et al.* 2015; Jofré *et al.* 2015) is that high precision abundances (uncertainties <0.15 dex) can be obtained for up to ten elements, including Al, Si, and Ca, while the least reliable abundances are obtained for Co, Ni, Zn, and Y. There are also problems for V at low metallicities. As a second example, the abundances obtained by the APOGEE survey (Holtzman *et al.* 2015; Mészáros *et al.* 2015; see also Holtzman *et al.* 2018; Jönsson *et al.* 2018) appear to achieve the highest precision (their uncertainties <0.05 dex) for α-elements, Fe, and Ni, and the lowest precision for V. Problems for Al, Ca, and Ti are encountered at low abundances.

The flags in the Gaia-ESO line list can be used to assess future data needs and to compile a wish list for new experimental gf-values in the *optical* wavelength region. Focussing on lines which are more or less unblended (*syn-flag* = Y or U), high priority should be given to species which have *gf-flag* = U or N for > 50% of these lines. This concerns ~240 Fe I lines, ~50 Ni I lines (whith high excitation energies), and some Fe II, Na I, Si I, and Ca II lines. However, there are a few species for which all of the few available "unblended" lines have uncertain gf-values, and these should be given even higher priority: Al I, S I, and Cr II.

In summary, we argue that accurate atomic data in the optical and IR are an important ingredient of large-scale stellar spectroscopic surveys. For recent progress in laboratory astrophysics see the latest report by the IAU Working Group on High-Accuracy Stellar Spectroscopy (Barklem *et al.* 2018). The Gaia-ESO survey provides a list of recommended lines for analysis of optical spectra of FGK stars. Abundances with typical precisions of ~0.1 dex are today being obtained on an industrial scale for ~10 chemical elements. Several elements with urgent need for better atomic data have been identified.

Acknowledgements

U.H. acknowledges support from the Swedish National Space Agency (SNSA/Rymdstyrelsen).

References

Barklem, P. S. 2016, *A&ARv*, 24, 9

Barklem, P. S., Nahar, S., Pickering, J., Przybilla, N., & Ryabchikova, T. 2018, Transactions of the IAU, *Vol. XXXA*, https://www.iau.org/static/science/scientific_bodies/working_groups/275/wg-hass-triennial-report-2015-2018.pdf

Blanco-Cuaresma, S., Soubiran, C., Jofré, P., & Heiter, U. 2014, *A&A*, 566, A98

Buder, S., Asplund, M., Duong, L., *et al.* 2018, *MNRAS*, 478, 4513

De Silva, G. M., Freeman, K. C., Bland-Hawthorn, J., et al. 2015, *MNRAS*, 449, 2604

Dubernet, M. L., Antony, B. K., Ba, Y. A., et al. 2016, Journal of Physics B Atomic Molecular Physics, 49, 074003

Gilmore, G., Randich, S., Asplund, M., et al. 2012, The Messenger, 147, 25

Heiter, U., Jofré, P., Gustafsson, B., et al. 2015, *A&A*, 582, A49

Holtzman, J. A., Hasselquist, S., Shetrone, M., et al. 2018, *AJ*, 156, 125

Holtzman, J. A., Shetrone, M., Johnson, J. A., et al. 2015, *AJ*, 150, 148

Jofré, P., Heiter, U., Soubiran, C., et al. 2015, *A&A*, 582, A81

Jönsson, H., Allende Prieto, C., Holtzman, J. A., et al. 2018, *AJ*, 156, 126

Kramida, A., Yu. Ralchenko, Reader, J., & and NIST ASD Team. 2018, NIST Atomic Spectra Database (ver. 5.6), [Online]. Available: https://physics.nist.gov/asd [Tue Oct 09 2018]. National Institute of Standards and Technology, Gaithersburg, MD. DOI: 10.18434/T4W30F

Lanzafame, A. C., Frasca, A., Damiani, F., et al. 2015, *A&A*, 576, A80
Majewski, S. R., Schiavon, R. P., Frinchaboy, P. M., et al. 2017, *AJ*, 154, 94
Mészáros, S., Martell, S. L., Shetrone, M., et al. 2015, *AJ*, 149, 153
Mikolaitis, Š., Hill, V., Recio-Blanco, A., et al. 2014, *A&A*, 572, A33
Randich, S., Gilmore, G., & Gaia-ESO Consortium. 2013, The Messenger, 154, 47
Ryabchikova, T., Piskunov, N., Kurucz, R. L., et al. 2015, scr, 90, 054005
Sahal-Bréchot, S., Dimitrijević, M. S., Moreau, N., & Nessib, N. B. 2017, in American Institute of Physics Conference Series, Vol. 1811, Atomic Processes in Plasmas (APiP 2016), 030003
Shetrone, M., Bizyaev, D., Lawler, J. E., et al. 2015, *ApJS*, 221, 24
Smiljanic, R., Korn, A. J., Bergemann, M., et al. 2014, *A&A*, 570, A122

Astronomy in Focus - XXX
Proceedings IAU Symposium No. XXX, 2018
M. T. Lago, ed.

doi:10.1017/S1743921319005106

The Stagger-grid: Synthetic stellar spectra and broad-band photometry

Andrea Chiavassa[1], L. Casagrande[2], R. Collet[3], Z. Magic, L. Bigot[1], F. Thévenin[1] and M. Asplund[2]

[1]Université Côte d'Azur, Observatoire de la Côte d'Azur, CNRS, Lagrange, CS 34229, Nice, France
email: andrea.chiavassa@oca.eu

[2]Research School of Astronomy & Astrophysics, Australian National University, Cotter Road, Weston ACT 2611, Australia

[3]Stellar Astrophysics Centre, Department of Physics and Astronomy, Ny Munkegade 120, Aarhus University, DK-8000 Aarhus C, Denmark

Resume

The STAGGER-grid (Magic *et al.* 2013) includes 3D stellar atmosphere simulations including hundreds of simulations with different metallicity, effective temperature, and surface gravity. Here we present the synthetic spectra computed for the grid with 3D

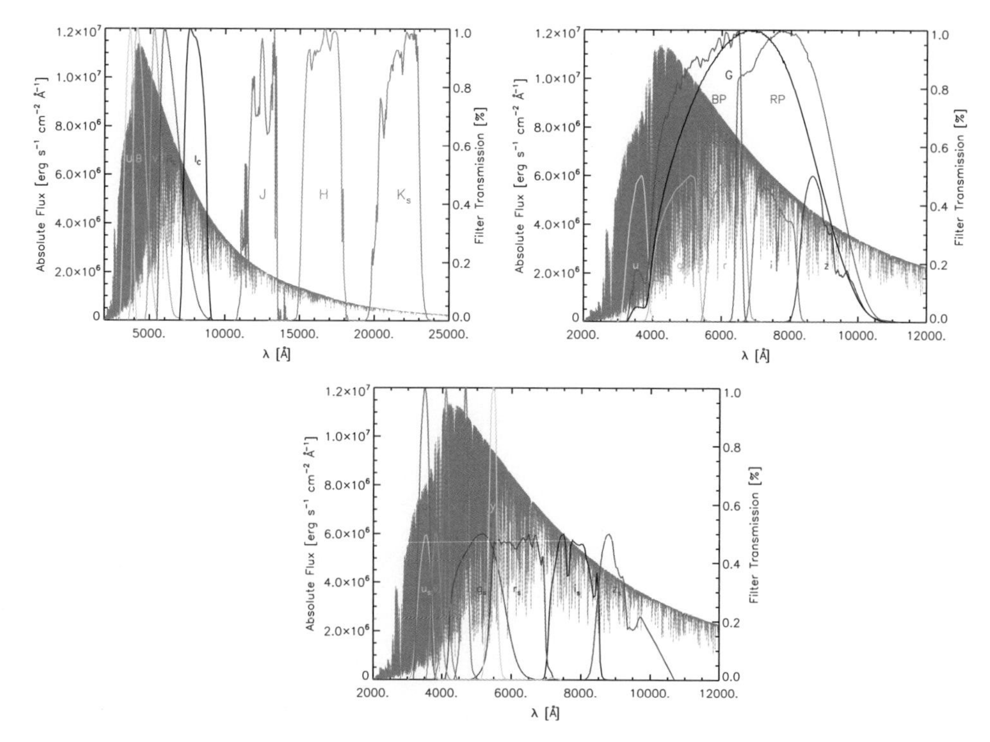

Figure 1. 3D synthetic spectrum of the solar simulation (grey) together with several system response functions: Johnson-Cousins system response functions (U, B, V, Rc, Ic in green), 2MASS in pink-violet, SDSS (u, g, r, i, z in yellow-red), Gaia (BP, RP, G), Strömgren (uvby in red-blue), SkyMapper (u_s, v_s, g_s, r_s, i_s, z_s, and HST-WFC3 (not shown here). For clarity, SDSS and SkyMapper functions are normalised to 0.5. Please refer to Chiavassa *et al.* (2018).

radiative transfer code OPTIM3D (Chiavassa *et al.* 2009). The spectra have been calculated with a constant resolving power of $\lambda/\Delta\lambda = 20\,000$ ($n_\lambda = 105\,767$ wavelength points) from 1000 to 200 000 Å and resolving power of 300 000 from 8470 to 8710 Å (Gaia RVS range).

In addition to this, we computed synthetic colours in the Johnson-Cousins, SDSS, 2MASS, Gaia, SkyMapper, Strömgren, HST-WFC3, and Gaiasystems (Fig. 1). We have made all the spectra publicly available for the community through the POLLUX† database (Palacios *et al.* 2010) and the bolometric corrections through CDS.

The reader should refer to the original paper for the details (Chiavassa *et al.* 2018).

References

Chiavassa, A., Plez, B., Josselin, E., & Freytag, B. 2009, *A&A*, 506, 1351
Chiavassa, A., Casagrande, L., Collet, R., *et al.* 2018, *A&A*, 611, A11
Magic, Z., Collet, R., Asplund, M., et al. 2013, *A&A*, 557, A26
Palacios, A., Gebran, M., Josselin, E., et al. 2010, *A&A*, 516, A13

† Available at http://pollux.oreme.org

Astronomy in Focus - XXX
Proceedings IAU Symposium No. XXX, 2018
M. T. Lago, ed.

doi:10.1017/S1743921319005118

All sky photometric zero-points from stellar effective temperatures

Luca Casagrande

Research School of Astronomy and Astrophysics, Australian National University, Australia
email: luca.casagrande@anu.edu.au

Abstract. I use SkyMapper DR1.1 to explore the quality of its *uvgriz* photometry, and zero-points. I introduce a formalism to derive photometric zero-points across the sky by benchmarking against stars with well known effective temperatures, bypassing the need for absolute spectrophotometry.

Keywords. surveys; techniques: photometric; stars: fundamental parameters, late-type; Galaxy: stellar content

---

Photometric zero-points in SkyMapper DR1.1 are not tied to spectrophotometric standard stars, but are obtained by predicting SkyMapper magnitudes of an ensemble of stars with photometry from other surveys (Wolf *et al.* 2018). If we wish to use SkyMapper photometry to study stellar populations across the Galaxy, and derive their parameters (most importantly metallicities), it is imperative to make sure that stellar parameters are not affected by variations of photometric zero-points.

With this goal in mind, we have conducted a thorough study of SkyMapper DR1.1 photometry. Since absolute flux standards are not available for SkyMapper, we have checked its standardisation by devising a new method based on the effective temperatures of a sample of reference stars (from Casagrande *et al.* 2010, 2011) to determine photometric zero-points across the sky. With our method we have recovered an offset of the *uv* zero-points that varies as a function of Galactic latitude. This variation is interpreted as a consequence of the reddening corrections currently employed to predict SkyMapper *uv* magnitudes from external photometry at longer wavelengths.

With a good control over photometric zero-points, we have then applied the InfraRed Flux Method (Casagrande *et al.* 2014) to compute effective temperatures for all stars in the GALAH spectroscopic survey, and provide empirical colour$-T_{\rm eff}$ relations. We have also used the GALAH spectroscopic metallicities to derive a calibration between them and SkyMapper v, g, and 2MASS K_S magnitudes. Our calibrations is validated down to approximately [Fe/H]$=-2$, and applies to late-type giants and dwarfs with $M_g < 7$. The reliability of our photometric metallicities is further checked against other spectroscopic surveys, confirming an overall precision of 0.2 dex. Finally, using ~ 9 million stars with Gaia parallaxes, we have produced a metallicity map in which we can clearly trace the mean metallicity decreasing as we move from the thin disc to the thick disc and then on into the halo, in agreement with what is expected from our knowledge of the Milky Way's structure.

References

Wolf, C., Onken, C. A., Luvaul, L. C., *et al.* 2018, *PASA*, 35, e010
Casagrande, L., Ramírez, I., Meléndez, J., Bessell, M., & Asplund, M. 2010, *A&A*, 512, A54
Casagrande, L., Schönrich, R., Asplund, M., *et al.* 2011, *A&A*, 530, A138
Casagrande, L., Portinari, L., Glass, I. S., *et al.* 2014, *MNRAS*, 439, 2060

Astronomy in Focus - XXX
Proceedings IAU Symposium No. XXX, 2018
M. T. Lago, ed.

doi:10.1017/S174392131900512X

Gaia Photometric Catalogue: the calibration of the DR2 photometry

D. W. Evans[1], M. Riello[1], F. De Angeli[1], J. M. Carrasco[2], P. Montegriffo[3], C. Fabricius[2], C. Jordi[2], L. Palaversa[1], C. Diener[1], G. Busso[1], C. Cacciari[3], E. Pancino[4] and F. van Leeuwen[1]

[1]Institute of Astronomy, University of Cambridge, Madingley Road, Cambridge CB3 0HA, UK

[2]Institut de Ciències del Cosmos, Universitat de Barcelona (IEEC-UB), Martí Franquès 1, 08028 Barcelona, Spain

[3]INAF–Osservatorio Astronomico di Bologna, via Ranzani 1, 40127 Bologna, Italy

[4]INAF - Osservatorio Astronofisico di Arcetri, Largo Enrico Fermi 5, 50125, Firenze, Italy
email: dwe@ast.cam.ac.uk

Abstract. Gaia DR2 was released in April 2018 and contains a photometric catalogue of more than 1 billion sources. This release contains colour information in the form of integrated BP and RP photometry in addition to the latest G-band photometry. The level of uncertainty can be as good as 2 mmag with some residual systematics at the 10 mmag level. The addition of colour information greatly enhances the value of the photometric data for the scientific community. A high level overview of the photometric processing, with a focus on the improvements with respect to Gaia DR1, was given. The definition of the Gaia photometric system, a crucial part of the calibration of the photometry, was also explained. Finally, some of the photometric improvements expected for the next data release were described.

Keywords. Catalogues, Surveys, Techniques: photometric

---

1. Introduction

The release of Gaia DR2 (Gaia Collaboration 2018) was a big step up from Gaia DR1 not only for the astrometry, but also the photometry in that colours for more than 1.3 billion sources were released. The presentation given described the processing and calibrations that went into generating the photometric data. See Evans *et al.* (2018) and Riello *et al.* (2018) for more details.

2. Internal calibrations

It is important to remember that the photometry in the different bands originates from two different sources on-board the satellite as this explains the different systematics. The G-band fluxes are the result of a fit to a set of corrected samples or pixels from the Astrometric Field (AF) CCDs. This fit consists of a Line or Point Spread Function (LSF/PSF) depending on whether the data is 1D or 2D. Only the brighter sources have 2D data transmitted to the ground for reasons of bandwidth limitation. The samples or pixels are corrected for bias and various background components e.g. charge release, straylight. These corrections are described in Hambly *et al.* (2018). The colour photometry comes from the prism-dispersed spectra on the BP (blue) and RP (red) CCDs. The photometric measurements are the results of summing up the bias and background-corrected samples and are equivalent to aperture photometry.

The basic principle of the calibration is one of a detailed iterative self-calibration followed by an external calibration which determines the zeropoints and the passbands that best represent the internal reference system (Carrasco *et al.* 2016). Since there is no photometric catalogue suitable for calibrating the Gaia photometry in detail, i.e. deep enough, all-sky, in the same photometric system and accurate at the mmag level, the main photometric calibration has to be carried out using the Gaia data itself. The task therefore is to calibrate the raw data onto a photometric system that represents the average instrument. By doing this, we ensure that a large number of sources can be used even for the rarest configurations of the satellite.

The general scheme of the iterative internal calibration is shown in Figure 1. The initial stage is to use all the raw photometric data for a source to form a crude average. Data is excluded from the periods of bad contamination for this initialization process. These averages are then used as reference values for calibrations which are calculated separately for each CCD, field-of-view and different configuration (gate and window class). The model for these calibrations includes the coordinate across the CCD and detailed colour information. The time variation is accounted for by calculating these calibrations per day of the mission. Following this initial calibration, calibrated photometry is used to form a new set of average fluxes for each source which then serve as reference fluxes to generate a new set of calibrations. This scheme is then iterated. A further set of calibrations gets carried out after this iterative scheme to account for a more detailed variation in response across each CCD. This is equivalent to a 1D flat field which is appropriate since the CCDs are read out in time-delayed integration mode.

The iterative scheme described here only works if the majority of sources are observed under many different conditions and the different observing configurations include many sources. In general, this is the case, but for the cases when this is a problem, special link calibrations are carried out. See Riello *et al.* (2018) for more details on these. Note that since this iterative initialization of the reference system is carried out for each processing cycle (which ends in a data release) and the data used for the initialization is different in the two cycles, the photometric system for DR2 is not the same as that for DR1.

Unexpected complexities have affected the photometric calibrations which were the result of unforeseen satellite problems. Soon after the mission started, it was seen that the photometric throughput was degrading at a rate higher than expected. This was caused by water from the spacecraft depositing as ice layers on the mirrors and the CCDs. This has been mitigated by periodically heating the mirrors and CCDs to remove this contaminant. During the period of the main mission, this decontamination procedure has been carried out three times. A consequence of this variation in throughput (by up to 0.5 mag) is to stress the photometric initialization process and to affect the detection limit at certain points of the mission. Another problem was one of additional straylight adding to the background level of the observations. This was caused by sunlight being scattered into the optical path by untrimmed fibres on the edge of the sunshield. This meant that the background-subtraction algorithms had to be improved to handle the large and rapid variations of this straylight component. However, at the faint end, there is a loss of performance and the G-band photometry is now sky limited. See Gaia Collaboration (2016) for more details about the satellite and the issues that affected the data.

3. Validation and testing

The precision of the G-band photometry is shown in Figure 2. This analysis is restricted to sources with about 100 CCD observations. By doing this, it is possible to compare the performance with nominal expectations. This nominal expected precision can also be convolved with a calibration floor to get an idea of the size of the uncalibrated effects that remain in the data at the individual CCD measurement level. For the G-band this is 2 mmag. At the bright end, the deviation from nominal is probably caused by saturation effects and problematic calibrations that have too few calibrators. The bumps at G=13

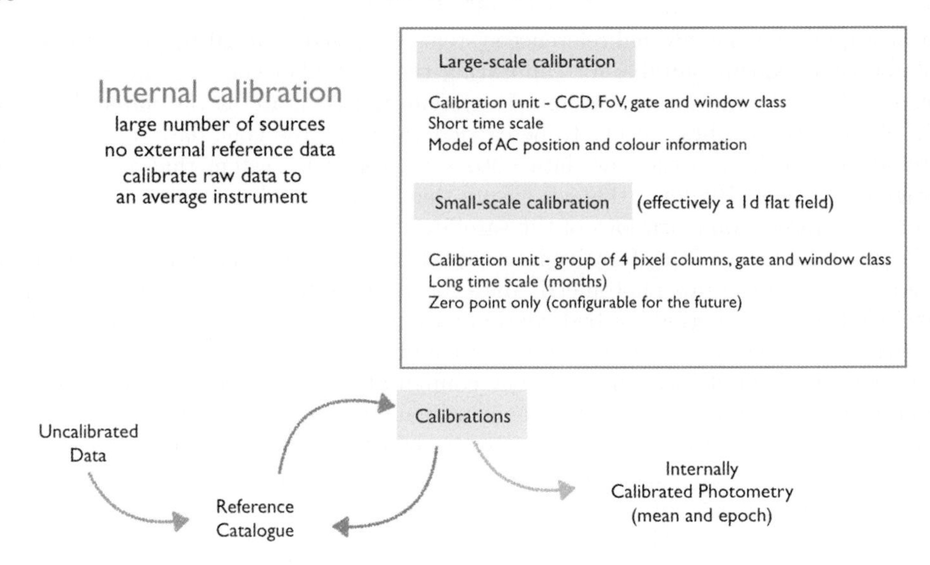

Figure 1. The general principle of the iterative internal calibration.

and 16 are linked to changes in the windowing configuration and some photometric inhomogeneities. At the faint end, it can be seen that the performance is poorer than expected due to the higher levels of straylight encountered in comparison to expectation. Whereas the expectation at the faint end of the G-band performance was that the observations would be source dominated, they are now sky dominated.

By restricting the analysis to sources with a fixed number of observations, a comparison between DR1 and DR2 was also possible and it shows that the performance has improved across the whole magnitude range.

Results were also shown for the BP and RP performances which had far fewer features in the uncertainty distribution than the G-band. This is due to fewer configuration changes and less saturation effects being present and to the overall larger uncertainties. The calibration floor for these passbands are 5 and 3 mmag for BP and RP respectively.

External catalogue comparisons have also been made, but suffer from the uncertainty in the source of any differences observed. In some cases it is clear that the difference has a Gaia origin due to the magnitude this occurs at corresponding to a Gaia configuration change. At G=13 and 16, small discontinuities remain at the 2–3 mmag level which is significantly smaller than those seen in the DR1 photometry. These comparisons also show possible background issues at the faint end.

Internal consistency checks can also be made between the three passbands. However, if differences are seen here, it is difficult to identify in which passband the features originate. An example of this can be seen in Figure 31 of Arenou *et al.* (2018). This shows a magnitude trend for sources brighter than $G = 16$ which is likely to be present in the G-band. This has been identified in a number of papers (Weiler 2018, Maíz Apellániz & Weiler 2018 and Casagrande & VandenBerg 2018). The likely cause of this magnitude trend is in the LSF/PSF fitting which generates the G-band fluxes where a non-time-varying calibration was used for the LSF/PSFs. Fainter than $G = 16$, the origin of the "hockey stick" feature seen can not be determined since it is likely that background issues are present in all three passbands.

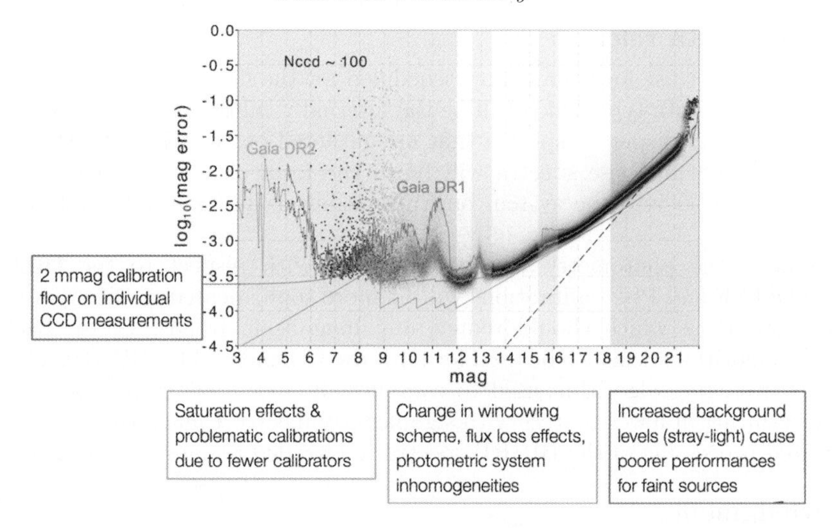

Figure 2. Precision of the G-band photometry. This analysis is restricted to sources with about 100 CCD observations. This enables a comparison to be made with nominal expectations (green line) and also facilitates the comparison between DR1 (magenta) and DR2 (orange). The red line is as the green one, but convolved with a 2 mmag calibration floor. The black dotted line has a gradient of 0.4 and shows the faint end to be sky dominated.

4. External calibration and the passbands

In addition to the G, BP and RP photometry, DR2 also released passbands that best represented the internal photometric system. These passbands were determined solely from the spectrophotometric data gathered for this specific determination. The version of this data used was SPSS V1. See Pancino *et al.* (2012) for more details on the SPSS catalogue.

Since the publication of DR2, two sets of alternative passbands have been published. Of these, the ones in Maíz Apellániz & Weiler (2018) are probably the best to use although note that there is no physical reason for the 2 passbands given for BP. The effects leading to this is likely to be a consequence of the photometric system not converging properly for sources with BP−RP<0.0. The main reason that these passbands are probably better than those released with DR2 is that additional information was used to help with the behaviour for very red sources, where the SPSS is lacking in data. Note that if you use these passbands, you must also use the zeropoints and G-band magnitude terms as formulated by the same authors. It is also important not to extrapolate the G-band magnitude terms fainter than G = 16 since the linear model used is not appropriate there and no comparison data was available.

An important point to consider is whether the magnitude term in the G-band determined in these papers is valid or not i.e. if the CALSPEC (Bohlin *et al.* 2017) magnitude scale is accurate at the mmag level. The clearest evidence for this comes from Figure 1 of Casagrande & VandenBerg (2018). At the very least, this shows that the BP and RP photometry are on the same magnitude scale as that of CALSPEC.

A final note of warning regarding the use of Gaia passbands in general. Even though much work has been done in building up a set of spectrophotometric standards, both SPSS and CALSPEC, they are not numerous, especially at the red end. The consequence of this is that the passbands are weakly constrained by this spectrophotometric data and that better results can be achieved by cross-matching photometric datasets with DR2 to derive colour-colour relationships.

5. The next data release

The next data release for Gaia is now scheduled for the first half of 2021. This release will contain a new data type which will be the internally calibrated mean source spectra. These are the prism-dispersed spectra that are gathered on the BP and RP CCDs. The external calibration of these spectra will be in the form of a utility that can convert a user-supplied SED into the system and resolution of the Gaia internally calibrated spectra.

Additionally, the photometry will be improved with respect to the DR2 performances. The LSF and PSF calibrations will be more sophisticated resulting in improved G-band photometry which should reduce any magnitude trends. Better background subtraction algorithms will also be used for both the G-band and for BP/RP. The robustness of the averaging algorithm used to compute the mean source photometry will be improved resulting in fewer sources with significantly incorrect magnitudes. Also, a more sophisticated processing of the BP/RP spectra in crowded regions will be used.

Acknowledgments

This work has made use of data from the European Space Agency (ESA) mission *Gaia* (https://www.cosmos.esa.int/gaia), processed by the *Gaia* Data Processing and Analysis Consortium (DPAC, https://www.cosmos.esa.int/web/gaia/dpac/consortium). Funding for the DPAC has been provided by national institutions, in particular the institutions participating in the *Gaia* Multilateral Agreement.

References

Arenou, F., Luri, X., Babusiaux, C. *et al.* 2018, *A&A* 616, A17
Bohlin, R. C., Mészáros, S., Fleming, S. W., *et al.* 2017, *AJ* 153, 234
Carrasco, J. M., Evans, D. W., Montegriffo, P. *et al.* 2016, *A&A* 595, A7
Casagrande, L., & VandenBerg, D.A. 2018, *MNRAS* 479, L102
Evans, D. W., Riello, M., De Angeli, F. *et al.* 2018, *A&A* 616, A4
Gaia Collaboration (Prusti, T., *et al.*) 2016, *A&A* 595, A1
Gaia Collaboration (Brown, A. G. A., *et al.*) 2018, *A&A* 616, A1
Hambly, N., Cropper, M., Boudreault, S., *et al.* 2018, *A&A* 616, A15
Maíz Apellániz, J. & Weiler, M. 2018, arXiv:1808.02820
Pancino, E., Altavilla, G., Marinoni, S., *et al.* 2012 *MNRAS* 426, 1767
Riello, M., De Angeli, F., Evans, D. W., *et al.* 2018, *A&A* 616, A3
Weiler, M. 2018, *A&A* 617, A138

Astronomy in Focus - XXX
Proceedings IAU Symposium No. XXX, 2018
M. T. Lago, ed.

doi:10.1017/S1743921319005131

The SkyMapper Southern Survey and its calibration

Christian Wolf

Research School of Astronomy and Astrophysics
Australian National University, Canberra ACT 2611, Australia
email: christian.wolf@anu.edu.au

Abstract. I discuss the photometric calibration of the SkyMapper Southern Survey, our adopted methods and what we learned from comparisons with external catalogues.

Keywords. surveys, catalogs, techniques: photometric

SkyMapper observes the whole Southern hemisphere in approx. 4,000 overlapping tiles with a field-of-view of 5.7°. It spans a dynamic range of 13 mag by combining shallow exposures with a deeper Main Survey. Data Release 1 is based on the Shallow Survey and world-accessible since Dec 2017 (Wolf *et al.* 2018, PASA 35, 10). Its six passbands are similar to SDSS filters except that the SDSS u band is split into two SkyMapper filters, an u(ultraviolet) and a v(iolet) band. Science goals include searching for extremely metal-poor stars and luminous high-redshift QSOs, to provide colour maps of nearby Southern galaxies, serve as a parent sample for the spectroscopic Taipan Galaxy Survey and as an optical complement to the radio surveys of the Australian SKA Pathfinders.

Photometric calibration can be done in three ways: (i) using airmass regression, but SkyMapper has a modest fraction of photometric nights, (ii) via frame overlaps, which require sufficient survey completion, and (iii) via external authority, which we adopted for now. A few challenges remain: (i) we are at the mercy of the external authority's own reliability, (ii) our exposures which can be as short as 5 sec freeze weather-related atmospheric throughput patterns across the wide field, and (iii) external all-sky catalogues currently don't cover the uv filter range, forcing us to extrapolate the calibration. Extrapolation from gri to uv depends on stellar metallicity, which has gradients across the sky; and on dust reddening, which is known for the integrated column along the line-of-sight to a calibrator star, but not for the star itself given its limited distance.

For DR1 we used 2MASS and APASS DR9; the latter has discontinuities across the sky, with residuals reaching over 0.3 mag. Given the exact APASS bandpasses are unknown, we fit an empirical transformation from APASS gri to Pan-STARRS1 gri and then apply a second transformation from PS1 to SkyMapper $uvgriz$ using the Pickles (1998, PASP 110, 863) spectral atlas. Calibrator stars are restricted to a colour range where transformations are close to linear. Calibration residuals between SkyMapper $griz$ and PS1 $griz$ have an rms scatter of 2%, but can be off by up to 10% in fat tails. For DR2 we replace APASS by Gaia DR2 (2018) and thus inherit its all-sky homogeneity.

Still, reddening at very low levels is not described with a precision of 1% yet, and at higher levels bandpass extrapolation still causes offsets. For DR2, Gaia allows us to restrict calibrator stars to a distance of less than 1 kpc and limit the amount of reddening.

Comparing Gaia DR2 and PS1 while creating the ATLAS Reference Catalog (Tonry *et al.* 2018, ApJ, in press), we find offsets of up to 10%, which can be modelled as a strong function of local star density and thus be suppressed to below 1%.

Astronomy in Focus - XXX
Proceedings IAU Symposium No. XXX, 2018
M. T. Lago, ed.

doi:10.1017/S1743921319005143

Passband reconstruction from photometry

Michael Weiler, Carme Jordi, Josep Manel Carrasco and Claus Fabricius

Institut de Ciències del Cosmos, Universitat de Barcelona, Spain
email: mweiler@fqa.ub.edu

Abstract. Photometric passbands are usually characterised through laboratory measurements and once in operations they are refined with true observations of reference sources with known spectral energy distribution. This paper revises the methods to determine those passbands and discusses the limitations encountered. The passbands are not fully constrained by the reference sources used and the method presented here allows to evaluate which is the constrained and the unconstrained component of the passband.

Keywords. standards, techniques: photometric, methods: analytical

---

1. Introduction

The interpretation of photometric observations in astrophysical terms requires knowledge of the photometric passbands the observations were performed in. These passbands are usually pre-defined by simulations and laboratory measurements on individual components of instrumentation, such as filter transmissivities and mirror reflectances. Once in operation, the passbands are refined using actual observations of astronomical objects with know spectral energy distributions. Weiler *et al.* (2018) have presented a detailed analysis of the constraints inherent to any reconstruction of the passband from photometric observations. In this work we summarise the important findings, provide a graphical interpretation of the results, and present examples for the reconstruction of the passband of ESA's HIPPARCOS mission ESA (1997).

2. Theoretical framework

2.1. *The mathematical formalism*

We consider the problem, that for a set of N calibration sources we have known spectral photon distributions (SPDs), as well as photometric observations in some passband. We want to determine the passband, i.e. derive the wavelength-dependent response curve $p(\lambda)$, λ being the wavelength. We assume $p(\lambda)$ to be the "photon response curve", i.e. the ratio of recorded photons (or, photo electrons) over the total number of photons entering the instrument. If c_i denotes the recorded number of photons per unit of time and aperture area for the calibration source i, $i=1,\ldots,N$, and $s_i(\lambda)$ the SPD of that source in terms of photons per unit of time, wavelength, and aperture area, then these two quantities are linked by the response curve via

$$c_i = \int_{\lambda_0}^{\lambda_1} p(\lambda) \cdot s_i(\lambda) \, \mathrm{d}\lambda. \tag{2.1}$$

This integral is evaluated over a finite wavelength interval $I = [\lambda_0, \lambda_1]$ which is chosen such that we can a-priori assume the response curve to be identical to zero everywhere outside I. To derive a suitable mathematical formalism in this context, we make use of the fact that both functions, $p(\lambda)$ and $s_i(\lambda)$, are for physical reasons square-integrable, i.e. the integral ofter the square of each of the functions has to be finite. All square integrable functions on the interval I form a vector space over the field of real numbers, which is endowed with a conical scalar product defined by

$$\langle f_1 \,|\, f_2 \rangle := \int_{\lambda_0}^{\lambda_2} f_1(\lambda) \cdot f_2(\lambda)\, \mathrm{d}\lambda. \tag{2.2}$$

Thanks to the scalar product, this vector space of square integrable functions on the interval I, denoted $\mathcal{L}^2(I)$, has properties which widely correspond to the properties well familiar from the more graphic Euclidian vector spaces. Elements of $\mathcal{L}^2$ can be developed in bases, and the existence of the scalar product ensures that bases can be chosen orthonormal. The scalar product also introduces a norm measuring the lengths of vectors, a metric measuring the distances between vectors, and angles between vectors. When we want to emphasise the vector nature of a function $f(\lambda)$ on the interval I, we conventionally write $|f\rangle$ for it. This convention corresponds to the convention of putting a little arrow on top of a letter to emphasise the vector nature of the corresponding quantity in a Euclidian vector space. Table 1 provides an overview of the correspondences between the well familiar Euclidian spaces and vector spaces formed by square-integrable functions.

Such a vector-based approach has first been introduced in photometry by Young (1994) for the problem of photometric transformations. Using this formalism also for the problem of passband reconstruction, we can write in short for Eq. (2.1)

$$c_i = \langle s_i \,|\, p \rangle \quad, i = 1, \ldots, N \tag{2.3}$$

and develop the N SPDs $|s_i\rangle$ in M orthonormal basis functions $|\varphi_j\rangle$, $j = 1, \ldots, M$ and $1 \leq M \leq N$,

$$|s_i\rangle = \sum_{j=1}^{M} a_{ij}\, |\varphi_j\rangle. \tag{2.4}$$

Taking $\mathbf{A}$ as the $N \times M$ matrix containing the coefficients a_{ij} of the development of the ith calibration SPD in the jth basis function, and combine the photometric observations c_i in a N–vector $\mathbf{c}$, we obtain

$$\mathbf{c} = \mathbf{A}\,\mathbf{p}_{\|}, \tag{2.5}$$

with $\mathbf{p}_{\|}$ the M–vector whose j–th element is

$$p_j = \langle p \,|\, \varphi_j \rangle. \tag{2.6}$$

Thus, $\mathbf{p}_{\|}$ is the vector containing the coefficients of the passband $|p\rangle$ developed in the basis $\{|\varphi_j\rangle\}_{j=1,\ldots,M}$. The optimal solution for the response curve $p(\lambda)$ can therefore be obtained by solving the simple linear system of equations that is Eq. (2.5) for $\mathbf{p}_{\|}$. Two observations are of importance in this respect.

First, in Eq. (2.4) we developed the N calibration SPDs $|s_i\rangle$ in a M–dimensional basis, with $1 \leq M \leq N$. Such a basis arises naturally from a set of N given vectors which contain M linearly independent vectors, where M is at least one and cannot exceed the number of vectors N. The optimal choice for the basis $\{|\varphi_j\rangle\}_{j=1,\ldots,M}$ is therefore derived from the calibration SPDs themselves. One may select the M linear independent vectors among the N vectors $|s_i\rangle$ and orthonormalize them. In practice, where the calibration SPDs may be available in tabulated form, this step can easily be done by a functional principal component analysis.

Second, the solution of Eq. (2.5) does not uniquely determine the passband. In fact, the solution for $\mathbf{p}_{\|}$ only provides the projection of the passband $|p\rangle$ onto the sub-space

Table 1. Comparison between Euclidian vector spaces and $\mathcal{L}^2$ over the field of real numbers.

Object	$\mathbb{R}^n$	$\mathcal{L}^2$
Notation for vector and basis vector	$\vec{r}$ and $\vec{e}_i$	$\lvert f\rangle$ and $\lvert\varphi_i\rangle$
Basis development	$\vec{r} = \sum_{i=1}^{n} a_i\,\vec{e}_i\,, a_i \in \mathbb{R}$	$\lvert f\,\rangle = \sum_{i=0}^{\infty} a_i\,\lvert\varphi_i\rangle\,, a_i \in \mathbb{R}$
Scalar product	$\vec{r}_1 \cdot \vec{r}_2$	$\langle\, f_1 \mid f_2\,\rangle$
Orthonormality of basis	$\vec{e}_i \cdot \vec{e}_j = \delta_{ij}$	$\langle\, \varphi_i \mid \varphi_j\,\rangle = \delta_{ij}$
Projection onto a basis vector	$(\vec{r}\cdot\vec{e}_i)\cdot\vec{e}_i\ [= a_i\,\vec{e}_i]$	$\langle\, f \mid \varphi_i\,\rangle \cdot \lvert\varphi_i\rangle\ [= a_i\,\lvert\varphi_i\rangle]$
Norm ("Length")	$\sqrt{\vec{r}\cdot\vec{r}}$	$\sqrt{\langle\, f \mid f\,\rangle}$
Metric ("Distance")	$\sqrt{(\vec{r}_1-\vec{r}_2)\cdot(\vec{r}_1-\vec{r}_2)}$	$\sqrt{\langle\, f_1 - f_2 \mid f_1 - f_2\,\rangle}$
Angle	$\mathrm{acos}\left(\frac{\vec{r}_1\cdot\vec{r}_2}{\sqrt{\vec{r}_1\cdot\vec{r}_1\ \vec{r}_2\cdot\vec{r}_2}}\right)$	$\mathrm{acos}\left(\frac{\langle\, f_1 \mid f_2\,\rangle}{\sqrt{\langle\, f_1 \mid f_1\,\rangle\,\langle\, f_2 \mid f_2\,\rangle}}\right)$

of $\mathcal{L}^2(I)$ that is spanned by the calibration SPDs. This will in general be a poor approximation to $|p\rangle$. We may however add any vector $|p_\perp\rangle$ which is orthogonal to the sub-space spanned by the SPDs, i.e. which is satisfying the conditions

$$\langle\, p_\perp \mid \varphi_j\,\rangle = 0 \quad , j = 1, \ldots, M, \tag{2.7}$$

to our solution $|p_\parallel\rangle$ without affecting the synthetic photometry of the calibration sources. We therefore write

$$|p\rangle = |p_\parallel\rangle + |p_\perp\rangle. \tag{2.8}$$

The component $|p_\parallel\rangle$ in this sum we refer to as the "parallel component" of the passband (with respect to the set of calibration sources used in the passband determination). This component is uniquely determined by the set of calibration SPDs available for passband determination and can be easily found by solving Eq. (2.5). The second component of the passband, $|p_\perp\rangle$, we refer to as the "orthogonal component". This component is entirely unconstrained by the set of calibration SPDs. We have to estimate this component, in such a way that the resulting passband $|p\rangle$ is satisfying the physical constraints (i.e., being non-negative, bound to unity in case of non-amplifying photon detectors) and in reasonable agreement with the available a-priori knowledge about the general shape of the passband.

The fact that the passband contains a component that is not constrained by the calibration SPDs may have important consequences for the synthetic photometry of SPDs which depend on the unconstrained orthogonal passband component. The need for eventually guessing $|p_\perp\rangle$ introduces a fundamental uncertainty to the passband $|p\rangle$. While for SPDs which lay within the sub-space of $\mathcal{L}^2(I)$ spanned by the calibration SPDs, i.e. which can be well described as a linear combination of the calibration SPDs, the synthetic photometry is reliable and unaffected by uncertainties in $|p_\perp\rangle$, this is not the case for SPDs that fall significantly out of this sub-space. For the later SPDs the uncertainty in $|p_\perp\rangle$ may introduce systematic errors in the synthetic photometry.

As contributions from $|p_\perp\rangle$, and the associated uncertainty, cannot be excluded in general, one may use the two passband components individually when computing the synthetic photometry for some SPD of interest. The ratio of the contributions of the two passband components to synthetic photometry, $\langle\, s \mid p_\perp\,\rangle / \langle\, s \mid p_\parallel\,\rangle$, then provides an estimate on how sensitive the photometric photometry is to the unconstrained component. A more graphical measure for this sensitivity is obtained if the ratio is weighted by the norms of the passband components, and converted into an angle by taking the inverse tangent,

$$\gamma = \text{atan}\left(\frac{\langle\, s \,|\, p_\perp \,\rangle}{\langle\, s \,|\, p_\parallel \,\rangle}\sqrt{\frac{\langle\, p_\parallel \,|p_\parallel \,\rangle}{\langle\, p_\perp \,|p_\perp \,\rangle}}\right). \tag{2.9}$$

This angle γ gives the orientation of some SPD $|s\rangle$ with respect to $|p_\parallel\rangle$, measured in the plane spanned by the two components $|p_\parallel\rangle$ and $|p_\perp\rangle$.

2.2. *Graphical interpretation*

The close analogy between Euclidian vector spaces and vector spaces of square integrable functions, as highlighted in Table 1, allows for a graphical illustration of the formalism described so far. For a truly graphical interpretation it is of course necessary to reduce the dimensionality of the problem. We replace the infinite-dimensional space of functions by a three-dimensional space, and we set $M = 1$, which corresponds to a situation where only calibration spectra of a single shape are available. In this case, the sub-space spanned by the calibration spectra is one-dimensional, which is graphically to be interpreted as a line. This case is shown in the left-hand panel of Fig. 1. The axes of the three-dimensional space are chosen arbitrarily in this figure, i.e. without any specific orientation with respect to the vectors of interest. The $(M = 1)$-dimensional sub-space spanned by the calibration SPDs is illustrated as a black line. The parallel component of the passband, shown as a red arrow, is oriented along this line. In a three-dimensional space, the orthogonal space to the one-dimensional line is a two-dimensional plane, oriented perpendicular to the line. The orientation of this plane is indicated by the blue circle. Any vector inside that plane can be added to $|p_\parallel\rangle$ without affecting the synthetic photometry for the calibration sources. The component of the passband within this plane is therefore unconstrained. Choosing the length and orientation of a vector within the plane for the orthogonal component of the passband has to be done such that the sum of the parallel and orthogonal component is in a position in agreement with the a-priori knowledge on the passband. This sum, the actual assumption of the passband, is shown as a green arrow in Fig. 1. The green shaded region indicates the plane spanned by the parallel and orthogonal component of the passband, being perpendicular to the blue plane of the orthogonal space.

The right-hand panel in Fig. 1 illustrates the situation when computing synthetic photometry. A SPD $|s\rangle$, shown as a magenta arrow, may lay somewhere in space. The contributions of the parallel and orthogonal components of the passband to the synthetic photometry for that SPD depend on the projections of $|s\rangle$ onto these passband components. These are shown as the orange on cyan lines, respectively. As the vector of the parallel component is constrained by the calibration sources, so is the projection of the SPD onto it. The projection onto the orthogonal component depends on the choice made for $|p_\perp\rangle$, illustrating the uncertainty in the synthetic photometry resulting from the orthogonal component. The dashed magenta line shows the projection of the SPD onto the plane spanned by the parallel and orthogonal passband components. The angle γ lays in this plane and is indicated in green. It measures from $|p_\parallel\rangle$ towards $|p_\perp\rangle$ and serves as a measure for the dependency of the synthetic photometry on the choice of $|p_\perp\rangle$. A value of zero for γ corresponds to a case where the projection of the SPD onto the plane spanned by $|p_\parallel\rangle$ and $|p_\perp\rangle$ coincides with the parallel component, and the synthetic photometry thus being independent of the selected orthogonal component. This situation corresponds to a reliable result for the synthetic magnitude. For $\gamma = 90°$, the opposite occurs. The synthetic photometry solely depends on the choice for $|p_\perp\rangle$, which may result in considerable systematic uncertainties on the synthetic magnitudes.

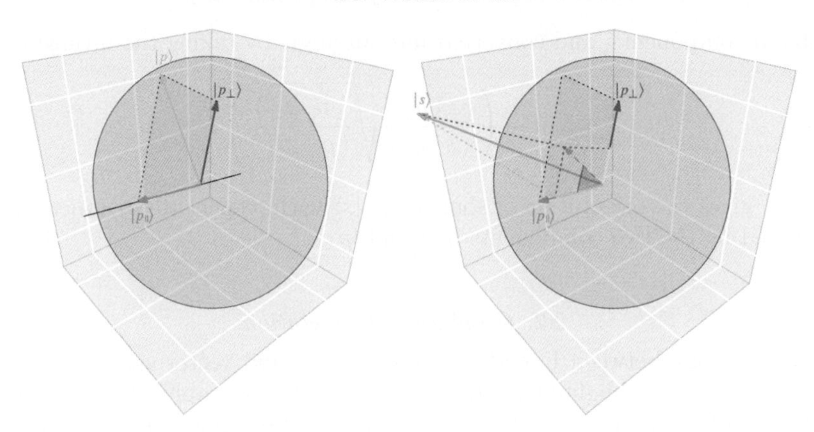

Figure 1. Sketch of the formalism described in this work. Left: Decomposition of the passband into parallel and orthogonal component. Right: Decomposition of a SPD with respect to the parallel and orthogonal passband components. For details see text.

2.3. *Estimating the orthogonal component*

The orthogonal passband component $|p_\perp\rangle$ has to be estimated such that its sum with $|p_\|\rangle$ satisfies the conditions one can a-priori impose on the passband, such as being non-negative and smooth. In particular the requirement for smoothness can be difficult to meet. The parallel component is determined by the shape of the calibration spectra, which are expected to contain many spectral features on many different scales. In order to easily find an orthogonal component compensating these features, it may be convenient to start from a smooth initial guess for the passband, $|p_{ini}\rangle$. One then may define a simple model for modifying the initial guess in a smooth way, and imposing the resulting modification to meet the parallel component derived by solving Eq. (2.5). We thus express the passband by a model

$$|p\rangle = \left(\sum_{k=0}^{K-1} \alpha_k \cdot \phi_k(\lambda) \right) |p_{ini}\rangle \tag{2.10}$$

with $\phi_k(\lambda)$ some basis functions for the smooth modification of the initial guess. Using Eq. (2.6), we obtain a linear system of equations for the coefficients α_k of the modification model, written in a K–vector $\boldsymbol{\alpha}$,

$$\mathbf{p}_\| = \mathbf{M} \cdot \boldsymbol{\alpha}. \tag{2.11}$$

The elements of the matrix $\mathbf{M}$ are given by

$$M_{n,m} = \langle\, \phi_m \, p_{ini} \,|\, \varphi_n \,\rangle. \tag{2.12}$$

For the K basis functions of the modification model, polynomials may be sufficient for simple passband shapes. For more complex passband shapes, where more freedom in modifying the initial guess may be desirable, B-spline basis functions may provide a convenient choice for the $\{\phi_k(\lambda)\}_{k=0,\ldots,K-1}$.

3. Examples from HIPPARCOS

To illustrate the effect of the unconstrained orthogonal component of the passband, we take as an example three different passband solutions for the HIPPARCOS passband, H_p, derived by Weiler *et al.* (2018). These three possible solutions for H_p were obtained using the Next Generation Spectral Library (NGSL, Heap & Lindler (2016)) as calibration spectra, and they only differ in their orthogonal component with respect to the calibration

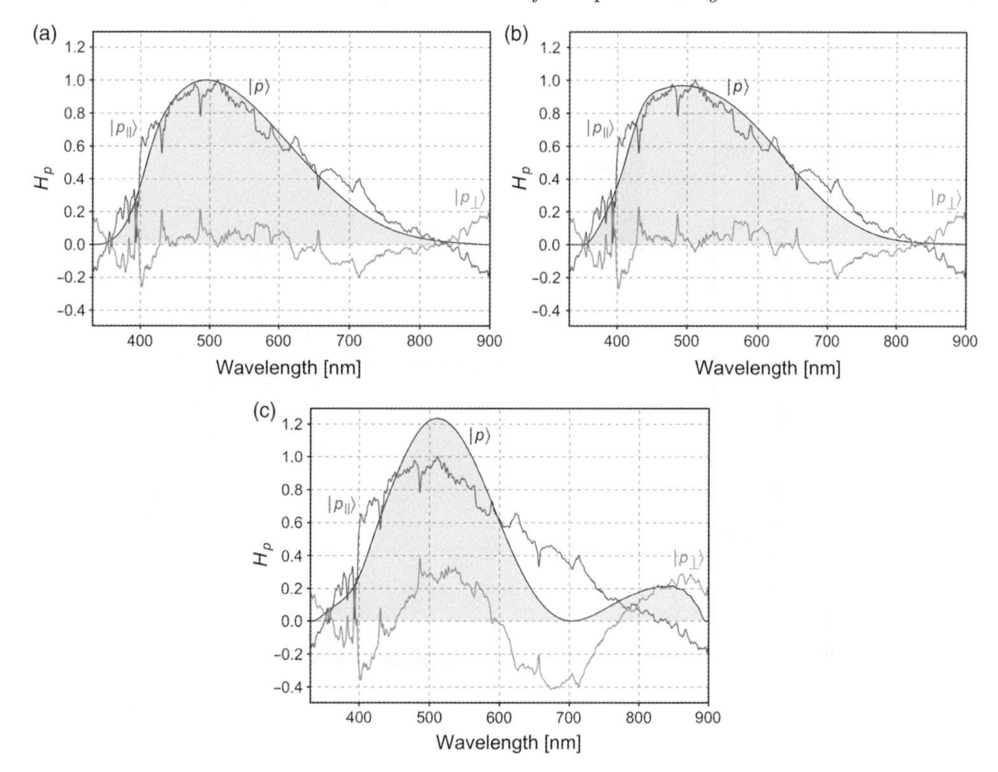

Figure 2. Three different solutions for the HIPPARCOS passband, which only differ in their orthogonal component (red lines), while the parallel component with respect to the calibration sources is kept fixed (blue lines). All passbands are normalised to the maximum of solution A.

spectra. The passbands, labeled A, B, and C, are shown in Fig 2. For each passband, the two components $|p_{\|}\rangle$ and $|p_{\perp}\rangle$ are provided, together with their sum $|p\rangle = |p_{\|}\rangle + |p_{\perp}\rangle$. While the solution A for H_p is the nominal solution from Weiler *et al.* (2018), solution B has been chosen to differ only slightly from A, while solution C has been deliberately chosen to be strongly different from any a-priori expectation for H_p. As $|p_{\|}\rangle$ is the same for all three solutions, they result in the same synthetic photometry for all calibration spectra used. However, the differences in $|p_{\perp}\rangle$ result in different synthetic photometry for sources whose SPDs are not well represented by a linear combination of the calibration SPDs.

To illustrate this effect, we take the empirical spectral library of Pickles (1998) and compute the synthetic HIPPARCOS magnitudes for the three different solutions. The difference between the synthetic magnitudes for solutions B and C with respect to solution A are shown in Fig. 3. The difference between the synthetic magnitudes from solution A and B (triangles in Fig. 3) is small, up to 57 mmag only. The differences however occur for sources with a large angle γ, which are the M-type spectra in the data set by Pickles (1998). Spectra of M-type sources are not included in the set of calibration sources, and these spectra are significantly different from the spectra included in the calibration set, in the sense that they are not linear combinations of the latter. The result is a large value for γ, and a strong dependency on the choice for $|p_{\perp}\rangle$. This effect is extreme for the difference between solution A and C. Although solution C results in essentially the same synthetic magnitudes for the Pickles spectra that are linear combinations of the NGSL calibration spectra (the sources with low values of γ), the differences for the M-type sources become extremely large, up to about 0.9 magnitudes. While this extreme case may be excluded

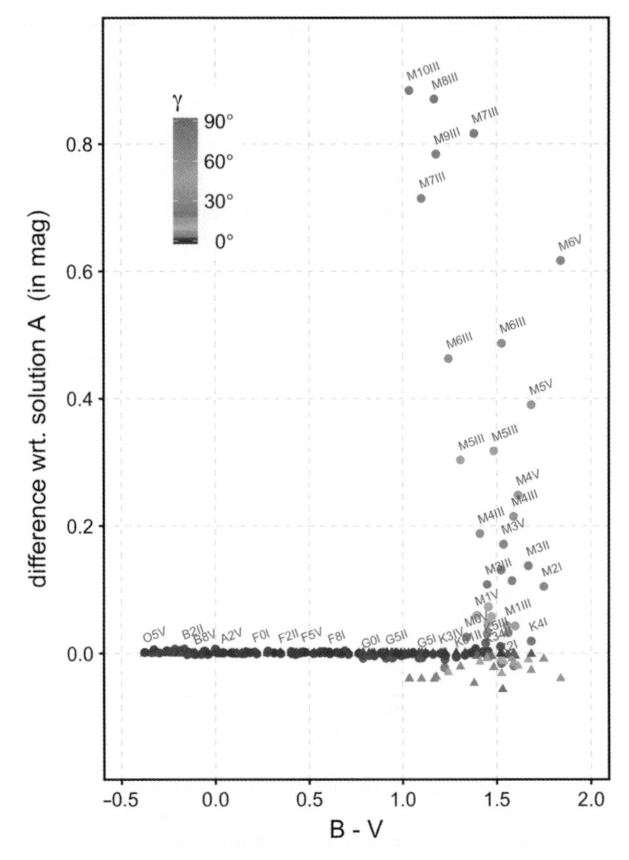

Figure 3. Difference between H_p solution B and A (triangles) and C and A (dots) for the spectra by Pickles (1998). The colour scale is indicating the γ angle according to Eq. (2.9). Selected spectral types are indicated.

in practice from the a-priori knowledge of the passband, less dramatic differences like the one between solution A and B may not be decided based on the available knowledge on how the passband should look like. Thus, systematic differences between the observed and synthetic magnitudes may result for sources not fully included by the vector space spanned by the calibration sources.

4. Summary and discussion

It has been demonstrated that deriving a passband from photometric observations of calibration sources with known SPDs is fundamentally limited by the choice of the calibration sources. A passband can be expressed as the sum of two functions. One of these functions, the "parallel component" of the passband, is a linear combination of the SPDs of the calibration sources and thus, up to uncertainties introduced by noise, uniquely determined by the calibration sources. The second function, the "orthogonal component" of the passband, is not contributing to the synthetic photometry of any of the calibration sources, and is in consequence unconstrained by the calibration spectra. The passband itself, i.e. the sum of the parallel and orthogonal component, is therefore not uniquely defined by any set of calibration spectra. A guess for the orthogonal component can be made based on any a-priori knowledge on the shape of the passband. Such guess however is intrinsically affected by uncertainties. While the synthetic photometry for SPDs which are a linear combination of the calibration SPDs only depends on the parallel passband component, and therefore being reliable, SPDs which are not well represented by a linear combination of the calibration SPDs may not result in a reliable synthetic photometry. In the latter case, the synthetic photometry depends on the unconstrained orthogonal component, and the more so the less well the SPD is represented by a linear combination of the calibration SPDs.

Currently available libraries of calibration spectra, such as CALSPEC (Bohlin 2007) or Stritzinger *et al.* (2005), have been compiled mainly for the calibration of spectrophotometric data. They therefore exhibit a rather limited range of different spectral shapes, being dominated by hot sources. As a consequence, the space spanned by linear combinations of these calibration SPDs is rather limited as well, leaving many SPDs poorly covered. Among the objects with poorly covered SPDs are M-type stars, strongly extinct stellar objects, or largely non-stellar spectra, e.g. those of quasi-stellar objects. The limitation resulting from the limited coverage of spectral shapes among calibration libraries therefore prevents reliable synthetic photometry for many scientifically interesting objects. The set of calibration sources which is produced for the photometric and spectrophotometric calibration of ESA's *Gaia* mission (Gaia Collaboration *et al.* 2016) is improving in covering different shapes of SPDs, including objects from O to early M type (Pancino *et al.* 2012). A recent work by Maíz Apellániz & Weiler (2018) however has demonstrated the remaining limitation by the orthogonal passband component for M-dwarfs in the *Gaia* Data Release 2 passbands. A stronger diversification of spectral shapes in calibration libraries, aiming for a larger space spanned by linear combinations of their SPDs, may therefore be desirable for improving the interpretation of photometric data.

References

Bohlin, R. C. 2007, *Astronomical Society of the Pacific Conference Series*, 364, 315
ESA 1997, *The Hipparcos and Tycho Catalogues*, ESA-SP 1200
Gaia Collaboration, Prusti *et al.* 2016, *A&A*, 595, A1
Heap, S. R., and Lindler, D. 2016, *ASP Conference Series*, 503
Maíz Apellániz, J., & Weiler, M. 2018, *A&A*, in press
Pancino, E., Altavilla, G., Marinoni, S., *et al.* 2012, *MNRAS*, 426, 1767
Pickles, A. J. 1998, *PASP*, 110, 863
Stritzinger, M., Suntzeff, N. B., Hamuy, M., *et al.* 2005, *PASP*, 117, 810
Weiler, M., Jordi, C., Fabricius, C., & Carrasco, J. M. 2018, *A&A*, 615, A24
Weiler, M. 2018, *A&A*, 617, A138
Young, A. T. 1994, *A&A*, 288, 683

Astronomy in Focus - XXX
Proceedings IAU Symposium No. XXX, 2018
M. T. Lago, ed.

doi:10.1017/S1743921319005155

Discovery of Blackbody Stars and the Accuracy of SDSS photometry

Masataka Fukugita

Kavli Institute for the Physics and Mathematics of the Universe, University of Tokyo, Kashiwa 277-8583 Japan

Institute for Advanced Study, Princeton NJ08540, U.S.A.

Abstract. We discovered stars that show spectra very close to the blackbody radiation without any line features. We found 17 such stars out of 0.8 million stellar objects in the SDSS archive. The blackbody temperature is approximately 10^4K. We identify these stars as DB white dwarfs with the helium atmosphere, possibly with a trace amount of hydrogen, that yields nearly perfect blackbody spectrum, which is also confirmed with our later study. These stars can be used to test the accuracy of the AB zero point across different colour bands, in particular including the NIR pass bands. The zero points of SDSS photometry are verified to < 0.01 mag.

Keywords. blackbody star, SDSS photometry, photometric calibration

---

1. Introduction

We report on the discovery of stars that exhibit spectra very close to the blackbody radiation without any line features (Fukugita and Suzuki 2017) and their use for calibration of the photometric system, especially a verification of the Sloan Digital Sky Survey (SDSS) photometric system. We noticed one such stellar object in the sample of quasar candidates in the SDSS that is selected on the basis of colour-colour diagrams. While this stellar object was eventually dropt from the quasar candidate, our scrutiny of this object reveals that the spectrum does not resemble a quasar but it is a star showing very close to the blackbody radiation. We find that it shows a significant proper motion, so it must be a star located nearby.

2. Blackbody stars

We have then searched for similar objects in the entire database of SDSS. We find 22 objects that do not show any line features and are consistent with the blackbody spectrum (χ^2/dof < 1.05) out of 800,000 objects that are classified as stellar. Among the 22, five show spectra in excess of black body in the near infrared obseved with the WISE W1 channel, so circumstellar objects, say a companion star or dust surrounding the star, are suspected and are dropt from our sample of blackbody stars. This 20% fraction is consistent with that of NIR excess known for white dwrafs (Debes *et al.* 2011). These objects all have brightness $r > 17$. We found none brighter than this brightness. They are quite rare objects in the sky. Four examples are shown in Figure. 1 (spectra in left panels and broad-band photometry in right panels). The bottom panels show fractional offset from the fit with positions of possible absorption lines indicated. The one in Figure 2 is an example where we detect an IR excess, while the optical spectrum is close to blackbody.

All stars, except one, are catalogued in Gaia DR2 (Gaia DR2 2018). They show parallaxes from 4.4 to 14 mas, so their distances are 70-230 pc. Their brightness together with these parallaxes suggests that these stars are white dwarfs with helium atmosphere with temperature too low to develop absorption line features. One may take them as DC white

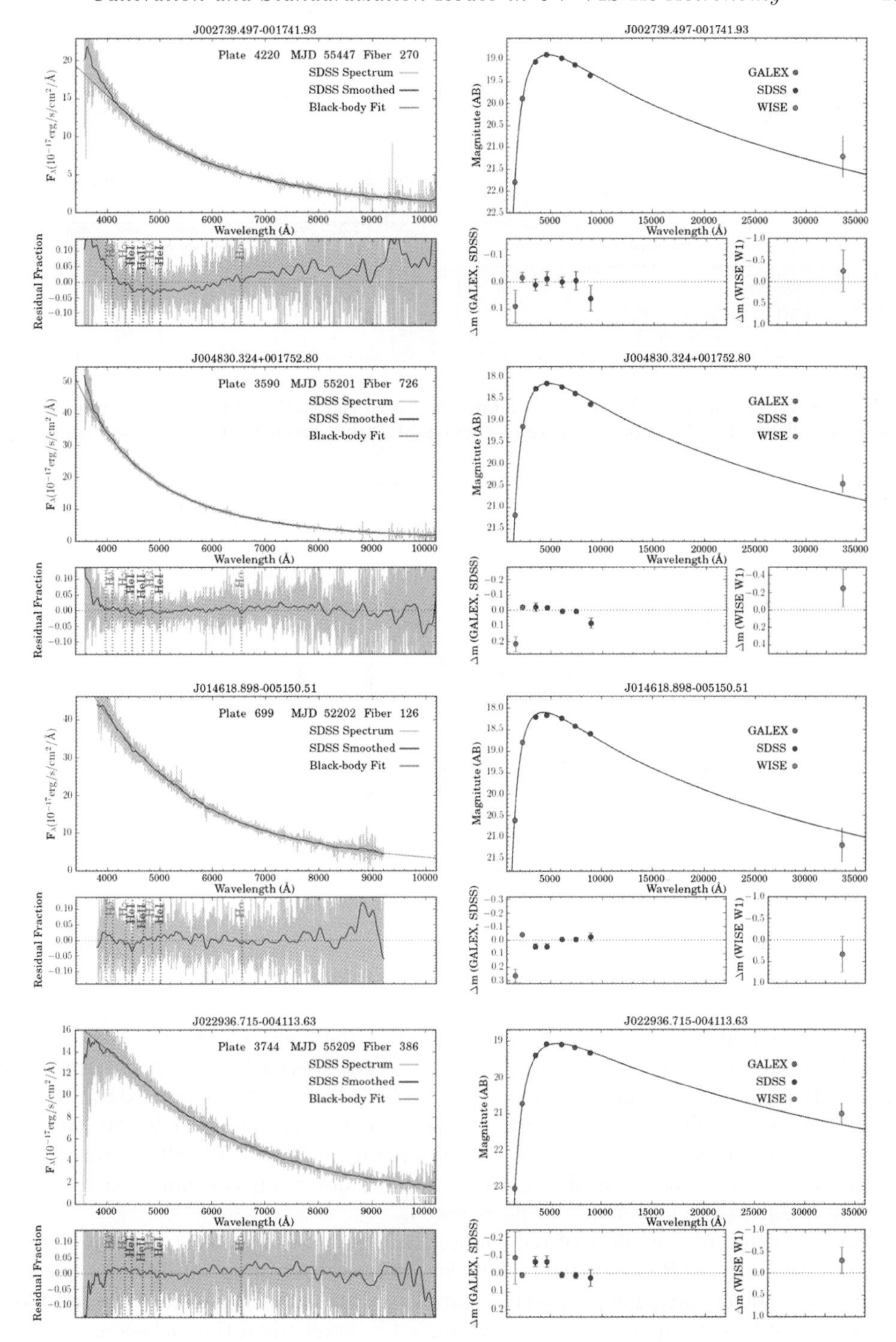

Figure 1. Four example spectra from the 17 blackbody stars we uncovered. In the left panels, observed spectra are indicated with grey and blue curves show smoothed spectra. Black-body fits are indicated with red. The right panels show the corresponding photometric data in the broad band from which our fit parameters are derived. In the bottom panels, the residuals (fractional values) from the fits are shown, with the positions of hydrogen and helium absorption lines indicated. Figures are taken from Suzuki & Fukugita (2017)

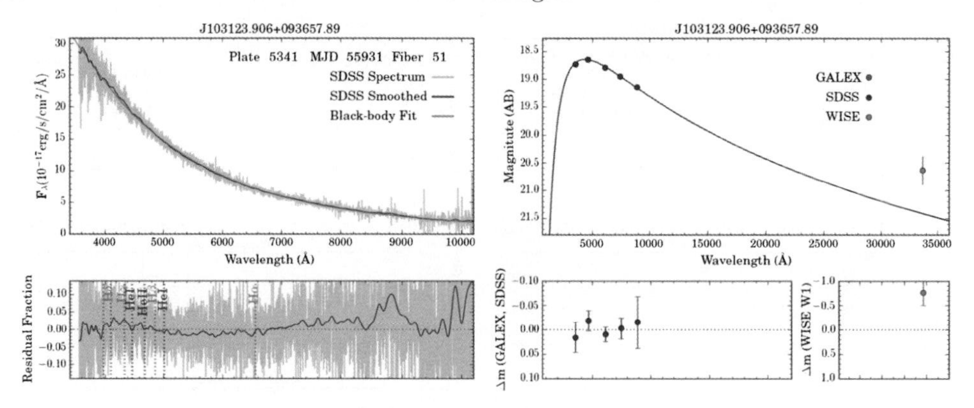

Figure 2. Example of the star that shows excess in the NIR, while the spectrum in the optical follows the blackbody spectrum. Figures are taken from Suzuki & Fukugita (2017).

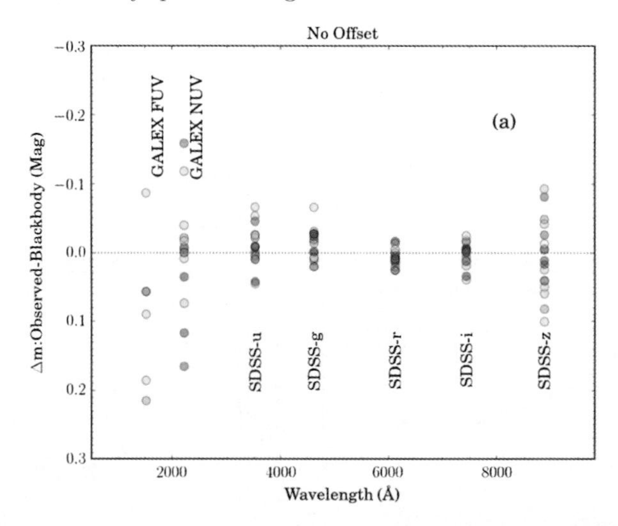

Figure 3. Residuals of the SDSS five-band photometric data and the GALEX NUV and FUV data from the blackbody fits in magnitude. Figure is taken from Suzuki & Fukugita (2017).

dwarfs by applying the observational criterion, but their physical nature are consistent with DB white dwarfs, as we confirmed in a later study (Serenelli *et al.* 2018). An accrate examination are carried out using more precise broad-band photometric data. The five band photometric data *ugriz* of SDSS, and also GALEX NUV plus FUV data, are fitted excellently with a blackbody spectrum with temperature 7400 to 12000K. Figure 3 shows the residual of our blackbody fit, showing that the deviation is small <0.03 mag and random in nature: no systematic trend is noted.

One may think of the effect of reddening. Reddening is expected to be small at this distance (typically $E(B-V)\approx 0.02-0.03$ is expected), but, if any, it does not affect the proximity to blackbody, for its effect is parallel to the change in temperature: it only shifts the resulting temperature by a few hunded K lower upon the inclusion.

We found that the helium atmosphere model can yield precisely these blackbody spectra when the temperature is 8000K to 11000K, and in particular when helium is contaminated with a trace amount ($\approx 10^{-8}-10^{-6}$) of hydrogen (Serenelli *et al.* 2018): see Figure 4. This hydrogen abundance is too small to detect its absorption features in current spectroscopic observations. We remark that the effective temperature of the atmosphere model turns out to be lower typically by 500 K than the temperature from

Table 1. Mean of residuals from the black-body fit (Δm=data − black-body fit) to the 17 blackbody stars with the SDSS zero point.

Data Name	SDSS-u	SDSS-g	SDSS-r	SDSS-i	SDSS-z
Mean Residuals	−0.006±0.031	−0.008±0.025	0.007±0.012	0.002±0.017	0.003±0.052

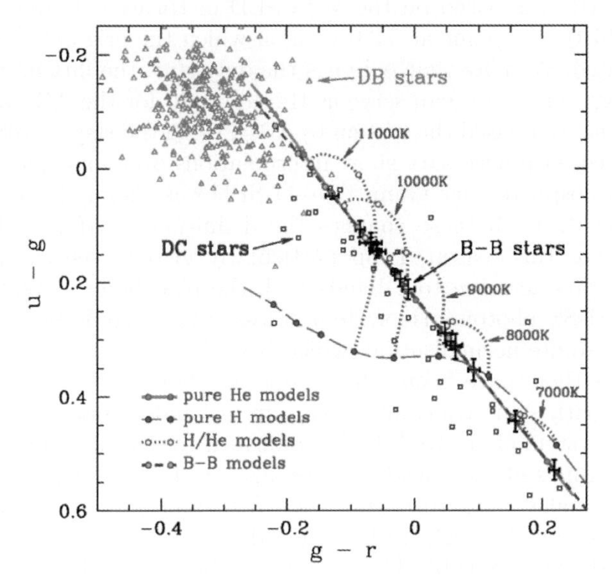

Figure 4. $u - g$ vs. $g - r$ colour-colour diagram of white dwarf stars. Our blackbody stars are shown with error bars. Green short-dashed line is the blackbody spectrum. Pure-He (red solid), pure-H (blue long-dashed) and H/He mixed atmosphere models (dotted) are also indicated (as in legends in the figure) in the range 12000-7000 K. Small circles on the curves show temperature in 1000K intervals. Those on the bridge between H and He models mark log $(N_{\rm H}/N_{\rm He}) = -2, -4, -6$. Figure is taken from Serenelli *et al.* (2018).

the blackbody fit. We note that all stars have temperature consistent with that in this range.

3. Accuracy of photometry

Irrespective of the physical nature, a significance of these blackbody stars lies in the fact that they can be used as photometric and spectrophotometric calibrators. As seen in Figure. 3 above the residual of our blackbody fit is typically a few times 0.01 mag, and is random in nature. When averaged over 17 stars, the departure from blackbody is smaller than 0.008 mag for all five colour bands (see Table 1). This means that these stars serve as excellent standards for photometry or spectrophotometry in optical bands. Their moderate faintness fits the standard to be used for 8-10 metre telescopes. GALEX photometry has larger errors (0.1 mag in NUV and 0.2 mag in FUV), but offsets are not detected.

We emphasise that the agreement of SDSS photometry of these stars (and GALEX photometry) with blackbody spectrum at this high accuracy, in turn, verifies the accuracy of the SDSS photomrtric system across the five colour bands. The SDSS photometric system, designated as AB_{95}, has been concocted from a number of elements Fukugita *et al.* (1996) and its accuracy is by no means evident but remained to be verified.

Briefly descibing, SDSS photometry is based on spectrophotometry of metal poor F-subdwarfs, BD+17°4708 and BD+26°2606, the spectra of which (F_ν) are relatively flat and show only very weak absorption features. This is in contrast to the prime calibrator α Lyr, which shows many conspicuous absorption lines and also the Balmer jump. Large parts of the spectrum of these SD stars is sufficently smooth. These BD spectra stand for, at the time, a handful of the most accurately measured SED, which were obtained by Oke & Gunn (1983) and further updated by Oke (1990). The Oke-Gunn standard, called AB_{79}, is based on the Vega SED of Hayes & Latham (1975) with the Oke & Schild (1970) zero point at 5480Å, upon which the original BD+17 and BD+26 SED were presented. We note that α Lyr is the only star, the flux of which is measured in the physical cgs units, thus can serve as the standard for the AB system.

In Fukugita *et al.* we revised the system to AB_{95}, using the Hayes (1985) compilation of αLyr supplemented with necessary short interpolations over some gaps using Castelli & Kurucz (1994) atmosphere. Oke-Gunn BD+17 SED was also updated with newer Oke's (1990) measurement. With these changes the definition of AB_{95} is by no means too straightforward. For this reason, I am particularly glad to see the agreement of the spectra of these stars across u to z bands with the blackbody stars, which verifies the accuracy of the SDSS photometric system across u to z bands being as good as ,0.01 mag, up to the absolute normalisation, which is yet to be verified.

As time passing there have been a few proposals that SDSS photometric zero points should be added with some constant varying from band to band from -0.04 to $+0.04$ mag to make the system closer to AB. We confirmed that such additions of offsets bring wiggles to the residuals of the blackbody fits, just by those added amounts. This means that the addition makes departure from the SB system only worse. This also includes the correction based on the CALSPEC standard.

Our final remark is that our blackbody stars can be used as flux calibrators, in particular in NIR photometry, for which good AB standards lack. An important feature is that the AB flux in NIR thus obtained should obviously be consistent with the AB system used in the optical bands.

Acknowledgement

I would like to thank Max-Planck-Institut für Astrophysik where this report was written, and Alexander von Humboldt stiftung for the support during his stay in Garching.

References

Castelli, F., & Kurucz, R. L. 1994, A&Ap, 281, 817
Debes, J. H., Hoard, D. W., Wachter, S., Leisawitz, D. T., & Cohen, M. 2011, ApJS, 197, 38
Fukugita, M., Ichikawa, T., Gunn, J. E., *et al.* 1996, AJ, 111, 1748
Gaia Collaboration, Brown, A. G. A., Vallenari, A., *et al.* 2018, A&Ap, 616, A1
Hayes, D. S. 1985, Calibration of Fundamental Stellar Quantities, 111, 225
Hayes, D. S., & Latham, D. W. 1975, ApJ, 197, 593
Oke, J. B. 1990, AJ, 99, 1621
Oke, J. B., & Gunn, J. E. 1983, ApJ, 266, 713
Oke, J. B., & Schild, R. E. 1970, ApJ, 161, 1015
Serenelli, A., Rohrmann, R., & Fukugita, M. 2018, arXiv:1804.01236
Suzuki, N., & Fukugita, M. 2017, arXiv:1711.01122, AJ in press (2018)

Astronomy in Focus - XXX
Proceedings IAU Symposium No. XXX, 2018
M. T. Lago, ed.

doi:10.1017/S1743921319005167

Testing of the LSST's photometric calibration strategy at the CTIO 0.9 meter telescope

Michael W. Coughlin[1], Susana Deustua[2], Augustin Guyonnet[3], Nicholas Mondrik[3,4], Joseph P. Rice[5], Christopher W. Stubbs[3,6] and John T. Woodward[5]

[1]Division of Physics, Math, and Astronomy, California Institute of Technology, Pasadena, CA 91125, USA,
email: `mcoughli@caltech.edu`

[2]Space Telescope Science Institute, 3700 San Martin Drive, Baltimore, MD 21218, USA,

[3]Department of Physics, Harvard University, Cambridge, MA 02138, USA,

[4]LSSTC Data Science Fellow

[5]National Institute of Standards and Technology (NIST), 100 Bureau Drive, Gaithersburg MD 20899

[6]Department of Astronomy, Harvard University, Cambridge MA 02138, USA

Abstract. The calibration hardware system of the Large Synoptic Survey Telescope (LSST) is designed to measure two quantities: a telescope's instrumental response and atmospheric transmission, both as a function of wavelength. First of all, a "collimated beam projector" is designed to measure the instrumental response function by projecting monochromatic light through a mask and a collimating optic onto the telescope. During the measurement, the light level is monitored with a NIST-traceable photodiode. This method does not suffer from stray light effects or the reflections (known as ghosting) present when using a flat-field screen illumination, which has a systematic source of uncertainty from uncontrolled reflections. It allows for an independent measurement of the throughput of the telescope's optical train as well as each filter's transmission as a function of position on the primary mirror. Second, CALSPEC stars can be used as calibrated light sources to illuminate the atmosphere and measure its transmission. To measure the atmosphere's transfer function, we use the telescope's imager with a Ronchi grating in place of a filter to configure it as a low resolution slitless spectrograph. In this paper, we describe this calibration strategy, focusing on results from a prototype system at the Cerro Tololo Inter-American Observatory (CTIO) 0.9 meter telescope. We compare the instrumental throughput measurements to nominal values measured using a laboratory spectrophotometer, and we describe measurements of the atmosphere made via CALSPEC standard stars during the same run.

Keywords. techniques: photometric, instrumentation: miscellaneous

---

Astronomy in Focus - XXX
Proceedings IAU Symposium No. XXX, 2018
M. T. Lago, ed.

doi:10.1017/S1743921319005179

New calibration of the Vilnius photometric system

M. Maskoliunas[1], J. Zdanavičius[1], V. Čepas[1], A. Kazlauskas[1], R. P. Boyle[2], K. Zdanavičius[1], K. Černis[1], K. Milašius[1] and M. Macijauskas[1]

[1]Astronomical Observatory of the Vilnius University,
Sauletekio 3, Vilnius 10257, the Lithuania. Email: marius.maskoliunas@tfai.vu.lt

[2]Vatican Observatory Research Group, Steward Observatory,
Tucson, Arizona 85721, U.S.A. Email: rpboyle@me.com

Abstract. The medium-band Vilnius photometric system with the mean wavelengths at 345 (U), 374 (P), 405 (X), 466 (Y), 516 (Z), 544 (V), and 656 (S) nm for many years was an important tool to determine interstellar reddenings and distances of single stars due to its ability to classify stars of all temperatures in spectral classes and luminosity classes in the presence of different interstellar reddenings. At present, Gaia DR2 presents distances to stars with an unprecedented accuracy at least up to 3 kpc. However, multicolor photometry, which allows the classification of stars as well as the preliminary determination of stellar temperatures, gravities, metallicities and interstellar reddenings, remains an important method for distant stars. Here we present an empirical calibration of the intrinsic color indices of the Vilnius system in terms of physical parameters of stars for dwarf and giant stars of spectral classes F-G-K-M. In any attempted photometric determination of physical parameters of stars it is important to have an extensive and homogeneous sample of spectroscopically determined parameters for stars for which there are also accurate photometric data. As a source catalogue for the Vilnius photometry the latest updated version of the Catalogue of Photoelectric Observations in the Vilnius System was used, which contains compilations from the published photometry for about 11 000 stars. The stars which had both the Gaia DR2 parallaxes and the determinations of stellar parameters from high-dispersion spectra were extracted from this catalogue. The final sample contains more than 1500 stars of spectral classes F-M. The majority of these stars (ca 70%) are not reddened, for others the values of interstellar reddening A_V were determined using the regular techniques of photometric classification in the Vilnius system. The absolute magnitudes M_V and consequently the luminosity classes were determined using Gaia DR2 parallaxes. We present the analytical expressions for the effective temperature T_{eff} and surface gravity log$g$ and evaluate the errors of solutions for dwarf and giant stars. To test the accuracy of the proposed method, we have compared our results with the stars observed by Gaia and with the stellar parameters available from the large spectroscopic surveys: APOGEE, Gaia-ESO, GALAH, LAMOST, RAVE and SEGUE. The results of comparison contain 5-6 % outliers.

The proposed method allows the fast and straightforward evaluation of stellar physical parameters for the stars observed in the Vilnius photometric system. Despite the fact, that the accuracy of determination is significantly lower than in the case of spectroscopic methods, the method described may be useful for distant faint stars, which are still inaccessible for spectroscopic observations.

Keywords. Techniques: photometric - Vilnius photometric system, stars: fundamental parameters, classification.

Astronomy in Focus - XXX
Proceedings IAU Symposium No. XXX, 2018
M. T. Lago, ed.

doi:10.1017/S1743921319005180

The Calibration of the UVIT Detectors for the ASTROSAT Observatory

Denis Leahy[1], J. Postma[1], J. B. Hutchings[2] and S. N. Tandon[3,4]

[1]Dept. of Physics and Astronomy, University of Calgary, Calgary, AB, T2N 1N4, Canada
email: leahy@ucalgary.ca

[2]Herzberg Institute of Astrophysics, 5071 West Saanich Road, Victoria, BC V9E 2E7, Canada

[3]Inter-University Center for Astronomy and Astrophysics, Pune, India

[4]Indian Institute of Astrophysics, Koramangala II Block, Bangalore-560034, India

Abstract. The UVIT ultraviolet and visual band detectors and electronics for the ASTROSAT observatory were calibrated in the vacuum laboratory at the University of Calgary. This work was supported by the Canadian Space Agengy and carried out prior to integration with the UVIT optical assembly and the ASTROSAT spacecraft. The multiband (X-ray, ultraviolet and optical) ASTROSAT observatory was successfully launched by the Indian Space Research Organization on Sept. 28, 2015, with subsequent in-orbit verification and ongoing calibration activities. Here we discuss the current issues of calibrating the UVIT data, such as distortion corrections, and how the laboratory data is being used to address these issues.

Keywords. instrumentation: detectors, techniques: image processing, techniques: photometric

1. Introduction

Astrosat is a multiwavelength space observatory of the Indian Space Research Organisation (ISRO), with contributions by the Canadian Space Agency and the University of Leicester. The satellite contains three pointed X-ray instruments and two ultraviolet (UV) and optical telescopes, all with fields of view that are aligned. An additional instrument is the all-sky X-ray scanning monitor. The full observatory details and capability are described in the ISRO World Wide Web site. The observatory constitutes a multiwavelength capability that allows simultaneous monitoring of targets from optical/UV wavelengths to 100 keV with high timing precision. The twin UV-optical telescopes (UVITs) have a 38 cm aperture and field of view of 28 arcmin, with 1.3 arsec spatial resolution. One telescope covers the far UV (FUV) and the other covers the near UV (NUV) and blue optical bands using a beam splitter.

The ground calibration of UVIT detectors was discussed in Postma *et al.* (2011). Custom software was developed to implement calibration procedures with data from the instrument in science operation mode. A description of the software package is given in Postma *et al.* (2017). The scientific performance of the UVIT instrument and its space calibration is discussed in Subramaniam *et al.* (2017) and Tandon *et al.* (2017). Some of the first scientific results with UVIT are described in e.g. George *et al.* (2018) and Leahy *et al.* (2018).

Here we describe briefly the calibrations and corrections that have to be applied to the raw centroid data from the UVIT telescopes. These are required to produce images which are useful for science analysis.

2. Discussion

Fig. 1 shows the flat field for the NUV which was measured by on-ground calibrations with an integrating sphere. The FUV channel flat field was similarly measured and is

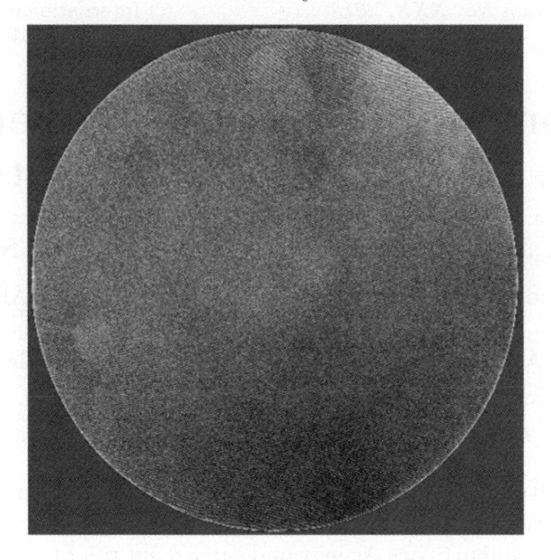

Figure 1. NUV flat-field image.

not shown, but is similar in quality to the NUV image figure. All centroids are thus given a weight corresponding to their position at which they were detected in the field. That is, a centroid nominally represents a unit detection of a photon event. However, at nominal high operating voltage for the detector there is a small variation in probability (10% across the field) that a photon will be detected when considering a large number of possible detections. Thus, the inverse of the detected flat field at a given location provides the weight of the centroid in that location: if the normalized flat field has a value of 0.9 where the photon falls, then relative to the normal that region is under-detecting and so a photon in that region is given a weight of 1/0.9. Because a centroid is a set of coordinates, the centroid list or table of photon event detection coordinates has a corresponding table list of coordinate weights. It is important to record the photon event weights at the initial ingestion of the data because later corrections for drift must move the photon event coordinates from their detected positions during the drift movements to a mean drift-corrected frame of coordinates which then no longer correspond to the detection coordinates.

The measurement of the flat field also provides for characterization of the fixed-pattern-noise (FPN) or centroiding bias introduced into the determination of the photon event centroids. The photon events are relatively under-sampled on the CMOS detector, having a FWHM (full width half maximum) of only approximately 1.3 ± 0.3 pixels. Sub-pixel resolution of photon event centroids is a design requirement of this system in order to achieve the desired 1" resolution, corresponding to 1/3 of a pixel of the CMOS chip. The photon event centroids are determined by a simple weighted mean of values and positions of the photon pulse over a 3x3 pixel kernel which captures $\simeq$95% of the photon event energy, and are recorded by the detector system at 1/32 pixel resolution. A 5×5 kernel was also explored in ground calibrations since this would capture the remaining 5% of signal of the photon event pulse, but it was found that the read noise in the outer 16 pixels prevented any improvement in the centroid determination. Thus, a weighted mean of an under-sampled approximately Gaussian signal provides centroids which have a preferential bias towards symmetry, that is, a centroid with a sub-pixel decimal value which pulls to zero. A truly symmetric photon event pulse on the CMOS results in a centroid with a decimal place of 0, while an asymmetric photon event pulse will have a non-zero decimal value but which is systematically biased towards zero. This can be seen in the left panel of Figure 2. This FPN bias can be measured and then corrected for by

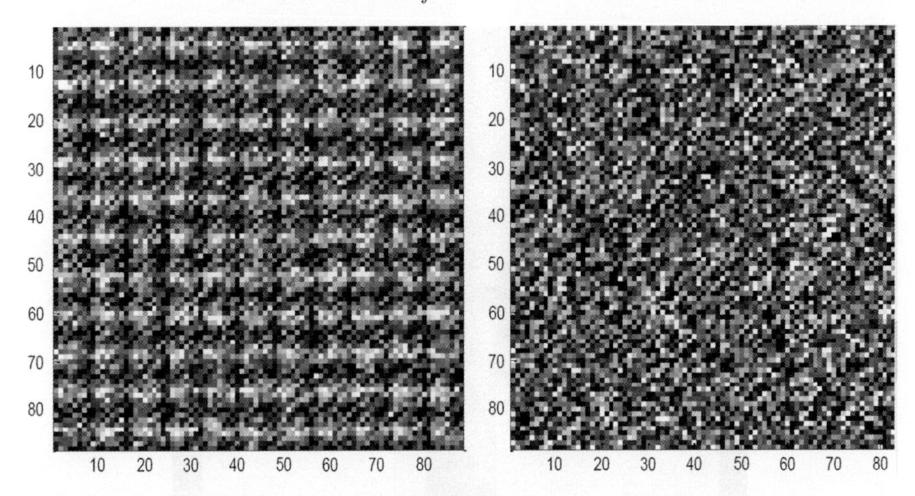

Figure 2. Closeup of a small region of the NUV image. Left panel: flat field image prior to fixed-pattern noise (FPN) correction. The FPN of the detector is seen in the image before correction. Right Panel: flat field image after the FPN correction.

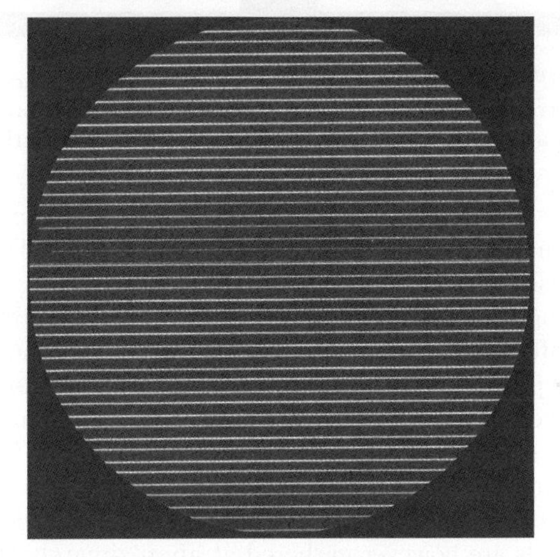

Figure 3. NUV image taken in laboratory with scanning of pinhole image for the purpose of distortion corrections. When examined in detail, the straight lines are imaged as curved lines. The distortion correction restores the lines to be straight.

determining the cumulative probability distribution of the decimal-places of all centroids detected from a flat field source, since a flat field will have no sub-pixel preferentiality. The FPN corrected flat field image can be seen on the right panel of Fig. 2.

For astrometric accuracy the fields must be corrected for distortions introduced by the fiber-optic taper (and other possible contributions). During ground-calibrations the detectors were continuously run while being scanned in either axis across their window with a micrometer stage and a multi-pinhole mask with rear-illumination, the result for the horizontal axis shown in Fig. 3. Measured at 1/8th subpixel resolution, after being corrected for the FPN, it is found that the scan lines are not straight lines but have structure resembling a quadratic curve exceeding several CMOS pixels variation from a straight line. Given that the detected lines were supposed to be straight lines, then the residuals of the detected line centroid coordinates from a fitted straight line provides the distortion across the field, and this is performed for each line and in both horizontal and

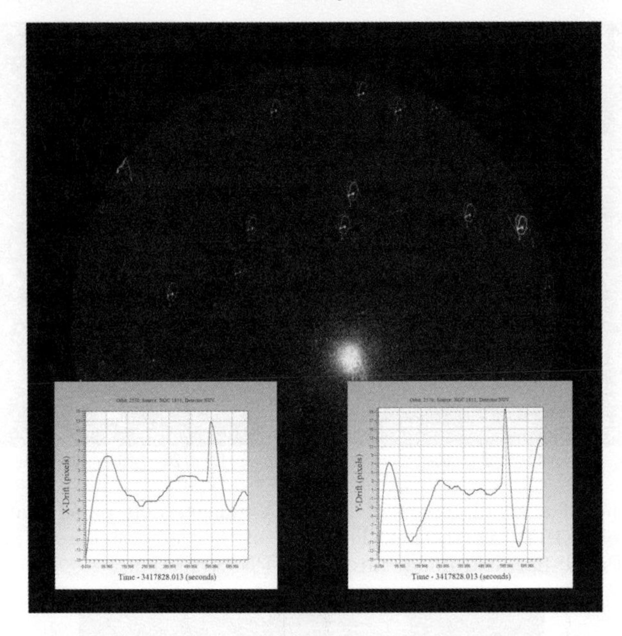

Figure 4. In-orbit NUV image of a field containing foreground stars and a globular cluster (below and right of centre). Star images are clearly trailed because of spacecraft pointing errors. The globular cluster image is also trailed which results in severe smearing.

vertical scan axes. The in-flight astrometry is thus improved to an accuracy approaching 1 arc second. Additional calibrations were done by imaging a 2-D grid of pin-holes. For details please see Girish *et al.* (2017).

For in-flight imaging of science targets, the ASTROSAT satellite is flown with an induced sinusoidal drift or dithering of nominal rate 1" per second and amplitude 80" in orthogonal axes. The purpose of this is to spread charge depletion of the microchannel plates inside the detectors evenly about their detector face, on average. In addition, there are random drifts up to $\sim$ 2"/s. The visible (VIS) channel is used for tracking the drift given that VIS-bright point sources (stars) are expected in any given observation, and the cadence of the VIS imaging for tracking is 1 Hz. The science data centroids in FUV and NUV channels are however gathered at approximately 30 Hz, and so the drift series determined from VIS at 1 Hz must be interpolated, typically with a simple cubic spline, to the times of the FUV and NUV centroids. A drift series can be seen in Fig. 4. along with the trailed-star image produced before drift-correction is applied, and Fig. 5 demonstrates the final science image after all corrections.

3. Summary

We have discussed briefly the various corrections to UVIT images required to put them in a form which is useful for imaging and photometry for science analysis. Corrections for flat-fielding, distortions and spacecraft pointing errors are essential to restore the images to the design spatial resolution, and to allow photometry, positions and shapes of astronomical images to be accurately determined in the NUV and FUV bands.

The Indian Space Science Data Centre (ISSDC) archives the final astronomer ready UVIT products (sky images in UV, exposure, etc) generated by the Payload Operation Centre (POC) at the Indian Institute of Astrophysics (IIA), using the UVIT Level-2 Pipeline developed by ISRO and the UVIT instrument team (http://uvit.iiap.res.in/Downloads).

Figure 5. NUV image of the same field containing foreground stars and a globular cluster (below and right of centre), after correction for trailing. The star images are now pointlike and the globular cluster is seen clearly.

Acknowledgement

This work was supported by the Canadian Space Agency.

References

George, K., Poggianti, B. M., Gullieuszik, M., *et al.* 2018, *MNRAS*, 479, 4126

Leahy, D., Bianchi, L. & Postma, J. 2018, *ApJ*, submitted

Girish, V., Tandon, S. N., Sriram, S., Kumar, A., & Postma, J 2017, *Experimental Astronomy*, 143, 59

Postma, J., Hutchings, J. B., & Leahy, D. 2011, *PASP*, 123, 833

Postma, J. E., & Leahy, D. 2017, *PASP*, 129, 115002

Subramaniam, A., Sindhu, N., Tandon, S. N., *et al.* 2017, *ApJ*, 833, L27

Tandon, S. N., Subramaniam, A., Girish, V., *et al.* 2017, *AJ*, 154, 128

Astronomy in Focus - XXX
Proceedings IAU Symposium No. XXX, 2018
M. T. Lago, ed.

doi:10.1017/S1743921319005192

WFC3/HST photometric calibration: color terms for the ultra-violet filters

Annalisa Calamida

Space Telescope Science Institute, 3700 San Martin Drive, Baltimore, MD 21218, USA
email: acalamida@stsci.edu

Color term corrections for magnitudes measured on the UVIS2 relative to the UVIS1 detector of the WFC3 camera on board Hubble Space Telescope are needed for three ultra-violet filters, namely $F218W$, $F225W$, and $F275W$. The two WFC3 detectors have different quantum efficiencies in the ultra-violet regime ($\lambda < 4{,}000$ Å), resulting in different count rate ratios as a function of the spectral type of the source. In the worst case, for cool red sources measured on UVIS2, there is a magnitude offset relative to UVIS1 up to ~ 0.08 mag, while the offset is negligible for hot ($T_{eff} \gtrsim 30{,}000$ K) blue sources.

We advise WFC3 users to apply the corrections provided in the Instrument Science Report Calamida *et al.* (2018, wfc.rept.8C) to UVIS2 magnitudes when calibrating photometry of stars cooler than ~30,000 K in the ultra-violet filters, i.e. when observing stellar fields that include stars of different spectral types, such as open and globular clusters, resolved local group galaxies, Galactic stellar populations. These transformations can also be used to correct photometry of other red sources, such as high-redshift galaxies, observed with the ultra-violet filters. For sources observed on the same detector the color term is smaller and no magnitude correction is required.

The color term transformations are provided as magnitude offsets as a function of color corresponding to different spectral types and are listed in lookup tables in Calamida *et al.* (2018).

Astronomy in Focus - XXX
Proceedings IAU Symposium No. XXX, 2018
M. T. Lago, ed.

doi:10.1017/S1743921319005209

Time-series photometry of a new set of candidate faint spectrophotometric standard DA white dwarfs

Annalisa Calamida

Space Telescope Science Institute, 3700 San Martin Drive, Baltimore, MD 21218, USA
email: acalamida@stsci.edu

Future facilities and deep surveys such as LSST, JWST and WFIRST, will require a network of standards faint enough to avoid saturation and homogenously distributed in both hemispheres. DA white dwarfs have almost pure hydrogen atmospheres and they are the simplest stars to model. The opacities are known from first principles, and for temperatures higher than $\sim$ 20,000 K, their photospheres are purely radiative and should be photometrically stable. DA white dwarfs are then the best candidates to establish a network of faint spectrophotometric standards. In order to provide standards in the dynamic range of large aperture (d $>$ 4m) telescopes, we collected Hubble Space Telescope WFC3 images and ground-based spectroscopy for 23 DA white dwarfs fainter than $r \sim$ 16.5 mag, distributed at equatorial and northern latitudes (see Saha *et al.* in these conference proceedings).

Las Cumbres Observatory time-series were also collected to monitor the stability of the candidate standard DA white dwarfs. Observations showed that most of our candidates are stable. However, two of them, namely SDSSJ203722.169-051302.964 and WD0554-165, show clear sign of variability in their light curves. The first star also shows emission features in the Balmer lines of the spectra implying the presence of a low-mass companion. We do not know the origin of the variability for WD0554-165. Two other DAWDs, SDSSJ010322.10-002047.7 and SDSSJ102430.93-003207.0 show hints of variability, but these results need to be confirmed with further data. SDSSJ20372.169-051302.964 and WD0554-165 will be excluded by our set of candidate standard DA white dwarfs.

The other white dwarfs will be established as standards and will provide a set of faint spectrophotometric standards to calibrate data from future facilities to a precision better than 1%.

Astronomy in Focus - XXX
Proceedings IAU Symposium No. XXX, 2018
M. T. Lago, ed.

doi:10.1017/S1743921319005210

SCALA: Towards a physical calibration of CALSPEC standard stars based on a NIST-traceable reference for SNIFS

D. Küsters[1,2] S. Lombardo[3,2], M. Kowalski[1,2], G. Aldering[4], K. Boone[4], Y. Copin[5], J. Nordin[2] and D. Rubin[6]

[1]Deutsches Elektronen-Synchrotron, D-15735 Zeuthen, Germany
[2]Institut fur Physik, Humboldt-Universitat zu Berlin, Newtonstr. 15, 12489 Berlin
email: kuesters@physik.hu-berlin.de
[3]Aix Marseille Univ, CNRS, CNES, LAM, Marseille, France
[4]Physics Division, Lawrence Berkeley National Laboratory, Berkeley, CA
[5]Universit de Lyon, Institut de Physique Nuclaire de Lyon
[6]Space Telescope Science Institute, Baltimore, MD

SCALA is a physical calibration device for the SuperNova Integral Field Spectrograph (SNIFS), mounted to the University Hawaii 2.2m telescope on Mauna Kea. For type Ia supernova (SN Ia) cosmology programs, an improved fundamental calibration directly translates into improved cosmological constraints. The aim of SCALA is to perform a fundamental calibration of the CALSPEC (Bohlin 2014) standard stars, which are currently calibrated relative to white dwarf model atmospheres.

SCALA transfers the calibration of a flux calibrated photodiode to SNIFS (Lombardo 2017). As the photodiode is sensitive over a broad wavelength range, a monochromator lamp setup is used to perform the transfer wavelength by wavelength. This calibration of SNIFS relative to a laboratory flux-calibrated broad-band detector can be compared with the usual SNIFS calibration relative to the CALSPEC standard stars, and consequently ties the laboratory detector standard with the models for the white dwarf atmospheres.

We report on the characterization of several systematics, which we correct our measurements for, most prominently internal reflections and out-of-band emission from our monochromator (Küsters 2018). We compare our results to the existing CALSPEC system (model atmospheres) and to a compilation of Vega flux calibrations using laboratory-calibrated light sources (Hayes 1985). Our measurements agree with the CALSPEC and Hayes 1985 results within $\pm 2\,\%$ in a wavelength range from 4500 Å to 9000 Å.

We plan to improve the baffling of our light source and detector to suppress reflections internal to SCALA. We further plan to use a double monochromator to suppress the out-of-band emission. Both improvements should lead to sub-percent uncertainties in the calibration of the standard stars and an increased wavelength range, from 3500 Å to 9900 Å.

References

Bohlin, R. C. 2014, *The Astronomical Journal*, 147.6, 127
Lombardo, S. et al 2017, *Astronomy and Astrophysics*, 607, A113
Küsters, D. 2018, *Humboldt Universität zu Berlin, Phd Thesis*
Hayes, D. S. 1985, *Symposium-International Astronomical Union*, 111, 225–252

FM13
Global Coordination of International Astrophysics and Heliophysics Activities from Space and Ground

Astronomy in Focus - XXX
Proceedings IAU Symposium No. XXX, 2018
M. T. Lago, ed.

doi:10.1017/S1743921319005222

FM13. Global Coordination of International Astrophysics and Heliophysics Activities from Space and Ground

David Spergel, Princeton University and Flatiron Institute (USA) on behalf of the IAU FM13 organizing committee

---

International collaboration has always been an important part of research in astronomy, astrophysics, and heliophysics. Over the past two decades, the increasing complexity and cost of new facilities, the constrained amount of funding available from individual sources, and the rapidly increasing volume of data produced by newer facilities have made international collaboration on large ground- and space-based facilities essential to moving the fields forward. As international cooperation becomes commonplace, data-sharing policies have become ever more important. All IAU members have a stake in the policy decisions made by nations and various scientific consortiums concerning data access and international collaborations. IAU FM13 aimed to provide a forum to discuss how to improve coordination of global strategic planning in astronomy, astrophysics, and heliophysics in order to maximize the scientific return from research facilities.

As astronomical projects grow larger, international collaboration has become essential for both ground and space-based astronomy and heliophysics. Europe -via multiple avenues like the European Commission and Horizon 2020 program- by its very nature has been a model of international cooperation for decades now, but a model that is not easily applied outside of the continent on a broader scale. The European Southern Observatory, CERN, and other projects demonstrate that there is much to learn from this regional space power and lessons to apply to other parts of the world separate from even the success of a fully functional space program (the European Space Agency). Major observatories such as ALMA, HST, SOHO, Planck, and Herschel are international projects with many partners making important technical, scientific and financial contributions.

Meanwhile, countries with starting, or increasing, investments in the space sciences are looking for ways to leverage precious resources by getting involved in international collaborative efforts. However, with a patchwork of national policies concerning data and facility access, as well as little in the way of international lessons learned or expectations for involvement in large-scale observatory projects, it is difficult to gauge the true value of involvement at the outset. At present, the U.S., Europe, Japan and China all conduct their own independent long-term planning process. In the United States, the planning is done primarily through its decadal survey process. ESA plans its major long-term projects as part of its its Cosmic Vision strategic plan, in particular the large L2/L3/L4/L5 missions. Japan has its own national planning process that spans much of its research program.

The 2010 U.S. National Academy of Sciences (NAS) astronomy and astrophysics decadal survey, New Worlds, New Horizons in Astronomy and Astrophysics, states that, "An important characteristic of contemporary astronomy, and therefore of this survey, is that most research is highly collaborative, involving international, interagency, private, and state partnerships. This feature has expanded the scope of what is possible but also makes assessment and prioritization more complicated." (p. xvi). Similarly, the 2012 U.S. NAS heliophysics decadal survey, Solar and Space Physics: A Science for a Technological Society states that "A comprehensive investigation in solar and space physics cannot take place in isolation but should be part of an international effort, with different countries able to bring to bear unique geographic advantages, observing platforms, and expertise", and "while participation in international solar and space research projects could be accomplished through numerous individual, bilateral initiatives and agreements, the overall impact would be increased by coordinated agency involvement".

How do we coordinate these international planning efforts? How can and should we share the data produced both by these international collaborations and by other projects? How should we provide access to these facilities? Furthermore, the huge volume of data produced by current and future observation systems necessitate modes of research that have not heretofore figured prominently in astronomical and heliophysical research enterprises. The Daniel K. Inouye Solar Telescope (DKIST) will collect 3.65 petabytes in its first year of science operations while the Large Synoptic Survey Telescope (LSST) alone will produce 30 TB of data per night.

The potential benefit of enhanced international coordination is high. Much can be learned in astrophysics by adopting a broad-scoped approach, in which ground and space-based facilities look at the same target with different wavelengths, timescales and technologies. Such an approach requires more resources than a single nation could maintain. Heliophysics has the added issue of coordinating truly global ground-based systems and space missions in various regions of the Sun-Earth system. In this context, Earth is an additional spacecraft embedded in its own space plasma environment. For the first time in history, we are capable of looking at a complicated coupled space system in its entirety, from the sun through the heliosphere, magnetosphere, ionosphere, and atmosphere down to the biosphere, in which we try to survive the present climate change. To study and understanding the system around us is the ultimate benchmark to be able to understand other star-planet systems.

Large progress towards combining international ground- and space-based assets has been made by, for example, the multi-spacecraft Cluster and THEMIS coordination with ground-based assets. Additional, truly global instrument networks have been developed by the community in international collaboration, through more-or-less grassroot activities, or in recent years through the International Living with a Star program on agency level. Nevertheless, we still have major gaps in our coverage and understanding of the Sun-Earth system as an entity.

At the XXXth GA, Focus Meeting 13 comprised six focused panels. Each of the panelists made initial remarks and then we had an engaged discussion led by the moderator and involving the panelists and the attendees.

The meeting opened with a panel moderated by Roger Davies on "Large international Space Projects: From Black Holes to Cosmology" at 10:30 on Monday August 30. Thomas Zurbuchen opened the session with a discussion of the role of decadal surveys in prioritization, the importance of independent costings in evaluating missions and the importance to NASA of international contributions. Günter Hasinger talked both about international collaboration and ESA and the excitement of the multi-messenger quest for high redshift black holes. Stefano Vitale provided an overview of the LISA project. He emphasized that unlike most astronomical projects, LISA's data is a time series. He discussed the potential role of multiple teams in the analysis. Zhan Hu provided an overview of the Chinese Space Station Optical Satellite (CSS-OS). He discussed the possibility of international contribution of new instrumentation. CSS-OS will be a serviceable mission with a long planned lifetime. He discussed pixel level co-processing with other upcoming great surveys such as LSST, Euclid and WFIRST. He ended with a discussion of the challenges of serving data to the community. Didier Barret provided an overview of the Athena mission and described its current status. Jeff Kruk followed with a discussion of the WFIRST mission. Ryan Hickox described the Lynx X-ray mission concept, a proposed mission that will be revised in the upcoming US decadal survey. He emphasized the search for the origin and growth of supermassive black holes and the connections with other project, particularly the upcoming major ground-based programs.

After the initial panel presentations, the attendees and the panelists discussed open data and the coordination of future missions. We then had an energetic discussion of the role of proprietary periods. Zurbuchen asked, "do we need a year proprietary period?" He noted that it doesn't create the sense of urgency and that scientists have "the right to make a fool of themselves". Alan Title commented on the solar physics community experience with rapid data release. Dave Silva noted that for Target of Opportunity observations, it is essential to get the data out very quickly. Ken Sembach, the Director of Space Telescope Science Institute, commented that there is no HST proprietary period for big programs and only six months for small program. He stated that there is no evidence for damage from short proprietary period from HST. The discussion then turned to whether models and software should be made public. Stefano Vitale noted the importance of the models for projects like LISA.

Athena Coustensis asked about the challenges for collaboration at the agency level. Tom Zurbuchen, the NASA Associate Administrator for Science, and Günther Hasinger, the ESA Director of Science, began by noting the importance of frequent informal communication on promoting and maintaining international collaboration Günter Hasinger described ESA plans for two medium class mission per decade, one ESA only and the other a likely collaboration with NASA on a big project, potentially one of NASA's big projects identified in next decadal survey. The discussion ended with comments on the importance of non-synchronicity in the prioritization process.

The meeting continued after lunch with a session on "Large international Space Projects: Opportunities for Studying Exoplanets, Planet and Star Formation". The incoming IAU President Ewine van Dishoeck chaired this session which brought together leaders of proposed long wavelength missions to discuss the science drivers for potential international projects. The presentations started with the exoplanet science case, summarizing and building on the dedicated Kavli-IAU workshop held in summer 2017 in Leiden, see arxiv.org/abs/1709.06992. Karl Staplefeldt introduced the big questions as well as an overview of the suite of new exoplanet missions on US side, from TESS to JWST and WFIRST, and on the European side from Gaia, to CHEOPS, JWST, Plato and Ariel.. These missions are steps towards answering "are we alone?". Jeff Kruk described the WFIRST exoplanet program, both coronagraphy (also as a technology demonstrator for future missions) and microlensing, with the latter program probing a unique part of the planet mass-orbital distance space . Karl Staplefeldt then presented the 4m HabEx mission, one of the study concepts for a future large US mission. Its main goal is to characterize our nearest planetary systems and detecting and characterizing a handful of ExoEarths. Starlight suppression at the 1e-10 level within 0.1 arc second needs to be achieved, either through advanced coronography or a free flying starshade. Thomas Henning then turned to the LUVOIR concept, after noting that ground-based ELTs and ALMA also will make important contributions to the scientific story. With 8-16m diameter and 4 scientific instruments, LUVOIR is more ambitious than HabEx and has a broader science case, including, for example, extragalactic UV spectroscopy and imaging. The larger aperture decreases the inner working angle so that significantly more exo-earth candidates are available for study. Thomas Henning stressed the importance of direct imaging spectroscopy, providing much more detailed exoplanetary atmosphere spectra than transit spectroscopy with its ppm depths can. The exoplanet discussion ended with a summary of the capabilities of the Origins Space Telescope by Ted Bergin. Origins operates in the mid-infrared wavelength range which includes strong, unique biomarkers for earth-like planets such as CH4 and O3. Ewine van Dishoeck stressed the need for broad wavelength coverage. The audience asked several questions of clarification and potential phasing of new missions. Also, critical technologies were discussed. Both HabEx, LUVOIR and Origins are study concepts that will be put forward to the US decadal committee.

The second part of this session centered on Star- and Planet formation. Doug Johnstone introduced the big questions, from the large scales of molecular clouds in our and other galaxies to the smallest scales of protoplanetary disks. Frank Helmich and Toru Yamada then jointly presented SPICA, a cooled 2.5m far-infrared mission and candidate for the ESA M5 slot with strong participation from JAXA. It would cover the wavelength range of 12-230 μm, in between JWST and Herschel/ALMA, with three instruments including a polarimeter to study magnetic fields in ISM filaments. Finally, Ted Bergin showed how Origins with its larger 6m aperture and 4 instruments covering 5-600 μm would be particularly powerful to address questions in star- and planet formation. He focused on the trail of water from clouds to disks and planets and the unique ability to measure gas disk masses through the HD line. With its higher spectral resolution and sensitivity compared with SPICA (which also highlights these science cases) it gains 1-2 orders of magnitude in limiting abundances. The importance of global collaboration and coordination for these missions, which is already happening, was stressed in the subsequent discussion with the audience.

On Tuesday, the focus meeting continued with a 10:30 AM panel on "Global Coordination of Ground-Based Astronomy" This session was a panel with leaders of the major next generation ground-based observatories, both the thirty-meter class telescopes and the large survey telescopes. Debra Elmegreen, the session chair, opened the session with a discussion of the importance of the global system and the history of the OIR system in the US. Xavier Barcons described the role of ESO, an intergovernmental organization that has now grown to 16 member states with the addition of Ireland, is building the ELT, a 39-meter telescope with built-in adaptive optics. The ELT is part of the broader ESO Paranal system. Ram Ramaprakash descried the efforts of the TMT partners (Canada, Caltech, China, India, Japan and University of California) to build the 30-meter telescope in Hawaii. He described the first light instruments, the infrared imager and the wide-field optical spectrometer, and the progress with the Hawaii lease. Pat McCarthy described the GMT partnership (Australia/Brazil/Korea/ASU/Texas/Arizona/Harvard-Smithsonian) to build its telescope on Las Campanas. The collaboration plans to complete the partial primary mirror array by 2024 and the full telescope by 2026. He noted that its GMT near infrared spectroscopy compares favorably with JWST (as does that of the other ELTs). Beth Willman introduced the Large Synoptic Survey Telescope and emphasized that it was a comprehensive system with 20% of the construction budget devoted to the data management system. She also introduced the National Center for Optical and Infrared Astronomy (NCOA), now in the planning stages, which will unify the U.S. share of Gemini, the National Optical Astronomy Observatories, and LSST for ease in management and coordination of operations and observations. Phil Diamond introduced the Square Kilometer Array, an international project involving 600 scientists from Australia, Canada, China, France, Spain, South Africa, the Netherlands and New Zealand through a proposed treaty organization similar to ESO and CERN. The first phase of SKA will run from 2020 to 2027. Phase 2 will be a multi-billion dollar project that will build 2000 dishes across Southern Africa. The SKA will primarily divide its time among scientists from member countries with a modest amount of time available to international scientists outside the collaboration. Alan McConnachie introduced the Maunakea Spectroscopic

Explorer, a project in an early stage and working towards first light in 2026. The MSE will use a 11.25 meter primary and will have 4300 fibers to enable low moderate and high resolution spectroscopy. The MSE repurposes the CFHT infrastructure/site and will be lighter than CFHT. McConnachie emphasized the importance of getting spectroscopy to complement the upcoming large imaging surveys. During the discussion, the panelists and the audience discussed the role of small institutions in these big projects, the lack of open data access and/or long proprietary period for some facilities and the pro/cons of international treaty versus limited liability company for building these large projects.

After lunch, Patricia Whitelock moderated the panel on "Engagement of countries with emerging astronomical communities in international efforts & Governance of International Projects". These panelists asked "Are there ways to better coordinate the major national and international funding agencies and also ground and space-based programs? What guidance for future international projects can be derived from studying the governance, fund-raising and project management strategies of past and current projects? What questions are so important and challenging that they can only be addressed by coordinating truly global resources?" This session also discussed possible ways to coordinate the engagement of astronomers in countries that do not have the resources to be major partners in large ground facilities and space missions. Ron Ekers began the discussion by noting the shared radio astronomy culture. He reviewed the OECD 1998 report and the role of the SKA pathfinders. Vanessa McBride described SALT, MeerKat and big projects in Africa. She contrasted South Africa's long tradition in optical astronomy which helped it build SALT with the lack of a long tradition in radio astronomy. She described the combination of the SALT experience, government support, and the MeerKat's role as a precursor is enabling South Africa's role in SKA. She described the government support of young astronomers and training programs and the development of the African VLBI network. Silvia Torres-Peimbert described the possible routes for small countries, such as Mexico, to participate in large projects. The options include participate in large projects directly: this approach requires money and expertise. An alternative is to construct small dedicated telescopes. While this approach is less expensive, it still requires expertise. She also noted the need for training options for students and scientists to work with the large data bases. She described two major Mexican projects, the Large Millimeter Telescope and the High Altitude Water Cherenkov (HAWC) Observatory. Peter Michelson described lessons learned from the FERMI space telescope involving emerging countries. Bob Williams discussed the role of the IAU working group and the need to involve more representatives from emerging countries. During the discussion, Beth Willman stressed the need to fund under-resourced communities and to make astronomy more inclusive. We also heard about opportunities for astronomy in Ethiopia and the possibilities of using Kilimanjaro as a site.

On Wednesday, FM13 turned towards the Sun with a session on "International Efforts in Heliophysics". Christina Mandrini moderated this panel at 15:30-17:00 on the science drivers on the ground and in space in the era of the Solar Orbiter and the Daniel K. Inouye Solar Telescope (DKIST). Valentin Pillet opened the session with an update on the construction of DKIST. DKIST has implemented a critical science plan and is coordinating its science program with ALMA. John Morgan described the Murchinson radio observatory, a multi-purpose science observatory that includes epoch of reionization and ionospheric physics. Masaki Fujimoto discussed post-Hinode space solar physics. He emphasized the key questions for future missions: the mechanism driving the solar cycle and irradiance variation. Future missions would need to probe the Sun at higher resolution so to understand the physical mechanism on the elemental scales. This will be enabled by high throughput spectroscopic telescopes with improved temporal resolution. Daniel Muller described the Solar Orbiter with its planned launch in Feb 2020. The mission will address "how does the Sun create and control the Heliosphere and why does solar activity change with time?" The mission will act in synergy with Parker Solar Probe and near-Earth assets such as SDO, DKIST and PROBA-3. Lika Guhathakurta concluded by discussing Heliophysics as a scientific discipline.

FM13's final session discussed "Gravitational Waves and Transient Science". Pietro Ubertini moderated this discussion of new opportunities for gravitational wave astronomy, Fast Radio Bursts, and opportunities for coordinated follow-ups in the LIGO/VIRGO/LSST era. He began the session by discussing the recent observations of GW 170817/GRB 170817A, a wonderful example of international collaboration. Fermi and Integral operating 100,000 km apart saw the same signal. Over 3500 scientists worked together to use a vast array of facilities to monitor the evolution of this event and produce transformative science. Matt Evans discussed the LIGO project. He noted the cultural changes from 2015 (before the detection of the binary black holes) to today. He described the next generation of detectors that will enable the detection of events out to cosmological scales.

Marisa Branchesi described the next run of LIGO/VIRGO. During this run, the detections will be announced to all. While the rates are uncertain, she anticipates that there will be 1-15 neutron star-neutron star mergers and tens of black hole/black hole mergers. How will the astronomy community follow-up on these events? In the discussion, David Silva noted that this upcoming set of events will be dimmer and will require 4m and 8m class telescopes. He stressed the need to prioritize the GW event follow ups. In the discussion of follow-up observations, Tim de Zeeuw, the former ESO Director General, noted that various pieces of the optical system have been in place for 20 years. For example, there are rapid response tools at VLT. Marisa Branchesi noted that the expertise of the GRB and SN community was essential for the follow-up and added that the challenge for future will be localization. Chryssa Kouveliotou emphasized the importance of instruments that will enable localization. Alvaro Gimenez, the former ESA Director of Science, discussed the role of space observatories in the localization. He emphasized the importance of the gamma ray burst detection of the counterpart, the dimmest and nearest GRB.

David Buckley's presentation asked, "How do we deal with transient follow-up in the GW/Multi-Messenger/LSST era?". There will be an enormous jump in the number of transients. This generates a need for tools to automatically follow-up on the fast transients with small telescopes. He described the aspirations at South Africa Astronomical Observatory (SAAO) to make the whole Sutherland site an integrated intelligent machine for transient follow-up.

Suijian Xue described the significant impact that the gravitational wave detection has had on Chinese astronomy. He described the plans for emerging observatories for multi-messenger astronomy on the "roof of the world", the Ali Observatories in Tibet. This site is located in the West of Tibet at the altitude of 5100 meters. Plans for this site include HinOTORI, a 40 cm robotic imager (with Hiroshima University) and a 2 meter telescope operating as part of the Las Cumbres Observatory Global Telescope (LCOGT) system. This site is also a potential location for CMB polarization experiments. The site is one of the four driest sites in the world and has the potential to be the "Atacama" of the North. He expressed an eagerness for international collaboration at the site.

The next several talks discussed plans for the next generation of gravitational wave detectors . Matt Evans described the U.S. roadmap of first upgrading LIGO and improving its sensitivity by factor of 2. Next, the program of building "Voyager" in the last 2020s. In the 2030s, the community aims to start construction of a detector capable of detecting events out to redshifts of 10-20. Marisa Branchesi described the formation of the Einstein Telescope consortium in Europe with possible sites in Hungary, the Netherlands and Italy. Alvaro Gimenez described the LISA pathfinder, its challenges in 2011, and its role as a precursor to LISA. Suijan Xue described China's plans to develop mission concepts that are broadly similar to LISA and to the Einstein Telescope.

Following the meeting, the FM13 working group met and made a number of recommendations for the upcoming year:

- We have opened the working group to all IAU members and encourage interested IAU members to contact David Spergel (dns@astro.princeton.edu) or Roger Davies (roger.davies@physics.ox.ac.uk).
- We have begun planning for a Kavli Meeting on transient science. The meeting will be held in Capetown in January 2020. The coming decade will be a very exciting time for transient science. We anticipate a rapid pace of transient discovery from LSST, LIGO, high energy missions, and radio observatories. The international astronomy community will need to be ready to follow-up on the most interesting of these events.
- We are planning on upgrading our website so that it can serve as a marketplace for international projects, particularly those seeking new partners.

FM14
IAU's Role in Global Astronomy Outreach - Meeting the Latest Challenges and Bridging Different Communities

Astronomy in Focus - XXX
Proceedings IAU Symposium No. XXX, 2018
M. T. Lago, ed.

doi:10.1017/S1743921319005234

FM14: IAU's Role in Global Astronomy Outreach - Meeting the Latest Challenges and Bridging Different Communities

Sze-leung Cheung[1] and William H. Waller[2]

[1]IAU Office for Astronomy Outreach, National Astronomical Observatory of Japan,
[2]The Galactic Inquirer, 243 Granite St., Rockport, MA, USA

The 2018 IAU General Assembly in Vienna provided an important opportunity for research astronomers and outreach professionals to jointly consider the IAU's role in global astronomy outreach. The organizers goals were to enable and coordinate astronomical outreach for the greatest good worldwide. Participants shared what was already done and discussed what could be done better going forward. Emphasis was placed on the role the IAU could actively play in better coordinating, catalyzing, and supporting global astronomy outreach efforts. As the largest body of professional astronomers in the world, the IAU has a recognized responsibility to be engaging with the public, providing access to astronomical information and widely communicating the science of astronomy (IAU Strategic Plan 2020 2030, p. 4). The structure of this Focus Meeting was designed to address these key points through its four oral sessions on (1) Bridging the Astronomy Research and Outreach Communities Recent Highlights, Emerging Collaborations, Best Practices, and Support Structures, (2) Communicating Astronomy in our Changing World, (3) The IAU National Outreach Contacts (NOC) Network Coordinating and Catalyzing Astronomy Outreach Worldwide, and (4) Outreach Action and Advocacy in the Context of IAUs 2020-2030 Strategic Plan.

Topics

- Developing and Sharing Big Data for Effective Astronomy Outreach
- Promoting Citizen Science Projects in Astronomy
- Brokering Pro-Am Partnerships in Astronomy
- Supporting Research Scientists in Astronomy Outreach
- Key Outcomes of the CAP 2018 Meeting in Fukuoka, Japan
- Effectively Navigating the Media Landscape
- Effectively Coordinating Museums and Planetaria Worldwide
- Supporting the Underserved and Dispossessed
- The IAU National Outreach Contacts (NOC) in East Asia and the Pacific, South Asia and Near East, Africa, Europe, and the Americas.
- Finding Common Purpose and Implementing Coordinated Action
- Addressing Key Aspects of the IAU 2020-2030 Strategic Plan
- Supporting IAU Research Astronomers in their Outreach
- Supporting Amateur Astronomers and other Volunteers in their Outreach
- Developing and Coordinating Astronomy and Information and Resource Dissemination Plans
- Working with UNESCO, ICSU/ISSC, for More Inclusive Outreach
- Formulating a Position and Statement on Terrestrial Climate Change and Other Astronomically-related Issues of Societal Concern

In a long day of invited talks, contributed talks, flash talks, panel discussions, and poster presentations, participants in FM 14 shared many success stories, challenges, lessons learned, best practices, and aspirations in our mutual quest to advance astronomy outreach worldwide. The proceedings that have been captured in the book and even in the on-line archive represent but a fraction of all the discussions and deliberations that transpired. Herein, some highlights are presented along with a compilation of potential action items that were recommended by participants for the IAU to pursue.

We outline below some potential actions for the IAU inspired by the speakers.

• From Sylvie Vauclair's talk on "Pro-Am Collaboration in Occitanie, Outreach and Consequences" — The IAU could help to coordinate and promote astro-tourism ventures worldwide.

• From Susanna Kohler's talk on "AAS Nova and Astrobites as Bridges between Astronomy Communities" — The emergent IAU Office of Astronomy Education (OAE) could help to increase student literacy by utilizing these sorts of well-vetted written resources.

• From Joseph Diamond's talk on "The New Age of Big Data Astronomy Digital Assets!" — Could the IAU serve as a clearinghouse of media products, and if so, what platform should it use?

• From the Panel Discussion on "Effectively Navigating the Media Landscape" — Rick Fienberg recommended that the IAU offer media training for its members, including help with translations.

• From the same Panel Discussion — Thiago Goncalves recommended that the IAU help to increase the visibility of Brazilian (and other nations') astronomy and astronomers, but how? Perhaps we could feature in our outreach and other newsletters research by astronomers from all member countries – one country at a time.

• From Mark SubbaRao's talk on "Effectively Coordinating Museums and Planetariums Worldwide" — The IAU, its OAO, OAD, and Commission 2 could work more closely with the IPS to better support and coordinate planetariums around the world.

• Also from Mark SubbaRao's talk — Rick Fienberg recommended that the IAU train scientists in outreach modalities, building on the progress made by the NASA Museum Alliance.

• From the Panel Discussion on "Supporting the Underserved and Dispossessed" Robert Massey recommended that the IAU, its OAO, OAD, and Commission C2 make similar partnerships with social agencies as the RAS for optimal outreach to the underserved.

• From Aswin Sekhar's presentation on "Challenges with Gender, Immigration and Diversity for Astronomers from Developing Countries" — The IAU could act as a nodal agency for engaging with policy makers involved in advancing gender equity, skilled immigration, and inclusive work environments.

• From Akihiko Tomita's presentation on "The 3D Map Astro/Geo Tour with your Fingertips" — The IAU could help to develop and share 3-D maps and other resources for engaging with the visually impaired.

• From Session 3 on "The IAU National Outreach Contacts (NOC) Network – Coordinating and Catalyzing Astronomy Outreach Worldwide" Samir Dhurde recommended that the IAU and OAO do a better job of coordinating activities within specific geographic regions such as South Asia.

• Also in Session 3 — Sze-leung Cheung encouraged the NOCs to find common purpose. They can reach more than 500,000 people, most of whom will be amateurs and members of the general public. About 10 percent will be students in schools. He asked, how can we work together better for the IAU 100th? We need more support and endorsements. Finally, we need to use the IAU 100th to celebrate engagement in astronomy, not the IAU.

• From the Panel Discussion on "Outreach Action and Advocacy in the Context of IAU's 2020-2030 Strategic Plan" — Rick Fienberg recommended that just as there are now IAU Ph.D. prizes for all the other divisions, let us make sure that there is a Ph.D. prize for Division C. Subsequent action during the General Assembly made such a prize an emergent reality by engaging the research communities related to Div C closer to the IAU.

• Also from this Panel Discussion — Bill Waller proposed that just as the IAU has weighed in on light pollution, we should seriously consider making a statement on terrestrial climate change. We astronomers have significant expertise on the subject, especially in making interplanetary comparisons (think Venus!).

This one-day Focus Meeting was held 23, August 2018 as part of the IAU XXX General Assembly in Vienna, Austria. Besides the invited and contributed talks, panel discussions occurred in 3 of the 4 oral sessions. In all 4 sessions, there was time allotted for audience participation which was recorded via distributed question sheets and the Slido online utility. Poster presentations were tended during the two coffee breaks and especially after the oral sessions ended. On the Wednesday prior to the Focus Meeting, participants enjoyed an evening reception and tour of the natural history museum in Vienna, where the largest collection of meteorites is curated.

Scientific Organizing Committee

• Sze-leung Cheung (co-chair) Tokyo, Japan
• William H. Waller (co-chair) Rockport, USA
• Lina Canas Tokyo, Japan
• Chenzhou Cui Beijing, China
• Nanjing Samir Dhurde Pune, India
• Richard Tresch Fienberg Watertown, USA
• Andrew Fraknoi Los Altos Hills, USA
• Beatriz Garcia Godoy Cruz, Argentina
• Kevin Govender Cape Town, South Africa
• Thilina Heenatigala Colombo, Sri Lanka
• Oana Sandu Bucharest, Romania
• Mike Simmons Calabasas, USA
• Sylvie Vauclair Toulouse, France

On behalf of the SOC, we would like to thank the IAU for providing this opportunity to shine a light on global astronomy outreach - and the many people who made our Focus Meeting so informative, lively, and forward looking.

Sze-leung Cheung and William H. Waller

Astronomy in Focus - XXX
Proceedings IAU Symposium No. XXX, 2018
M. T. Lago, ed.

doi:10.1017/S1743921319005246

FM14 Session 1: Bridging the Astronomy Research and Outreach Communities - Recent Highlights, Emerging Collaborations, Best Practices and Support Structures

Sze-leung Cheung[1], Sylvie D. Vauclair[2], Chenzhou Cui[3], Shanshan Li[3], Yoichiro Hanaoka[4], Sharon E. Hunt[5], Shio K. Kawagoe[6], Nobuhiko Kusakabe[7], Shigeru Nakamura[8], Grigoris Maravelias[9,10,11], Emmanouel Vourliotis[9,12], Krinio Marouda[9], Ioannis Belias[9], Emmanouel Kardasis[9], Pierros Papadeas[9], Iakovos D. Strikis[9], Eleftherios Vakalopoulos[9], Orfefs Voutyras[13], Lucia Marchetti[14,15], Thomas H. Jarrett[14], Franck Marchis[16,17,18], Arnaud Malvache[16], Laurent Marfisi[16], Antonin Borot[16], Emmanuel Arbouch[16], I. Villicaña-Pedraza[19], F. Carreto-Parra[20], S. Prugh[19], K. Lopez[19], J. Nuss[19], D. Cadena[19], V. Lopez[19] and Priya Shah[21]

[1]IAU Office for Astronomy Outreach, National Astronomical Observatory of Japan,
email: cheungszeleung@iau.org,

[2]IRAP, Toulouse, France,

[3]National Astronomical Observatories, CAS, Beijing, China,

[4]National Astronomical Observatory of Japan, Japan,

[5]National Optical Astronomy Observatory, USA,

[6]Institute of Industrial Science, The University of Tokyo, Japan,

[7]Astrobiology Center, NINS, Japan,

[8]Chiyoda Ward's Kudan Secondary School, Japan,

[9]Hellenic Amateur Astronomy Association, Athens, Greece,

[10]Foundation for Research and Technology-Hellas, Greece,

[11]University of Crete, Heraklion, Greece,

[12]National and Kapodistrian University of Athens, Zografou, Greece,

[13]National Technical University of Athens, Zografou, Greece,

[14]Department of Astronomy, University of Cape Town, South Africa,

[15]Department of Physics and Astronomy, University of the Western Cape, South Africa,

[16]Unistellar, France,

[17]Carl Sagan Center, SETI Institute, USA,

[18]LESIA, Observatoire de Paris, Meudon, France,

[19]DACC, New Mexico State University, USA,

[20]Physics Department, New Mexico State University, USA,

[21]Department of Physics, Maulana Azad National Urdu University, India

Abstract. Section 1 of the FM14 focus on bridging the astronomy research and outreach communities - recent highlights, emerging collaborations, best practices and support structures. This paper also contains supplementary materials that point to contributed talks and poster presentations that can be found online.

Keywords. astronomy outreach, big data outreach, pro-am activities, citizen sciences

---

1. Supplementary materials

The following contributed talk of section 1 can be found online in the format of supplementary material.

• Vauclair, S. *et al.* Pro-Am collaboration and outreach in French Occitanie. Supplementary material 1-01.

The following poster presentations of section 1 can be found online in the format of supplementary materials.

• Cui, C. *et al.* Data Driven Astronomy Education and Public Outreach. Supplementary material 1-02.

• Hanaoka, Y. Professional-Amateur Collaboration in the Scientific Observations of Total Solar Eclipses. Supplementary material 1-03.

• Hunt, H. Digital archives as an outreach tool: The role of the observatory librarian. Supplementary material 1-04.

• Kawagoe, K. *et al.* Development of stargazing party for local elementary school students. Supplementary material 1-05.

• Kusakabe, N. Outreach activities of Astrobiology Center of Japan. Supplementary material 1-06.

• Maravelias, G. *et al.* A paradigm to develop new contributors to Astronomy. Supplementary material 1-07.

• Marchetti, L. *et al.* The Iziko Planetarium 8K Digital Dome: Big Data visualisation & public engagement. Supplementary material 1-08.

• Marchis, F. *et al.* Democratizing Astronomy with the Unistellar eVscope Network. Supplementary material 1-09.

• Villicaña-Pedraza, I. *et al.* Citizen Science and the Virtual Observatory with college students: Characterization of exoplanets. Supplementary material 1-10.

• Shah, P. The Open Universe and Data-driven Astronomy. Supplementary material 1-11.

Astronomy in Focus - XXX
Proceedings IAU Symposium No. XXX, 2018
M. T. Lago, ed.

doi:10.1017/S1743921319005258

VO for education and outreach

Giulia Iafrate on behalf of the education group of IVOA

INAF - Astronomical Observatory of Trieste,
Via G.B. Tiepolo 11, I - 34143, Trieste, Italy
email: giulia.iafrate@inaf.it

Abstract. The Virtual Observatory (VO) is an international astronomical community-based initiative. VO aims to allow global electronic access to the available astronomical data archives of space and ground-based observatories and other sky survey databases. *VO for education* is a project developed within the framework of the European Virtual Observatory (EuroVO) with the aim of diffusing VO data and software to the public, in particular students, teachers and astronomy enthusiasts. *VO for education* offers use cases, pedagogical units, and simplified professional software that will allow a taste of the emotion of scientific research even to those approaching astronomy for the first time or simply wishing to wander between stars.

Keywords. miscellaneous, astronomical data bases: miscellaneous, atlases, catalogs, surveys

1. Introduction

Billions of Bytes are written every night in the computers of astronomical observatories. Such a wealth of information is often underused because data are not distributed widely enough, or because formats are very different from place to place, or because it is difficult to retrieve data once found. In order to improve on this situation several teams of astronomers have created the Virtual Observatory (VO). The goal is to have a virtual instrument that to an astronomer is, or is going to be, just another telescope. There are familiar interfaces and tools and, in principle, nothing set the astronomer know that his/her observations are retrieved from digital archives rather than collected directly at a telescope. Of course, only data of observatories that join the project are available, but almost all major data centers have already adhered to the VO and most new telescopes and instruments will be VO compliant.

The International Virtual Obervatory Alliance (IVOA - Hanisch R. J. *et al.* 2010) now coordinates 21 national or super-national VO projects. Among these projects there is the European VO (EuroVO). EURO-VO aims at deploying an operational VO in Europe. It supports the utilization of VO tools and services by the scientific community, technology take-up and VO compliant resource provision, and building of the technical infrastructure. The specific project ASTERICS† (Astronomy ESFRI & Research Infrastructure Cluster), is leading VO into operations.

Among the many tasks of the EuroVO, an important part deals with service activities for higher education and outreach (Demleitner, M. *et al.* 2018). It is certainly important that some of the benefits that VO offers to professional astronomers, also become resources for teachers, students and people who are simply curious about celestial objects. Researchers of many European countries contribute to the development of software tools, documentation and use cases that constitute the main product of VO for education.

† ASTERICS is supported by the European Commission Framework Programme Horizon 2020 Research and Innovation action under grant agreement n. 653477

In the present contribution we outline the VO for education project, we briefly describe its development, then we give information on products and on how to retrieve them. Finally we present the educational activities we perform in Trieste combining real, remote and virtual observations and we describe the IVOA *education interest group* activities.

It is worth pointing out that all our products can be freely downloaded and used. If you will use them, we will be glad to hear from you: our possibility to keep developing our educational material also depends on its documented use by as many people as possible. Of course, also bug reports, suggestions, and complaints are most welcome.

2. *VO for education* products

In the framework of *VO for education* we want, through the VO resources, to bring science to students, get them engaged and give students an involving glimpse of the professional world of astronomy, including a perception of the infrastructure. We want also to make them realize they could become scientists and teach some elements of astronomy also.

In order to reach our purposes we adapted and simplified key tools (with users help), we provide a library of use cases and propose activities that mix "serious work" with "fun". We organize workshop for teachers training also.

VO for education offers a software to visualize the sky (Stellarium) and a software to access VO data (Aladin). Both software come from a professional version and have been modified to be used by everybody. Several use cases and pedagogical units are available, each one focused on a specific astrophysical problem. These examples include user guides on how to use the software and have been developed with the help of teachers and students (Freistetter, F. *et al.* 2010).

VO for education software tools and use cases have been revised in the last school year during an intense program with high school students, in the framework of the H2020 ASTERICS project. They are available in several languages and can be free downloaded from the web site http://vo-for-education.oats.inaf.it/.

2.1. *Software tools*

We identified existing professional software tools for the retrieval, visualization and analysis of VO data that could be efficiently and successfully adapted for educational and/or outreach purposes. We selected Aladin (Bonnarel, F. *et al.* 2010), a tool to retrieve and display VO images, and Stellarium (http://stellarium.org/), a sky browser. These tools were already stable and in use among professional astronomers, their developers working in Institutions of the EuroVO collaboration.

Aladin is an interactive celestial atlas for the visualization of digital images available in VO format, together with a selection of large area sky surveys. It also displays the nicest images taken with the NASA Hubble Space Telescope. Aladin allows to mark on images the positions of the celestial objects contained in the astronomical catalogs available in international archives like Simbad and VizieR at the Centre de Donnes astronomiques de Strasbourg (CDS). Aladin has several built-in functions that allow image handling and a quick analysis. All professional applications developed for VO are available to Aladin.

Stellarium simulates a realistic sky on the screen of the PC, like it could be seen by naked eye or with a telescope from anywhere on the earth and at any time. Stellarium may be used by anybody, but in particular it may be useful to teachers for exploring the night sky and the basics of the celestial sphere. It is also a useful tool for amateur astronomers who wish to plan an observing session.

Both Aladin and Stellarium have been modified and simplified according to students' and teachers' feedback, new features have been added also (e.g. search objects by class in Aladin and information about star spectral type in Stellarium).

2.2. *Use cases*

Usage examples aim at familiarizing the users with Aladin and Stellarium and at stimulating further interest and activities in astronomy. Usage examples are in the form of pedagogic modules consisting of two main parts. The first part presents a typical astronomical problem with a short introduction and a description of the solution found by astronomers, or, in some cases, an expanded treatment of the problem. The second part is a step-by-step guide to the commands needed to reach the solution of the problem with Aladin or Stellarium. Some of our usage examples include exercises that are proposed for teachers' activities in the classroom. Solutions are provided separately.

Use cases have been developed with the support of teachers, are dedicated to middle and high school students and are divided into three difficult levels: *basic* - requires no astronomical knowledge or ability to solve math computations, age group 12-15; *intermediate* - requires basic astronomy and math knowledge and the capability to understand a plot, age group 15-18; *advanced* - requires good astronomy and math knowledge and the capability to build and understand a plot - age group 18+.

- ***Basic***
  - **Introduction to Aladin**: Aladin's main features, each presented with examples taken from its most common use.
  - **Stellarium for beginners**: Stellarium's main features, each presented with examples taken from its most common use.
  - **The sky**: celestial coordinates, Earth's rotation and revolution, constellations, light pollution.
  - **The shape of galaxies**: galaxy morphology and classification according to the Hubble diagram.
  - **Planetary conjunctions**: motion of planets both around the Sun and in the sky, planetary conjunctions and the Star of Bethlehem.
  - **Introduction to Stellarium for preschoolers**: easiest and most appealing functions of Stellarium for a presentation to preschoolers, as the constellations and their variation between different cultures, light pollution.
  - **The constellations of the Zodiac**: the orbital motion of the Earth, history of astronomy and the precession of the equinoxes.
  - **The Messier catalog**: examples of the most interesting categories of celestial objects, from stellar clusters to galaxies, belonging to the Messier catalog that includes some of the most viewed objects of the deep sky.

- ***Intermediate***
  - **The stars**: basic observational parameters of stars, color, magnitude, temperature, Herzsprung-Russell diagram and how stars work and evolve.
  - **Proper motion of the Barnard's star**: motion on the celestial sphere of stars that seem "fixed" on the sky, estimation of displacement on the sky of the Barnard's star.
  - **Confirmation of an apparent supernova**: supernova search and discovery, astrometrical solution of images.
  - **Distance of the Crab nebula**: supernovae, exploding or exploded stars, the Crab Nebula (the 1054 AD supernova registered by Chinese astronomers) and its distance.

◦ **Asteroids in the Solar System**: main characteristics of asteroid orbits and their distribution within the Solar System.
◦ **The Moon**: geometry of the orbit of the Moon and the nature of its phases, eclipses of Moon and Sun.
◦ **The mass of Jupiter**: determination of the mass of Jupiter by observing the orbits of the Galileian moons and by inserting these data into Kepler's laws.
◦ **Star clusters**: intrinsic linear size and apparent angular size, basic facts about star clusters.
◦ **The disk of the Milky Way**: shape and thickness of the disc of our own Galaxy by counting stars within and around the Milky Way.

• ***Advanced***

◦ **The Pleiades open cluster**: distance of Pleiades open cluster derived from parallax, Herzsprung-Russell diagram of stellar clusters.
◦ **Distance of the Andromeda galaxy**; distance of the Andromeda Galaxy by identification of variable stars of the Cepheid class and the determination of the relation between their period and their intrinsic luminosity.
◦ **Planetary motion**: Kepler's laws, a cornerstone of astronomy and a fundamental brick of both Newton's and Einstein's theories of gravitation.

3. Combining real, virtual and remote observations for kids

At the astronomical observatory of Trieste (part of the Italian National Institute for Astrophysics) we developed a completely remoted educational telescope, called "Stars go to school" (SVAS - Baldini *et al.* 2010), and set up a interactive laboratory (Esploracosmo) where students can deal with real, virtual and remote observations.

SVAS offers to schools and teachers a remote laboratory with which carry out real observation sessions, managed in real time by the students under the supervision of the teacher, in the classroom, and of an astronomer, at the Astronomical Observatory of Trieste, "Osservatorio Astronomico di Trieste (OATs)" in Italian, thanks to the telematic link between the school and the observatory. Students and teachers experience real astronomical observations, through the interactive participation to the different steps of planning, observing and acquiring the data.

The project is addressed to 13-18 yr students. Every observation is previously planned together with the teachers, according to age and curriculum of the students, with the aim to maximize the results. The observing activity, lasting about 90 minutes and led by an astronomer of the OATs, can be done during the morning (observation of the Sun) or in the evening (observation of stars, nebulae, clusters and galaxies). SVAS is a member of the Italian Remote Network of Educational Telescopes (IRNET).

Esploracosmo is the interactive computer laboratory in operation at the Observational Branch of Basovizza of the Astronomical Observatory of Trieste. It proposes a modern tool to support teaching of astronomy, through the study and experimentation of its observation methods. Esploracosmo can host 25 students and has an educational network that connects the PCs equipped with software expressly developed for schools. Projection systems and TV monitors complete its equipment. Thanks to the data access, coming from both the OATs archives and the archives of the major professional telescopes of the world through the VO, and the possibility to perform remote observations with educational telescopes available in the world, Espploracosmo allows the exploration of the Solar System, stars, nebulae and galaxies, as well as the study of the gravitational laws, one of the building blocks of the structure and evolution of the Universe. Esploracosmo allows to perform the observations with the telescopes of the SVAS project and the activities of the *VO for education*. The activity proposed in Esploracosmo is managed by an astronomer and lasts about 90 minutes. To whom who desire to match astronomy with a tour in a research institute, Esploracosmo offers the emotion of the scientific research in contact to the instruments and, thanks to the joint between real and virtual observations, assures educational activities also under a cloudy sky.

4. IVOA education interest group

The *education interest group* (eduIG) of the IVOA has been founded in 2012 with the assignment to coordinate the widest global distribution of VO tools, data and practices in support of astronomy teaching in schools and universities.

The eduIG is a a two-way communication channel between the public and VO. The main activities developed by the eduIG are:

- identify scientists/educators interested in working within eduIG;
- collect information on educational activities that would immediately benefit from VO resources or that are already using them;
- compile lists of general educational resources in astronomy that may be useful in the VO environment (either as they are, or as a starting point for possible VO developments);
- open communication channels with projects active in astronomy and education worldwide;
- evaluate and discuss educational requirements on tools, data, and guides to be submitted to IVOA, providing information for possible actions (e.g. simplification of tools or publication of data). Feedback to the relevant IVOA WG;
- provide IVOA endorsement in support of national VO actions aimed at obtaining recognition and funding from high level national government education structures and/or from nation-wide organizations of teachers;
- create and maintain an online repository where to store documentation, lists of available resources and projects for educators as well as VO developers and astronomers.

If you are interested in joining the eduIG and subscribing to the IVOA education mailing list visit: http://www.ivoa.net/members.

4.1. *VO doc registry*

As argued above, disseminating text-like material like tutorials, worked-out use cases, or larger introductions, is an important task in facilitating VO exploitation. Within the VO community, there is a large body of educational material for a wide variety of audiences ranging from pre-school to researchers.

To date, such material has been collected informally by the various projects on plain web pages as the *VO for education* web site. The VO already has a registry extension for standards, which of course are also text-like. This extension, however, focuses on metadata important for standards e.g., vocabularies and status that is not pertinent for educational material. Conversely, it is not concerned with document language (which can safely be assumed to be English for standards), or education levels, and it disregards the issue of locating formatted and source versions, which for educational material is important.

In order to improve upon this situation a dedicated VO registry to keep record of educational material has been created.

4.2. *SVN repository for edu resources and Virtual Observatory Text Treasures*

Registering text documents as VO resources allows searching for tutorials and similar material through standard registry interfaces. But keeping tutorials up to date, in their master form and also in their translated versions, is an obviously important management issue not really addressed by the registry. For tracking changes and versions, the standard tool is a version control system. Therefore, a versioned repository (SVN) has been set up at GAVO (German Astrophysical Virtual Observatory). It collects part of the already existing VO tutorials with the goal of preserving them and letting users update and translate them.

The Virtual Observatory Text Treasures VOTT is a formatted list of VO educational and outreach texts: use cases, tutorials, courses, etc. VOTT contains material for all settings, from pre-school to graduate. It is generated from the documents known to the VO registry and available at the url http://dc.zah.uni-heidelberg.de/VOTT.

4.3. *VAPE web application*

VAPE (http://ia2-edu.oats.inaf.it:8080/vape) is an application for the publication of educational data in the VO, developed by IA2 (Italian center for Astronomical Archive). Thanks to VAPE, institutes managing educational telescopes can publish their data in the VO without facing the complexity of (more complete) professional tools. The creation of an educational data archive a) provides teachers performing observations with an educational telescope to easily store and access their observations, b) makes observations available to teachers that have otherwise no access to a telescope. An added advantage of the publication in VO is the availability of many free tools for displaying and/or analyzing data.

5. Conclusions

Many remote educational telescopes are now available around the world and the VO is growing also. We think VO and remote observing, together with dedicated examples and use cases, is the winning combination for astronomy education and teachers' and students' feedback confirm it. So we aim to a wide diffusion of *VO for education* software tools and use cases along with an increasing number of educational remote telescopes.

References

Baldini, V., Calderone G., Cepparo F., Cirami R., Coretti I., Di Marcantonio P., Iafrate G., Londero E. & Zorba S., *A complete automatization of an educational observatory at INAF-OATs*, Proc. SPIE conference 10707, 2018

Bonnarel, F., Fernique, P., Bienaymé, O., Egret, D., Genova, F., Louys, M., Ochsenbein, F., Wenger, M. & Bartlett, J. G. 2000, *A&AS*, 143, 33

Demleitner, M., Molinaro, M., Ramella, M., Iafrate, G. & Heinl, H. 2018, *IVOA note*, http://www.ivoa.net/documents/Notes/EDU

Freistetter, F., Iafrate, G., Ramella, M. & the AIDA-Wp5 Team 2010, *CAP Journal*, 10, 18

Hanisch R. J., Quinn P. J., De Young D., Padovani P. & Pasian F. 2010, *IVOA note*, http://www.ivoa.net/documents/Notes/IVOAParticipation

Astronomy in Focus - XXX
Proceedings IAU Symposium No. XXX, 2018
M. T. Lago, ed.

doi:10.1017/S174392131900526X

Transforming research (and public engagement) through citizen science

Samantha Blickhan[1], **Laura Trouille**[1] **and Chris J. Lintott**[2]

[1]Department of Citizen Science, The Adler Planetarium
1300 South Lake Shore Dr. Chicago, IL 60605, USA
emails: samantha@zooniverse.org, trouille@zooniverse.org

[2]Department of Physics, The University of Oxford
Keble Road, Oxford, OX1 3RH, UK
email: chris.lintott@physics.ox.ac.uk

Abstract. Processing our increasingly large datasets poses a bottleneck for producing real scientific outcomes and citizen science - engaging the public in research - provides a solution, particularly when coupled with automated routines. In this talk we will provide a broad overview of citizen science approaches and best practices. We will also highlight in particular recent advances through Zooniverse, the world's largest platform for online citizen science, engaging more than 1.7 million volunteers in tasks including discovering exoplanets, identifying features on Mars' surface, transcribing artist's notebooks, and tracking resistance to antibiotics.

Keywords. citizen science, public engagement, machine learning

---

1. Introduction

As data rates and volumes grow, many research communities are turning to online citizen science as a method of coping with large datasets. Citizen science typically refers to collaborative research involving professional scientists and volunteer participants (Marshall *et al.* 2015). By involving the public in the research process through citizen science activities such as data collection, classification or evaluation, and analysis, we can process massive datasets in a reasonable amount of time while simultaneously engaging the public and promoting a greater understanding of the scientific process.

Zooniverse (www.zooniverse.org) is the largest platform for online citizen science in the world. It began with a single project called Galaxy Zoo, launched in July 2007. The goal of the project was to process just under one million images of galaxies from the Sloan Digital Sky Survey into single morphological categories. Several hundred thousand volunteers participated, and not only did they complete the task in a matter of months, but each image was classified an average of 38 times, instead of the single classification that would have been produced by an individual researcher. The results of the initial Galaxy Zoo project suggested that the general public can reliably classify large sets of galaxies with a similar accuracy to professional astronomers (Lintott *et al.* 2008). The success of Galaxy Zoo led to the foundation of the Zooniverse platform, which currently hosts more than 90 active projects and a volunteer community of 1.7 million registered participants in 234 countries. Among online citizen science platforms Zooniverse is unique, due to its 1) shared open-source software, communication and participation among inter- and multi-disciplinary users; 2) reliable, flexible, and scalable Application Programming Interface (API), which can be used for a variety of development tasks; 3) the Project Builder, a do-it-yourself (DIY) tool which allows anyone to build their own project for free (described further below); and 4) the size and scale of its audience. Zooniverse partners with hundreds of researchers across the disciplines, and has become a compelling option for research framework as citizen science gains popularity around the world. Since the launch of the Galaxy Zoo project in 2007, data from Zooniverse projects have been used in over 150 peer-reviewed publications across many disciplines, including astronomy (Fortson *et al.* 2012, Johnson *et al.* 2015, Lintott *et al.* 2008, Marshall *et al.* 2016, biomedical (dos Reis *et al.* 2015), climate science (Hennon *et al.* 2015, Rosenthal *et al.* 2018), ecology (Matsunaga *et al.* 2016, Swanson *et al.* 2016), and the humanities (Grayson 2016, Williams *et al.* 2014). These publications and the projects from which they stem have set a precedent for online citizen science producing significant, high-quality data.

2. Types of discoveries

With so many eyes on project data, combined with intuitive design and a supportive culture, Zooniverse is in a unique position to support multiple kinds of discovery: the Known Unknowns and the Unknown Unknowns. The former refers to making discoveries on the Zooniverse platform through the classification task; when the intended engagement with the data results in discovery. For example, in April 2017, the Exoplanet Explorers project (Schwamb *et al.* 2012) built on the work of the previous Planet Hunters project, in which citizen scientists examined time series photometry from *Kepler* to identify potential planetary transits. In Exoplanet Explorers, researchers at UC Santa Cruz and Caltech used processed *K2* time series data for candidates identified by machine learning. The project launch was set to coincide with a television program called Stargazing Live, on the Australian Broadcasting Corporation, during which the presenters of the program encouraged viewers to participate online, and gave live updates on the project results. Thousands of citizen scientists volunteered, and on the final night of the three-day program, researchers announced the discovery of a four-planet system by participants in the project. Since then, the near-resonant system has been named K2-138 and the research team have found that has a fifth planet, and possibly a sixth, although additional verification is needed (Christiansen *et al.* 2018). This example illustrates the enormous potential for discovery through crowdsourced data processing, and is one of the main reasons why the Zooniverse platform has been steadily growing for more than 10 years.

From its beginnings, the Zooniverse platform design has not only supported discovery through the classification task, but has also enabled *serendipitous* discovery; a result of having multiple independent examinations of the data, but also taking advantage of our innate human ability to notice things that are out of the ordinary. According to Luczak-Roesch *et al.* (2014), it is not uncommon for citizen scientists to go to great lengths to alert researchers to the existence of an unusual object of interest and/or work on independent research questions based on something they noticed while performing the prescribed main classification task. The Zooniverse 'Talk' discussion forums have been central to these volunteer-led investigations. The forums were intentionally designed to facilitate serendipitous discovery: when a volunteer finishes a classification on any Zooniverse project, they can choose to go directly to the Talk board to discuss the subject they have just seen. An example of the Talk board from the Gravity Spy project can be seen below, in Fig. 1.†

Talk provides a space for a wide range of information sharing and exchange of ideas between volunteers and researchers, enabling collaboration and community building. The interactions in Talk range from the purely social to goal-driven and benefit the project directly, as a means of responding to inquiry, and indirectly, as a way of ensuring long-term engagement (Mugar *et al.* 2014).

† https://www.zooniverse.org/projects/zooniverse/gravity-spy/talk

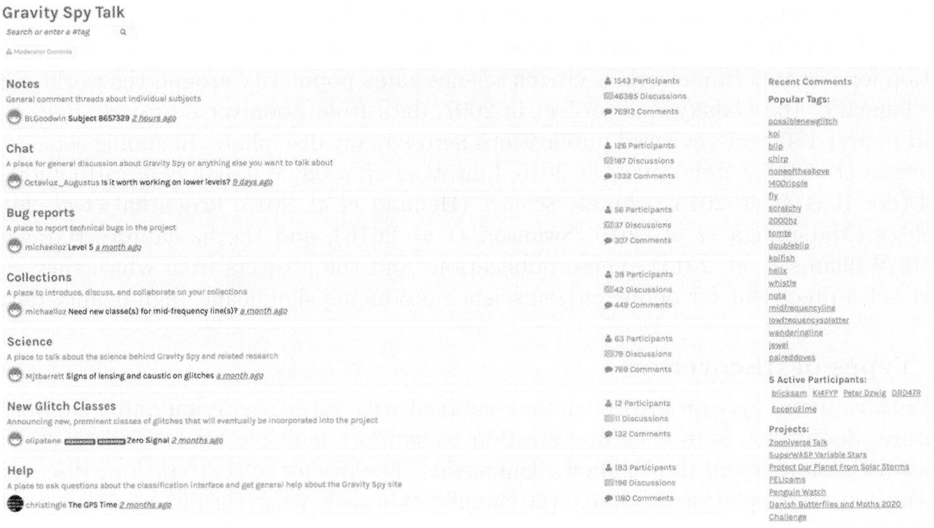

Figure 1. 'Talk' board for the Gravity Spy project.

Over the years, Zooniverse volunteer effort has led to a number of serendipitous discoveries that have been transformative in their respective fields. One thread in particular, posted on the Galaxy Zoo Talk board, led to a discovery which highlights the power of including message boards with each project. While classifying, several volunteers had noticed strange, compact green blobs which looked like peas, and subsequently started a Talk discussion thread titled 'Give peas a chance'. The researchers worked alongside the volunteers (who called themselves the Peas Corps) to refine the collection of objects, ultimately identifying 250 'green peas' in the million-galaxy data set. We now know that these compact green galaxies are low mass, low metallicity galaxies with high star formation rates (Cardamone *et al.* 2009). Yet when the Galaxy Zoo project began, this type of galaxy was not part of the taxonomy of galaxy types, let alone something that the research team was actively searching for with the classification task – a true 'Unknown Unknown'. According to Dr. Carolin Cardamone, one of the lead researchers for Galaxy Zoo, "No one person could have done this on their own. Even if we had managed to look through 10,000 of these images, we would have only come across a few Green Peas and wouldnt have recognized them as a unique class of galaxies."†

The 'Green Peas' are just one example of serendipitous discovery on the Zooniverse platform. Other examples include Hanny's Voorwerp (Lintott *et al.* 2009) and 'Boyajian's Star', KIC 8462852 (Boyajian *et al.* 2016), as well as the re-discovery of a group of 18th century female scientific illustrators on the Science Gossip project‡.

3. Project growth

As rates and volumes of data steadily climb, the number of research teams wanting to build crowdsourcing projects has similarly grown. As noted above, Zooniverse has grown from a single, custom-built project in 2007, to a platform with more than 90 active projects in which volunteers can choose to participate. The accelerated expansion of Zooniverse is a result of the launch in July 2015 of our free Project Builder tool,¶

† https://news.yale.edu/2009/07/27/galaxy-zoo-hunters-help-astronomers-discover-rare-green-pea- galaxies

‡ https://talk.sciencegossip.org/#/boards/BSC0000004/discussions/DSC00004s8

¶ https://www.zooniverse.org/lab

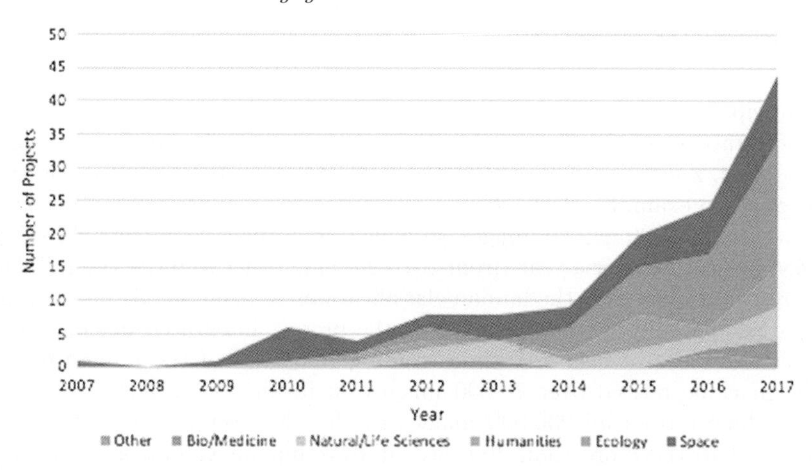

Figure 2. Zooniverse projects launched per year, 2007-2017.

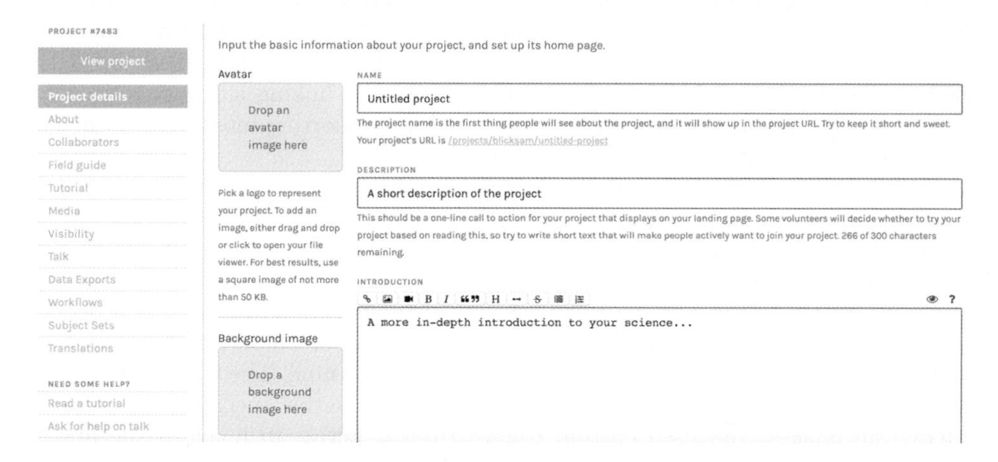

Figure 3. 'Project details' tab of the Zooniverse Project Builder.

which enables anyone to create an online citizen science project using a web browser-based toolkit. Prior to the development of the Project Builder, a typical online citizen science project was entirely custom-built, and required months to years of professional web development time. Fig. 2 shows project growth by year, from 2007 to 2017.

The Project Builder supports common types of interaction including classification, multiple-choice questions, comparison tasks, text entry, marking and drawing tools, or any combination thereof. The Project Builder front-end is a series of forms and text boxes a researcher fills out to create the project's classification interface and website (Fig. 3). All Project Builder projects come with a landing page, classification interface, discussion forum, and 'About' pages for content about the research, the research team, and results from the project.

Since the Project Builder launch, over 1,800 people have attempted building their own project. Some projects go through beta review and are launched publicly, while other projects remain private, for internal use among research teams, community-based projects, etc. One new public project is launched per week. Though we have over 1.7 million registered volunteers, we must acknowledge that this rate of project growth will either require renewed efforts toward recruiting even more volunteers, or finding other ways to affect project efficiency, as described below.

4. Future developments

As we plan to provide support for the growing number of Zooniverse projects, we also need to anticipate huge amounts of data from future surveys, such as LSST, which will produce 30 terabytes of data per night (Ivezic *et al.* 2018). In preparation for massive amounts of data, Zooniverse has explored the use of human-machine systems across several projects. For example, Beck *et al.* (2018) details an experiment with the Galaxy Zoo 2 dataset of 200,000 galaxy images, in which Zooniverse volunteers classified on the Galaxy Zoo project for over a year (from 2010-2011) in order to retire the full dataset. In 2017, the research team used the human classifications to train an off-the-shelf algorithm to simulate what would have happened had the project integrated machine learning in 2010. After four days of near real-time training (days 4-8 of the human classification data), the machine retired over 70,000 images on its first run. Through active learning, the machine retired all 200,000 images by day 30, over a factor of 8 decrease in time compared to the more than 400 days it took human volunteers to complete the dataset.

Zooniverse machine learning efforts are closely aligned with an auxiliary service known as Caesar; a secondary Rails module that monitors classifications in real time, supporting aggregation and subject retirement and promotion. Caesar can set rules and actions based on those rules, such as responsive retirement and linking subjects retired from one workflow to the next logical workflow, as well as supporting integration of machine learning algorithms into the Zooniverse system.

A number of projects have used volunteer classifications to generate training sets for automated methods to efficiently classify remaining data. For example, an early Supernova Hunters project on Zooniverse, using data from the Palomar Transient Factory, retired from the system after the volunteer classifications provided a large enough training set for the researchers' machine learning algorithms to automatically process the remaining data with confidence. Since the implementation of Caesar, projects have been able to utilize human effort concurrently with machine learning. The Camera CATalogue project trained models to provide predicted classifications for species identification in camera trap images, and used custom Caesar rules to retire an image if the first two human volunteers agreed with the model's prediction. Ultimately, this has allowed the project to maintain classification accuracy rates while reducing human effort by 43% (Willi *et al.* 2018).

As another example, the current Supernova Hunters project† uses machine learning to flag Pan-STARRS telescope images as containing potential supernovae candidates. Subjects which the machine deems unlikely to contain supernovae are retired automatically; the remaining subjects are uploaded to the project for volunteer classification. Wright *et al.* (2017) found that the human classifications and machine learning results were complementary; by combining the aggregated volunteer classifications with the machine learning results, they are able to create the most pure and complete sample of new supernovae candidates.

The integration of machine learning into Zooniverse may at first seem antithetical to the existence of a platform dedicated to crowdsourced participation in research. However, we have shown that machine and human *collaboration* can produce superior classification results than either one alone. Integration of machines can also greatly increase the efficiency of a citizen science system, but tensions exist when designing for engagement, inclusion and serendipity. Trouille *et al.* (2018) discuss these tensions in detail, and offer potential solutions, such as the Snapshot Wisconsin team experiment about including blank images in ecology camera-trap projects (Bowyer *et al.* 2015).

† zooniverse.org/projects/dwright04/supernova-hunters

5. Concluding remarks

Zooniverse is a transformative tool for research and public engagement in science. In the eleven years since the launch of Galaxy Zoo, the Zooniverse platform has grown exponentially, a result of steadily-growing data rates and volumes. Recent experiments with machine learning have shown that human-computer collaboration will be essential to the continued health of the platform - and, arguably, the field of citizen science - in years to come.

6. Acknowledgements

This publication uses data generated via the Zooniverse.org platform, development of which is funded by generous support, including a Global Impact Award from Google and by a grant from the Alfred P. Sloan Foundation. The development of Caesar was supported in part by support from STFC under grant ST/N003179/1. We gratefully acknowledge our amazing Zooniverse web development team, the research teams leading each Zooniverse project, and the worldwide community of Zooniverse volunteers who make this all possible.

References

Anderson, TM, *et al.* 2016, *Philosophical Transactions of the Royal Society B: Biological Sciences* 371(1703):20150314
Areta, C, *et al.* 2016, in *European Conference on Computer Vision* 483–98
Barr, AJ, *et al.* 2016, arXiv:1610.02214 [physics.soc-ph]
Beck, MR, *et al.* 2018, *MNRAS* 476(4): 5516–34
Bowyer, A, *et al.* 2015, in *Human Computation and Crowdsourcing: Works in Progress and Demonstrations. An Adjunct to the Proceedings of the Third AAAI Conference on Human Computation and Crowdsourcing*
Boyajian, TS, *et al.* 2016, *MNRAS* 457(4): 3988–4004
Cardamone, C, *et al.* 2009, *MNRAS* 399(3): 1191–1205
Christiansen, JL, *et al.* 2018, *AJ* 155(2): 57
Fortson, L, *et al.* 2012, in *Advances in Machine Learning and Data Mining for Astronomy* (CRC Press): 214-33
Grayson, RS 2016, *British Journal for Military History* 2(2): 160–85
Hennon, CC, *et al.* 2015, *Bul. Am. Met. Soc.* 96(4): 591–607
Ivezic, Z, *et al.* 2018, arXiv:0805.2366v5 [astro-ph]
Johnson, LC, *et al.* 2015, *ApJ* 802(2): 127
Lintott, CJ, *et al.* 2008, *MNRAS* 389(3): 1179–89
Lintott, CJ, *et al.* 2009, *MNRAS* 399(1): 129–40
Luczak-Roesch M, *et al.* 2014, in *ICWSM* aaai.org, http://www.aaai.org/ocs/index.php/ICWSM/ICWSM14/paper/viewFile/8092/8136
Marshall, PJ, *et al.* 2015, *ARAA* 53: 247–78
Marshall, PJ, *et al.* 2016, *MNRAS* 455: 1171–90
Matsunaga, A, *et al.* 2016, *Future Generation Computer Systems* 56: 526–36
Mugar, G, *et al.* 2014, in: *Proceedings of the 17th ACM Conference on Computer Supported Cooperative Work and Social Computing* (New York, NY, USA: ACM): 109–19
dos Reis, FJC, *et al.* 2015, *EBioMedicine* 2(7): 681–89
Rosenthal, IS, *et al.* 2018, *in press*
Schwamb, ME, *et al.* 2012, *apj* 754(2): 129–146
Swanson, A, *et al.* 2012, *Conservation Biology* 30(3): 520–31
Trouille, L, *et al.* 2018, accepted to *PNAS*
Willi, M, *et al.* 2018, submitted to *Methods in Ecology and Evolution*
Williams, AC, *et al.* 2014, in *IEEE International Conference on Big Data* 100–105
Wright, D, *et al.* 2017, in *Monthly Notices of the Royal Astronomical Society* 472(2): 1315–23
Zevin, M, *et al.* 2017, *Classical and Quantum Gravity* 34(6): 064003

Astronomy in Focus - XXX
Proceedings IAU Symposium No. XXX, 2018
M. T. Lago, ed.

doi:10.1017/S1743921319005271

AAS Nova and Astrobites as Bridges Between Astronomy Communities

Susanna Kohler, the AAS Publishing Team, and the Astrobites Collaboration

American Astronomical Society,
1667 K Street NW, Suite 800, Washington, DC 20006, United States
email: susanna.kohler@aas.org

Abstract. Education and outreach in astronomy often focuses on communicating broad astronomical concepts. But how can educators and outreach practitioners also share current astronomical research results with students and the public, conveying both the process of science and the excitement of new discoveries? AAS Nova and Astrobites are two resources freely available to the astronomy community and the general public, intended to help readers learn about the most recent research published across the field of astronomy. Both supported by the American Astronomical Society, these two daily astrophysical literature blogs provide accessible summaries of recent publications in AAS journals and on the arXiv. As both AAS Nova and Astrobites directly distill original studies, these resources constitute a critical bridge between astronomy researchers and educators, outreach practitioners, and the broader astronomy community. The material on these two websites — which includes a total archive of more than 2,500 research study summaries — is written accessibly while still providing access to the original sources and outcomes. As a result, AAS Nova and Astrobites can be used by educators and outreach practitioners to easily introduce the latest in astronomical research studies into classrooms and outreach events.

Keywords. education, outreach

---

1. Introduction

The topic of science communication encompasses multiple channels; scientific researchers must be able to communicate not only with each other, but also, ultimately, with the public. Communication between scientists and the public is important both from a financial perspective — as much of scientific research is publicly funded — and from the perspective of encouraging education and engagement of the public with science.

Traditionally, communication between scientists and the public occurs through intermediary channels, like the media, outreach practitioners, and educators. Under this model, scientific researchers publish academic papers, and intermediary channels must then understand these outcomes and communicate them to the public.

There are a number of places where this process can break down, however. The first is in getting the information from the researchers — who are generally not trained in science communication (see e.g. The Royal Society 2006) — to the intermediaries. Use of field-specific jargon, complex methodology descriptions, a lack of provided context, and additional challenges can often obscure the results of a study to the point where they can't be easily understood by someone outside of the field. A bridge is needed to make this information more accessible.

Another point of communication breakdown can occur between the intermediaries and the public. Science can be misrepresented in the process of being shared with the public,

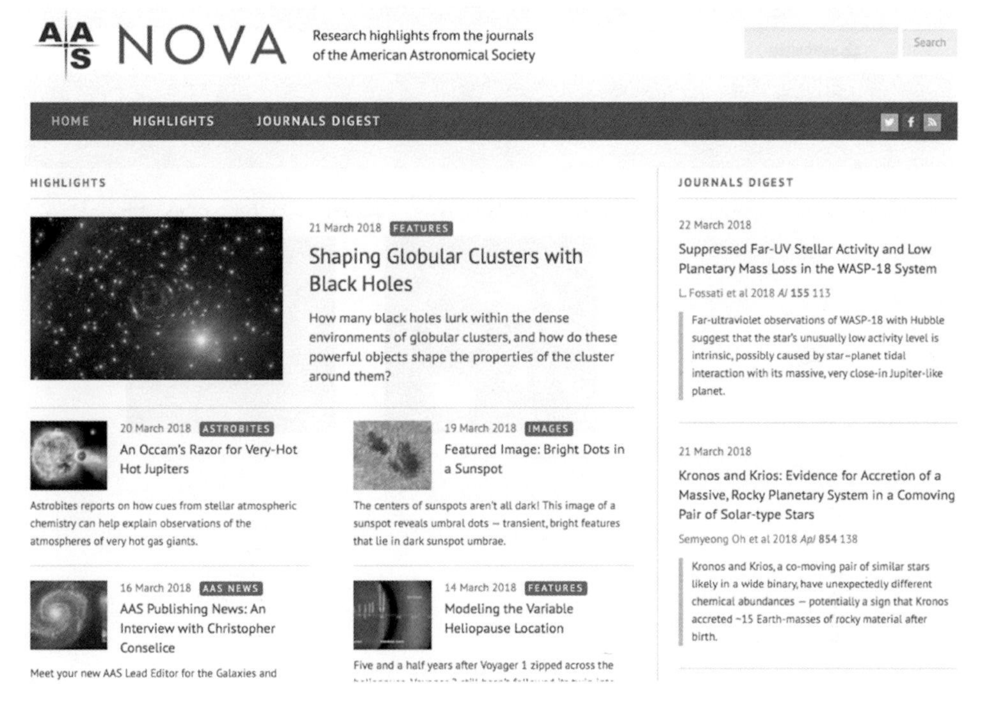

Figure 1. The AAS Nova homepage.

whether by a sensationalized headline or by an oversimplified reporting of a result (see e.g. Caulfield & Bubela 2004). As we caution the public to be more skeptical of the news they consume, it would be beneficial if we could point them to tools that allow a reader to dig deeper when they see a headline they're skeptical of, or even just to find out more about a scientific study that interests them. This requires a further bridge between the astronomy researchers and the public.

Fortunately, these bridges exist. In the following proceeding, we introduce two of them: AAS Nova and Astrobites.

2. Two Resources Available

2.1. *AAS Nova*

AAS Nova (https://aasnova.org; Figure 1) is a website developed by the American Astronomical Society (AAS) in 2015. The main goal of the site is to present curated summaries of recent astronomy research that has been published in the journals of the AAS, with the goal of making this work more accessible to a broad variety of groups. Articles featured on the website are selected by AAS journal editors as research of especially large impact or likely to be of interest to a broad audience. While highlighted articles sometimes include major research results that receive independent press attention, they are more often studies with important but less-flashy results, which might not have otherwise been noticed without the attention brought to them via AAS Nova. AAS Nova highlights are published three times per week and are typically around 400–500 words long. Highlights always link back to the original study, they include and explain figures from the original paper, and they add context and background, emphasizing the main results of the study and why they are important.

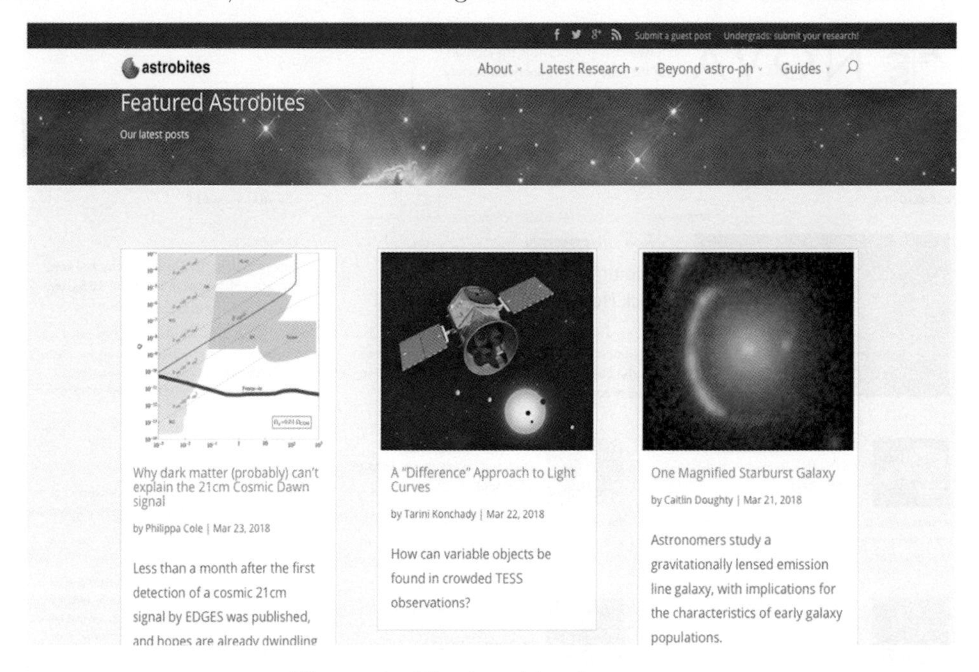

Figure 2. The Astrobites homepage.

2.2. *Astrobites*

Astrobites (https://astrobites.org; Figure 2) is a website that provides summaries of recent astronomy research published on the arXiv astro-ph preprint server. Astrobites was founded in 2011 and is run by a rotating group of graduate students with the specific goal of making astronomy research easier to learn about for undergraduates first starting out in the field. Astrobites is supported by the AAS since 2016. Astrobites articles are published five times per week and often go into greater depth than an AAS Nova highlight; Astrobites posts typically are 1,000–1,500 words in length.

3. Bridging Between Astronomy Communities

Astrobites and AAS Nova can be used in a variety of ways to bridge between various astronomy communities. Here we discuss just a few examples.

3.1. *Media*

The media can use the latest AAS Nova and Astrobites posts to discover new astronomical research and obtain information at an accessible level. To facilitate this process, AAS Nova publishes a biweekly tip sheet of recent AAS Nova posts via the American Astronomical Society press-release distribution service.

3.2. *Education and Outreach*

AAS Nova and Astrobites have been used in outreach activities and undergraduate and graduate classes in a variety of ways. In particular, Astrobites has published multiple studies exploring different approaches to integrating Astrobites articles into undergraduate and graduate astronomy classes (Sanders *et al.* 2012; Sanders *et al.* 2017). The studies include detailed lesson plans for guided reading assignments (Figure 3), literature research projects, and writing projects in which students write their own Astrobites-style

Sample: Astrobites Guided Questions for 'Settling the Proxima Centauri Question'

Read the Astrobite here: https://astrobites.org/2016/11/16/settling-the-proxima-centauri-question/

Provide 1–3 sentence responses to each of the following questions.

* Required

Name / Student ID *

Used to assign grades

Your answer

Question 1: Proxima Centauri is currently the closest known star to our solar system. According to the information in the paper, was this true 300,0000 years ago?

This is a sample question for an entry level astronomy course.

Your answer

Figure 3. Sample questions from a guided reading assignment in the Astrobites lesson plans.

posts. All lesson plans are based on Astrobites resources and are targeted at three levels: lower-level undergraduates, upper-level undergraduates, and graduate students.

3.3. *Preparing Future Science Communicators*

Astrobites and AAS Nova provide an additional bridge of preparing current young scientists to become future science communicators. Through the American Astronomical Society, AAS Nova offers a rotating media fellowship for astronomy graduate students, intended to provide training and experience in science communication. At present, Astrobites has around 100 current and former authors who are astronomy graduate students from around the world, and their tenure writing for Astrobites provides them with experience and peer feedback — elements that have been shown to improve science communication skills (see e.g. Liang & Tsai 2010). In these ways, Astrobites and AAS Nova are helping to further bridge the gap between astronomy researchers and the public in the future, helping to increase the readability of future scientific literature.

4. Conclusion

AAS Nova and Astrobites are two resources — each with a slightly different model — that provide bridges between astronomy researchers and educators, outreach practitioners, and the broader astronomy community. Both sites hope to continue to grow and collaborate, and they welcome feedback from the community. Please reach out to aasnova@aas.org or astrobites@gmail.com to contact the teams.

References

Caulfield, T., & Bubela, T. 2014, *Health Law Rev.*, 12, 2, 53–61

Liang, J.-C., & Tsai, C.-C. 2010, *The Internet and Higher Ed.*, 13, 4, 242–247

Sanders, N., Newton, E., Kohler, S., & The Astrobites Collaboration 2012, *Astron. Educ. Rev.*, 11, 010201

Sanders, N., Kohler, S., Faesi, C., Villar, A., Zevin, M., & The Astrobites Collaboration 2017, *Am. J. Phys*, 85, 10

The Royal Society 2006, *Factors affecting science communication: A survey of scientists and engineers*

Astronomy in Focus - XXX
Proceedings IAU Symposium No. XXX, 2018
M. T. Lago, ed.

doi:10.1017/S1743921319005283

FM14 Session 2: Communicating Astronomy in our Changing World

William H. Waller[1], Lina Canas[2], Hidehiko Agata[3], Hitoshi Yamaoka[3], Shigeyuki Karino[4], Davide Cenadelli[5], Andrea Ettore Bernagozzi[5,6], Jean Marc Christille[5], Matteo Benedetto[5], Matteo Calabrese[5], Paolo Calcidese[5], Richard Gelderman[7], Saeko S. Hayashi[8,9,10], Donald Lubowich[11], Thomas Madura[12], Carol Christian[13], David Hurd[14], Ken Silberman[15], Kyle Walker[16], Shannon McVoy[17], Robert Massey[18], Bogumił Radajewski[19], Maciej Mikołajewski[20,21], Krzysztof Czart[20,21], Iwona Guz[19], Adam Rubaszewski[19], Tomasz Stelmach[19], Rosa M. Ros[22], Ederlinda Viñuales[23], Beatriz García[24], Yuly E. Sánchez[25], Santiago Vargas Domínguez[26], Cesar Acosta[27], Nayive Rodríguez[28], Aswin Sekhar[29,30], Maria Sundin[31], Petra Andersson[32], Christian Finnsgård[33], Lars Larsson[34], Ron Miller[35], Akihiko Tomita[36] and Yogesh Wadadekar[37]

[1]*The Galactic Inquirer*, 243 Granite St., Rockport, MA, USA
email: `williamhwaller@gmail.com`,

[2]IAU Office for Astronomy Outreach / National Astronomical Observatory of Japan, Mitaka, Japan,

[3]National Astronomical Observatory of Japan, Mitaka, Japan,

[4]Kyushu Sangyo University, Fukuoka, Japan,

[5]Fondazione C. Fillietroz-ONLUS Astronomical Observatory of the Autonomous Region of the Aosta Valley and Planetarium of Lignan, Italy,

[6]UNICAMearth Working Group, Geology, University of Camerino School of Science and Technology, Italy,

[7]Western Kentucky University Physics & Astronomy, USA,

[8]TMT-Japan Project Office, National Astronomical Observatory of Japan, National Institutes of Natural Sciences,

[9]Department of Astronomy, SOKENDAI (Graduate University for Advance Studies),

[10]Astronomical Society of Japan,

[11]Department of Physics and Astronomy, Hofstra University, USA,

[12]San José State University, USA,

[13]Space Telescope Science Institute, USA,

[14]Edinboro University, USA,

[15]NASA Goddard Space Flight Center, USA,

[16]South Carolina Commission for the Blind, USA,

[17]Michigan Bureau of Services for Blind Persons, USA,

[18]The Royal Astronomical Society, UK,

[19]TVP Bydgoszcz, Telewizja Polska S.A., Poland,

[20]Polish Astronomical Society, Poland,

[21]Urania - Postpy Astronomii, Toruń, Poland,

[22]Universdad Politécnica de Cataluña, Spain,

[23]Universidad de Zaragoza,
[24]ITeDAM-CONICET-CNEA-UNSAM, UTN Mendoza, Lab. Pierre Auger,
[25]Departamento de Física, Universidad Nacional de Colombia, Bogotá, Colombia,
[26]Observatorio Astronómico Nacional, Universidad Nacional de Colombia, Bogotá, Colombia,
[27]Universidad Nacional de Colombia, Bogotá, Colombia,
[28]Colegio San Agustín IED, Bogotá, Colombia,
[29]CEED, Faculty of Mathematics and Natural Sciences, University of Oslo, Norway,
[30]Armagh Observatory and Planetarium, Armagh, UK,
[31]Department of Physics, University of Gothenburg, Sweden,
[32]Department of Philosophy, Linguistics, Theory of Science, University of Gothenburg, Sweden,
[33]Department of Mechanics and Maritime Sciences, Chalmers University of Technology, Sweden,
[34]SSPA Sweden AB and Chalmers University of Technology, Sweden,
[35]Black Cat Studios, Virginia, USA,
[36]Faculty of Education, Wakayama University, Japan,
[37]National Centre for Radio Astrophysics, Pune, India

Abstract. As the IAU heads towards its second century, many changes have simultaneously transformed Astronomy and the human condition world-wide. Amid the amazing recent discoveries of exoplanets, primeval galaxies, and gravitational radiation, the human condition on Earth has become blazingly interconnected, yet beset with ever-increasing problems of over-population, pollution, and never-ending wars. Fossil-fueled global climate change has begun to yield perilous consequences. And the displacement of people from war-torn nations has reached levels not seen since World War II.

Keywords. astronomy outreach, astronomy education, astronomy for development

---

1. Summary

In this session on "Communicating Astronomy in our Changing World," we endeavored to reconcile the latest challenges in astronomy communication and addressed the role of astronomy outreach within the new social context. We began with a brief invited talk by Oana Sandu of the European Southern Observatory on "Conclusions from the Communicating Astronomy with the Public Conference" that occurred in Japan this past March. We then held a panel discussion on "Effectively Navigating the Media Landscape" featuring Lars Lindberg Christensen from the European Southern Observatory, Rick Fienberg from the American Astronomical Society, and Thiago Goncalves from the Federal University of Rio de Janeiro. The panel discussion was followed by an invited talk on "Effectively Coordinating Museums and Planetariums Worldwide" by Mark SubbaRao from the Adler Planetarium. His talk was followed by another panel discussion on "Supporting the Underserved and Dispossessed" that featured Mike Simmons from Astronomers without Borders, Robert Massey of the Royal Astronomical Society, and Ramasamy Venugopal from the IAU Office of Astronomy for Development. Olayinka Fagbemiro from Astronomers without Borders and the Universe Awareness program in Nigeria was unable to make the focus meeting, and so Sze-leung Cheung gave a brief summary of her important work with Nigerian children. We ended this session with a brief invited talk on "Astronomy and Host Communities" – Considerations of Science,

Culture, Environment and Relationships with Host Communities" by Gordon Squires of the Thirty Meter Telescope project.

2. Supplementary materials

The poster presentations of section 2 can be found online in the format of supplementary materials

- Canas, L. *et al.* Communicating Astronomy with the Public 2018: Efforts on Bringing Together the International Astronomy Communication Community. Supplementary material 2-01.
- Cenadelli, D. *et al.* Light, Water, Life: The Search For New Worlds In The Galaxy. An educational project at Regional level in Italy. Supplementary material 2-02.
- Gelderman, R. Interactive Planetarium Presentations that Support a Personal Understanding of the Cosmos. Supplementary material 2-03.
- Hayashi, S. Bridging the Gap of How One Feels about Large Facilities. Supplementary material 2-04.
- Lubowich, D. Methods to increase the audiences, promote citizen science projects, and include women/underserved groups. Supplementary material 2-05.
- Madura, T. *et al.* Astronomy for Students with Visual Impairments: Development of a Hands-on Career Exploration Lab. Supplementary material 2-06.
- Massey, R. ARAS200: Sky & Earth: Engaging diverse partners and diverse audiences with astronomy: a new approach to public engagement. Supplementary material 2-07.
- Radajewski, B. *et al.* Astronarium TV series as an example of cooperation between astronomers and the media. Supplementary material 2-08.
- Ros, R. *et al.* NASE and Cultural Astronomy: the rescue of the "every day" Astronomy. Supplementary material 2-09.
- Sánchez, Y. *et al.* Astronomy as a tool for inclusion of blind and visually-impaired students at university. Supplementary material 2-10.
- Sekhar, A. Challenges with Gender, Immigration and Diversity for Astronomers from Developing Countries. Supplementary material 2-11.
- Sundin, M. *et al.* Two studies using space sports in education and outreach - Sailing on Titan and Equestrian Sports on Mars. Supplementary material 2-12.
- Tomita, A. The 3D map astro/geo tour with your fingertips. Supplementary material 2-13.
- Wadadekar, Y. Learnings from multilingual astronomy outreach in eight Indian states. Supplementary material 2-14.

Astronomy in Focus - XXX
Proceedings IAU Symposium No. XXX, 2018
M. T. Lago, ed.

doi:10.1017/S1743921319005295

Report on Communicating Astronomy with the Public (CAP) Conference 2018

Oana Sandu[1] **and Sze-leung Cheung**[2]

[1]European Southern Observatory
Karl-Schwarzschild-Strasse 2 D-85748 Garching bei Mnchen Germany
email: osandu@partner.eso.org

[2]Office for Astronomy Outreach, International Astronomical Union,
Office: 310 South Building, National Astronomical Observatory of Japan 2-21-1 Osawa,
Mitaka, Tokyo, 181-8588, JAPAN
email: cheung.szeleung@nao.ac.jp

Abstract. Since 2003, the Communicating Astronomy with the Public (CAP) Conference has facilitated the exchange of ideas and best practices among professionals in the field. This paper reports on the latest edition, CAP 2018, organised in Fukuoka, Japan. It presents a few quantitative outcomes of the conference, the programme and a selection of ideas that were presented and discussed during the meeting. For further details, please consult the Book of Proceedings Communication Astronomy with the Public Conference 2018 2nd Edition, available at: https://www.communicatingastronomy.org/cap2018/

Keywords. communication, outreach, CAP Conference, challenges.

---

1. Presentation of the CAP Conference

Since 2003, the Communicating Astronomy with the Public (CAP) Conference has facilitated the exchange of ideas and best practices among professionals in the field. The CAP 2018 programme included: 5 keynote talks, 24 plenary talks, 141 talks in parallel sessions, 4 unconference slots, 111 posters and 24 workshops.

The local organisation of the 2018 edition was led by the National Astronomical Observatory of Japan (NAOJ) and Fukuoka City, supported by a very strong national and local team of astronomy communicators, city officials and other partners. The Scientific Programme of the conference was led by the IAU Commission C2 CAP Conference Working Group.

2. Data from CAP 2018

In our evaluation of the conference, we gathered data from an online survey among participants, as well as by looking at registration data. The online survey was implemented for the last three editions (2011, 2016 and 2018). The response rate of the 2018 edition was 144 out of 430 participants. The reported percentile gender balance was 33% male, 47% female, and 20% NA.

CAP 2018 was the most well attended edition from the CAP Conference series (see Figure 1). The number of participants almost doubled in 2018 compared with the previous most attended edition. While we did not record data on diversity over time, if we look at the participants' country distribution, the 2018 edition went from 25 to 53 countries represented.

We have observed an increased interest in submitting content. Abstract submissions for talks have increased from 149, for the 2016 edition to 283, for the 2018 edition (see Figure 2).

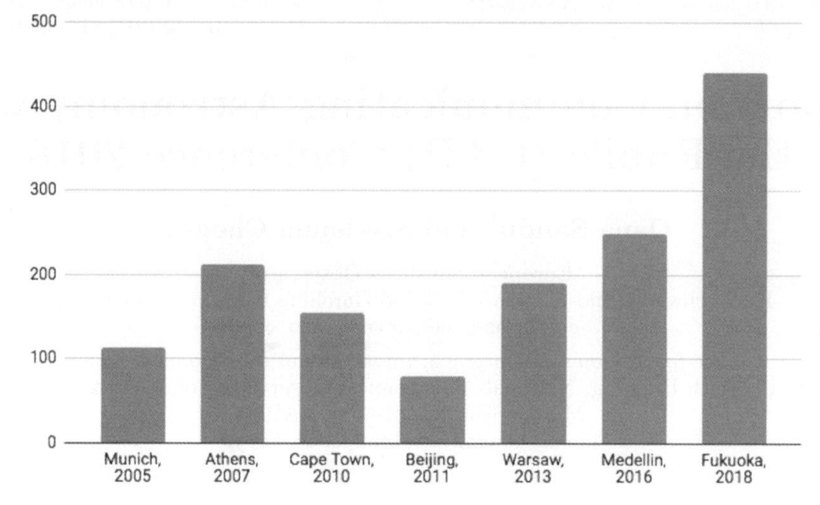

Figure 1. Number of participants per CAP Conference edition. Data gathered by Lucas Ellerbroek.

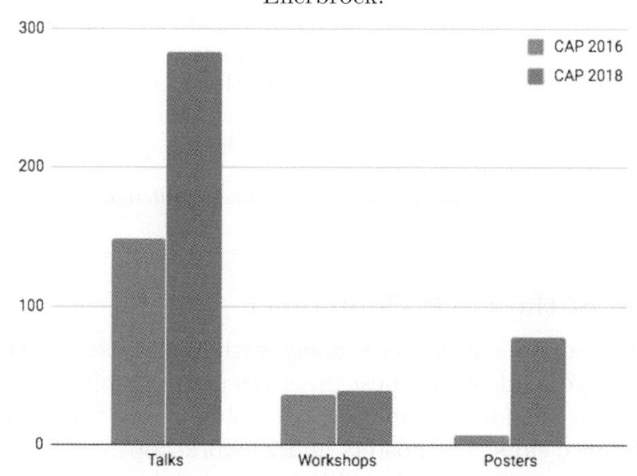

Figure 2. Comparison of content submission between CAP 2016 and CAP 2018. Content is divided by type.

We observed good results on delivering practical content to the community. Over 60% of respondents in the 2018 survey said they left the conference with more than three key learnings (see Figure 3).

3. CAP 2018 Programme

The main theme of the conference was Communicating Astronomy in Today's World: Purpose & Methods, with sub-themes focusing in topics such as: • Current Challenges in Astronomy Communication; • Best Practices in Public Outreach; • Inclusion, Diversity, Equity, and Empathy in Communicating Astronomy; • Astronomy Communication for a Better World; • The Media's Role in Astronomy Communication; • Using Multimedia, Social Media, Immersive Environments, and other Technologies for Public Engagement with Astronomy; • Special Topic: Public Engagement Opportunities during the IAU Centennial Celebration.

We had five invited speakers: • Norio Kaifu, Professor Emeritus of the National Astronomical Observatory of Japan NAOJ, Advisor to the IAU, writer and lecturer, Japan; • Wanda Diaz Merced, Postdoctoral Researcher at the IAU Office of Astronomy

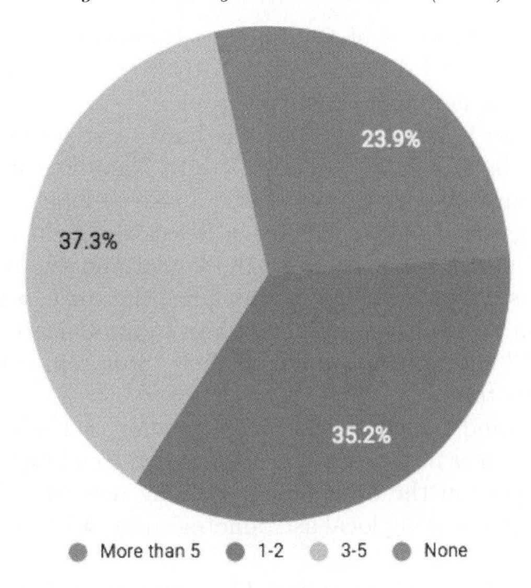

Figure 3. Answers to survey question: "With how many key learnings have you left CAP 2018?"

for Development, Cape Town, South Africa; • Hitoshi Murayama, PhD theoretical physicist, Professor at the University of California, Berkeley, and Director of the Kavli Institute for the Physics and Mathematics of the Universe at the University of Tokyo, Japan; • Dominique Brossard, Professor and Chair in the Department of Life Sciences Communication at the University of Wisconsin-Madison, USA; • Jennifer Ouellette, science writer and author, former science editor of Gizmodo, USA.

4. Challenges in Communicating Astronomy in Todays World—a selection of ideas presented at the conference

(1) Changing media environments that promote narrowcasting, false facts & other toxic behavior

Data from 35 countries show that younger groups are using social media as main sources of news, instead of websites or organizations, TVs or newspapers.

There is increased likelihood of sending around unreliable science news, stories based on a grain of truth, but totally distorted to be more appealing to the public.

The problem does not lie with the social platform itself, but with human psychology. From this perspective, it is important to understand why we share things on social media, some of the reasons being: •We share stories that confirm our beliefs (confirmation bias); •We share stories because they give us hope (even though not realistic); •We share stories because they are funny (not necessarily accurate).

Narrowcasting — the idea that media will make sure that you see things that you are likely to check out based on your previous online behavior— seems to be a technique that will be more and more used. This approach creates the filter bubbles, a major problem for science communication as it is hard to reach the non-converted. In trying to answer the question of who is responsible for identifying false facts and stop their spreading, a few suggestions were made: • Readers could follow a few simple steps to check information; • Press officers could check the media coverage and make sure it is correct; • Scientists could have public engagement as part of their job and they could be trained to do communication; • Journalists could use some of the tools developed for fact checking.

Refer to Brossard (2018) Book of Proceedings Communication Astronomy with the Public Conference 2018 2nd Edition, Japan: NAOJ, "Communicating Science in New Media Environments" for more details on this topic.

(2) Social media noise is increasing and it is hard to stand out. Storytelling can be more persuasive than lists of facts, because stories communicate concepts and values in a format that humans are wired to recognize and remember. Therefore, a few pieces of advice in managing social media include: • Select the right platform and the right amount; • Develop a narrative (a protagonist, a goal and the development); • Adapt your language; • Focus on one message; • Humanize the story; • Evaluate and adjust.

Refer to Heenatigala (2018) Book of Proceedings Communication Astronomy with the Public Conference 2018 2nd Edition, Japan: NAOJ, "Storytelling through Social Media" for more details on this topic.

Other methods to stand out on social media include: • Keeping a balance between astronomical content such as news, discoveries, facts and trending topics, on the one hand, and promotional content, on the other hand; • Using non-related topics to talk about astronomy (sports etc.); • Sharing local astrophotography; • Finding and promoting local heroes; • Using comics.

Refer to Riaza (2018) Book of Proceedings Communication Astronomy with the Public Conference 2018 2nd Edition, Japan: NAOJ, "Astronomy and its Digital Sex appeal: the Art Behind Making People Fall in Love through the Social Networks" for more details on this topic.

(3) Our own communication is sometimes faulty

Concerns were raised that science communication itself has its own problems: • The way we visualize scientific results: Astronomical observations are mainly made of numbers representing the incoming data, which are often converted into images, in order to be better understood. The display process implies a set of rules for encoding the information in a visual form, but this code is not always acknowledged and can lead to misinterpretation. • Misuse of concepts and words: Media reports on exoplanetary findings are often over-hyped, making comparisons with Earth and habitability that current data cannot support. We should be paying more attention to using terms such as "Habitable Zone", "Earth-like" etc.

Refer to Varano (2018) Book of Proceedings Communication Astronomy with the Public Conference 2018 2nd Edition, Japan: NAOJ, "Far From Reality: Scientific Visualization," S. Varano and **?** "We Have Not Found Earth 2.0: Debunking the Media, for more details on this topic."

(4) We face global challenges, but we also face local challenges

Language continues to be a barrier in spreading existing good content. Efforts to set up a global Astronomy Translation Network should not underestimate aspects such as: motivating volunteer translators, finding good content and evaluating translations.

Refer to Shibata *et al.* (2018) Book of Proceedings Communication Astronomy with the Public Conference 2018 2nd Edition, Japan: NAOJ, "Astronomy Translation Network: the Challenges of Translating Astronomy Resources Globally" and the same authors' contribution to the FM 14 proceedings for more details on this topic.

There is a vacuum in science communication in developing countries, due to lack of investment in outreach. This gap is filled with an inflow of information from press offices in wealthier nations. The end result is a lack of awareness and/or interest in science, or alternatively an impression of monopoly in scientific discoveries by developed economies. In Brazil, a project was started to address this problem by: • Ceating a working group acting as liaison between the academic community and the press; • Focusing on the local aspect of astronomical research, highlighting national contribution and Brazilian scientists; • Collaborating with popular social media users by offering them advice to increase science accuracy.

Refer to Goncalves *et al.* (2018) Book of Proceedings Communication Astronomy with the Public Conference 2018 2nd Edition, Japan: NAOJ, "Astronews: Scientific Journalism in Developing Countries" for more details on this topic.

(5) Astronomy has the opportunity to become more inclusive and bring different communities together.

Numerous examples were given at the conference on how to reach different communities. Among such activities and recommendations are:

Moving beyond the visual by developing hands-on activities to communicate astronomy through other senses: • smelling and tasting different molecules that have been found in the Universe; • touching meteorites and through them learning about physical characteristics of objects; • hearing and feeling vibrations of energy released by astronomical objects at different wavelengths.

Refer to De Leo-Winkler *et al.* (2018) Book of Proceedings Communication Astronomy with the Public Conference 2018 2nd Edition, Japan: NAOJ, "Sensing the Universe: Outreach Activities for Inclusion" for more details on this topic.

• Considering adopting universal design for products and content by making them suitable for people, regardless of age, ability, or status in life. • Tackling challenging problems. The unfamiliar and overstimulating environment of a planetarium can be turned to a friendly space for visitors with Autism Spectrum Disorder. • Expanding the reach towards less common communities: people at borders in conflict areas, prisoners, long-term hospitalised people etc.

Refer to Book of Proceedings Communication Astronomy with the Public Conference 2018 2nd Edition, Japan: NAOJ for more details on all of these topics.

5. Final remarks

The book of proceedings for the Communicating Astronomy with the Public Conference 2018 2nd Edition, Japan: NAOJ was made possible by the CAP 2018 LOC team with support from the SOC team. It is available at: https://www.communicatingastronomy.org/cap2018/ where more information on all of the topics presented here and more can be found.

Further reflections and lessons learned on developing and implementing the CAP 2018 conference can be found in the presentation by L. Canas *et al.* on "Communicating Astronomy with the Public 2018: Efforts on Bringing Together the International Astronomy Communication Community" in the supp. materials for the proceeding of FM 14.

Astronomy in Focus - XXX
Proceedings IAU Symposium No. XXX, 2018
M. T. Lago, ed.

doi:10.1017/S1743921319005301

Effectively Coordinating Museums and Planetariums Worldwide

Mark U. SubbaRao[1]

[1]Adler Planetarium,
1300 S, Lake Shore Drive, Chicago IL, USA
email: msubbarao@adlerplanetarium.org

Abstract. Informal science educators at museums and planetariums face the challenging task of engaging a diverse public audience in contemporary science. To do this they need a solid background in the science itself, educational pedagogy, and modern practices in science communication. The task has gotten even more challenging in the era of big data. Interpreting and visualizing these datasets in planetarium shows and museum exhibits requires specialized technical skills. Furthermore, the increasing pace of discovery means that informal science educators have less time to accomplish these tasks. This presentation will summarize a variety of museum and planetarium community efforts to address these challenges through worldwide collaboration and coordination among museums and planetariums. Solutions include content sharing and distribution mechanisms as well as networking museums and planetariums together to create global worldwide events.

Keywords. Keyword1, keyword2, keyword3, etc.

---

1. Introduction

Museums and planetariums have a tremendous potential for communicating astronomy with the public. These free-choice learning environments are unbound by the content restrictions inherent in formal education. The nature of the museum and planetarium experience allows for more in depth exploration than is often possible in other informal education formats (e.g. traditional media and social media). The presence of trained facilitators and presenters allows for the creation of an experience that is both exploratory and explanatory. This facilitated exploration has been termed "Exploranation" (Ynnerman *et al.* 2018). As the pace of scientific discovery accelerates, and as the flow of new data increases the role of science museums and planetariums needs to change. They should transform themselves from presenters of well established scientific truths, to places where the community can come together with experts in a facilitated 'exploranation' of the latest discoveries.

Museums and planetariums reach a large worldwide audience. There are approximately 3,000 science centers in the world serving an audience of roughly 310 million people each year (Mechelen Decleration 2014). A rough estimate gives 4,000 planetariums in the world, with an annual attendance of 150 million (Petersen, M., 2018). These institutions are scattered all over the world. (Fig. 1) Progress is being made in establishing new institutions in the under represented areas of the world. At the 2018 International Planetarium Society Biennial Congress the first meeting of the African Planetarium Association took place (Padavatan 2018). This group will help nurture and grow the African planetarium community.

Informal science institutions generally benefit from a high level of public trust. This is especially important in an era when trust in the traditional sources of information such

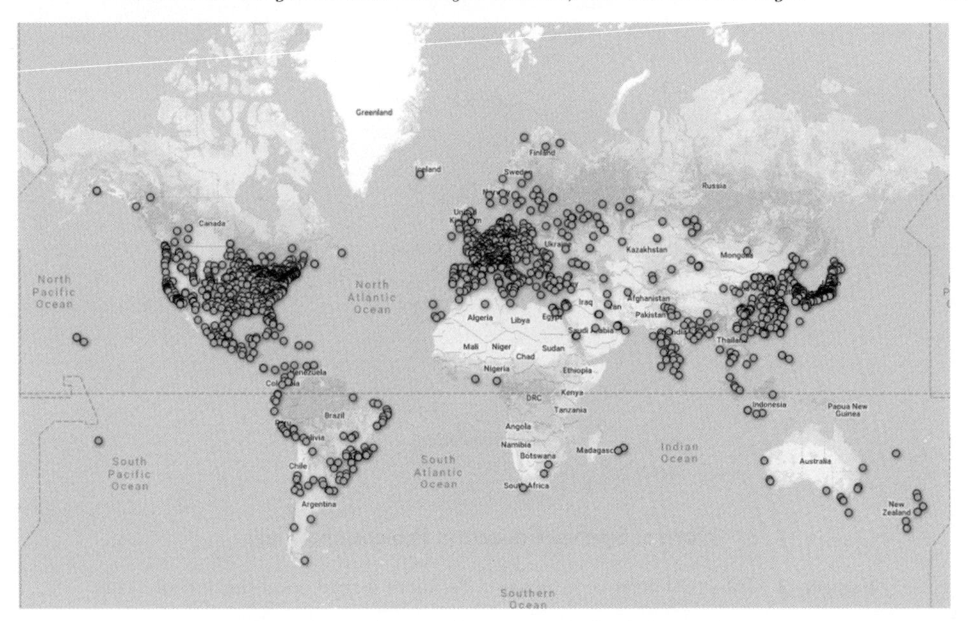

Figure 1. A map of the worlds planetariums.

news media and politicians are low (Leiserowitz *et al.* 2010). This high level of public trust is especially important when dealing with potentially controversial topics such as the Big Bang, evolution, or climate change.

2. Challenges

This article lays out an optimistic vision of the potential for sciences museums and planetariums to radically change the public's understanding, appreciation, and interest in science. There are some serious challenges in reaching that vision. Chief among them are the burdens placed on museum and planetarium staff. These positions require a combination of skills, a solid background in the science itself, educational pedagogy, and modern practices in science communication. In addition 'data science' skills are increasingly important as we shift towards communicating more immediate and more data intensive scientific results.

Professional development is critical to combating this challenge. Since 2002, JPLs Informal Education group has run NASAs Museum Alliance (Sohus 2006), providing museums and other informal education institutions with access to NASA staff, resources, and professional development. More than 700 organizations around the world are Museum Alliance members. The Museum Alliance's primary product is professional development teleconferences with NASA scientists and other experts.

While certain trends, like an increasing pace of scientific discovery, are making the job of the Informal Science Educator more difficult, there are others making it easier. In particular there has been development in our understanding of how effective certain science communication strategies are.

Figure 3 shows the relationship between the traditional 'deficit model' of science communication, the 'engagement model' and the 'participatory model'. These should not be thought of as a hierarchy, and in general an institution should employ all three models. However when possible museums should try and move from the one way flow of information in the 'deficit model' to the two–way flow of information in the 'engagement model'. The 'participatory model' is enabled by citizen science projects such as the Zooniverse

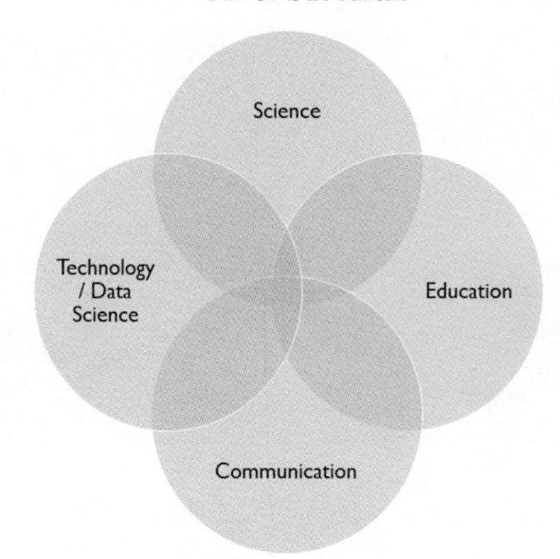

Figure 2. Informal science education requires a rare combination of skills.

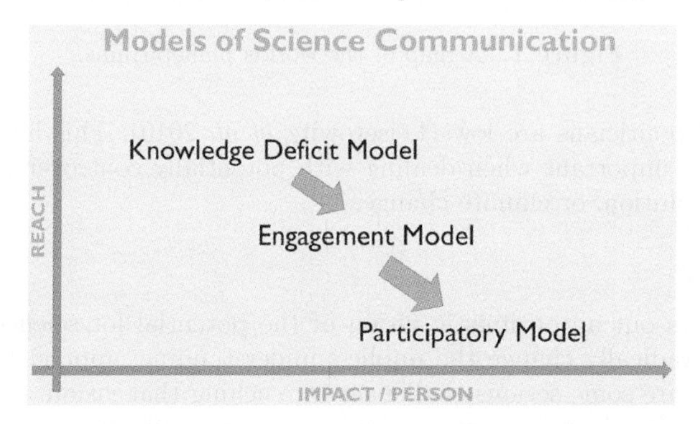

Figure 3. Science communication models

(Smith *et al.* 2013). These projects provide opportunities for museum and planetarium visitors to contribute to scientific projects.

3. Technology

Modern museum exhibits have been transformed by technology. Traditional optomechanical planetariums have largely been replaced by digital projection systems. No longer limited to the night sky, the modern digital planetarium is capable of presenting any astrophysical phenomenon. The open data revolution means that more and more of this astronomical data is publicly available, much of it streamed almost instantaneously. As the LSST comes on line and ushers in an era of time domain astronomy. Planetariums and museum exhibits will be capable of pulling in, processing, and visualizing these data streams. Figure 4 illustrates a planetarium (the Grainger Sky Theater at the Adler Planetarium) being used for big data visualization. At the core of realizing the potential of this data rich future is a philosophical shift in the role of the planetarian or museum facilitator. From that of a curator of astronomical data to that of an 'astronomical weatherperson' - an interpreter of the continuous flow of information coming from telescopes, space missions, and computer simulations. Hopefully this will also inspire a shift in how the public views a planetarium or science center from a place to be visited

Figure 4. The modern digital planetarium has become a world class immersive data visualization facility. (Photo Nick Ulivieri)

once in a lifetime, to a place that one should visit frequently to keep abreast of our growing understanding of our place in the universe.

4. Data to Dome

In the planetarium community it has been recognized that the field needs to evolve to thrive in the big data era. This evolution requires both the development of technical infrastructure as well as the professional development of planetarium staff. To accomplish this the International Planetarium Society established the Science and Data Visualization Task Force. The mission of the Science and Data Visualization Task Force is to streamline the process of going from data to dome, increasing the potential for scientific communication and storytelling in the planetarium.

The task force has undertaken initiatives aimed at:

• Preparing planetaria for the big data streams that will come from next generation telescopes, satellites, experiments and computational simulations.

• Creating professional development opportunities aimed at developing more 'data savvy' planetarians.

• Developing and promoting best practices for data visualization in the dome.

• Connecting data suppliers with vendors and planetarium end-users by setting, and recommending, standards for real-time scientific content distribution.

• Encouraging the visualization of a wide range of scientific data in the dome (moving beyond astronomy).

• Advocating for the inclusion of dome visualization tools in standard scientific analysis and visualization packages.

• Encouraging planetaria to make their facilities available to researchers from their communities to use as a visualization tool.

As a result of these initiatives a new data streaming standard was developed (data2dome). Spearheaded by the European Southern Observatory, this standard provides a mechanism for content providers to deliver content directly to planetariums. On the same day as the press release of a major discovery, planetarium operators will find planetarium content related to that discovery on their console. This includes not only the visual assets to display on the dome, but also the contextual information necessary to interpret them effectively.

Figure 5. Domecasting synchronizes planetariums around the world, enabling the creation of global science communication events.

5. Domecasting

This article has outlined how museums and planetariums can coordinate through the sharing of resources and professional development opportunities. However a more direct form of coordination involves linking museums and planetariums together through networked events.

Domecasting is the synchronizing of planetariums around the world. This technology was first demonstrated at the 2008 International Planetarium Society conference by the planetarium company SCISS, where they connected a portable planetarium in Chicago with the Ghana Planetarium in Accra. One decade later the technology is now quite mature.

The largest domecasting event is the Kavli Fulldome Lecture Series. This lecture happens twice a year featuring a prominent astronomer. They are produced by the Adler Planetarium and simulcast to a network of approximately 30 planetariums around the world.

One challenge of a worldwide presentation are timezones. Currently the Kavli Fulldome Lectures are performed twice to accommodate North and South America, Europe and Africa. Language is also a challenge, the Adler is experimenting with crowdsourcing the simultaneous translations among different museums.

Simulcasting allows the presentation to reach far more people than it would in a single planetarium. This justifies more developmental effort in the lecture, which leads to a higher quality experience. Domecasting also provides an opportunity for people to participate in the lecture in rural areas, or parts of the world that are infrequently visited by top scientists. Perhaps most important is the sense of global community that is created by these live connections.

6. Conclusions

Because of the number of people they reach and the public trust they hold, museums and planetariums are in a unique position to elevate the public's understanding and interest in science globally. To accomplish this it is imperative that museums and planetariums develop both their technical infrastructure to display and visualize scientific data streams, and the scientific communication skills of the staff that help the public interpret these data. To accomplish this we need to effectively coordinate our resources and expertise across the global museum and planetarium community.

Discussion

W. Waller: How could the IAU best partner with the IPS for optimal outreach?

M. SubbaRao: We are actively looking for ways to strengthen the relationship between the IAU and the IPS. My invitation to give this invited talk at this focus meeting is a reflection of that. This involves working with the OAO and IAU Comission C3 to include planetarium content at the CAP meetings. Increasingly I hope to work with the OAD to better support planetariums in emerging communities around the world.

References

Leiserowitz, Anthony and Maibach, Edward and Roser-Renouf, Connie and Smith, N 2010, *Yale and George Mason University. Yale Project on Climate Change*
Mechelen Decleration, 2014 Science Center World Summit 2014
Padavatan, J. 2018, *Planetarian, The Journal of the Internationl Planetarium Society*, 47, 3
Petersen, M. 2018, *Tallying the Worlds Planetariums, Loch Ness Productions*
Sohus, A. 2006, *Bulletin of the American Astronomical Society*, 39,1306
Smith, Arfon M and Lynn, Stuart and Lintott, Chris J 2013, *First AAAI conference on human computation and crowdsourcing*
Ynnerman A., Lowgren J., and Tibell L. 2018, *IEEE Computer Graphics and Applications*, 38, 3

Astronomy in Focus - XXX
Proceedings IAU Symposium No. XXX, 2018
M. T. Lago, ed.

doi:10.1017/S1743921319005313

FM14 Session 3: The IAU National Outreach Coordinators (NOCs) Network – Coordinating and Catalyzing Astronomy Outreach Worldwide

Sze-leung Cheung[1], Prospery C. Simpemba[2], Zouhair Benkhaldoun[3], Martin Aubé[4], Ismael Moumen[5], Raid M. Suleiman[6], Krzysztof Czart[7], Tomasz Brudziński[7], Paweł Grochowalski[7], Agnieszka Nowak[7], Dawid Pałka[7], Krzysztof Pęcek[7], Artur Sporna[7], A. B. Morcos[8], Zara Randriamanakoto[9,10], L. Randrianjanahary[9,11], N. Randriamiarinarivo[9,11], Andrea Sosa[12], Fernando Albornoz[13], Adrian Basedas[13], Carlos Fariello[13], Daniel Gastelu[13], Fernando Giménez[22], Andrea Maciel[15], Oscar Méndez[16], Katyuska Motta[20,17,18], Valentina Pezano[12,13], Reina Pintos[14,19,13], Jorge Ramírez[19], Daniel Scarpa[21] and Hitoshi Yamaoka[23]

[1]IAU Office for Astronomy Outreach, National Astronomical Observatory of Japan,
email: cheungszeleung@iau.org,

[2]Copperbelt University, Kitwe, Zambia

[3]Oukaimeden Observatory, LPHEA, FSSM, Caddi Ayadd University, Marrakech, Morocco

[4]Cégep de Sherbrooke, Sherbrooke, Canada

[5]Université Laval, Québec, Canada

[6]Harvard-Smithsonian Centre for Astrophysics, Boston, USA

[7]Urania - Postpy Astronomii, Toru, Poland

[8]National Research Institute of Astronomy and Geophysics, Helwan, Egypt and KCSCE

[9]Malagasy Astronomy & Space Science, Antananarivo, Madagascar

[10]Department of Astronomy, University of Cape Town, South Africa

[11]Department of Physics, University of the Western Cape, South Africa

[12]Centro Universitario Regional del Este (CURE), Universidad de la República (UdelaR)

[13]Consejo de Educación Secundaria (CES)

[14]Inspección Nacional de Astronomía

[15]Observatorio Astronómico Los Molinos (OALM), D2C2 - Ministerio de Educación y Cultura (MEC)

[16]Planetario de Montevideo *Agrimensor Germán Barbato*

[17]Consejo de Formación en Educación (CFE)

[18]Consejo de Educación Inicial y Primaria (CEIP)

[19]Instituto de Profesores *Artigas* (IPA)

[20]Instituto de Formación Docente de Rocha (IFD Rocha)

[21]Planetario Móvil *Kappa Crucis*

[22]Productora radial *Telecosmos*

[23]National Astronomical Observatory of Japan

Abstract. Section 3 of the FM14 focus on the The IAU National Outreach Contacts (NOC) Network. This paper also contains supplementary materials that point to poster presentations that can be found online.

Keyword. astronomy outreach

---

1. What is NOC?

The National Outreach Coordinator (NOC) is the principal national representative appointed under the IAU structure responsible for the implementation of the proposed IAU outreach initiatives at the national level. The NOCs form a global network and is managed by the IAU Office for Astronomy Outreach (OAO).

Responsibilities of the NOCs
• To coordinate the IAU outreach initiatives (such as the IAU 100th anniversary activities) at the national level, to develop plans and action lists.
• To build and maintain a database of outreach experts mapped with their skills.
• To disseminate IAU communications and distribute them nationally.
• To maintain a database of national astronomical organizations at all levels.
• To maintain the relationship with the national communities of amateur astronomers/outreach professionals with the IAU.

2. Panel discussions

In this session, the NOCs break into small group panel discussions, namely - (1) Asian and Pacific Region, (2) Middle East and African Region, and (3) American and European Region, reporting their activities.

3. Supplementary materials

The following poster presentations of section 3 can be found online in the format of supplementary materials
• Simpemba, P. Panel discussion of NOC activities in Middle East and Africa Regions. Supplementary material 3-01.
• Benkhaldoun, Z. *et al.* Atlas Dark Sky Reserve. Supplementary material 3-02.
• Czart, K. *et al.* AstroGPS - database of all astronomy and space related events in Poland with mobile app. Supplementary material 3-03.
• Morcos, A. Activities in Egypt in the Frame Work of NOC. Supplementary material 3-04.
• Randriamanakoto, Z. *et al.* Growing Astronomy Outreach in Madagascar. Supplementary material 3-05.
• Sosa, A. *et al.* Astronomy Outreach Activities in Uruguay. Supplementary material 3-06.
• Yamaoka, H. Astronomical Activities in Japan: NOC report. Supplementary material 3-07.

Discussion

W.Waller: Several NOCs are involved in National Astronomy Olympiads. Can you provide more information on this, and how they could be more coordinated internationally?

NOCs: Members of the audience noted that there are two National Astronomy Olympiads, and there have been problems coordinating them.

Aniket Sule: The International Olympiad on Astronomy and Astrophysics (IOAA) has always desired close cooperation with the IAU. Presently, 45 national teams participate in the IOAA. On the other hand, the International Olympiad on Astronomy (IAO) is down to about 15 teams. The fact that the IAO chairman does not want to discuss a possible merger or other coordination of the two events should not come in the way of cooperation between the IAU and the IOAA.

Astronomy in Focus - XXX
Proceedings IAU Symposium No. XXX, 2018
M. T. Lago, ed.

doi:10.1017/S1743921319005325

FM14 Session 4: Outreach Action and Advocacy in the Context of IAU's 2020-2030 Strategic Plan

Sze-leung Cheung[1], William H. Waller[2], Yukiko Shibata[3], Kumiko Usuda-Sato[3], Berenice Himmelfarb[1], Lina Canas[1], Hidehiko Agata[3], Nuno R. C. Gomes[4] and Rosa Doran[4]

[1]IAU Office for Astronomy Outreach, National Astronomical Observatory of Japan, email: cheungszeleung@iau.org,

[2]Rockport Public Schools and *The Galactic Inquirer*, Rockport, MA, USA, [3]National Astronomical Observatory of Japan, Mitaka, Japan, [4]NUCLIO – Ncleo Interativo de Astronomia, Portugal

Abstract. Section 4 of the FM14 focus on the outreach action and advocacy in the context of IAUs 2020-2030 Strategic Plan. This paper also contains supplementary materials that point to contributed talks and poster presentations that can be found online.

Keywords. astronomy outreach, astronomy education, astronomy for development

---

1. Summary

In this session on "Outreach Action and Advocacy in the Context of IAUs 2020-2030 Strategic Plan," was started by an overview of the the the IAU 2020-2030 Strategic Plan by the Incoming IAU President Ewine van Dishoeck, followed by a few related talks corresponding to the action items on the strategic plan such as translations and dark-sky protection. A panel discussion was followed with panelist represented the Division C, Commission C2, IAU Office for Astronomy Outreach and IAU Office of Astronomy for Development.

2. Discussions

These reflections were aired in the last Panel Discussion on "Outreach Action and Advocacy in the Context of IAU's 2020-2030 Strategic Plan."

• Incoming IAU President Ewine van Dishoeck emphasized the theme of our home planet as the Pale Blue Dot, from which we make our cosmic explorations while taking better care of our home planet.

• Sylvie Vauclair recommended that our outreach efforts promote critical thinking. Bill Waller agreed with Sylvie that advancing critical thinking provides an important lens for focusing our outreach efforts.

• Rick Fienberg noted that Commission C2 and the OAO are very close in their missions. There continue to be the CAP Journal and CAP Conference Working Groups. He sees a need to sort out the purviews of these activities.

• Kevin Govender urged us to make major contributions to the UN's sustainability goals and to develop measures of success.

• Sze-leung Cheung said we need to plan to set measurable goals. For example, he likes the prospect of coordinating and developing translation glossaries in astronomy.

• Rick Fienberg opined that outreach differs from education in terms of intent. Are you trying to build awareness and engagement, or you trying to teach content and skills?

• Kevin Govender reminded us that the OAD provides seed funding for projects that build capacity.

• Thilina Heenatigala wanted to see more synergies between the IAU offices.

• Kevin Govender responded by stating that we will need to combine our opportunities and communicate them together.

• Incoming Division C President Susanna Deustua agreed that we have yet to coordinate the offices and so address key opportunities. In closing, she urged us to "apply skills to our passions."

3. Supplementary materials

The following contributed talk of section 4 can be found online in the format of supplementary material

• Shibata, Y. *et al.* The Astronomy Translation Network: Outreach Action and Advocacy in the Context of IAU's 2020-2030 Strategic Plan. Supplementary material 4-01.

The following poster presentations of section 4 can be found online in the format of supplementary materials

• Canas, L. *et al.* IAU and the Public: IAU Office for Astronomy Outreach Communications. Supplementary material 4-02.

• Gomes, N. *et al.* Dark Skies Rangers: a flagship to bridge the gap between schools and communities. Supplementary material 4-03.

• Waller, W. Towards an IAU Position and Statement on Terrestrial Climate Change. Supplementary material 4-04.

Astronomy in Focus - XXX
Proceedings IAU Symposium No. XXX, 2018
M. T. Lago, ed.

doi:10.1017/S1743921319005337

The IAU Strategic Plan for 2020-2030: OAO

Ewine F. van Dishoeck[1] **and Debra Meloy Elmegreen**[2]

[1]Leiden Observatory,
P.O. Box 9513, NL-2300 RA, Leiden, the Netherlands
email: ewine@strw.leidenuniv.nl

[2]Dept. of Physics & Astronomy, Vassar College,
Poughkeepsie, NY 12604 USA
email: elmegreen@vassar.edu

Abstract. The IAU Strategic Plan for 2020-2030 presents an overview of all of the activities of the IAU along with priorities, key goals, mandates, and specific actions. Here future plans and goals are outlined for the Office of Astronomy for Development (OAO).

Keywords. editorials, notices

---

1. Introduction

The first formal Strategic Plan of the International Astronomical Union for the decade 2010-2020, was focused on the then-new Office of Astronomy for Development (OAD), which uses astronomy in developing countries to impact UN Sustainable Development Goals (SDGs). The IAU Executive Committee (EC) decided that the new Strategic Plan (SP) for 2020–2030 should encompass all activities of the IAU. The EC working group charged to write the new SP included Ewine van Dishoeck, Debra Elmegreen, Piero Benvenuti and Renée Kraan-Korteweg and received extensive input from its membership and the various Offices, in particular from Sze-leung Cheung and his team on the OAO part.

The final Strategic Plan 2020-2030 was presented for adoption as Resolution A1 at this XXXth GA, where it was approved at the second Business Meeting (Fig. 1). The revised mission of the IAU is to "promote and safeguard astronomy in all its aspects (including research, communication, education, and development) through international cooperation," where the additional parenthetical aspects make explicit that the IAU has branched out beyond its original purpose of fostering scientific communication and exchange of ideas among professional astronomers.

2. OAO and other Offices of the IAU

The 3 current offices of the IAU were formed within the last decade to carry out additional aspects of IAU activities beyond the advancement of astronomy. The Office for Astronomy Outreach (OAO) is a partnership with the National Astronomical Observatory of Japan, located at the Mitaka campus in Tokyo, and focuses on engaging with the public, providing access to astronomical information and astronomy communication. It maintains a network of National Outreach Contacts (NOCs) and amateur groups. The OAO interacts closely with the OAD, hosted in Cape Town, South Africa, which focuses on the use of astronomy for development by capitalizing on the field's scientific, technological and cultural links and its impacts on society. The Office of Young Astronomers, hosted by the Norwegian Academy of Science and Letters, focuses on the training of young astronomers at University level, and organizes the International School for Young Astronomers (ISYA). A new office is proposed: the Office of Astronomy for

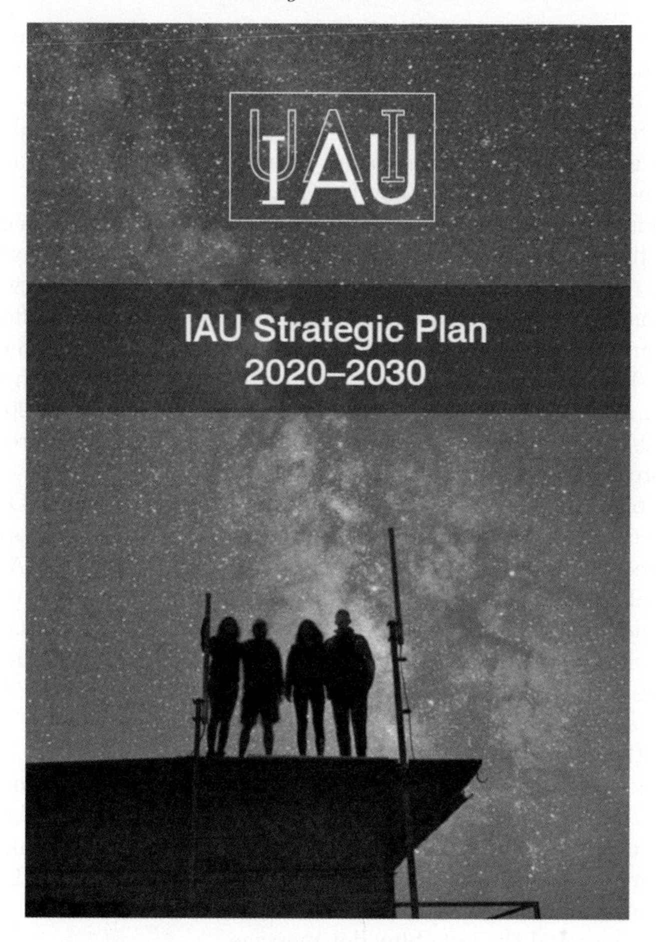

Figure 1. Cover of the new IAU strategic plan, illustrating the importance of outreach, engaging with the general public and dark skies.

Education (OAE) will focus on providing training and resources for using astronomy as a stimulus for teaching and education from elementary to high school level (astronomy and science education). Each of these offices has its own strategic plan, and each office has a liaison on the EC to ensure that the strategic plans align with the overall goals of the IAU. Sometimes activities span more than one office, so having connections with the EC helps coordinate overlapping activities.

3. Office for Astronomy Outreach (OAO)

The Office for Astronomy Outreach (OAO) coordinates public outreach activities and communication of science to the public. Its long-term vision is: that all people throughout the world will have access to knowledge of frontline astronomy; that all countries will have good access to astronomical research, culture and experiences to help build a literate society; and that astronomers are a strong part of the global citizenship.

The central OAO function is *accessibility*. The OAO generally does not create outreach material itself, but works with the IAU and other organisations to increase the impact of its activities. Outreach is a strong component of many different units within the IAU, and the OAO interfaces with all of them. For example, Commission C2, 'Communicating Astronomy with the Public' (CAP), addresses astronomy communication and outreach issues, with the OAO editing the CAP journal as well as an Astronomy Outreach Newsletter. Also, the OAO works with the OAD to conduct numerous actions related to public outreach in order to achieve SDGs. By working together, both offices can maximise their synergies in the same area but with different goals. The OAO also takes part in EPO-like (Education and Public Outreach) activities, thus providing an interface with the newly proposed Office of Astronomy for Education.

To reach its long-term goals, OAO activities will include the provision of easily accessible public-friendly information on astronomical terminologies and objects in the Universe. To help reach larger audiences, the OAO will expand its translation network in the coming decade to manage and distribute astronomical results in several different languages. Activities to connect professional and amateur astronomers will be strengthened in the coming decade. The OAO also coordinates some of the worldwide citizen-science projects and campaigns, such as the public exoplanet naming competition in 2015 2016. Taken together, the OAO activities contribute to providing inclusive resources globally and achieving a more diverse astronomical community.

The network of National Outreach Coordinators, maintained by the OAO, can coordinate and advance projects in their country, as was done with great success during the 2009 International Year of Astronomy, and is being done in 2019 for the IAU100 activities. The NOCs are also the formal point of contact for engagement with amateur astronomy groups within each country. Furthermore, NOCs can play a role in helping to spread the word about the dark and quiet sky initiatives and to promote citizen science projects. In the coming years, the NOC system will be redefined, restructured and enlarged to ensure its effectiveness across the world.

In summary, the goals of the OAO for the next decade are:
• Increase the network of NOCs; restructure and ensure their effectiveness.
• Facilitate international communication through exchanges and translations.
• Provide open databases and public-friendly access to astronomical information.
• Encourage communication of science and critical thinking through IAU member public engagement, professional-amateur, and citizen science activities.
• Promote dark skies and the pale blue dot message

The IAU is grateful to the OAO staff and the many people who have contributed to the achievements of the OAO so far, especially the National Outreach Coordinators. We encourage colleagues and early career astronomers to join the IAU and contribute to the OAO goals.

References

Astronomy for Development: Strategic Plan 2010-2020,
https://www.iau.org/static/education/strategicplan_2010-2020.pdf
IAU Strategic Plan 2020-2030,
https://www.iau.org/static/education/strategicplan-2020-2030.pdf

FM15
Astronomy for Development

Astronomy in Focus - XXX
Proceedings IAU Symposium No. XXX, 2018
M. T. Lago, ed.

doi:10.1017/S1743921319005349

Summary: FM15 Astronomy for Development

Vanessa A. McBride[1,2]

[1]IAU Office of Astronomy for Development, c/o South African Astronomical Observatory, PO Box 9, Observatory, 7935, South Africa

[2]Dept. of Astronomy, University of Cape Town, Private Bay X3, Rondebosch 7700, South Africa
email: vanessa@astro4dev.org

Abstract. This summary captures, in the broadest sense, some of the achievements, challenges and spirit of the astronomy for development community at the 30th General Assembly of the IAU.

Keywords. miscellaneous, sociology of astronomy, astronomy for development

1. Introduction

The two-day Focus Meeting 15 (FM15): Astronomy for Development comprised two poster sessions, a series of invited and contributed talks, panel discussions, and a brainstorming session. In particular the meeting aimed to bring together science communicators, experts from development and science policy backgrounds, and astronomers in the various Divisions of the International Astronomical Union. A key theme of FM15, and of the 30th IAU General Assembly in a broader sense, was the pronounced inclusion of astronomy for development in the 10 year strategic plan of the International Astronomical Union. This is articulated as Goal #3: "The IAU promotes the use of astronomy as a tool for development in every country".

It became clear through the submissions to FM15 that the landscape of astronomy for development is evolving rapidly. While the astronomy community has converged on a shared understanding of development where the United Nations Sustainable Development Goals are central, we are cognisant that "development" itself is a fluid concept, being rapidly redefined at global, regional and and domain-specific levels.

2. Selected Highlights

This meeting covered a great many aspects of astronomy for development, and one can't do them justice in this short summary. I discuss a few short examples below:

- Astro-tourism is emerging as a strong focus in the astronomy for development community. It has the potential to contribute to the economic empowerment of societies that live around observatories or other astronomy-related attractions. (See submissions by Patatanyan, Jiwaji and El Yazidi).
- The United Nations sustainable development goals provide a broad, international focus for development efforts over the coming decade. Chinigò's submission points out that it is possible to lose focus on small communities and individuals when adhering to

a global definition, and care must be taken to conceptualise development at different scales: international goals, regional and community-based.

• Contributions from the astronomy for education community resulted in a renewed understanding of the topology of the astronomy education landscape (Bretones, Eriksson, Alves-Brito, Gutiérrez and many others). Specifically, the need for an astronomy teaching practitioners journal was highlighted. These developments are foundational for the establishment of the Office of Astronomy for Education in the forthcoming triennium. This new office is expected to have a close, synergistic relationship with the Office of Astronomy for Development and Office for Astronomy Outreach. Specific implementation possibilities include existing working groups, such as those on solar eclipses or magnetic activity (Division E), where clear opportunities for aligning science, outreach and education are available.

• A panel discussion led by Division presidents or representatives suggested the establishment of an inter-divisional working group or think-tank that may interface with other disciplines to address development imperatives. Areas such as data and related techniques (Divisions B and H), high energy detector technology (Division D) and citizen science and astronomy education research (Division C) would benefit from and contribute to such a cross-disciplinary forum.

• Accessibility and inclusion featured prominently in the programme, with discussion of specific interventions (Voelker, Spuck, Gastrow & Diaz Merced) as well as systemic changes that would be required to allow broader participation in astronomy. This was complemented by similar focus from the IAU through the Inspiring Stars exhibition.

• In various sessions of FM15, it was noted that recognition and respect for the value that is added by social science (López, Gastrow & Diaz Merced) and the progress that has already been made by other disciplines in the science for development narrative. Such awareness is essential in a field where progress is so strongly dependent on nurturing cross-disciplinary relationships.

• Many excellent contributions were made through poster presentations. Some of these are available as part of the supplementary material to the FM 15 proceedings.

3. Challenges

As at 2018, there are now ten regional offices of astronomy for development. While these regional offices subscribe to a common vision, there is significant variation in focus areas and implementation of the astronomy for development vision. This is a strength, because regions can accommodate localised needs, but also a challenge because one can get complacent with one's own understanding of development and progress. It is clear we need to work harder to engage in the local and global definitions of development, using the network of regional offices to drive this.

Work on sustainable development is interdisciplinary by nature, and there are many science-for-development initiatives from other disciplines. The cross-disciplinary nature of development work requires humility, funding and mutual respect. While we keep looking to redefining and updating our understanding of development, it is worth noting that the concept of astronomy for development is now embedded in the IAU. The time is ripe for an audacious vision that will allow astronomers to mobilise, alongside other scientists, economists and development agencies, in an effort to tackle the big socioeconomic and environmental issues facing the planet.

Astronomy in Focus - XXX
Proceedings IAU Symposium No. XXX, 2018
M. T. Lago, ed.

doi:10.1017/S1743921319005350

Overview of the OAD: Achievements, Challenges and Plans

Vanessa McBride[1,2] and Ramasamy Venugopal[1]

[1]South African Astronomical Observatory, 1, Observatory Road Cape Town, South Africa
email: vanessa@astro4dev.org rv@astro4dev.org

[2] Dept. of Astronomy, University of Cape Town,
Private Bay X3, Rondebosch 7700, South Africa

Abstract. The Office of Astronomy for Development (OAD) aims to use astronomy, including its tools, practitioners and skills, to benefit society. The OAD, a joint project of the International Astronomical Union and the South African National Research Foundation, has the vision of using 'Astronomy for a better world'. Since 2013, the OAD has funded more than 120 projects that use astronomy to address developmental issues as defined under the United Nations Sustainable Development Goals (SDG).

Keywords. sustainable development, SDG, astronomy for development, capacity building, inclusion, social sciences

---

1. Introduction

Astronomy is an exciting and popular topic because it connects exotic science, cutting edge technology and a sense of cultural connection or inspiration. Taking advantage of this appeal, astronomy can be used as a vehicle to draw attention to, and address issues of sustainable development. At the OAD, we consider how the tools, skills and methods of astronomy can be used to encourage astronomers and other scientists to work towards the SDGs. In addition to the OAD's global coordinating office in Cape Town, there are ten regional or language offices across the globe, where the SDGs are afforded regional priority and implementation. The OAD favours a dual approach to maximise the potential impact of astronomy for development - i) grassroots projects responding to local needs, run by stakeholders in the community which are funded annualy through an open call and ii) special projects driven by members of the OAD staff or other collaborators.

2. Annually Funded Projects

Every year, the OAD coordinates a global call for proposals for projects that aim to use astronomy to promote sustainable development. Since 2013, the office has disbursed IAU grants totalling 628,025 Euros to over 120 projects around the world. Figure 1 illustrates some of the ways in which past projects have tried to influence the SDGs. Below are specific examples of astronomy for development: i) Astronomy for Capacity Development: Several OAD funded projects have focused on capacity building in education by conducting workshops, schools, trainings etc. targeting communities and regions which are disadvantaged or under-represented. These actions directly impact on SDG 4 Quality Education and SDG 10 Reduced Inequalities. Projects have used astronomy at school and university level to improve skills in scientific inquiry, programming, data science, mathematics etc. ii) Astronomy for Economic Stimulation: Astro-tourism projects, such as those funded by the OAD, are based on the idea that astronomical sites can be systematically promoted as points of interest and together with the tourism industry

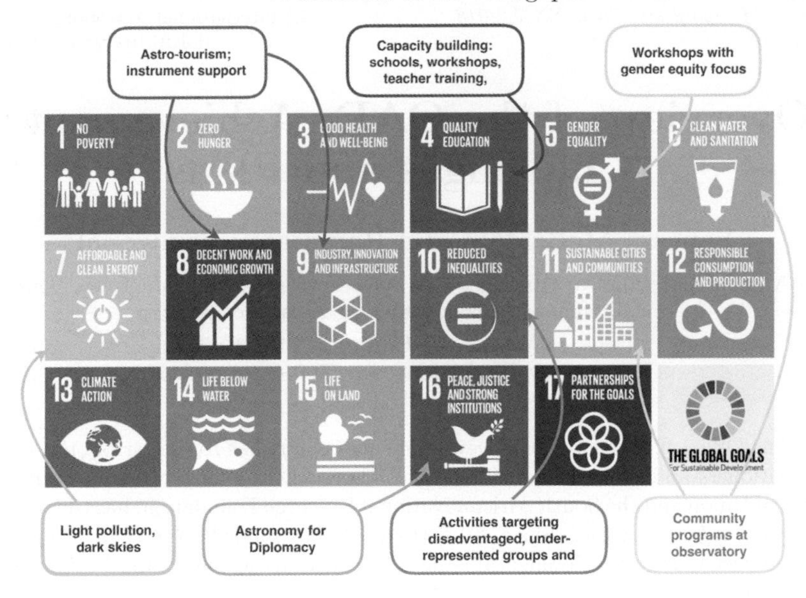

Figure 1. Mapping OAD funded projects to the Sustainable Development Goals

can contribute to the local economy. This relates to SDG 8 Decent Work and Economic Growth and SDG 9 Industry, Innovation and Infrastructure. iii) Astronomy for Inclusion and Equity: Several OAD projects have used astronomy to bring science to audiences traditionally under-represented or excluded. For example, developing tactile astronomy resources for visually impaired children.

3. Special Projects

Astronomy for development is concerned with activities that seek to affect human development, hence the projects are "social interventions". Working with other disciplines, especially the social sciences, is critical to cover the development aspects. At the OAD, this is achieved through special projects and partnerships. Two such examples are listed below: i) Applying astronomy tools in the field of development economics: OAD Fellow Tawanda Chingozha is a development economist who is collaborating with the OAD to explore the application of astronomy skills in economics. In his study of changes in urban informality patterns in developing countries, Tawanda is relying on citizen science methods similar to those used in popular astronomy projects such as Galaxy Zoo. ii) A research collaboration between the OAD and Human Sciences Research Council in South Africa to improve access to science for visually impaired audiences.

4. Conclusion

The OAD, its Regional Offices, collaborators, partners, and large volunteer community have successfully implemented a number of for-development actions globally. Although these actions by the community have influenced key SDGs, there is a need to work across disciplines and explore specific inter-disciplinary collaborations in order to understand and contribute better to sustainable development. The OAD is also grateful to its global volunteer community and invites ideas and feedback on astronomy for development.

Astronomy in Focus - XXX
Proceedings IAU Symposium No. XXX, 2018
M. T. Lago, ed.

doi:10.1017/S1743921319005362

Overview of IAU OAD Regional Offices and Language Centres

Rosa Doran[1], German Chaparro[2], S. V. Farmanyan[3], Jaime E. Forero-Romero[4], Angela Patricia Perez Henao[5], Wichan Insiri[6], Awni Khasawneh[7], M. B. N. Kouwenhoven[8], Joana Latas[1], A. M. Mickaelian[3], G. A. Mikayelyan[3], George Miley[9], Lenganji M. Mutembo[10], Bonaventure Okere[11], Pedro Russo[9,12], Prospery C. Simpemba[10], Michelle Willebrands[12] and Alemiye Mamo Yacob[13]

[1]NUCLIO Nucleo Interactivo de Astronomia, Sao Domingos de Rana, Portugal
[2]Vicerrectoría de Investigación, Universidad ECCI, Calle 19 No. 49-20, Bogotá, Colombia
[3]Byurakan Astrophysical Observatory (BAO), Byurakan 0213, Aragatzotn Province, Armenia
[4]Departamento de Física, Universidad de los Andes, Calle 18A No. 1 - 10, Bogotá, Colombia
[5]Planetario de Medellín, Carrera 52 No. 71 - 117, Medellín, Colombia
[6]Southeast Asia ROAD, National Astronomical Research Institute of Thailand
[7]Director General, Royal Jordanian Geographical Centre, Jordan
[8]Department of Mathematical Sciences, XJTLU, 111 Ren'ai Rd, SIP, Suzhou 215123, China
[9]Leiden Observatory, Leiden University, the Netherlands
[10]Southern African ROAD, Copperbelt University, P.O. Box 21692, Kitwe, Zambia
[11]West African ROAD, NASRDA Centre for Basic Space Science, Nsukka, Nigeria
[12]Department of Communication and Society, Leiden University, the Netherlands
[13]East African ROAD, Ethiopian Space Science & Technology Institute, Ethiopia
Contact email: `info@astro4dev.org`

This paper presents a very brief overview of the 10 Regional Offices (ROADs) and Language Centres (LOADs) established as part of the IAU's Astronomy-for-Development effort. Due to space constraints here, longer 2-page papers on each office are available as Supplementary Material in the electronic proceedings as well as on the website of the IAU's Office of Astronomy for Development (www.astro4dev.org). Authors are listed in alphabetical order of last name.

Astronomy is the science that connects the world; it is the science that aims to answer the most fundamental questions of the Universe we live in. The benefits of astronomy to society extend far beyond those of scientific knowledge. Astronomy can be used to help achieve the Sustainable Development Goals outlined by the United Nations. The IAU East-Asia Regional Office of Astronomy for Development (EA-ROAD) and the Chinese Language Expertise Center (LOAD) were established in 2012, and operate in the People's Republic of China, Mongolia, and the Democratic People's Republic of Korea and encompassing roughly 20% of the world's population. The consortium uses astronomy as a tool for development in the East-Asia region, with the aim of utilising all aspects of astronomy and astrophysics to promote development in the entire East-Asia region.

Armenia hosts the South West and Central Asian (SWCA) ROAD. So far, 6 countries have officially joined (Armenia, Georgia, Iran, Kazakhstan, Tajikistan, and Turkey). The SWCA ROAD plays an important role in maintaining contacts between the region's countries with various cultures, and regularly organizes regional astronomical workshops and summer schools. The International Conference "Astronomical Heritage of the Middle East" sponsored by UNESCO was devoted to the role of astronomy in many fields of human activities. The SWCA ROAD's Astro Tourism project perfectly fits to the IAU Strategic Plan's goals as a tool for development of society. Up-to-date information about IAU SWCA ROAD is available on the website†.

The Andean ROAD started its activities in 2013 and was officially signed into existence in 2015. Its most important success has been keeping a conversation with the central OAD office and the IAU members interested in development activities in the Andean region. The networks created in such conversations have helped us to keep motivated and define new strategies. The main challenge has been running the activities through volunteers and without permanent funding. Trying to define what development means for those in the region is another dilemma; the mainstream development concept used by the OAD is influenced by global north ideologies distanced from post-development and decolonial concepts, which are highly relevant to our realities.‡

The European ROAD was established in February 2018, jointly hosted by Leiden University and the European Astronomical Society. Its mission is to use astronomy to help accomplish the UN Sustainable Development Goals 4 (Lifelong Education), 16 (Peaceful Societies), 13 (Climate Action) and 5 (Gender Equality). Focusses are on young people in migrant and disadvantaged communities and contributing European partnerships and resources to global activities. Although we seek structural funds for a ROAD coordinator, the office will initially be supported by relevant European Commission projects. Following EU Universe Awareness (2011 – 2013) and EU Space Awareness (2014 – 2017), Leiden University will start coordinating SPACE.EU in 2019. This will contribute to the mission of the ROAD with a portfolio of activities relevant to SDGs 4 and 16 for young people in disadvantaged communities.

Since the official launch of West African Regional Office of Astronomy for Development (WAROAD), in November 2015 at Nsukka, Nigeria, the West African Region has witnessed increased activities in Astronomy outreach, education and research. Among these activities are: (1) Formation of Astronomy clubs in many institutions across the region including Burkina Faso, Ghana, Nigeria, Senegal, among others. (2) Hands-on Basic Space Science Workshop for primary and secondary school science teachers in Nigeria. (3) Astronomical outreach/workshops at the schools and public levels especially the AstroBus activity by Senegal among others such as eclipse observations. (4) Astro Camp for girls (5) The West African International Summer School for young Astronomers. (6) Dunlap Institute for Astronomy and Astrophysics, Canada and University of Toronto, Canada. Funds for these activities are locally sourced with support from OAD, SKA, ICTP, DARA Project, etc., though getting enough funds to run our projects has been the greatest challenge. In terms of astronomy research facilities, the region can confidently boast of a 1m optical telescope in Burkina Faso and the converted 32 m radio telescope in Kutunse, Ghana.

In the past decade, astronomy and space science showed significant development in the East African Region. The establishment of space agencies and institutes in Kenya and Ethiopia, formulation of space policies in Ethiopia and the introduction of astronomy and astrophysics both in the curriculum and as a specialty on both undergraduate and postgraduate levels in universities in Ethiopia, Kenya, Uganda and Rwanda, contribute a lot towards the region's success. Moreover, the establishment of East Africa Astronomical Society (EAAS), East African Astrophysics Research Network (EAARN) and East Africa Regional Office of Astronomy for Development (EAf-ROAD) all played a role for the

† http://iau-swa-road.aras.am/eng/index.php

‡ Grosfoguel, R. (2002), Colonial Difference, Geopolitics of Knowledge and Global Coloniality in the Modern/Colonial Capitalist World System, *Review* 19, 2, pp. 131-154

development of the field in the region. Though there are promising landscapes to flourish astronomy and space science in the region, there are also bottlenecks in coordinating and synergizing the region to reach the point where astronomy should be. Thus, EAf-ROAD will take the lead and create a platform that can be used to engage government and policy makers, science educators, advocators and professional societies to be aware about the role of astronomy for development and to inspire and attract young people to the field in line with the SDGs and IAU 2020–2030 strategic plan.

The Southern African Regional Office of Astronomy for Development (SAROAD) came into being on the 14th of August 2014, after the signing of the Memorandum of Understanding between the Copperbelt University (CBU) and the IAU in Kitwe, Zambia. Since then there have been both achievements made by the office and challenges faced. Among the notable activities of SAROAD are the hosting of the annual regional Astronomy workshops (AstroLab), outreach to Schools, public engagement and dissemination of information about new developments and upcoming events in astronomy and space science. Three AstroLab workshops have already been held and several outreach activities carried out. The regional office has been facilitating the involvement of member states in most astronomy cornerstone projects.

The Portuguese Language Expertise Centre for the Office of Astronomy for Development (PLOAD) is hosted by NUCLIO in Portugal since 2015. The vision is to use astronomy as a tool for development in this specific region and language, namely in the Portuguese speaking countries and communities at a global level. PLOAD acts as a link between institutions with common objectives regarding Astronomy, in a concerted action following the three tasks forces: Astronomy for Universities and Research, Astronomy for Children and schools, Astronomy for the Public. The PLOAD mission is to fulfil its vision by following the guidelines of the IAU strategic plan and to build on existing challenges and opportunities towards strong collaborative and active structures. This mission is being accomplished by several steps, starting from a careful research of resources and existing needs and the design of an effective implementation strategy of aid and support.

The Southeast Asia Regional Office of Astronomy for Development (SEA-ROAD) is hosted at the National Astronomical Research Institute of Thailand. SEA-ROAD aims to strengthen the already existing Southeast Asia Astronomy Network's (SEAAN) ties among the active national members as one of the driving forces behind SEA-ROAD. It aims to integrate SEA-ROAD with the International Training Centre in Astronomy under the auspices of UNESCO (ITCA) via trainings and workshops such as winter and summer schools in all levels covering schoolteachers, young researchers, university students, university lecturers, etc. SEA-ROAD aspires to be the ultimate human resource database and excellence centre in Astronomy of the region that also taps on value chains of astronomy and related sciences in the region. SEA-ROAD activities 2017 to August 2018 at a glance: 21 Schools/trainings/workshops; 25 Countries participated; 766 Participants; 722 Southeast Asian participants.

The Arab Regional Office of Astronomy for Development (Arab-ROAD) and the Arabic Language Expertise Center (Arab-LOAD) is hosted in Jordan by the Arab Union of Astronomy and Space Science (AUASS) since December 2015. Since its official inauguration, the office has been organizing local and regional activities in collaboration with the Royal Jordanian Geographic Center (RJGC), Jordanian Astronomical Society (JAS), the Syrian Astronomical Society (SAS), the Sharjah Center for Astronomy and Space Science (SCASS), the Oman Astronomical Society (OAS), the Sirius Astronomical Association (Algeria), Ibn al-Haytham Association for Science and Astronomy (Algeria), the Tunisian Astronomical Society, and the Sudan Astronomical Society.

These 10 offices, together with the OAD in South Africa, form the core of the IAU's efforts towards especially Goal 3 of its 2020–2030 Strategic Plan, namely "The IAU promotes the use of astronomy as a tool for development in every country".

Astronomy in Focus - XXX
Proceedings IAU Symposium No. XXX, 2018
M. T. Lago, ed.

doi:10.1017/S1743921319005374

Science and the Sustainable Development Goals

Heide Hackmann

CEO, International Science Council
email: Heide.Hackmann@council.science

1. Introduction

In September 2015, the 193 member nations of the UN adopted Agenda 2030 consisting of a set of 17 Sustainable Development Goals (SDGs), including 169 targets. The SDGs replaced the Millennium Development Goals (MDGs), agreed in 2000, after their 2015 deadline. Building on the principles put forward in the outcome document of the Rio+20 Conference, held in 2012, the SDGs represent: a global agenda relevant to all countries in all parts of the world; an integrated agenda in which environmental sustainability, social inclusion, and economic development are equally valued; an inclusive agenda, calling on multi-sectoral, multi-stakeholder collaboration and "whole of government" approaches; and a potentially transformative agenda, challenging and changing systems, institutions and the values underlying them. Ban Ki Moon, UN Secretary General at that time coined the narrative of an agenda for "people, planet, peace, prosperity, and partnerships".

Agenda 2030 also embraces other major post-2015 agendas for action, notably the Paris Climate Agreement under the UN Framework Convention on Climate Change (UNFCCC), the New Urban Agenda in the context of UN Habitat and the Sendai Framework representing the 2015 – 2030 update of the UN International Strategy for Disaster Reduction (UNISDR).

The Rio+20 outcome document affirmed that the development of the global goals and their implementation at national, regional and global levels must be evidence based and the scientific community must be made a key partner. To this end, several science based initiatives have been launched after 2015 in the context of the UN and individual UN agencies. The essentials of this new UN landscape in science for the SDGs include: first, science-based initiatives of individual UN agencies/programmes, generic or SDG specific, such as the UNESCO sustainability science initiative or the UNESCO-IOC led UN Decade of Ocean Science. Second, the UN is mandated to prepare in four-year intervals a Global Sustainable Development Report (GSDR) aimed at "assessing" the science behind the set of 17 SDGs, with special attention to the information needs of decision-makers, and thus strengthening the science-policy interface. Third, UN member states established equally in 2015 the Technology Facilitation Mechanism (TFM), a dedicated structure to advance science, technology and innovation (STI) for the SDGs.

The TFM consists of an Inter-Agency Task Team (IATT) on STI (35 UN bodies, including the World Bank) advised by the multi-stakeholder 10 Member Group appointed by the UN Secretary General. A good number of scientists from a range of disciplines and from all parts of the world are currently members of this advisory group. The IATT and the 10 Member Group are jointly responsible for organising an annual STI Forum at the UN in New York, an event during which the STI community engages directly with UN member states to discuss strategies and priorities for their collaboration on SDG implementation.

The main political body for monitoring and guiding the implementation of the SDGs is the UN High-Level Political Forum (HLPF) on sustainable development. It meets annually under the auspices of ECOSOC, the UN's Economic and Social Council. While being the main platform providing political leadership and guidance on sustainable development issues at the international level, the core element on its agenda annually relates to reviewing implementation of SDGs. This review process is undertaken in two ways, by annually focusing on the state of implementation of a selected cluster of SDGs and by inviting country-led "Voluntary National Reviews" covering all SDGs and allowing to address challenges and opportunities, as well as best practice in SDG implementation at the national level.

Promoting science for sustainable development is one of the main areas of activities of the International Science Council. This is done by ISC focusing on four functions. The first one consists of ISC representing and advocating international science in UN policy formulation, implementation and review. The formal mechanism for ISC to be able to sit at the table is the UN Major Groups System, with the Scientific and Technological Community being one of the nine Major Groups invited by the UN to participate in UN policy processes. Promoting international research relevant to the SDGs represents the second function of ISC supporting science for the SDGs. ISC is the key scientific sponsor of the leading SDGs related international research programmes: Future Earth, the World Climate Research Programme, Integrated Research on Disaster Risk, and Urban Health and Wellbeing, as well as Comparative Research on Poverty. The third way for ISC to promote science for SDGs consists of funding SDG-relevant research through two funding schemes made possible by grants from the Swedish International Development Agency (SIDA): (i) Transformations to Sustainability, a scheme with world-wide coverage, and (ii) Leading Integrated Research for Agenda 2030 in Africa (LIRA). Finally, the fourth function is focused on preparing expert reports providing scientific input and advice related to SDG implementation. A first report presented a review of SDG targets, while the second report demonstrated a methodology to address interactions between SDGs. A third project aimed at providing scientific advice for national-level SDG interactions analysis, prioritization and implementation is being developed in cooperation with the International Network of Government Science Advisors (INGSA).

Within the broader international scientific landscape related to science for the SDGs, the International Science Council works with different partners. There are those covering specific sectors and niches of the global scientific and technological community such as the World Federation of Engineering Organisations, the InterAcademy Panel or the Sustainable Development Solutions Network. Other international interdisciplinary partners such as the International Institute for Applied Systems Analysis focus on specific methodological approaches. A third category of potential partners in action aimed at strengthening science for the SDGs consists of the broad array of our international scientific union members cutting across most scientific disciplines in the natural and social sciences. The International Astronomical Union (IAU) with its Flagship Programme "Astronomy and the Sustainable Development Goals" represents a fine example in this respect.

Astronomy in Focus - XXX
Proceedings IAU Symposium No. XXX, 2018
M. T. Lago, ed.

doi:10.1017/S1743921319005386

The IAU Strategic Plan for 2020-2030: OAD

Ewine F. van Dishoeck[1] and Debra Meloy Elmegreen[2]

[1]Leiden Observatory,
P.O. Box 9513, NL-2300 RA, Leiden, the Netherlands
email: ewine@strw.leidenuniv.nl

[2]Dept. of Physics & Astronomy, Vassar College,
Poughkeepsie, NY 12604 USA
email: elmegreen@vassar.edu

Abstract. The IAU Strategic Plan for 2020-2030 presents an overview of all of the activities of the IAU along with priorities, key goals, mandates, and specific actions. Here future plans and goals are outlined for the Office of Astronomy for Development (OAD).

Keywords. editorials, notices

1. Introduction

The first formal Strategic Plan of the International Astronomical Union for the decade 2010-2020, was focused on the then-new Office of Astronomy for Development (OAD), which uses astronomy to impact UN Sustainable Development Goals (SDGs). At the XXIXth General Assembly in Honolulu, Hawaii, USA in August 2015, the Executive Committee agreed to present for approval at the XXXth General Assembly in Vienna, Austria in August 2018 an extended plan for future activities of the OAD. At the 98th Executive Committee (EC) meeting in Mexico City in May 2016, a working group was appointed, including Ewine van Dishoeck, Debra Elmegreen, Piero Benvenuti, and Renée Kraan-Korteweg. The EC further decided that the new Strategic Plan for 2020–2030 should encompass all activities of the IAU. The working group received extensive input from its membership and the various Offices, in particular from Kevin Govender and his team on the OAD part.

The final Strategic Plan 2020-2030 was presented for adoption as Resolution A1 at this XXXth GA, where it was approved at the second Business Meeting. The revised mission of the IAU is to "promote and safeguard astronomy in all its aspects (including research, communication, education, and development) through international cooperation," where the additional parenthetical aspects make explicit that the IAU has branched out beyond its original purpose of fostering scientific communication and exchange of ideas among professional astronomers.

2. OAD and other Offices of the IAU

The 3 current offices of the IAU were formed within the last decade to carry out additional aspects of IAU activities beyond the advancement of astronomy. The OAD, hosted in Cape Town, South Africa, focuses on the use of astronomy for development by capitalizing on the field's scientific, technological and cultural links and its impacts on society. It interacts with the other Offices. The Office of Young Astronomers, hosted by the Norwegian Academy of Science and Letters, focuses on the training of young astronomers at University level, and organizes the International School for Young Astronomers (ISYA). The Office for Astronomy Outreach (OAO), hosted by the National

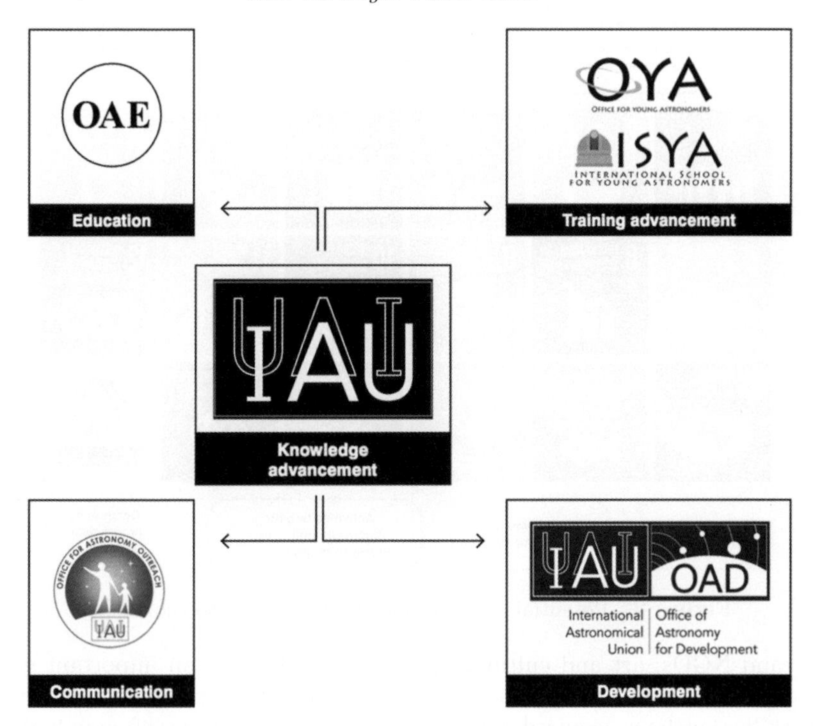

Figure 1. Interconnection of the Offices of the IAU.

Astronomical Observatory of Japan, focuses on engaging with the public, providing access to astronomical information and astronomy communication. It maintains a network of National Outreach Contacts (NOCs) and amateur groups. A new office is proposed: the Office of Astronomy for Education (OAE) will focus on providing training and resources for using astronomy as a stimulus for teaching and education from elementary to high school level (astronomy and science education). Each of these offices has its own strategic plan, and each office has a liaison on the EC to ensure that the strategic plans align with the overall goals of the IAU. Sometimes activities span more than one office, so having connections with the EC helps coordinate overlapping activities. Figure 1 is a schematic to emphasize the relations between the offices and the EC.

3. Office of Astronomy for Development (OAD)

The overarching goal of the OAD is to promote the use of astronomy as a tool for development in every country. More specifically, the OAD uses the UN Sustainable Development Goals (SDGs) as the global definition of development in calling annually for proposals. Fig. 2 below illustrates some potential contributions of astronomy to the SDGs. A goal for the 2020 - 2030 strategic plan is that – depending on the region – about half of the 232 SDG indicators should have been positively affected by OAD projects on all of the populated continents.

The Regional (ROADs) and Language (LOADs) Offices form the global core structure of the OAD and a goal is to further solidify and expand this network. Interactions between them stimulate synergies among different geographic and cultural regions. The OAD aims to identify a number of global 'signature' projects that can be expanded and regularly carried out worldwide in the next decade.

Partners from fields other than astronomy should also be included in this scheme, as astronomy connects with the space sector, ICT, branches of social sciences, relevant

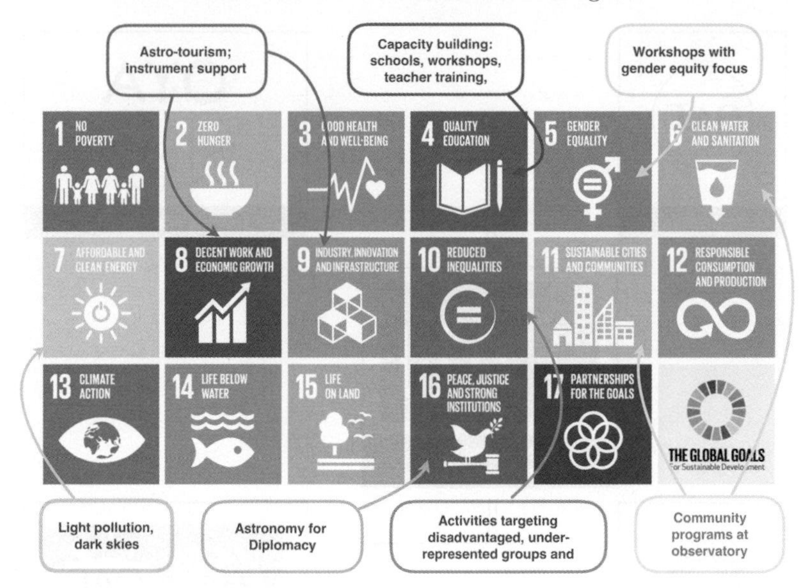

Figure 2. Potential contributions of astronomy to the SDGs.

industries and NGOs, art and cultural organisations. Indeed, an important goal of the OAD is to position young people for non-academic job opportunities in which they can apply the skills they have acquired in their career so far, as they are often in high demand. The field of astronomy can lead other sciences in terms of the societal benefits of blue skies research, and also with regard to the value that science brings to challenges facing humanity. The OAD can be a 'working space' for collaborations across sciences. The recent merger of the ICSU and ISSC clearly promotes such an approach.

4. Conclusions

We are grateful to the many people who have contributed to the achievements of the OAD in its first decade. Moving forward, the IAU depends on its members for the successful execution and implementation of the plan. Members are encouraged to continue to be involved in IAU activities through Working Groups, Commissions, Divisions, and Offices, and to encourage colleagues and early career astronomers to join the IAU and contribute to the OAD goals.

References

Astronomy for Development: Strategic Plan 2010-2020,
https://www.iau.org/static/education/strategicplan_2010-2020.pdf
IAU Strategic Plan 2020-2030,
https://www.iau.org/static/education/strategicplan-2020-2030.pdf

Astronomy in Focus - XXX
Proceedings IAU Symposium No. XXX, 2018
M. T. Lago, ed.

doi:10.1017/S1743921319005398

Synergies among the IAU Offices

Kevin Govender[1], Sze-Leung Cheung[2], Itziar Aretxaga[3] and Oddbjørn Engvold[4]

[1]IAU Office of Astronomy for Development, c/o South African Astronomical Observatory, PO Box 9, Observatory, 7935, South Africa. Email: kg@astro4dev.org

[2]IAU Office for Astronomy Outreach, c/o National Astronomical Observatory of Japan Email: cheungszeleung@gmail.com

[3]Instituto Nacional de Astrofísica, Óptica y Electrónica, Luis Enrique Erro 1, Sta. María Tonantzintla, 72840 Puebla, Mexico. Email: itziar@inaoep.mx

[4]Rosseland Centre for Solar Physics, Institute of Theoretical Astrophysics, University of Oslo, Oslo, Norway. Email: oddbjorn.engvold@astro.uio.no

---

The IAU Strategic Plan 2020-2030 envisages four offices (Office of Astronomy for Development; Office for Astronomy Outreach; Office for Young Astronomers; Office for Astronomy Education). The IAU's three current offices (OAD, OAO, OYA) have distinct mandates but need to work in synergy to ensure maximum effectiveness.

The IAU established the Office of Astronomy for Development (OAD) in 2011 in partnership with the National Research Foundation (NRF) of South Africa, with strong support from the South African Department of Science and Technology. The OAD is hosted at the South African Astronomical Observatory (SAAO) in Cape Town, South Africa, and aims to use astronomy to impact on the United Nations Sustainable Development Goals (SDGs). Its activities revolve largely around an annual open call for proposals, the establishment and oversight of regional structures, and the building of collaborations and networks between astronomy and development fields. As of 2018 the OAD has supported 122 "astronomy-for-development" projects reaching over 85 countries; established 10 regional offices around the world; and registered over 600 volunteers.

The IAU Office for Astronomy Outreach (OAO) is the IAU's hub for coordinating its public outreach activities around the world. It is based at the National Astronomical Observatory of Japan (NAOJ) with funding contributions from several astronomical institutes. The OAO coordinates and supports worldwide efforts to enhance public knowledge, appreciation and education of astronomy and related sciences. The OAO promotes public awareness of the IAU activities and coordinates the IAU international outreach campaigns. The OAO maintains and coordinates the IAU network of National Outreach Coordinators (NOC), who are responsible for the implementation of IAU outreach initiatives at national level and for maintaining the relationship with the national communities of amateur astronomers. The OAO also works with the IAU's Commission C2 to support the astronomy communication communities through the CAP Journal (Communicating Astronomy with the Public) and CAP conference.

The three-week International Schools for Young Astronomers (ISYA) program of Office for Young Astronomers (OYA) are intensive postgraduate schools in regions where students have limited opportunity to be exposed to full extent, up-to-date lectures in astronomy. They provide basic introductory courses led by an international panel of $\sim 10-12$ professors, that cover theoretical and observational lectures and lab work with local observing facilities (when available), data reduction labs, virtual observatory and/or data mining labs. The selection of topics for each school is tailored to regional needs. The schools accommodate $\sim 30-50$ students, ensuring a lively interaction with the lecturers throughout the school, where professional longer-term interactions often develop. In recent years, collaboration of ISYA and regional OAD offices and OAO NOCs has given some fruitful examples of a path to reinforce in the future. In the Ethiopia 2017 ISYA, for instance, the Square Kilometre Array South African Office offered a communication workshop supported by the regional OAD, where students designed development projects that could be submitted to the OAD calls for proposals. The Colombia 2018 ISYA was co-organized by an alliance of Colombian universities and the regional OAD, and additional curriculum modules for high performance computing, machine learning, and big data handling were included into the program. Additionally, many schools include science communication workshops led by prominent local outreach agents, with links to OAO. Activities such as how to design a good outreach talk/paper/science activity or use social media are part of these workshops.

These complementary activities strengthen the notion that the schools have to be flexible in their curricula and adapt to local needs, opportunities and conditions, and that collaboration on development (OAD), outreach (OAO) and education (OAE) have a natural fit into the young astronomers (OYA) ISYA curricula. One of the major partnerships among the offices is the sharing of networks, for example the OAD regional offices have the know-how at regional levels that advise the OAO to identify the key person to serve as the OAO network of National Outreach Coordinators (NOCs). This is also true in reverse, since the OAO NOCs network covers regions that do not have a regional OAD office. The OAO NOC network was a useful tool to reach out to people beyond the OAD reach. In the same way, the OAO can also support OYA and ISYA needs of identifying potential future hosts. Given the fact that in many cases those who are conducting activities related to development, outreach, education and professional development are the same group of people, the sharing of human networks and cross-advertisement of activities are particularly useful among the offices.

Astronomy in Focus - XXX
Proceedings IAU Symposium No. XXX, 2018
M. T. Lago, ed.

doi:10.1017/S1743921319005404

Hands on the Stars

Amelia Ortiz-Gil[1], Beatriz García[2] and Dominique Proust[3]

[1]Astronomical Observatory, University of Valencia, Spain
Email: amelia.ortiz@uv.es
[2]ITeDAM (CNEA-CONICET-UNSAM)
UTN Facultad Mendoza, Lab. Pierre Auger, Argentina
Email: beatriz.garcia@iteda.cnea.gov.ar
[3]Observatoire de Paris-Meudon, France
Email: dominique.proust@obspm.fr

Abstract. Hands on the Stars is a long-term project developed by the IAU Commission C1 Education and Development of Astronomy and its WG3 Astronomy for Equity and Inclusion with the goal of creating the first international comparative list of astronomical words in as many sign languages as possible.

Keyword. Accessibility in Astronomy

---

1. Introduction

This project is motivated by the fact that technical words in Astronomy are absent in many sign languages. To alleviate this problem, we proposed the creation of a comparative list of signs that already exist in different languages so they can be used as the starting point to create these signs in other languages in which they have not yet been defined. The ultimate goal would be to try to converge as much as possible to a universal sign, although this is quite difficult as signs are developed with a strong link to each particular cultural heritage.

2. An OAD funded project: Sign Language Universal Encyclopedic Dictionary

Our starting point was the dictionary in French sign language *Les mains dans les étoiles: Dictionnaire encyclopédique d'astronomie pour la Langue des Signes Française (LSF)* by D. Proust *et al.*, 2009. This dictionary contains approximately 300 signs describing several classical celestial bodies such as planets, asteroids, galaxies and quasars, as well as technical terms such as telescope, spectrograph and photometry. Many of them were developed by the authors of this book as these terms did not exist previously.

The OAD project had therefore two goals. First, we translated the French Encyclopedic Dictionary into English and Spanish. The translations are available online at the WG3 website (http://sion.frm.utn.edu.ar/iau-inclusion/). Second, we developed a first comparative list comprising 47 astronomical terms in 31 languages. The list is available at http://bit.ly/2HHsXul. We are currently working on an expansion of this list.

3. Acknowledgements

The authors wish to acknowledge the efforts of a large international collaboration with many individual volunteers and support from schools and institutions for the deaf mainly from Argentina, Canada-Québec, France, Italy, Japan, Spain, UK-Scotland and the USA. A. Ortiz-Gil acknowledges financial support by the Spanish Ministry of Science project AYA2016-81065-C2-2.

Astronomy in Focus - XXX
Proceedings IAU Symposium No. XXX, 2018
M. T. Lago, ed.

doi:10.1017/S1743921319005416

The Quality Lighting Teaching Kit: Utilizing Problem-Based Learning in Classrooms

Constance E. Walker and Stephen M. Pompea

National Optical Astronomy Observatory, 950, N. Cherry Ave., Tucson, AZ 85719 USA
Email: cwalker@noao.edu, spompea@noao.edu

Abstract. The U.S. National Optical Astronomy Observatory's Education and Public Outreach group has produced a Quality Lighting Teaching Kit. The kits are designed around problem-based learning scenarios. The kit's six activities allow students to address real lighting problems that relate to wildlife, sky glow, aging eyes, energy consumption, safety, and light trespass. The activities are optimized for 11-16 year olds. As part of the IAU100 celebration, the kits will be manufactured and made available to observatories and communities around the world.

Poor quality lighting not only impedes astronomy research and our right to see a starry night sky, but creates safety issues, affects human circadian sensitivities, disrupts ecosystems, and wastes billions of dollars/year in energy consumption. It also leads to excess carbon emissions. How do you change the mindset of society that is used to turning night into day? You educate the next generation on quality lighting.

As an outcome of the International Year of Light 2015, the U.S. National Optical Astronomy Observatory's (NOAO's) Education and Public Outreach (EPO) group has produced a Quality Lighting Teaching Kit. The kits are designed around problem-based learning scenarios. The kit's six activities allow students to address real lighting problems that relate to wildlife, sky glow, aging eyes, energy consumption, safety, and light trespass. The activities are optimized for 11-16 year olds but can be expanded to younger and older. All materials are in both English and Spanish. Most of the activities can be done within in a few minutes during class or afterschool and as stations or as stand-alones. Everything you need for the six activities is included in the kit. Tutorial videos on how to do the activities can be found at www.noao.edu/education/qltkit.php. 92 out of 100 kits have been distributed in 32 countries through SPIE-The International Society for Optical Engineering, CIE-International Commission on Illuminations, OSA-The Optical Society, IDA-the International Dark Sky Association, and the IAU OAD-Office of Astronomy Development. Successful feedback has NOAO's EPO group on the cusp of commercializing the kit. The aim is to have kits available to observatories and communities around the world, as part of the Dark Skies for All flagship project during the IAU100 celebration.

Astronomy in Focus - XXX
Proceedings IAU Symposium No. XXX, 2018
M. T. Lago, ed.

doi:10.1017/S1743921319005428

The United Nations Open Universe Initiative for Open Data in Space Science

Ulisses Barres de Almeida[1], **Paolo Giommi**[2] **and Jorge Del Rio Vera**[3]

[1]Centro Brasileiro de Pesquisas Físicas (CBPF)
Rua Dr. Xavier Sigaud 150, URCA 22290-180, Rio de Janeiro, Brazil
Email: ulisses@cbpf.br

[2]Agenzia Spaziale Italiana (ASI)
Via del Politecnico snc 00133, Roma, Italy
Email: paolo.giommi@asi.it

[3]United Nations Office for Outer Space Affairs (UNOOSA)
Vienna International Centre, Wagramerstrasse 5, A-1220, Vienna, Austria
Email: jorge.delriovera@un.org

Keywords. Space Sciences, Astronomical Data, Open Universe, UNOOSA, SDGs

---

The information revolution is rapidly becoming the major force in social, economic and cultural transformation worldwide, and the internet is today an asset capable of globally achieving the long sought, fundamental goals of transparency, availability and accessibility to information. The benefits of openness and transparency, for both users and providers of information, have been widely emphasised in the most diverse areas of society, and are no exception in the case of the space sciences.

Much has been done in recent years, especially in space astronomy, to offer open access, user-friendly and integrated platforms and services. However, there is still a considerable degree of unevenness in such services. Further efforts are therefore necessary to consolidate, standardise and expand them, aiming to promote a significant leap in an inspirational data-driven surge in training, education and discovery in space sciences. Such a process, leading to a much larger level of availability and integration of space science data, should be expanded beyond the scientific community, to non-scientific sectors of society, as a driver for development.

The Open Universe is an initiative under the auspices of the Committee on the Peaceful Uses of Outer Space (COPUOS) with the objective of stimulating a dramatic increase in the availability and usability of space science data, extending the potential of scientific discovery to new participants in all parts of the world, and empowering global educational services, especially in developing and underdeveloped countries. The far-reaching vision of the Initiative – which is carried out in cooperation with, and under the leadership of the United Nations Office of Outer Space affairs (UNOOSA) – and its potentially global reach, call for a wide international cooperation. Initially developed as part of the activities in preparation for UNISPACE+50, it is an initiative in response to the UN Sustainable Development Goals (SDGs), in particular Sustainable Development Goal 4, as a tool for Quality Education.

Open Universe will ensure that space science data will become gradually more openly available, easily discoverable, and free of bureaucratic, administrative and technical barriers, and therefore usable by the widest possible community, from space professionals, to the common interested citizen. By doing so, it seeks to trigger a major evolution in the culture of space science through data availability, by fostering the publication of all existing open space science data, and promoting its immediate usability, thus responding to the global demand for information and transparency. Through the integration and interoperability of existing services, data sets and software tools, provided by many actors worldwide, it will promote VO-based technology and protocols in the multi-messenger era of astronomical research, and improve international cooperation in space between governments, academia and with the third sector.

Astronomy in Focus - XXX
Proceedings IAU Symposium No. XXX, 2018
M. T. Lago, ed.

doi:10.1017/S174392131900543X

The DARA Big Data Project

Anna M. M. Scaife[1] **and Sally E. Cooper**[2]

[1]Jodrell Bank Centre for Astrophysics, University of Manchester,
Alan Turing Building, Oxford Road, Manchester M13 9PL, UK
Email: anna.scaife@manchester.ac.uk

[2]Jodrell Bank Centre for Astrophysics, University of Manchester,
Alan Turing Building, Oxford Road, Manchester M13 9PL, UK
Email: sally.cooper@manchester.ac.uk

Abstract. The DARA Big Data project is a flagship UK Newton Fund & GCRF program in partnership with the South African Department of Science & Technology (DST). DARA Big Data provides bursaries for students from the partner countries of the African VLBI Network (AVN), namely Botswana, Ghana, Kenya, Madagascar, Mauritius, Mozambique, Namibia and Zambia, to study for MSc(R) and PhD degrees at universities in South Africa and the UK. These degrees are in the three data intensive DARA Big Data focus areas of astrophysics, health data and sustainable agriculture. The project also provides training courses in machine learning, big data techniques and data intensive methodologies as part of the Big Data Africa initiative.

Keywords. Radio Astronomy, Data Intensive, Big Data

The Development in Africa with Radio Astronomy (DARA) Big Data project is a sister program to the original DARA Project (www.dara-project.com), which provides training in radio astronomy to support the African partner countries of the Square Kilometre Array (SKA) project develop additional STEM capacity in high tech areas, ahead of the SKA and African VLBI Network (AVN) telescopes. The DARA Big Data project (www.darabigdata.com) builds on this capacity development to encourage translation of STEM skills from radio astronomy into other data intensive fields.

DARA Big Data provides bursaries for students from the AVN countries to study for MSc(R) and PhD degrees at universities in South Africa and the UK. These degrees are in the three data intensive DARA Big Data focus areas of astrophysics, health data and sustainable agriculture, and include co-supervision from academics in AVN countries. In addition to providing studentship bursaries, DARA Big Data also works in partnership with the South African SKA project, now incorporated into the South African Radio Astronomy Observatory (SARAO), and the Institute for Data Intensive Astronomy (IDIA; www.idia.ac.za) on the broader Big Data Africa training program. Big Data Africa provides training workshops in machine learning, big data techniques and data intensive methodologies across the three DARA Big Data focus areas. These workshops and training courses currently take place in South Africa, are open to students from across the AVN country network, and provide financial support for attendees.

Astronomy in Focus - XXX
Proceedings IAU Symposium No. XXX, 2018
M. T. Lago, ed.

doi:10.1017/S1743921319005441

Overview of the Astronomy Education Research landscape

Paulo S. Bretones
Departamento de Metodologia de Ensino/UFSCar, Brazil
Email: bretones@ufscar.br

1. Introduction

When the landscape of a research area is analyzed, its scientific production is mainly identified by the following publications: theses, conference proceedings and journal papers. To investigate this output the works may be classified and analyzed by the categories: year, country, institution, grade level, topic/content, focus of study in education, type of academic research, theoretical framework. The present work deals with review articles published in Astronomy Education Research (AER), their results and suggestions. Examples of surveys and summary reviews are: Wall (1973); Bishop (1977); Bailey & Slater (2004); Bailey, Prather & Slater (2014); Slater (2008); Lelliott & Rollnick (2010); Franknoi (2014). Some articles also published reviews about students' conceptions of astronomy concepts as Trumper (2001).

About reviews of proceedings, as an example, Bretones & Megid Neto (2011) analyzed 283 papers dealing with astronomy education published in the IAU proceedings from 1988 to 2006. The results show the majority of each category as: country (USA-35.6%); grade level: university education (37.8%) and public outreach (27.6%); focus: non-school-programs (28.3%), curricular programs (26.3%) and teaching materials (18.4%); content: General (83.4%); type of academic research: Reports of Education Experience (R&D, Reports of Practices etc.): (67.5%). 87.6% did not show any theoretical frameworks.

Concerning journals Lelliott & Rollnick (2010) reviewed 103 peer-reviewed journal articles from 1974 to 2008 using a conceptual framework of "big ideas" in astronomy, five of which accounted for over 80% of the studies: conceptions of the Earth, gravity, the day-night cycle, the seasons, and the Earth-Sun-Moon system. Most of the remaining studies were of stars, the solar system, and the concepts of size and distance. About published papers in journals, an example of analysis can be seen in Bretones, Jafelice & Horvath (2016) about the first ten years of the Latin-American Journal of Astronomy Education (RELEA), with 103 articles published in 24 editions (2004 – now) and some discussions about the Teaching of Astronomy in Asian-Pacific Region which published 20 issues featuring 171 articles (1990-2003) and Astronomy Education Review which published 19 issues featuring 255 articles (2001-2013).These surveys show trends and gaps, already discussed in the literature of the area and point towards the less addressed contents and recommendations for further work.

The answers to a questionnaire of IAU CC1 and WG on Theory and Methods in Astronomy Education members about the achievements and challenges of AER are discussed. About the achievements and impacts of AER mentioned in the last decades were: effective techniques for teaching astronomy; construction of a variety of concept knowledge inventories; strategies for alternative conceptions assessment and development of classroom techniques to overcome them; development and evaluation of active learning; creation of journals; publications of theses or dissertations, conference proceedings and journal papers. Among the objectives and challenges of AER for the next decades, our colleagues mentioned: consolidating the achievements; deeper treatments dealing with epistemological questions; increasing the methodological rigor; development of models to connect new technologies in a variety of contexts and instruments to probe student attitudes; astronomy to improve science education and to link other branches of culture; investigate the roots of astronomy in each nation and respect multiculturalism. Regarding the efforts of the WG, recent surveys of publications from some countries are also shown evidencing the dispersion of AE literature. The role and goals of astronomy teaching should be discussed considering contents, methods, levels, resources and purposes. Given the needs and complexity of education nowadays and the role of astronomy in this context, the potential of education research is also evaluated, considering knowledge, practices, policies and the training of teachers.

An important issue is about the formation of a community in an area where the astronomers are trained as scientists and the need of training of education researchers. About this, the different approach of hard sciences, that advance from their points of arrival differently from the social sciences in which education is inserted that advance from their starting points is pointed out and discussed. Because of this, educational sciences have lack of memory and for their researches, it is necessary to know about what have been done and previous results as mentioned by Charlot (2006). Considering the memories of the publications on astronomy education research and accepted astronomy teaching practice, Franknoi (2014) mentions the importance of a journal for the community.

Finally, collaborations for surveys, literature reviews and the advertisement of such materials, aiming to strengthen the training of researchers and the practitioners as well are also encouraged.

Acknowledgements

I would like to express my deepest thanks to Jorge E. Horvath for his suggestions. I also thank the IAU and FAPESP for the travel grants and financial support (grant 2018/07912-9, São Paulo Research Foundation (FAPESP)).

References

Bailey, J. M., Prather, E. E. & Slater 2014 *T. F. Advances in Space Research*, 34, p. 2136–2144

Bailey, J. M., & Slater, T. F. 2004 *Astronomy Education Review*, 2, p. 20–45

Bretones, P. S.; Jafelice, L. C. & Horvath, J. E. 2016, *Journal of Astronomy & Earth Sciences Education*, v. 3, n. 2, p. 111–124

Bretones, P. S. & Megid Neto, J. 2011, *Astronomy Education Review* v.10, n. 1

Bishop, J.E. 1977, *Science Education* 61, p. 295–305

Charlot, B. 2006, *Revista Brasileira de Educação* v. 11 n. 31. p. 7–18

Fraknoi, A 2014, *Journal of Astronomy & Earth Sciences Education* 1, 37–40

Lelliott, A. & Rollnick, M. 2010, *International Journal of Science Education*, 32, 1771

Slater, T. F. 2008, *Astronomy Education Review* 7, 1

Trumper, R. 2001, *International Journal of Science Education* v. 23, n. 11, p. 1111–1123

Wall, C. A. 1973, *School Science and Mathematics*, 73, p. 653–669

Astronomy in Focus - XXX
Proceedings IAU Symposium No. XXX, 2018
M. T. Lago, ed.

doi:10.1017/S1743921319005453

AstroAccess: Creative Approaches to Disability Inclusion in STEM

Anna Voelker[1]

[1]The Ohio State University, Columbus, OH 43210, USA
Email: voelker.30@osu.edu

Disability-based exclusion is a pervasive issue in the fields of astronomy and STEM as a whole. In the United States, nearly 20% of the population has a disability [1] and yet according to the National Science Foundation, people with disabilities make up only 8.4% of the country's employed scientists and engineers [2]. Furthermore, approximately 1% of U.S. doctorate degrees in science and engineering are held by people with disabilities [3].

From an underrepresentation of researchers with disabilities to a deficit of educational materials available for students with specialized needs, there are a multitude of accessibility barriers that need to be acknowledged and addressed. AstroAccess is a yearlong astronomy outreach initiative designed to tackle this problem.

This project is being supported by The Ohio State University's President's Prize [4]. It began at the International Astronomical Union General Assembly (IAU GA) in Vienna with a presentation on science accessibility during Focus Meeting 15. This overview highlighted Sensory Friendly Day (SFD), an inclusive outreach event that took place at the Center of Science and Industry (COSI), Columbus, Ohio's science centre. SFD was designed to make COSI more accessible for children with developmental disabilities and featured a series of science theatre games that I modeled after Kelly Hunter's Shakespeare and Autism Program [5].

After the IAU GA, the AstroAccess initiative began a two-month collaboration with the IAU Office of Astronomy for Development, in Cape Town, South Africa. I worked with Dr. Wanda Díaz Merced to develop astronomy accessibility guidelines and organize local outreach programming for blind students in the Cape Town community. The AstroAccess project culminated in SciAccess, a science accessibility conference that took place at The Ohio State University in Columbus, Ohio on June 28 and 29, 2019. This interdisciplinary conference drew approximately 250 people from around the world to address barriers to science faced by people with disabilities. It featured over 60 speakers, including keynote presentations by Dr. Temple Grandin, renowned autism advocate, and Anousheh Ansari, the first female private space explorer. To learn more, please visit sciaccess.org or email sciaccess2019@gmail.com.

References

[1] US Census Bureau Public Information Office. "Nearly 1 in 5 People Have a Disability in the U.S., Census Bureau Reports." Census.gov, United States Census Bureau, 25 July 2012

[2] "Accountability for Broadening Participation in STEM: CEOSE 2015-2016 Biennial Report to Congress." NSF.gov, Committee on Equal Opportunities in Science and Engineering, 2017

[3] Sevo, Ruta. "Recommended Reading: Disabilities and Diversity in Science and Engineering" In B. Bogue & E. Cady (Eds.). Apply Research to Practice (ARP) Resources, 2012

[4] The Ohio State University President's Prize: https://presidentsprize.osu.edu/about/

[5] Shakespeare and Autism: https://shakespeare.osu.edu/autism

Astronomy in Focus - XXX
Proceedings IAU Symposium No. XXX, 2018
M. T. Lago, ed.

doi:10.1017/S1743921319005465

A 3D Universe? Students' and professors' perception of multidimensionality

Urban Eriksson[1] **and Wolfgang Steffen**[2]

[1]NRCF, Department of Physics, Lund University, Box 118, 221 00, Lund, Sweden
Email: urban.eriksson@fysik.lu.se

[2]Instituto de Astronomia, OAN, UNAM, Ensenada, Mexico. Email: wsteffen@astro.unam.mx

Abstract. This paper discusses the importance of learning to understand the three-dimensionality of astronomical objects, in particular nebulae. After collecting data from students' and professors' discernment of 3D we finds that this is difficult for both students and professors, which highlights the importance of addressing extrapolating three-dimensionality in astronomy education.

Keywords. Spatial thinking, 3D, Astronomy Education Research

The competency to be able to extrapolate three-dimensionality (E3D) in one's mind from 1D and 2D representations has been identified as an important factor for success in learning astronomy and understanding the Universe. However, only little research has been done in investigating this competency (Eriksson *et al.* 2014, Heyer *et al.* 2013), while at the same time there is a growing interest for what and how 3D representations can contribute to learning astronomy (Cole *et al.* 2018). This paper discusses the competency to E3D in one's mind and reports on the preliminary findings from an investigation concerning students' and professors' perception of three-dimensionality when looking at 2D representations, images and simulations, of a sample of nebulae. Images are 2D and simulations can offer a sense of 3D by offering parallax motion–this is often referred to as psudo-3D. Through an on-line questionnaire university students and professors are exposed to various images and simulation of nebulae (Steffen *et al.* 2007) and asked for their perception of depth, both by numbering certain features and in explaining their reasoning. The data collection, which is still ongoing, will be analysed using a standard qualitative research methodology. The preliminary results indicate that the competency to E3D vary significantly between the participants and in particular many students struggle to see nebulae as 3D objects, which confirms earlier studies (see, for example, Eriksson *et al.* 2014, Heyer *et al.* 2013). An awareness of these findings by astronomy educators may have great importance for how teaching and learning astronomy are viewed and also how curricula development could be enhanced for optimizing astronomy education at university level. From this, and previous research (Eriksson *et al.* 2014), we recommend astronomy educators to consider, and take into account, students' difficulties in E3D regarding astronomical objects when teaching. Finally, we recommend educators helping students discern relevant 3D aspects and features of the representations used.

References

Cole, M., Cohen, C., Wilhelm, J., & Lindell, R. 2018, *Physical Review Physics Education Research*, 14, 1

Eriksson, U., Linder, C., Airey, J., & Redfors, A. 2014, *Science Education*, 98(3), 31

Heyer, I., Slater, S., & Slater, T. 2013, *Revista Latino-Americana de Educação em Astronomia - RELEA*, (16), p. 45-61

Steffen, W., Koning, N., Wenger, S., Morisset, C., & Magnor, M. 2007, *IEEE Transactions on Visualization and Computer Graphics*, 17(4), p. 454-465

Astronomy in Focus - XXX
Proceedings IAU Symposium No. XXX, 2018
M. T. Lago, ed.

doi:10.1017/S1743921319005477

Evaluating Quality in Education: NASE new metrics

Rosa M. Ros[1] **and Beatriz García**[2]

[1]Universdad Politécnica de Cataluña, Spain.
Email: rosamariaros27@gmail.com
[2]ITeDAM-CONICET-CNEA-UNSAM, UTN Mendoza, Lab. Pierre Auger

Abstract. The Learning Services Management System of the Network for the Education of Astronomy in the School (IAU-NASE) has been developed following the guidelines of the ISO 29990: 2013 Standard, which understands on the "Learning services for non-formal education and training", and which aims to improve quality of learning services and facilitate comparison on worldwide basis.

Keywords. NASE Program, ISO 29990:2013

---

1. Introduction

The Network for Astronomy School Education (NASE) has performed more than 110 courses between 2010 and 2017 on the Globe and reaches more than 5000 professors. The benefits arising from the work under a International Quality Management System (QMS) ensure not only quality in teaching-learning processes, but also a method to evaluate the service and produce continuous improvement. Based on specific indicators we could analyse quantitatively the impact of the activity under different contexts and cultures.

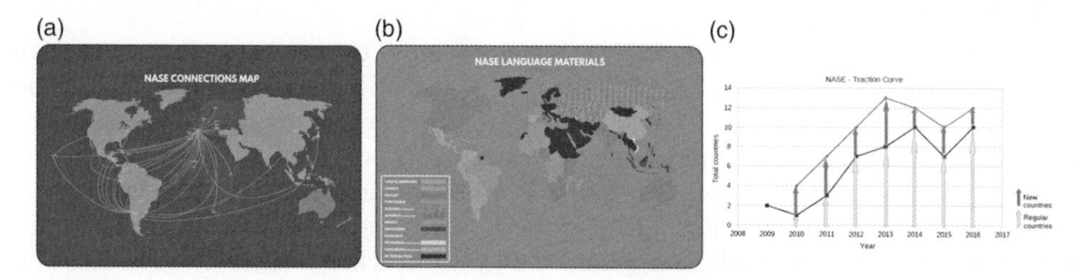

Figure 1. NASE connections: ambassadors(a); NASE translations(b); NASE traction curve(c)

2. QMS New Metrics

The evaluation of the process of Quality, which means make plans, evaluate performance and set indicators was followed to analyse the impact of NASE. In this sense, evaluations of the assistants pre and post workshops, satisfaction poll and mid and long term evaluation as well us other techniques like traction curves (Fig. 1c) were used to conclude that NASE is one of the most successful IAU programs on Education.

3. Conclusion

The progress of NASE can be summarise in the following topics: expansion (110 courses since 2010, in 4 continents), translations (didactic material in 7 languages), connection (NASE ambassadors teaching in different countries), communication (books, newsletters).

Astronomy in Focus - XXX
Proceedings IAU Symposium No. XXX, 2018
M. T. Lago, ed.

doi:10.1017/S1743921319005489

A Pilot Project to Evaluate the Effect of the Pale Blue Dot Hypothesis

Ramasamy Venugopal[1] **and Kodai Fukushima**[2]

[1]South African Astronomical Observatory,
1,observatory road Cape Town, South Africa
Email: rv@astro4dev.org

[2]The University of Tokyo, Tokyo, Japan
Email: kodai.fukushima417@gmail.com

Abstract. Astronomy and Space topics are perceived as holding universal fascination. It is widely considered that exposure to such topics inspires people, changes their perspective and leads to an uptake in science and STEM subjects. But very rarely is the impact of such communication evaluated rigorously and scientifically. There is a need for more rigorous evaluation methods which would reveal the successes and failures of current methods and tools of astronomy communication and whether they might lead to any inadvertent harm. The IAU Office of Astronomy for Development (OAD) and Hosei University together with the South African Astronomical Observatory conducted a randomised controlled trial (RCT) in Cape Town, South Africa to test whether exposure to an astronomy intervention affects empathy and altruism in children. The pilot demonstrated that it is possible to use such methods to evaluate impact of science communication in an inexpensive manner.

Keywords. randomized controlled trial, evaluation, pale blue dot

1. Experiment

From October to November 2015, the astronomy outreach project One World Experiment was carried out among 938 secondary school students in Cape Town, South Africa. The main objectives of the study were a) to test whether exposure to an astronomy intervention affects empathy and altruism in children b) prove feasibility of evaluating an astronomy intervention in a low-cost manner. Participants were randomly assigned (as a class group) into experimental and control groups. The experimental group received an astronomy intervention where a qualified Physics teacher taught the students about their place on Earth and in the solar system. The students were taken on a tour of the solar system to foster the idea of One Common Humanity. The intervention was followed by two measurements: 1) a voting process intended to test the helping behaviour of the children toward children from other groups. 2) questionnaire to test their feelings towards children from ingroup and outgroup. There were two sets of five questions, one for the home country and the other for the chosen foreign country.

2. Conclusion

The RCT was implemented as a pilot project to test the feasibility of adding and running a low-cost evaluation component to a typical educational intervention at the school level. We hope this pilot initiative will guide others interested in repeating this experiment. Full analysis of the data, which will compare the control group and experimental group results and focus on the impact of the astronomy intervention, will be published in the future.

Astronomy in Focus - XXX
Proceedings IAU Symposium No. XXX, 2018
M. T. Lago, ed.

doi:10.1017/S1743921319005490

Astronomy as entrance to STEAM capacity building

Premana W. Premadi[1]

[1]Bosscha Observatory, Institut Teknologi Bandung, Lembang 40391, Indonesia
Email: premadi@as.itb.ac.id

The reasons that make astronomy appealing to all are precisely the ones that make astronomy a good entrance to Science Technology Engineering Arts Mathematics (STEAM) education. Astronomy is universal and inspirational, relates to all sciences, pushes technology forward, and invites people to reflect. Those reasons are woven to construct the goals of learning STEAM via astronomy, which are: To raise awareness and comprehension of the physical world and natural causal relation; To promote rational perception and respect towards abiotic and biotic environments, including human being; To promote curiosity, knowledge, skill, creativity in STEAM; and To promote positive participation and mindfulness. Those four goals resemble and emphasize the key spirit of the Sustainable Development Goals.

We apply this approach in an area surrounding the site for a new astronomical observatory in Timor island, Indonesia. The sky quality is excellent, yet the national development is still far behind, with minimal infrastructure and very low living condition. The challenge is to ensure the coexistence of a modern observatory and thriving villages. We ask ourselves how astronomy, with its high STEM level, assists a community to develop. The key is to find a common ground for the observatory and villages to grow together and be clear about what each aspires to be and the requirement for success. STEAM has the following elements in its embodiment of knowledge and skill learning: empowering rational thinking, encouraging life-long learning, expanding horizons, promoting creativity, and fostering teamwork. For an observatory to optimally function and produce high quality science in a region shared with a community eager to move forward, it must have a clear goal and strategy that embody sustainable development ideas. To engage and to empower rational thinking it is mandatory that we first learn and acknowledge the existing way of life and relation with the nature. This is important to understand the local mindset and identify obstacles in the logic that could hinder development. Then we together identify the challenges and construct a roadmap towards the future. Astronomy comes in from a wider perspective and then zooms in towards some chosen STE(A)M challenge. Since electricity and clean water are not yet available, they are the centre topic. Most astronomy material to be introduced eventually lead to water and energy, and are prepared according to learners' cognitive level. Deliverable topics are: Earth and the Universe, Living on Earth, Water, Energy, Human and STEM, and Good data. The strategy for proper deliverance is by building teamwork, optimizing communication, and using training and as empowerment modes.

Acknowledgement

We are grateful to: our collaborators: UNAWE Indonesia, Nusa Cendana University, the Indonesian Institute for Energy Economics, Kupang Polytechnic Institute; the IAU OAD grant 2016, grants from Institut Teknologi Bandung and The Indonesia Science Fund; and conference travel grant from the Leids Kerkhoven Bosscha Fonds.

Astronomy in Focus - XXX
Proceedings IAU Symposium No. XXX, 2018
M. T. Lago, ed.

doi:10.1017/S1743921319005507

The Columba-Hypatia Project: Astronomy for Peace†

Francesca Fragkoudi[1]

[1]Max-Planck-Institut für Astrophysik, Karl-Schwarzschild-Str. 1, 85741 Garching, Germany
email: ffrag@mpa-garching.mpg.de

Abstract. "Columba-Hypatia: Astronomy for Peace" is a joint astronomy outreach project by GalileoMobile and the Association for Historical Dialogue and Research (AHDR) which takes place on the divided island of Cyprus. The project aims to inspire young people, through astronomy, to be curious about science and the cosmos, while also using astronomy as a tool for promoting meaningful communication and a Culture of Peace and Non-violence. We conduct educational astronomy activities and explore the cosmos with children and the public, bringing together individuals from the various communities of Cyprus 'under the same sky' to look beyond borders and inspire a sense of global citizenship.

Keywords. miscellaneous

---

1. Implementation

The main phase of the project began in 2017, and ran throughout the entire year. The project involved mono-communal school visits – where trainers visited Greek-Cypriot (GC) and Turkish-Cypriot (TC) schools separately in Cyprus – and bi-communal activity days, were children from GC and TC schools came together in the United Nations-controled 'buffer zone', to meet each other and participate in astronomy activities.

During the mono-communal visits we focused on introducing the children to the project, and to basic astronomical concepts. The activities were chosen to give the children an idea of the place of the Earth in the context of modern astronomy and to introduce them to the vast scales and sizes of the Universe, as well as introducing the concept of the Earth as a 'Pale blue dot'‡.

During the bi-communal days, children from the GC and TC communities came together in the buffer zone to participate in astronomy activities, such as the "Building a Cyprus Golden Record" activity. During this activity the children were split into mixed groups and were given the opportunity to discuss between them what they would send to an alien civilisation as a representation of the whole of the island of Cyprus. This allowed the children to get to know each others' cultures, and to discover the many similarities between them, while also appreciating the differences and rich diversity of the island.

2. Concluding remarks

Through this project, we experienced how effective astronomy is in promoting a feeling of global citizenship and a culture of peace, enabling children to broaden their views and interactively explore together their place on Earth and, specifically for our project, on the island of Cyprus. The project was implemented successfully during 2017, and current and future instalments are ongoing and planned for the coming years.

† https://www.columbahypatia-project.org/
‡ http://www.planetary.org/explore/space-topics/earth/pale-blue-dot.html

Astronomy in Focus - XXX
Proceedings IAU Symposium No. XXX, 2018
M. T. Lago, ed.

doi:10.1017/S1743921319005519

Considering the Astro-tourism Potential in Indonesia using GCIS-MCDA

Dwi Y. Yuna[1] **and Premana W. Premadi**[2]

[1]Department of Astronomy, Institut Teknologi Bandung, Jln. Ganesha no.10, Bandung, Indonesia email: dwiyyuna@gmail.com

[2]Bosscha Observatory, Lembang, Indonesia email: premadi@as.itb.ac.id

Abstract. We developed a method to identify potential astro-tourism sites by considering parameters characteristically relevant to astronomical observation such as air quality, dark sky quality, annual average cloud coverage, as well as terrain feature. Applying this method on Indonesia by perusing data from Geographic Information System and applying Multi Criteria Decision Analysis we identify a number of potential astro-tourism sites. We cross correlate this with Indonesia Tourist Destination to produce a list of recommended sites. Fulfilling the astrotourism criteria is one sure way towards sustainable tourism.

Keywords. Astro-tourism, Geographic Information System (GIS), Multi-criteria Decision Analysis (MCDA)

Indonesia has numerous tourist destinations which makes tourism one of Indonesias economic leading sectors. However, sustainability aspect is a concern, mainly due to various environmental issues yet to be addressed. We introduce astro-tourism – as a form of sustainable tourism – in several potential existing tourist destinations.

We examine the astro-tourism potential in Indonesia, in particular the 80 points of strategic natural tourist destinations, which includes 6 geoparks. We utilized Geographic Information System (GIS) (Malczewski (2006)), considered four astro-climatogical factors (Graham *et al.* (2005)) such as artificial light, aerosol optical thickness (AOT), altitude and cloud coverage. We set up a classification model for the astro-tourism potential sites over those 80 destinations. as an approach to prioritize the development strategies for astro-tourism in potential location.

About 73% of Indonesian area has good night sky with low light pollution and nearly half of area has good air quality suitable for astronomical observation. With nearly 90% of the area having more than 70% annual cloud coverage a tourism strategy needs to be developed. This leads to 14 out of the 80 investigated existing tourist destinations recommended as astro-tourism sites.

Indonesia has considerable astro-tourism potential. The success of astro-tourism will not only enhance local economy development but also be beneficial for the advancement of astronomical works in Indonesia, public science education, whilst ensuring sustainable development through environmental protection.

References

Graham et al. 2005, *Meteoroligal Aplications*, 12, 77-81
Malczewski, E. 2006, *Int. J. Geogr.Inform. Sci*, 20, 703-726

Astronomy in Focus - XXX
Proceedings IAU Symposium No. XXX, 2018
M. T. Lago, ed.

doi:10.1017/S1743921319005520

Cultural astronomy perspectives on "development"

Alejandro Martín López

CONICET, Sección de Etnología, ICA, UBA, Bs. As., Argentina
Email: astroamlopez@hotmail.com

Keywords. Development, interdisciplinarity, social sciences, cultural astronomy

---

1. Introduction

"Astronomy for development" has been a priority for the IAU in recent years, as is shown by the strategic plans 2010–2020 and 2020-2030. These plans insist on the interdisciplinary nature of this goal. But, the asymmetric relations among academic disciplines is an obstacle to this effort. It implies strong preconceptions about the social sciences among astronomers that are even reflected in the mentioned strategic plans. For this reason, it is crucial to include the perspective of cultural astronomy. This is an interdisciplinary area that deals with understanding the systems of knowledge and practices about the sky of different cultures. Many members of the IAU have devoted themselves to this field. They have important experience and knowledge about interdisciplinary work, especially with the social sciences.

As a brief example of the possible cultural astronomy contributions, we can mention the necessity of analyzing the very idea of "development", a matter of strong debate in social sciences. We see the implications for the three areas recurrently mentioned in the debates on astronomy for development: education, heritage and economics.

Regarding education, cultural astronomy has very important contributions to do, because a true scientific education in the contemporary World must be an intercultural education. An education for a World with a great cultural diversity in a context of inequality. Cultural astronomy also plays a crucial role in the joint initiatives of the IAU and UNESCO on astronomical heritage. Here we need to break a static vision of heritage and also take into account that heritage has become a language for a great variety of conflicts (López 2016). Finally, in reference to economy, we must be especially attentive to the impact on local populations of the great international astronomical facilities. Many recent cases (Swanner 2013) show that we are following the ways of the large extractive industries instead of setting agenda to them. In a world in which imposition and authoritarianism are often the easiest way out, the astronomical community has the opportunity to show that reason, dialogue and listening to the other are the right path.

Supplementary Materials

*Fig 1: Example of misconceptions about social sciences.
*Table 1: Terminology and preconceptions about social sciences among astronomers.
*Extended References.

References

López A. M. 2016, *Astronomical Heritage and Aboriginal People: Conflicts and Possibilities. In: Benvenuti, P. (Ed.), 2016. Astronomy in Focus. As presented at the IAU XXIX General Assembly 2015, Cambridge: Cambridge University Press.* 142-145.

Swanner, L. A. 2013, *Mountains of Controversy: Narrative and the Making of Contested Landscapes in Post-war American Astronomy. Ph.D. Thesis. Cambridge: Harvard University.*

Astronomy in Focus - XXX
Proceedings IAU Symposium No. XXX, 2018
M. T. Lago, ed.

doi:10.1017/S1743921319005532

Scope for Citizen Science and Public Outreach Projects in the Developing World

Aswin Sekhar[1,2]

[1]CEED, Faculty of Mathematics and Natural Sciences, University of Oslo, Norway
[2]Armagh Observatory and Planetarium, Armagh BT61 9DG, United Kingdom
emails: aswin.sekhar@geo.uio.no, aswin.sekhar@armagh.ac.uk

Abstract. Both science public outreach and citizen science projects have gained immense momentum in the western world in recent years. This work aims to outline some ongoing and possible future projects in the developing world on this front. Cooperation and collaboration with IAU-OAD is envisaged.

Keywords. Public Outreach, Citizen Science, Developing World, Science Projects

1. Brief Introduction

The potential and scope for more science public outreach projects in the developing world is simply astronomical! There are various possibilities for such projects at multiple levels namely global, national, state and local levels.

2. Engagement with Public - Multiple Levels

2.1. *Mega Science CERN India Exhibition - Global Level*

An initiative was taken from the Indian side to bring CERN travelling exhibition to our country. This exhibition aims to reach 100,000s of Indian students and general public next year. Top agencies of Govt of India have agreed to support this project.

2.2. *Remote Telescope Observations - Global Level*

Las Cumbres Observatory Global Telescope Network provides opportunities to school students in EU and US to do short remote observing projects. In near future, the idea is to extend this opportunity to many school students in India.

2.3. *Citizen Science and Data Mining - National Level*

Embracing citizen science projects through Indian partners is aiming for young students, amateur astronomers and ordinary Indian citizens to identify and discover new astronomical objects using extensive observational databases and data mining techniques.

2.4. *School Visits to Observatories and Research Institutes - Local Level*

Opportunities are being given to students from tribal and rural school backgrounds to visit world class observatories and research institutes in order to instill scientific spirit and excitement in them.

3. Summary and Discussion

Such projects at multiple levels eventually lead to creation of budding stars in astronomy! Moreover such efforts aim to address the substantial gender and diversity gap issues of STEM subjects in the long run.

Astronomy in Focus - XXX
Proceedings IAU Symposium No. XXX, 2018
M. T. Lago, ed.

doi:10.1017/S1743921319005544

The Digital Revolution, Open Science and Development

Geoffrey Boulton

University of Edinburgh & the International Science Council
emai: G.Boulton@ed.ac.uk

Global society is in the throes of a revolution in the means by which information and knowledge are acquired, stored and communicated, which have always been powerful drivers of human material and social progress. Cost reduction and increased flexibility as consequences of replacement of analogue by digital systems, that permit any device with its own power source to collect non-trivial information about its environment, have been the drivers of this revolution. It has been mediated through four key capabilities: high rates of data acquisition; ever increasing data storage capacity; ever increasing computational power; and the growth of the world-wide-web and the ubiquitous communication devices that interrogate it.

This phenomenal growth in the amount and variety of available data has brought the AI learning algorithms that were developed decades ago into their own, for we are now able to feed their voracious appetites for data at rates that were formerly impossible. They are able to reveal hitherto inaccessible but profound relationships in nature and society on all scales from the molecular to the cosmic and in all areas of human concern, from fundamental business practices to the efficient delivery of health systems to global sustainability. There are few areas of individual, commercial, social or political action to which these developments are not relevant. The high rates at which data can be acquired by sensors and ingested and analysed by machine learning algorithms is now also the basis of autonomous systems able to make judgements once thought to be the exclusive prerogatives of humans, and indeed to act on those judgements.

The greatest novel dimension for science lies in enhancing understanding of complex systems. During the 50-60 years since the development of the modern high performance computer, we have become increasingly adept at analysing the dynamics of complex, coupled systems, such that integrated modelling is now the norm. In principle we now also have the capacity to characterise complexity by integrating data from a variety of sensors that contribute data from different perspectives that relate to the same complex phenomenon. The capacity to model complex system dynamics coupled with identification of deep patterns in system states is a powerful new tool in the scientific armoury. However, except for across relatively narrow disciplinary ranges, such as in modern weather forecasting, this is currently impossible because of the different standards and inadequate or incompatible vocabularies and ontologies used in compiling data. The International Science Council's (ISC) Committee on Data for Science and Technology (CODATA) is currently developing a decadal programme, together with ISC unions and associations, to work on this crucial priority for interdisciplinary data integration. These capacities will also be vital in addressing the UN Sustainable Development Goals (SDGs), such as in understanding the complexities of cities as organisms, where most global population growth is located, in understanding the vectors of infectious disease or in evaluating sustainable routes to increased agricultural productivity.

If science is to exploit this potential, it must break out of the siloes within which individuals and groups tend only to have access to a limited range of data reflecting their own disciplinary focus. We have the choice, either to maintain a system were we only have access to that data which we "own", or to develop one where we have access to a much wider range of the data from many disciplines, and which is the necessary pre-requisite to address many of the complexities of this world. I argue that the latter offers the greatest global public good. However, it is a route that is inimical to deeply engrained practices amongst scientists and their institutions, where data that is generally acquired from publicly funded research is deemed to be "owned" by scientists or by their institutions. The public good argument should take precedence, with scientists and institutions moving from a model of ownership to one of access, as has been argued by Science International†.

As the new paradigm of "Open Science" develops, it begs the question: open to whom? Open data and open access publishing still only represent, scientists talking to other scientists, albeit more efficiently. I argue that science must become a public enterprise that engages actively with business, policymakers, governments, communities and citizens, as a knowledge partner, not as a unique source of wisdom, in jointly framing questions and jointly seeking solutions rather than an enterprise conducted behind closed laboratory and library doors or one that is only found in the publicly incomprehensible pages of scientific journals. Without this development science will fail to contribute its unique insights to a changing society.

The web is a crucial part of the infrastructure for open science. It has created a globally networked society that enables unprecedented access to data and information by a great diversity of public and private actors in the creation and use of knowledge. However, although the web may be ethically neutral, its uses are not. A decade ago, it was expected that the web would be a democratising force, enabling debate and engagement between different social actors and inhibiting the creation of knowledge monopolies. The reality has been different. Major web platforms and social media have fostered the development of "echo chambers" in which we are confronted only with material to which we are personally sympathetic. The web has proven to be indifferent to falsehood and honesty, and a powerful enabler of "alternative facts" in a so-called "post truth society". It has been the essential weapon in an environment where interest groups energetically question the credibility and authority of scientific evidence and the trust that science should be afforded.

Further profound issues for society arise from the digital revolution in the reality and potential of autonomous systems powered by AI for: the future of work, cyber-crime and cyber-warfare, the imminent possibility for brain implants, the dangers of privatisation of data and knowledge as data is increasingly valued as feedstock for voracious machine-learning systems, whether and how we create global governance for the data needed to support management towards planetary sustainability, and in analysing the potential existential threat AI might pose to the human role on the planet. The temptation for many scientists is to take the ethically neutral stance of focussing on discovery without considering use. It is now fundamentally important that we do not take the position that these problematic issues are for government and their agencies alone. We in the natural sciences should be working with our social science colleagues to address these issues in a holistic fashion. The opportunity to do so lies in the frame created by the new International Science Council, and should be grasped.

Scientific exploitation of the digital revolution is inevitably faster in those states with well-funded science systems, with the potential to create yet another global north-south

† Note the principles set out in the Science International Accord, Open Data in Big Data World, endorsed by over 120 major scientific bodies worldwide.

knowledge divide. But there are promising moves to avoid this outcome. Collaboration between ISC, CODATA and major African institutions has led to the concept of an African Open Science Platform, to be launched at Science Forum South Africa in December 2018. The Platform's mission is to put African scientists at the cutting edge of contemporary, data-intensive science as a fundamental resource for a modern society. Its building blocks are:

- a federated hardware, communications and software infrastructure, including policies and enabling practices, to support open science in the digital era;
- a network of excellence in open science that supports scientists & other societal actors in accumulating and using modern data resources to maximise scientific, social and economic benefit.

This may only be the beginning. A parallel development to create a similar platform for the ASEAN group of countries† is being led by the ISC regional office for the Asia-Pacific, and discussions are under way about possible creation of a Latin America and Caribbean Platform, with profound and exciting possibilities of a creative and influential south-south scientific enterprise.

† ASEAN – Indonesia, Thailand, Singapore, Malaysia, Philippines, Vietnam, Myanmar, Cambodia, Brunei, Laos

Astronomy in Focus - XXX
Proceedings IAU Symposium No. XXX, 2018
M. T. Lago, ed.

doi:10.1017/S1743921319005556

How publication and peer review are evolving in the life sciences: implications for astronomy and development

Liz Allen[1,2]

[1]F1000,
[2]Policy Institute, King's College London
email: Liz.Allen@f1000.com

The world of scholarly publishing is in flux. Many current and legacy publishing systems are known to be outmoded and dependent upon models of peer review that are burdensome and largely non-transparent; perhaps most importantly, the established system of scholarly publishing is thought to be a significant cause of inefficiency and research waste, an issue thought to be particularly acute for the life and biomedical sciences.

In a world of the web, where space restrictions largely disappear, there is a demand for more rapid and fuller access to research findings. In addition, research funders and institutions are being increasingly directive in their requirements for publicly-funded research findings to be made openly available; most recently exemplified by 'Plan S', announcing that a number of major European public research funders (including the EC and UKRI) will fund only the publication of articles that are made immediately (gold) open access (OA) from 2020.

A number of scholarly publishing initiatives have emerged that provide opportunities for researchers to share research faster and more fully. Perhaps most notably, signalling the researcher demand for more rapid models of publishing, there has been a massive growth in the number of 'preprint' articles posted across discipline specific pre-print servers, with *Crossref* noting an impressive 20% growth in the number of preprints being posted between 2016 and 2018 compared to growth of 2–3% of traditional research articles for the same period. And in 2013, F1000 launched F1000Research†, the world's first open research publishing platform, combining the ability to publish rapidly with functionality to ensure greater transparency, robustness and reproducibility of research; importantly, and uniquely, F1000Research provides a post-publication, open peer review model of scientific publishing.

Explaining F1000's publishing model

F1000Research combines the benefits of pre-printing (rapid publication) with expert, invited peer review (quality assurance). While open peer review is not an essential feature of a post-publication peer review model F1000Research chose to adopt perhaps the purest version of open peer review being used by scholarly publishers today‡ – requiring full disclosure and naming of reviewers. The purpose of peer review in our model is to validate and improve the research being shared, helping authors to improve the quality of their work, and providing readers with context to support potential use and re-use. To do this most effectively requires full transparency and openness and it is for this reason that open peer review is now commonly described as a 'pillar' of open science¶.

† https://f1000research.com/
‡ https://f1000research.com/articles/6-588/v2
¶ See for example: Foster Open Science - https://www.fosteropenscience.eu/learning/open-peer-review/

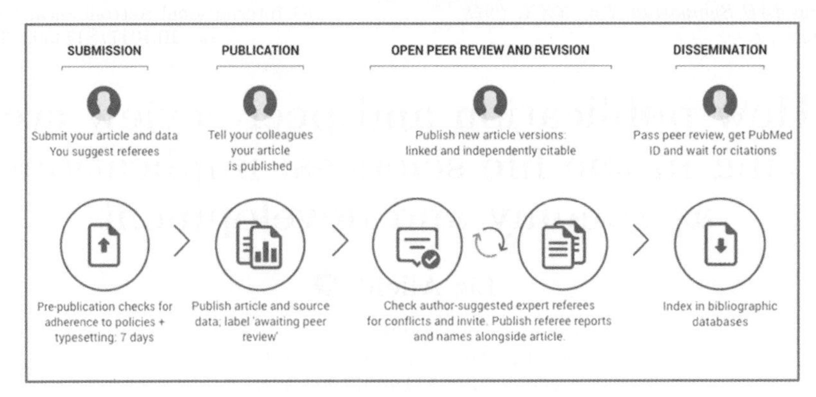

Figure 1. Overview of F1000's post-publication, open peer review publishing model

F1000 editors are not employed to screen content for interest, they instead conduct basic policy and author credential checks, and if the content adheres to our requirements, the work is published openly on the platform (see Fig. 1 for overview of process). Researchers can share a wide range of output types, including those that can be difficult to publish via traditional routes (e.g. case reports; datasets; research protocols; negative & confirmatory studies). And while the approach makes things easy for researchers from across the career spectrum, by removing the barriers and simplifying the process of publishing research, there are likely to be particular benefits for early career researchers who often find it difficult to build a portfolio of work at the start of a career.

As noted, peer reviewers are *invited* experts, whose affiliation is made public. Reviewers are required to provide a narrative report, which appears alongside the article, and to make one of three determinations: *approved*, *approved with reservations*, or *not approved* – if an article is 'not approved' it remains published. Once content is 'approved' it is indexed and fully discoverable on a range of international bibliographic databases. Full transparency of the peer review, together with an ability to cite reviewer reports independently of the article, provides a route for reviewers to gain credit and recognition for their contribution to the work of others and to science more broadly.

All content published is immediately (gold) OA and licenced by the authors under Creative Commons licenses (CC-BY for articles or CC0 for research data) without charge for others to view and access. Peer reviewer reports are also published under a CC BY license.

Working alongside research funding agencies and institutions

The volume of articles published on F1000Research continues to grow year on year and F1000 is now working with a number of high profile funding agencies and research organisations across the world to provide open research and data publishing services for the researcher community they support (including the Gates Foundation, and the African Academy of Sciences†). **In working with us, F1000's clients share our common high-level goals: to remove unnecessary delays and barriers that researchers face when sharing their research, to build research and researcher capacity, and to accelerate the use and potential impact of that research.**

Importantly, F1000 also provides a *transparent* and *cost-effective* route to achieving immediate open access (OA), removes the burden of article payment and management from the publishing authors, their research institutions and libraries, as well as removing subscription costs from the whole system.

† https://wellcomeopenresearch.org/; https://gatesopenresearch.org/; https://aasopenresearch.org/

Astronomy in Focus - XXX
Proceedings IAU Symposium No. XXX, 2018
M. T. Lago, ed.

doi:10.1017/S1743921319005568

Applying astronomy tools in the field of development economics

T. Chingozha[1] and D. von Fintel[1,2]

[1]Department of Economics, Stellenbosch University, Private Bag X1, 7602 Matieland, South Africa email: tchingozha@gmail.com
[2]Research affiliate, Institute of Labor Economics (IZA), Bonn, Germany

Abstract. Astronomy has largely relied on imagery and innovative data mining techniques. This article briefly illustrates the use of machine learning to create agricultural production data sets from satellite imagery to answer different development questions. Of late, astronomy has also relied on citizen science in the identification of new galaxies (a case in point is the Galaxy Zoo project). The use of citizen science in the examination of changes in urban informality patterns in developing countries (another development question) is also discussed.

Keywords. astronomy, development, multi-disciplinary, very high resolution, citizen science, machine learning

---

1. Introduction

The Sustainable Development Goals (SDGs) are an important yardstick through which the world's different societies can evaluate progress towards bettering the lives of their citizens. Within sub-Saharan Africa (SSA) and other developing nation contexts, the development of the agricultural sector is of paramount importance given that the majority of people live in rural areas and depend on land. Against the background of limited availability of data at sub-national level in SSA, remotely sensed data and related information extraction and analysis techniques become indispensable. This emphasises the importance of multi-disciplinary research and skills sharing amongst disciplines. Astronomy has an important part to play because it has traditionally centred on image analysis and has developed techniques and infrastructures for handling computationally intensive problems.

Devarajan (2013) observes the phenomenon of a "statistical tragedy in Africa" where governments are unable or do not have the political will to provide accurate, objective and accurate statistics that can be used by researchers to inform relevant policy. This therefore calls for multi-disciplinary approaches to data gathering and empirical analysis. Employing machine-learning techniques (borrowed from geographers and astronomers) to classify Landsat imagery and generate agricultural data illustrates benefits of multi-disciplinary research. Aside from the rural (agricultural) sector, citizen science may help identify and measure phenomena such as urban informality from Very High Resolution (VHR) imagery.

Urban poverty and high informality characterise developing countries and citizen science can help create a novel dataset on informality out of VHR satellite imagery so that policy makers may have an idea about the size and growth patterns of the shadow economy. The Zooniverse.org platform (originally "Galaxy Zoo") was developed by astronomers and physicists to harness the power of internet crowdsourcing and help in the discovery of new galaxies from images. The Zooniverse platform has successfully hosted different projects, and it could also help in creating a dataset on informality from VHR image classification as discussed in the next section.

2. Investigating size and growth patterns of the informal sector in Developing countries

Efforts to effectively reduce poverty and achieve inclusive growth largely depend on the availability of data on small-holder agriculture, informality and migration as these are intricately related to development. For instance, Bhattacharya (1995) criticises dualistic development models (Harris & Todaro 1970; Lewis 1954) in that they simplistically assume that ancient societies were solely engaged in food production yet there existed an informal economy in which they produced different goods and services. This highlights the informal sector as a pre-requisite for industrialization. Yet, by definition, the informal sector is unrecognized and there are no reliable estimates on its size.

We conduct a pilot study (with Stellenbosch University students) on the Zooniverse.org platform to test whether citizen science can be relied upon to accurately measure the size of the informal sector in developing countries. After viewing some tutorials, participants were asked to identify different land use types from VHR satellite imagery over Harare, the capital of Zimbabwe. Classification accuracy depends on a number of parameters. Thus, various parameters were manipulated in order to identify the best settings under which higher classification accuracy may be achieved. In citizen science, the end decision depends on consensus, thus the number of participants classifying a single image was varied in order to achieve the best threshold. The study takes advantage of a 2005 clean-up operation that destroyed informal structures in Harare. Thus, we are able to test whether or not volunteers can accurately identify *no changes* for 2004 and 2006 images for areas not affected (placebo) and accurately identify *changes* in areas affected (treatment) by the operation. This study is ongoing.

3. Conclusion

As development questions become complex, or where there are challenges, this study argues that multi-disciplinary approaches and collaboration are key to unlock value. It discusses techniques that have traditionally been employed by geographers, astronomers and others and makes an important case for multidisciplinary research. This is important if the SDGs can be achieved and the world can be a better place for everyone.

References

Bhattacharya, P. C.,1995. The economics of development: A review article. *Journal of Economic Studies* 22(2), 59-74

Devarajan, S., 2013. Africa's statistical tragedy. *Review of Income and Wealth*, 59(S1)

Harris, J. R., & Todaro, M. P. (1970). Migration, unemployment and development: a two-sector analysis. *American Economic Review*, 126-142

Lewis, W. A. (1954). A model of dualistic economics. *American Economic Review*, 36(1), 46-51

Astronomy in Focus - XXX
Proceedings IAU Symposium No. XXX, 2018
M. T. Lago, ed.

doi:10.1017/S174392131900557X

The Data Observatory, a vehicle to foster digital economy using natural advantages in astronomy in Chile

Demián Arancibia[1], Amelia Bayo[2], Guillermo Cabrera-Vives[2], Francisco Föster[2], Roberto González[2], Mario Hamuy[3], Juan Carlos Maureira[2], Peter Quinn[3], Juan Rada[3], Gabriel Rodriguez[3], Juande Santander-Vela[1], Massimo Tarenghi[3], María Teresa Ruiz[3], Mauro San-Martin[2] and Robert Williams[3]

[1]Astroinformatics Program Project Team
[2]Astroinformatics Program Advisor
[3]Astroinformatics Program Management Committee

Abstract. The Astroinformatics Program is funded by the Chilean Economy Ministry's (FIE Grant FIE-2016-V022, CORFO Grant 16IFI6626) with the mission to identify and initiate investments to foster Chilean Digital Economy, using Astronomy data-centric tools (known as astroinformatics). Over 2017 we worked with communities across sectors identifying opportunities to achieve the program mission, the Data Observatory vision emerged from that work and will guide design activities throughout 2018.

Keywords. large facilities, capacity building through astronomy, digital transformation, innovation policy

---

1. Astronomy and Chile

Astronomy Digital Transformation. Over the last century the technology of astronomical observatories has improved dramatically. The understanding of the origin and destiny of our universe has evolved accordingly, now we know how limited is our ability to explain most of it; there is so much more to discover. Progress in technology ignited a transformation in astronomy: knowledge that emerged from individual's minds, now flows from multi-disciplinary teams using data-centric tools. On one hand, data blooms from observatories; on the other, data bursts from cosmological simulations on computing clusters. Telescopes will produce zetta-scale datasets over next decade (Quinn *et al.* 2015), and theoretical astrophysics will generate similar data volumes and challenges.

The Capital of Ground Astronomy. Over the last decades, collaboration between the Chilean government and international observatories has brought 40% of Earth's telescopes to our territory. That share will grow to up to 55% in 2021 (Catanzaro 2014). The inauguration of instruments in the next decade will further enshrine the Atacama Desert as a capital of ground astronomy. According to our work, the volume of astronomical data acquired in Chile will go from ∼1 $^{Pb}/$year today, to ∼20 $^{Pb}/$year in 2021.

2. The Data Observatory

Our work shows that Chile has two main strategical advantages to achieve the astroinformatics program mission: a) the ability of the Chilean Government to establish a neutral-broker institution at the heart of astronomy, academy, and productive sector collaboration, embedded in our astronomy-interested-society and its natural environment, and b) the proximity of this institution to observatories, their teams and its users.

The Data Observatory (DO) vision emerges as a vehicle to exploit these advantages and achieve the program mission. The DO will honor the legacy of telescopes in our territory and will be a cross-continental system for both current and historical data and data-centric technology for storage, access, analysis, exploration, and visualization of the data produced here. The DO will operate under a quintuple-helix model (Carayannis *et al.* 2012), to coordinate the relationships between diverse actors and their contexts, and foster digital economy in Chile.

The DO & Astronomy. The DO will work with the global astroinformatics community and international observatories in Chile to be involved in the data-centric challenges of astronomy in the era of multi-messenger and large survey instruments.

The DO & the Virtual Observatory. The DO will foster and grow IVOA capacities, helping in the application of those standards for Chilean observations.

The DO & Academy. The DO will offer a unique data-centric fellowship system to form the talent of the future, providing hands-on training in challenges related to sophisticated analysis on high-value and volume datasets.

The DO & Cloud Computing Companies. An international consulting firm conducted research in the productive sector and showed that even though astronomy has obvious differences with industry, it's similar to it in the data-centric tasks (see next section 3). The DO will work with cloud providers to improve the region's industry capacity for data-centric challenges, through a bi-directional technology transfer system between astronomy and the productive sector.

The DO & Society. The DO will work for a very enthusiastic general public. Astronomy has a great position in Chile, it is by far the most productive scientific field, and also the field that produces more interest from the general public considering publications both at national and international level (Fundación MarcaChile 2018).

The DO & Natural Environment. The Atacama desert has the highest levels of solar irradiance on Earth, due to the same reasons that make it great for astronomy (Rondanelli *et al.* 2014). As of May 2018, the government has awarded generation contracts that will produce power at $24/MWh for photo-voltaic systems, and at $48/MWh for concentrated solar power systems, world's best prices for solar energy, motivated by the green mining demand. The DO will work with the emerging Solar industry in the Atacama desert to create green infrastructure.

3. Astronomy data-centric tasks

Working with the international community, we identified a set of data-centric tasks critical for the production of knowledge in astronomy. A component diagram of these tasks is introduced in Fig. 1, and we introduce each set of tasks in the following paragraphs.

Data acquisition and generation tasks. Data-centric challenges demand data that comes from scientific experiments, curated data archives or simulations. Quality assured data has to be processed enough for conversion to appropriate physical units for it to be used.

Data access and governance tasks. To use and re-use data, it has to be standardized, stored and indexed, enabling to be searched and filtered by either the initial project that demanded it, or others that require it.

Data analysis tasks. Data obtained from curated systems or from simulations or observations is analyzed (in real-time or not) to get new knowledge from it. The process does not end with this task, new data and insights may be ingested in curated systems and made available for other users.

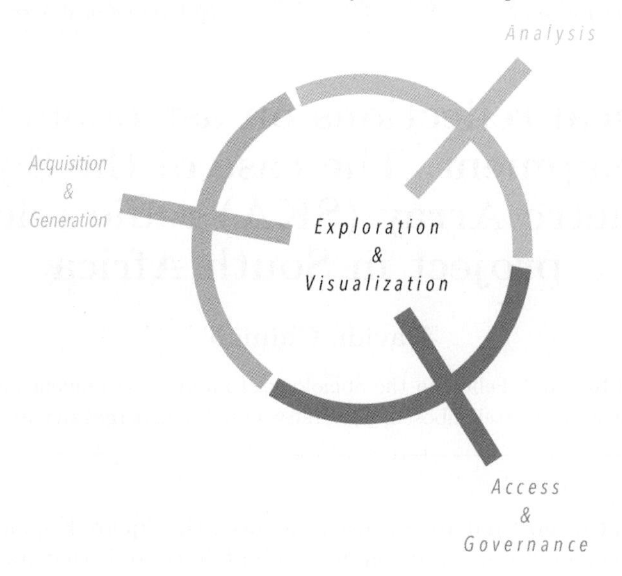

Figure 1. data-centric tasks in astronomy

Data exploration and visualization tasks. In performing each of the tasks above, exploration and visualization of the data is required, at minimum to enable humans work with it, at more sophisticated levels to enable discovery.

4. Conclusion

As mentioned before, over 2017 we worked with communities across sectors identifying opportunities to achieve the program mission of growing Chilean Digital Economy, using Astronomy data-centric tools. The DO vision emerged as an investment able to achieve this mission, and we will proceed to a next step of design of the DO throughout 2018, aiming to start its implementation in 2019.

References

P. Quinn, T. Axelrod, I. Bird, R. Dodson, A. Szalay, and A. Wicenec, Proceedings of Science (2015).

M. Catanzaro, Nature (2014).

E. G. Carayannis, T. D. Barth, and D. F. Campbell, Journal of Innovation and Entrepreneurship **A Systems View Across Time and Space** (2012).

Fundación MarcaChile, "Ciencia y tecnología de chile en la prensa internacional 2017," (2018).

R. Rondanelli, A. Molina, and M. Falvey, Bulletin of the American Meteorological Society (2014).

Astronomy in Focus - XXX
Proceedings IAU Symposium No. XXX, 2018
M. T. Lago, ed.

doi:10.1017/S1743921319005581

Critical reflections on astronomy and development. The case of the Square Kilometre Array (SKA) radio telescope project in South Africa

Davide Chinigò[1]†

[1]Postdoctoral Research Fellow in the Sociology of Land, Environment and Sustainable Development, Stellenbosch University. email: dchinigo@sun.ac.za

---

Perhaps one of the most remarkable features about the Square Kilometre Array (SKA) project, when completed the world's largest radio telescope, is that its main host site is located in Africa. When I began my sociological study of the semi-arid Karoo region of South Africa in 2016 where recently the South African SKA precursor MeerKAT has been inaugurated, I thought this enabled a reversal: for the first time in history Africa was going to be represented for what it is, and will eventually become, rather than for what it is not, or has not yet achieved, as many analysts depict the continent's developmental trajectory. I was therefore very intrigued to hear first hand how people in the small towns around the telescope viewed such a significant endeavour, expected to answer some of the most fundamental questions about the past and the future of humanity. Even more surprising was to realize soon after I started spending time in Carnarvon, the most prominent of these small towns, that for many people the SKA was a highly controversial issue, raising a number of concerns and public critiques. This backlash has been despite the fact that the South African SKA, now SARAO, runs a significant number of local development initiatives across different fields, including education, the upgrading of infrastructure, and support to local businesses.

To explain why the SKA became a controversial topic I will briefly explore two key moments illuminating the complex relationship between the SKA and Carnarvon. The underlying argument is that the 'development of astronomy' the undertaking of astronomy for scientific progress – and 'astronomy for development' – the societal impact resulting from the undertaking of astronomy – are two propositions reflecting priorities that are not always aligned and, for this reason, need careful examination. The first phase of this interaction dates back to the early 2000s when the intention to build the two SKA precursors (KAT-7 and then MeerKAT) was first announced. This was a moment in time characterised by the setting of high expectations about what the project could deliver for the small towns around the core site. This moment culminated in 2012 when South Africa won the international bid to host the broader international SKA project. Inter alia, then President Zuma visited Carnarvon, an exceptional event for local people, and declared that the "SKA will put Carnarvon on the world map". Local people were excited and honoured by the opportunities that would seemingly arise from the project. All this took place in a context characterised by significant marginality and social challenges. Small towns in this region of South Africa are characterised by high unemployment and school

†This work is based on the research supported by the South African Research Chairs Initiative of the Department of Science and Technology and National Research Foundation of South Africa (grant number 98765).

dropout rates, dependence on government social grants, and significant substance abuse. These towns are still characterised by strongly racialised social hierarchies, a legacy of the apartheid period, involving a tiny, relatively wealthy white elite and a large underclass of impoverished "coloured" people. In this context, the SKA seemed to constitute a once-in-a-lifetime opportunity to change the course of a history of marginality. Effectively, it represented an opportunity to make, not remain outside of, history. This was also seen as fulfilling the promises of the post-apartheid transition, in particular the values of pan-Africanist humanism of the African Renaissance proposed by President Mbeki in his term in office (1999 – 2008).

However, soon afterwards a second and more turbulent phase opened in the relationship between Carnarvon residents and the SKA. Relationships started deteriorating from 2016, when the SKA implemented a land acquisition programme for the core site of the infrastructure. The acquisition programme was in line with the regulations set by the 2007 Astronomy Geographic Advantage Act that provides the Minister of Science and Technology with special powers to prioritize astronomy in the region, primarily to minimize radio frequency interference around the telescope. Initially, protests came from the white commercial farmers who owned the farms that were being bought, notwithstanding the payment to them of generous compensation. Later, protests extended to important segments of the 'coloured' majority. Key concerns they expressed included their lack of involvement in local decision-making processes, questions about the distribution of resources for local development, and the lack of transparency in communications between the community and project management. Assessing the validity of these claims is a complex task that is beyond the scope of this short article. But what is important to stress here is another paradox. Most complaints about the SKA come from Carnarvon which is also the local town that has received the lion's share of development support. The way in which anti-SKA sentiment has unfolded suggests two complementary readings. Firstly, it is how the initially high local expectations of the project have unfolded over time in a very marginalised context. Secondly, it is how the expectations intersect with historically rooted concerns around land and forge the ways in which people relate to each other and to external interventions. Another important aspect is that while from the perspective of the SKA project the Karoo is often portrayed as a desert region ripe for new developments serving the future of South Africa and humanity at large, it is certainly not unpopulated, even if it is sparsely populated region. Current concerns that people raise about the major shift from farming to astronomy intersect with a long and complex history of land dispossession in the region; they also overlap with other land use changes, including renewable energy, mining, and conservation. This complexity generates considerable anxiety in people about their individual futures and the future of the region.

Two main lessons that can be drawn from this case study are firstly, that the two propositions 'development of astronomy' and 'astronomy for development' are not neatly aligned, and secondly, that by implication we need to unpack what we mean by 'development'. Can large astronomy projects that are about the advancement of science be simultaneously about science for development? Who should take responsibility for delivering on the promises of astronomy for development? What does science for the benefit of humanity mean when we narrow our focus to a small town in the Karoo? The case of Carnarvon suggests that the relationship between astronomy for development and the development of astronomy is much more complex than it appears at first glance. There is a disconnect between the national and global benefits that are set to flow from the investment in astronomy and its local development impacts. Policy-makers need to take this disconnect seriously by designing policy options and possible mitigation mechanisms for those who stand to lose the most.

Astronomy in Focus - XXX
Proceedings IAU Symposium No. XXX, 2018
M. T. Lago, ed.

doi:10.1017/S1743921319005593

Astronomy and inclusive development: access to astronomy for people with disabilities

Wanda Diaz Merced[1] **and Michael Gastrow**[2]

[1]IAU Office of Astronomy for Development, c/o SAAO, PO Box 9, Observatory 7935
email: wdm@astro4dev.org

[2]Human Sciences Research Council, South Africa email: mgastrow@hsrc.ac.za

---

1. Introduction: Astronomy and inclusive development

Globally, the institutions of astronomy have prioritized inclusive development within their strategic visions. The International Astronomical Union (IAU), and its Office of Astronomy for Development (OAD), have a clearly articulated strategy for social inclusion in the domain of astronomy.

In order to achieve the goals of the IAUs strategy, it will be necessary to enhance access to astronomy knowledge for people with disabilities (PWD). However, there is a paucity of knowledge in this particular domain. This proposed pilot study aims to develop new knowledge to contribute towards filling this gap. In particular, we propose to undertake a small-scale study focusing on access to astronomy knowledge for the visually impaired. The study will test the efficacy of a software tool designed to translate astronomical data into audio signals that can be interpreted by people with visual impairment. By using baseline data (prior to the intervention) and post-intervention data, we will assess the manner and extent to which the software package, and the attendant didactic techniques and classroom activities, facilitate changes in knowledge about astronomy, interest in astronomy, attitudes towards astronomy, and aspirations towards astronomy careers.

The longer-term objective of the study is to develop a knowledge base upon which the intervention can be scaled up, both geographically and to other modalities of disability. At the same time, PWD have great potential to contribute towards astronomy as a discipline, and greater access for this group will therefore increase the global pool of available talent for astronomy.

2. Astronomy as a gateway to science for people with disabilities

Astronomical topics such as weightlessness, black holes, and asteroid impacts have been identified as the most interesting topics among physics students. Astronomy also integrates chemistry, mathematics, biology, computational thinking with multiple STEM disciplines. Science has been considered one of the most valuable subjects taught to students with disabilities, and teachers identify science as the subject most suited for mainstreaming special needs students.

The World Report on Disability (2011) by the World Health Organisation and the World Bank estimates that more than a billion people across the world, or approximately 15% of the worlds population, live with some form of disability. However, in the sciences, professionals and students who are classified as disabled are considered to be a significantly under-represented population. This may be because mathematics and science often rely on one modality of learning to convey key aspects of the content. While computers can make the world more accessible to the disabled community, the visual nature of computing as it is usually implemented, a lack of resources, and a lack of role models, make it difficult for students with disabilities to participate in science courses and to see themselves as potential. Among the main reasons why disabled learners do not participate in sciences and mathematics is the absence of appropriate access technologies, and the attitudes of teachers and lecturers.

3. Methodology

A *literature review* will assess the state of knowledge with respect to: access to science knowledge for the visually impaired; the use of specialized software to facilitate learning for the visually impaired; and the mechanisms and dynamics of science learning for the visually impaired. The literature review will provide an informed basis on which to develop appropriate final research instruments, and guide the development of a detailed analytical framework.

Research instruments will include baseline measurements (pre-intervention), intervention observations, and post-intervention measurements. Instrument development will also include the development of a fieldwork plan for the deployment of the software package. The *location of the study* will be the Athlone School for the Blind, located in Cape Town, South Africa. *Baseline measurements* will capture demographic data. *Pre-intervention interviews* will include questions to elicit participants perceptions of their capabilities as a result of their academic preparation and access to information. The *intervention* will take the form of the didactic deployment of a specialized software package 'SoundOne'. This sonification (text-to-sound translation) tool has been designed to present astronomy telemetry from text files. *Post-interview questions* will consist of a series of scripted open-ended questions. *Analysis* will draw on the collective research data, both quantitative and qualitative, in order to: assess changes in perceptions of astronomy and science in terms of interest, access, and potential career paths; theorise mechanisms to explain changes in perceptions; identify changes in astronomy knowledge as a consequence of the intervention; identify mechanisms to explain changes in knowledge; reflect on the efficacy of the software package, and the efficacy of the didactic methods used to apply the software in the classroom setting; analyse demographic variables in relation to changes in knowledge and perceptions; and identify lessons for future application and scaling up.

Author index